Coasts are some of
Understanding the n
landforms and the pı
better management of

Coasts provides a
those studying coastal
the coast, their respo
pattern of evolution t
over thousands of ye
are used to illustrate t
paid to coastal morp
form, on rocky, reef, s

This valuable tex
dents is clearly illustra
will also be of great in
ners and coastal man

COLIN WOODROFF
University of Cambri
of mangrove swamps
positions at the Ur
Louisiana State Uni
before taking up his
an active member of
Landscape Change R
research on the geom
on islands in the Pac
co-edited another bo

Coasts: form, process and evolution

Colin D. Woodroffe
School of Geosciences,
University of Wollongong,
NSW 2522, Australia

PUBLISHED BY THE PRESS SYNDICATE OF THE UNIVERSITY OF CAMBRIDGE
The Pitt Building, Trumpington Street, Cambridge, United Kingdom

CAMBRIDGE UNIVERSITY PRESS
The Edinburgh Building, Cambridge CB2 2RU, UK
40 West 20th Street, New York, NY 10011–4211, USA
477 Williamstown Road, Port Melbourne, VIC 3207, Australia
Ruiz de Alarcón 13, 28014 Madrid, Spain
Dock House, The Waterfront, Cape Town 8001, South Africa

http://www.cambridge.org

First published 2003

Printed in the United Kingdom at the University Press, Cambridge

Typeface Monotype Times 9.5/13 pt *System* QuarkXPress™ [SE]

A catalogue record for this book is available from the British Library

Library of Congress Cataloguing in Publication data

Woodroffe, C.D.
Coasts: form, process, and evolution / Colin D. Woodroffe.
p. cm.
Includes bibliographical references (p.).
ISBN 0 521 81254 2 (hb) – ISBN 0 521 01183 3 (pb)
1. Coasts. 2. Coast changes. I. Title.

GB451.2 W65 2002
551.45′7–dc21 2002017418

ISBN 0 521 81254 2 hardback
ISBN 0 521 01183 3 paperback

To
David Stoddart

Inspirational teacher, mentor, scholar and friend

Contents

Preface

This book outlines the way that coasts operate. It is written for students of coastal geomorphology, coastal environments, and coastal geology, and for all those with an interest in coastal landforms or who seek insights into the way the coast behaves. It brings together studies of process operation and studies of coastal evolution concerned with longer-term landform development, into morphodynamic models. Coastal morphodynamics involves the mutual co-adjustment of process and form. It provides a framework from which to generalise across space and time scales. The book introduces these concepts and outlines geological setting, materials and coastal processes. Although there are physical principles which govern the response of sediment to forcing factors such as wave energy, the complexity of non-linear interactions means that it is generally difficult to scale up to explain behaviour over time scales that are relevant to human societies.

The book is based heavily on my own research experiences in Australia, Britain, the United States, New Zealand and on many islands in the West Indies, Pacific and Indian Oceans. It also draws extensively on the scientific literature and pays particular tribute in terms of historical perspective to those coastal scientists who have built the foundations of what we know. I hope that it instils something of the sense of wonder that I feel about the coast, and offers new perspectives on how the coast is shaped.

I have been privileged to learn about coasts at several of the most active centres of coastal research, including the University of Cambridge, University of Auckland, the Australian National University, Louisiana State University and, most recently, the University of Wollongong, where I am an active member of two research centres, the Oceans and Coastal Research Centre and the Landscape Change Research Centre. My interests and studies on coasts have been stimulated by four outstanding coastal geomorphologists, and it is a pleasure to acknowledge the considerable debt that I owe to each of them: David

Stoddart, Roger McLean, Bruce Thom and John Chappell. Many of the ideas and concepts that I describe in this book have been derived from the innovation, dedication and generosity of these and other friends and colleagues. I thank them all for many memorable times.

In writing this book I am deeply indebted to Bruce Thom and Scott Smithers for reading and commenting on the entire first draft. I am also grateful to Roger McLean, Brian Jones, Tom Spencer and Mark Dickson for reading and commenting on sections. The figures were drawn by David Martin, with assistance from Richard Miller. I am grateful to the Oceans and Coastal Research Centre for funding towards the cost of diagram production. For photographs, and many exciting discussions, I am grateful to Colin Murray-Wallace, Bob Young, John Morrison, Scott Smithers, David Kennedy, Brendan Brooke, Peter Cowell, Sandy Tudhope, Toshio Kawana, Gary Griggs, David Stoddart, and Bill and Clare Carter. I also appreciate the opportunity to use photographs from Mark Hallam, Environment Australia, and South West Water (UK).

I thank my father, David Woodroffe, for photographs, and for the encouragement to write this book. I extend my thanks to all my family for their support during the writing of this book, and for the time and space to devote to it.

I hope that the reader finds some of the exhilaration and fascination that I have found when reading about, and researching about, the coast.

Chapter 1
Introduction

> The scenic features of the coast – its ragged scarps, its ever-changing beaches and bars, its silent marshes with their mysterious past – all excite the imagination, and tempt the wanderer by the shore to seek an explanation for these manifestations of Nature's handiwork. (Johnson, 1925)

Coasts are often highly scenic and contain abundant natural resources. The majority of the world's population lives close to the sea. As many as 3 billion people (50% of the global total) live within 60 km of the shoreline. The development of urbanised societies was associated with deltaic plains in semiarid areas, and the first cities appeared shortly after the geomorphological evolution of these plains (Stanley and Warne, 1993a). The coast plays an important role in global transportation, and is the destination of many of the world's tourists.

The shoreline is where the land meets the sea, and it is continually changing. Coastal scientists, and the casual 'wanderer by the shore', have attempted to understand the shoreline in relation to the processes that shape it, and interrelationships with the adjacent shallow marine and terrestrial hinterland environments. Explaining the geomorphological changes that are occurring on the coast is becoming increasingly important in order to manage coastal resources in a sustainable way.

This book examines the coast as a dynamic geomorphological system. Geomorphology is the study of landforms, and coastal geomorphology is concerned primarily with explaining the many different types of coastal landforms, and understanding the factors that shape them. Physical processes, including both hydrodynamics and aerodynamics, are influenced by substrate and the biota (plants and animals) growing on it. Landforms provide habitats within which coastal ecosystems function. In many cases, the biota themselves contribute to shaping the landforms.

Coastal geomorphologists were one of the first groups of scientists to examine coastal landforms, with interests in the form, process and dynamics of the coast. A wide range of other scientists, however, also study the coast. The physical setting, including rocks and sediments, is examined by geologists, physical geographers and geophysicists. The living organisms are described by ecologists and natural scientists. Processes are studied by oceanographers and climatologists. Economists, human geographers, planners, and other social scientists have separate, and at times conflicting, interests in human use of the coast; and engineers and coastal managers are concerned about the stability of the coast and use of coastal resources. The challenges posed at the land–sea interface by the impact of global environmental change, and by intensive human use of coastal resources, will require further cross-disciplinary collaboration, and interdisciplinary consultation and study. This book focuses on coastal geomorphology, recognising the need to take a multidisciplinary and holistic approach to the diverse factors that influence the present form of the coast and the geomorphological processes that continue to shape it.

1.1 The coastal zone

The coast comprises the interface between the land and the sea. The 'shoreline' is the actual margin of the two, whereas the term 'coast' is much broader and includes areas below and above the water line, such as shoals, dunes and cliffs (Figure 1.1a). The 'coastal zone' is a broad transitional area in which terrestrial environments influence marine environments and in which marine environments influence terrestrial environments (Carter, 1988). Although a statutory definition of the coastal zone may be needed for planning and legislative purposes, it is difficult to identify precise landward and seaward boundaries to the mutual interaction between land and sea. It is often more appropriate for physical, biological or administrative purposes to use a definition relevant to the particular management issue (Kay and Alder, 1999).

Coastal geomorphology focuses on explaining landforms in the coastal zone by examining the form, sediments and depositional history at the modern shoreline. It can include study of the shallow marine environment that is influenced by terrestrial factors and the land where the influence of the sea is felt if this is appropriate to understand the longer-term evolution of the coast. Migration of the shoreline as a result of sea-level fluctuations during the ice ages (the Quaternary Period) means that, in many cases, it is necessary to broaden the area of study beyond the modern coastal zone to include older landforms. Understanding changes at the coast can require examination of processes well outside the coastal zone. A delta is likely to respond to changes within the river

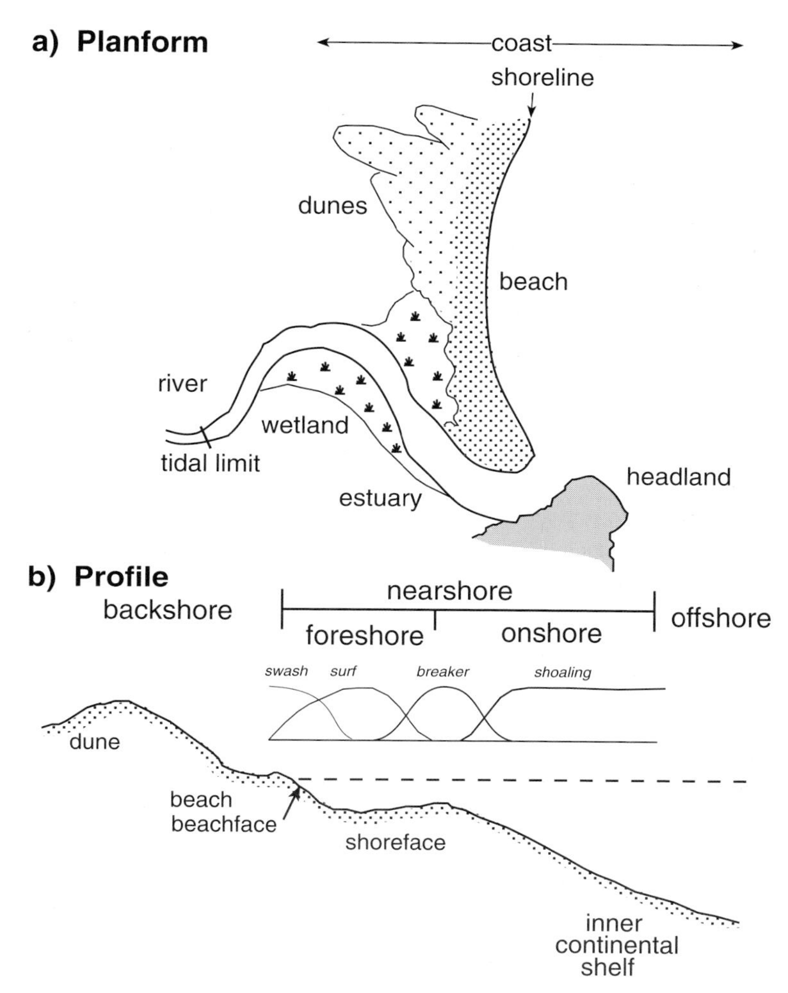

Figure 1.1. A schematic coastal embayment in (a) planform and (b) profile. Some of the key definitions based on proximity to the shore, wave characteristics, and substrate are shown.

drainage basin; for example, construction of the Aswan Dam on the River Nile has decreased the volume of sediment reaching the Nile Delta causing serious coastal erosion (Milliman *et al.*, 1989). Elsewhere, the coast may respond to events seaward of the coastal zone; for instance, mass slumping of delta fronts may carry sediment into deeper water.

1.2 Coastal geomorphology

The shape of coastal landforms is a response of the materials that are available to the processes acting on them. The geomorphology of the coast can be examined in planform (also called shore-parallel, or longshore) or in profile (also called cross-section, cross-shore, shore-normal, or orthogonal) as shown schematically in Figure 1.1. In planform, a coastal embayment can be divided into relatively distinct landforms based on factors such as topography and lithology, or sediment texture and resistance. The most resistant are rocky coasts, characterised by cliffs. Sand and gravel are more mobile in the coastal zone and form coastal barriers on which there are beaches and associated landforms. Deltas or estuaries occur where rivers reach the sea. The coastal zone extends upstream as far as tidal influence is felt, which can vary over time in response to tidal or river flow characteristics. Muddy substrates occur in the more sheltered areas and can support coastal wetlands.

Figure 1.2 shows an example of a coastal embayment in south-eastern Australia. The Minnamurra River drains from the Southern Highlands of New South Wales, and its mouth is constrained by a headland to the south. An estuary has developed behind a sandy barrier on which there is a beach and dune. This coastal embayment shows many of the characteristics of barrier estuaries that have been incorporated in an evolutionary model described by Roy and examined further in Figure 1.7 (Roy, 1984; Roy *et al.*, 1994).

Figure 1.2. The estuary of the Minnamurra River in New South Wales, southeastern Australia is an example of a coastal embayment in which the landforms represented schematically in Figure 1.1 can be seen (photograph R.J. Morrison.

1.2.1 Coastal landforms

The broad zones shown in planform in Figure 1.1a, or discernible in Figure 1.2, can be subdivided into smaller-scale landforms which are described in separate chapters in this book. Cliffs and shore platforms typical of rocky headlands and other bold coasts are described in Chapter 4. Beaches and dunes that form on sand and gravel barriers are described in Chapter 6. Estuaries (and deltas) are described in Chapter 7, and muddy shorelines, and the wetlands that characterise them, are described in Chapter 8. Coral reefs contain a distinctive group of landforms, occurring on tropical and subtropical coasts, and are examined in Chapter 5. It is the distribution and interdependence of these landforms, as shown in Figure 1.1a, which determines the topographic character of a particular stretch of coastline. The resistant headlands are obstacles that influence the pattern of wave energy alongshore and consequently affect the shape of sand and gravel barriers. The supply of sediment from a river, filtered through an estuary, can determine changes in the total volume of sediment in the barrier (beach and dune) and nearshore. Muddy shorelines and associated wetlands generally occur where conditions are sheltered within the rocky, barrier or estuarine suite of landforms.

In profile it is also convenient to divide coasts into zones based on morphology and the processes that operate. The terms used to describe these zones differ slightly between geomorphologists and other scientists, such as engineers, and between European and North American researchers. A general scheme is shown in Figure 1.1b, but usage is not always consistent. The backshore is the zone above the water line. The nearshore comprises shallow water in which waves interact with the seafloor, and consists of foreshore where waves break, and onshore. Deeper water areas are termed offshore. The terms can be applied to most coast types, including rocky and muddy coasts, but are illustrated in Figure 1.1b with reference to a sandy embayment, because the subdivisions are generally distinct in a beach setting. The subaerial beach and dune are part of the backshore. The seafloor is termed the shoreface; whereas that part of the beach which is subject to variations in water line (intertidal) is termed the beachface. The nearshore can be further subdivided on the basis of the behaviour of waves in terms of shoaling, breaker, surf and swash zones (see Chapters 3 and 6). Offshore, the shoreface merges into the inner continental shelf.

Coasts are among the most dynamic parts of the earth's surface. The land and the sea rarely meet at a constant boundary. The shoreline migrates daily with the tide; it can change seasonally, and varies over longer time scales as the coast erodes or deposits, or as sea level changes.

Coastal sediment deposits are shaped and reshaped by wave and current processes, which in turn vary through time. The coastal geomorphologist is concerned with the way in which coastal morphology changes with time, both in terms of its present-day dynamics and its longer-term evolution. The techniques by which these morphological adjustments can be determined include monitoring landform changes as they take place, reconstructing past events by comparison of surveys (including maps and aerial photographs), and inferring changes from erosional landforms and depositional sedimentary sequences.

Comparison of coastal landforms from different places can indicate a sequence of modern landforms in different stages of development. The description and interpretation of the sequence of superimposed sedimentary units is called stratigraphy, and stratigraphic units can be examined using subsurface coring or seismic studies. Former shorelines, such as raised terraces or beaches on uplifted coasts or beach and nearshore sediment deposits, give an insight into past coastal conditions. In some cases the shape of a body of sediment can be more diagnostic of environment of deposition than sedimentological characteristics alone, and separate depositional episodes are often distinguished on the basis of surface or subsurface form. Where stratigraphy is examined in association with the shape of former landforms it is called morphostratigraphy. Stratigraphic and morphostratigraphic units provide a sequential, though often incomplete, sedimentary record of past depositional and erosional changes.

Coastal evolution is the study of how and why the characteristics and position of the shoreline have altered (Carter and Woodroffe, 1994). Reconstructions of past coastal landforms are of considerable relevance in view of the dynamic nature of most coastal environments. It is possible to develop 3-dimensional models of the coastal landscape based on sedimentary units (facies) and habitats (paleoecology) determined on the basis of biological remains or activity (e.g. trace fossils). A series of dating techniques have become available which allow a firmer time control on the formation of coastal landforms and sediment deposits (chronostratigraphy). Coastal features may be placed in temporal context based on relative ages, for instance, using archaeological evidence such as artefacts (e.g. pottery, stone tools, midden deposits or other cultural remains), or absolute ages, based on the radiometric decay of a range of isotopes. These analyses reveal that the shoreline has varied markedly in its location through time. For instance, it is now clear that repeated cycles of shoreline adjustment have occurred as a result of sea-level changes (see Chapter 2). Particular sequences of coastal landforms characterise periods of sea-level rise (transgression) and sea-level fall (regression), and complex landforms have developed as a result of erosion and deposition during periods in which sea level has been relatively constant (stillstand).

1.2.2 Coastal morphodynamics

The mutual co-adjustment of coastal form and process is termed coastal morphodynamics (Wright and Thom, 1977). The processes that operate on a section of coast are a function of oceanographic and climatic factors, the nature of the hinterland, and the pre-existing topography. Wave and current processes lead to deposition in some places and erosion elsewhere, modifying form, which in turn alters the rate of operation or effect of the processes themselves. Coastal morphodynamics provides a framework for understanding the interaction of forces and topography that gives rise to the distinctive character of the world's shorelines. Morphodynamic models, based on empirical field and laboratory studies, and supplemented by computer simulations, synthesise these relationships to generate a greater understanding of how coasts operate (Cowell and Thom, 1994).

Models play an important part in the study of the coast, and will play an even more significant role in future, as they become more rigorous and more complex. Models are useful because they are simplifications of real world situations in which it is possible to control variables and look at one thing at a time (see Chapter 9). They also offer a conceptual framework within which patterns of coastal evolution can be extrapolated from one site to another (Carter and Woodroffe, 1994). For example, Figure 1.1a represents a 'model' of the planform of a section of coast, whereas the photograph in Figure 1.2 shows that a real coast is much more complex than the model implies. Nevertheless, several of the key features of the relationship between landforms in an estuary or embayment are portrayed schematically in the 'model' in Figure 1.1a, and this simplification will be used several times in subsequent chapters to generalise about the way that such systems operate. Quantitative models can be developed where there is a good understanding of the interrelationships between form and process. Computer simulations can be based on patterns of change inferred over long time scales and consistent with processes known to operate over shorter time scales (see Section 1.4 below). Simulation modelling is not intended to reconstruct past or predict future coastal evolution exactly, but provides a tool for experimentation and extrapolation within which broad scenarios of change can be modelled and sensitivity to variables examined.

In Chapters 4–8, emphasis is placed on morphodynamic models and their ability to help understand the way that particular coasts function and change with time, synthesising process studies on modern coastal landforms with reconstructions of the way in which coasts have evolved. The response of coastal landforms to rare extreme events or to changes in controlling variables such as oceanographic conditions needs to be understood. The way in which coasts respond to human impacts is less clear, and has become a major challenge for coastal geomorphologists to address (see Chapter 10). The threat of future anticipated sea-level rise as a result of the enhanced greenhouse effect has focused effort towards applied studies of coasts, a trend that will increase in the 21st century. However, it needs to be emphasised that it has only been in recent decades that understanding of the timing and magnitude of Quaternary sea-level changes has improved, and many of the fine details, and the way in which the coast responds, remain to be resolved.

Our understanding of how coasts operate is based on an ever-increasing body of scientific knowledge. The most widely accepted interpretations of today will be revised as that body of knowledge increases further. It is important to recognise that modern models and interpretations are based on the research and insights of many past coastal scientists. The development of ideas in coastal geomorphology, as in other disciplines, comprises a web of interwoven and recurring conjecture, theory and empirical (field and laboratory) study. Many elements of contemporary theory and models can be seen to have a firm basis in early concepts of coastal geomorphology, such as shoreline development through juvenile and mature stages, or the operation of a cycle of change. Consideration of the history of how ideas have changed provides an important perspective on modern theories and models. For this reason, each of Chapters 2–8 also begins with a brief historical perspective on the development of scientific knowledge as background.

This chapter expands on these themes in coastal geomorphology. First, a brief historical perspective of the development of ideas in coastal geomorphology is given, recognising some of the key researchers and research schools that have contributed to ideas which are developed further in subsequent chapters. Second, time scales and space scales appropriate for the study of coastal geomorphology are discussed. Third, a systems approach to the study of coasts is outlined, and some of the fundamental concepts such as feedback and equilibrium are defined. It is important to recognise that human influence now plays a role in most coastal systems and needs to be considered in the longer-term management of coasts. In the chapters that follow, individual coastal landforms are discussed in terms of their morphodynamics. These morphodynamic concepts are re-evaluated in Chapter 9 drawing together themes that emerge from the discussion of various types of coast. The final chapter considers briefly the influence that humans have, or have had, and recognises that this role will become increasingly prominent in study of coasts in the future.

1.3 Historical perspective

People have always been fascinated by the coast; study of the coast is probably as old as any study. The Greeks described the delta at the mouth of the River Nile; Herodotus remarked on the similarity of its shape to the Greek letter 'delta'. The arcuate curvature of the shoreline was accurately portrayed by Leonardo da Vinci in his plan for draining the coastal Pontine Marshes in Italy in the 15th century. However, the first true scientific interest in coasts began post-Renaissance in the context of navigation beyond the Mediterranean. Surveys by hydrographers (e.g. Bellingshausen, Wharton) were essential components of global exploration; geological and biological studies were often extensions of such survey work, undertaken by naturalists.

1.3.1 Pre-20th-century foundations

Coastal geomorphology has its foundations in the geological studies that followed from the recognition of the destructive work of the sea along coastlines (e.g. Lomonosov, 1759). Voyages of discovery, particularly those of Captain James Cook, had increased the known shores, and the naturalists who followed described landscapes in detail. Exploration and survey enabled recognition of the operation of modern-day processes over long periods of time to achieve substantial erosion or deposition, the concept of uniformitarianism (Hutton, 1788; Playfair, 1802). The geological ideas which were derived from this principle (Lyell, 1832; de Beaumont, 1845) matured into the study of physiography, which emphasised the interconnectedness between causal processes and resulting landforms (Huxley, 1878).

The 19th century was a time of increased scientific awareness of the diversity of the world. Charles Darwin's voyage on the Beagle, and James Dana's travels as part of the United States Exploring Expedition led to numerous observations by these great geologists as to how the coast had come to look as it now does. Darwin realised the significance of tectonic activity, not only in terms of the uplift that he observed along the coast of South America, but also in terms of subsidence that he could infer as a process in reef-forming seas (Darwin, 1842). Darwin's remarkable deduction that fringing reefs might become barrier reefs which in turn might form atolls as a result of gradual subsidence of volcanic islands combined with vertical reef growth (Figure 1.3) provides one of the earliest examples of a conceptual model of coastal

a) Fringing reef

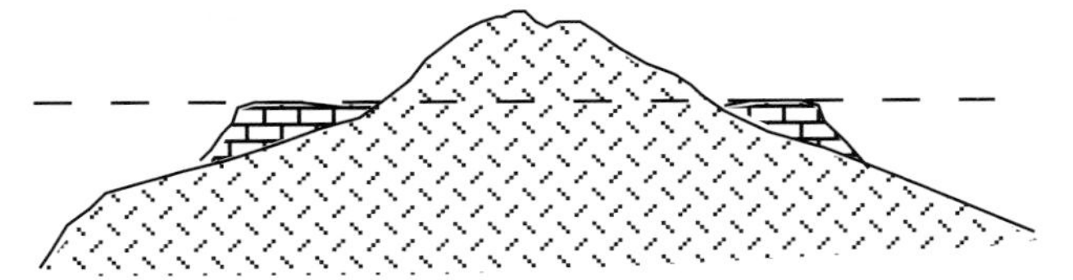

b) Barrier reef

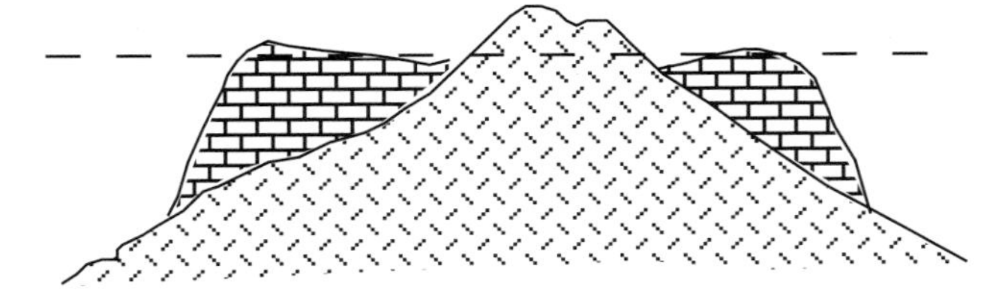

c) Atoll

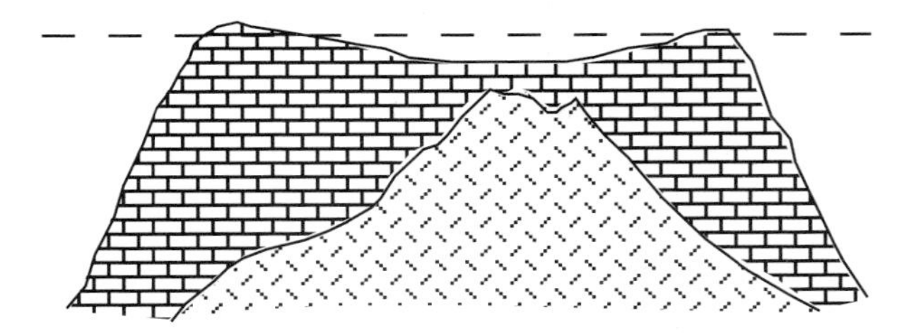

Figure 1.3. Schematic representation of the progression deduced by Charles Darwin showing how (a) fringing reefs around a volcanic island would develop into (b) barrier reefs and barrier reefs would develop into (c) an atoll, as a result of subsidence and vertical reef growth (based on Huxley, 1878).

evolution (preceding publication of his views on evolution of species). It has been largely substantiated by subsequent studies (see Chapter 5), and has stimulated development of other models of coastal change. Dana enthusiastically promoted Darwin's theory recognising that the embayed nature of the shoreline on reef-encircled islands in the Pacific Ocean supported the concept of subsidence, and he described shore platforms (see Chapter 4) and discussed their formation (Dana, 1849).

Detailed interpretations by Grove Karl Gilbert of the stratigraphic relationships of sedimentary units associated with terraces in Lake Bonneville marking abandoned Pleistocene shorelines provided a foundation upon which modern concepts of shoreline equilibrium would become based (Gilbert, 1885). An early classification of coasts recognised straight (Pacific) coasts where the structure of the landscape runs parallel to the shoreline, and irregular (Atlantic) coasts where the structure is at an angle to the shoreline (von Richthofen, 1886; Suess, 1888).

1.3.2 Early-20th-century geographical context

William Morris Davis played a pivotal and charismatic role in physical geography at the close of the 19th century, and his idea of a 'geographical cycle' of uplift and erosion was widely adopted during the early 20th century (Davis, 1899, 1909). Davis was primarily concerned with the development of terrestrial landscapes, recognising the importance of structure, process and time. He emphasised time as the prime factor, and thought of it as progressive, irreversible and orderly. Davis believed that the cycle of erosion, punctuated by periodic uplift rejuvenating an eroded landscape, would have implications for the coast. Figure 1.4a indicates how Davis believed a coast would evolve in planform becoming less indented over time, with erosion of promontories and infill of embayments (Davis, 1896). It shows a remarkable similarity to his concept of planation of the landscape in profile (with reduction of relief and peneplanation of the landscape).

Figure 1.4b shows a detailed study of the evolution of the Provincelands spit at Cape Cod, Massachusetts, by Davis (1896). The initial morphology was hypothesised, and subsequent change was

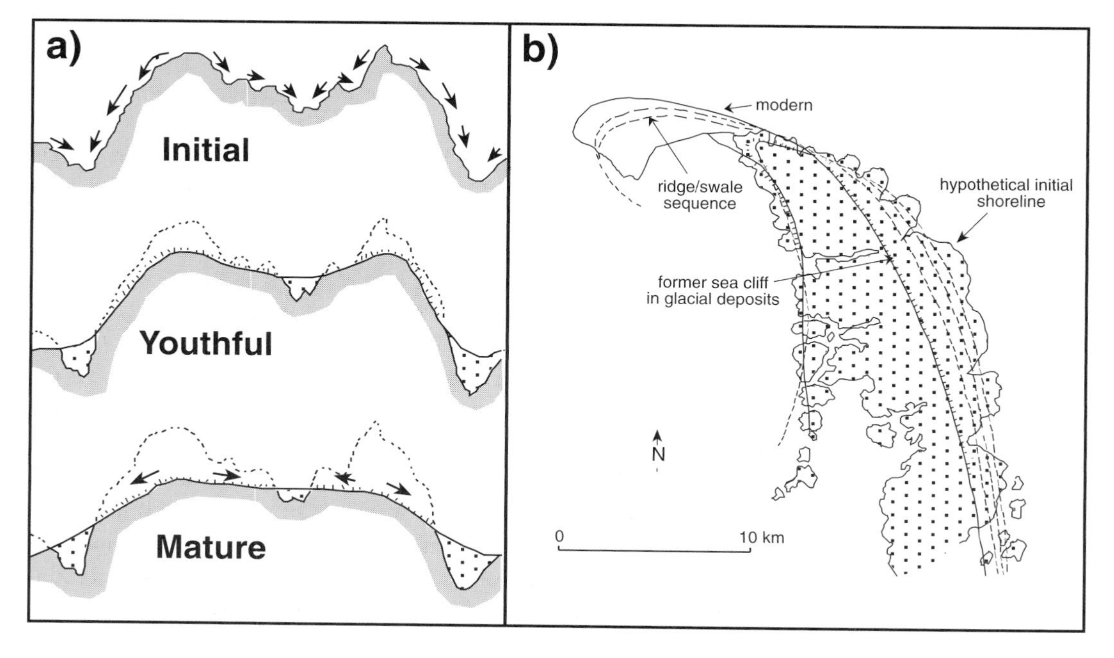

Figure 1.4. The Davisian cycle as applied to the coast. (a) Schematic representation of planform stages as conceived by W.M. Davis. A shoreline progresses from an initial rugged form, through maturity, to a regularised shoreline that is cliffed and infilled with sand (stippled). The arrows indicate direction of sediment movement. He envisaged this in parallel to the geographical cycle of erosion by which mountains (like the initial shoreline form) would be reduced to a peneplain (like the ultimate mature shoreline). (b) Schematic evolution of Cape Cod, Massachusetts (based on Davis, 1896).

accommodated into the progression from youth to maturity. Davis reconstructed the previous positions of the spit's shoreline pivoting around a fulcrum of spit development, and suggested an equilibrium profile developed offshore. Whether the northern end of the spit has really been characterised by a sequence of ridges each of the same length but building seaward (Davis, 1896), or whether the whole spit has remained constant in width but grown longer (Zeigler *et al.*, 1965), is still unresolved (FitzGerald *et al.*, 1994).

The Davisian cycle of erosion was 'by intention a scheme of the imagination and not a matter of observation' (Davis, 1909, p. 281). The concept was extended and applied to coasts by a series of successors (Gulliver, 1899; de Martonne, 1906, 1909), most notably by Douglas Johnson. Johnson included in the study of coastal evolution not only the description of landforms, and the historical reconstruction of the origin of a particular shore, but also a framework in which to view the systematic evolution, from initial to sequential forms. Figure 1.5 shows progressive stages of coastal evolution as proposed by Johnson (1919). He believed that initial form exerted a control on the first stages of coastal erosion. The coast passed through a youthful stage in which there was the greatest diversity of beaches barriers and spits. On the coast of New England, reworking of glacial deposits provided episodic

Figure 1.5. Schematic stages in the evolution of a shoreline of submergence as envisaged by Johnson (1919, 1925), (a) initial, (b) youthful and (c) mature.

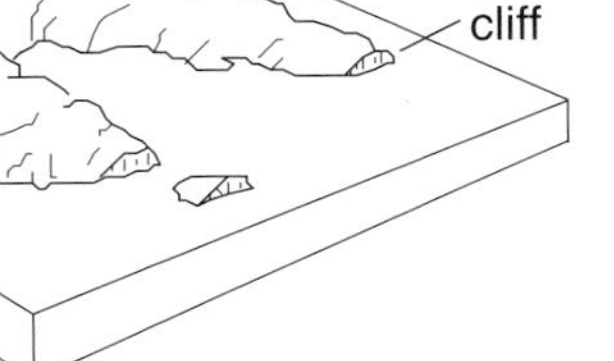

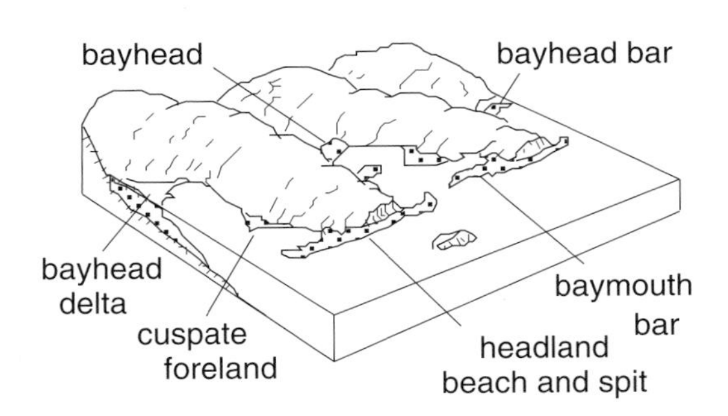

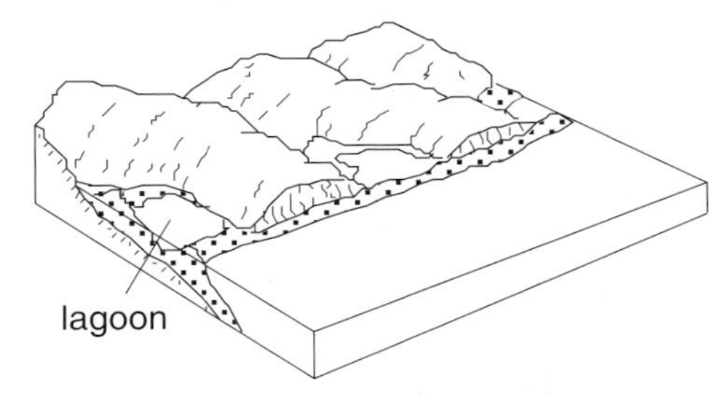

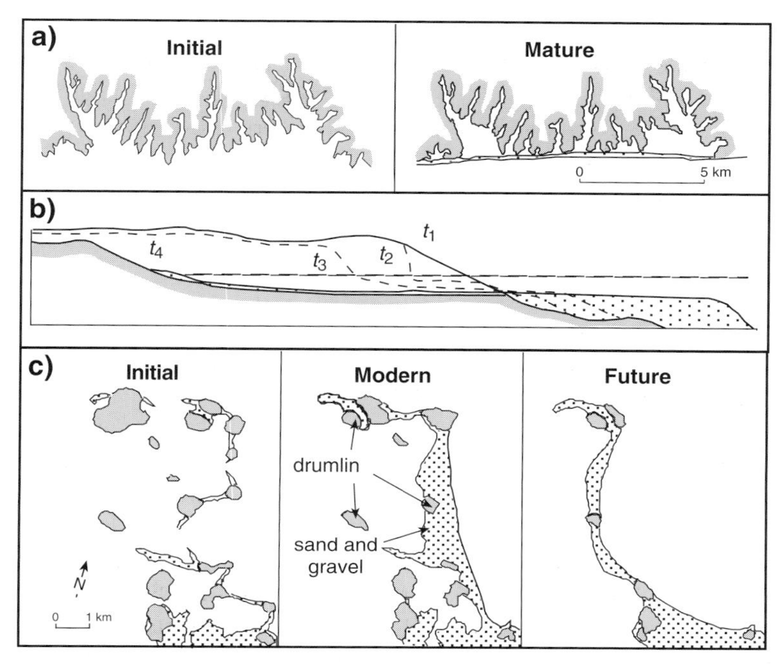

Figure 1.6. Evolution of the coast as envisaged by Johnson. (a) Schematic evolution for the Martha's Vineyard area of New England, showing the hypothesised initial and the modern regularised, mature form of the shore. (b) Stages in the development of an equilibrium shore profile as envisaged for a shoreline of submergence by Johnson (based on Johnson, 1919). (c) Schematic evolution of Nantasket Beach, Massachusetts, showing inferred initial, modern and hypothesised future stages (simplified from Johnson and Reed, 1910).

supply of a range of sediment sizes (Johnson, 1925). In mature stages the coast became less complex. These ideas provided a framework within which to view coastal evolution, and a sequence by which it might be possible to reconstruct former, or future, coastal configurations. Johnson's book on the formation of the Acadian–New England coast is a magnificent example of the use of a selection of landforms in space to infer a sequence through time (Johnson, 1925). Johnson adopted the idea that the coast would become less embayed with time; Figure 1.6a shows his reconstruction of initial form of the coast around Martha's Vineyard, Massachusetts, in comparison to its modern form (Johnson, 1919).

The concept of equilibrium is an important one in coastal geomorphology, and one which is central to much of this book. Johnson believed that the offshore profile formed in soft sediments would tend towards an equilibrium shape within the context of the geographical cycle. The nature of erosion and sedimentation on continental shelves was hypothesised on the basis of very sparse field data, and led to the concept of a nearshore equilibrium profile (Figure 1.6b). This wave-formed terrace was considered to develop as a balance between oscillatory wave action and undertow versus onshore currents (Fenneman, 1902). This early concept of an equilibrium profile applied to continental shelves has not been upheld in the context in which it was initially proposed (Dietz, 1963; see also Chapter 4), although it remains an issue of debate in relation to beach and barrier coasts (Dean, 1977, 1991; see also Chapters 6 and 9).

Figure 1.6c is a detailed coastal reconstruction of Nantasket Beach, on the southern side of Boston Harbor, Massachusetts (Johnson and Reed, 1910). Drumlins, elongate or oval mounds of glacial deposits formed beneath ice sheets, are eroded by wave action and their sediments become incorporated into beaches and spits. Spits connect existing, eroding drumlins, and a sequence of prior drumlins can be envisaged before retreat of the coast to its present position. Drumlins anchor beaches and spits; when they are finally eroded, the beaches retreat landward until they intersect the next drumlin. Johnson and Reed (1910) used this logic to project how a future stage might look (Figure 1.6c). Their interpretations of shoreline reworking of drumlins remain important in understanding this coastline, although it does appear that well-sorted sand has also been derived from an offshore source, and not solely from the drumlins (FitzGerald *et al.*, 1994). This approach to the explanation of coastal evolution, adopted by Johnson, has been the foundation for the development of many conceptual models in coastal geomorphology. The drumlin coast of New England and Nova Scotia has also continued to play a central role in the development of morphodynamic models (Carter and Orford, 1993; Forbes *et al.*, 1995).

The distinction between relative fall (emergence) or rise (submergence) of the sea in relation to the land (Davis, 1909) was extended by Cotton (1916) and particularly by Johnson (1919), who recognised coasts of emergence and coasts of submergence, as well as compound and neutral coasts. However, although Johnson's approach placed relative sea-level changes central to the classification of coasts, it preceded a clear understanding of the way that sea level had fluctuated. Evidence from numerous shores indicating that the sea level had varied, and the realisation that there had been considerable changes in ocean volume associated with the ice ages (Daly, 1910, 1925, 1934) became a central theme for much of the rest of the century.

The Davisian concept of cycles of erosion, progressing through youth, maturity and old age, was heavily influenced by Darwinian evolutionary theory (Stoddart, 1966). It was an approach based on explan-

Table 1.1. Three mid-20th-century approaches to the classification of coasts

Shepard, 1948	Valentin, 1952	Cotton, 1954
I Primary or youthful coasts	I Advancing coasts	I Coasts of stable regions
(a) shaped by terrestrial erosion	(a) emerged coasts	(a) recently submerged
(b) shaped by terrestrial deposition	(b) constructional coasts	(b) previously emerged
(c) shaped by volcanic activity	(i) organic (mangrove, coral)	(c) miscellaneous
(d) shaped by diastrophism	(ii) inorganic (marine, alluvial)	II Coasts of mobile regions
II Secondary or mature coasts	II Retreating coasts	(a) recently submerged
(a) shaped by marine erosion	(a) submerged coasts	(b) recently emerged
(b) shaped by marine deposition	(i) glacial (eroded, deposited)	(c) fault and monoclinal
	(ii) fluvial (folded, flat-lying)	(d) miscellaneous
	(b) retrograded (cliffed)	

atory description, whereby generalising principles were inferred, and observations were interpreted within a preconceived cycle (Rhoads and Thorn, 1996). Meanwhile, the inductive empiricism of geologists such as G.K. Gilbert, in which field observational data were collected based on theoretical premises with the objective of testing between competing or multiple working hypotheses (Gilbert, 1885), received relatively little attention until the second half of the 20th century.

1.3.3 Mid-20th-century coastal studies

During the mid 20th century the development of coasts was considered within the context of complex denudation chronologies that were based on geographical cycles of erosion. Examples include the Weald in south-eastern England (Wooldridge and Linton, 1938, 1955) and the landscapes of Wales (Brown, 1960). These and other interpretations were constrained by presupposition that the landscape had undergone a series of cycles of uplift and gradual erosion. In the absence of techniques to date land surfaces, studies were inevitably conjectural and subject to varying interpretations. Many such landscape narratives have been refuted now that geochronological techniques are available to provide a partial time frame.

Considerable effort was directed towards the classification of coasts. Classifications were either descriptive, concerned with features visible in the modern shoreline, or genetic, concerned with how the coast had evolved. Table 1.1 outlines the principal divisions within three of the more popular schemes (Shepard, 1948; Valentin, 1952; Cotton, 1954). The basic variables which were used to classify different types of coast included the structure of the coast, stage of development, relative movements of land and sea, and progradation and retreat of the coast.

Several classifications expanded Johnson's ideas, either extending the concept of youthful versus mature, or the distinction between submerging and emerging coasts. Although several different coastal classifications were proposed, none received universal acceptance (McGill, 1958; Tanner, 1960; Russell, 1967; King, 1972; Shepard, 1976).

Major descriptive studies were undertaken, with inferred time frames for development; an outstanding example is the history of the Mississippi River valley and delta in the United States by Fisk (1944). Similar detailed coastal studies were undertaken by extremely active individual researchers such as Schou (1945) in Denmark, Kuenen (1950) in Holland, Steers (1953) and King (1959) in Britain, Guilcher (1958) in France, Zenkovich (1967) in Russia, Russell (1967) in the United States, and Cotton (1974) in New Zealand. These, and studies by other researchers, provided the material from which it was possible to develop generalisations from field-based observational facts, drawing upon similarities (or, in some cases, dissimilarities) from wide geographical areas.

During the 1950s and 1960s geomorphology as a whole became more focused upon process operation within the landscape (Strahler, 1952a; Hack and Goodlett, 1960). Process studies had already become an important element of the study of sediment movement in coastal environments (Reynolds, 1933; Saville, 1950; Rector, 1954), with strong traditions established at the University of Cambridge (Lewis, 1931) and in London (Bagnold, 1940, 1946), on which further studies could be based. The growing acceptance of the unifying concepts of plate tectonics provided a framework within which the distinction between Pacific (active plate margin) and Atlantic (passive plate margin) coasts (see Chapter 2) could be placed (Inman and Nordstrom, 1971). Coastal studies provided the foundations from which marine geology and oceanography were to develop.

The poor level of understanding of beach environments had been highlighted by several situations during World War II where landing craft and military personnel had experienced difficulties with beach landings because of the dynamic nature of coastal landforms (Williams, 1960). Research programmes funded through the Geography Program of the US Office of Naval Research in the United States, many based at the Coastal Studies Institute at Louisiana State University or Scripps Institute of Oceanography in California, played a central role (Shepard and Wanless, 1971). The initiation of the Beach Erosion Board in the US Corps of Engineers, later the Coastal Engineering Research Center (Krumbein, 1944), the Wallingford Hydraulics Research Laboratory in Britain, and Delft Hydraulics in the Netherlands reflected these directions of research effort. They have given rise to an extensive body of knowledge on beach behaviour (Bascom, 1964; Komar, 1976), discussed in Chapter 6, with recognition of a series of beach types responding to the energy conditions to which the beach has been subjected.

1.3.4 Late-20th-century process and historical geomorphology

While some coastal geomorphologists were focusing on process studies using laboratory and field experimentation, historical geomorphology was also continuing to expand, especially in the context of sea-level history. The development of dating techniques had a major impact. Availability of radiocarbon dating, especially its application to interpretation of the behaviour of the Mississippi River and the Gulf of Mexico coast, led to some classic studies (Gould and McFarlan, 1959; Frazier, 1967). Geochronology brought a new rigour to studies of coastal evolution and, although not without problems associated with interpretation, provided a means of calibrating rates of development that had previously been conjectural. An approach based on dating the past began which continues to provide a strong focus for coastal geomorphology.

It became clear that sea level, which has been close to its present level during interglaciations and 80–150 m below present during glaciations, had been an important constraint on coastal evolution. Sea-level curves, indicating a presumed global eustatic sea-level trend, were compiled by Fairbridge (1961) and many others. It soon became apparent that sea-level history was different at different places in the world (see Chapter 2), and INQUA (International Quaternary Association) and IGCP (International Geological Correlation Program) projects began to document this variability (Bloom, 1977). Identification of relative sea-level history from individual locations, establishment of regional variability and compilations of global syntheses have been objectives of much recent coastal research (Pirazzoli, 1991, 1996). A further incentive to understand coastal response to sea-level variation in the final decades of the 20th century has been the anticipation of accelerated sea-level rise as a result of the enhanced greenhouse effect (Bird, 1993a).

Coastal studies up until the 1960s had been heavily focused on the shores of northwest Europe, North America and the Soviet states. These coasts are not particularly representative of the world's shorelines as a whole (Davies, 1980); many are still experiencing isostatic adjustment to the melting of Pleistocene ice sheets (see Chapter 2), and are shaped by winter storms (see Chapter 3). Coastal studies were considerably broadened as a result of a series of major overseas research projects, many funded by the US Office of Naval Research. There also developed strong research schools within other countries, including Japan (Horikawa,

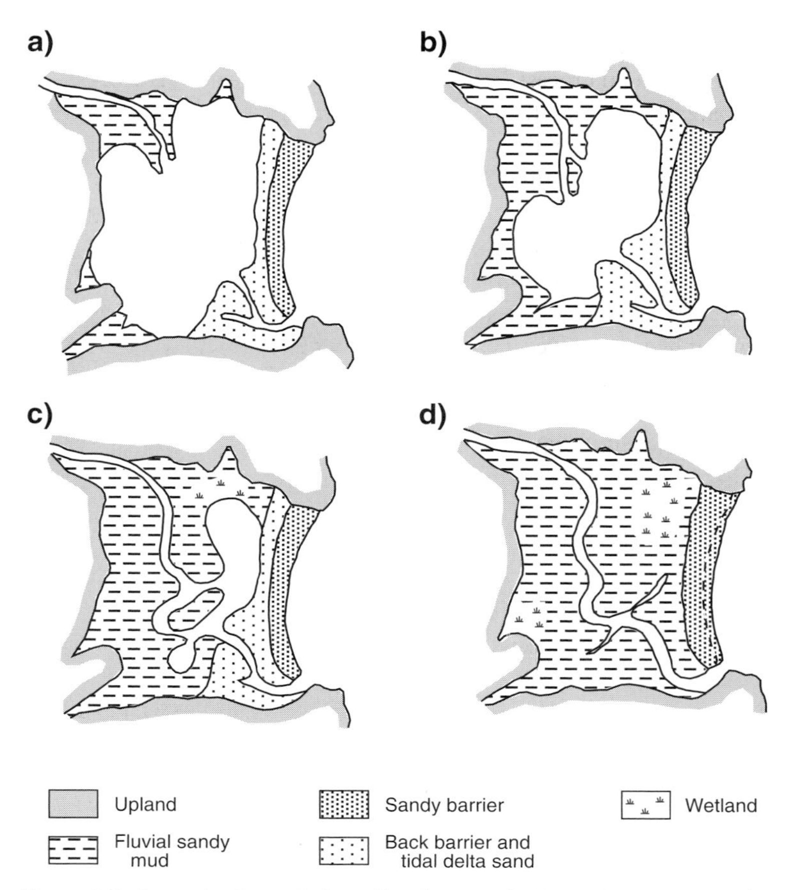

Figure 1.7. Stages in the evolution of barrier estuaries in southeastern Australia. Individual estuaries can be seen that are in each of these different stages (after Roy, 1984). See detailed discussion in Chapter 7.

1988) and Australia, where individuals, such as Trevor Langford-Smith and Joe Jennings, inspired coastal geomorphology research groups at the Coastal Studies Unit of the University of Sydney and at the Australian National University. Research studies combined an understanding of variation in process operation with geomorphological reconstructions of the evolution of landforms (McCann, 1977). These syntheses led to the development of a morphodynamic approach, emphasising the co-adjustment of form and process (Wright and Thom, 1977), and this morphodynamic emphasis has provided the focus for many of the studies described in this book.

Figure 1.7 is an example of an evolutionary model that developed from this collaborative research. It records the stages of Holocene evolution of barrier estuaries in southeastern Australia as a result of sediment input by the contributing river and landward reworking of sand by wave and tidal processes. Sharp gradients in transport potential lead to progressive infill of the muddy estuarine embayment until the river becomes channelised through wetlands and sediment bypasses the estuary and is carried directly to the sea (Roy, 1984; Roy *et al.*, 1994).

Sea-level change has also been shown to be a prominent control on the longer-term stratigraphy of coastal and nearshore sediments in large sedimentary basins as part of development of seismic and other geological studies, called sequence stratigraphy (Vail *et al.*, 1977). Sequence stratigraphy involves recognition of major depositional cycles (termed systems tracts) when the sea was low (lowstand systems tracts) and when it was high (highstand), as well as transgressive systems tracts when it was rising (Miall, 1996). The sequence stratigraphical approach of sedimentologists and the morphostratigraphic approach of coastal geomorphologists can be combined. For example, the classification of sedimentary environments in relation to a continuum of wave and tidal processes proposed by Boyd *et al.* (1992) provides a framework which is described in detail in later chapters (see Figure 7.5).

As a result of conceptual developments during the 20th century, coastal landforms were regarded as very dynamic, responding to frequently changing external factors (boundary conditions) including climate and sea level. This increasing body of coastal studies and literature made possible a series of global syntheses and comparisons of shoreline and coastal types (Bird, 1976; Davies, 1980; Bird and Schwartz, 1985; Kelletat, 1995). It became apparent that coastal sedimentary environments not only responded passively to external stimuli but are also capable of a series of internal morphodynamic adjustments (Chappell and Thom, 1986; Cowell and Thom, 1994).

In the 1990s, concerns about global change, particularly sea-level rise (Barth and Titus, 1984), have given a new impetus to the study of coasts, re-uniting historical and process geomorphology (Bird, 1993a). Major international initiatives, such as the International Geosphere Biosphere Program (IGBP), combine patterns of past behaviour with process studies to address issues relating to future coasts at scales relevant to planning and coastal management (see Chapter 10). In order to be able to extrapolate across scales in time or in space it is necessary to examine the space and time scales over which processes operate and coastal landforms evolve.

1.4 Temporal and spatial scales

The study of coasts is concerned with a range of scales in both space and time. In the case of space, the width of the coastal zone can vary substantially. In some cases it is necessary to include the continental shelf and much of the coastal catchment, for instance where sediment flux is traced from river catchment across a delta and down a submarine canyon. In other cases, such as high cliffs plunging into deep water, there may be little more than the cliff face itself and some salt-affected cliff-top vegetation communities, which are relevant to coastal studies.

The scale of study can also vary in relation to the lithological and sedimentary characteristics of the coast. Featureless cliffs can extend for hundreds of kilometres and change little in either time or space. On beaches, however, subtle sorting of grain size and microtopography of bars can result in variations along the shoreline over a few metres, which change over the progress of a tidal cycle or in response to storms or seasonal events.

There is a hierarchy of time scales relevant to coastal geomorphology. Geologists are generally concerned with events that occur over geological time. Rocky coastal topography can have originated from events that occurred millions of years ago, formed or shaped in the Tertiary or earlier. Other shoreline features have been shaped during the pronounced climate changes of the Quaternary (the past 2 million years, characterised by a sequence of glacial and interglacial phases, termed the Pleistocene, and the past 10000 years of postglacial, termed the Holocene). Engineers and coastal managers may be more concerned with providing an estimate of the amount of erosion by individual events through observation of processes of sediment movement on a beach during one storm or one season. Coastal geomorphology needs to study landforms at each of these scales; it may be appropriate to consider several time scales to understand adequately modern coastal management issues.

Geomorphological study of landscapes has recognised three time scales: geological time (often termed cyclic time, reflecting the long-term geographical cycle of W.M. Davis); graded time at which a river tends toward a dynamic equilibrium with landscape denudation; and steady-state time, a short-term perspective within which external factors (boundary conditions) are considered not to change (Schumm and Lichty, 1965). There have been attempts to extend these time scales to coastal systems. For instance, the adjustment of soft-rock cliffs composed of glacial deposits in eastern Britain has been considered within cyclic (several thousand years), graded (2–100 years), and steady-state (<1 year) scales (Cambers, 1976). It is convenient to recognise relative scales such as megascale, macroscale, mesoscale and microscale adjustments. However, these terms have been used to cover different scales for different coastal types. For instance, beach adjustment has been described at megascale (10–20 ka), macroscale (7 days – 6 months), and microscale

(10–30 min) (Schwartz, 1968). In the case of dunes, similar terms have been used to refer to different time scales, with macroscale (greater than decades), mesoscale (months–decades) and microscale (seconds–months) (Sherman, 1995).

In view of the coast's sensitivity to external factors and internal morphodynamic readjustments within the system, different rates of process operation occur on different types of coast. Cliffs generally respond to external factors more slowly than beaches. The long-term evolution of coral atolls, as envisaged by Darwin (see Figure 1.3), operates over geological time. An atoll, such as that shown in Figure 1.8, is a function of the rate at which corals grow and at which reef framework is built. Although sediment movement occurs in response to wave and tide processes, accretion of framework occurs over decades to centuries (see Chapter 5). Reef growth can occur over different time scales to the erosion of drumlin fields and the adjustment of sandy bars and spits, as described by Johnson in his study of the New England shoreline (see Figure 1.5). Modern process studies examine coastal change under typical conditions, but it may require study of past coastal changes to put those into a temporal or spatial perspective, or to provide insights into rates of operation of processes that cannot be observed directly, or occur infrequently.

Figure 1.8. Elizabeth Reef, the southernmost coral atoll, occurring in the southwest Pacific Ocean. The rim of the atoll is built of reef framework and is nearly continuous, but there are only two small sandy islands, termed cays (photograph M. Hallam, Environment Australia).

1.4.1 Hierarchy of time scales

There are advantages in viewing individual coasts over a range of different time scales. A convenient framework for examining time and space scales in coastal geomorphology has been described by Cowell and Thom (1994) and is shown in Figure 1.9. The smallest scale is an 'instantaneous' time scale over which the principles of fluid dynamics apply, governed by the laws of physics, but operating stochastically (randomly). At any point in time, such as during the passage of a wave, there is tight coupling between fluid flows and sediment transport brought about by these physical processes. It might seem possible to monitor in 'real-time' the details of sediment entrainment and the complexities of turbulent flow, including processes such as the deposition of individual bedform laminae, with a view to understanding the coastal system perfectly at this scale. Under fieldwork conditions, however, studies can rarely, if ever, be undertaken with this degree of exactitude, and the stochastic nature of the processes precludes such detailed knowledge. More controlled investigations may be possible in laboratory studies, but questions then arise as to how representative these experiments are of field conditions.

The 'event' time scale is concerned with recurrent sequences of processes, such as a tidal cycle with rising, slack and ebbing conditions, storm or flood events, and seasonal variations. At the event scale, 'instantaneous-scale' processes are averaged to gain an understanding

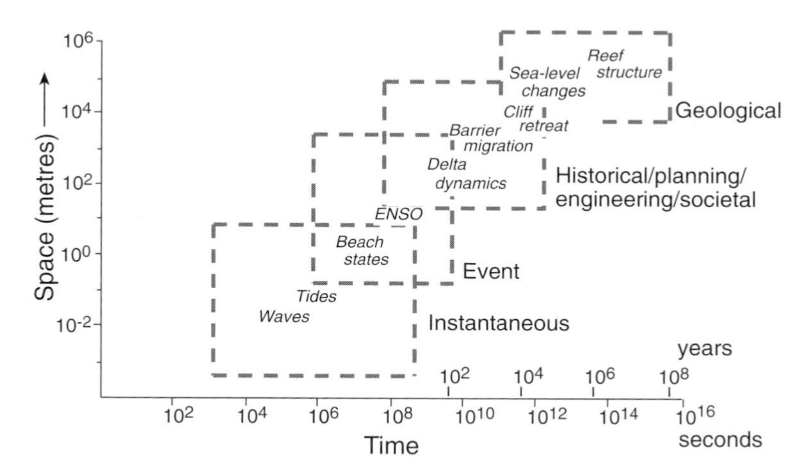

Figure 1.9. The representation of space and time scales appropriate for the study of coastal morphodynamics and the schematic representation of examples of coastal systems or patterns of variation and their general position in terms of space–time scales (based on Cowell and Thom, 1994).

of morphological response to single events or perturbations, such as a single storm. It is convenient to think of the coast in terms of the operation of 'normal conditions' (or 'non-events'), and 'events' which are generally higher-energy conditions. The coast appears to be more or less balanced during 'non-events', but adjusts to a perturbation or event. However, such a distinction is artificial; coastal landforms are the cumulative product of numerous events (and 'non-events').

Extreme events represent perturbations to the coastal system. There has been considerable geomorphological discussion as to the role of high-magnitude, low-frequency events as opposed to events of more regular occurrence (Wolman and Miller, 1960). Particularly large events may exceed thresholds that more regular processes cannot achieve. Storms move large boulders and floods cause delta distributaries to switch channel position. In some cases, particularly extreme events can have a disproportionate impact on the landscape. Volcanic eruptions, such as the 1883 explosive eruption of Krakatau, result in widespread deposition of volcanic products and reshape entire islands. Associated tsunami (tidal waves) may be experienced on coasts on the other sides of oceans and can have significant geomorphological impacts (Bryant, 2001). It is important to assess the relative significance of events of different magnitudes and frequencies in terms of their impact on particular coastal systems (see Chapter 9).

The mechanistic linkages between form and process which are incompletely understood at the instantaneous scale are generally scaled up in a deterministic way to describe morphological changes at the event scale. For instance, although the physics of sediment movement under waves cannot be fully determined, empirically derived relationships in terms of sediment transport enable some generalisation about changes in nearshore morphology. As an example, the pattern of beach erosion during a storm and the more gradual recovery and build up of the beach after the storm can be partially understood on the basis of these extrapolated generalisations, together with observational data from experience of past events (see Chapter 6).

The next level of time scale is termed 'engineering' or 'historical' by Cowell and Thom (1994), covering several decades and combining many 'events'. The term 'engineering' appears to presuppose that the coast has been or should be manipulated by engineers to allow human use. The term 'historical', in contrast, implies that concern is only with the past. On the contrary, the most important issues at this time scale relate to the future, and to the future coexistence of human activity within coastal systems. The terms 'societal' or 'planning' may be more appropriate in recognition that this time scale, of decades to centuries, is that over which humans have interacted, and will continue to interact, with the coast. This is a time scale that is difficult to study because it transcends the length of research projects, political commitment and, in some cases, human generations. More particularly, it is problematical because non-linearities which are present in the system but discounted by empirical generalisations at the event scale make behaviour at the societal scale chaotic. It has been the focus of recent computer simulation modelling, particularly that called 'large-scale coastal behaviour' (List, 1993; List and Terwindt, 1995). Although this often remains small-scale in comparison with much historical geomorphology, it offers an approach within which the probable patterns of landform change can be considered, and is examined in more detail in Chapter 9.

Geological time scales are longer term (Cowell and Thom, 1994). Studies of the way that coasts have evolved over geological time, particularly the Quaternary, are primarily descriptive, reconstructed on the basis of paleoenvironmental evidence. It is never possible to prove historical causes, because they occurred in the past and cannot be revisited. Instead, the course of inferred change in conditions can be tentatively determined if several pieces of evidence are consistent with one of several competing hypotheses (see Chapter 9). Advances in paleoecological and geochronological (dating) techniques have considerably improved the resolution possible. Nevertheless, it remains the case that in many coastal depositional environments, and still more so in relation to erosional landforms, the past is incompletely preserved.

The scale at which a system is studied should be relevant to the question that is to be answered. It will comprise the entire coast in one case but focus on a particular beach in another. The spatial scale over which it is appropriate to examine a coastal landform is generally related to the temporal scale of study (Figure 1.9). For instance, the study of whether a beach is accreted or eroded, focused over the event time scale, generally looks at a local scale of the individual beach, a part of a beach, or some closely related beaches. A study of delta distributary dynamics or delta lobe changes will need to consider a longer time scale of decades to millennia, generally focusing on a larger area. A study of reef structure, such as evolution of the atoll shown in Figure 1.8, involves regional and Quaternary or longer time scales, although aspects of surface morphology such as erosion and transport of sand or individual patch reefs focus on part of a reef at shorter time scales.

The chapters of this book illustrate these points. However, it is convenient to summarise the concept of time and space scale in terms of a coastal embayment like that illustrated in Figures 1.1 and 1.2. Wave interactions and the pattern of refraction influence sand on the shoreface at 'instantaneous' time scales, requiring study at representative locations throughout the bay to understand completely. The event scale of study is appropriate to understand the way that the beach responds to an individual storm, or a seasonal change in wave characteristics. At the engineering/historical or societal/planning scale, the variation in beach shape and volume over decades is studied in order to understand the movements and net accretion of bars and the shifting of beach berms and exchanges between beach and dune or estuary and near-shore. Over the geological time scale, the long-term build up of sand within the barrier is studied, including how this has changed with the rise of sea level from the last glaciation, and the relationship with sand bodies deposited during previous sea-level conditions.

1.5 Coastal systems

Studying the coast as a system provides a framework within which interactions between the many elements of coastal geomorphology can be considered (Haslett, 2000). It enables the interrelationships between different components of complex landforms and the processes that shape them to be expressed, based on general systems theory (von Bertalanffy, 1952, 1968). Elements of the real world (termed entities) are described by a series of variables (also termed attributes or parameters) such as the size and shape of landforms, or wave and tide characteristics.

Factors outside the system being studied (external factors) are termed independent variables; they are sometimes referred to as exogenous, extrinsic or allogenic. They represent boundary conditions, or forcing factors, because when they alter they can cause changes within the system. Variables within the system (endogenous or intrinsic) are termed dependent variables and they vary in response to changes in independent variables (boundary conditions). There are also adjustments between dependent variables which come about as a result of intrinsic changes within the system, without a change in boundary conditions. Recognition of these internal morphodynamic adjustments is important because they might otherwise be interpreted as a response to an external factor (such as human impact) whereas they may be natural (see Chapters 9 and 10).

1.5.1 Landform morphology

A system is in a particular state at any time, that state being defined (or measured) by its variables. The scale at which it is most appropriate to define a system depends on the questions that are to be addressed. For instance, the whole earth can be thought of as a system, and the set of variables that apply to it at a particular time characterise the 'state' that it is in. There have been two distinct states that have been of particular significance in the Quaternary, glaciation and interglaciation. Orbital variables associated with the eccentricity and obliquity of orbit and precession of equinoxes (discussed in more detail in Chapter 2) are the external or independent variables that appear to determine in which of these states the earth is. Several variables characterise the earth in the glaciated state, most obviously lower temperature and the greater extent of the ice sheets, but also the lower sea level. These global states clearly have implications for the much smaller-scale study of individual coastlines; for instance, very different coastal environments in terms of position and supply of sediment were associated with low sea level during glaciation (termed lowstand systems tracts in the terminology of sequence stratigraphy). Glaciation and interglaciation can be expressed as distinct states.

An example at a smaller scale involves two contrasting states based on climatic variables within the Pacific Basin, El Niño and La Niña (Philander, 1990). The El Niño–Southern Oscillation (ENSO) phenomenon can be characterised by several variables (sea surface temperature, sea level across the Pacific), but is generally described by the Southern Oscillation index, a measure of the pressure difference between Darwin and Tahiti (see Chapter 3). The factors that drive the fluctuation between these states remain the subject of research, but oceanographic circulation and atmospheric variables associated with wind fields are important. ENSO is expressed to varying extents in the coastal zones of much of the Pacific region, including patterns of beach variation in eastern Australia and of cliff retreat in California.

It is often convenient to define states associated with the endpoints of a continuum; they may be the most distinct and contrasting states (e.g. glaciation and interglaciation; El Niño and La Niña). However, it is often the case that the system spends much time in less clearly defined intermediate states between those extremes. The Pacific Ocean is not in either El Niño or La Niña for much of the time, but adopts an intermediate (and transitional) state and also responds to the longer-term Pacific Decadal Oscillation that appears to influence ENSO (Mantua *et al.*, 1997). Similarly, in terms of glacials and interglacials, it will be shown in Chapter 2 that the earth only spent about 10% of the Quaternary in either of these polarised states, and that for the majority of time it was in some intermediate state. In terms of sea level, highstands and lowstands occurred for relatively short periods and, for much of the time, sea level was intermediate, as when it was rising – a transgressive state. The transition from rising sea level (transgressive) to relatively stable sea level (stillstand or highstand) has been a particularly significant event in terms of the geomorphology of coasts globally. Many coasts are still

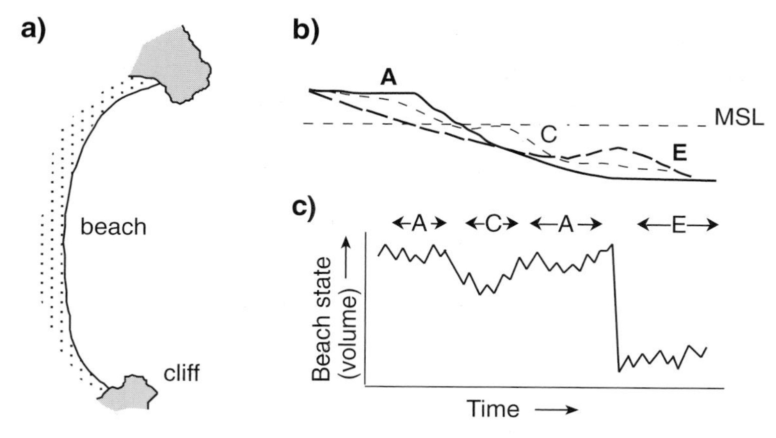

Figure 1.10. The morphological state of a simple beach. (a) The beach in planform. (b) The beach states, shown schematically in profile, consist of an accreted (A) and an eroded (E) state. It will be seen in Chapter 6 that these correspond to reflective and dissipative beaches respectively. An intermediate state (C) is also shown. (c) The beach fluctuates between these states as shown schematically; it can be represented using some surrogate variable, such as beach volume. (MSL, mean sea level).

adjusting to this transition; its importance will become clearer in the context of individual coastal systems in later chapters.

Morphology is an especially appropriate measure of the state of a coastal system. Figure 1.10 portrays a simple wave-dominated beach profile. It contains two distinct endpoint morphological states, accreted (A) for the steep beach and eroded (E) for the flatter beach. On mid-latitude northern hemisphere beaches, the accreted profile (constructional) has been termed the summer profile, because it is common during summer months and the eroded profile (destructional) has been called the winter profile, because it occurs in winter, as a result of storms (Lewis, 1931; King, 1972). However, it will be shown in Chapter 6 that such profiles are common extremes, and robust morphodynamic models of beach state have been developed independently of seasonal variation (e.g. Wright and Short, 1984). The extremes have been termed reflective (predominantly reflecting wave energy) and dissipative (generally dissipating wave energy).

In Figure 1.10, the extremes and an intermediate state have been described by their overall morphology. It would also be possible to characterise them by some quantitative variable, such as beach slope or beach volume (both being greater in the case of A). As with the variation between glacials and interglacials, or the ENSO phenomenon in the Pacific, the state of a beach varies within a continuum. It need not reach the extreme endpoints, but varies over time in response to driving forces.

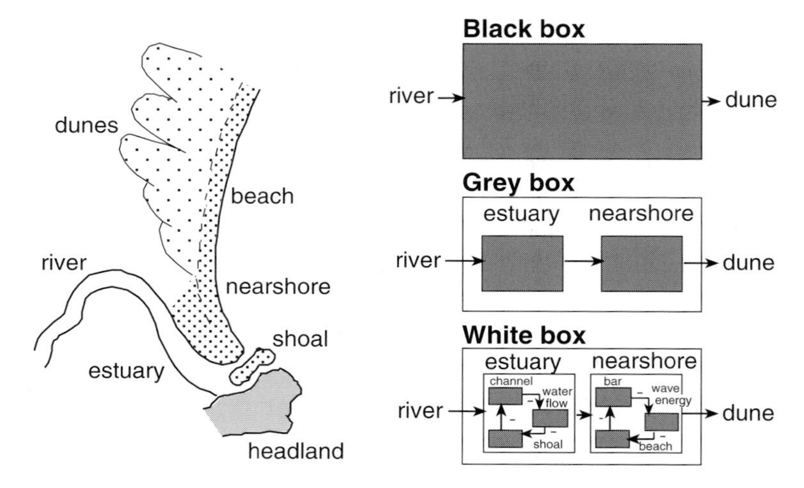

Figure 1.11. A coastal system, like that shown in Figures 1.1 and 1.2. The links between river, estuary, shoal, nearshore, beach and dune can be envisaged in terms of various systems approaches. A white-box, grey-box and black-box approach are shown schematically.

1.5.2 System interrelationships, feedback and equilibrium

A system is composed of many subsystems (Chorley and Kennedy, 1971; Haslett, 2000). Figure 1.11 illustrates schematically the interrelationships between several coastal subsystems, including a river draining into an estuary like that schematised in Figure 1.1 or illustrated in Figure 1.2, with an estuarine shoal at its mouth linked to nearshore, beach and dune subsystems. In the most detailed studies, the complexities of the interactions between variables must be known, but at higher levels of generalisation it may suffice to integrate these smaller-scale relationships.

The linkages within a system can be represented in different ways. For instance, in Section 1.2 it was shown that the different landforms are spatially related in terms of process operation, whereas in Figure 1.11 beach and dune subsystem variables are related to other factors such as the size of the contributing river catchment. A cascading system is one characterised by flows and a process–response system is one in which a given input generates a particular output as a result of a specific process or transfer function. The relationship is generally assumed to be linear (although it can be logarithmic, exponential, or otherwise transformed); for example, the volume of sediment added to the beach or shoal subsystems might be related to the total annual river discharge. The flux of sediment can be represented schematically as in Figure 1.11, with sediment storage in the river, estuary, beach or dune. For instance, there might be a maximum storage volume for the shoal, and the extent to which sediment is transferred from estuary to nearshore relates to the extent to which the storage volume of the shoal is full.

System and subsystem components

The spatial scale of study will vary according to the boundaries that are defined around a coastal system. It may be appropriate to study the river mouth shoal in isolation, or the river mouth and the beach within a single system, concentrating on the interchange of sediments between the two. In either case, the supply of sediment down the river would be an external variable (an independent boundary condition); in a broader study encompassing the entire catchment, river sediment load would be an internal variable.

The detail in which the subsystems are defined can vary according to level of generalisation. A black-box approach is where no components of the subsystems are identified (Figure 1.11). In the case of the simplified example, increased river sediment supply might result in additional dune building. Where all components of the subsystems are identified, this is called a white-box approach; a grey-box approach is intermediate. The intricacies of interrelationships are closely related to scale. A broad understanding of sediment movement may be sufficient to reconstruct the evolution of coastal stratigraphy on a geological time scale without recourse to the details of fluid dynamic relationships in the nearshore. However, on the event scale, the calculation of sediment volumes moved will need much closer assessment of these smaller-scale variabilities. Scientific studies which attempt to offer comprehensive explanations of one element of a subsystem (white box), reducing it to all its minor interrelationships, are called reductionist. Those studies which concentrate on the system as a whole (black box), believing that the entire system functions in a way which is more than a sum of its component parts, are called holistic.

There are a number of cases in the coastal zone where discrete biotic states can be recognised. In the case of coral reefs, discussed in Chapter 5, distinctive zones occur, and zonation is often dynamic as distribution of organisms changes with time. Co-adjustment of physical and ecological systems, termed biogeomorphology (Viles, 1988), is observed on salt-marsh and mangrove muddy coasts and on sand dunes. An ecosystem is an example of a cascading system, with energy transfer through the trophic levels.

It is sediment transport that provides the coupling between landform and process, with sediment properties influencing which processes occur, and sediment availability controlling the extent to which potential transport is realised (Cowell and Thom, 1994). The physical rules of fluid dynamics which govern sediment transport can be extrapolated upwards from instantaneous time to event time (de Vriend *et al.*, 1993a,b), but they are usually not well enough understood to extrapolate to societal or geological time scales. Sedimentation can be examined by inference from stratigraphy, scaling down from geological time. However, many of the physical principles are based on uniform uncohesive sediment, whereas in field situations minor effects, such as microbial influence on sediment properties or binding of sediment by plants, introduce complexities and non-linearities of sediment dynamics that can have far-reaching consequences.

Initially, landscapes were treated as closed systems without interactions with external variables, and tending towards maximum entropy (Langbein and Leopold, 1966). However, there are clearly external controls, such as climate, and it is more appropriate to treat landscapes as open systems (Haslett, 2000). An open system has inputs and outputs of mass and energy. A coastal embayment may be a closed system within which sediment movement and mass budget can be determined, termed a sediment cell (see Chapter 6). However, more often it is an open system, receiving inputs (of water, energy and sediment), losing outputs, and involving a series of processes concerned with changes in matter and energy. The estuary and beach system shown in Figure 1.11 functions as an open system with sediment input from the river, as well as other inputs and losses offshore and alongshore.

Feedback

Systems are more complex than these simple relationships suggest because they are often non-linear. Linear relationships are those in which an increase in one variable results in a proportional response of another variable. Many coastal systems do not function linearly, but exhibit positive or negative feedback mechanisms. Negative feedback tends to maintain balance in a system; positive feedback reinforces change until a threshold is reached beyond which a different type of response occurs. If natural non-linear systems are described using linear generalisations, their behaviour may be approximated, but will incorporate a degree of distortion as a result (Bennett and Chorley, 1978).

In the case of the estuary-mouth shoal in Figure 1.11, the size of the shoal has a negative relationship to estuary channel cross-sectional area (the larger the shoal, the smaller the channel area). This in turn can accelerate flow, increasing sediment entrainment. Deposition on the shoal increases flow velocity leading to erosion; conversely, if the shoal gets smaller, channel area increases and velocity decreases, promoting sedimentation. This is an example of negative feedback that tends to

Figure 1.12. A tidal creek system in northern Australia. There is a large tidal range and the different levels reached by successive tides can be seen around the creek marked by salt crystallisation on the mudflat and wetting zones. Mangroves occur along those parts of the creek that are inundated frequently. Sedimentation rate is likely to slow with distance from the creek because the frequency of inundation is less; an example of negative feedback.

maintain the shoal close to an equilibrium size that reflects optimum hydrodynamic and sedimentary conditions.

Negative feedback is common in coastal systems and maintains an equilibrium through self-regulation (also termed self-maintenance and homeostasis). For instance, as an intertidal mudflat surface builds up vertically through sedimentation, substrate elevation is raised with respect to the tides, and it is inundated less frequently, slowing the rate of sediment accumulation. A tidal creek system in northern Australia is shown in Figure 1.12. It is clear that tides of different heights inundate the mudflats to differing extents, and that these inundation levels influence the distribution of mangroves and other salt-tolerant vegetation. Those areas that are infrequently inundated will undergo much slower accretion (vertical accumulation) of sediment because they rarely receive tidal waters.

Positive feedback augments systems and leads to self-organisation, or to the system crossing a threshold. Development of a dune blowout represents an example of positive feedback where, after initial breaching of a foredune during one storm event, an area of vulnerability can grow into a dune blowout, because the wind exploits the absence of vegetation, smothering adjacent vegetation with the mobile sand.

The state of an estuary, expressed by its shape, determines the velocity of flow and hence relates to rates of erosion at one point and deposition at another. However, morphological changes as a result of deposition and erosion alter the shape and hence, in turn, modify the flows and rates of process operation. Coastal morphodynamics provides many examples of such feedback interrelationships in the mutual co-adjustment of form and process (Wright and Thom, 1977).

Equilibrium

Equilibrium is a time-dependent concept, generally inferred from some state, such as morphology, and viewed at a particular time scale. Whether or not a coastal system adopts equilibrium is an important component of coastal morphodynamics. It is generally interpreted in one of several ways: stability of form, correlation between form and process, balance over time in sediment budgets (where inputs and outputs are balanced), or relatively even distribution of energy over space (Chorley *et al.*, 1984). The concept is simple in principle, but often hard to define in practice, and is examined in greater detail in Chapter 9. It is rarely the case that all inputs and outputs involved in a coastal system can be defined, let alone measured.

Where external boundary conditions do not change, they are termed stationary and a coast may adopt a steady-state equilibrium. More often there are variations in boundary conditions that cause the system to change (termed non-stationarity), and coastal landforms adjust in an effort to maintain a dynamic equilibrium (see Chapter 9). The detailed response of coastal systems is likely to be unpredictable because of the wide range of variables and the non-linear interrelationships between them. However, in common with other chaotic systems, coastal systems tend to vary within a specific domain with certain key states being recurrent.

The morphology of the coast often exhibits geological inheritance. Previous coastal landforms survive, and features derived from antecedent conditions persist. For instance, cliffs retain the evidence of previous shorelines (see Chapter 4), and beach shape, although responding to incident wave processes (the contemporaneous wave conditions), is also dependent on the former shape of the beach that was in the process of adjusting to previous (antecedent) wave conditions (see Chapter 6). The recurrence of particular morphology is also typical of coasts because coastal landforms are often resistant, built of rocks or sediments that require a threshold to be exceeded before they change. Different coastal systems exhibit different response times to external, or internal, changes or perturbations. These issues are examined in detail in Chapter 9, after the various coastal types have been discussed.

1.6 Human impact on the coast

Almost all coastal systems have been subject to influence by human actions, and it is becoming increasingly inappropriate to consider the geomorphology of coasts without factoring in human influence. Although the coastal geomorphologist has an interest in the natural dynamics of coasts, there are many pressing reasons to study coasts to assist coastal management. The aims of coastal management are to facilitate human use of the coastal zone, minimising or otherwise influencing impacts of such human use, and protecting human interests at the coast from natural and human-related processes.

Human impact has been both exploitative, where resources have been extracted and land reclaimed, and manipulative, where some degree of control of the direction of process operation has been sought. Often inadvertent interference with coastal sediment transport has had far-reaching implications. Indirect interruption of sediment input occurs where the rate of sediment yield from catchments has been altered. Landuse change often leads to significant increase in sediment supplied to the coast, resulting in progradation at river mouths, augmentation of littoral drift, infilling of estuaries, construction of updrift strandplains and dune building. It can lead to silting of harbours and the need for dredging or training works. For example, the Huanghe

(Yellow) River flows through thick loess deposits in central China and its sediment load of over 1×10^9 t a^{-1} of mainly fine silt has been significantly increased by forest clearing in the catchment, resulting in a rapidly changing delta (Milliman *et al.*, 1989). In contrast, damming of river systems has decreased sediment loads on other coastlines with consequent problems of erosion. This is demonstrated by the Nile River where an active delta, adjusted to high seasonal sediment load, is now depleted as a result of construction of the Aswan Dam, and the delta is undergoing erosion (Stanley and Warne, 1993b). It will be shown in Chapter 7 that erosion and deposition are components of all deltaic regions, but the magnitude in these extreme cases is greatly modified by human activity.

It is not possible to predict in detail how coasts will respond in future but the morphodynamic approaches described in this book can reduce the uncertainty about what adjustments will occur. The challenge for coastal geomorphologists is successfully to disentangle secular climate change, natural variability generated by external forces, internal dynamics, the role of extreme events, and human impact on the coast (Slaymaker and Spencer, 1998). This challenge is examined in more detail in Chapter 10.

1.7 Summary and outline of following chapters

The coast, or the coastal zone over which land and sea meet, is dynamic, has a history of change, and will continue to change in the future. Coastal geomorphology offers a way of studying coastal landforms as a system, composed of various subsystems, with interactions through the exchange of energy and mass. Morphology of the coast is one expression of the state that the system is in. It responds to independent boundary conditions, such as wave energy or sea level. Feedback between components of the system constrains the shape of the coast, although complex adjustments to external boundary conditions and non-linearities in interactions mean that coastal landforms are unlikely to achieve or maintain equilibrium. Morphodynamics is the co-adjustment of form and process, and morphodynamic models based on these principles provide a framework within which to view coastal change.

In this chapter, the concepts of equilibrium and feedback have been introduced. These will become increasingly apparent as coastal landforms are described, over a range of temporal and spatial scales. Although it may not be possible to predict specific outcomes, conceptual models of morphodynamic behaviour provide some indication of the probable response of coasts to changes in boundary conditions and perturbations. The most significant boundary conditions are the nature of the geological and topographical setting of the coast and the nature and abundance of material (lithology and sediments) and sea-level history. These are examined in Chapter 2. The frequency and magnitude of processes operating on the coast are examined in Chapter 3.

Coasts are set within a series of spatial and temporal scales, from hard rock coasts which change slowly, responding to events on a geological time scale, especially sea-level changes over the Quaternary, to beaches which respond to wave conditions on a much shorter time scale, and muddy low-energy environments in which tidal and biotic factors are important. These coastal settings are examined in turn: rocky coasts (Chapter 4); coral reefs (Chapter 5); beach and barrier coasts (Chapter 6); deltas and estuaries (Chapter 7); and muddy coasts (Chapter 8). The morphodynamics of coastal systems and the development of models are re-examined in Chapter 9.

Sustainable human use of the coast and successful coastal management depend on understanding the processes operating and the landforms resulting in each of these coastal systems. Chapter 10 examines human influence on the coast and the way that coasts may change in the future in the context of coastal morphodynamics and modelling.

Chapter 2
Geological setting and materials

> Understanding comes only when, standing on a beach, we can sense the long rhythms of earth and sea that sculptured its land forms and produced the rock and sand of which it is composed. (Carson, 1955)

The coastal system operates within a series of boundary conditions. Geological setting and materials exert primary controls on the range of landforms that can be formed by coastal processes. The plate-tectonic setting is important in terms of vertical and horizontal movements over the long-term evolution of a particular coast. Climate affects the rate at which terrestrial and oceanographic processes operate, and is linked to sea-level change. Relative changes of sea level constrain shoreline position and geomorphological process operation, and influence how the coast is shaped at a range of scales. Lithology and the materials from which the coast is composed are also important.

This chapter examines factors that constitute boundary conditions, adopting a geological time-scale perspective, and then examining conditions at increasingly smaller temporal and spatial scales. Plate-tectonic setting influences long-term vertical and horizontal displacements of land relative to sea level. Fluctuations in the level of the sea influence the way in which the land is abraded by marine processes. In some cases, these can be deciphered from terraces or shelves or from sequences of sediments that have been deposited. The significance of sea-level changes at geological time scales, especially through the Quaternary, has been widely accepted. The issue of sea-level change at shorter time scales, as a result of anthropogenic actions, is of growing concern in relation to global environmental change that is anticipated to result from the enhanced greenhouse effect.

Vertical displacement between land and sea involves three components. First, there are tectonic movements, the result of compressive stresses and strain in or between lithospheric plates. Second, there are sea-level variations superimposed on these, particularly as a result of ocean-volume changes associated with extension and contraction of ice sheets during the Quaternary. Third, there are flexural adjustments of the crust, termed isostasy, in response to imposition and redistribution of water, ice, rock and sediment. Although all three components operate simultaneously, they become most apparent at successively shorter time scales. Plate tectonics operate and are expressed at geological time scales of millions of years. Sea-level fluctuations have dominated the Quaternary period; oscillations with a periodicity of thousands of years (typically 10^5 years) were associated with the ice ages. Isostatic flexure is apparent on a shorter time scale, particularly over the past few thousand years during which the ocean volume has varied little. These processes are examined in this order in the first part of this chapter.

If the relative positions of land and sea remain unchanged, coasts evolve constrained primarily by the nature and supply of material and the oceanographic processes that shape them. The role of coastal lithology (the type of rock) and sediments (fragmented material derived primarily from rocks) is examined in the second part of this chapter; it is sediment transfer in response to processes that shapes coastal morphodynamics. A major distinction is drawn between clastic sediment and carbonate sediment. As sea level changes, antecedent landforms also exert an influence on the morphology of the coast. Paleoenvironmental reconstruction provides insights into past coastal morphology and the way that coastal landforms have changed.

2.1 Historical perspective

It is important to recognise that the first attempts to explain the way in which coasts evolved preceded either an understanding of plate tectonics or the way in which sea level had changed. Although evidence for relative movements of land and sea was recognised, it was not until the second half of the 20th century that plate tectonics or the details of Quaternary sea-level fluctuations became widely accepted. The oceans and the coasts of the world provided much of the evidence leading to acceptance of these ideas.

2.1.1 Global tectonics

The ruins of the so-called Temple of Serapis (in the Phlegraean Fields caldera near Naples, Italy) were described by Charles Lyell (1832) who inferred that the level of the sea had changed through time. The columns of this Roman market show a zone of borings by the marine mollusc, *Lithophaga*, indicating that they have been partially submerged since construction in the 2nd century BC. Geomorphological, archaeological and historical evidence is now interpreted to indicate alternating subsidence and uplift in response to local volcanic activity (Pirazzoli, 1996). This, and other European evidence, such as widespread submerged forests around the coast of southern Britain, indicated changes in the relationship between land and sea (Huxley, 1878).

Exploration of more remote parts of the world indicated additional evidence for vertical movements of land and sea. Darwin (1842) witnessed rapid uplift on the coast of South America and described marine fossils at several hundred metres above sea level indicating episodic uplift. The contrast between the relatively straight and mountainous Pacific coast of the Americas and the Atlantic coast, with its more irregular shoreline, was recognised by Suess (1888), and became the basis for a classification of coasts, described below, that was extended in a plate-tectonic context by Inman and Nordstrom (1971). Sea level was recognised as the lower limit to which rivers could erode their valleys, termed 'base level' by Powell (1875). There was considerable debate about the extent to which rivers grade to base level (Gilbert, 1877; Mackin, 1948). The view proposed by Davis (1909), and widely accepted, involved a geographical cycle in which landscapes underwent uplift and planation (see Chapter 1).

Continental drift was proposed by Alfred Wegener (1915), who presented a strong case for a large former continental mass that he called Pangaea based on the shape and continuity of geological features across continents. The concept that continents have undergone horizontal movement was initially ridiculed by geologists, in contrast to indisputable evidence that there had been changes in the vertical relationship between land and sea. It became increasingly clear, however, that not only was the distribution of mountain ranges on continents consistent with plate-tectonic theory, but bathymetry and early gravity readings at plate boundaries from the ocean also provided support (Vening Meinesz *et al.*, 1934).

The distribution of volcanoes, mapped in association with coral reefs, had provided the first clues to plate tectonics (Darwin, 1842). Mid-ocean volcanoes are almost exclusively basaltic, whereas those around the margin of the oceans are more acidic. An 'andesite line' occurs around the margin of the Pacific Plate, connecting island arcs (for example the Kermadec Islands, Tonga, Solomon Islands, Marianas Islands, Kurile Islands, and Aleutian Islands), characterised by continental-type andesitic volcanoes (Hess, 1948).

Evidence for a time sequence of erosion along chains of mid-oceanic basaltic islands was recognised by James Dana (1849), based on the eroded surfaces of Maui and Molokai, and the deeply dissected surfaces of Oahu and Kauai within the Hawaiian Islands. Assuming the

Table 2.1. Alternative names for glacial, interglacial and postglacial stages based upon regional compilations and deep-sea core marine oxygen isotope boundaries

General description	Britain	NW Europe	Alps	North America	Marine oxygen-isotope stage	A (ka)	B (ka)	C (ka)
Holocene	Flandrian	–	–	–	1	13–0	12–0	12–0
Last Glacial	Devensian	Weichselian	Würm	Wisconsin	2–4	75–13	71–12	74–12
Last Interglacial	Ipswichian	Eemian	–	Sangamon	5	128–75	128–71	130–74
Penultimate Glacial	Wolstonian	Saalian	Riss	Illinoian	6	195–128	186–128	190–130
Penultimate Interglacial	Hoxnian	Holsteinian	–	Yarmouth	7	251–195	245–186	244–190
	Anglian	Elsterian	Mindel	Kansan	8	297–251	303–245	
	Cromerian	Cromerian	–	Aftonian	9	347–297	339–303	
	Beestonian	Menapian	Günz	Nebraskan	10	367–347	362–339	

Notes:
A, Shackleton and Opdyke, 1973; B, Imbrie *et al.*, 1984; C, Martinson *et al.*, 1987.

same intensity of erosion, Wentworth (1927) interpreted the progressive stages of extinction of active volcanism and subsequent erosion to indicate that greater time had been available on islands to the northwest of the chain. Subsequently, not only were further chains recognised in which there was a similar progression (Chubb, 1957), but the ability to date the time of volcanic activity by potassium/argon dating (K/Ar) has indicated that islands were active sequentially in many of these chains. A linear chain of islands appears to have developed as a consequence of gradual movement of the plate over a 'hot spot' (Wilson, 1963; McDougall, 1964; McDougall and Duncan, 1980).

2.1.2 Quaternary variations

The concept of several successive glaciations during the Quaternary was initially based on moraines in the Alps where four ice advances called Günz, Mindel, Riss and Würm were identified (Penck and Brückner, 1909). Detailed sequences of glacial advances were later identified from North America and elsewhere. The Günz–Mindel–Riss–Würm sequence was believed to be global, but because there were problems of dating and correlation, independent chronologies had to be derived from each location. The Würm glaciation was known as the Wisconsin in North America, the Weichselian in northern Europe, and the Devensian in Britain (see Table 2.1).

It is now realised that the Quaternary Period (approximately the past 2 Ma) has been characterised by many alternate extensions and contractions of the polar ice sheets. These appear to be correlated with temporal changes in insolation that result from variations in the earth's orbit, as proposed by the Serbian mathematician Milutin Milankovitch (1941). The quasi-periodic variations in eccentricity of the earth's orbit (~100 ka), obliquity (~41 ka), and precession (~21.7 ka) combine in a manner which appears to drive the gradual build up of polar ice to a state of glaciation and its more rapid termination by ice melt (Hays *et al.*, 1976). There have been a series of interglacials recurring at a periodicity of around 100–150 ka, although it remains unclear why the 100 ka periodicity has been so dominant over the past 900 ka (Imbrie *et al.*, 1992, 1993).

Expansion of ice sheets resulted in lowering of sea level, and Reginald Daly attributed significant geomorphological roles to lowered sea level, particularly in relation to coral reefs (Daly, 1910, 1915). He also recognised the important role of isostasy, flexure of the crust as a result of the extension of ice sheets over the northern continents depressing the land, with rebound on its removal (Daly, 1934). Sea-level stages that were presumed contemporary with interglacials between each of the Günz–Mindel–Riss–Würm glaciations were recognised from the Mediterranean, and it was presumed that the oldest were the highest as a result of falling sea level throughout the Quaternary (Zeuner, 1945).

The oceans have provided further insights into climate change during the Quaternary. Although mapping and discrimination of moraines indicating ice extent on land demonstrates recurrent glaciation on Northern Hemisphere landmasses, the most continuous record of periodic advance and retreat of polar ice sheets and associated fall and rise of sea level is preserved in the ocean. The record on the land is very patchily preserved because it has been reworked by successive glacial advances, whereas the record from selected deep-sea cores is one of almost continuous accumulation (Bowen, 1978). Oxygen-isotope analyses of foraminiferal tests that have accumulated on the ocean floor provide an insight into ice- and ocean-volume variations (Shackleton and Opdyke, 1973).

The record of sea-level changes has been increasingly refined for the Quaternary. Postglacial sea-level rise, reconstructed for coasts around the world, was initially compiled into one 'eustatic' curve, until it became apparent that the pattern of sea-level change differed from place to place (Fairbridge, 1961). These patterns are examined below (Section 2.3.2) because the relative adjustments between land and sea provide the setting in which the coastal system operates. These relationships are examined at a range of scales, from global variation over geological time scales relevant to plate tectonics, to future ongoing sea-level rise over coming decades which it has been suggested may result from global warming (Gutenberg, 1941).

2.2 Plate-tectonic setting

The earth's surface is covered by 10 large lithospheric plates, and several smaller ones, which gradually move over the denser asthenosphere (Figure 2.1). The plates are rigid lithosphere, comprising upper mantle and crust, and move tangential to the surface of the earth (Cox and Hart, 1986). The crust beneath the ocean, the oceanic crust, is composed of sima (basaltic material in which silica and magnesium are relatively abundant, density 2.9–3.0), whereas the continents are continental crust, composed of sial (lighter rocks, in which silica and aluminium are relatively abundant, including granites with densities of 2.7–2.8). Both vertical and horizontal movements of crust are possible, and are most pronounced where the continental margin abuts the plate boundary.

New oceanic crust is formed by passive upwelling of basaltic magma at 'mid-ocean ridges'. In some cases, as in the Atlantic, these are actually in the middle of the ocean; in other cases they are offset, for instance the East Pacific Rise, the ridge from which the Pacific plate is spreading westwards and the Nazca plate is spreading eastwards (Figure 2.1). Mid-ocean ridges are broad topographic features (1–2 km high, and >1000 km wide), marking divergence of plates that can spread at rates of up to 170 mm a^{-1}. The spreading centres are characterised by tensional cracks at which earthquakes are experienced, confined to shallow depths in the weak crust. Rarely is land associated with the mid-ocean

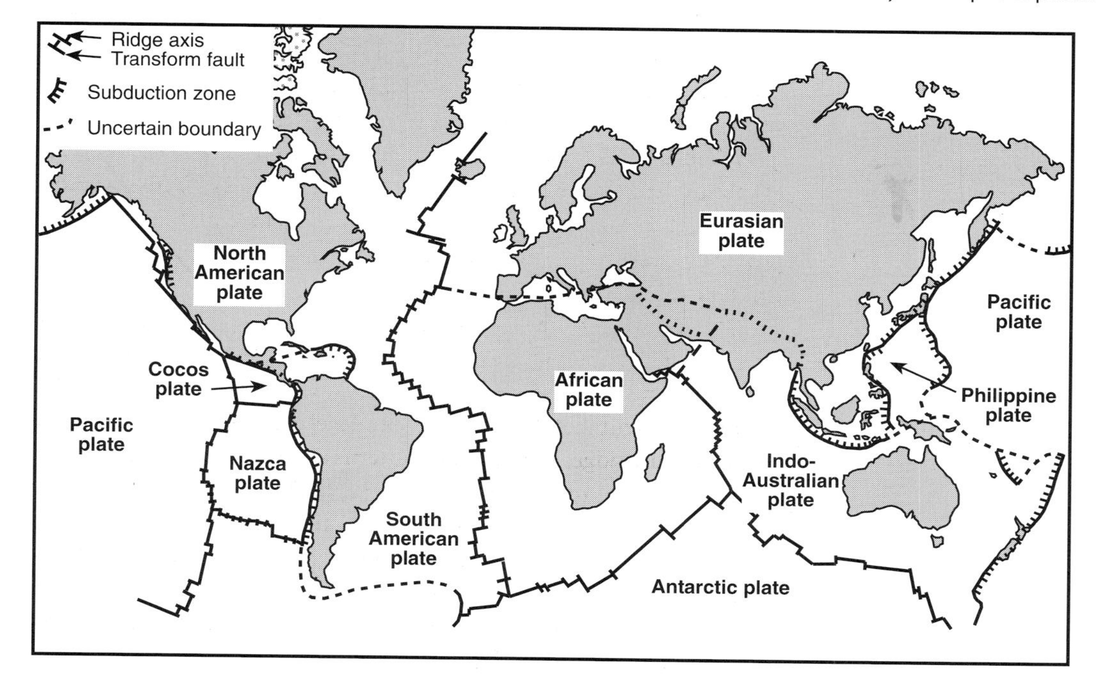

Figure 2.1. Distribution of the major lithospheric plates.

ridge, although in the case of Iceland a hot spot (shown by the Iceland–Faeroe ridge) coincides with the mid-ocean ridge, and volcanic activity occurs, as evidenced by the abrupt eruption of the island of Surtsey in 1963. St Paul and Amsterdam Islands in the southeast Indian Ocean are associated with mid-ocean ridges, and Reunion may have initially formed on such a ridge (Nunn, 1994).

The ocean floor is composed of a series of parallel zones of alternate reversed and normal polarity on opposing flanks of the mid-ocean ridge, within which magnetic minerals demonstrate polarity typical of that operating at the time of formation (Vine, 2001). These alternate zones enable dating of the sea floor (Figure 2.2). As the oceanic crust moves away from the spreading centre it cools and contracts and, as a result, the depth to the sea floor increases, reflecting the relative rate of spreading. There is a well-established relationship between ocean depth and time, with depth increasing with the square root of time, relative to the height of the mid-ocean ridge at which it formed. The thickness of the crust varies from 10–20 km at spreading centres, to perhaps 100 km beneath older crust (>100 Ma) over which ocean depths are generally around 4000 m.

Subduction zones occur where the plate collides with another plate and one is subducted beneath the other. Oceanic crust is generally subducted for hundreds of kilometres beneath less-dense continental crust, being carried at an angle of 30–45° in what is called the Benioff zone. This zone is marked by earthquakes, which occur at progressively greater depths with distance, and the overriding plate is generally characterised by volcanic activity. Subduction zones are associated with deep ocean trenches; depths in the Puerto Rico Trench in the Atlantic Ocean exceed 9000 m, whereas depths in the Kuril, Tonga, Mindinao and Marianas Trenches in the Pacific Ocean exceed 10 000 m.

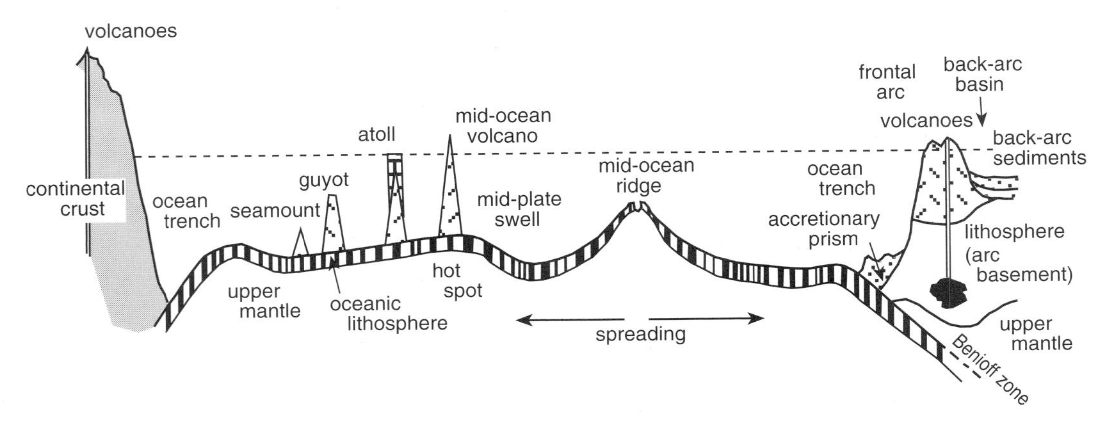

Figure 2.2. Schematic cross-section of oceanic plate showing nature of collision at plate margins. Note mirror-image pattern of polarity reversals, indicated by black and white shading, with distance from mid-ocean ridge.

It is convenient to consider the implications of plate-tectonic setting for continental and island coasts separately.

2.2.1 Continental coasts

Plate-tectonic setting enables a broad classification of continental coasts proposed by Inman and Nordstrom (1971). At the largest scale, coasts are classified as collision coasts or trailing-edge coasts, depending on whether they are active plate margins or passive mid-plate locations. This follows the division into Pacific and Atlantic coasts, shown to be especially appropriate to the west and east coasts of the United States by Suess (1888), and extended using the terms accordant and discordant (see Chapter 4).

Collision coasts are also termed leading-edge coasts; they are typically rocky, and steep, with only a narrow continental shelf. They are coasts of convergence where plates meet, and can be divided into continental collision coasts, such as western North and South America, and island arc collision coasts, such as the Indonesian island arc. The compressive forces associated with convergent plate margins result in uplift of the land, as demonstrated by mountain ranges such as the Andes and Himalayas, together with gliding, elastic rebound, faulting, folding and tilting of crustal blocks. Reheating of the subducted oceanic crust causes magma to liquefy at depth and steep explosive volcanoes composed of viscous lavas occur along the overriding plate, for instance the Cascade Mountains of Oregon and Mount Fuji in Japan.

The coasts that characterise convergent plate margins are often rocky with a series of uplifted terraces dissected by steep, short and actively downcutting rivers, particularly clearly seen on the arid coast of Chile (Figure 2.3). The sediment that these rivers bring to the coast is either carried into deeper water, or reworked by vigorous wave activity that results in longshore drift into small bays within an otherwise relatively straight coast. Sudden seismic events cause vertical displacement and rapid drowning or emergence that can be preserved in the stratigraphy of sediments. For example, the earthquake experienced in Napier in eastern North Island, New Zealand, in 1931 caused local uplift of over 1 m. An earthquake in Prince William Sound, Alaska, in 1964 caused a tsunami (see Chapter 3) that deposited a distinct layer of sediment in many coastal embayments along the western North American coast (Atwater and Moore, 1992).

Trailing-edge coasts are low-gradient, with a broad continental shelf and typically thick sedimentary accumulation. Trailing-edge coasts have been divided into three types that reflect the period of time over which marine planation and the accumulation of sediments have been occurring. Neo-trailing coasts, such as those flanking an active rift (spreading centre), for example in the Red Sea, retain topography reflecting their rock structure. Afro-trailing coasts, for example southern Africa, are composed of cliffs that are bevelled and they have greater sediment accumulation than neo-trailing coasts. On Amero-trailing coasts, such as the eastern coasts of much of North and South America, there are large accumulations of deltaic sediment.

Sedimentary basins have accumulated particularly on subsiding trailing-edge coasts as a result of continued fluvial input of sediment to the shoreline and its redistribution across the shallow continental shelf. Relative sea-level variations have been major controls on sediment deposition over the millions of years that these have accumulated. Sequence stratigraphy is based upon interpretation of deep seismic reflection profiles of sedimentary basins together with borehole data (Vail *et al.*, 1977; van Wagoner *et al.*, 1990). It recognises sedimentary strata and the bounding surfaces between them (Vail, 1992; Miall, 1996). Limited age control may be possible based on planktonic foraminifera, but otherwise sediment sequences are placed in relative time sequences based on regional unconformities and coastal onlap of marine sediments over those unconformities (marine sediment is said to onlap where it overlies terrestrial sediment indicating relative sea-level rise). Figure 2.4 shows schematically how this might occur. The marine and coastal deposits

Figure 2.3. A Late Pleistocene marine terrace at Cobija Village, 130 km north of Antofagasta on the coast of Chile. This arid coast occurs on a plate margin at the convergence of the South American and Nazca plates, and has been rapidly uplifted. The terrace at about 15 m above present sea level is inferred to be of Last Interglacial age (photograph C.V. Murray-Wallace).

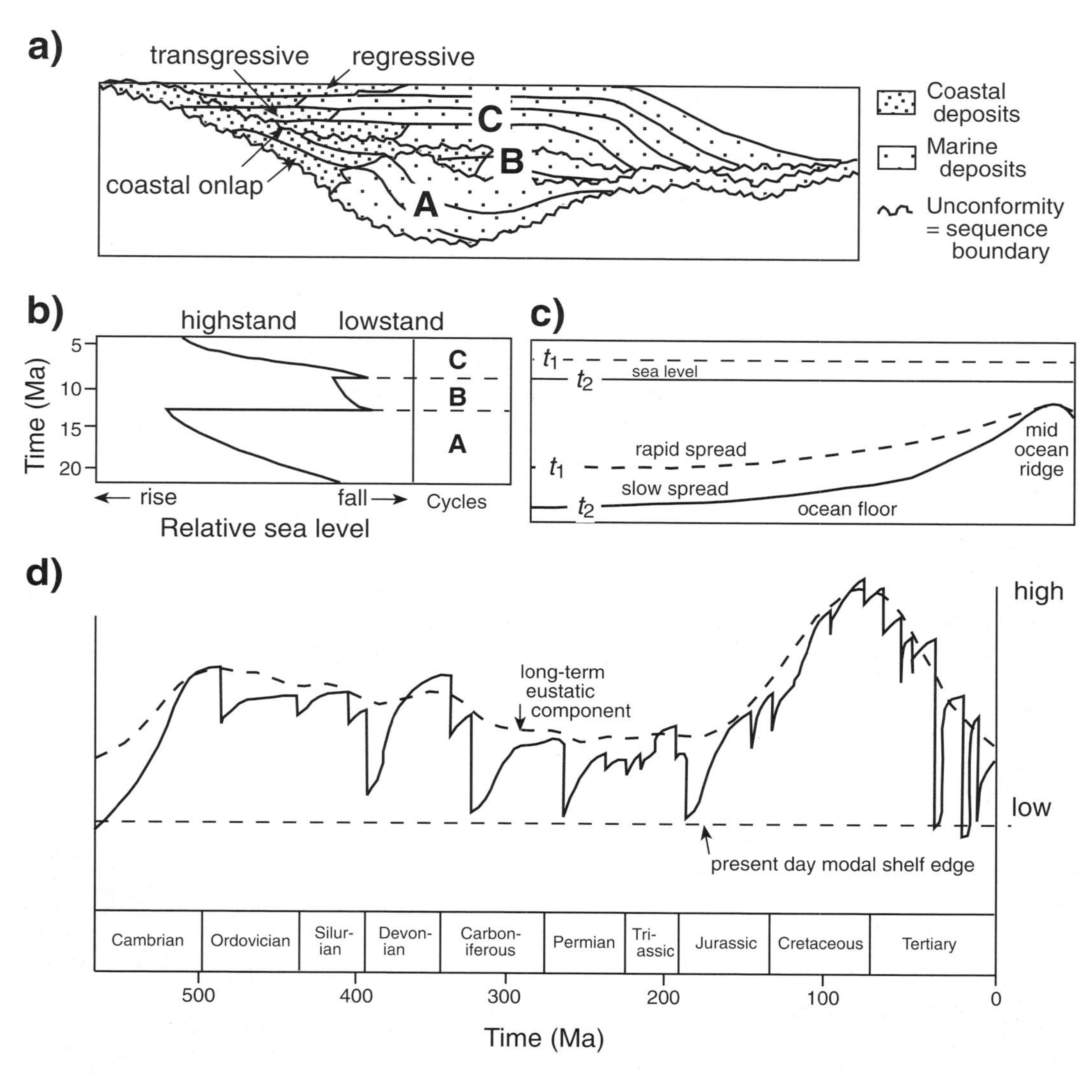

Figure 2.4. The principles of sequence stratigraphy. (a) A sequence of sedimentary strata containing coastal and marine deposits and unconformities. Where marine sediments overlie an unconformity coastal onlap indicates transgression; where coastal sediments grade laterally into marine sediments regression is indicated (based on Vail and Hardenbol, 1979). (b) Compilation of sequences into relative sea-level histories based on relative rise (onlap) and fall. (c) Schematic cross-section of ocean floor showing rapid spreading (t_1) and slower spreading (t_2) which alters the shape of the ocean floor and consequently the height of sea level. (d) Inferred sea-level history for the Phanerozoic (based on Vail *et al.*, 1977).

within sequence A have been deposited (onlap) over the previous surface and this must have occurred during relative sea-level rise (relative sea-level rise involves a rise of sea relative to the land and could be due to subsidence of the land). The unconformity across the upper surface of sequence A (a ravinement surface) indicates a relative fall of sea level, whereas sequence B onlaps over these sediments indicating another relative sea-level rise but of smaller magnitude. Sequence B is topped by an unconformity which is onlapped by sequence C. The relative sea-level sequence is shown schematically in Figure 2.4b.

The procedure recognises the importance of the availability of space in which sediment can be deposited, termed 'accommodation space' (Jervey, 1988). This refers particularly to the space available below sea level (often referred to as base level) within which marine sediments can accumulate, and demonstrates the significance of sea level, which is considered in the next section, as a boundary condition. The response of sedimentation to sea level and subsidence, which combine to represent relative sea-level change, generates groups of sedimentary units, called systems tracts in the sequence stratigraphy literature (Vail *et al.*, 1977; Posamentier and Vail, 1988).

The classification of continental coasts by Inman and Nordstrom (1971) included several other types of coast. They identified marginal sea coasts which occur on continental margins where the plate boundary is marked by an island arc. Examples of these include the Gulf of Mexico and the South China Sea. Other distinctive coasts occur along transform faults. Transform faults occur where the crest of the mid-ocean ridge is offset, and are major fractures along which there is movement of adjacent sections of the plate, but in opposite directions (see Figure 2.1). These are characterised by shallow earthquakes. The San Andreas fault, which runs through San Francisco south to the Gulf of California is one such ridge–ridge transform (from a spreading ridge off the coast of northern California to another ridge in the Gulf of California).

Classification into collision (active) and trailing-edge (passive) coasts is particularly appropriate for North and South America, each of which has distinct Pacific and Atlantic coasts. The Pacific coast consists of a high mountain range, characterised by active uplift and volcanism, and provides a substantial source of sediment which is carried by large rivers from the continental interior and delivered to the Atlantic coast. This contrasts with the African continent, which has large rivers but does not supply a large sediment load down those rivers (see Chapter 7). Inman and Nordstrom's classification is less appropriate for other coasts, such as the southern coast of Australia which is steep with a narrow continental shelf but is not an active plate margin, and 'it seems clear that the important distinction is between collision coasts lying along a zone of convergence and all the others which do not' (Davies, 1980, p. 12).

2.2.2 Island coasts

Islands can also be divided into those at active plate margins and those in passive mid-plate settings. They show many similarities in terms of the processes that are operating to collision and trailing-edge continental coasts respectively, but oceanic crust responds in a simpler way than continental crust.

An island arc occurs where two oceanic plates converge. Most island arcs are conspicuously arcuate, although the Tonga–Kermadec island arc in the central Pacific north of New Zealand is linear. It has a forearc or frontal arc that is non-volcanic, then a forearc basin behind which is the volcanic arc (Tofua ridge). Elsewhere, as in the Japanese island arc, there is not a prominent frontal arc. There are situations in the Philippines where two island arcs appear to be colliding. The inactive former topographic features associated with transform faults are known as fracture zones, characterised by undersea cliffs and high ridges. Rarely are islands found on such features, although Cikobia in Fiji could be on a transverse plate boundary of the Fiji Fracture Zone, and Clipperton Atoll could be on a fracture zone (Nunn, 1994). Islands within an island arc are often characterised by relatively rapid uplift or subsidence. In tropical seas the coastline usually supports coral growth and many islands consist of a sequence of uplifted reef terraces, which provide insights into past sea-level history (see Section 2.3.2).

Mid-plate islands are usually either volcanic, or consist of a volcanic foundation subsequently veneered by reef limestone (see Chapter 5). Mid-ocean volcanism builds massive shield volcanoes predominantly of basalt upwelled through structural weaknesses in the crust. Low-angle fluid lavas form into ropey, billowy 'pahoehoe', or angular, clinker-like 'aa', as can be seen on active volcanoes such as Kilauea on Hawaii (Macdonald *et al.*, 1970). There are many weaknesses in the oceanic crust and volcanic edifices are very abundant in the oceans (Ollier, 1988). The vast majority of mid-plate volcanoes do not reach the sea surface but remain under water as seamounts. Some of those that do reach the surface are subsequently truncated by marine abrasion and form a 'guyot'. Island or seamount building continues for a relatively short period in geological time, perhaps 1 Ma or less.

Island chains are interpreted to have developed where oceanic crust has migrated over a persistent or intermittently active volcanic 'hot spot', and islands track the motion of the plate (Clouard and

Bonneville, 2001). Submerged seamounts and guyots often continue the trend of the chain; for instance, the Emperor seamount chain extends northwest beyond the Hawaiian Islands and indicates older episodes of volcanic activity (Batiza, 2001). However, there are chains that do not show the anticipated progression of ages and groups, such as the Cape Verde, Samoan and Galapagos Islands where the direction and rate of migration appear reversed or islands are not neatly aligned (Nunn, 1994).

The oceanic crust gradually deepens with age, but it also adjusts to any load placed on it. If the crust is rigid enough, which generally means if it is old enough, a volcanic island extruded onto the crust will be regionally supported. If the crust is not rigid enough it will be locally supported. Menard has used the analogy of a skater on ice. Regional compensation occurs when the ice around the skater is depressed slightly and that further away is slightly bowed upwards; a skater who breaks through the ice will sink until floating buoyantly, and in this case is locally supported (Menard, 1986). A volcanic island, such as Iceland or the Galapagos, which represents a considerable load on young plate is initially locally supported because the crust is not rigid enough to support the mass. The majority of islands, however, are regionally supported and do not break through the crust. Massive islands like Hawaii, which rises from water depths of around 4000 m, to heights in excess of 9000 m from the sea bed, depress the plate at the point where they impose a load. This forms a moat near the island and produces an arch 150–330 km from the load, depending on the age and rigidity of the crust (McNutt and Menard, 1978). This flexure of the lithosphere is called volcano-isostasy and is one example of a series of isostatic adjustments that the crust makes to a change of load. Whereas the Hawaiian example has received considerable attention (Walcott, 1970; Watts and Ribe, 1984), there are also impressive cases associated with the young volcanics of Tahiti in the Society Islands and Rarotonga in the Cook Islands (Lambeck and Nakiboglu, 1980, 1981; Spencer *et al.*, 1987). One of the advantages of the latter is the presence of reefal limestones on islands on the arch which provide the opportunity to assess the extent of differential movement experienced (Woodroffe *et al.*, 1991a).

2.3 Sea-level variations

Movements of land or sea level lead to complex relative sea-level changes that are an important boundary condition for coastal systems. Sea level reflects variations in ocean volume, which are usually considered global, termed 'eustatic' by Suess (1888), because the oceans are interconnected. The truly eustatic component of sea level is influenced by the addition of new water (juvenile water) from within the earth's rocks, which occurs at a negligible rate, and the volume of water that is locked up in, and subsequently released from, the ice sheets. A minor component of this water may be stored at other places in the hydrological cycle (such as in lakes or groundwater).

In many cases, the sea-level curves are based on fossil shoreline evidence. The aggregation of data from many sites enables the pattern of sea-level change to be determined which in turn can provide information on the way in which a shoreline responds to a particular pattern of sea-level change. It is increasingly appreciated that minor vertical adjustments of lithosphere are occurring ubiquitously and there is no overall eustatic sea-level curve that can be derived; eustasy (at any scale) is a quaint oversimplification (Sloss, 1991). Nevertheless, it is useful to separate sea-level influences into those that affect the absolute volume of water in the seas (the eustatic component) and those which result in a redistribution of the water already in the ocean basin (Church and Gregory, 2001).

2.3.1 Pre-Quaternary ocean volume variations

Sequence stratigraphy and the interpretation of sedimentary basins indicate that relative sea level changes over geological time scales. The extent to which relative sea-level change reflects regional sea-level curves or basin subsidence history has been subject to debate. The variations in relative sea-level history that are generated over millions of years result primarily from alterations in the ocean basin volume that can occur as a result of differences in the spreading rate from mid-ocean ridges (Pitman, 1978). If the plate cools and migrates slowly, then deeper depths occur close to the ridge than in the case where the plate migrates rapidly (Figure 2.4c). The slower the spreading rate the larger the ocean volume and the lower sea level is (because crust near the ridge will have had longer to cool and so the sea floor will have contracted to a lower level and the ocean will be deeper). The pattern of sea-level variation through the Tertiary is poorly known; it seems likely that it has declined from a level 270–350 m above present over the past 100 Ma (Kominz, 2001). Compilation of a series of relative sea-level curves from seismic data (many of them in the files of the petroleum industry) indicates a long-term pattern of sea-level variation (Haq *et al.*, 1987; Hallam and Wignall, 1999), indicated as the long-term eustatic component in Figure 2.4d.

Studies of subsurface relationships over geological time scales have been developed for particular geological purposes (especially hydrocarbon exploration) and offer little insight into the present geomorphology

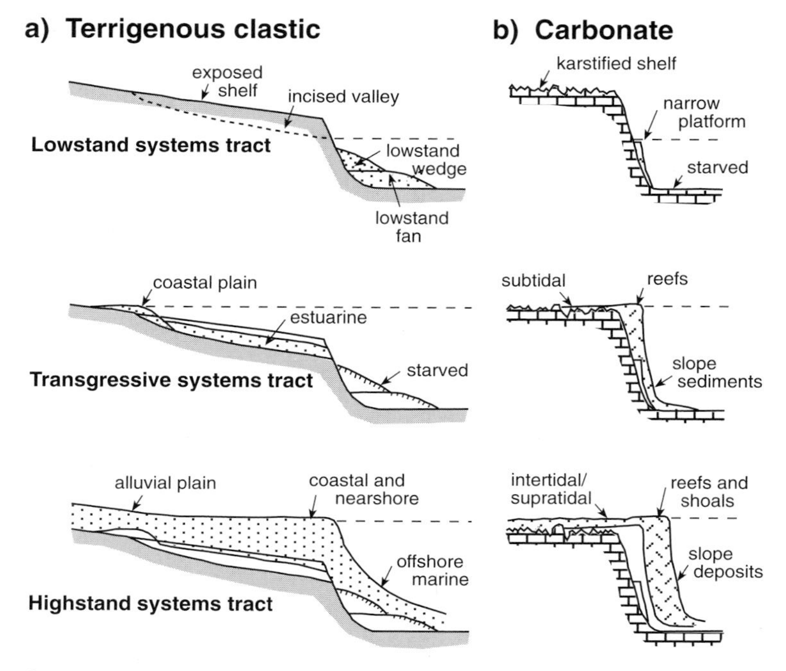

Figure 2.5. Contrasts in sedimentary sequences in terrigenous clastic and carbonate shelf-slope systems, showing lowstand systems tract, transgressive systems tract and highstand systems tract (based on James and Kendall, 1992).

of the coast on these passive margins. However, the systems tract approach does offer a useful framework to consider how a coast responds to relative sea-level changes. These relationships are explored in Figure 2.5 in the context of the two most significant sedimentary environments, terrigenous clastic (siliceous sediments derived primarily from weathering of rocks on land) and carbonate (calcareous sediments derived primarily from biological production of shallow-water marine organisms). The significance and distribution of these types of sediment province are examined in Section 2.4 below.

In terrestrial clastic settings it is generally interpreted that a river valley incises into the exposed continental shelf during early stages of sea-level fall (lowstand) with widespread fluvial deposition forming a lowstand wedge and lowstand fan off the shelf margin (Figure 2.5). During the subsequent sea-level rise (termed a transgression) the transgressive systems tract consists of estuarine and coastal plain sediments deposited over (onlapping) the previously exposed valley surface. When the sea is stable at its peak (highstand) the space in which the coastal highstand systems tract can be deposited is limited and the shoreline builds seaward. A reduction in accommodation space leads to progradation, and the shoreline is said to be regressive.

Sequence stratigraphy implies that siliciclastic depositional systems shed most sediment into deep basins during lowstands of the sea, at which time rivers deposit directly onto the upper continental shelf and shelf sediments of the preceding highstand are eroded (Figure 2.5). By contrast, on carbonate platforms, particularly those with reefs around their rim, most sediment is shed during highstand, at which time there are extensive carbonate producing shallow-water environments (James and Kendall, 1992; Schlager *et al.*, 1994; see also Chapter 5).

The sequence stratigraphy approach is a large-scale overview of the sedimentary responses during periods when the sea is low, high, rising or falling. In later chapters in this book, the more detailed geomorphology of landforms, sediments and erosion surfaces is examined. Coastal geomorphology often uses morphostratigraphy to determine these relationships in a very similar, but smaller-scale, approach to that used in sequence stratigraphy.

Late Quaternary geomorphological reconstructions of coastal environments could be used to examine to what extent the relationships postulated by sequence stratigraphy have occurred during the most recent sea-level cycle. Where this has been attempted, there have been some interesting refinements to the sequence stratigraphic approach. For instance, it appears that not all rivers follow the self-cannibalisation cut-and-fill sequence of incision during lowstands and subsequent infill during transgressive and highstand systems tracts (Galloway, 1989). In addition, rivers may adjust their channel pattern, rather than incise, to reflect altered conditions (Schumm, 1993; Wright and Marriott, 1993; Wescott, 1993). Some incision of the lower Mississippi River and other rivers draining into the Gulf of Mexico has been shown (Thomas and Anderson, 1992). The significance of variations in sediment supply has been emphasised on the Atlantic continental shelf of the United States (Swift and Thorne, 1991; Thorne and Swift, 1991a). Incision and mass sedimentation did not correspond to the typical terrigenous clastic model when sea level was low in the region of the Great Barrier Reef off northeastern Australia (Woolfe *et al.*, 1998a; Dunbar *et al.*, 2000).

2.3.2 Quaternary sea-level variations

The pattern of ice–ocean volume changes over the Quaternary can be most effectively reconstructed using oxygen-isotope analyses from deep-sea cores through sediments that have been accumulating on the floor of the ocean. Oxygen-isotope analyses on foraminifera from deep-sea cores were initially undertaken to use temperature-based fractionation, with more of the lighter ^{16}O isotope being incorporated into carbonate skeletons (relative to ^{18}O) under warmer temperatures, as a

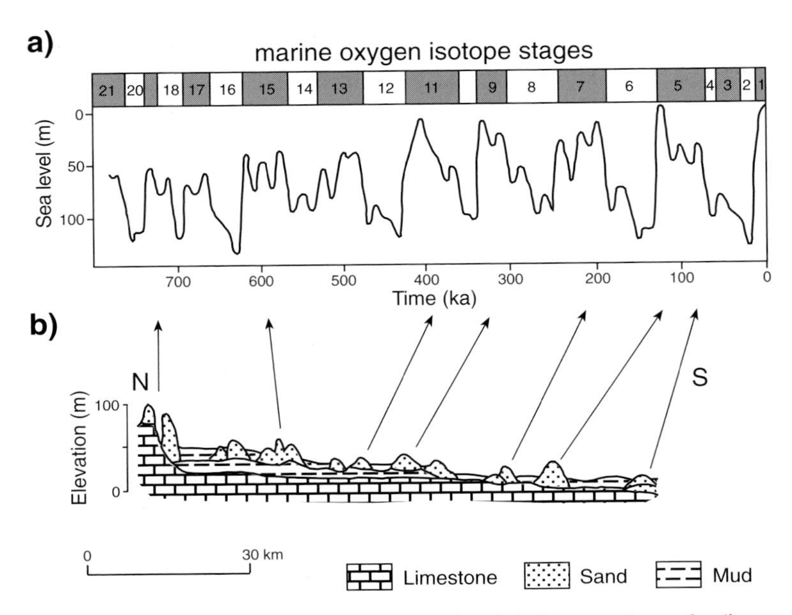

Figure 2.6. Successive highstands of sea level and their expression as fossil shorelines. (a) Sea-level variation over the past 800 ka determined from oxygen-isotope analyses on planktonic foraminifera from five deep-sea cores (based on Imbrie *et al.*, 1984). (b) A sequence of dune-topped beach ridges from Naracoorte in the north to Robe in the south, on the gradually uplifting coastal plain of southeastern South Australia, and the correlation of ages with particular highstands (based on Huntley *et al.*, 1993; Murray-Wallace *et al.*, 1998).

paleo-thermometer. In fact the isotope ratio in the foraminifera also depends on the ratio in the ocean itself, and this has varied considerably. During ice build up, the lighter ^{16}O is preferentially evaporated from the oceans and subsequently precipitated in the snow that contributes to the ice sheets. The ocean, therefore, tends to have a relatively higher ratio of ^{18}O during glaciations, and this effect accentuates, and largely masks, the direct temperature effect. Of the 1.7‰ change in foraminifera $\delta^{18}O$ between glacial and interglacial extremes, about 1.3‰ is considered to relate to ice volume change (Shackleton, 1987).

Figure 2.6a shows a generalised record of sea-level change for the past 800000 years (800 ka) based on oxygen-isotope analyses of planktonic foraminifera from five deep-sea cores (Imbrie *et al.*, 1984). The time scale, back to the Brunhes/Matuyama magnetic reversal, has been orbitally tuned; marine oxygen-isotope stage numbers refer to successive highstands (odd numbers, except stage 3) and lowstands (see Table 2.1). The level of the sea approximately coincided with present sea level during recurrent interglacials. The variations in sea level have resulted in substantial movements of the shoreline. Fossil shorelines relating to individual highstands are sometimes preserved in areas where sediment is abundant and erosion is slow. Figure 2.6b shows a cross-section of a sequence of quartz and calcareous sandy ridges, interfingered with estuarine and lagoonal muds, running parallel to the shore for more than 300 km in southeastern South Australia (Cook *et al.*, 1977). The sequence between Robe and Naracoorte comprises beach ridges topped by dune deposits, preserved because of gradual uplift of the coast and calcrete cementation. The age of the ridges has been constrained using thermoluminescence dating and whole-rock aminostratigraphy and can be seen to relate to sea-level highstands, although several of the ridges are composite in age (Huntley *et al.*, 1993; Murray-Wallace *et al.*, 2001a).

Reef terraces

Late Quaternary sea-level variations have been deciphered by comparison of deep-sea core isotope records and fossil shorelines. Recognition of former shorelines from rapidly uplifting sites where flights of terraces are preserved has been combined with observations of the height of former interglacial shorelines in areas considered stable. Veeh (1966) showed that the Last Interglacial shoreline (marine oxygen-isotope substage 5e, dated at 120 000 years BP) occurred at around 6 m on several mid-plate oceanic islands (such as Oahu in the Hawaiian Islands, and Mangaia in the Cook Islands) and on stable continental sites (such as along the coast of Western Australia). Slightly warmer conditions than present are indicated by the more southerly extension of coral on the western and eastern coasts of Australia during the Last Interglacial (Marshall and Thom, 1976; Szabo, 1979). Further dating from the Bahamas (Chen *et al.*, 1991), Seychelles (Israelson and Wohlfarth, 1999), Hawaii (Ku *et al.*, 1990), and Western Australia (Stirling *et al.*, 1995, 1998) has provided evidence that sea level at the peak of the Last Interglacial reached 2–10 m above present.

Oceanic islands are not necessarily stable. As indicated above, atolls are generally subsiding, whereas other islands can show flexural responses (see Section 2.2.2) to plate subduction or volcanic loading (i.e. Oahu is on the arch adjacent to the young volcanic island of Hawaii; Mangaia is on the arch around Rarotonga) or to hydro-isostasy (Nakada, 1986). Lambeck and Nakada (1992) investigated the possibility that the sea may have been at the same level as present, and produced modelling results reflecting the difference between Atlantic and Pacific sea-level curves. The discrepancy between the elevation of the shoreline formed during the peak of the Last Interglacial and the modern shoreline can be used to give a first-order approximation of whether the coast is undergoing long-term rise or fall.

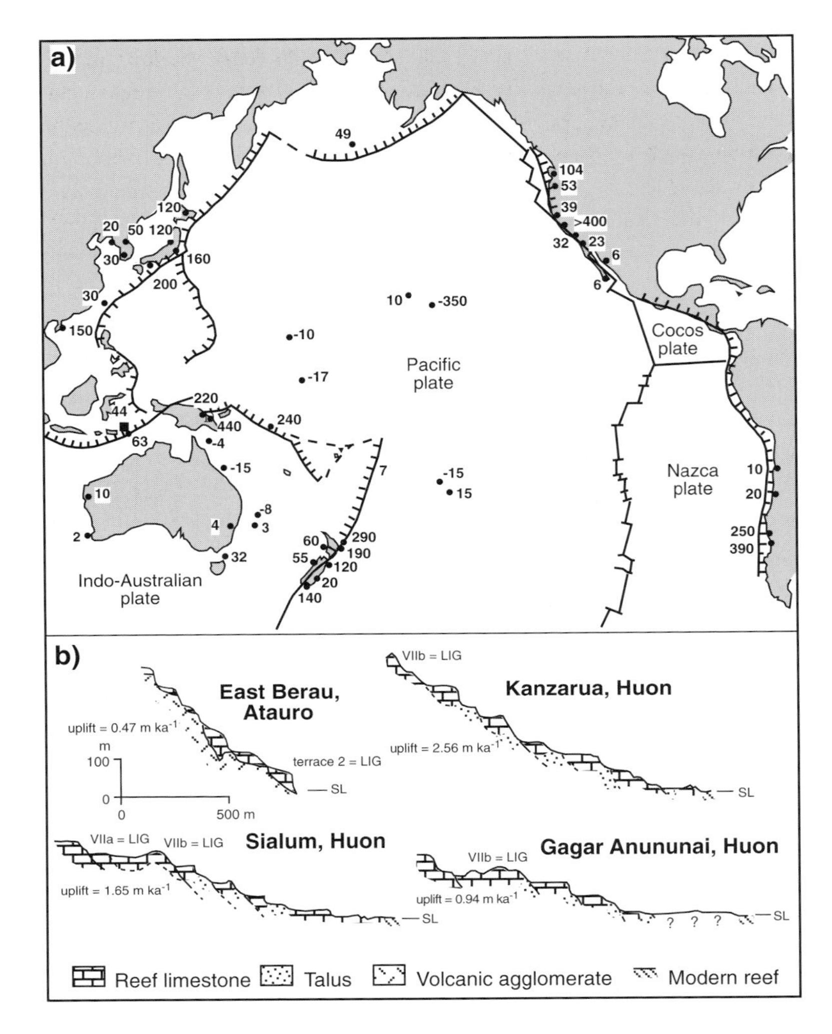

Figure 2.7. Late Quaternary shorelines in the Pacific. (a) Height at which Last Interglacial terrace is found around the Pacific (based on Ota and Kaizuka, 1991). At plate margins the terrace is generally raised above the level at which it formed. Although vertical movements are less marked in mid plate, there is still variability. A Last Interglacial terrace is found well below sea level on the Big Island of Hawaii (isostatically submerging), at shallow depths on atolls (gradually submerging), and raised above sea level on Makatea islands (undergoing lithospheric flexure). (b) Coral terraces from the rapidly uplifting coast of Huon Peninsula (triangle on (a)), indicating the sequence of terraces with more terraces on those coasts which uplift more rapidly, i.e. Kanzarua section (based on Aharon and Chappell, 1986). Section from Atauro (square on (a)) shown for comparison to illustrate sequence of higher terraces (based on Chappell and Veeh, 1978). LIG, Last Interglacial; SL, sea level.

Stable regions, paradoxically, are not as suitable for sea-level reconstructions over the past 120000 years as uplifting shores, because evidence can be eroded, reworked or buried. Stable regions can indicate the relative heights of several successive interglacials, for example on Bermuda (Hearty and Vacher, 1994), Grand Cayman (Vézina *et al.*, 1999) and Bahamas (Kindler and Hearty, 2000). Collision coasts that are being rapidly uplifted contain a sequence of former shorelines that are spread out and which can be deciphered if the terraces can be dated and the rate of uplift estimated. Initially dating of raised reefs on Barbados in the West Indies provided evidence for several sea-level highstands (Broecker *et al.*, 1968); two terraces associated with interstadials (substages 5a and 5c) are found below the Last Interglacial Rendezvous Hill terrace (Mesolella *et al.*, 1969). Further terraces are found at other uplifting sites around the world and several other sequences of raised reefs have been dated and interpreted based on assumptions about rate of uplift (Bloom *et al.*, 1974; Dodge *et al.*, 1983; Pirazzoli *et al.*, 1991).

The Last Interglacial shoreline can be seen at varying elevations at locations around the margin of the Pacific (Ota and Kaizuka, 1991). More terraces are likely to be preserved within a time period on a rapidly uplifting coast than a more gradually uplifting one (Figure 2.7a). Numerous terraces occur in Taranaki, New Zealand, on Sumba, Indonesia, and on Atauro, East Timor. Figure 2.7b shows a series of cross-sections of uplifted sequences of coral terraces marking former reefs; the Last Interglacial (marine oxygen-isotope substage 5e) is shown in each case. There are up to 19 individual reef terraces along the Huon Peninsula on the northern coast of Papua New Guinea (Figure 2.8). This has become one of the most important sites because of its rapid rate of uplift at the point of collision between the Pacific and Indo-Australian plates (Chappell, 1974a). In Figure 2.7b the rapidly uplifting Kanzarua section, the Sialum section and the slowly uplifting Gagar Anununai sections at Huon Peninsula are shown (Aharon and Chappell, 1986) and are contrasted with the sequence on Atauro, East Timor which extends back before the Last Interglacial (Chappell and Veeh, 1978).

Figure 2.9 shows an interpretation of sea level based on uranium-series dating of Huon terraces compared with the oxygen-isotope record from a Pacific deep-sea core V19–30 (Chappell and Shackleton, 1986; Shackleton and Chappell, 1987). The isotope record has been constrained by further dating of the sequence of terraces on the Huon Peninsula (Chappell *et al.*, 1996) and Figure 2.9 shows details of this cross-correlation. The details of the fall in sea level from 80000 years BP to the peak of the last glaciation around 20 000 years BP remain controversial because this period is largely beyond the range of radiocarbon dating (Thom, 1973). There is increasing evidence for a divergence between the isotope record and reef terraces around 50–40 ka BP (Shackleton, 2000; Cabioch and Ayliffe, 2001).

Last glaciation

During the last glaciation, the ice sheets reached a maximum extent at a time that has been radiocarbon dated to around 18–17 ka BP. Uranium-series calibration of the radiocarbon time scale reveals this to have been 22 000–21 000 years BP (Bard *et al.*, 1990a). Table 2.2 records evidence for lowering of sea level at this time (Yokoyama *et al.*, 2000). In addition to the dated evidence, based on shallow-water sediments, shells or corals, there is further morphological evidence such as distinct notches or breaks of slope on forereefs (Land and Moore, 1977; Anderson, 1998), or submerged deltas (Emmel and Curray, 1982; Kudrass and Schlüter, 1994).

Ice melt occurred rapidly after the peak of the last glaciation and sea-level curves from several parts of the world are in general agreement (Figure 2.10). For instance, postglacial sea-level curves from Barbados (Fairbanks, 1989; Bard *et al.*, 1990a), Huon Peninsula, New Guinea (Chappell and Polach, 1991), Abrolhos Islands, Western Australia

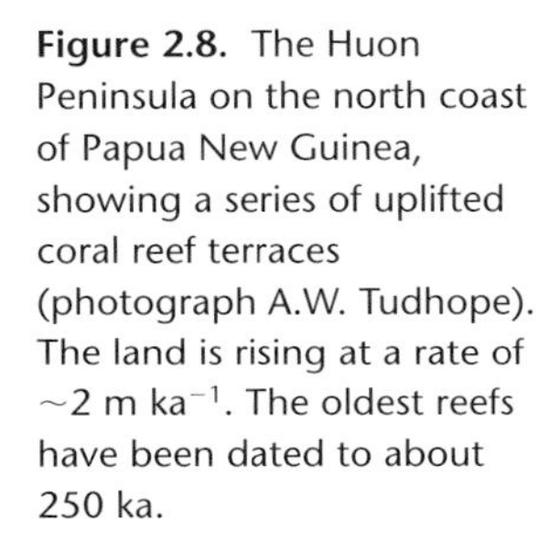

Figure 2.8. The Huon Peninsula on the north coast of Papua New Guinea, showing a series of uplifted coral reef terraces (photograph A.W. Tudhope). The land is rising at a rate of ~2 m ka^{-1}. The oldest reefs have been dated to about 250 ka.

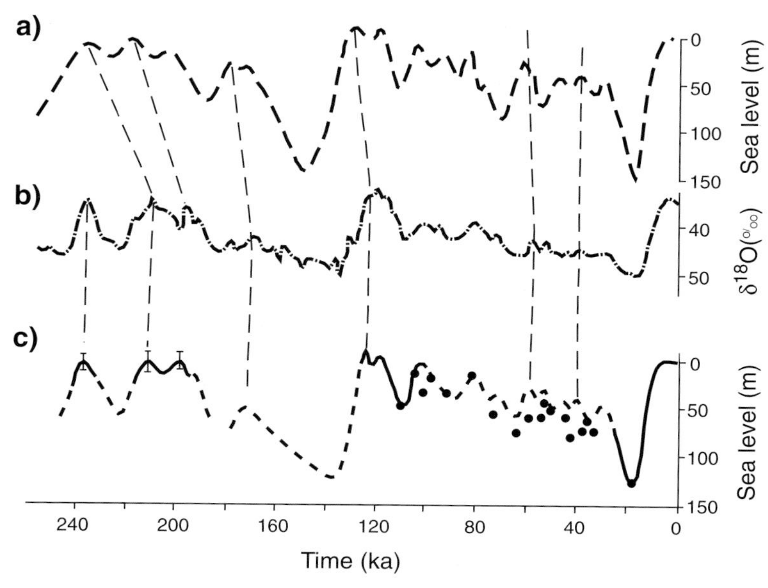

Figure 2.9. Sea-level variation over the past 240 000 years. (a) Reconstruction of sea level based on ages and uplift-corrected heights of reef terraces on Huon Peninsula. (b) Oxygen-isotope record from core V19–30 from the equatorial Pacific Ocean. (c) Reconciliation of those curves (based on Chappell and Shackleton, 1986), and revision of details of further dating (black dots) from Huon terraces (based on Chappell *et al.*, 1996).

Table 2.2. Dated indicators of last glacial maximum lowstand sea level

Location	Material	Depth (m)	Ages (^{14}C years BP)	Reference
Mazatlan, Mexico	Shallow-water organisms	119	19 300 ± 300	Curray, 1961
Timor Sea, Australia	Shallow-water shell, *Chlamys*	130	16 910 ± 500	Van Andel and Veevers, 1967
Atlantic coast, United States	Shallow-water sediments and oyster shells	125	c.15 000	Milliman and Emery, 1968
Arafura Sea, Australia	Beachrock	130–175	18 700 ± 350	Jongsma, 1970
Great Barrier Reef, Australia	Shallow-water coral, *Galaxea*	160–175	17 000 ± 1000[b]	Veeh and Veevers, 1970
Bonaparte Gulf, Australia	Microfossils in cores	121–129	18 000–16 000	Yokoyama *et al.*, 2001
Guinea–Sierra Leone	Algal rock	103–111	18 750 ± 350	McMaster *et al.*, 1970
Barbados	Shallow-water coral, *Acropora palmata*	116–126[a]	c.17 000	Fairbanks, 1989
Southeastern Australia	Shallow-water shells, *Pecten*	110–130	17 320 ± 220	Ferland *et al.*, 1995
Mayotte Island, Indian Ocean	Coral, *Acropora*	152	18 400 ± 500	Pirazzoli, 1996
Sardinia	*In-situ* vermetid gastropods	126–150	18 860 ± 170	Pirazzoli, 1996

Notes:
[a] uplift-corrected; [b] U-series disequilibrium age.

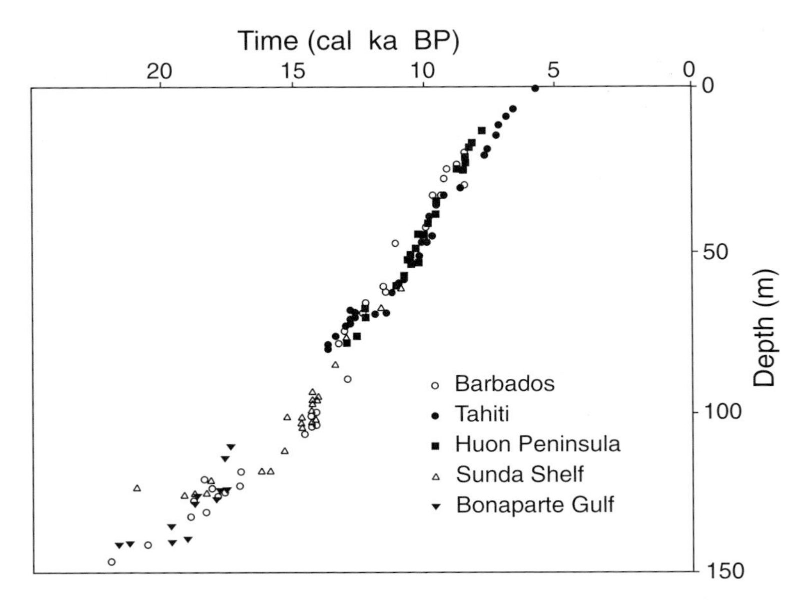

Figure 2.10. Record of postglacial sea-level rise (in calibrated years BP) from peak of last glacial until apparent cessation of ice melt around 6 ka (based on coral data from Barbados, Fairbanks, 1989; Bard *et al.*, 1990a; New Guinea, Chappell and Polach, 1991; Tahiti, Bard *et al.*, 1996; Montaggioni *et al.*, 1997; and mangrove sediments from the Sunda Shelf, Hanebuth *et al.*, 2000; and Joseph Bonaparte Gulf, Yokoyama *et al.*, 2001).

(Eisenhauser *et al.*, 1993), Tahiti (Bard *et al.*, 1996; Montaggioni *et al.*, 1997), and Joseph Bonaparte Gulf, northwestern Australia (Yokoyama *et al.*, 2001) are remarkably consistent with data on mangrove sediments from the Sunda Shelf (Hanebuth *et al.*, 2000). The Barbados curve shows slow rise after 20 ka BP at a rate of around 5 m ka^{-1}, until around 14 ka when rise as rapid as 37 m ka^{-1} and around 12 ka when rise of about 12 m in <1000 years occurred. Several events during which sea level rose at >45 m ka^{-1} have been interpreted from reef cores around Barbados (Blanchon and Shaw, 1995), drowning the reefs (see Chapter 5).

Postglacial sea-level rise inundated a series of land bridges, including the English Channel isolating the British Isles from Europe, opening the Bering Straits and Torres Strait, and flooding the Sunda Shelf (Hanebuth *et al.*, 2000). The rapid sea-level rise must have had dramatic effects on the broad continental shelves. In northern Australia, rapid sea-level rise would have resulted in horizontal rates of shoreline movement across the area now covered by the Timor Sea of up to 20 m a^{-1} (over a metre a month!). This shoreline instability must have limited the type of coastal landforms, habitats and ecological assemblages that could survive (Chappell and Thom, 1977).

Holocene sea-level variation

The pattern of relative sea-level change during the Holocene is particularly important (Figure 2.11), and shows considerable variation from sites around the world in contrast to the primarily eustatic far-field record of early postglacial sea level shown in Figure 2.10. There has been a history of disagreement between researchers attempting to reconstruct Holocene sea level. For example, Godwin in Britain considered that the sea had reached a position close to present in mid Holocene and varied little over recent millennia (Godwin *et al.*, 1958). However, researchers in the United States and northwestern Europe inferred that the sea had risen but at a decelerating rate up until present (Jelgersma and Pannekoek, 1960; Shepard, 1963b).

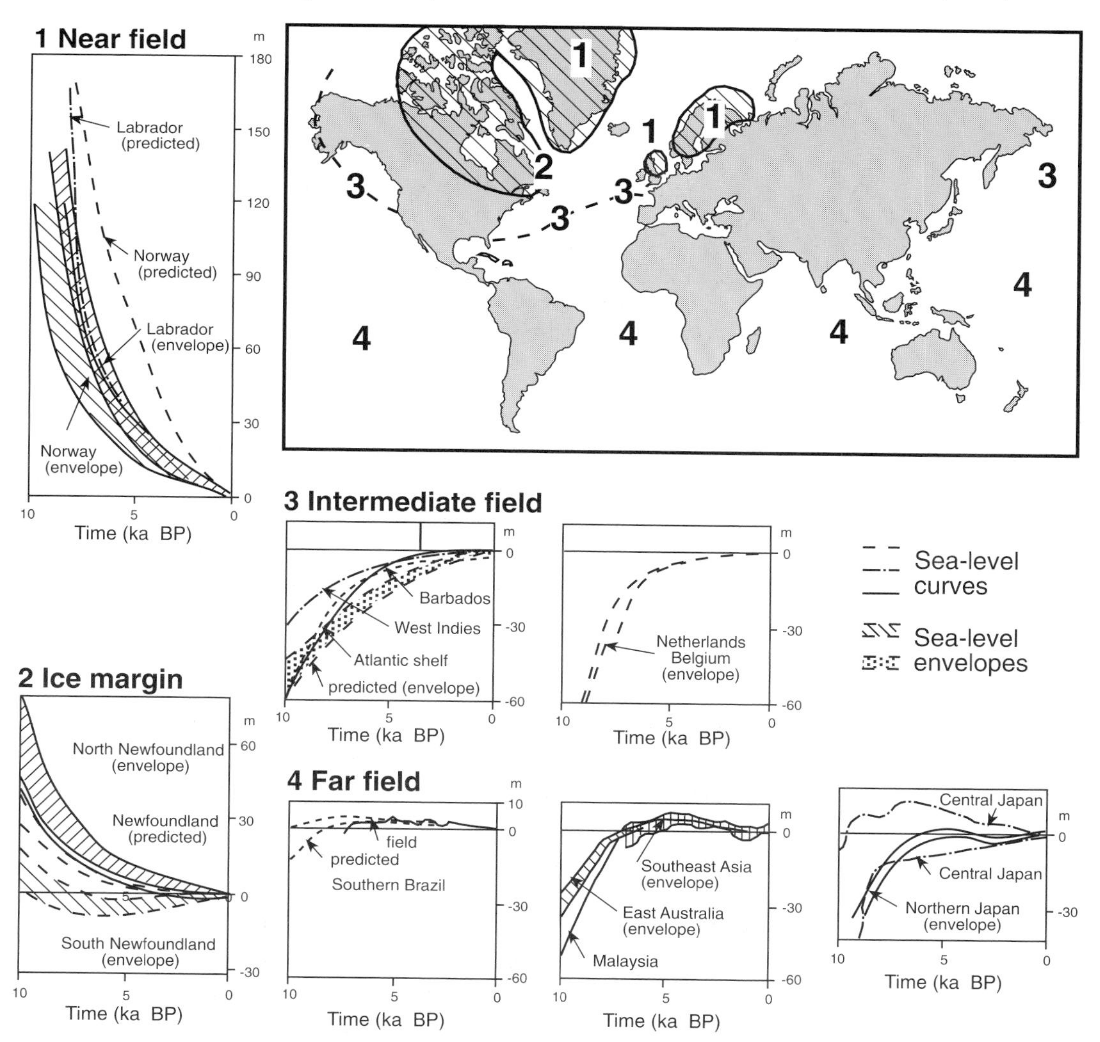

Figure 2.11. Variation in Holocene sea-level curves, illustrated with examples from near field, ice margin, intermediate field and far field (based on sea-level data in Pirazzoli, 1991, 1996).

In 1961, Fairbridge produced a global compilation of radiocarbon dating results indicating a series of oscillations of sea level during the Holocene (Fairbridge, 1961). The compilation comprised periods during which sea level appeared to have been below present, based on evidence from the Bahamas and Florida, and periods during which it seemed to have been higher than present, based on evidence from Western Australia. The data implied different sea-level histories for the Americas and the Australian region. Atlantic sites, such as the Caribbean and Florida, demonstrated decelerating rise up until present (Adey, 1978; Lighty *et al.*, 1982). By contrast, in the Australian region sea level appeared to have reached a level close to present around 6500–6000 years BP and to have varied little since (Thom and Chappell, 1975; Thom and Roy, 1985).

Not all coastlines preserve evidence of former shorelines, and it is generally necessary to exercise caution in determining past sea level from a set of indicators. Some lines of evidence are direct, whereas others are indicative, implying that sea level was above or below a particular elevation (van de Plassche, 1986). As an example, it is worth noting that within the Pacific Basin Clark *et al.* (1978) indicated a region that had experienced a decelerating (Atlantic-type) sea-level curve. They based this on studies of the reefs of Hanauma Bay by Easton and Olson (1976). However, it is likely that the dated corals were not at sea level, and that the curve produced is a reef growth curve rather than a sea-level curve (see Chapter 5).

The sea surface is not flat, but is an equipotential surface determined by the gravity field, called the geoid, and influenced by the distribution of mass of underlying water and rock (Gaposchkin, 1973). There are bumps and depressions; the lowest point, −200 m, occurs in the Maldives, while high points are seen in some plate-margin locations, such as the Huon Peninsula in Papua New Guinea. Variations in the distribution of mass that effect gravity therefore have an influence on the relative shoreline position (Mörner, 1976). Gravitational adjustments to sea level are likely to have been particularly large in the near field and ice margin which were depressed and have subsequently rebounded with melting of the ice sheets, and which are affected by the gravitational attraction of the mass of ice.

2.3.3 Isostatic adjustments

Isostasy is the buoyant equilibrium maintained (or sought) between the crust and the load on it. Glacial isostasy, the redistribution of mass associated with the extension and melt of polar ice sheets, was described by Daly (1934). The load of an ice sheet deforms the earth's crust, and the resulting subsidence beneath the ice is accompanied by highly viscous deformation of material in the mantle (Smith and Dawson, 1983). As the ice melts, the crust responds elastically and relatively quickly, whereas the upper mantle has a viscous response that can take several thousand years, which means that a delayed response occurs to return the crust towards its pre-load configuration. The response to changed load is not immediate as can be seen from Scandinavia and the Hudson Bay region of northern Canada (Pirazzoli, 1991; Fjeldskaar, 1991). Fossil shorelines up to 300 m above present indicate that these areas have uplifted after melt of the ice; uplift rates of 9–10 mm a^{-1} are still continuing (Eronen *et al.*, 2001). A forebulge appears adjacent to the ice and collapses as, and after, the ice melts. The region around the ice sheet (ice margin) shows a complex pattern of movement (Andrews, 1970).

The visco-elastic response of the globe to the changing mass of ice involves responses that are observed globally, with the mass of water added to the oceans causing a small hydro-isostatic response. This hydro-isostatic adjustment was modelled by Walcott (1972) who envisaged that the centre of the ocean basins would respond elastically to the addition (and subtraction) of a volume of water, and that there would also be a longer-term visco-elastic response. There have been further attempts to model this (Clark *et al.*, 1978; Clark and Lingle, 1979), with different models based on different assumed continental ice loads and their melting histories. The visco-elastic flexural response of the earth is non-linear, and therefore non-additive and non-repeatable; it cannot therefore be assumed that over repeated glacial cycles the shoreline returns to the same elevation.

The collapse of the forebulge ice margin region results in draining of water from the equatorial ocean in the far field, called equatorial ocean siphoning by Mitrovica and Peltier (1991). This appears to be an important reason, together with hydro-isostasy, that many islands in the mid-Pacific Ocean show evidence of shorelines above present, dating from the mid–late Holocene (Dickinson, 2001), and why this higher sea level is only a relative feature and does not indicate a truly eustatic sea level (Nakada, 1986).

The continents might be anticipated to react more slowly because of the greater rigidity of continental crust. The margins of continental collision coasts show abrupt transitions in plate thickness and rigidity across subduction zones, and are complicated by local tectonism. Preliminary modelling of this layered earth implied that the continental shelf would flex in response to the load of water, termed hydro-isostasy. If there is a broad shelf, the water flooding over it is likely to downwarp the outer part of the shelf and to uplift the shoreline itself.

Consequently, on continental coasts a Holocene shoreline is likely to be elevated above present, and this is likely to vary at the head of embayments (where it will be higher) in relation to promontories, where the contemporaneous shoreline will be lower (Chappell, 1974b). Field evidence from across the Great Barrier Reef and along the Queensland shoreline appears to confirm this pattern (Chappell *et al.*, 1982, 1983). The evidence of higher sea level reported from the Spencer Gulf region may indicate flexure in response to hydro-isostasy (Chappell, 1987).

Geophysical models of the isostatic response of the earth's surface to changes in load have been developed by Peltier (1988, 1999, 2001) and involve iterations to generate values for rigidity and viscosity, and tend to use the sea-level data from fossil shorelines as a constraint. These geophysical models depend on a good understanding of ice-volume history, and model lithospheric deformation of a visco-elastic earth to changes in surface load (Lambeck, 2001). Nevertheless the viscous relaxation is continuing, and modern sea-level changes contain this ongoing response (Mitrovica and Davis, 1995).

Similar models have been used to generate regional patterns of isostatic response. In particular Lambeck has run models covering Australia (Lambeck and Nakada, 1990), northwestern Europe (Lambeck *et al.*, 1990; Lambeck, 1993a, 1993b; Lambeck, 1995a; Lambeck *et al.*, 1996), the Mediterranean (Lambeck, 1995b, 1996a), and the Persian Gulf (Lambeck, 1996b). Different patterns of relative sea-level response occur in near field (defined as sites within the limits of the former ice sheets), ice margin (near former ice margins, forebulge regions), intermediate field and far field (Lambeck, 1993c). In the near field, the dominant contribution to sea-level change comes from ice-load effects, with rapid uplift resulting in a general exponential fall of relative sea level to its present level. At ice-margin sites, the relative sea-level change was characterised by an initially rapid fall during the late glacial followed by stability and a more recent rise to present position. Intermediate field sites correspond to the forebulge, with late-glacial and postglacial subsidence at decelerating rates. Far-field sites appear to have experienced a sea-level curve similar to the pattern of ice melt, rising to near present around 6000 years BP. However, they generally show a slight fall since then as a result of hydro-isostatic adjustment or equatorial ocean siphoning.

Figure 2.11 indicates relative sea-level curves for each of the major regions, near field, ice margin, intermediate field and far field (based on Pirazzoli, 1991, 1996). There are unlikely to be sharp transitions between the regions for which a particular pattern of sea-level variation applies. Near-field sites are those that were covered by ice. There is generally a broad envelope of change that is predicted by geophysical modelling, but field evidence, such as emerged shorelines, is consistent with rapid uplift of Norway and other Scandinavian countries (Hafsten, 1983) as also in Labrador (Quinlan, 1985), and the Hudson Bay region of Canada (Dredge and Nixon, 1992).

Relative sea-level history is particularly complex around the margin of former ice sheets as a result of the interplay of isostatic rebound and eustatic sea level. Even around the island of Newfoundland observed and predicted curves vary along the coast (Figure 2.11). Initial rapid uplift of the land (emergence) means that shorelines dated to before 10000 years BP are elevated above modern sea level (Newman *et al.*, 1980), but as the rate of uplift has declined, the rising sea has flooded over some areas that were initially emerged, resulting in further submergence (Scott and Medioli, 1980).

Intermediate-field sites include those places where it had been inferred that the sea level was rising at a decelerating rate until present. There is considerable similarity between the relative sea-level curves from the Netherlands (Jelgersma, 1979; van de Plassche, 1982) and Belgium (Denys and Baeteman, 1995), and those from the western Atlantic which are influenced by the forebulge of the Laurentide ice sheet that covered North America (Milliman and Emery, 1968). Geophysical modelling indicates decelerating sea-level rise along these coasts and field data from Florida and the West Indies support this interpretation (Lighty *et al.*, 1982; Fairbanks, 1989).

A relative sea-level history in which sea level has been close to its present level for most of the past 6000 years is observed around much of the coastline of Australia (Thom and Chappell, 1975; Thom and Roy, 1985). This pattern, shown in Figure 2.11, or a similar pattern, appears appropriate for much of the Pacific (Hopley, 1987; Pirazzoli *et al.*, 1988), Southeast Asia (Tjia, 1996; Hesp *et al.*, 1998; Scoffin and Le Tissier, 1998) and South America (Martin *et al.*, 1985; Angulo and Lessa, 1998). There are regional discrepancies. For instance, since 6000 years BP much of the Australian coastline appears to have experienced a slight fall of sea level from an elevation of around 1 m above present as a result of hydro-isostatic adjustments as the shelf has been inundated (Chappell, 1983a; Lambeck and Nakada, 1990). Possible sea-level oscillations of 1 m or more above present are still the subject of debate (Baker *et al.*, 2001). Curves for Japan indicate that sea level has been close to present for several thousand years but there is considerable variability which results from local tectonism (Fujii and Fuji, 1967; Fujii and Mogi, 1970; Ota and Omura, 1991). Other areas with a significant tectonic influence also deviate from the broad pattern (e.g. New Zealand, Ota *et al.*, 1988, 1992).

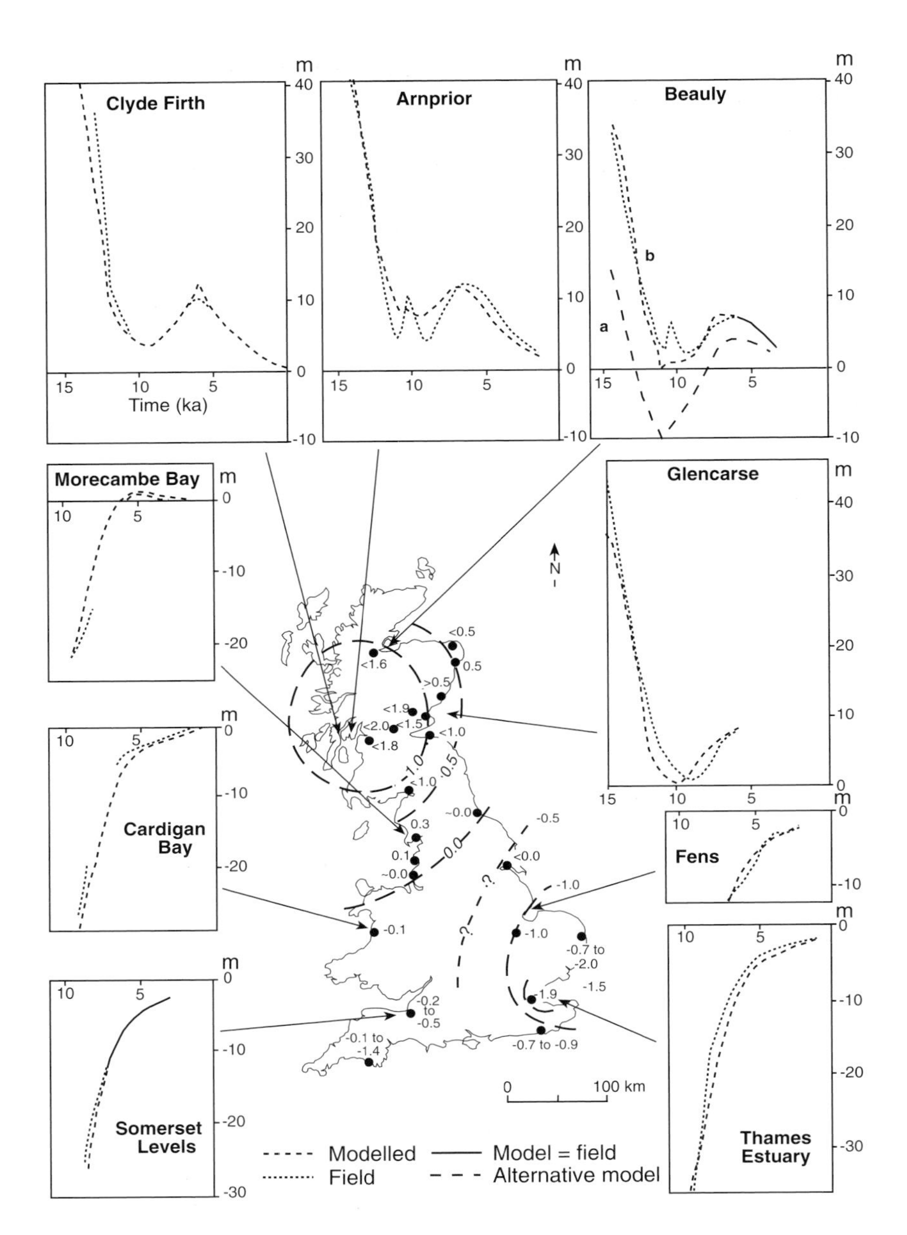

Figure 2.12. Variation in Holocene sea-level curves around Great Britain (based on Lambeck, 1995a). The curves show modelled (dashed line) and field reconstructions (dotted line) of sea levels and where these coincide (solid line). The map shows estimated current rates of crustal movement (mm/radiocarbon year) around Great Britain (after Shennan, 1989). Note modelling shows a particularly large discrepancy for Beauly Firth; the ice model used for other sites is shown by curve b; curve a shows simulation if 20% more ice is assumed over Scotland.

Figure 2.12 examines a range of modelled and observed sea-level curves around Great Britain. Modelling of the geophysical response of Great Britain to the melting of an ice load over Scotland (Lambeck, 1993a,b, 1995a) shows some agreement with the pattern of ongoing crustal movement (Figure 2.12) as deduced from local studies of sea-level trend (Shennan, 1989; Shennan and Woodworth, 1992). Scotland has undergone near-field response and is still uplifting. Biostratigraphy and radiocarbon dating of sediments in isolation basins (bedrock hollows that at one time were inundated by marine waters, but which are now isolated beyond marine influence as a result of uplift) enable further constraint on the pattern of uplift (Hafsten, 1983; Shennan *et al.*, 1994; Long *et al.*, 1999).

Although there appears to be good agreement between modelled and observed sea-level history for much of Great Britain as shown in Figure 2.12, there are several regions where field data are not consistent with the model predictions (D.E. Smith *et al.*, 2000). For instance, the apparent history of sea-level changes in the Beauly Firth area of eastern Scotland would be better modelled if ice thickness at this location were increased by 20% from that used in the ice model (Lambeck, 1995a; Johnston, 1995; Lambeck *et al.*, 1996; Shennan *et al.*, 1999, 2000a). In order to minimise discrepancies in the regional relative sea-level response it is possible to alter the ice- or earth-model parameters. A different pattern is generated if a different ice load is assumed, or if the values for rigidity of the crust are varied, but it is not possible to tell which is the more appropriate of these to adjust. For example, the pattern of sea-level change in northeastern England can be correlated with several alternative combinations of ice- and earth-model parameters because individual iterations of the geophysical models do not produce unique solutions (Shennan *et al.*, 2000b,c).

There can be no doubt that geophysical modelling, both at global and regional scales, has added a powerful tool in the ever-finer resolution of relative sea-level history. However, it seems likely that it will require further refinements before the fine details of sea-level variation are entirely described by models. There is a range of external factors which complicate the modelling. On continental shores, and at plate boundaries, additional local and regional geodynamic factors are superimposed on the glacio-isostatic adjustment. Load responses to sediment mass will also occur, as can be seen where deltas load the continental shelf (see Chapter 7). For instance, whereas the northern Gulf of Mexico is interpreted as undergoing rapid subsidence in relation to delta loading, the sea-level history of the Texas coast remains controversial (Morton *et al.*, 2000a; Blum *et al.*, 2001; Otvos, 2001).

2.3.4 Present and future sea-level trends

The significance of sea level as a boundary condition acting as a constraint on the coastal system has taken on a new significance in the light of perceived ongoing, and anticipated future, sea-level rise as a result of the enhanced greenhouse effect (Barth and Titus, 1984). The issue of the impact of future sea-level changes on coastal systems is examined in Chapter 10. Below, the background to future sea-level change is considered.

Changes of sea level in future will retain a component of each of the longer-term trends demonstrated to occur through plate-tectonic processes, ice-volume changes, and lithospheric deformation by isostasy, operating at its characteristic frequency. However, particular concern has been expressed about a sea-level rise that is driven primarily by the thermal expansion of the upper oceans (termed steric), with lesser contributions from other factors such as groundwater diversion and damming of rivers (Manabe and Stouffer, 1994; Sahagian *et al.*, 1994). Anthropogenic increases in the atmospheric concentration of greenhouse gases (particularly carbon dioxide, nitrous oxide and methane) through burning of fossil fuels (coal, oil and natural gas) appear to be leading to a gradual human-induced warming of the globe. Globally averaged surface air temperatures appear to have increased by 0.6 ± 0.2 °C over the 20th century (Houghton *et al.*, 2001). Although warming will lead to some further melting of grounded ice and melt from small glaciers, sea level will increase primarily through expansion of the upper ocean (Douglas *et al.*, 2001).

The most detailed data on present trends in sea-level change come from tide-gauge records. Tidal levels have been measured in Amsterdam since 1682, in Stockholm since 1704, in Brest since 1807, in Swinoujscie since 1811 and in Venice since 1825 (Pirazzoli, 1996). A recording tide gauge has been in operation for over 165 years at Sheerness in England and since 1854 at San Francisco in the United States. Tidal measurements were initially made for navigational or other nautical purposes and the methods, accuracy and frequency of recording varied considerably. Where trends have been sought these are best examined where there is 40–50 years of data (Pirazzoli, 1986a; Emery and Aubrey, 1991; Douglas, 2001).

There have been several predictions of the global pattern of future sea-level change. However, early predictions of sea-level rise of the order of 3–5 m by the year 2100 (Barth and Titus, 1984) have now been revised downward to a more modest range of 0.09–0.88 m by 2100 (Wigley and Raper, 1992; Douglas, 2001; Houghton *et al.*, 2001).

Tide gauges show a complex pattern of change (Woodworth, 1993) and individual gauges can show variable patterns (Belperio, 1993). Along much of the eastern coast of the United States there has been a trend of sea-level rise of 2–4 mm a^{-1}, ranging up to 6.3 mm a^{-1} at Galveston. Much of this area is in the forebulge region and is continuing to experience isostatic adjustment to melting of the Laurentide ice sheet. The particularly high rate at Galveston seems to reflect a response to groundwater pumping in addition to isostatic effects (Paine, 1993). Tidal gauges can also show idiosyncratic records where they are subject to tectonic influences. The Kanto earthquake of 1923 uplifted the gauge at Aburatsubo in Japan by 1.37 m and other events have had lesser effects (Baker, 1993). An earthquake at Antofagasta in Chile in 1995 uplifted the gauge there by 40 cm, whereas changes in the pattern of mean sea level recorded by the tide gauge at Manila in the Philippines appear to relate to harbour extensions (Douglas, 2001).

Attempts to determine average global trends which do not make allowance for tectonic or isostatic factors tend to give rates of 1.0–1.5 mm a^{-1}, and those corrected for such factors using geophysical modelling indicate rates of 2–3 mm a^{-1} (Peltier and Tushingham, 1989; Tushingham and Peltier, 1991; Peltier, 2001). A problem with interpreting trends from tide gauges is that they are spatially clustered. Not only are they absent from the middle of most oceans, but 70% of the sites with records of more than 50 years are sited in the northern hemisphere in or close to forebulge regions (Pirazzoli, 1986a).

More recently satellite altimetry has become a powerful way to assess sea-level changes (Cheney, 2001). Initially, Skylab in 1973, GEOS 3 in 1975, SEASAT in 1978 and GEOSAT in 1985 provided data with some opportunities for intercomparison (Cheney *et al.*, 1989; Périgaud and Delecluse, 1992). More recent satellites such as TOPEX-Poseidon enable a much closer scrutiny of sea-surface behaviour (Nerem, 1995, 1999). They show patterns of sea-level variation that correspond to sea-surface temperature changes associated with periodic El Niño–Southern Oscillation phases (see Section 3.5.3), implying that records of several decades will be necessary before trends can be detected (Cabanes *et al.*, 2001; Nerem and Mitchum, 2001).

2.4 Materials

The geophysical setting and geological history of a region determine the materials that are available to be shaped by coastal processes. The rate of landform development depends on the strength or resistance of rocks or material strength of sediments which determines how susceptible they are to reshaping by the processes that they experience. Sediments can also be important because they contain a record of past coastal landforms and their pattern of development.

The broad plate-tectonic setting exerts a control on the coastal landforms that occur along a particular stretch of continental coast. These terrigenous settings are composed of rocks associated with continental crust and the sediments derived from them, which are termed clastic sediments (these are generally siliceous sediments and the term siliciclastic is also used). In addition to these coastal settings, there are significant areas that are dominated by carbonate sediments, most of which are derived from calcareous organisms. In the sections below, coastal lithology and terrestrial sediments, termed clastic, are examined briefly. Carbonate sediments and carbonate platforms are discussed in more detail.

2.4.1 Lithology

The type of rock that makes up a coast, termed lithology, is an important constraint on coastal geomorphology. Rocks can be divided into igneous, metamorphic and sedimentary. Igneous and metamorphic rocks are generally more resistant than sedimentary rocks which include a series of soft-rock lithologies, such as alluvium, talus, and glacial deposits. Lithology may be of differing density (i.e. basalt, 3.0–3.3 g cm^{-3} is harder than granite, 2.7–2.9 g cm^{-3}) with differing resistance to various forces on the shoreline. Unconsolidated, or soft-rock, coasts are the least resistant and may not have sufficient rock strength to maintain vertical cliff faces in response to erosion (see Chapter 4).

Structural characteristics of the rock are also important. Rock mass strength and resistance to shearing depend on the strength of the intact rock and its state of weathering, and the nature of fissures or weaknesses in the rock. These are reflected in the spacing, dip, width, continuity and infill of joints and fissures, together with movement of water into or out of the rock (Selby, 1985). Table 2.3 summarises the relative rock strengths of rock and unconsolidated sediments. Rock mass can be weakened (or strengthened) by intrusions (i.e. dykes), or by jointing and bedding patterns. Rocks, even where heavily jointed, are generally more resistant than unconsolidated sediment.

Lithology is a fundamental constraint on coastal geomorphology, where bedrock forms the shoreline. There are distinctive landforms associated with some lithologies. For instance, limestone coasts contain karst landforms reflecting the solubility and rock strength of limestone (see Section 4.3.2), and to a lesser extent granite is characterised by a suite of distinctive landforms.

Table 2.3. Rock mass strength classification (after Selby, 1985)

Strength class	Description	Examples of rock types
1	Very strong rock: requires many blows with geological pick to break	Quartzite, dolerite, gabbro
2	Strong rock: requires single blow with geological pick to break handheld sample	Marble, dolomite, andesite, granite, gneiss
3	Moderately strong rock: firm blow with geological pick makes shallow indentation	Slate, shale, ignimbrite, sandstone, mudstone
4	Weak rock: surface can be cut or scratched by knife	Coal, schist, siltstone
5	Very weak rock: crumbles when hit with geological pick, can be cut with knife	Chalk, lignite, rocksalt

The extent to which the geology of an area is expressed on the coast depends on whether the underlying rocks are veneered with more recent sediments. It has already been shown that there are sedimentary basins accumulating on trailing-edge continental margins. Sediments are less likely to accumulate on island margins which are more likely to have crenulate coasts reflecting the control of rock type and structure (Mitchell, 1998). Many tropical islands are fringed by coral reefs that protect rocky coasts from the direct impact of waves (see Chapter 5). In the humid tropics, a depositional swathe of coastal sediment of marine or fluvial origin often flanks the coast, whereas in higher latitudes lithological control may be more direct (Tricart, 1972). For instance, the Precambrian shield areas of high latitudes (North America and northern Europe) outcrop on the coast, whereas many of those in Brazil, tropical Africa, India and Australia are flanked by coastal plains (Davies, 1980), although the rugged Kimberley coast of Western Australia (see Figure 2.13) is an exception. Rocky coasts are examined in detail in Chapter 4.

Along much of the world's coastline, there are muds, sand and gravel derived primarily from river discharge. The range of grain size and mineral composition depend on physical and chemical weathering processes in the source area which are determined by climate and tectonic factors (Ollier, 1984). In these cases the lithology of the hinterland may be more important than the immediate underlying rock in influencing the composition of sediments. Worldwide, rivers supply around 90% of coastal sediments whereas cliff erosion accounts for only about 5%.

2.4.2 Clastic sediments

The rocks of a catchment are broken down by weathering processes and are then more amenable to erosion and reworking. The supply of coastal sediment will reflect source; for instance, the coloured sands of Alum Bay on the Isle of Wight, southern England, or the green olivine-rich sands of southern shores on Hawaii (derived from the phenocrysts of the surrounding volcanic rocks) are locally sourced. Most beaches in temperate areas are composed of quartz and feldspar grains, derived ultimately from the weathering of granitic rocks, gneisses and schists typical of the continents. There can be small quantities of heavy minerals such as hornblende, garnet and magnetite, reflecting lithology of the source rocks; these can be concentrated by coastal processes into commercially viable 'placer' deposits (Komar and Wang, 1984). For example, zircon and rutile migrate along beaches in western and eastern Australia and several beaches have been mined for the small concentrations of these minerals that they contain. Less resistant minerals, such as mica and chlorite are rare because they are broken down or remain in suspension and are lost seawards.

Coastal sediments are important in relation to geomorphology in two respects. First, these are the materials from which landforms are constructed and by which the coastal system adjusts. This is most clearly demonstrated on sandy beaches, where the waves expend energy redis-

Figure 2.13. Whirlpool Passage, on the Kimberley coast of northwestern Australia consists of resistant lithology which controls the shape of the coast. In this picture, mangroves line the coast (photograph B.P. Brooke).

tributing sediments so that the beach is reshaped towards equilibrium with the waves (see Chapter 6). Second, the sediments are important because their deposition and preservation provides evidence of past coastal geomorphology. It is not always possible directly to observe the coast evolving; this is because processes that shaped the coast can have changed, or much of the work may be done by rare, extreme events. In this case, reconstruction and inference from the coastal sedimentary record provides an invaluable additional tool.

There are important sedimentary properties, referred to as petrological characteristics, that are used to describe sediments. These include grain size and shape. Grain-size description is generally based on the Wentworth scale, a negative logarithmic relation, to base 2, called the phi scale (Table 2.4). Sand comprises grains that are >62 μm and <2000 μm (2 mm). A mix of sand and mud (<62 μm, comprising silt and clay) is described as a sandy mud when it contains 25–50% sand. Grain or clast (individual grains are also referred to as clasts) size can be measured with calipers, sieved, or analysed using hydrometer, settling or optical techniques. Bulk sediment size properties are measured using several different methods that assess different properties; sieving determines size on the basis of intermediate axis; optical techniques assess the equivalent spherical characteristics of particles, whereas hydraulic settling assesses the likelihood of settling in particular flows.

The grain-size distributions of coastal sediments are often expressed as a cumulative plot drawn on probability scale paper, from which percentiles can be read off to derive the size statistics indicated in Table 2.4 (Folk, 1964). The mean or modal grain size gives a measure of central tendency that is often used to infer the energy field. The extent to which a sediment sample is sorted, or composed of similar-sized grains, is indicated by the standard deviation. Sorting can be used to infer the maturity of a sediment deposit, or to infer sediment transport pathways if the sediment has been moved. Sedimentary processes tend to concentrate size fractions; skewness is a measure of the extent to which the distribution is asymmetrical, which can also indicate the history of movement, for instance, differentiating dune sand from beach sand. Dune sand tends to have a fine tail (to the distribution), but lacks coarse sediment so can be positively skewed, whereas the beach sediment, in which the material too coarse for transport to a dune concentrates as a lag, is often negatively skewed. Kurtosis is a measure of the degree to which the curve is flattened or peaked (Folk, 1964).

Additional sediment properties include the packing of grains, degree of imbrication (alignment or stacking), and porosity. Shape of individual clasts (the term is often used for the larger grains) can be characterised in terms of roundness and sphericity. Roundness is a measure of

Table 2.4. Sediment size characteristics and classes

Size			Wentworth size class	Friedman size class
millimetres	microns	phi		
				Very large boulders
2048		−11		
				Large boulders
1024		−10	Boulders	
				Medium boulders
512		−9		
				Small boulders
256		−8		
				Large cobbles
128		−7	Cobbles	
				Small cobbles
64		−6		
				Very coarse pebbles
32		−5		
				Coarse pebbles
16		−4	Pebbles	
				Medium pebbles
8		−3		
				Fine pebbles
4		−2		
			Granules	Very fine pebbles
2	2000	−1		
			Very coarse sand	Very coarse sand
1	1000	0		
			Coarse sand	Coarse sand
0.5	500	1		
			Medium sand	Medium sand
0.25	250	2		
			Fine sand	Fine sand
0.125	125	3		
			Very fine sand	Very fine sand
0.063	63	4		
				Very coarse silt
0.031	31	5		
				Coarse silt
0.016	16	6		
			Silt	Medium silt
0.008	8	7		
				Fine silt
0.004	4	8		
				Very fine silt
0.002	2	9		
			Clay	
				Clay

attrition and pebbles lose their angularity as a result of abrasion during transport. Sphericity is the degree to which a clast resembles a sphere. Settling velocities are often calculated in relation to spherical grains. Clastic sediment derived from siliceous material, as opposed to calcareous sediment derived from the skeletons of organisms and described in the following section, is usually relatively spherical, but pebbles do seem to abrade to become preferentially disc-shaped. Sand grains may become smaller by abrasion; but grains less than 0.25 mm diameter do not seem to abrade because they do not attain great enough collision velocities.

Gravel beaches are composed of pebbles and cobbles of a range of rock types. In Britain, these are often termed shingle. They are generally closely linked with source material which influences the range of sizes and angularities. Glacial moraines yield a wide variety of sediment sizes, but beach deposits become progressively sorted, and individual grains more rounded as they are reworked by beach processes (Carter, 1988). Gravel is often sorted on a beach with disc-shaped clasts at the top of the beach and imbrication (stacking) of gravel on the beachface. Particular assortments of gravel clasts relate to source. For instance, flint can be weathered out of chalk cliffs and accumulate on beaches at the foot of chalk cliffs. An assortment of coral boulders and shingle occur on coral reefs, and their size often reflects the skeletal characteristics of the contributing coral colonies (see Section 2.4.3). Figure 2.14

Figure 2.14. The Moeraki boulders, 40 km south of Oamaru on the east coast of the South Island of New Zealand, are large spherical boulders that appear to be calcareous concretions weathered out of Paleocene mudstones. The boulders are derived from a local source.

shows a highly local influence where concretions (called the Moeraki boulders) are weathered from rock on a beach on the South Island of New Zealand, and contribute to the beach material.

2.4.3 Carbonate sediments

Carbonate depositional environments are composed predominantly of the skeletal remains of organisms (termed bioclastic or biogenic), with a minor proportion of calcareous sediment resulting from direct precipitation of calcium carbonate from seawater. Carbonate can occur in two different crystalline forms: calcite and aragonite. Carbonate sediments are extensive, particularly within the tropics where they are dominated by coral reefs in which coral and coralline algae are prominent (called chlorozoan assemblages). Chapter 5 examines the geomorphology of reefs in detail; this section introduces the distinctive calcareous sediments and outlines the variation seen on carbonate banks. The geological record contains large carbonate platforms the centre of which consisted of low-energy, fine-grained carbonate sediments. The Great Bahama Bank and Florida Bay as well as coastal salt flats called sabkhas, such as those along the Trucial coast (Persian Gulf), provide modern analogues for some of these environments.

There are also extensive temperate carbonate depositional environments (Nelson, 1988; Fórnos and Ahr, 1997). These are often dominated by molluscs and foraminifera (called foramol assemblages). Within the subtropics there is a transition zone between the coral-dominated associations of warm waters, and the cool-water foramol associations typical of temperate regions. For instance, in the Ryukyu Islands in the northwestern Pacific Ocean, there is a transition from inshore coral reefs to an outer shelf and upper slope dominated by coralline algae (Tsuji, 1993). In the southern hemisphere, transitional subtropical carbonate shelves occur along the coast of Brazil (Carannante *et al.*, 1988) and on both the west and east coasts of Australia (Marshall *et al.*, 1998; James *et al.*, 1999).

Southern Australia is dominated by bioclastic temperate carbonates (James *et al.*, 1992). The most extensive cool-water carbonate shelf occurs in the Great Australian Bight (James *et al.*, 2001). Calcitic carbonates, dominated by coralline algae, bryozoans, foraminifera and molluscs, occur in southwestern Australia, south of Perth (Collins, 1988), and on the shelf of southern New South Wales (Marshall and Davies, 1978). Carbonate sediments can be important on other temperate coasts, either because, in the absence of major river input, *in situ* production of biogenic (organically produced) sediment becomes proportionally significant, or because the wave-energy conditions selectively concentrate carbonate such as shell from other sediment. For example, the significance of molluscan carbonate varies along the eastern United States coast. The beaches of New England to the north contain little carbonate (<1%), because shell production is low compared with supply of other sediment, but carbonate dominates beaches in Florida, because shell production is high and other sediment supply is limited (Giles and Pilkey, 1965).

In this section, the organisms that contribute skeletal material to the bioclastic sediments are described (Table 2.5), then the example of the Great Bahama Bank, on which inorganic precipitation is important, is outlined in greater detail.

Calcifying organisms

Coral reefs comprise a framework built primarily by reef-building (hermatypic) corals, also called scleractinian corals because they belong to the class Scleractinia. Other corals (soft corals such as alcyonarians and gorgonians) play a minor role in contributing to the reef carbonate; for example, the octocoral *Heliopora* is locally important in reef-crest environments of Indo-Pacific reefs, and the hydrozoan *Millepora* occurs on high-energy reef crests, particularly in the Caribbean. In several reef settings, however, corals are rare or absent, and other calcareous organisms, especially coralline algae and foraminifera, assume a major role (see Chapter 5).

Algae are a diverse group comprising coralline (red), codiacean (green) and coccoid (blue-green) algae. Coralline algae are a prominent component on coral reefs, often thriving in conditions of high wave energy, such as the reef crest (see Chapter 5). Coralline algae can also occur as spherical free-living organisms, termed rhodoliths, which are significant in subtropical regions (Bosence, 1983). The codiacean algae *Halimeda* forms segmented calcareous platelets, and is highly productive in many reef settings. *Halimeda* platelets, and the fine sand and aragonite needles that they break into, make a significant contribution to many reef sediments. *Halimeda* can also form large mounds, called *Halimeda* banks (or bioherms), for example at water depths of about 30 m in the Java Sea and on the Nicaraguan Rise (Roberts *et al.*, 1987; Hallock *et al.*, 1988). Filamentous blue-green algae (Cyanophyta) form algal mats and can be important in binding reef sediment and sediment in temperate areas. Stromatolites, built by cyanobacteria, are prominent in the geological record. Modern stromatolites occur in Shark Bay in Western Australia (Playford, 1990), but are otherwise a minor component of modern carbonate sedimentology.

Other contributors to the production of carbonate include molluscs and foraminifera. Molluscs range in size from the giant clams, *Tridacna*,

Table 2.5. Calcareous organisms contributing to reef-building or carbonate coastal sediments

Taxon	Taxonomic grouping	Typical genera	Role on reef	Sedimentary characteristics	Mineralogy
Coral					
Scleractinian	Hermatypic 'stony' hexacorals (madrepores) Zoantharia = Hexacorallia	*Acropora* (branching) *Porites*, *Diploria* (massive) *Agaricia* (platy, encrusting)	Framework builders encrusters and binders	Massive to branching boulders to gravel, breaking down to grit	Aragonite
Alcyonarian	'Soft' octocorals, Alcyonaria, including gorgonians (sea fans) and alcyonarians (soft corals)	*Heliopora* (coenothecalia) *Tubipora* (stolonifera)	Minor reef-top colonies	Friable skeleton	Calcite
Milleporoid	Hydrozoan coral	*Millepora* (firecoral)	Reef-crest encruster	Friable skeleton	Aragonite
Algae					
Coralline (red)	Crustose Melobesioideae (Rhodophyta; nullipores)	*Porolithon*, *Goniolithon*, *Lithothamnium*	Encrusting on high-energy reef-rim, or as rhodoliths	Massive, branching, encrusting or nodular	Calcite
	Articulated Corallinoideae (Rhodophyta)	*Amphiroa*, *Jania*, *Corallina*	Benthic, articulated	Break into sand-sized rods	Calcite
Codiacean (green)	Chlorophyta – Codiaceae	*Halimeda*, *Penicillus*, *Rhipocephalus*, *Udotea*	Benthic, calcify as tubular filaments	Segments (blades) break to aragonite needles	Aragonite
Coccoid (blue-green)	Cyanophyta – (~Cyanobacteria) Charophyta	*Lyngbya*	Filamentous mucus-secreting algae that bind sediment	Laminated algal mats and stromatolites	–
Foraminifera	Unicellular, chambered Protozoa, benthonic and planktonic	*Marginopora*, *Calcarina*, *Homotrema*	Benthic epibionts	Tests contribute directly to sand	Calcite

Molluscs	Mollusca, includes bivalves (pelecypods) and gastropods	*Tridacna* (clams), *Strombus*, *Crassostrea* (oyster), *Cerithidea*	Benthic infauna, grazers, borers	Robust shell, shell hash; coiled vermetid gastropods minor reef-building role	Aragonite (except oysters)
Echinoderms	Echinodermata, comprising echinoids (sea urchins) asteroids (starfish) and holothurians (sea cucumbers)	*Acanthaster*	Benthic grazers, herbivores and sediment-feeders	Plates, segments, spines and sclerites minor contribution to sediment	Calcite
Crustaceans					
Decapods	Decapod crustaceans include crabs, shrimps and lobsters	*Uca*, *Callianassa*	Wide ranging infauna and highly mobile	Minor skeletal remains	Calcite
Ostracods	Minute bivalved crustaceans		Sediment dwellers	Minor contributors to sediment	Calcite
Barnacles	Cirripedia		Colonise firm substrate	Minor role	Calcite
Bryozoans	Sessile colonial Bryozoa	*Domopora*, *Porella*	Benthic dwellers on stable substrate	Tubular branching cryptic or binding other substrates, occasionally reef-building	Calcite
Brachiopods	Bivalved Brachiopoda		Cavity dwellers on shallow reefs	Cryptic, minor role	Calcite
Sponges	Porifera include siliceous demosponges and calcareous sclerosponges	*Cliona*	Benthic, encrusting and boring	Spicules, minor contributor to sediment	Rarely calcareous
Worms	Annelida, include polychaete marine segmented worms	*Spirorbis*, *Galeolaria*	Secrete tubular exoskeleton; rare serpulid or sabellarid reefs	Minor encrusting role	Aragonite or calcite

that produce solid shells that are anchored to tropical reefs, to the many cryptic bivalves and gastropods, including those that bore into coral and rock, such as *Lithophaga*. Oysters can build subtidal to intertidal reef-like structures up to a kilometre long in temperate to subtropical estuarine environments (Bartol *et al.*, 1999). Foraminifera and ostracods are generally smaller than molluscs. Foraminifera, growing on seagrass and turf algae on the reef flat, contribute tests that can account for >50% of lagoonal and reef-island sediments. They also contribute to calcareous and clastic sediments in temperate regions and their tests are common in coastal sediments and can be used as ecological indicators and tracers (Haslett *et al.*, 2000). Lesser contributions are made by echinoderms, bryozoans, and sponges (see Table 2.5).

Although corals and algae are the prime reef-building organisms (biohermal), some other organisms such as oysters and vermetid gastropods, bryozoans and serpulid and sabellariid worms can build mounds and minor reefs in temperate waters. Vermetid gastropods are filter feeders that attach to a hard substrate in subtropical intertidal to shallow subtidal settings (Jones and Hunter, 1995); *Dendropoma corrodens* is a vermetid that forms small reefs together with coralline algae in Bermuda (Thomas and Stevens, 1991). Serpulid worms (e.g. *Ficopomatus enigmaticus*) are benthic suspension feeders that occur in low-energy settings and can form mounds (Fórnos *et al.*, 1997). Sabellariid worms (e.g. *Phragmatopoma californica* and *Sabellaria cementarium*) can form mounds at the foot of a beach in higher energy settings, such as in Florida, New Zealand and in the Baie du Mont St Michel in northwestern France (Ekdale and Lewis, 1993).

The sediment size characteristics of calcareous sediments reflect the biological organisms that contribute, each breaking down through a discrete series of size classes, termed the Sorby principle. In a classic study of the Isla Perez, on Alacran Reef, in the western Caribbean, Folk and Robles (1964) demonstrated that corals break down through boulders, cobbles, sticks (30 mm) and then coral grit (0.25 mm), rarely forming grains of intermediate size (Figure 2.15). Coral shingle (fine gravel) composed of sticks of *Acropora* is a significant element of reef sediments. Similarly, *Halimeda* platelets, although variable in size themselves, break down from sand to fine dust. The tests of foraminifera contribute directly to coarse sand and may constitute more than 70% of reef sediments on some Pacific reefs (McKee, 1959; Weber and Woodhead, 1972; Woodroffe and Morrison, 2001).

The prime biological contributors also determine the mineralogy (Table 2.5). Corals, *Halimeda* and molluscs secrete aragonite, whereas coralline algae and foraminifera are largely calcitic (Bathurst, 1975; Scoffin, 1987). Grain shape is also a function of contributing organisms.

Rarely are grains spherical; more often they are segmented (e.g. *Halimeda*), massive (e.g. coralline algae and large molluscs), branching (e.g. branching corals and bryozoans), spicular (e.g. sponge spicules), or small hollow tests (e.g. foraminifera). Shape affects the way in which the grains respond to hydrodynamic forces, in contrast to clastic sediments which may more closely resemble the ideal spherical form (Maiklem, 1968; Kench and McLean, 1996). The sedimentary characteristics of calcareous sediments are often poorly sorted, reflecting the mixture of source organisms; they can be reworked and sorted to various degrees by processes of sediment transport (see Chapter 5).

Carbonate platforms

On carbonate platforms (or banks) almost all sediment is calcareous, but it can be derived both from organic (bioclastic) sources and from inorganic precipitation. The Great Bahama Bank is the most extensive modern carbonate bank. Much of the bank surface is covered by water that is less than 10 m deep and extensive areas are intertidal or supratidal (Figure 2.16). The distribution of sedimentary environments across the bank reflects physical and ecological parameters, particularly distance from the bank margin and shelter provided by islands (Table 2.6). A similar succession of environments to that seen in the Holocene occurs in the Pleistocene sediments of South Florida (Hoffmeister *et al.*, 1967).

Some of the margin of the Great Bahama Bank is characterised by coral reefs. Elsewhere along the margin there are shoals composed of oolites (or ooids) which are chemically precipitated spherical grains,

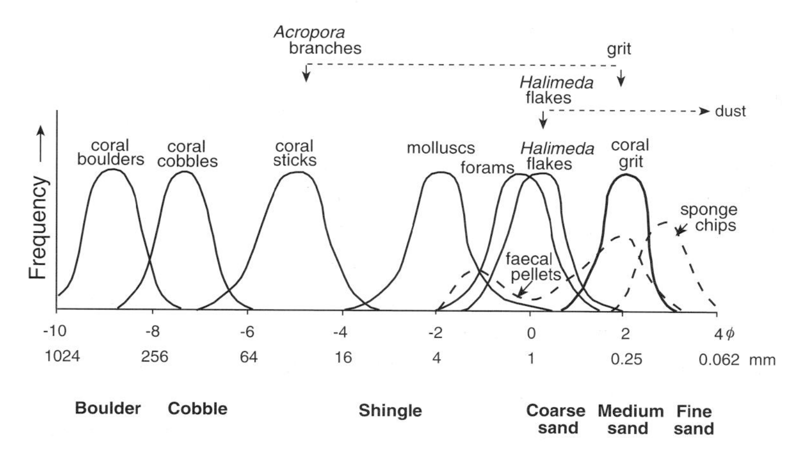

Figure 2.15. Reef sediments break down via a series of discrete size classes depending on the biological skeletal elements and the processes of breakdown which are operative (based on Folk and Robles, 1964; Scoffin, 1987).

comprising concentric layers of aragonite. The oolite shoals are up to 1 km wide and are dominated by strong tidal currents that move the grains, giving rise to cross-bedding and ripples, but inhibiting most biological activity (Ball, 1967). Lower energy environments occur with distance onto the platform. Oolitic environments occur that are less mobile than the oolite shoals; biota is abundant stabilising the sediment and bioturbating it (Figure 2.16). Grapestone comprises aggregates of particles which, because they are not agitated, become lightly cemented by cryptocrystalline cement (Winland and Matthews, 1974). The platform interior contains non-skeletal mud and pellet mud (the faecal pellets of the organisms such as polychaete worms that graze the surface), with a tidal range of <0.8 m (Purdy, 1963). The poor circulation and salinities in excess of 39‰ appear to favour precipitation of mud (Scoffin, 1987).

Carbonate platforms like the Great Bahama Bank have been extensive in the past and their productivity is related to the pattern of sea-level change. Such banks appear to have been most productive of sediment during highstands (see Figure 2.5). The Holocene history of the Great Bahama Bank and adjacent banks changed as postglacial sea level rose. The extent to which the platform is rimmed by carbonate deposits influences the rate at which it sheds carbonate sediment to the adjacent deep-water environments, and the pattern of Holocene

Figure 2.16. Supratidal carbonate sediments on the leeward side of Joulters Cays on the Great Bahama Bank. The white sediments are composed of a range of bioclastic material and oolites. The mangrove, *Avicennia germinans*, has colonised this environment where it is subject to infrequent inundation by seawater. The lateral roots of the mangrove beneath the sediment surface extend a series of vertical breathing pencil roots called pneumatophores. The root system has the effects of both stabilising the sediment, and locally mixing (bioturbating) it.

Table 2.6. Major depositional environments of the Bahama banks

Environment	Location	Energy setting	Biota	Sediment composition and structure	Mineralogy
Reef	Bank margin, low turbidity	High wave energy, unidirectional	Diverse coral and algal communities	Coralgal, biothermal reef growth and lithification	Aragonite > high Mg calcite
Oolite shoal	Bank margin, high turbidity	High tidal energy, bi-directional	Little organic activity	c.100% ooids, actively precipitated and formed into rippled and cross-bedded shoals	Aragonite (c.100%)
Oolitic	Bank surface, exposed	Medium energy, some grain mobility	Seagrass, algae and molluscs	Ooid-dominated, moderately skeletal sand, bioturbated	Aragonite (minor high-Mg calcite)
Grapestone	Bank surface, mildly exposed	Medium energy, fines winnowed, but sand cemented	Seagrass and algae	Oolitic sands cemented into grapestone aggregates	Aragonite (minor high-Mg calcite)
Pellet mud	Bank surface, sheltered	Low energy, poor circulation	Few biota, sparse seagrass, polychaete infauna	Mud, dominated by faecal pellets	Aragonite > high-Mg calcite
Mud	Bank surface, sheltered	Low energy, poor circulation	Few biota	Mud, aragonite needles with minor pellets	Aragonite ≫ high-Mg calcite

bank-margin sedimentation on Abaco is related to sediment (aragonite) accumulation in the deep-sea basin (Dix and Kyser, 2000).

2.5 Paleoenvironmental analysis and sedimentary history

Landforms can provide evidence about past environments that enables a reconstruction of coastal evolution. In the case of rocky coasts, the evidence is erosional and generally difficult to set into a time frame. However, on sedimentary coasts the stratigraphy of coastal sediments provides important geomorphological clues. This section summarises approaches to examining coastal sediments in relation to environment and time of deposition. The sequence of sedimentary deposits can be described in one of five ways: (i) in terms of the petrological characteristics of sediment (lithostratigraphy); (ii) in terms of the ecological changes that occur (biostratigraphy); (iii) in terms of the time at which sedimentary units were deposited (chronostratigraphy); (iv) in terms of shape of landforms or sediment units (morphostratigraphy); or (v) in terms of cycles of deposition that can be detected in the sedimentary column or core (allostratigraphy). The methods available for dating fossil material in coastal sediments are also discussed briefly.

2.5.1 Paleoenvironmental analysis

Determining the environment of deposition is important in geomorphological reconstruction. This can be undertaken on the basis of physical characteristics of the sediment, for instance grain size and sorting, or in terms of sedimentary structures. Sedimentary structures include features such as bedforms, including ripples and swash marks, laminations and various biologically formed features (Collinson and Thompson, 1982). The preservation of ripple patterns can serve to denote the hydrodynamic conditions in operation at the time of deposition and current bedding can indicate the velocity and direction of flow. For instance, deposition of sequences of sand and mud laminae recording a 14-day neap–spring tidal cycle, called tidal bundles, reflect tidal periodicity and indicate intertidal or subtidal deposition in a tide-dominated environment.

Environment of deposition can often be reconstructed on the basis of ecological evidence. In some cases, it may be physical preservation of some form that results from biological activity. For instance, trace fossils are the physical preservation of the disturbance features left by organisms whose bodies are too soft to have been preserved. Burrowing patterns and other signs of reworking of sediment by organisms, called bioturbation, can destroy sedimentary structures and mix sediments. The form of tree trunks, or root structures (rhizoliths) can be preserved, although none of the original biological material remains. Where fossil organisms are preserved, ecological assemblages are generally compared with extant communities to reconstruct past environments. For instance, relationships between sediment composition, organisms (biota) and energy setting are demonstrated for the Great Bahama Bank in Table 2.6. Understanding the patterns of modern environments enables sedimentary units, termed facies, to be interpreted in subsurface sediment samples and cores. Facies analysis can be especially detailed in the case of reefs where much of the past community may be preserved, and can be compared with coral and algal reef growth typical of modern or Holocene reef. In the Caribbean, it seems that similar facies characterised Last Interglacial reefs which show little change in coral assemblages (Pandolfi, 1996).

Corals in reefs can be preserved in their position of growth, although there are many cases where coral framework has been broken, transported and redeposited. Other organisms, however, make more subtle contributions to the deposits. Shells, or broken shells (shell hash) can be transported from their position of growth into death assemblages. Death assemblages rarely resemble life assemblages at the same location (Pandolfi and Minchin, 1995). Microfossil analysis is particularly valuable because small sediment samples can contain a large number of individual specimens enabling a greater quantification and degree of resolution. There have been several pollen reconstructions specifically on coastal vegetation. For instance, changes relating to saltmarshes (Clark, 1986) and mangroves have been analysed using pollen analysis (Grindrod, 1988) and can enable broad changes of coastal positions to be discerned (Grindrod *et al.*, 1999).

Foraminifera have played a very important role in coastal paleoecology. In some cases they can be related specifically to a particular part of the tidal range (Scott and Medioli, 1978, 1986; Gehrels, 1994). In areas of large tidal range, discrimination of paleotidal range is possible where the significance of particular species has been demonstrated in the modern range (Haslett *et al.*, 1998; Horton *et al.*, 1999).

Ostracods and diatoms also allow definition of detailed habitats (Devoy, 1982). Such reconstructions are based on the assumptions that environmental conditions constrain the occurrence of these organisms, their tolerances have remained the same, and assemblages are in equilibrium with optimal conditions. Greater confidence is possible where several indicators are used; for example combined use of diatoms, foraminifera and macrophytes indicates that elevation is a key factor in intertidal zones (Patterson *et al.*, 2000).

2.5.2 Dating coastal landforms

Coastal geomorphologists could only speculate on the history and rate of change of coastal landforms before the advent of geochronological dating. A range of dating techniques are now available and can be used in conjunction with morphological parameters, such as degree of dissection, horizon development in coastal soils, archaeological and other evidence that was traditionally used to interpret coastal landform sequences.

Table 2.7 summarises a series of dating techniques with wide application in reconstructing the dynamics of coastal environments during the Quaternary, outlining the method, its basis, age range, applicability, precision and principal limitations (after Williams *et al.*, 1998). Each of the dating procedures is based on assumptions, and caution needs to be exercised in interpreting age estimates in terms of the evolution of landforms (Aitken, 1990; Geyh and Schleicher, 1990; Noller *et al.*, 2000).

Dating methods are generally divided into those that are relative, indicating that a particular sedimentary feature, or fossil, is older than another, and those that produce an absolute age estimate. There are four approaches to dating: (i) there are techniques that are based on the rate of radioactive decay of a radioisotope (such as radiocarbon); (ii) there are techniques based on the relative balance of parent and daughter radioisotopes (such as uranium-series dating); (iii) there are techniques based on accumulation of electron traps (such as luminescence and electron spin resonance); and (iv) there are techniques based on a chemical change over time (such as amino-acid racemisation). Radiometric techniques are at a more advanced state of refinement than other methods, and hence produce results that can generally be interpreted with greater confidence than the others (Noller *et al.*, 2000).

Different dating techniques can be applied to different materials and are most suitable within various age ranges. Significant advances in processing have broadened the opportunities for application of radiometric dating techniques, in particular accelerator mass spectrometry (AMS) in the case of radiocarbon, and thermal ionisation mass spectrometry (TIMS) in the case of uranium-series dating. These analyses enable much smaller samples to be analysed, considerably extending the range of applications.

Radiocarbon dating

Radiocarbon dating uses the radioactive decay of ^{14}C (formed in the atmosphere by cosmic radiation bombarding ^{14}N) which is taken up as carbon dioxide by a wide range of organisms, or precipitated in sedimentary environments. Most carbon is in the form of ^{12}C (98.9%), a

Table 2.7. Principal Quaternary dating methods; the range over which they are most appropriate and major limitations and cautions in their use

Method	Basis	Range	Best materials	Precision (%)	Principal cautions
Conventional radiocarbon	^{14}C decay by gas-proportional or liquid-scintillation radiometry	0–40 ka	Wood, charcoal, shell, coral, peat etc. (several grams)	~1	System closure or contamination Reservoir effects Calibration Background
AMS radiocarbon	^{14}C ion mass spectrometry	0–40 ka	Wood, charcoal, shell, coral, peat etc. (milligrams)	~1	System closure or contamination Reservoir effects Calibration Background
Uranium-series	U parent/daughter isotope radiometry	0–250 ka	Coral, speleothem, eggshell	~1	System closure or contamination
TIMS U-series	U parent/daughter spectometry	0–500 ka	Coral, speleothem, eggshell	<0.5	System closure or contamination
Potassium-argon	^{40}K decay to ^{40}Ar mass spectrometry	>500 ka	Igneous and metamorphic rocks	0.5	Background argon
Thermoluminescence (TL)	TL accumulation since deposition	0–500 ka	Quartz, feldspar, tephra	10–15	Dosimetry of sample Bleaching before deposition
Optically stimulated luminescence (OSL)	OSL accumulation since deposition	0–500ka	Quartz, feldspar	7–10	Dosimetry of sample
Electron spin resonance (ESR)	Electron traps filled since deposition	0–1000 ka	Coral, teeth, calcite	10–20	Dosimetry of sample
Amino-acid racemisation (AAR)	L- and D- isomer ratio, chromatography	0–500 ka	Shells	>15	Temperature effect
^{210}Pb	Decay of ^{210}Pb	0–100 a	Mud	–	Supported/unsupported ratio
Cosmogenic radioisotopes	^{10}Be, ^{26}Al, ^{36}Cl mass spectrometry	various	Bare rock	variable	Exposure and erosion

Source: Based on Williams *et al.*, 1998.

lesser amount is ^{13}C (1.1%), and a very small amount is ^{14}C (1×10^{-10}%). Radiocarbon, which is likely to have been in equilibrium with atmospheric concentration, commences radioactive decay at the time of death of an organism. Radiocarbon dates are a measure of the extent of decay, and are conventionally reported in terms of years before present (BP), where present is defined as 1950. Since bomb testing, which occurred especially in the 1950s, the amount of ^{14}C in the atmosphere has been artificially increased.

It is necessary to find appropriate material, such as wood, shell, coral, or other biogenic carbonate, suitable for radiocarbon dating in order to determine an age relating to a coastal event. Laboratories generally measure ^{14}C either by gas proportional or liquid scintillation counting, and report the date with one (or sometimes two) standard deviation errors (±67% or 95%). Marine carbonate, where the reservoir that the living organism has taken carbon from is the ocean, also requires a correction for reservoir effects because the ocean is depleted in ^{14}C giving it an 'apparent' age; the value is generally about 400 years.

Radiocarbon dates are usually reported as conventional ages in radiocarbon years, corrected for fractionation effects. Paired radiocarbon and uranium-series dating has indicated that radiocarbon years do not correspond exactly to calendar years. Using corals from Barbados, the timing of the last glacial maximum that seems to have occurred around 18 000 radiocarbon years BP, corresponds to an absolute age of around 22 000 years BP by uranium (U)-series (Bard *et al.*, 1990b). A calibration curve is available, but variations in atmospheric carbon isotope concentration mean that one radiocarbon age can actually correspond to more than one period in calendar years (Stuiver *et al.*, 1998).

In the coastal environment, radiocarbon dating has been widely used to determine chronology. Radiocarbon should be reported as conventional ages, with laboratory code, with details of environmental correction or calibration clearly identified to assist subsequent researchers. In this book, many studies of the past dynamics of the coast are examined and the ages used are those reported in the relevant literature. In some cases, geomorphological reconstructions based on numerous dates have been generalised using lines that connect points of presumed similar age, termed isochrons, and ages have been rounded to thousands of years (ka). This can be adequate to understand the longer-term morphodynamics of coasts. However, conversion to calibrated ages would be necessary if more precise correlation with archaeological or historical evidence is intended.

Radiocarbon dating has been used over the past 50 years to determine the rate of development of coastal landforms and the pattern of sea-level change. There have been significant advances in laboratory practice and recent development of accelerator mass spectrometry (AMS) radiocarbon dating enables the analysis of samples of only a few milligrams of carbon, significantly extending the range of materials that can be aged. However, it remains the case that there are underlying assumptions in using an age determined as the time of death of an organism or organic material to constrain the time of development of a coastal landform (see Williams *et al.*, 1998). There are issues related to whether the sample might have been contaminated by extraneous radiocarbon and whether it was contemporaneous with the landform or has been reworked and subject to transport since it died (post-mortem). Although radiocarbon dating can be used back until about 40000–50000 years BP, the amount of ^{14}C is close to background and minor contamination can yield an erroneous date, meaning that the practical limit of dating is likely to be considerably less (Chappell, 1987).

Other dating techniques

Uranium-series dating has been widely applied in coastal geomorphological studies, particularly to Pleistocene reef deposits. U-series dating involves determining the extent to which there is equilibrium between uranium isotopes and their daughters (generally thorium) on the decay chain towards lead. Methods have been developed that can be applied with great precision to coral. Corals incorporate uranium from seawater into the carbonate they secrete, but thorium (Th) is not incorporated into the skeleton and appears primarily as a result of decay of the uranium. Thorium accumulates until it occurs in equilibrium with the levels of uranium. The age of the coral determined by traditional measurements (radiometry of alpha particles) is based on the degree to which the coral is out of equilibrium between uranium and its daughter isotopes in the decay chain. Considerably improved precision is possible using thermal ionisation mass spectrometry (TIMS); ages on corals from the Last Interglacial have been reported with error ranges of less than a thousand years (Edwards *et al.*, 1987; Chen *et al.*, 1991; Stirling *et al.*, 1998). Similar approaches can be applied to speleothems (stalactites and stalagmites), but are less effective with molluscs, which do not appear to remain closed systems after death.

There are several luminescence and related dating techniques which are more complex than radiometric dating but measure the progressive accumulation of trapped electrons within minerals (Williams *et al.*, 1998). The first is thermoluminescence (TL), which measures the luminescence given off when these trapped electrons are released by heating. TL dating is very effective at determining pottery ages where the heating process is known to zero TL, so that the TL which is measured is known to have accumulated after firing of a pot. However, sediment grains, generally quartz, are harder to date because there is less certainty that TL has been set to zero by bleaching under sunlight (Price *et al.*, 2001).

TL ages are determined on the basis of the equivalent dose (TL acquired) that the sample appears to have received divided by the natural dose rate. The rate at which the sample is being bombarded (dose rate) can be determined, either through direct dosimetry, or by measuring the uranium, thorium and potassium in surrounding sediment. Variations in water content over the period of burial can have effects on the dose rate. Recently, measurement of trapped electrons by optically stimulated luminescence (OSL), including single-grain analyses, have greatly extended the range of application of luminescence dating techniques (Murray and Roberts, 1997; Murray-Wallace *et al.*, 2002). Nevertheless, issues associated with sample collection, determination of dose rates (radiation by which the defects accumulate), and many technical issues involved with the physics of the analysis, make application of TL and OSL dating an issue requiring specialist input. Electron spin resonance (ESR) is also based on accumulation of imperfections in the lattice structure.

Amino-acid racemisation (AAR) is based on the chemical change of mirror image carbon bonds (isomers) occurring in amino acids (Lowe and Walker, 1997). In living organisms, these are almost wholly left-handed L-(levo), but after death of the organism racemise (or epimerise if there are two asymmetrical carbon bonds) to half L- and half D-(dextro). The rate of racemisation differs for different amino acids, and is also highly dependent on temperature. This technique was effectively used to discriminate fossil beaches on the Gower Peninsula of Wales (Davies, 1983), and has recently been extended to whole-rock analyses enabling development of age sequences (aminostratigraphy) for calcareous biogenic rocks on eolianite islands (fossil dune, see Section 6.5.5) of Bermuda and the Bahamas (Hearty *et al.*, 1992; Hearty and Kaufman, 2000), and southern Australia (Murray-Wallace *et al.*, 2001).

2.6 Summary

The tectonic and geophysical setting, particularly the relationship of the land to the sea, has varied in the past. On the geological time scale of millions of years, it has been shown that the variable rate of plate-tectonic movements exerts a control on ocean basin volume which can be expressed in the development of abrasion terraces, platforms, and sequence stratigraphy. During the Quaternary, sea-level variations occurred with a periodicity of about 100 000 years driven by glacio-eustatic fluctuations, reflecting the exchange of water between the ice caps and the oceans. Sea-level changes in the Holocene show considerable spatial variation driven by glacio-isostatic and hydro-isostatic responses to the redistribution of mass over land and oceans. Sea level is affected by astronomical effects, meteorological effects, oceanographic effects (hydro-isostasy, steric effects), and isostatic responses to ice (glacio-isostasy), volcanic load (volcano-isostasy), sediment load (sedimento-isostasy) and earth's interior (thermo-isostasy, density changes).

These land–sea variations, particularly Quaternary sea-level fluctuations, have affected all coasts, and most coasts are undergoing some continued readjustment, at a gradual and largely imperceptible rate, to redistribution of ice and water masses. Inheritance, based on geological conditions or topographic expression, plays an important role in coastal geomorphology. On many coasts, the modern shoreline is revisiting a part of the topography that has seen previous sea levels. For those tropical or subtropical shorelines that are distant from any ice sheets the shoreline may roughly coincide with the Last Interglacial shore. This can have significant consequences in terms of gradient or sediment availability. Elsewhere the coast is adjusting to other tectonic factors and can be rapidly uplifting or subsiding.

It has been shown in Section 2.2 that collision coasts are more likely to consist of steep rocky shorelines than are trailing-edge coasts which are more likely to have been centres of accumulation of sedimentary basins. These differences reflect several features, including the steepness of the shoreface and continental shelf, the supply of sediment from the hinterland and availability of sufficiently sheltered coastal embayments in which that sediment can be deposited.

The lithology and weathering present the raw materials that either form the coast directly or contribute to the sediments from which landforms are constructed. Clastic shorelines dominate many of the continents, but carbonate sediments can be significant where organic production is high, and there are extensive tropical carbonate environments, such as coral reefs and carbonate banks. The rocks and sediments not only form the materials that are worked on by a variety of coastal processes, described in the next chapter, but they also contain a paleoenvironmental record which can provide insights into the pattern and timing of deposition.

Chapter 3
Coastal processes

> The real waves look and act nothing like the neat ones that endlessly roll down the wave channel or march across the blackboard in orderly equations. These waves are dishevelled, irregular, and moving in many directions. No alignment can be seen between a series of crests; some of the crests actually turn into troughs while we are watching them. (Bascom, 1964)

Coastal landforms are shaped by a range of processes. This chapter examines the principal sources of energy by which these materials are moved. The processes are considered individually, although in practice a stretch of coast is likely to be influenced by several processes, acting simultaneously or in sequence. A modern coastline is the outcome of processes that operate today and those that have operated over time, and many landforms have been partially or wholly inherited from former conditions.

Coastal landforms can be composed of rocks or sediments, as indicated in the previous chapter. The movement of sediment is a primary control on coastal morphodynamics and this chapter begins with a consideration of the sedimentary processes involved with erosion, transport and deposition of this material.

The principal source of energy for most coastal systems comes from waves. The generation of waves and their transformation as they approach the shore are described. Although incident waves (the waves arriving at a beach at the time of study) can be seen to move sediment, it has become clear that various longer-period events also play a very significant role in shaping the coast, including infragravity waves and extreme storms and other disturbances, and these are also examined.

On some coasts, tidal processes are significant and tides and their variability are examined in the latter part of this chapter. Tidal processes become particularly dominant in estuaries or other embayments, and the adjustment of tidal flows within these systems is introduced, with further discussion in Chapter 7. Other oceanographic processes are associated with ocean currents, sea ice, and ocean–atmospheric circulation changes, such as the El Niño–Southern Oscillation phenomenon.

The coast is also shaped by several subaerial processes. Included in this category is river action that can supply sediment to the coast especially in delta environments (discussed in detail in Chapter 7). Wind can be important, particularly in generating waves, but also in moving dry sediment and building dunes. In the case of rocky or other bold coasts, hillslope processes can also shape coastal landforms.

Biological processes are important on many coasts. Organisms can have influences varying from minor roles that alter sediment properties or affect the surface of substrates, to boring and rasping, and in some cases secretion of sediment and construction of framework. Coral reefs are especially significant and are considered in detail in Chapter 5. Plants play substantial roles in trapping or binding sediments on dunes, and the upper intertidal areas of muddy coasts (see Chapter 8). Finally the implications of extreme events are examined, placing the incident processes in the context of the longer-term pattern of process operation.

3.1 Historical perspective

Coastal landforms are built through the erosion or deposition of sediment and links between sediment entrainment and the physics of fluid dynamics have been important themes in coastal geomorphology. Hjulström (1935) derived an empirical relationship between the velocity of flow and movement of sediment particles of different sizes. He demonstrated a non-linear relationship between flow velocity and grain diameter which is shown in Figure 3.1, and which still forms a useful framework within which to examine sedimentary processes. Sediments are deposited when the flow decreases below the fall velocity; the larger grain sizes are deposited before the smaller grain sizes which continue to be transported. Clays can flocculate and are then held cohesively together by electrolytic forces that make them more difficult to entrain than fine sand. The critical threshold for initiation of sediment motion was examined on a flat-bed surface by Shields (1936) who derived a dimensionless parameter in relation to median grain size. Movement of sediment occurs initially as traction, then as saltation, and finally as suspended load (Bagnold, 1940, 1941). In practice, sediment is rarely uniform and a series of factors influence cohesiveness, including sediment sorting as well as biota. Although these sediment principles were developed for sediment transport in rivers, they have been applied to other situations in which sediment is eroded, transported and deposited including wave and tidal processes.

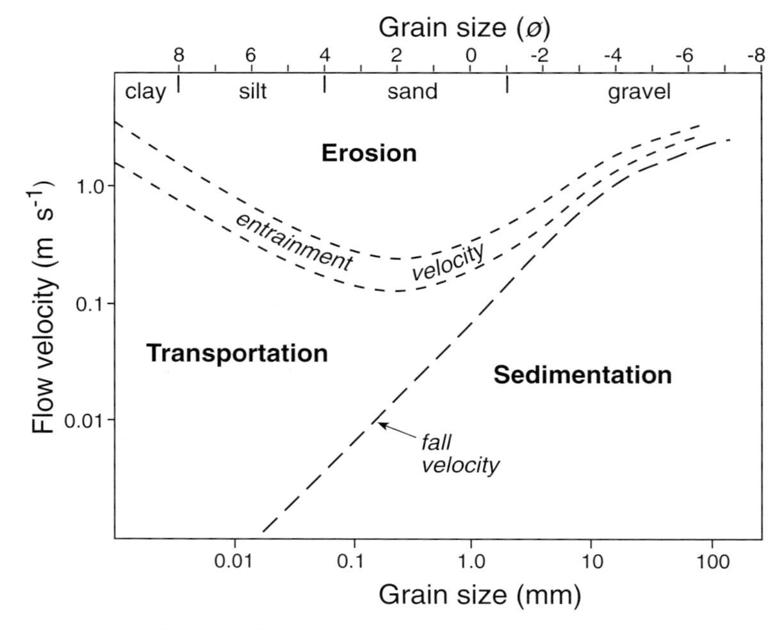

Figure 3.1. Relationship between sediment entrainment and deposition and grain size and fluid velocity (based on Hjulström, 1935).

The physics of waves and tidal processes has a long history of study (Bascom, 1964). Aristotle recognised that wind created waves on the ocean. The basics of deep-water wave theory have been understood for well over a century, but waves generally undergo significant transformations as they move into shallow water. By the 1850s, Airy had developed what is called linear wave theory (Airy, 1845), which treats waves as a sinusoidal form. This is an oversimplification but it works well for many purposes. In the 1880s, Stokes developed a more appropriate second-order theory of waves, differing slightly from that produced by the simpler Airy theory (Stokes, 1847). Further refinements to wave theory were made by Rayleigh (1876, 1877), Reynolds (1877) and McCowan (1894).

The wind transfers energy to waves, which appear to propagate as a result of pressure differences caused by the sheltering effect of successive crests (Jeffreys, 1925). Waves are often described in terms of significant wave height, which is the mean height of the highest one third of waves. It was conventional to use this before it was possible to measure the entire spectrum of water-level variation. Methods were developed to forecast significant wave height and period in relation to weather conditions based on correlation of wind velocity, fetch and storm duration from a series of observations (Sverdrup and Munk, 1947). The approach was extended by Bretschneider (1958), and became known as the S–M–B (Sverdrup–Munk–Bretschneider) method (CERC, 1984).

It has been recognised that wave action can be constructional on some coasts and destructive on others (Cornish, 1898; Lewis, 1931). Coastal geomorphology has incorporated the physics of wave dynamics since the mid 20th century (Shepard, 1948; Kuenen, 1950; King, 1972). Laboratory studies have played a central role in linking physical properties and the potential to do work in a whole series of areas of geomorphology (Bagnold, 1940, 1941, 1966). It has become increasingly recognised, however, that wave theory inadequately describes the complexities of what actually happens (Bascom, 1964; see quotation at beginning of chapter). Although waves exert a primary control on the geomorphology of many coastal landforms, particularly beaches (Komar, 1976), it has become apparent that longer-period variations are also significant, and these are described in more detail below.

Tidal variations in water level have been recognised for a long time. Basic tidal theory derives from Newton's theory of gravity, and was further refined by Laplace in the 18th century. Tidal records for various ports have been kept for navigational purposes for much of the 20th century and the problem of predicting astronomical tides within the oceans has, in essence, been solved (Pugh, 1987; Cartwright, 1999). Tidal records have recently also become significant in relation to whether the level of the sea is changing (see Chapter 2). Other periodic changes in ocean conditions, such as the El Niño–Southern Oscillation phenomenon, have also been recognised as significant in terms of oceanographic conditions (Bjerknes, 1969), with implications for geomorphological processes. These and other coastal processes are examined in more detail in the remainder of this chapter.

3.2 Sedimentary processes

The processes of sediment erosion, transport and deposition are discussed in this section. On the coast, rock fragments and sediment are generally transported in a fluid, usually water, although occasionally also air or ice. The rate at which these processes operate depends on the rock or sediment, including both the characteristics of individual grains and bulk properties (such as density, porosity, shape, size, sorting, fall velocity and angle of repose) and on the physical characteristics and mechanics of the fluid. Application of fluid dynamic principles to the transport of coastal sediments involves a high degree of mathematical simplification in relation to how idealised grains might act under steady flow of incompressible fluids (termed inviscid). In the field, conditions rarely behave as these simplified assumptions presume. For instance, flows are generally not ideal and there are fluid–grain and grain–grain interactions which can occur over a range of spatial and temporal scales, influencing the initiation of motion of grains by currents, the behaviour of grains during transport, and their settling (Bagnold, 1966).

3.2.1 Sediment entrainment

Whether or not a fluid (water or air) entrains sediment grains depends on the density and viscosity of the fluid and the velocity of flow. Density is a measure of the weight per unit volume of the fluid, often expressed in relation to water by the dimensionless term 'specific gravity'. Density influences the magnitude of forces and the effectiveness with which disturbances such as waves travel through the fluid. Fresh water has a density of 1000 kg m^{-3}, whereas sea water can be 1028 kg m^{-3} (specific gravity 1.028). Viscosity is a measure of the resistance of the fluid to deform or to flow. It is measured as a ratio between the shear stress (the shearing force applied per unit area) and the rate of deformation, and is temperature dependent. Velocity is the rate at which a fluid flows; it is a vector having a magnitude and direction.

Coastal geomorphology is concerned with a range of water movements, including the backward and forward motion beneath waves, river-influenced flow in delta channels, and flood and ebb movements in tidal creeks and estuaries. Flows are often complex in the boundary layer over the surface of a substrate. It is especially important whether flow is laminar or turbulent (Figure 3.2). Laminar flow, in which streamlines are parallel, occurs where velocity is low and viscosity is high, whereas turbulence, in which there are random eddies, occurs where flow is rapid and viscosity is low. Many flows are turbulent; for instance, a strong wind demonstrates gustiness, and the surface of a river or estuary may appear to 'boil'. Velocity of turbulent flows is variable over time, but can be expressed as time-averaged 'mean' velocity. The threshold between these states is expressed by the Reynolds number, the ratio between inertial forces (mean velocity over a defined distance or depth) and viscous forces. Laminar flow occurs where the Reynolds number is <500, turbulent flow occurs where it is >2000, and transitional flow occurs in the range 500–2000 (Figure 3.2a). Flow in most coastal waters is turbulent, and turbulent eddies are often set up as flow crosses obstacles. Even in turbulent flows there is generally a viscous sub layer in the boundary layer directly over the surface of the substrate below turbulent flow, although in many cases this is irregular and is termed 'streaked'. The roughness of the bed, expressed either as grain roughness, or as form roughness, is also important in determining whether or not flows are turbulent (Hardisty, 1994).

a) Reynolds number

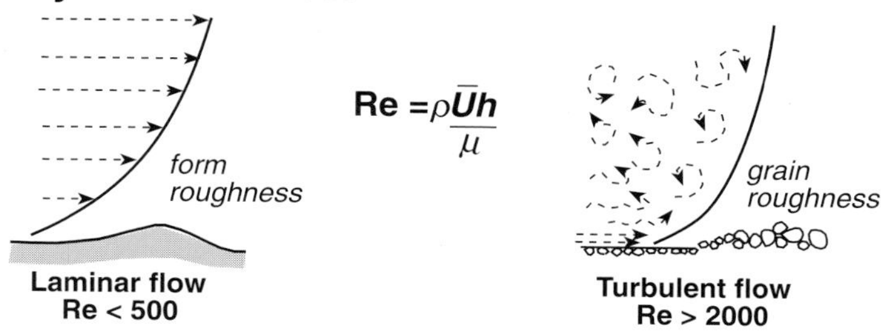

b) Froude number

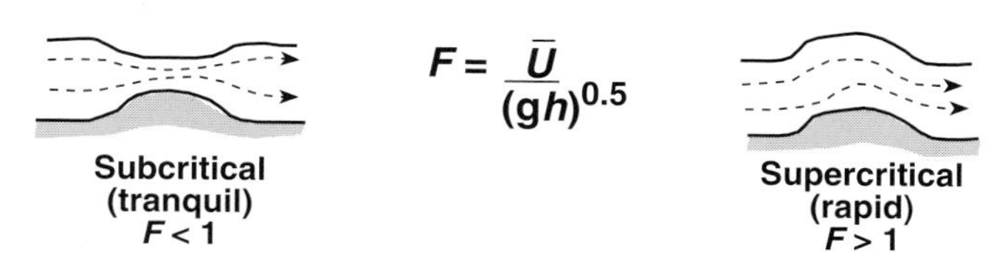

c) Forces on particles

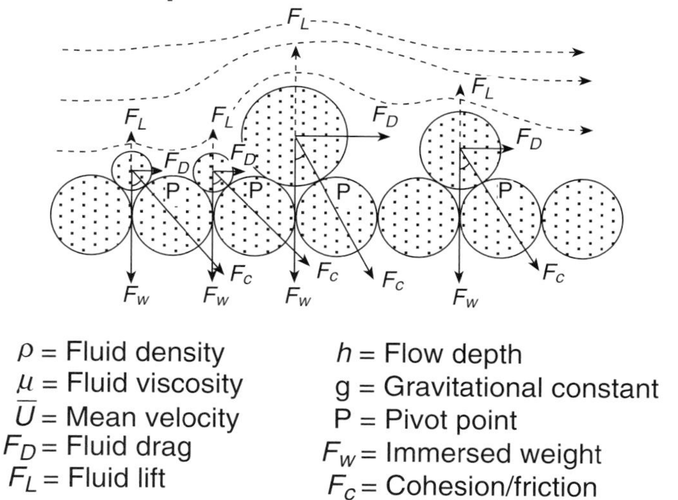

Figure 3.2. Fluid properties and the entrainment of sediment. (a) Definition of laminar and turbulent flow in terms of Reynolds number; (b) definition of subcritical and supercritical flow in terms of Froude number, and (c) the forces involved with sediment grain entrainment.

The flow behaves differently, on the basis of the ratio of inertial forces (characterised by the velocity of flow) to gravitational forces (characterised by the square root of the product of depth of the fluid and the gravitational constant, see Figure 3.2b). This is expressed by the Froude number (Figure 3.2b). The boundary between supercritical (or rapid) flow and subcritical (or tranquil) flow occurs at a Froude number of 1 when the total flow energy is at a minimum. If the Froude number is >1, flow is supercritical with inertial forces exceeding gravitational. At Froude numbers <1, flow is subcritical, and has a smooth and glassy surface because gravitational forces predominate. Abrupt changes from one state to the other are termed a hydraulic jump or hydraulic drop (Allen, 1994).

A fluid at rest exerts a static force due to its mass (termed hydrostatic in the case of water); its potential energy is a function of its mass and gravity. A fluid in motion has kinetic energy, a measure of the work done getting into that state of motion. It loses energy through friction, and the sum of potential and kinetic energy is a constant.

Material strength is generally defined as the ability to resist the principal erosional stresses; tensile stresses involve stretching, compressive stresses involve crushing the material, and shear stresses involve deformation by sliding. Whereas compressive strength can be important in relation to wave impact on rock surfaces (see Chapter 4), it is the stresses set up by shear of moving fluids over the sea floor that are the primary forces of geomorphological significance in the coastal zone. Rock generally has considerable material strength, and shear forces rarely manage to pluck or quarry material from a rock surface except where there are structural weaknesses or where the rock has been weathered by physiochemical processes, such as corrosion. Other erosional processes on rock include abrasion (or corrasion) which is the mechanical wearing, grinding and scraping by sediment already entrained (see Chapter 4). Cavitation occurs under very rapid water movement where pressure falls to vapour pressure and air bubbles form and collapse.

Sediments can be divided into cohesionless and cohesive. Sand and gravel are cohesionless, whereas fine-grained muddy sediments (comprising silt and clay) are cohesive with electrostatic forces holding grains together more effectively than gravitational forces separate them. Chemical bonds (intermolecular attractive forces) and organic films increase cohesion, and mean that greater forces are necessary to entrain them. In the case of cohesive muddy sediments, the stress applied results in a strain (change in unit length), which increases until the yield strength (plastic limit) or until they undergo liquefaction (van Rijn, 1993).

Figure 3.2c indicates the forces exerted on grains on a sediment bed. The bed shear stress, related to velocity of flow, is the force per unit area parallel to the bed and exerts fluid drag across the projected area of the grain. It is not possible to measure the shear stress directly; it is likely to be very variable in space and in time. Time-averaged velocity measurements at two or more depths, usually plotted as logarithmic height above the bed, indicate the slope of the velocity profile, and provide an estimate of shear stress. A sediment entrainment threshold occurs at a critical shear velocity, depending primarily on grain size. For turbulent flows over a plane bed, a dimensionless critical shear (termed Shields' parameter) relates flow velocity and grain diameter, but the nature of the relationship is complex where the substrate is irregular and grain sizes are mixed with individual grains protruding. If flow is constricted over a grain protruding from the bed, streamlines are accelerated and exert a vertical lifting force, termed fluid lift. Determining the degree of lift is difficult in turbulent flows because of the variable eddies. The resistance to movement comes from the gravitational force and depends on the immersed weight of the grain and the cohesion between grains. Movement occurs where the forces cause an individual grain to pivot over adjacent grains (Figure 3.2c).

The comparison of grain size between the grain subject to the drag and the bed on which it lies is important; it is possible for a coarse grain on a bed of finer grains to show less resistance around the pivotal point than shown by smaller grains (Komar, 1987). Differences of this type can lead to selection of grains in terms of size, and contribute to sediment sorting, the construction of bedforms such as ripples, and the concentration of heavy minerals, called placers (Komar and Wang, 1984). Other issues, such as grain shape, including roundness and sphericity, surface texture (microrelief or polished surfaces), and sediment bulk properties, such as density, moisture content, grain-size variability, packing or imbrication, are also important.

Variation in the threshold of motion for different grain sizes is indicated in the original empirical curve derived by Hjulström (1935) and shown in Figure 3.1. Although there has been more variability demonstrated for fine-grained sediments than the curve shows, it indicates the greater cohesion of muds. Above a grain diameter of 0.6 mm there is a more regular relationship between grain size and the velocity or critical shear stress needed to entrain sediment. This is often expressed as a dimensionless ratio called the Shields' entrainment function (van Rijn, 1993).

There are many reasons why these theoretical relationships do not adequately describe conditions on the majority of coasts. The generalisations are derived for ideal sediments, in which grains are uniform, spherical and of similar density. Most coastal landforms are composed of bimodal or irregular sediment-size distributions, and grains are of variable shape and density. In addition to initiation of sediment movement by fluid drag, motion may occur at slightly lower velocities where impact from grains already in transit increases entrainment, known as the impact threshold. In mixed sediment types stress–strain relationships are often non-linear and bedforms such as ripples further influence thresholds to sediment mobility. Under waves it is common for there to be a flurry of sediment grains moved forward under one wave, and then moved back again before the next wave. Ripples influence shear velocities and can result in entrainment and transport as suspended load. Under higher velocities the flurries of sediment movement can become sheet flow which obliterates bedforms (Haff, 1991).

3.2.2 Sediment transport

Once sediment is entrained, its transport depends on several factors. Sediment grains may be moved in water by sliding or rolling along the seafloor; at greater velocities impact from grains settling can trigger a burst of new grains to be suspended into the water column, a process termed saltation. Under the most rapid flows sediment is transported in suspension (Figure 3.1). Very fine silt and clay material (<50 μm) can be carried from the catchment by rivers as washload. Which mode of transport occurs is a function of grain size, and it is possible to recognise discrete sections of the sediment-size distribution which relate to suspension, saltation and rolling (traction) or surface creep (Visher, 1969). The fluid viscosity and the degree of turbulence are very important, but concentration of sediment in the fluid and the viscosity of the fluid itself alter as sediment is entrained.

Estimates of the amount of sediment transported by a fluid are limited theoretically in terms of derivation of suitable equations, and empirically in terms of field measurement of transport rate (van Rijn, 1993). There are several predictive formulae with the same general structure that enable bedload to be estimated, most based on the work of Bagnold (1966), but they do not take many of the complexities of flow into account (Allen, 1994). There are a number of methodological problems. First, it is assumed that the instantaneous rate of sediment transport is a function of the velocity at that point in time, although it requires a finite time for sediment to be entrained or deposited. It is rarely possible to record continuous velocity measurements at sufficient locations to calculate sediment movement. It is generally preferable to use integrated flow formulae that predict net mass transport over the passage of a single wave, tidal cycle, or longer, rather than instantaneous estimates of transport rate at maximum flow velocities (Hardisty, 1994). Estimation is also based on the assumption that there is unlimited supply of sediment for entrainment, which is rarely the case.

There are various approaches that can be adopted to measure suspended sediment transport. It is possible to measure suspended sediment concentration using optical or acoustic backscatter devices to get near-continuous records. This is a great advantage in comparison to bedload transport where sediment traps provide only an aggregate measure of sediment movement. However, attempts to use suspended concentrations and mean velocity to calculate mass transport generally yield poor results because the vertical concentration profile is highly variable (Black and Rosenberg, 1991).

Instantaneous measurements of sediment transport, even if they are possible, reveal little about the dynamics of landforms at longer time scales, such as the event scale. Often, the rate of process operation is very variable, and the net transport is not a linear function of instantaneous rates. For instance, under low-energy conditions incident wave processes dominate cross-shore sediment transport with net onshore transport, whereas under storm conditions, infragravity motions dominate, carrying sediment either onshore or offshore (Hardisty, 1994). Time-averaged patterns of transport can be estimated by more traditional methods (see Chapter 6). It is possible to use tracers, either fluorescent or radioactive markers to track sediment pathways. It can be more appropriate to determine transport rates by determining changes of volume over time, for instance comparing a series of beach profiles, giving an indication of net longer-term transport. The issue of extrapolating beyond an instantaneous or individual event scale to longer scales is examined in more detail in Chapter 9.

3.2.3 Sediment deposition

The hydraulic behaviour of sediment depends on individual grain characteristics. The settling or fall velocity is the rate at which a grain settles through fluid. It can be measured experimentally. After introduction into a fluid, a grain accelerates until its rate of settling reaches a constant rate, called the terminal fall velocity. The rate at which sediment settles out of suspension in a fluid of any particular viscosity is governed by Stokes' law, which depends on the ratio of particle and fluid density, but varies with the square of sediment diameter. It is also a function of grain shape, and is generally determined on ideal sediment (quartz spheres). Real sediment differs in many ways. For instance, fragments of shell differ in density and shape, and are likely to settle much more slowly than quartz spheres. Settling depends on the fluid viscosity that is partly a function of suspended sediment concentration; when this is high there are more frequent grain–grain contacts that alter settling characteristics.

In the case of clays, settling is greatly assisted by aggregation and flocculation. Clay particles are generally negatively charged and attractive forces are significant, especially in water, because clays have a large surface area relative to their volume. In seawater, a high electrolyte concentration encourages flocculation. This is especially important where fluvial waters discharge into estuaries (see Chapter 7).

3.3 Wave processes

The principal source of energy in the coastal zone, both for erosion and deposition, comes from waves. Waves are undulations on the surface of a liquid that move across the liquid, transferring energy but generally not resulting in a net transfer of mass. Waves can occur in many different media; those on the ocean surface can be generated from several sources, but all consist of a crest and a trough, and wave height and length can be measured as shown in Figure 3.3a.

Waves occur across a broad spectrum of sizes. Wave height is a measure of the elevational difference between a wave crest and the adjacent wave trough; it is twice wave amplitude. Wave heights can be measured in the open ocean (H_0) where they are believed to occur up to heights of about 35 m, or at the point where the wave breaks (H_b). However, height is independent of other wave variables which are interrelated. The time interval between passage of corresponding points on the wave profile, such as wave crests, at a fixed point in space, is termed the wave period (T). The horizontal distance between those points on successive wave profiles is termed wave length (L). The speed at which waves travel, termed phase velocity or celerity (C), is the ratio of wave length to wave period (in open ocean). Waves can also be described by frequency, which is the reciprocal of wave period and is measured in cycles per second or Hertz (Hz). Waves of up to 20 s period (0.05 Hz) are formed by wind on the surface of the ocean.

3.3.1 Types of waves

Some confusion can arise between the different types of wave. Most of the waves that can be seen on the surface of the ocean have been generated by wind. The term 'wind waves' is often used, but waves can travel outside the area where a storm is generating choppy conditions, and are then called 'swell'. Waves on the surface of the ocean are maintained by gravity and so have been called 'gravity waves'. The term 'incident waves' applies to those waves reaching a shore, consisting of wind waves (also called a 'sea') and swell.

The tide is a long-period wave; it is a result of the gravitational attraction of sun and moon (see Section 3.4). Although gravity is a cause of this wave form, it is not a gravity wave. The term 'tidal wave' has generally been used for a solitary and catastrophic deep ocean wave resulting from submarine eruption or earthquake, but the Japanese term

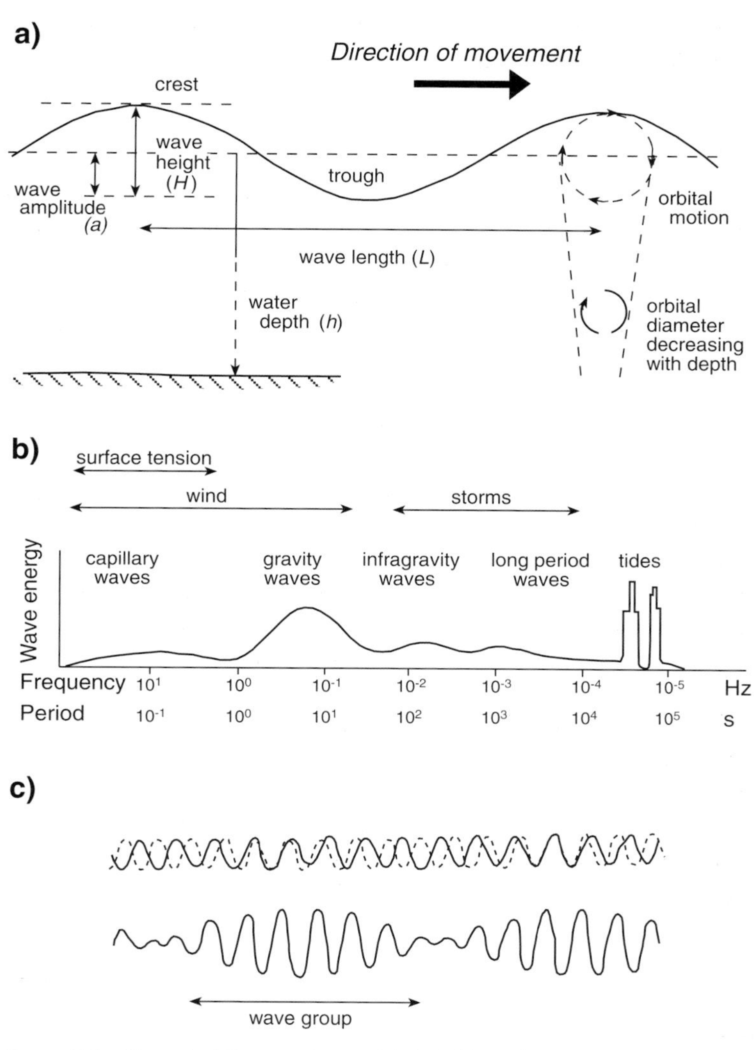

Figure 3.3. Characteristic wave forms. (a) Definition of terms; (b) wave period (or frequency) and the major wave forming factors; and (c) the schematic amalgamation of two regular wave trains of similar period to give groupiness.

'tsunami' is more appropriate for this type of wave, which is not related to the tide (see Section 3.5.4).

Waves can be divided into progressive waves, that move across the ocean, and standing waves, which fluctuate up and down but do not appear to move. Progressive waves radiate out from the area where they are generated, for instance ripples spread across the surface of a lake into which a pebble is dropped. Frictional stress between the wind and the sea transfers energy to the surface of the ocean and waves move in the direction that the wind is blowing. The more that the wind speed exceeds the wave speed, the steeper the wave will become (steepness is defined as wave height divided by wave length). The wave form moves progressively, but the water particles themselves do not undergo any net landward movement (in deep water), consisting of a series of circular orbits. The diameter of orbits decreases with depth (Figure 3.3a).

Standing (or stationary) waves occur where the water surface appears to rise and fall, but the crest does not move across the surface. If a bucket of water is agitated from side to side it develops a standing wave with its water surface slopping backwards and forwards. A standing wave often occurs where a wave is reflected off a solid vertical cliff or seawall; this is called 'clapotis' (see Chapter 4). The wave form is reflected off the wall and meets the next incoming wave. The water surface is broken into nodes, at which no change in surface elevation occurs, and interspersed antinodes, where it fluctuates up and down. Although the distinction between progressive and standing waves is very important, some situations combine elements of both types. For instance, the progressive swell waves that reach a beach may interact with standing waves that are trapped in the surf zone (see Section 3.3.5), and the way in which the tide moves up estuaries can show both progressive and standing wave characteristics (see Section 3.4.2).

Two other types of waves are significant. Seiches are standing waves that occur within closed or partially closed basins. They develop a resonance where the wave length is twice the effective basin length, or in the case of open-ended basins where the wave length is four times the basin length. These surface disturbances can then slosh backwards and forwards independently of the forcing mechanism that began the water motion. A bore differs from a wave in that the mass of water propagates forward in a bore, also called a 'soliton'. If wave height (H) is less than a fourth of water depth (h) ($H/h < 0.25$), an undular bore, comprising a series of secondary waves, occurs. If wave height is larger in comparison to water depth ($H/h > 0.75$), a fully developed turbulent bore occurs. The movement of mass of water, rather than just wave form, results in rapid dissipation (loss) of energy through turbulence and friction and a turbulent bore loses energy more rapidly than an undulatory bore (Komar, 1998).

There are many aspects of wave theory that have application to the coastal zone and these are examined in detail in Komar (1998). In the following sections, wave processes are described for a classic beach topography in which shoaling, breaking and surf zones can be recognised, with swash and backwash on the beachface (see Figure 1.1). Wave generation in the open ocean and transformation of waves through a shoaling zone to the point where they break is of great significance to processes of beach sediment movement and deposition. However, there are many other situations in which waves exert an influence on the coastal zone, and it should not be assumed that all of these processes occur on all shorelines. In Section 3.4, it will be shown that tidal processes often operate as a wave, and that aspects of wave theory are important for understanding rates of tidal process operation. Waves may encounter coasts that are steep, such as cliffed coastlines, or the steep forereef of coral reefs, in which case there is relatively little nearshore transformation as waves approach the shore. Local wind waves can be generated in estuaries or coastal lagoons and large lakes, in addition to the storm and sea-breeze waves that impinge on exposed beaches.

3.3.2 Wave generation and movement

Most ocean waves are created by tangential drag exerted on the surface of the ocean by winds of speeds in excess of 1 m s^{-1}. Waves grow in size, depending on wind speed, and their upper limit is set by gravity and surface tension. Small scale capillary waves with typical sizes <2 mm high and <20 mm long result from, and are maintained by, surface tension. There are a series of oscillations which occur at frequencies longer than 20 s termed 'infragravity waves' (see Section 3.3.5), and storms can also develop longer-period waves (Figure 3.3b). The tide is a large wave driven by the moon and the sun (see Section 3.4); earthquakes can generate solitary rapidly travelling tsunami waves (Komar, 1998).

The energy in a wave consists of potential energy (based on static forces) and kinetic energy (based on wave motion). Wave energy is related to the square of wave height (Table 3.1); it is also called 'energy density' because it represents the sum of energy integrated over one wave length, per unit length of wave crest. Wave energy varies as wave height changes during shoaling. The rate at which energy is carried along by the waves per unit length of wave crest, or energy flux, also called 'wave power' (power per unit wave-crest length), remains relatively constant as waves approach the shore. It is a function of wave height and wave group speed (Table 3.1).

The incident waves observed at a shore are unlikely to conform to the simple ideal form shown in Figure 3.3a for a number of reasons. First, there are significant transformations associated with the coastal boundary zone in which waves interact with the topography of the seafloor. These interactions become particularly important as water depth (h) decreases towards the value of wave height (H) and the wave shoals and breaks (see Section 3.3.4). Second, waves are rarely developed from a single storm or impulse. Although it is convenient to treat a train of

Table 3.1. Equations for Airy or linear wave theory

	Deep water ($h/L>0.25$)	Transitional water depth (general expressions) ($0.25>h/L>0.05$)	Shallow water ($h/L<0.05$)
Wave phase velocity (C)	$C=\frac{gT}{2\pi}$	$C=\frac{gT}{2\pi}\tanh\frac{2\pi h}{L}$	$C=(gh)^{0.5}$
Wave length (L)	$L=\frac{gT^2}{2\pi}$	$L=\frac{gT^2}{2\pi}\tanh\frac{2\pi h}{L}$	$L=T(gh)^{0.5}$
Wave group velocity (C_g)	$C_g=0.5\left(\frac{gT}{2\pi}\right)$	$C_g=0.5\left(\frac{gT}{2\pi}\right)\left[1+\frac{2\sigma d}{\sinh 2\sigma d}\right]$	$C_g=C=(gh)^{0.5}$
Energy density (E)	$E=0.5\rho gL\left(\frac{H}{2}\right)^2$	$E=0.5\rho gL\left(\frac{H}{2}\right)^2$	$E=0.5\rho gL\left(\frac{H}{2}\right)^2$
Energy flux (Ef)	$Ef=0.25\rho g\left(\frac{H}{2}\right)^2\frac{gT}{2\pi}$	$Ef=0.25\rho g\left(\frac{H}{2}\right)^2\left(\frac{gT}{2\pi}\tanh\frac{2\pi d}{L}\right)\left(1+\frac{2kh}{\sinh 2kh}\right)$	$Ef=0.5\rho g\left(\frac{H}{2}\right)^2(gh)^{0.5}$
Radiation stress (S_{xx}) (onshore)	$S_{xx}=\frac{1}{2}E$	$S_{xx}=E\left[\frac{1}{2}+\frac{2kh}{\sinh 2kh}\right]$	$S_{xx}=\frac{3}{2}E$
Radiation stress (S_{yy}) (longshore)	$S_{yy}=0$	$S_{yy}=E\left[\frac{kh}{\sinh 2kh}\right]$	$S_{yy}=\frac{1}{2}E$
Radiation stress (S_{xy}) (x flux of y momentum)	$S_{xy}=\frac{1}{2}E\sin\alpha\cos\alpha$	$S_{xy}=\frac{1}{2}E\left[1+\frac{2kh}{\sinh 2kh}\right]\sin\alpha\cos\alpha$	$S_{xy}=E\sin\alpha\cos\alpha$

Notes:
g = 9.81 m s^{-2}; T, wave period; $\pi=3.1416$; $E=2.7183$; H, wave height; h, water depth; $\sigma=2\pi/T$; $k=2\pi/L$; ρ = density; α, angle of wave approach; x, y, z, cartesian coordinates.

waves as a simple parcel of wave forms, a single storm will generate a whole spectrum of waves of differing heights and lengths. Third, their character changes as they travel, and waves from different areas, and of different sizes, interact. Wave trains of similar frequency combine to give an irregular surface as shown schematically in Figure 3.3c. When crests from the two wave groups coincide, the crest height is increased, but when they are offset, crest height is diminished. This effect results in water-level variations at infragravity frequencies termed wave groupiness, discussed in Section 3.3.5. Wave groupiness can be observed in the surf zone, as sets of several larger wave crests that recur with a period of several minutes, referred to as surf beat.

In the area where waves are generated, the sea surface becomes very choppy as individual waves are shaped into sharp angular crests with smaller wavelets and ripples superimposed on them. The height of the wind waves that are generated depends on the wind speed, the fetch, which is the distance over which the wind is affecting the sea surface, and the duration of the wind event. When fetch length is sufficient or the storm has been blowing for sufficient duration, the choppiness reaches a state that is called a fully arisen (or fully developed) sea (Figure 3.4a). At this point, energy is dissipated by waves at the same rate as they receive energy from the wind, and after this state is reached the size and characteristics of the waves do not change. It appears that wave characteristics may briefly overshoot this equilibrium state, but they rapidly return to it (Barnett and Sutherland, 1968).

The wave form, though not the water mass, travels out of the area of generation, and is then termed swell. In the area of wave generation, there is a broad spectrum of wave periods from 0.1 s ripples to 15 s or larger waves. The smallest waves disappear over short distances, and larger crests are formed by summation of smaller waves. Swell shows a smaller range of wave periods than the fully arisen sea (Figure 3.4a). The speed (velocity or celerity) at which the wave travels and its period are related to its length (see Table 3.1). The largest waves travel considerably faster then the smaller waves. As swell crosses an ocean there is more time for the smaller waves to become absorbed. A relatively small range of wave periods is likely to be experienced within the surf zone when groups of swell waves reach the shore. Although the wind and current fields of the northern and southern hemispheres are discrete, swell can cross from one hemisphere to another; for example, swell generated in the Southern Ocean can travel across the equator and reach the western coast of North America.

Wave length and period are maintained as a wave group travels beyond the area of wave generation, but wave height can decrease slightly as waves travel. Although the wave group is shown schematically

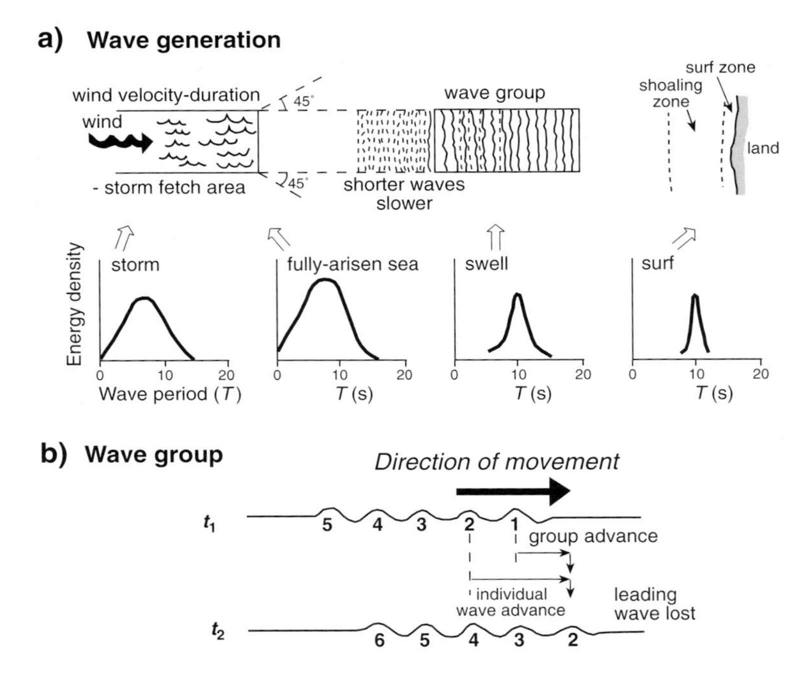

Figure 3.4. (a) Transformation of waves from the area of generation into wave groups, the longer of which travel faster, and thus reach the shore first. Wave period becomes narrower as waves move from area of storm through fully arisen sea, through swell to the shore where the narrowest range of periods is seen in the surf zone. (b) Individual waves travel at twice the speed of the group of waves. Individual waves appear to form at the rear of the group, travel through the group, and disappear at the front.

in Figure 3.4a as of uniform width, the waves disperse radially from the storm area, with about 90% of wave energy contained within an angle of 45° from the direction of travel. In deep water, the speed of travel of the wave group (group velocity) is half the speed of the individual waves (phase velocity). The first wave in the group acts to initiate movement of the surface of the water and it disappears from the front of the group and a new wave forms at the rear. Therefore, as the individual waves advance twice the length of a wave, the group only advances one length (Figure 3.4b). Individual waves travel through the group, starting at the back, moving through to the front where they are lost.

Wave height is an important variable unrelated to the other parameters but depending on wind speed, fetch length and wind duration. Figure 3.5 shows an empirical determination of the significant wave height that is generated under combinations of wind speed, duration and fetch length. The chart is called a nomogram, expanding the Sverdrup–Munk–Bretschneider method, based on the Jonswap experiments in the North Sea (Hasselmann *et al.*, 1976). To use the nomogram, observed wind speed is identified on the left-hand side of the chart and extrapolated across to the right until either the appropriate fetch or wind duration is reached, whichever is nearest the left of the chart. This determines whether the set of waves are fetch- or duration-limited, enabling determination of the expected wave height (CERC, 1984). This approach can be useful for determining probable maximum wave conditions for enclosed seas and embayments of limited fetch.

Wave theories provide mathematical approximations of the form and movement of waves. Figure 3.6 shows the form of different waves and where each wave theory is most appropriate in terms of the relationship of wave height and length to water depth (Komar, 1998). Wave form was initially considered trochoidal by Gerstner, the shape followed by a point within a circle, such as a single wheel nut or tyre valve, as a wheel travels along. This was revised by Airy to a sinusoidal shape involving simple harmonic motion. Airy theory does not take into account terms to the second or higher order, which are incorporated within the refinement undertaken by Stokes. Stokes' theory relates to waves that have flatter troughs, but the more complex mathematics means that Airy theory is more often adopted. Cnoidal theory is appropriate in shallow water and accounts for the distortion of wave shape by

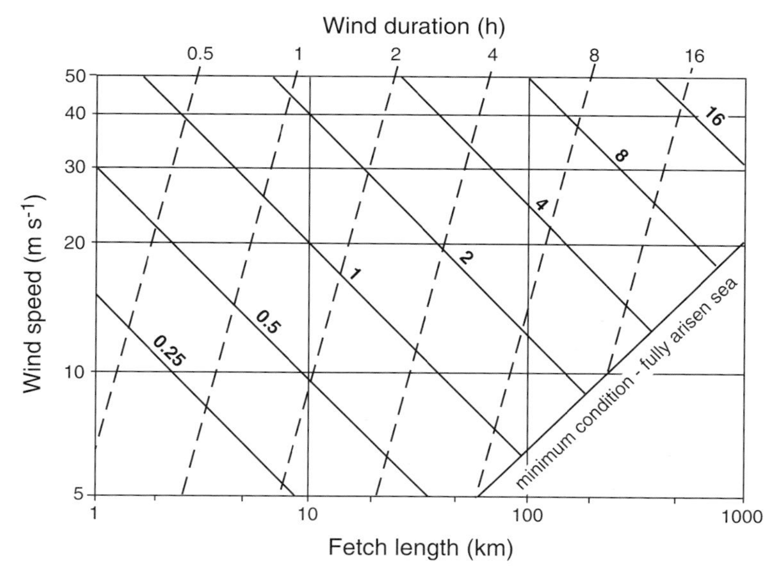

Figure 3.5. Nomogram for prediction of wave heights (metres) from Jonswap data on wind speed and fetch length or wind duration (based on CERC, 1984).

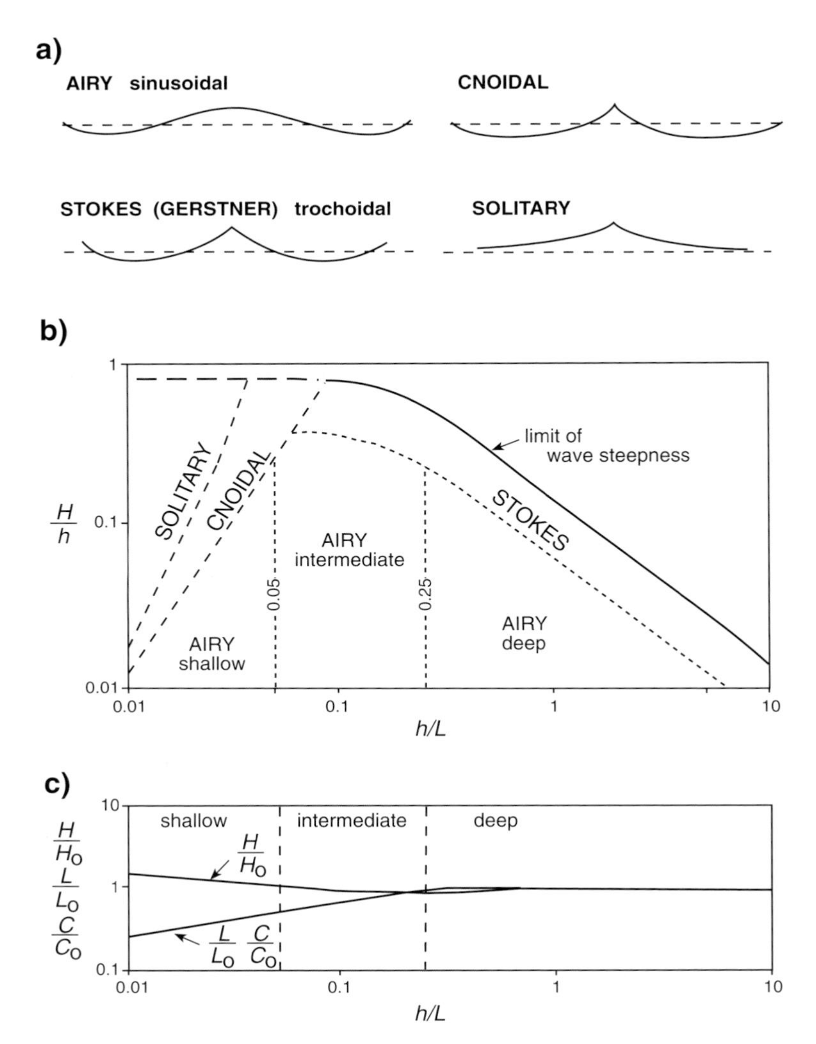

Figure 3.6. Open water waves and their transformation in shallow water. (a) The characteristic wave shapes on which major wave theories are based. (b) The area in which each of the wave theories is most appropriate based on relative wave height (H/h) and water depth (h/L). (c) The dimensionless expression of wave transformation as it moves from deep into shallow water (Airy theory); note that height (H) increases relative to open water height, but length (L) and celerity (C) decrease (based on Komar, 1998).

bottom interference, and solitary wave theory has also been developed where individual wave crests are of interest (see Komar, 1998). For most purposes, Airy theory, also called linear or small-amplitude wave theory, provides a good approximation of wave characteristics. The principal equations of Airy theory are given in Table 3.1.

3.3.3 Wave transformations

Deep-water waves transform as they move into shallower water (see Figure 1.1). Wave speed decreases as the wave interacts with the seafloor in the shoaling zone, attenuating wave power as a result of transfer of energy into sediment mobilisation. The waves may also refract so that the pattern of wave crests becomes increasingly parallel to the shore. In Airy theory, the diameter of wave orbits is directly related to wave height. The orbital motion of water particles involves movement in the direction of wave movement at the wave crest, but return movement in the trough. The diameter of orbits decreases with water depth and, at a depth equivalent to half of the wavelength, there is negligible movement, and this is termed 'wave base', and in shallow water waves transform because they 'feel the bottom'. Water depth is expressed in relation to wave height, distinguishing deep ocean from shallow water (Table 3.1). The deep-water equations apply when the water depth is more than one quarter of wave length. When it is less than one twentieth of wave length, shallow-water relationships apply, and intermediate conditions apply between these deep and shallow water waves (Komar, 1998).

Deep-water wave length increases proportionally to wave period squared, whereas the velocity of the wave increases linearly with wave period (Table 3.1). In shallow water it is water depth which is the main control on the velocity at which a wave travels. Figure 3.6c shows the non-linear transformation of wave parameters as waves shoal; as water depth decreases, wave length and velocity (speed of travel of individual wave, as opposed to wave group) both decrease, but wave height increases (Balsillie *et al.*, 1976). The fact that wave characteristics such as velocity relate to size of wave in deep water, but are depth-limited in shallow water has important implications in terms of sediment movement in shallow water. It will be demonstrated below that a series of infragravity variations become increasingly significant in comparison to incident wave conditions.

Waves entering shallow water are likely to refract. Their wave crests 'bend', becoming increasingly parallel to seafloor contours and until they are almost parallel to the shore; for example, Davies (1958a) has indicated how waves transform to parallel the numerous beaches in complex embayments such as Frederick Henry Bay, Tasmania. This occurs because water depth varies along a wave crest that approaches the shore at an oblique angle. The speed at which the wave travels in shallow water is related to water depth (see Table 3.1). Those parts of a wave which enter shallow water move forward more slowly than those parts in deeper water, causing the wave crest to bend towards alignment with the bottom contours (Figure 3.7a). Wave rays, or orthogonals, can

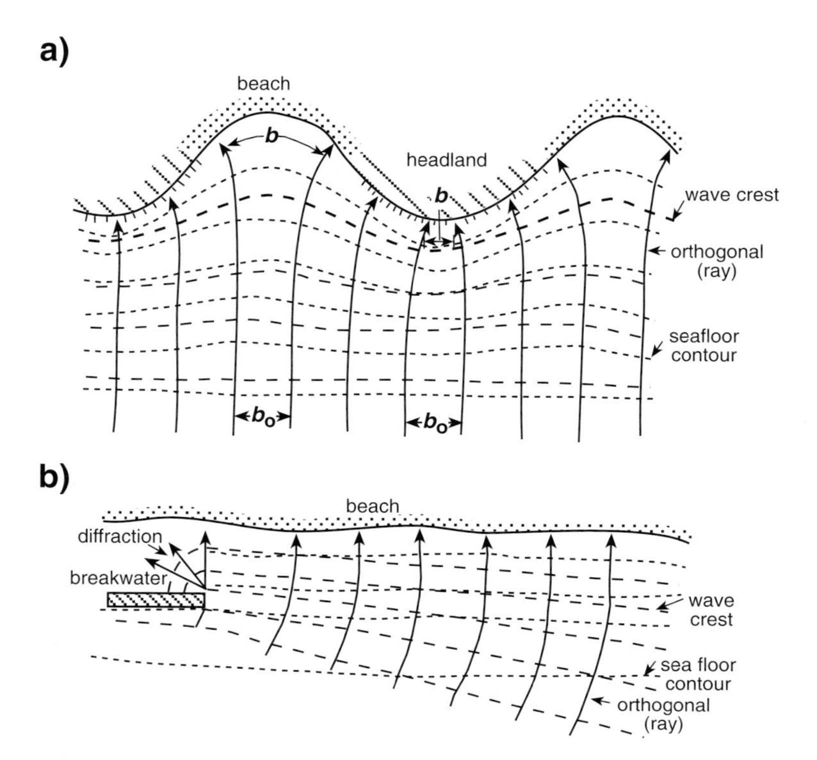

Figure 3.7. Wave refraction. (a) Waves approaching normal to shore. Convergence of rays indicated where $b<b_o$ and divergence $b>b_o$. (b) Waves approaching oblique to shore. Diffraction occurs behind breakwater.

be drawn at right angles to the wave crest, marking the path of imaginary objects carried on the waves and brought in to shore. The orthogonals tend to converge on headlands, indicating a concentration of wave energy and an increase in wave height. In the intervening bays, the orthogonals diverge, indicating dispersion of energy. Where there are obstacles, such as a breakwater, waves bend around them, a process termed diffraction (Figure 3.7b).

Swell waves are more likely to refract than locally generated wind waves which can therefore play an important role in the longshore transport of sediment. As the waves interact with the seafloor they also lose energy that is transferred to the bed and results in sediment movement. The angle that waves make with seafloor contours and wave velocity can be described by Snell's law (see Figure 3.9c) which provides the basis for several techniques to construct wave refraction diagrams for particular directions and periodicities of wave approach (Kirby and Dalrymple, 1986; Dalrymple, 1988). Various models are available to interpolate wave refraction across a grid by non-linear finite difference techniques (Komar, 1998; Maa *et al.*, 2000).

3.3.4 Breaking waves, reflection and dissipation

Waves break in water approximately as deep as the waves are high because they oversteepen (Komar, 1998). As waves progress into shallower water, they become steeper because wave height increases but wave length decreases (Figure 3.6c). Steepness, the ratio of wave height over wave length, increases until around 0.147, at which point the wave becomes unstable. At a critical point the orbital velocity of water particles exceeds the velocity of the wave form and this may also contribute to the breaking process because water particles break out of the wave form. It can also be shown that, as velocity of the wave is dependent upon water depth, the rate of travel of the crest is faster than that of the trough, because water depth is greater and so the crest catches up and 'overtakes' the trough (Thorpe, 2001).

None of these factors completely explains wave breaking. Low waves can run into shallower water than high waves before they break. Although on average waves break when the ratio of wave height to water depth is at a value of 0.78 (McCowan, 1894), on steep beaches this may reach 1.2, whereas on flatter beaches it may be 0.6 (Galvin, 1972). On this basis, a wave of height 1.2 m could move into water depths of 1 m on a steep beach, but would break in water depths of 2 m on a flat beach. In practice, incident waves contain a spectrum of periods and a variety of wave heights. Waves seem to break when the root-mean-squared wave height is around 0.42 of the water depth (Thornton and Guza, 1982; Bowen and Huntley, 1984). Much of the energy that the wave acquired from the wind, and which has been carried across the sea, is expended in wave breaking.

At the break point, waves adopt one of a continuum of four forms: spilling, plunging, collapsing or surging (Figure 3.8). Spilling waves occur where the seafloor is gently sloping; the top of the crest breaks and spills down the face of the wave as turbulent water with foam and bubbles. It is typical for spilling waves to break some distance from the beachface, and transform into a bore with the water moving forward across the surf zone. Energy dissipation occurs across the surf zone and there can be several breaker lines over bars (see Chapter 6). If the seafloor deepens, with a trough or channel between the breakpoint and the beach, the bore can reform into much smaller secondary waves which surge up the beachface. Plunging waves are typical where the seafloor is steeper; the front face of the wave is nearly vertical as the crest curls over

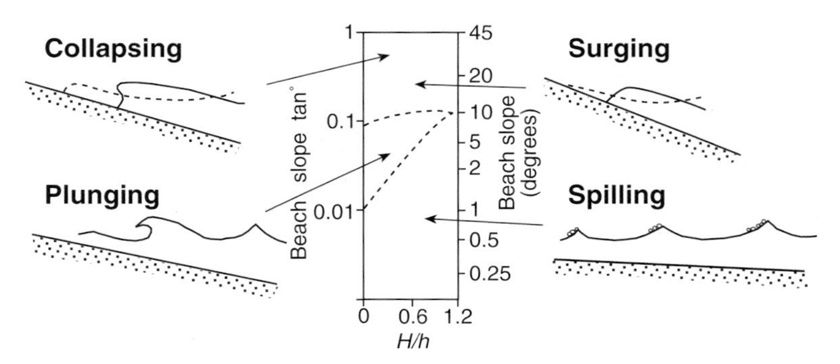

Figure 3.8. Breaking wave forms and the conditions of beach slope and relative wave height (*H*/*h*) under which they occur (based on Galvin, 1968).

and falls forward. They typically dissipate their energy in a concentrated zone shoreward of the breaker line. Surging waves occur on the steepest beaches; the front face and crest are relatively smooth with little foam or bubbles, but the base slides up the beachface and disintegrates as swash. These waves break where water depth is the same as wave height and there is no surf zone. Collapsing waves are intermediate between plunging and surging. The occurrence of these different types of breaking waves in relation to beach slope and the relative water depth is shown in Figure 3.8 (Galvin, 1968).

On the beachface, the uprush of water is called swash and the return is called backwash. The relative strength of swash and backwash depends on wave conditions, but it is also a function of the permeability of the beach. Run-up of swash on the beachface is related to deep-water wave height ($R_s = 0.7\ H_o$), but most of the energy of swash is in the infragravity frequency range (Guza and Thornton, 1982). In the case of gravel beaches there is often almost no backwash because water penetrates into the coarse porous beachface.

When waves break at a breakpoint bar, the wave often decomposes into four to six smaller secondary waves. Energy is then transferred across the surf zone as an undulatory bore within which there is relatively little further dissipation of energy where there is pronounced shallowing of the surf zone, or as a turbulent bore which does lose energy as it propagates where depth remains unchanged (Masselink, 1998a).

If a wave reaches the shore without breaking, there is a high potential for wave energy to be reflected seaward off the land. This can be most clearly observed when watching waves hitting a vertical cliff face or seawall where a standing wave develops (see Chapter 4). Reflection of wave energy is also likely where waves reach a steeply sloping beachface, as is often the case on coarse sand and gravel beaches. If the nearshore is gradually sloping, however, the wave interacts with the bottom and energy is dissipated, as is more common on fine sandy beaches. The extent to which wave energy is reflected or dissipated has become a powerful assessment of the geomorphological function of wave processes (Wright and Short, 1984). A continuum of beach types from highly reflective to highly dissipative is discussed in Chapter 6; individual beaches tend to experience a range of broken and unbroken waves (Thornton and Guza, 1983). Other landforms can also be considered in terms of this dissipative–reflective continuum. Steep cliffs are reflective; a reef crest is dissipative; and mud flats may also rapidly attenuate any wave energy they receive. These concepts are examined in greater detail in several of the following chapters.

3.3.5 Infragravity waves

Wave height increases during a storm but, on a broad dissipative beach, waves break further offshore with minimal increase in incident wave energy at the shore. Wave height in the surf zone is primarily depth-limited (see Table 3.1) rather than a function of incident swell, so that other factors must be involved in relation to why it is that storm conditions are so much more effective geomorphologically than normal sea and swell conditions. The pattern of energy dissipation in the breaker and surf zones results in significant nearshore circulation patterns and these appear to respond to water-level fluctuations dominated by wave set-up and infragravity waves. Detection of the more subtle currents requires simultaneous water-level observations both shore-normal and longshore which can be achieved by sophisticated monitoring equipment and spectral analysis (Holman, 1983, 2001).

Wave set-up represents a rise in mean water level above open-water still-water level as waves approach the shore in the surf zone (Bowen *et al.*, 1968). This occurs because waves need to balance energy, momentum and mass. Set-up produces a pressure gradient or force that balances the onshore component of radiation stress (Table 3.1). Whereas energy is lost through breaking, momentum is preserved. Radiation stress (or wave thrust) is the excess flow of momentum due solely to the presence of waves, and can exist in onshore–offshore (xx), longshore (yy) and oblique (xy) directions (Longuet-Higgins and Stewart, 1964). It has been shown that radiation stress in Airy waves is related to energy transmission, and in deep water it approximates half the energy of the wave crest, whereas in shallow water it is $1.5 \times$ energy of wave crest shore-normal, but $0.5 \times$ energy in longshore direction in shallow water (Carter, 1988).

In the shoaling zone energy flux (wave power) varies little as the shore is approached; however, momentum increases inshore as a result of changes in particle velocity and pressure associated with increasingly asymmetric orbits (see Table 3.1). In the shoaling zone, radiation stress increases shorewards and, to compensate, the water level is set down to conserve the momentum flux. In the surf zone, energy is rapidly dissipated and there is a marked decrease in wave height (and hence energy), and momentum declines. This leads to a decline in radiation stress and water level is set up to balance momentum. The degree of wave set-up depends on beach slope; it is greater in storms, and can be 15–30% of wave height (Aagaard and Masselink, 1999).

Infragravity waves are waves of frequencies of 20 s to 200 s. These are too long to be dissipated in the shoaling zone, in contrast to gravity waves, and so they cause significant variations in water level in the breaker and surf zones. Waves of 20 s give rise to infragravity surf beat or wave groupiness of up to several minutes in frequency (see Figure 3.3c). This kind of oscillation is a bound wave (it is not 'free' to travel independently of the incident waves), lowering the average water surface during the passage of a group of high waves. Typically such infragravity waves in the surf zone can be 20–60% of the offshore wave height, and their influence is felt most pronouncedly towards the shore; for instance, as standing waves within the surf zone. Infragravity waves in the surf zone are related to the generation of rips, crescentic bars, and other longshore bars (Bowen and Inman, 1969).

There are three sorts of currents in the surf zone, undertow, rip currents and longshore currents (see Figure 3.9). Undertow or bed return currents occur as part of the balance of stresses and represent a transfer of water back along the bed (Bagnold, 1940). They have received relatively little study (Stive and Wind, 1982; Masselink and Black, 1995). Rip currents are conspicuous on many beaches, and have received more attention particularly because of the hazard that strong rips can represent. Rip currents are generally a part of a cell circulation. Their occurrence is indicated by an increase in turbulence, an interruption in the line of breakers, or local plumes of sediment being carried offshore (McKenzie, 1958). Rip currents can flow at velocities of greater than 1 m s^{-1}, and often pulse at infragravity frequencies. Rip-current spacing increases with wave height, and because higher waves break in deeper water further offshore, it is therefore also related to surf-zone width (Brander, 1999; Brander and Short, 2000a,b).

Rip currents form in response to longshore variations in wave height and set-up particularly in relation to edge waves as indicated in Figure 3.9a. Longshore variations in the height of the water surface occur where refraction of orthogonals leads to convergence. They also result from topographic control, and rip currents can be fixed over canyons, or close to headlands or reefs, and can act on beach topography to fix their positions. In other cases, incoming waves interact with edge waves which

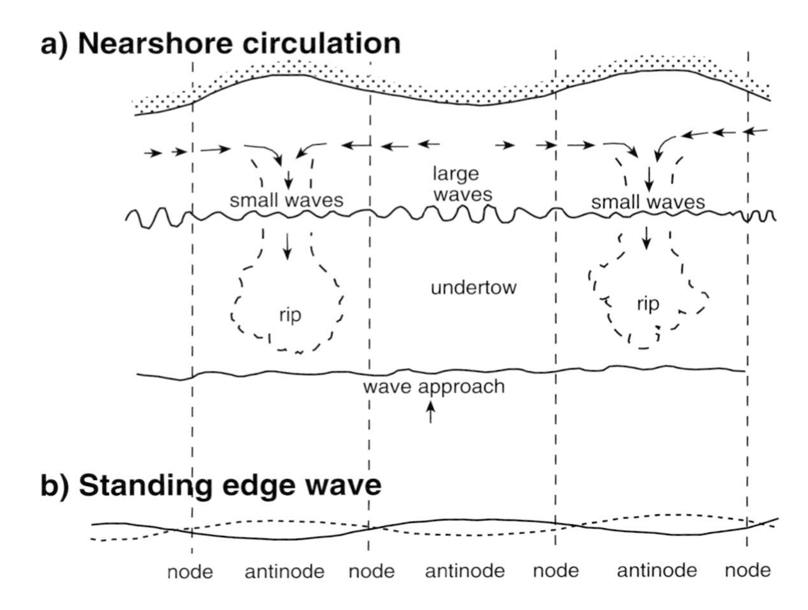

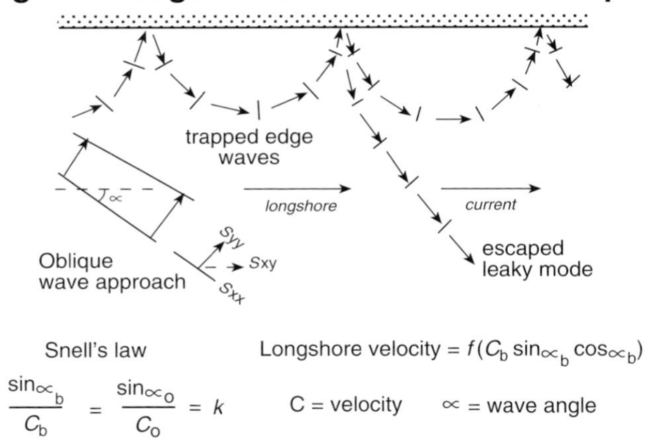

Figure 3.9. Development of rip currents, undertow and longshore currents in the surf zone. (a) The coincidence of incident waves with water level variations related to standing edge waves results in rip currents in association with antinodes. (b) Development of a standing edge wave along the shore; no variation occurs in water level at nodes. (c) Longshore current where waves approach the shore obliquely, and the reflection of a trapped edge wave (based on Komar, 1998).

are trapped in the nearshore region. The edge waves show no discernible water level fluctuation at their nodes, but oscillate with maximum variation at antinodes, with the same period (or subharmonic) as incident waves. Consequently, at alternate antinodes the water level surface coincident with a wave crest is amplified by the edge wave, whereas at the intervening antinodes the wave crest is diminished because the edge wave is at a minimum (see Figure 3.9b). A circulation consequently develops with flow of water from the high water surface, with rip currents existing seaward of the breaker zone through the low antinodes (Komar, 1998). Rip currents occur at an angle to the shore if the angle of wave approach to the shore is small. Where the angle of wave approach is large, rip circulations are more likely to be replaced by longshore currents.

Edge waves are very-low-amplitude variations that occur along the beach. They have been hypothesised, but are generally difficult to detect because they are obscured by the run-up of incident waves (Huntley and Bowen, 1973; Bowen and Guza, 1978). A standing edge wave develops if two progressive waves of the same period are moving alongshore in opposite directions, and can occur on steep reflective beaches. Where waves approach a beach at an angle they are reflected from the beachface and travel off at an angle such that they are refracted and bend back into shore again to be further reflected. They are trapped in the surf zone (Figure 3.9c) and progress alongshore. Leaky waves are infragravity waves that are free waves, and that are reflected off the shore and propagate back into the ocean as escaped leaky mode waves (Figure 3.9c). As incident waves often occur in two or more frequencies, edge waves occur at some longer frequency that is an integer multiple of the incident frequencies (Holman, 1983).

Longshore currents give rise to longshore sediment transport, also called littoral drift. At the shore, there can be movement of sediment within the swash zone as a result of uprush of swash at an angle, but return of backwash by gravity, normal to the beach. On gravel beaches, longshore transport occurs primarily in the swash zone, whereas on sandy beaches it occurs across the entire surf zone (Carter, 1988). Longshore drift can generally be inferred from a series of indicators, such as the direction in which spits have built across estuaries, or river mouths have been diverted, or the side on which sand builds against groynes, breakwaters or other obstructions, including natural headlands and rocky outcrops. Longshore drift can also be determined using tracers, either natural indicators where distinctive lithologies offer the opportunity to track sediment movement, or through artificial tracers, such as fluorescent or radioactive labelling of sand grains to tag sediment movement. Alternatively, there are several methods available to attempt to calculate the potential rate of sediment transfer by longshore currents (Schoonees and Theron, 1993; Van Wellen *et al.*, 2000). The methods calculate the submerged weight of sediment as a function of the longshore component of wave power which depends on the angle of wave approach (see Figure 3.9c). Although several equations are available (e.g. CERC, 1984; Kamphuis, 1991), there are large uncertainties associated with the likely error terms attached to each attribute (Kamphuis, 2000). Longshore transport rates are a function of the oblique radiation stress (S_{xy}), and maximum longshore transport rates occur where wave approach is around 45°. In practice, it remains complex to calculate reliable values for longshore transport (Allen, 1988) and calculations are rarely tested empirically (Sherman, 1988).

3.3.6 Wave measurement

Waves can be measured in a series of ways (Horikawa, 1988; Komar, 1998). Wave frequency can be measured by timing wave passage relative to a fixed point; wave length is hard to measure in the field, but can be determined from aerial photographs. Wave height is often measured using capacitance probes placed in beaches where wave energy will not be great enough to damage them. Wave rider buoys anchored in open ocean can be used to record long-term records. Short-term analysis is undertaken on wave records of 10–20 minutes duration because to generalise about a particular wave regime it is important that the characteristics are stationary, and not undergoing gradual change over time. Pressure transducers are effective methods of measuring water-level variations but do not generally record at frequencies sufficient to detect short-term water-level variations associated with the passage of waves, particularly in shallow water. These are more generally used to measure tides, often installed in stilling wells which are intended to smooth out the oscillations of surface variations associated with waves. Wave recorders include non-directional measurements by acoustic sea-bed sensors, wave orbital velocity sensors, wave staffs, pressure sensors, current meters and accelerometers that measure movement of vibrating masses. Remote sensing methods provide data on mid-ocean waves and wave data are increasingly available from such remote sensing techniques as satellite radar altimetry, synthetic aperture radar (SAR), and radar backscatterance as well as airborne laser and radar (Krogstad *et al.*, 1999).

Figure 3.10 shows the analysis of an analogue record of water-level variation. Using the zero-upcrossing method (recording a wave every time the water level rises across the zero line), individual waves can be detected from a continuous time-series record of water-level variation (Figure 3.10a). They can be expressed as a histogram of wave heights (Figure 3.10b), or statistical approaches such as spectral analysis can be used to determine a graph of energy against frequency from a time-series (see Figure 3.10c). Mean wave height and maximum wave height are often identified, and significant wave height, based on the highest one-third of waves, is still often used. Mean wave height tends to be 0.63

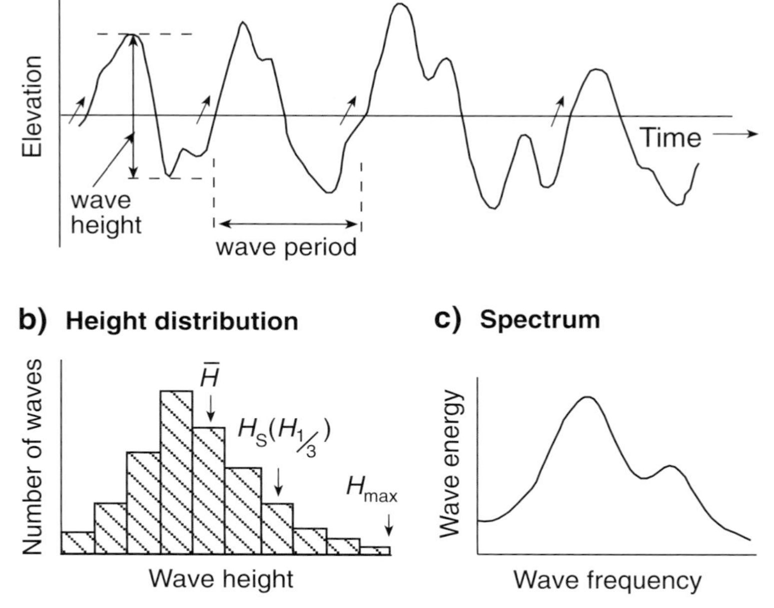

Figure 3.10. Analogue wave record and the zero-upcrossing technique by which height distribution or spectra are defined (based on Sunamura, 1992). (a) Time-series of water surface elevation; (b) frequency distribution of wave heights; and (c) wave spectrum.

times significant wave height (Horikawa, 1988). More complete descriptions include directional wave spectra (Komar, 1998).

3.3.7 Wave climate

There are significant global patterns in terms of wave climate. The largest waves are associated with gale force winds of temperate latitudes (Table 3.2), characterised by the occurrence of westerly winds and frontal activity. In the southern hemisphere, particularly in the Southern Ocean, these occur especially in latitudes around 60° S, and in the northern hemisphere in the 'roaring forties', 40–60° N. Strong persistent winds characterise the tradewind belts. Strong winds are rare in the doldrums of equatorial latitudes. Intense tropical storms, called hurricanes in the Atlantic and typhoons in Asia, and cyclones in the western Pacific are significant influences on many tropical coasts, but do not occur in the region 5° N – 5° S of the equator. The major area of influence of these is shown in Figure 3.11, and they are discussed below.

Five broad wave environments have been recognised by Davies

Table 3.2. Beaufort scale of wind and significant wave characteristics under conditions of fully arisen sea

Beaufort wind force	Wind speed m s^{-1}	Description of wind	Description of sea	Approx. H_s (m)	Approx. T_s (s)
0	0–0.2	Calm	Like a mirror	0	1
1	0.3–1.5	Light air	Rippled	0.025	2
2	1.6–3.3	Light breeze	Small wavelets	0.1	3
3	3.4–5.4	Gentle breeze	Large wavelets	0.4	4
4	5.5–7.9	Moderate breeze	Small waves	1	5
5	8.0–10.7	Fresh breeze	Moderate waves	2	6
6	10.8–13.8	Strong breeze	Large waves	4	8
7	13.9–17.1	Moderate gale	Sea heaped up with white foam breakers	7	10
8	17.2–20.7	Fresh gale	Moderately high waves with spindrift	11	13
9	20.8–24.4	Strong gale	High waves with streaks of foam	18	16
10	24.5–28.4	Whole gale	Very high waves with overhanging crests	25[a]	18[a]
11	28.5–32.7	Storm	Exceptionally high waves with extensive foam patches	35[a]	20[a]
12	>32.7	Hurricane	Sea completely covered with foam, driving spray	40[a]	22[a]

Notes:
[a] Seldom attains fully-arisen sea because required duration and fetch rarely achieved. H_s, significant wave height; T_s, significant wave period.

(1980). Storm-wave environments occur where gales generate short, high-energy waves of varying direction. West coast swell occurs on coasts experiencing low long swell waves of consistent direction, but rarely experiencing storms. East coast swell consists of rather weaker swell, amplified by local wave generation (as in the Tasman and Coral Seas off the east coast of Australia; see Chapter 4), and occasional tropical storms. Areas of monsoon influence experience a reversal of principal wind and wave directions during a year; they are common in southeast Asia and northern Australia. Protected seas are large water bodies in which wave generation is local rather than swells from the large ocean basins (Davies, 1980). Figure 3.11 combines the mapping of these broad wave zones with some characteristic examples of wave height variation throughout the year (Short, 1999a). The significance of wave climate is re-examined in relation to individual coastal types in several of the following chapters; for instance, it plays an important role in relation to beach morphology which is discussed in Chapter 6

3.4 Tides and tidal influence

On almost all coasts there is a rhythmic tidal rise (flood) and fall (ebb) of water. The tides are long waves; generally the period is 43000 s, or 12 h 25 min (semidiurnal). The wave that constitutes the tide is driven by the gravitational pull of the moon and, to a lesser extent, that of the sun. Newton showed that gravitational forces arise from the product of the masses of two objects divided by the square of the distance between them. Therefore, although the sun is many times the mass of the moon, because it is much further away its gravitational influence is only 0.46 times that of the moon (Pugh, 2001).

3.4.1 Tidal oscillations

The moon remains in orbit around the earth because the gravitational force of attraction is equal to the centrifugal force. This produces the orbital motion of the moon and the earth around a common centre of mass (this centre of mass is actually within the earth because of its con-

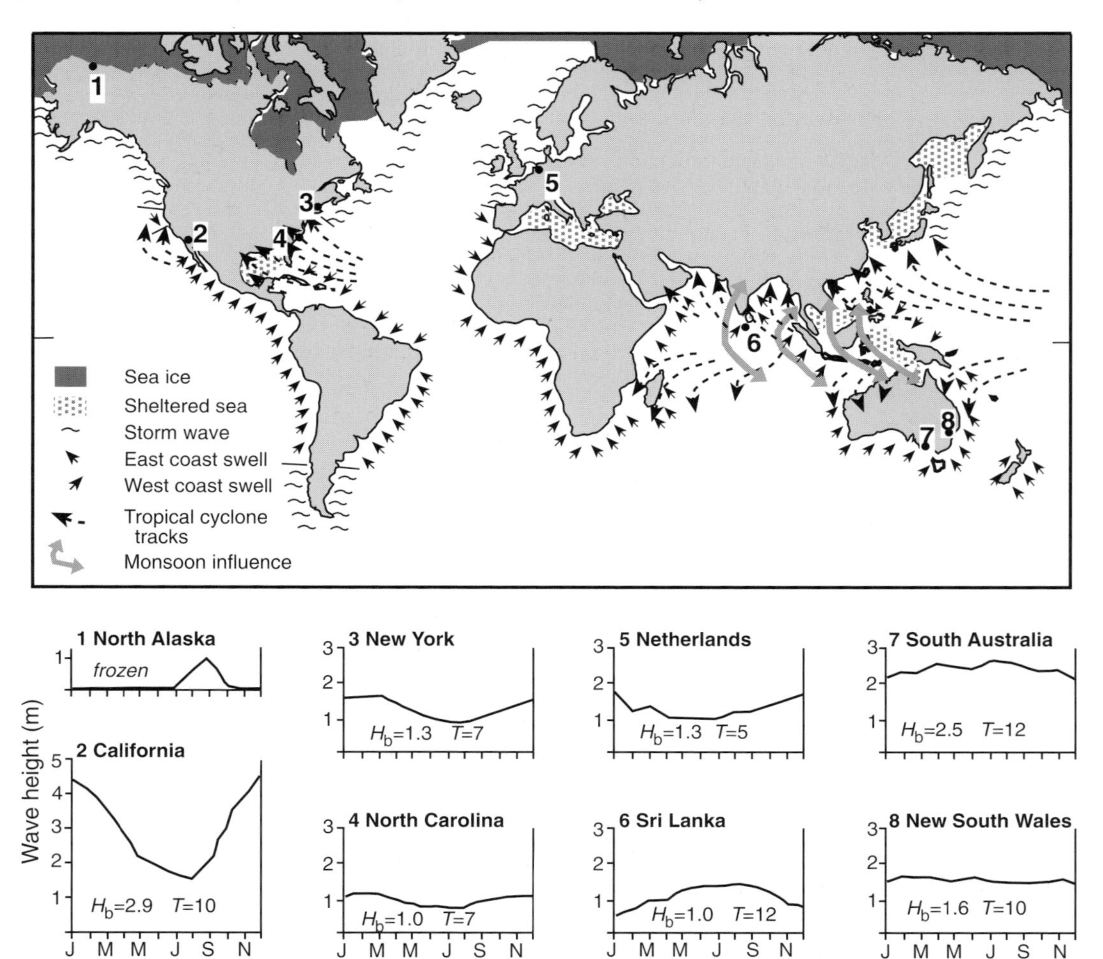

Figure 3.11. Major wave climate of worlds coasts (based on Davies, 1980) and sample wave climates based on monthly mean wave height for selected locations (based on Short, 1999a).

siderably larger mass). The water in the oceans is pulled towards the moon by the gravitational force; the land is also attracted, though to a lesser extent, resulting in scarcely perceptible earth tides. The gravitational attraction and the centrifugal force are only equal at a mid point within the body; the gravitational attraction is greater than the centrifugal force on the side of the earth close to the moon, while the reverse is true on the side further away (Pethick, 1984; Komar, 1998).

Theoretically this inequality of forces results in the water surface having a bulge both on the side of the earth closest to the moon (gravitational force > centrifugal force), and on the side away from the moon (centrifugal force > gravitational force). As a consequence, if the earth were entirely covered with water, its surface would bulge towards and away from the moon. The spin of the earth means any point on the earth would pass beneath each of the two bulges in a 24-hour rotation. However, in that period of 24 hours, the moon would also move in the same direction as the earth's rotation, so that the period over which the two bulges are experienced is actually 24 hours and 50.47 minutes. The bulges and therefore the tidal height would be of similar magnitude when the moon is overhead at the equator, but as the moon moves towards maximum declination (28°) the height of the two daily tides diverge, referred to as a diurnal inequality (Pethick, 1984). If one of the semidiurnal peaks is absent, the tide is diurnal; if there is a pronounced inequality, it is called mixed (Figure 3.12).

The lesser attraction of the sun also results in a pair of small bulges associated with it. The combination of bulges associated with the moon and with the sun depends on the relative positions of sun and moon with respect to the earth. When they act in the same direction, as they do at full moon and new moon, the bulges accentuate each other causing spring tides, whereas when the moon and sun are at right angles, the effects serve to partially cancel each other, and neap tides are experienced (Figure 3.12).

This simple picture of an earth with bulges on it is referred to as the equilibrium theory of tides (Cartwright, 1999). If the surface of the earth were only covered with water, then the earth would rotate under these bulges, producing long-period (semidiurnal) waves. However, the presence of land impedes the path of this wave, except in the Southern Ocean where that is effectively what happens and one tidal wave propagates right around the globe (Figure 3.13). Elsewhere the tidal wave is limited by the distribution of land masses. In view of the long wavelength, even the deepest ocean basin represents shallow water. The equilibrium theory of the tides supposes the wave would behave as a linear wave in shallow water, with velocity equal to the square root of water depth multiplied by the gravitational constant (just as with wind waves

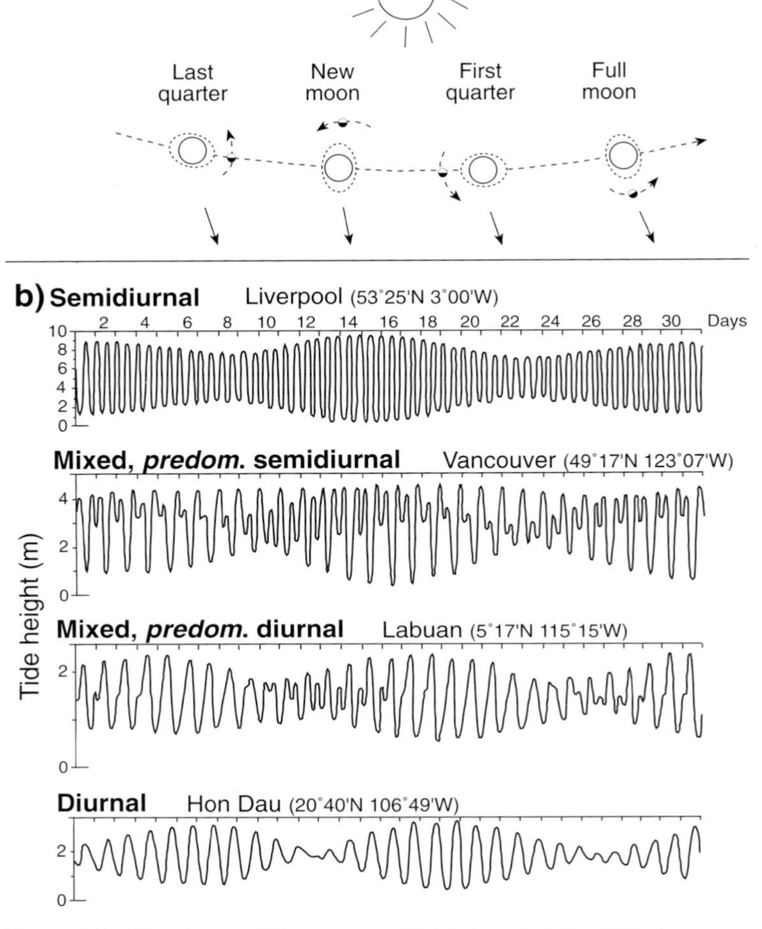

Figure 3.12. The phases of the moon and tidal characteristics; (a) bulge generation in relation to moon and sun's position (not to scale), and (b) an example of tidal response showing semidiurnal, mixed and diurnal tides.

in shallow water, see Table 3.1). The tide can reach velocities of up to 230 m s^{-1} in deep ocean. However, the wave is not a free progressive wave like a wind wave, but a forced wave driven by the moon, and it is reflected off landmasses (Pugh, 1987).

In a small ocean basin (or a large lake), the equilibrium-theory tidal wave would move across the basin from east to west, and then it would be reflected back again off the western shore, forming a standing wave. Water level would oscillate around a node in the middle of the basin (at which no change in water level would be seen), reaching its maximum elevation at antinodes on either shore. Wave speed is dependent upon the depth of the basin (as for shallow-water waves). If the length of the basin coincides with a multiple of wavelength, it is called resonant, and a seiche is established. These factors are incorporated into what is called the dynamic theory of tidal behaviour.

The Coriolis force (due to the rotation of the earth) causes the standing wave to rotate counterclockwise in the northern hemisphere and clockwise in the southern hemisphere, so that the standing tide (a Kelvin wave) rotates around a node that is called an amphidromic point. The tide has no amplitude at the amphidromic point. Much of the open ocean has a tidal range of <0.5 m. However, amplitude increases away from amphidromic points, and can reach several metres on the margin of wide oceans (Figure 3.13). The dynamic tide appears to rotate around lakes or bays, whereas on open-ocean coasts it appears progressive, occurring later at successive points along the coast. The tide is actually a complex summation of a series of constituents caused by the various tide-generating forces. Magnitudes of tide-generating forces can be precisely resolved to give tidal predictions. However, in practice the height of the water is also subject to atmospheric and meteorological factors and tidal predictions only approximate actual water levels.

Tidal range is divided into a series of levels shown in Table 3.3. The most extreme water levels which result from tides are termed Highest and Lowest Astronomical Tide (HAT and LAT), although meteorological

Figure 3.13. Global distribution of principal amphidromic points and of megatidal, macrotidal, mesotidal and microtidal regimes (based on Davies, 1980). Cotidal lines join places experiencing high tide simultaneously and the number represents hours within the 12-h passage of tides.

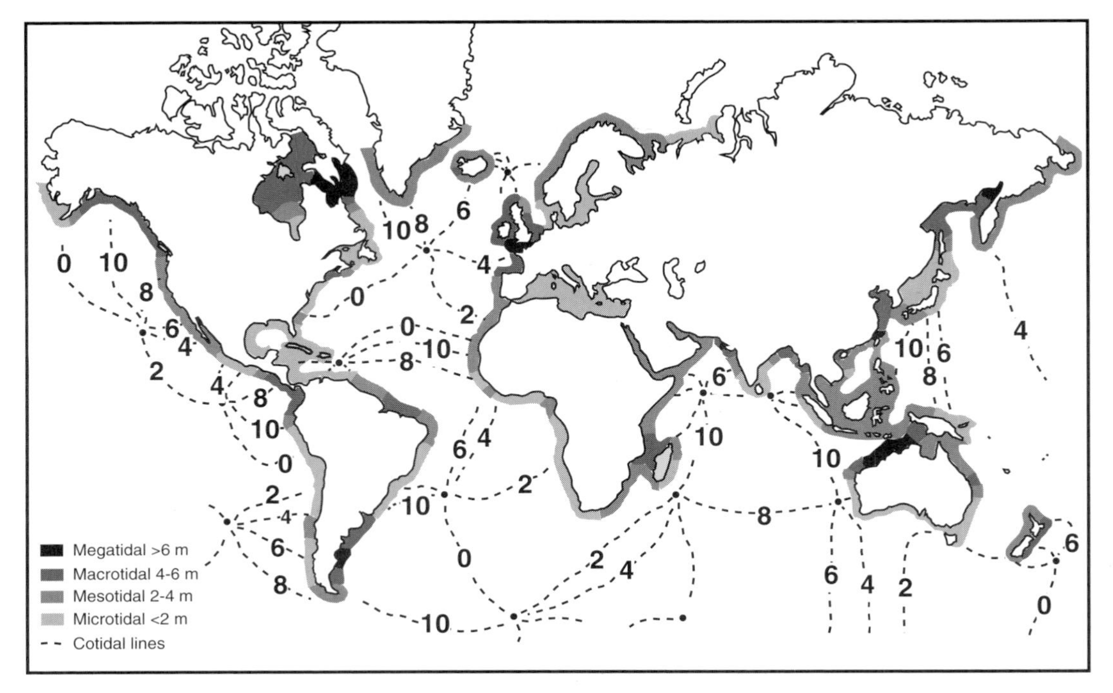

Table 3.3. Definition of major tidal heights

Abbreviation	Tidal height	Definition of level
SS	Statistical storm	Maximum storm surge expected in a specified number of years
AS	Annual storm	Maximum expected annual storm surge
HAT	Highest astronomical tide	Highest which can be predicted to occur under any combination of astronomical conditions
MHWS	Mean high water springs	Average (19-year) of high water occurring on spring tides (full moon +2 days, and new moon +2 days)
MHHW[a]	Mean higher high water	Average (19-year) of higher tides each day
MHW	Mean high water	Average (19-year) of high water
MSL	Mean sea level	Average (19-year) of hourly water-level heights
MTL	Mean tidal level	Average of MHW and MLW (a plane midway between MHW and MLW)
MLW	Mean low water	Average (19-year) of low water
MLLW[a]	Mean lower low water	Average (19-year) of lower low tides each day
MLWS	Mean low water springs	Average (19-year) of low water on spring tides (full moon +2 days, and new moon +2 days)
LAT	Lowest astronomical tide	Lowest which can be predicted to occur under any combination of astronomical conditions, often used to define Chart Datum

Notes:
[a] Mixed tidal regime only (when daily highs are pronouncedly unequal).

and other factors can result in higher or lower water levels. Mean sea level is the long-term average; it is generally necessary to have at least 18.6 years of record to be able to observe the variations that occur in a lunar cycle and determine a precise average. The terms mean higher high water (MHHW) and mean lower low water (MLLW) are applied to stations which experience predominantly mixed tides. On other coasts it is customary to distinguish spring and neap tide levels.

Tidal range is increased and the velocity of the tidal wave is slowed over the shallow bathymetry of broad continental shelves (Davies, 1980). Coasts can be divided into microtidal where tidal range is <2 m, mesotidal where it is 2–4 m, macrotidal where it is 4–6 m, and megatidal where it is >8 m (Figure 3.13). Macro- and megatidal ranges are common within gulfs and embayments where the volume of the tidal wave is constrained, and tides reach a maximum spring range of 16.8 m in the Bay of Fundy in eastern Canada.

Tidal range influences the broad distribution of coastal landforms (Davies, 1980). River deltas are well-developed in microtidal areas (see Chapter 7), as are barrier islands (see Chapter 6) and spits (Hayes, 1975). There are also several non-tidal basins, for example the Mediterranean and the Baltic Seas. Tidal processes also tend to be less pronounced in those seas which experience primarily diurnal tides (one high per day, as opposed to semidiurnal which experience two tides per day), such as the Gulf of Mexico or the Gulf of Thailand. There are many funnel-shaped estuaries in areas which experience a macrotidal regime (see Chapter 7), and these are generally flanked by extensive mudflats often with broad expanses of mangroves or salt marshes (see Chapter 8). The reasons for increased or distorted tidal range in these embayments is examined in the next section.

3.4.2 Tidal processes in embayments, estuaries and creeks

The velocity of tidal currents depends on tidal range and the topography of the coast. Tidal currents can be the dominant water movements in embayments, estuaries and tidal creeks, resulting in distinctive tide-dominated landforms. In large gulfs or embayments the tide can resonate. For instance, the pattern of tides in the North Sea has been shown to have three amphidromic points, displaced as a result of friction to the east (Pethick, 1984; Komar, 1998). Resonance results in large tidal ranges experienced along the east coast of Britain, further from the amphidromic points, with amplification of the tide into estuaries such as the Humber and the Wash (>5m). Amplification of tides frequently occurs into bays and gulfs (see Figure 3.13).

Distortion of tides occurs along major estuaries, in terms of amplitude, symmetry, and duration of flood and ebb tides. Many of the principles of river hydrology apply to estuaries, although a significant difference is that tidal flows in estuaries are bi-directional. Whereas river discharge is determined by runoff from the catchment and is independent of channel size, tidal discharge is dependent on channel dimensions and tides are forced flows that are 'not going anywhere'. The volume of water between the high-tide surface and the low-tide surface is called the tidal prism. In small tidal inlets in which a system is full at high tide and empty at low tide, discharge is determined primarily by the tidal prism, and flows consist of this volume of water passing through the inlet on rising and falling tides. In larger estuaries and tidal creeks, the tide progresses as a wave, and the timing of high (and low) tide varies along the system.

There are two types of flow in estuaries: tidal currents which are the most important physical or advective processes, and residual currents which are chemical or diffusive processes set up by density differences resulting from mixing of fresh and salt water (see Chapter 7). In the open ocean, tides occur as long waves, attaining maximum velocities (tidal currents) at high water and low water, but with no residual transport of water (comparable to orbital fluxes beneath waves). In estuaries, however, flows become modified as a result of frictional dampening, landward constriction of the channel (termed convergence), and reflection from channel banks, shoals and the channel head. Maximum velocities are experienced at mid tide, or just before and after high tide.

The tide moves as a shallow-water wave in the coastal zone, because the seafloor is shallow compared to tidal wavelength, and undergoes similar transformations to waves in shallow water. Tides in an estuary can be either standing or progressive. An ideal progressive wave will move upstream, instead of high water occurring simultaneously throughout the estuary. In the case of a standing wave, high water occurs simultaneously throughout the estuary and there is no flow at high (and low) water. It is generally the case that flows in estuaries are negligible at high and low slack water, but also that high tide occurs later with distance along an estuary (Figure 3.14). Estuarine flows experience friction and reflection from channel boundaries causing energy loss, which is likely to be expressed as a loss of tidal amplitude upstream. Consequently, neither classic progressive wave nor standing wave characteristics dominate, but estuaries show elements of both standing and progressive wave behaviour, depending on channel geometry (Hunt, 1964; Ippen and Harleman, 1966). For instance, pronounced tapering of the estuarine funnel of the Ord River in northwestern Australia and variability in its width/depth ratio coincides with a tendency towards standing wave behaviour, whereas the adjacent highly sinuous but less tapered King River appears to have more progressive wave characteristics (Wright *et al.*, 1973).

The process–form interactions responsible for distinct tidal channel morphology have been investigated by Wright *et al.* (1973), who demonstrated that longitudinal gradients of tidal discharge, rate of water-level change, and channel width mutually co-adjust along macrotidal estuaries. They suggested that depth and width convergence upstream balances frictional dissipation, such that equal work per unit area of channel is performed. Estuarine channel morphology is both a product of, and a determinant of, the hydrodynamics, and tidal velocity, ampli-

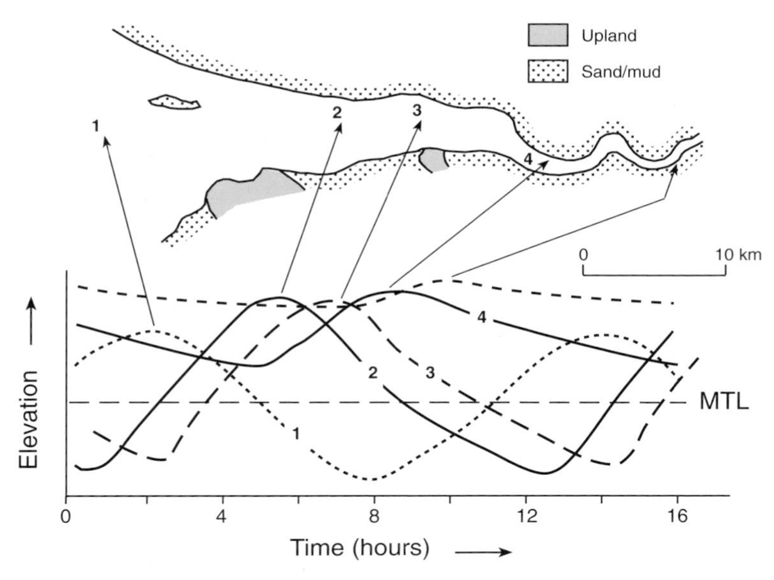

Figure 3.14. Distortion of tidal wave propagating up a schematic estuary. The tidal amplitude varies as a result of changes in width. See text for details.

tude and tidal harmonics are linked by morphodynamic feedbacks (Friedrichs and Aubrey, 1994). The rate at which estuarine channels taper (converge) determines the tidal amplitude along a system (Nichols and Biggs, 1985). If the rate of convergence is in equilibrium with tidal flows (generally adopting an exponential decrease in width with distance) tidal range remains relatively constant along the channel (termed 'synchronous' by Nichols and Biggs, although this term is confusing because it appears to imply that high tide occurs at the same time throughout the system, which is not the case). If channel convergence is rapid, energy is concentrated and tidal range increases upstream, whereas if the channel widens rapidly tidal range decreases. Slight variations in tidal range are shown schematically in the system in Figure 3.14, in relation to constrictions of the channel imposed by bedrock outcrops.

In a long estuarine channel, the water surface in the open ocean and at the head of the embayment will not adjust simultaneously, but it will take a finite time for the tidal wave (progressive) to travel along the estuary. The rate of travel of the tidal wave varies with the square root of water depth (see Table 3.1; velocity, C, is the velocity of a wave in shallow water). Since water depth is deeper at high water than at low water, the crest (high tide) travels more rapidly up the estuary than the trough (low tide). Consequently, the duration of the flood limb gets progressively shorter upstream, whereas the ebb limb gets longer (Figure

3.14). Flows become increasingly asymmetric, both in duration and velocity, further upstream. In some macrotidal estuaries, this velocity asymmetry becomes oversteepened and the tide propagates as a bore. Impressive bores, forming a wall of water up to 5 m high, occur on large tides in some estuaries. For example, there are bores on the Severn and Trent Rivers in Britain, the Elbe River in Germany, the Colorado River draining into the Gulf of California, the Daly River in northern Australia, the Changjiang (Yangtse) River in China, and the Amazon River in South America. The implications of velocity asymmetry in terms of sediment movement within estuaries are examined in Chapter 7.

3.5 Other oceanographic processes

There are several other oceanographic processes that are significant in terms of coastal geomorphology. Ocean currents have an effect on the global distribution of organisms and wave climates. Sea ice is a significant factor for some high latitude coasts. Broad global circulation patterns, such as deep-water circulation and the El Niño and La Niña phenomena in the Pacific Ocean, can result in long-term variations in geomorphological processes on coasts that are affected by these variations, and the role of extreme events such as storms and tsunami can also be expressed in terms of coastal landforms. These processes are examined briefly in this section.

3.5.1 Ocean currents

The oceanography of coastal seas is concerned with the boundary layer between the oceans and land (Yanagi, 1999). Physical, chemical and biological processes, including tidal currents, residual currents set up as a result of density differences, and wind-driven currents are influenced by the larger ocean, but are also affected by the proximity of land. These usually have only a minor influence on coastal processes, for instance, minor fluctuations in sea level are associated with operation of the Florida current as it feeds the Gulf Stream (Maul *et al.*, 1985).

Ocean currents are the result of wind stresses on the surface of the open ocean, driven by horizontal pressure gradients, and differences in density as a result of variations in temperature and salinity, termed thermohaline. Atmospheric circulation and thermohaline patterns are influenced by the distribution of land masses, with deflection of currents by the Coriolis force, as a result of the earth's rotation, resulting in the broad circulation patterns shown in Figure 3.15. Warming of equatorial waters, together with excess heat in the tropics results in expansion of air which rises, whereas at the poles there is deficit and air sinks.

The broad ocean circulation pattern has evolved on plate-tectonic time scales, as the land masses have adopted their present positions (see Chapter 2). Some aspects of ocean circulation have changed as sea level has fluctuated during the Quaternary, for instance straits such as the English Channel and Torres Strait have only re-opened as Holocene sea level flooded previous land bridges. Inter-ocean connections, such as the Indonesian throughflow connecting the Pacific and Indian Oceans, would have operated differently when the sea was lower. There have been other variations in ocean circulation; for instance, the pattern of deep-ocean circulation (shown in Figure 3.15). The North Atlantic loses heat to the atmosphere and the dense water sinks and feeds deep-water circulation; this deep-water 'conveyor', which can be traced through the Indian and Pacific Oceans, appears to have undergone major switches from one pattern (or state) to another at several times during the Late Quaternary (Broecker, 1997).

Ocean circulation exerts an influence on the distribution of biological communities and wave climate. The clockwise circulation in northern hemisphere oceans and anticlockwise circulation in southern hemisphere oceans, results in warm currents on the western side of oceans, and significant upwelling of cold, nutrient-rich waters along the

Figure 3.15. Global ocean currents, and the deep-water circulation.

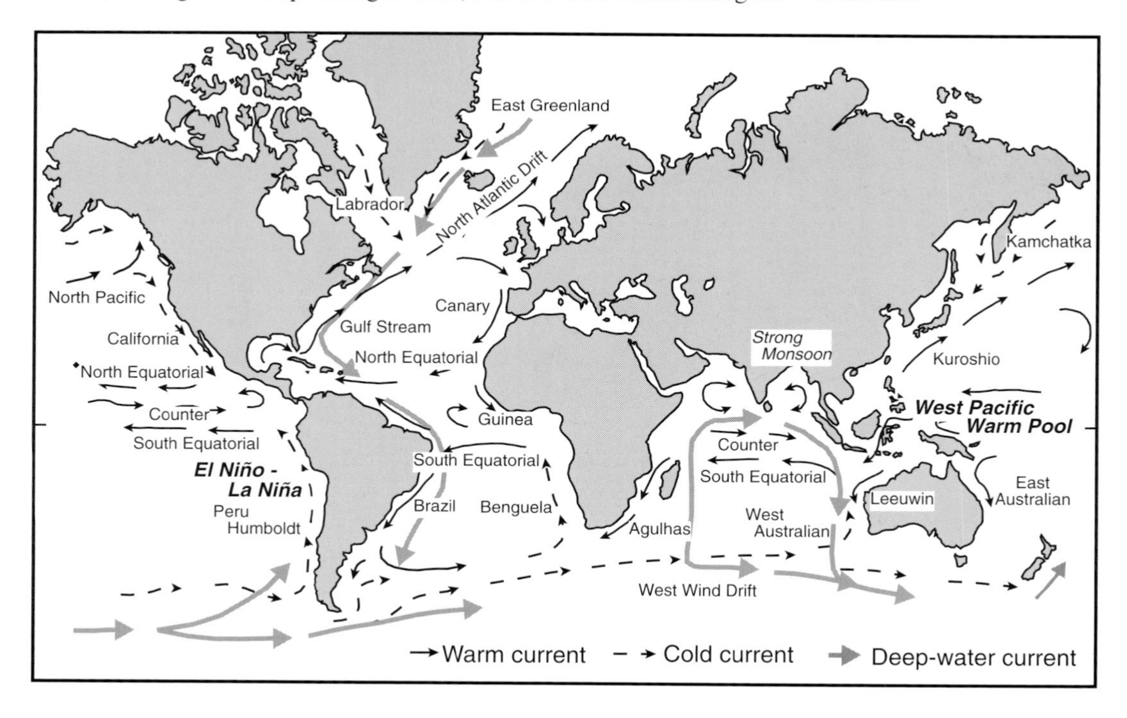

eastern margin of oceans. These broad circulation patterns have indirect effects on coastal processes, especially in terms of the climatic influences, such as the pronounced aridity of western coasts of the major continental landmasses. The distribution of organisms such as corals and other shallow-water communities is influenced by ocean circulation, not only in terms of sea-surface temperatures, but also in terms of the supply of larvae and recruitment. Circulation patterns are also seen as significant in terms of the influence they have on ocean–atmospheric interactions associated with oscillations such as the El Niño–Southern Oscillation described below.

3.5.2 Sea ice

Processes associated with ice and frost action are restricted to high-latitude coasts, such as the margins of Antarctica, northern Canada and Greenland, and other areas around the Arctic Ocean. In these areas, ice is perennial or seasonal, and the length of time that waters are ice-free is especially significant. The freeze-up occurs in autumn and 'frazil' or 'slush' ice, an unconsolidated mass of water and ice crystals, gradually becomes pan ice. Arctic waters are relatively constricted, meaning that ice floes (pan ice) collide, overriding each other. In the Antarctic, floating ice can reach greater thicknesses (3–4 m), but the waters are more open and ice floats away as icebergs. Floating ice comprises pack ice, and goes through the stages of freeze-up and break-up. In contrast ice-foot is fast ice that is anchored onto the shoreline (John and Sugden, 1975; Forbes and Taylor, 1994).

When there is ice cover, then the role of waves and tides and the impact of storms are reduced, because the waters are frozen and the sediments are less mobile (see North Alaska, example 1 in Figure 3.11). After the peak of the winter, the ice breaks up particularly as a result of tide-induced cracks in the ice cover, and 'dirty ice' with sediment inclusion often melts first. When ice no longer covers the water, wave processes can become important again. During the spring thaw there are also large sediment loads brought down by rivers draining to the Arctic Sea. If the sea ice has not melted, the material brought down by these turbid rivers is deposited on the surface of the ice, often in layers that are incorporated into the ice.

Ice can lift sediments with it when floated, resulting in a series of boulder deposits when the ice melts along the shoreline. These include ice-pushed ridges up to 20 m high, scour marks and other bulldozed landforms (Trenhaile, 1997). Ice ride-up occurs where ice floes slide over one another, whereas ice pile-up results from buckling of ice as it is pushed up onto the shore (Hill *et al.*, 1994). Boulders can be entrained where they are frozen into the ice, and they can also serve to gouge the substrate when the ice moves. For instance, several million tons of ice is ice-rafted along the St Lawrence Estuary each year, accumulating in the high turbidity region. Ice also has effects on the marshes and mudflats, scouring the surface and leaving gouges, termed ice wallows (Dionne, 1999).

The sediment carried by ice includes coarse boulders that are frozen into the ice and plucked away from the bottom by anchored blocks of ice, as well as finer sediment layers incorporated into the ice. These are dumped when the ice melts and, when wave action is again sufficient, these dumped deposits are reworked, the coarse fraction remains as a lag deposit but waves winnow the finer material. Usually, the changes to the seafloor are minor from year to year, with little adjustment to any equilibrium profile as a result of ice grounding on the shoreface (Héquette and Barnes, 1990).

3.5.3 El Niño–Southern Oscillation

El Niño and La Niña phases involve recurrent variations in the wind strength and direction and movement of warm water in the central and eastern Pacific. El Niño, meaning the 'Christmas child', is a South American term to describe the occurrence of a series of conditions that accompany a decline in upwelling along the South American coast (Trenberth, 2001). It appears that reduced tradewinds are associated with less pronounced circulation from east to west across the Pacific, and warm water remains along the South American coast. The cold water that usually upwells brings nutrients with it and supports the anchovy fishery along the coast, which fails during the warmer El Niño. The variation in the eastern Pacific is mirrored in the western Pacific, where there are temperature fluctuations in what is called the West Pacific Warm Pool (Figure 3.15), warm waters that are more extensive during La Niña than during El Niño phases. During El Niño phases, heavy rainfall is experienced on islands in the central Pacific, such as Tarawa in the Gilbert chain, Canton in the Phoenix Islands and Christmas Island in the Line Islands (Bjerknes, 1969). In La Niña phases, when the Southern Oscillation Index (SOI) is positive, warm water occurs in the western Pacific, and a tongue of cool water crosses the central Pacific. Islands near the equator experience drought, whereas abundant rainfall is received in the western Pacific (Philander, 1990). There are also variations in sea level associated with the ocean circulation patterns (Philander, 2001).

El Niño and La Niña phases are strongly associated with the Southern Oscillation, which is a change in atmospheric pressure conditions across the Pacific. It is generally measured as the SOI, the pressure difference between Tahiti and Darwin. A negative SOI is associated with El Niño, and brings drought to areas of the western Pacific, such as eastern Australia and Indonesia. The El Niño–Southern Oscillation (ENSO), as the combined variation is called, is quasi-cyclic with an average period over recent decades of about 4 years, varying from 2 years to 7 years. Longer-term records indicate decadal variation, termed the Pacific Decadal Oscillation (PDO), which influences the intensity of ENSO (Mantua *et al.*, 1997). Although the driving mechanisms remain poorly understood, it is clear that ENSO has far-reaching effects and climatic responses can be seen in regions well beyond the Pacific.

Although the ENSO phenomenon has only been well documented in recent years, it is becoming increasingly clear that it has important ramifications for coastal geomorphology. It has effects on water level, storm frequency and intensity, river discharge and sediment delivery, and wind and wave climate, each of which can result in adjustments to coastal landforms. For instance, westerly winds associated with El Niño result in different patterns of beach accumulation or erosion and shifts in island orientation in the central Pacific (Solomon and Forbes, 1999). At the same time, tropical cyclones are more frequent in French Polynesia, impacting islands that do not experience such storms at other times. Accretion and erosion of beaches in eastern Australia (Bryant, 1983), and patterns of cliff erosion along the coast of California (Storlazzi and Griggs, 2000) have been related to ENSO. Some of these are examined in the chapters that follow.

3.5.4 Storms and extreme events

The geomorphological significance of an event is governed by the amount of energy it expends on the landscape relative to the resistance of that landscape to change. Over a period of time, there will generally be a large number of moderate events and a smaller number of large events. If the period of observation is long enough, then the record may be more-or-less representative of all events, but this is rarely the case. The magnitude–frequency relationship is generally drawn as a probability plot from which recurrence can be determined and extrapolated beyond the observed data set to estimate the magnitude of larger events that have not been experienced. Wave data can be described in terms of annual exceedence probability; for instance, the 1% (one in 100 years) storm event on the northern coast of New South Wales consists of wave heights of 12.3 m, whereas on the southern New South Wales coast it consists of waves reaching 9–10 m.

There has been a long debate as to whether more geomorphological work is done by recurrent events of moderate intensity, or events of high magnitude but low frequency. For instance, a storm that occurs once in 100 years is likely to have a greater impact than a 10-year storm, but it is less clear whether it will have a greater effect than the several 10-year events that will occur over a longer term.

Recurrence intervals are probabilistic; they do not indicate when events will occur, but merely indicate the probability associated with an event of a particular size. Events of greater magnitude have a lower probability of occurring; their recurrence is longer on average. Recurrence intervals are determined presuming that climate is constant over the long term (stationarity). If climate change occurs, the expected frequency of events will also change, and this particularly affects the ability to extend the recorded data to extrapolate the magnitude of infrequent events.

The term storm is defined as wind speeds of above 8 on the Beaufort scale (see Table 3.2), but it is used more generally to include relatively frequent high-energy events. The movement of water during a storm depends on water-level changes caused by lateral variations in barometric pressure, wind-driven surface currents, coastal set-up (a piling up of water in the nearshore zone), and mass transport and set-up by waves. A storm surge is generated where atmospheric pressure is particularly low and the sea surface responds by raising in elevation (Flather, 2001). If a surge, which consists of just one crest, coincides with high tides at perigee (when the moon is closest to the earth) the sea surface can be increased by several metres (Hubbert and McInnes, 1999). For example, the storm surge of 31 January and 1 February 1953 in the North Sea had a water-level 3.3 m above expected.

High-latitude storms result from low-pressure systems, called cyclones, which are formed at 3-to-4-day intervals. Cyclones move slowly, and storm conditions can influence a coast, raising water level for several days (McInnes and Hubbert, 2001). In the tropics, violent low-pressure systems, termed hurricanes in the Caribbean, typhoons in Asia, and tropical cyclones in the Pacific, affect those coasts which lie in their paths (see Figure 3.11). These storms are formed over the ocean through heat exchange, particularly latent heat, from the ocean, within belts either side of the equator (Simpson and Riehl, 1981). They move along unpredictable tracks but within well-defined parts of the ocean (see Figure 3.11), maintaining intensity over warm water (>27 °C), but dissipating rapidly over land or cooler water.

Figure 3.16 is a schematic summary of the major effects that can be anticipated in relation to typical storms affecting hypothetical continents. As a storm passes a coast, wind speeds and directions change and the pressure results in surges or set-down. On the north

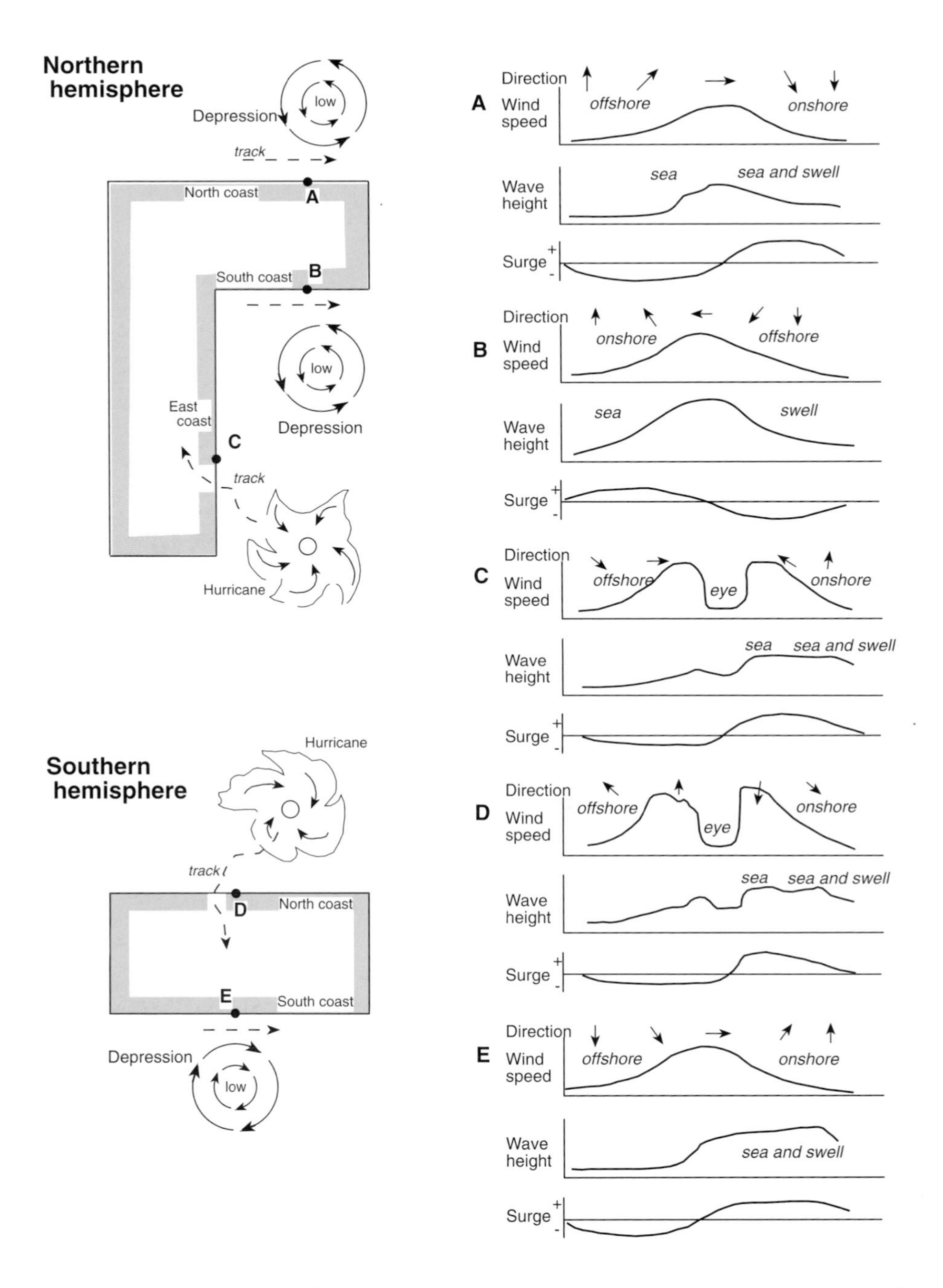

Figure 3.16. Schematic representation of storm passage in relation to coasts of generalised land masses, showing the variation in wind direction, wind speed, wave height and surge characteristics. Passage of northern hemisphere depressions (based on Carter, 1988). Note the different track and speed of tropical cyclone travel and the pronounced eye of the storm.

coast of a land mass in the northern hemisphere, a storm moving from west to east will show a progression from offshore winds generating only small inshore waves which gradually turn westerly (longshore) and then south (onshore). Wave heights will rise sharply, and plateau as the cyclone passes and sea and swell combine. Wave period reaches a maximum after the storm as swell waves arrive at the shore, and a negative surge will change into a positive one after the storm passes (Carter, 1988).

The sequence in which peaks of wind and sea/swell or surge occur will be different for other settings (Figure 3.16). Tropical cyclones are characterised by a central calm, termed the eye of the storm, with pronounced wind shifts. Storms are generally accompanied by major rainfall events that are also likely to have significant impacts. Single independent or isolated events can have an unduly significant impact in shaping coasts. Two events occurring in succession can have accentuated geomorphological impacts, and the effects of storms on particular coastal landforms are discussed in the chapters that follow.

Rarer and more extreme events include tsunami. Tsunami, from the Japanese term for harbour wave, are large-amplitude waves created by a disturbance, such as an earthquake or submarine landslide. They are progressive waves with large wave lengths, several 100 km, that travel very rapidly (>200 m s^{-1}) across ocean basins because relative to wave length these represent shallow water (Komar, 1998). Seismic activity, or submarine landslides on the steep shelves along collision coasts, create a sequence of several tsunami waves. For instance, a tsunami generated by an earthquake with its epicentre deep off Unimak in the Aleutian islands hit the Hawaiian Islands on 1 April 1946. The first three waves got progressively higher, from around 4 m to >6 m above high water mark, being especially damaging at closer to 10 m above still-water level at Hilo on the island of Hawaii (Shepard *et al.*, 1950).

The geomorphological work done by tsunami can be considerable. They generate long period waves, beneath which high peak orbital velocities can persist for several minutes (Bryant, 2001). The net onshore flow associated with these can carry large volumes of sediment onto the shore. Although tsunami behave as shallow-water waves with amplitudes of <1 m in the open ocean, they increase considerably in height as they cross the continental shelf and approach the shore. The damage that such waves can do was powerfully demonstrated by the sequence of three large tsunami waves of 17 July 1998 that overtopped a sand barrier and swept away several villages, leaving a trail of devastation along 30 km of the coast of West Sepik province, Papua New Guinea, (McSaveney *et al.*, 2000). The geomorphological role of tsunami and other extreme events is considered further in Chapter 9.

In terms of determining how significant high-magnitude, low-frequency geomorphological events are in shaping the landscape, it has been recognised for some time that the work done is not a linear function of the magnitude of process operation. For instance, sediment yield from a catchment increases with rainfall but not in a linear fashion. Sediment runoff from semiarid catchments shows a rapid increase as rainfall increases, but this is countered to some extent by the more extensive vegetation in wetter climates (Langbein and Schumm, 1958). Although large floods carry tremendous sediment loads, their infrequency means that over a long period most sediment movement is achieved by events of moderate magnitude and frequency (Wolman and Miller, 1960). If the threshold of sediment movement is high, as where sediment comprises boulders, then maximum sediment movement could be associated with extremely rare events (Baker, 1977).

In the case of changes on the coast, it is clear that movements of very large rocks, or other events needing a major energy source, occur only during extreme events, but that moderate events that recur frequently can achieve a lot of work over a prolonged period of operation. For instance, one storm experienced once in 20 years in Alaska appeared to have had a more major effect in terms of coastal change than smaller events achieved in intervening periods (Hume and Schalk, 1967). The passage of cold fronts causes moderate storms along the Gulf of Mexico coast which achieve most of the geomorphological work of building landforms (Nummedal *et al.*, 1980). Extreme events, such as hurricane Gilbert in 1988, cause rapid retreat of the shoreline but, a short time after a major storm, the system readjusts and regular rates of retreat are re-established (Dingler *et al.*, 1993). Whether or not they are responsible for more work overall, there can be little doubt that storms are powerful forces in shaping the coast, and their effects will be considered for different coastal landforms in the following chapters.

3.6 Terrestrial and subaerial processes

There are many other processes that have an effect on the coast, but only brief mention of a few will be made here. Wind is significant, not only in relation to wave generation, but also in terms of dune building. In high latitudes, frost action can play a role in shattering rock. Rivers are important in terms of the water and the sediment that they bring to the coast; their role is examined in relation to deltas and estuaries in Chapter 7 but some general observations are made in the section below.

Hillslope processes are also discussed briefly, but are considered in more detail in Chapter 4.

3.6.1 Wind action

Wind is an important factor responsible for the generation of waves as discussed in Section 3.3. Wave characteristics typical of a fully arisen sea under particular wind conditions (defined on the Beaufort scale) are shown in Table 3.2. Although many coasts are influenced by swell, generated by waves associated with storms that are beyond the immediate influence of the coast, there are also significant locally generated wind waves which can have important effects on the coast. There are winds that are particularly associated with the coast. For instance, a sea breeze occurs on many coasts because of the different thermal conductivities of land and sea. During the day, the land heats more than the sea, causing air to rise over the land, with development of an afternoon sea breeze (blowing onshore). At night, the land cools whereas the sea remains warm and a land breeze blows from land to sea (offshore). Sea breezes of around 2.5 m s^{-1} are typical, but in some cases they can be 15 m s^{-1} or stronger (Abbs and Physick, 1992). A summer sea breeze, locally called the Fremantle Doctor, is experienced around Perth, Western Australia, blowing parallel to shore, generating wind waves of up to 0.9 m height, and eroding beach cusps (Pattiaratchi *et al.*, 1997).

Wind also plays a major role in the formation of dunes. Early attempts to determine net direction of sand transport involved construction of resultants of wind direction (Bagnold, 1941; Schou, 1952; Landsberg, 1956; Jennings, 1957). These generally involved winds over a threshold velocity of around 5 m s^{-1} that was considered necessary to entrain sand. Sand movement was then considered proportional to the cube of wind speed; often only onshore winds were considered. Sand movement by winds blowing along a beach can also be significant (Bourman, 1986).

Sand grains are moved when wind velocity exceeds the critical shear stress. Wind sorts sediment, preferentially moving sand in the range 0.1–0.3 mm, with the coarser fraction being left as surface lag. Entrainment thresholds occur at higher velocities than in water. Surface roughness leads to retardation and the development of turbulence. A logarithmic decrease in velocity near the ground has generally been demonstrated in laboratory experiments, with a zone of no movement (Z_0) which is dependent on surface roughness (a function of grain diameter) or vegetation (Figure 3.17). This has been presumed also to apply

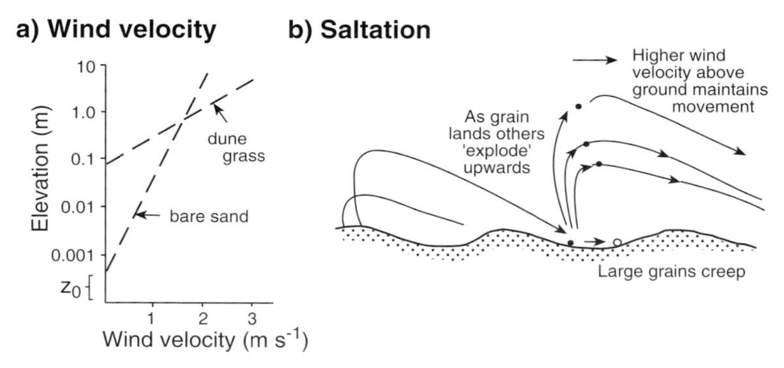

Figure 3.17. Processes operating on dunes. (a) Wind velocity with elevation above a dune vegetated with dune grass and a bare dune. The value Z_0 is defined by roughness, being related to vegetation height or grain size respectively (based on Carter, 1988). (b) The process of saltation (based on Pethick, 1984).

in the field. Typical wind-velocity distributions would differ beneath dune grass and over bare sand as shown in Figure 3.17a.

There are various theoretical, empirical and experimental equations to calculate the volume of sand transported by wind (Table 3.4). Many of these are based on the initial studies by Bagnold (1941), assuming that there is a steady and uniform wind field, that the surface is flat and does not impede transport, that sand grains are dry and non-cohesive, and that transport is primarily by saltation. Saltation is the process by which impact of wind-borne sand grains on the floor ejects further grains into suspension (Figure 3.17b). Such movement is highly episodic, with rapid response of bedforms to variable wind conditions. It may be necessary to generate sufficient lift to get particles into suspension; alternatively, the impact of particles falling from suspension can flick many other particles into the air. The higher velocity just above the surface is sufficient then to maintain the particle in suspension while it is carried forward. Deviations from the best-fit logarithmic relationships shown in Figure 3.17, arising because transport depends on aspects of wind energy, direction, wind resultants, tidal range and sediment supply, can cause major discrepancies in calculations of eolian sand transport. Moisture, temperature, surface crusts, surface roughness and beach morphology can all affect eolian transport, as well as armouring by coarse lag material that is left over the surface by the winnowing of finer material (Pye, 1983). Observed transport rates also differ from predicted because of development of microtopography (Sherman and Hotta, 1990). The types of dune and their formation are considered further in Chapter 6.

Table 3.4. Eolian sand transport formulae

Equation	Source	Comments
$Q_a = A\frac{\rho_a}{g}\left(\frac{d}{D}\right)^{0.5} U_*^3$	Bagnold, 1941	A = 1.5 for uniform sand A = 1.8 for graded sand A = 2.8 for mixed sand
$Q_a = C\frac{\rho_a}{g} U_*^3$	Chepil, 1945	C is between 1 and 3.1
$Q_a = z\frac{\rho_a}{g}\left(\frac{d}{D}\right)^{0.75} U_*^3$	Zingg, 1952	$z = 0.83$
$Q_a = H\left(\frac{U}{(gd)^{0.5}}\right)^3$	Hsü, 1971	$H = \exp(-0.42 \times 4.91d) \times 10^{-4}$
$Q_a = B\left(\frac{\rho_a}{g}\right) u^n \left(\frac{d_{50}}{D_{50}}\right)(u_* + u_{*c})z\,(u_* - u_{*c})$	Horikawa *et al.*, 1986	

Notes:
U, wind velocity; ρ_a, density of air ($\approx 1.2 \times 10^{-6}$); g, gravity (9.8); d, grain size; D, standard grain size (0.25mm).
Subscripts: c, critical or threshold velocity; *shear velocity; u = uniformity coefficient for grain size (d_{60}/d_{10}), d_{10} d_{50} d_{60} represent grain size percentiles. B, n = empirical constants.
Source: Based upon Horikawa *et al.*, 1986; Carter, 1988.

3.6.2 Frost action

Freezing and thawing of rocks and sediments, called cryogenic processes, can affect high-latitude coasts, and include the action of permafrost and ground ice as well as freeze–thaw processes. The effect of these frost processes depends on the lithology of rocks, the supply of water, and the extent to which the products of freeze–thaw action are removed. Permafrost heaves sediments and causes ice mounds. Ice lenses are near-circular depressions. On the shore, the upper intertidal zone can be frozen and inactive, whereas inundation means that the lower intertidal zone is unfrozen and still active. Frost shattering of rocks is also an important process in the formation and shaping of cliffs (see Chapter 4) resulting in angular talus, especially in mid latitudes. For instance, talus or scree has accumulated at the foot of many cliffed shorelines in Scotland, such as Ailsa Craig, or Drumadoon Point on the Isle of Arran (Steers, 1973). The legacy of frost action during colder past phases can still be seen in paraglacial regions (Forbes and Syvitski, 1994). Thaw of permafrost can lead to important flow slides as a cliff or hillslope retreats. Buried snow or ice produces elongated pits in the beach surface known as kettle holes. When the temperature is below freezing point, tidal waters can serve to melt frozen sediments, especially in mid latitudes, and the freeze–thaw processes can be directly associated with frequency of inundation.

3.6.3 Fluvial processes

The role of rivers is particularly important in relation to estuaries and deltas. Rivers bring fresh water and sediment to the coast (Milliman, 2001). In the case of deltas, the sediments are likely to accumulate at the mouth of the river unless nearshore processes are sufficient to redistribute them. Estuaries can be a sink for fluvial sediments and for nearshore sediments.

The range of landforms that develop at the mouths of rivers is the subject of Chapter 7. River-borne clastic sediments can be an important addition to the littoral sediment budget (see Chapter 2), and their contribution can have important consequences for beach and dune systems downdrift of the mouth. If a system has adjusted to the addition of river sediment, the disruption of that supply of sediments can also have significant consequences (see Chapter 10).

3.6.4 Weathering and hillslope processes

There are also other subaerial processes that can be important in some coastal settings. Climatic factors have an influence on the rate of weathering which determines the resistance of the rock or sediment (Ollier, 1984). In the case of rocky coasts, subaerial hillslope processes such as creep, slopewash, rockfall and mass movement can be significant. These processes depend on the nature of the slope and the material of which it is made, and include a range of gravity-controlled failures and falls, such as spalling, toppling, sliding, slumping and flowing, which are discussed further in Chapter 4. Other subaerial processes include the weathering of rock that can occur on coasts that are subject to salt spray. Salt can influence the breakdown of rock, pitting its surface. There are other physiochemical influences that become important in some coastal settings, as in the case where seawater ponds on rock surfaces. Similar issues relate to the effects of solubility in the case of limestone. These are also considered in Chapter 4.

3.7 Biological processes

Plants and animals can also have important effects on coastal geomorphology. The clearest example is the case of coral reefs, where entire landforms result from the constructional activities of organisms. Not only do corals form skeletons that compose the framework of the reef (see Chapter 2), but other organisms also contribute to the sediments that infill and bind the reef (see Chapter 5).

Biota play less central roles but also are significant in other settings. There are many organisms that produce calcareous sediments (see Table 2.5). The nature of the organisms contributing to sands determines the skeletal components which influence the size characteristics of the sediment (see Figure 2.15). For instance, coral breaks into distinctive sizes of shingle, and powders to a coral grit.

Biota play other biomechanical and biochemical roles. They veneer and protect surfaces. They bind sediments as in the case of algal mats. Bioerosional roles are important. Organisms, particularly gastropods graze and bore rock. In softer sediments, the activity of infauna turns over the sediment (bioturbation), while a range of other organisms ingest the surface sediments (Spencer, 1988a). In the case of dunes, vegetation acts to bind sand, and on muddy coasts, the role of halophytic (salt-tolerant) vegetation in trapping fine-grained sediment is examined in Chapter 8.

Perhaps one of the most important factors is human impact. Humankind has drastically modified many coasts and exerts an influence, however subtle, on almost all coasts. The role of human activity in terms of coastal geomorphology is considered in Chapter 10.

3.8 Summary

Processes of sediment erosion, transport and deposition are fundamental to the geomorphology of coastal landforms. The nature of the material, considered in Chapter 2, including individual grain and sediment bulk characteristics, influences how effectively landforms are reworked by the processes that operate on them.

Of the many processes that affect the coast, wave and tide processes have been emphasised in this chapter, because they are the principal sources of energy. Although the physics of how waves travel and the forces generated by the tides have been understood for some time, the relationship between the shear stresses that are exerted on coastal rock or sediment and the degree to which landforms are reworked remains difficult to quantify over time scales that are useful in understanding the morphodynamics of coastal landforms. The lack of uniformity, in terms of sediment characteristics, bimodality of sediment size distributions, and the role of cohesion and biological activity, complicate the responses and frustrate real predictive capacity in terms of sediment movement. The highly variable nature of process operation over time, particularly the bi-directional nature of flows and hence of sediment transport beneath waves or in tidal channels, also thwarts attempts to derive an estimate of integrated sediment transport rates. Longer-term variations in the rates at which processes operate mean that it is still unclear to what extent it is normal incident conditions which shape the coast, or the impact of extreme events.

Although some processes have been quantified at the instantaneous time scale, it is clear that feedbacks between morphology and process mean that the deposition of sediment on the bed, or erosion of sediment from the bed modifies the morphology in a way that impacts on the rate of operation of the processes responsible for further erosion or deposition. The co-adjustment of form and process is central to coastal morphodynamics, and these issues will be examined in relation to particular types of coasts in the chapters that follow.

Chapter 4
Rocky coasts

> If the coast is bold and rocky, it speaks a language easy to be interpreted. Its broken and abrupt contour, the deep gulfs and salient promontories by which it is indented, and the proportion which these irregularities bear to the force of the waves, combined with the inequality of hardness in the rocks, prove, that the present line of the shore has been determined by the action of the sea . . . It is true, we do not see the successive steps of this progress exemplified in the states of the same individual rock, but we see them clearly in different individuals; and the conviction thus produced, when the phenomena are sufficiently multiplied and varied, is as irresistible, as if we saw the changes actually effected in the moment of observation. (Playfair, 1802)

Rocky coasts occur where rugged or relatively resistant terrestrial lithology abuts the ocean, forming a distinct or abrupt transition between land and sea. They are typically high-energy coasts, primarily of sea cliffs and other steeply inclined shorelines, on which the influence of underlying rock type is plainly apparent. Many of these coasts evolve at slow rates; in places, resistant pre-Tertiary rocks appear to have changed little over millions of years. However, it is clear that tectonic activity and fluctuations of sea level, particularly during the Quaternary, have caused shoreline position to adjust over time. The sea has periodically reoccupied former positions, and modern rocky coastal morphology is often partially inherited.

The most spectacular rocky coasts are cliffed. However, the principles that are discussed in this chapter also apply to a series of other coastlines that are less bold. For instance, resistant lithology can form isolated headlands separated by embayments that contain other types of coastal landforms (such as beaches or estuarine and deltaic deposits). Some coasts consist of bouldery slopes that shelve gradually seaward, and others comprise 'soft rocks' (poorly consolidated sediments such as glacial and alluvial deposits) that are scarped by the action of the sea.

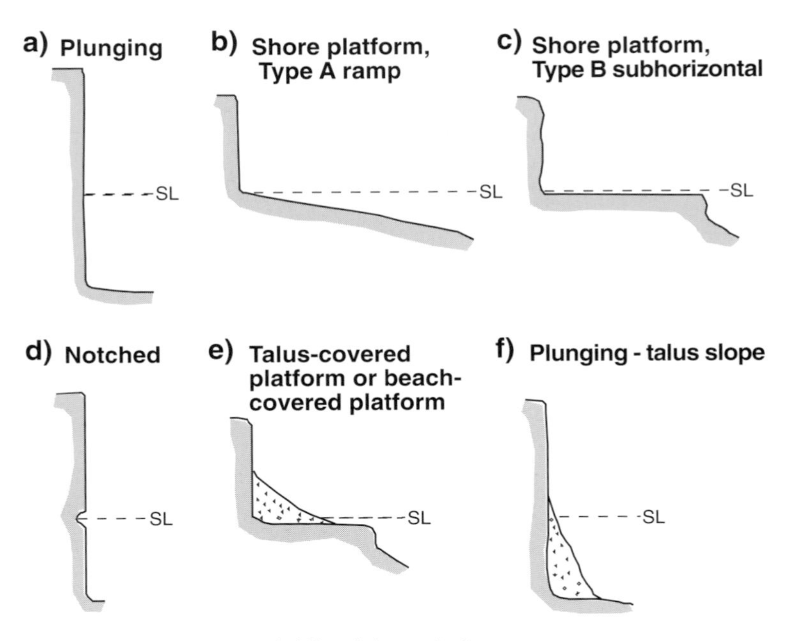

Figure 4.1. Major types of cliff and shore platform.

Six distinct types of rocky coast morphology are shown in profile in Figure 4.1. Shorelines cut into resistant rock generally adopt one of these forms, whereas less resistant lithology adopts a broader range of forms including intermediate profiles. Plunging cliffs are defined as near-vertical slopes that continue below the water line; they are typically high (>100 m), developed in resistant rocks, and have bases which extend into deep water. Steep cliffs of this type reflect wave energy in the same way that waves rebound off vertical seawalls (see Chapter 3). In rare cases, particularly on rocks that can be dissolved by seawater such as limestone, a plunging cliff can be notched around sea level. At the base of many cliffs there are near-horizontal intertidal or supratidal platforms which are called shore platforms. These usually adopt one of two morphologies; they either slope seawards, termed a ramp (Type A according to Sunamura, 1992), or they form a subhorizontal platform with a seaward low-tide scarp (Type B according to Sunamura, 1992). The talus that is produced by the weathering and erosion of rocks can accumulate at the foot of cliffs when supply exceeds removal by wave processes. A talus slope of coarse material, or a beach (sometimes ephemeral), protects the rockface from direct wave impact. These morphologies are examined further in Section 4.3.

The term 'cliff' is a morphological term denoting a steep slope. It is not genetic and does not presuppose an origin; not all cliffs have been eroded primarily by direct wave attack. It is convenient to distinguish

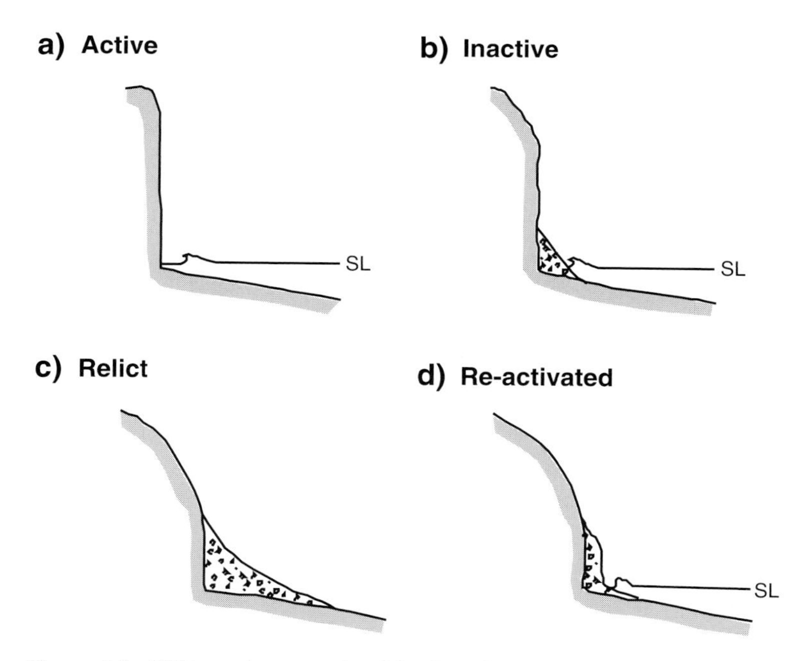

Figure 4.2. Cliff types in terms of activity (based on Emery and Kuhn, 1982).

several types of cliffs in terms of their 'activity' (Figure 4.2). Cliffs that are subject to wave action at their base are active. When talus has accumulated at the base and cliffs are protected from wave action they are inactive, although they can still be shaped by subaerial processes. Cliffs are considered relict (termed 'former' by Emery and Kuhn, 1982) where inactivity is prolonged, when for example the influence of the sea has been removed because sea level has dropped. In these cases, subaerial slope processes dominate. Relict cliffs are reactivated when subject to wave action again. Oscillations of sea level have meant that the majority of cliffs have had an episodic life history, punctuated by periods of inactivity. Cliff form may often have been initiated during a previous highstand, and reworked and extended during subsequent highstands of the sea. Many cliffs, therefore, are not solely the product of modern processes but have been subject to a sequence of different formative processes and are termed 'polygenetic' (Russell, 1963).

This chapter sets the historical perspective within which rocky coasts have been viewed. Rocky coasts are examined within the plate-tectonic setting of the continental or island shorelines on which they form. Characteristic planform and profile morphology is described in terms of lithology and structure, and the processes that are operating. The morphodynamic development of shore platforms and soft-rock cliffs are discussed in greater detail.

4.1 Historical perspective

It has long been appreciated that wave action is an effective agent of erosion which, over time, cuts back the shoreline (Playfair, 1802; see the quotation at the beginning of this chapter). Waves were considered to abrade both broad submarine shelves and narrower intertidal or supratidal shore platforms at the foot of cliffs, but controversy arose as to the precise role and relative significance of wave action. In the early literature it was believed that cliffs were eroded to wave base (the depth at which there is still detectable water motion beneath waves as described in Chapter 3). The product of wave planation was considered to be a broad submarine platform or shelf (Lyell, 1832). However, it was recognised that the ability of waves to cut back cliffs would be limited when an initial bench had been cut because of the attenuation of wave activity across that bench.

4.1.1 Darwin, Dana and Davis

Darwin described cliffs around St Helena and suggested that waves did not cut to wave base, but to a level close to sea level. He proposed that cliffs only continued to erode, bevelling a shelf or eventually truncating islands, where subsidence enabled large waves to reach the cliff 'with fresh and unimpaired vigour' (Darwin, 1846). At a similar time, Dana described impressive near-horizontal shore platforms, initially referred to as 'wave-cut' platforms, from the Bay of Islands in New Zealand and along the shoreline near Sydney, Australia (Dana, 1849). Dana considered that these platforms were cut into resistant rock by waves at the level at which wave energy was most concentrated in a similar manner to that in which reef flats developed (see Chapter 5). He also recognised platforms that had developed in weathered material and in lower-energy settings.

Gilbert (1885) and von Richthofen (1886) adopted Darwin's suggestion that it was necessary for the land to subside (or sea level to rise) for marine abrasion to be effective in cutting back the shoreline. In his review of the coral reef problem, Davis undertook a comprehensive assessment of tropical coasts and regarded cliffed shorelines as compelling evidence for subsidence of reef-encircled islands (Davis, 1928). Recognition of broad shelves backed by plunging cliffs around volcanic oceanic islands, such as St Helena, Auckland Islands, Campbell Islands and Lord Howe Island, but also in other settings such as Banks Peninsula in New Zealand, further supported this view (Cotton, 1969a, Bal, 1997). These truncated shelves experienced substantial sea-level changes over the Quaternary, which must have influenced the cutting of the cliffs. Bradley (1958) and Cotton (1963, 1969b), for instance, considered that plunging cliffs must have been cut when sea level was lower and wave action could attack their base.

As a part of the geographical cycle proposed by Davis (see Chapter 1), wave action across the continental shelf was believed to form a wave-built terrace (Gulliver, 1899; Fenneman, 1902). The view was supported by numerous bathymetric profiles (see Figure 1.6b) that appeared to support the concept of an equilibrium offshore profile (Johnson, 1919, 1925). The issue of an equilibrium profile is still contentious in relation to sandy shorelines (see Chapter 6), but the concept of wave-built terraces on continental shelves has been comprehensively dismissed (Dietz, 1963).

Davis also proposed as part of the geographical cycle that initial and 'youthful' shorelines would be cut back and become 'mature' and less indented with time (see Figure 1.4). Identifying the stage that cliffs were in led to differing interpretations; for instance, the impressive chalk cliffs flanking the English Channel were considered to be mature by Davis (1909) and de Martonne (1909), but were viewed as youthful by Johnson (1919). The recognition of regular Quaternary sea-level fluctuations (Daly, 1925, 1934) made it possible to envisage more frequent phases of cliff-cutting over shorter periods of time. Davis himself came to recognise several cycles of cliff erosion on uplifted coasts. He attributed terraces along the Californian coast (see Section 4.2.1) to rapid cliff erosion during times when positive movements of sea level overtook and outpaced the rate of gradual uplift (Davis, 1933).

Uplifted, fossil shorelines were interpreted from terraces elsewhere, for instance in Scotland, and led to the view of a cycle of active cliff erosion and abandonment on uplifting coasts (Jamieson, 1908; Wright, 1914; Putnam, 1937). Recognition of former shorelines on rocky coasts became a prominent component of coastal geomorphology (e.g. King, 1930; Balchin, 1941; King, 1963). Morphological correlation of these raised shorelines, together with stratigraphic interpretation of associated sedimentary deposits, became a focus of regional compilations and formed a basis for morphostratigraphical studies (Sissons *et al.*, 1966).

4.1.2 Lithology and weathering

The significance of lithology and its influence on the rate at which processes operate has been a central issue on rocky coasts. Cliff morphology along the north coast of Devon was shown to be correlated with dipping strata and intense folding of the hinterland by Arber (1911). The distinctive hogs back, or 'slope-over-wall' cliffs, on which a more gentle upper slope occurs above a steep lower cliff, were shown to result from re-excavation of interglacial cliffs buried by periglacial deposits (Arber, 1949, 1974). Reactivation of these and other relict cliffs was shown to be an important constraint on their shape (Guilcher, 1958; Orme, 1962; Cotton, 1963, 1974). When cliffs are abandoned, for example after sediment is deposited in front of them cutting off the effects of wave action, the cliff profile degrades as a result of subaerial processes (Savigear, 1952, 1962). The concept that cliffed shorelines might be polycyclic or polygenetic, not only in formerly glaciated areas, but also in tropical areas, was further extended by Russell (1963).

Resistant rock types that retain vertical cliffs in cold climates are likely to experience more rapid weathering in the tropics, with the consequence that cliffs are often more degraded. For instance, dolerite headlands protrude from the intervening strongly weathered granite and gneiss in Liberia (Tricart, 1962) and, in Yampi Sound in semiarid Western Australia, bold promontories remain in quartzite, whereas the metamorphic rocks have been more intensively chemically weathered and are eroded into shore platforms (Edwards, 1958). Whereas it has been suggested that gradually sloping, well-vegetated coastal bluffs, or 'versant-falaise', are more common than rugged cliffs in the humid tropics of South America (Tricart, 1972), vertical cliffs with basal talus are common in parts of Brazil (Guilcher, 1985).

Cliff erosion has also been studied on soft-rock cliffs. Historical records were used to assess rapidly retreating cliffs along the Californian coast (Vaughan, 1932; Emery, 1941; Kuhn and Shepard, 1984) and soft-rock cliffs in Europe, such as the glacial deposits of eastern England (Valentin, 1954). Mass movement and other subaerial slope processes are significant on these cliffs, and the relative balance between marine and subaerial erosion processes explains aspects of cliff morphology (Emery and Kuhn, 1980, 1982) and is examined further in Section 4.3.2.

Subaerial processes of a different kind, physiochemical action resulting from the alternate wetting and drying of intertidal or supratidal rock surfaces, appear to be significant factors in the lowering of shore platforms. Distinctive 'Old Hat' islands in sheltered parts of the Bay of Islands, New Zealand, have shore platforms of similar width on all sides. These were interpreted to indicate that waves were capable of removing weathered rock from the back of cliffs, but did not have sufficient energy to erode the platform directly (Bartrum, 1916, 1926). Bartrum believed that weathering occurred on the platform surface down to a level termed 'saturation level' (the level to which the rock is saturated by seawater) corresponding to high tide level, rather than the level of maximum wave activity proposed by Dana (Bartrum, 1935, 1938).

The role of platform weathering was emphasised in relation to lowering of horizontal shore platforms (Bartrum and Turner, 1928), and gained support from studies on volcanic tuff on the island of Oahu, Hawaii, with recognition of water layer levelling (Wentworth, 1938). Water layer levelling encompasses physiochemical action by water ponded on the platform (see Section 4.4.1) and has been implicated in the formation of platforms along the coast of southeastern Australia (Johnson, 1931; Jutson, 1939; Hills, 1949; Bird and Dent, 1966) and New Zealand (Healy, 1968). These ideas remain the subject of ongoing debate and the relative balance of wave action and subaerial platform processes is reviewed by Trenhaile (1987) and Sunamura (1992), and is re-examined in Section 4.3.

4.2 Plate-tectonic setting and wave planation

Rocky coasts are found throughout the world. They have been described in most detail from the temperate shorelines of North America, Britain and Europe, Russia and the former Soviet states, New Zealand and Australia. There are several factors that influence the distribution of rocky coasts at global scales, including plate-tectonic setting. Long-term wave abrasion of shorelines can lead to planation of the continental shelf and shelves around mid-oceanic islands. These issues are examined in this section.

4.2.1 Plate-tectonic setting

Rugged shorelines can develop in any setting where resistant lithology outcrops at the coast. However, rocky coasts are prominent on active plate margins. They occur along Pacific-type collision coasts dominated by vertical tectonic movements, particularly long-term uplift (Inman and Nordstrom, 1971). In these circumstances the coast is parallel with the structural trend (concordant), and is often characterised by volcanism and seismic activity with a deep-sea trench offshore, narrow continental shelf, and a shoreline composed of steep cliffs backed by uplifted marine terraces (Griggs and Trenhaile, 1994). A suite of uplifted marine terraces is characteristic of much of the shoreline on the western coast of North and South America (see Figure 2.3). The rate of recession of cliffs on this uplift-dominated collision coast is imperceptible in the short-term decadal–century time scale (Komar and Shih, 1993). Relict cliffs occur at the back of these terraces and persist for several glacial cycles (Figure 4.3). However, their slopes undergo degradation as a result of subaerial slope processes and dissection by rivers and streams, and the marine terraces and associated relict cliffs have a limited life (of the order of a million years), on geological time scales (Anderson *et al.*, 1999).

Uplift, whether on tectonically active uplifting coastlines, or resulting from isostatic rebound (e.g. Scotland; Steers, 1973; Sissons, 1974), exposes new rock surfaces to the erosive forces of the sea. The formation and erosion of these terraces can be simulated in computer models (Cinque *et al.*, 1995). However, cliffs are rare on coasts that are uplifting very rapidly because there is insufficient time for their development, (for instance, in parts of Scandinavia such as the Baltic basin; Eronen *et al.*, 2001).

Rocky coasts on active plate margins in tropical seas are generally suitable for coral-reef development, and distinct uplifted marine terraces of constructional reef facies are typical (see Figure 2.8). These flights of raised fossil reefs have provided evidence on the basis of which Pleistocene sea levels have been reconstructed because corals within the reefs can be dated (see Chapter 2). The elevation of individual reefs, when corrected for the rate of uplift, indicates highstands of sea level. Sequences of reef terraces have been dated from a series of key plate-margin sites, such as Barbados (Broecker *et al.*, 1968), Huon Peninsula in Papua New Guinea (Chappell, 1974; Chappell *et al.*, 1996), and Sumba in Indonesia (Pirazzoli *et al.*, 1991).

Figure 4.3. Marine terraces along the coast of California (photograph G.B. Griggs).

There are distinctive sequences of marine terraces on other uplifted coasts, particularly at plate margins around the Pacific (Bull, 1985; Bloom and Yonekura, 1985; Ota, 1986). Figure 4.3 shows a sequence of terraces from the coast of California. Sequences of marine terraces without coral are less amenable to dating, unless there are other means of dating them, such as the presence of distinct tephra layers, as in New Zealand (Ota, 1986). However, uplift rates have been assumed constant in several situations and it has been possible to link reefal and non-reefal terraces to determine tectonic history (Yoshikawa, 1985; Ota and Omura, 1991; Muhs *et al.*, 1994). The height at which the Last Interglacial terrace occurs at a series of sites around the Pacific is shown in Figure 2.7a.

On passive margins, greater tectonic stability and the more sustained supply of sediment leads to more subdued topography. The underlying basement rock is much more likely to be cloaked in sedimentary deposits or to be part of a drowned embayed coast. Resistant interfluves persist as rocky headlands that alternate with embayment fills (Bishop and Cowell, 1997). With the continued supply of sediment, broad coastal lowlands accumulate.

4.2.2 Shelf abrasion and island planation

Erosion of rocky coasts over sufficient time can form a near-horizontal submarine platform. For example, there is an abrasion surface, which is of the order of 1000 km long and 10 km wide along the east Australian coast, reflecting the concentration of wave energy along this coastline since the Miocene. The process can be particularly effectively viewed in the case of mid-oceanic islands of different ages, offering a perspective on different stages of abrasion.

Mid-plate volcanic islands occur in linear island chains resulting from hot-spot activity (see Chapter 2). The islands migrate passively with plate motion, and are progressively truncated by wave abrasion, ultimately forming a guyot (see Figure 2.2). Mid-latitude oceanic islands outside reef-forming seas are subject to vigorous wave abrasion. Many, such as St Helena and Tristan da Cunha in the Atlantic, sit in the middle of broad near-horizontal submarine shelves and are flanked by high plunging cliffs indicating the efficiency of marine abrasion (Cotton, 1969a). The near-horizontality of the shelf implies that these islands have not experienced subsidence or that subsidence has been countered by isostatic readjustment to the loss of material by erosion (Menard, 1983). Islands in tropical seas are protected from wave abrasion by coral reefs, discussed in the next chapter. A coral reef attenuates wave energy and reef-encircled islands are not truncated, and appear to be subsiding in association with plate migration (Menard, 1986).

Lord Howe Island is an example of a volcanic island that has been gradually migrating northwards on the Indo-Australian plate and which is just at the threshold at which reefs can form. It sits in the middle of a broad shelf, 8–10 km across and 30–50 m deep, which forms an abrasion surface that has been cut in the 6 million years since volcanism. Steep plunging cliffs (with cliff base in deep water) occur around much of the island (Figure 4.4). The coral reef that fringes part of the western shore of the island is the southernmost coral reef in the Pacific, and the hillslopes that are protected behind that reef are gradual convex slopes, in contrast to the plunging cliffs along unprotected shoreline.

Truncation of abrasion surfaces occurs as the integrated outcome of sea-level variations through time. Truncation does not occur at uniform average rates; as the platform widens, wave energy will be attenuated across it, and it will be cut at progressively slower rates with time (Trenhaile, 1989). Young volcanic islands are susceptible to very rapid initial rates of erosion; for instance, the pahoehoe lava flows of Kilauea on the Big Island of Hawaii, undergoing cooling and fracturing where they flow into the sea, are rapidly being cut back by erosion (Wentworth, 1927; Mogi *et al.*, 1980). On the island of Surtsey, which formed from a submarine eruption in 1963 off the coast of Iceland, recession of the tephra averaged 30–70 m a^{-1} (Norrman, 1980). Similarly, Krakatau in Indonesia, which erupted in 1883, appears to have retreated at a rate of

Figure 4.4. Lord Howe Island, southwestern Pacific. This basaltic island is the partially truncated remnant of a volcano and horizontally bedded lava flows dominate Mount Lidgbird (centre) and Mount Gower (far right). There are impressive plunging cliffs around Mount Gower rising to 900 m, in places with talus at their base. A fringing reef (the southernmost reef in the Pacific) provides protection for the shoreline and basalt slopes draining into the lagoon are convex, shaped by subaerial slope processes.

33 m a^{-1} over the period 1883–1928, but slowed to 5–7 m a^{-1} in the period 1981–83 (Sunamura, 1992).

In the early post-eruptive stages of volcanic island evolution, there can be significant mega-landslides, often generating large tsunamis (Holcomb and Searle, 1991; Masson, 1996). The evolution of the Hawaiian Islands appears to have been punctuated by large subaerial and submarine slides that have accounted for a large proportion of the loss of material. The largest cliff in this archipelago, the sea cliff on northeastern Molokai, appears also partly maintained by such massive block fall (Moore *et al.*, 1989, 1994b).

The history of truncation of guyots, shelves around islands outside reef-forming seas, and continental abrasion surfaces remains undecipherable. It is likely to have been intermittent, punctuated by major sea-level fluctuations, with frequent reoccupation and reactivation of former shorelines. The chronology of truncation is probably as complex as it has been on uplifted shorelines, but the record is less clearly preserved.

4.3 Cliff and shore platform morphology

The morphology of cliffs and shore platforms can be viewed in planform and in profile. The processes that operate on these rocky coasts are generally too slow to observe, and they have usually been inferred from morphology. In this section, the planform topography of rocky coasts is discussed. Sequences of erosional landforms such as caves and arches appear to offer insight into the pattern of gradual change on cliffed coastlines. The importance of a series of factors, such as lithology and structure, wave energy, and subaerial processes are considered, with reference to particular examples. The nature and morphology of shore platforms is examined. Finally, the significance of inheritance and the polygenetic shorelines that develop where processes change with time are described.

4.3.1 Planform of rocky coasts

The planform of rocky coasts is highly variable. It is useful to distinguish between concordant coasts that run parallel to geological structure, and discordant coasts that cut across the structure, extending the early recognition of these using the terms Pacific and Atlantic respectively (Suess, 1888; Johnson, 1919). The distinction can be viewed at a series of scales. At the largest scale much of the western (Pacific) coast of the Americas is concordant; at a smaller scale individual outcrops

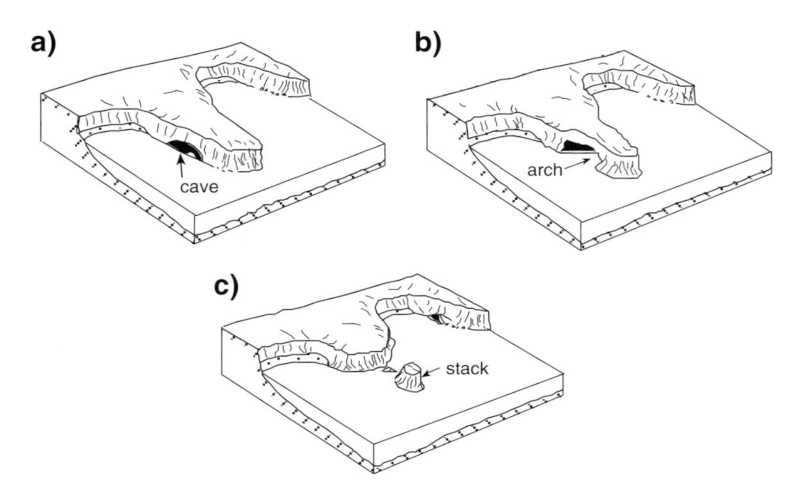

Figure 4.5. Progressive formation of sea cave, arch and sea stack on an eroding cliff shoreline (based on studies of the Port Campbell coast of Victoria by Bird, 1993b).

can be concordant (also termed accordant). Discordant coasts, with structure perpendicular, or at a high angle, to the shoreline are generally highly embayed, for example the Kimberley coast in northwestern Australia, or County Kerry in Ireland. The role of structure is clearly expressed in planform on the coast of Dorset in southern England where resistant Jurassic limestone runs along the shoreline to seaward of more erodible Wealden clays. The southern Dorest coast is a concordant (Pacific) type of coast and a series of features, such as a remnant sea arch (Durdle Door), small embayments (Stair Hole, Lulworth Cove), and a larger bay (Mupe–Worbarrow Bays), appear to represent successive stages of erosion (Horsfall, 1993). This contrasts with a discordant east coast near Swanage that truncates the lithology and where there are resistant headlands, rocky outcrops and sea stacks (Old Harry Rocks).

Vertical cliffs can form uniform smooth shorelines, for instance the coast of the Nullabor Plain in southern Australia (Bird, 1976). The vertical chalk cliffs of Sussex contain a cliff-top sequence of undulations corresponding to dry valleys (locally termed coombes) that were cut by periglacial activity during the Pleistocene (Wood, 1968; Bird, 2000). The significance of dipping strata can be seen along the north Devon coast; cliffs are relatively stable where the dip is away from the coast, but if the dip is towards the sea they are much less stable, with landslides along structural bedding planes (Arber, 1911). Embayed coastlines can result where there are pronounced lithological differences with the more resistant lithology forming the headlands, or where persistent interfluves between adjacent catchments remain as headlands. Embayment size and the distribution of rocky headlands in southeast Australia are related to catchment characteristics rather than lithology (Bishop and Cowell, 1997).

Sea caves, arches and sea stacks are some of the more spectacular erosional landforms that characterise rocky coasts. These landforms represent a progressive sequence in terms of wave attack on headlands (Johnson, 1919). They are best developed on coasts which comprise rocks that are strong enough to stand in near-vertical slopes, supporting the roof of caves or arches, but contain sufficient zones of weakness (such as cracks, fissures, cleavage planes, joints, faults or folds and stratifications) that they can be excavated. Exploitation of weak zones can puncture a headland to form an arch, and the collapse of the roof of an arch leaves a sea stack which, in turn, is weathered away (Figure 4.5). The sequence depicted in Figure 4.5 is well illustrated in the Port Campbell region of the coast of Victoria, which contains arches, such as London Bridge (Figure 4.6) and stacks, such as the Twelve Apostles (Figure 4.7), cut in the soft Tertiary Port Campbell limestone.

Sea caves are caves at the water line that have been cut by wave action. They can reach 50 m or more in length (e.g., Geographers' Cave at Cape Gilyanly in the Caspian Sea; Tsutendo Cave, San-In coast of Japan; Fingal's Cave on the Isle of Staffa, Scotland). Caves form preferentially

Figure 4.6. London Bridge, an arch on the coast of Victoria, Australia is eroded from soft Tertiary Port Campbell limestone. It consisted of a double arch, the stack being connected to the mainland, until 15 January 1990 when the inner arch collapsed.

in response to structural weakness such as jointing in sedimentary rocks or through wave-induced erosion of weathered cliff material (Moore, 1954). In the first phases of cave development, fractures or openings are widened by the hydraulic action of waves supplemented by the pneumatic action of trapped air, which acts as the positive feedback to enlarge a cavity. Subsequently, the rate of cave extension will slow. Blowholes occur where hydraulic and pneumatic action puncture the roof of a cave. Spectacular spurts of water through the confined opening have made blowholes a tourist attraction, for instance, at Kiama on the southern coast of New South Wales.

Arches are openings through a headland and are ephemeral features. They have been referred to as sea tunnels when their length exceeds their width, as in the case of Merlin's Cave near Tintagel, Cornwall, which is 100 m long. The example of the double arch called London Bridge illustrated in Figure 4.6 shows a stage in the progression; the inner arch collapsed on 15 January 1990 (Bird, 1993b). An arch with a life history of only a little over a year has been described from Black Rock Point near Melbourne (Bird and Rosengren, 1987). The 13-year history of an arch at Table Head, Cape Breton Island, Nova Scotia was recorded in a sequence of photographs (Johnson, 1925). Other arches

Figure 4.7. The Twelve Apostles, Victoria, Australia. These are stacks left by the retreat of the Port Campbell limestone.

can be seen to have undergone little change, such as Jump-off-Joe near Newport, Oregon, which changed imperceptibly in photographs from 1880 to 1930, but collapsed in 1935 (Sayre and Komar, 1998). An arch termed the Giant's Eye Glass in basalt near Giant's Causeway in Northern Ireland collapsed in 1949. The destruction of arches and the subsequent removal of the stacks have been recorded for San Diego County, California (Shepard and Kuhn, 1983). Cathedral Rock Arch near La Jolla, for instance, collapsed in 1906, one buttress remained until 1963 and had finally disappeared by 1968 (Kuhn and Shepard, 1983, 1984).

Stacks are isolated pinnacles or outliers of rock; they can form as buttresses remaining after an arch has collapsed (see Figure 4.5), but can also result from other causes. The Needles at the western end of the Isle of Wight are cut in steeply dipping resistant chalk (Steers, 1969). The Old Man of Hoy in the Orkney Islands is a stack in resistant Old Red Sandstone; it is 140 m high, with prominent horizontal bedding. The life expectancy of a stack is related to rock strength. It is not necessarily the case that the rock of which stacks are composed is more resistant than that previously around it; in some cases their persistent morphology may result from dissection along intervening joint patterns (Trenhaile *et al.*, 1998). The Twelve Apostles (Figure 4.7) represent a sequence of stacks left by retreat of the Tertiary Port Campbell limestone on the coast of Victoria, Australia.

4.3.2 Cliffs in profile

Cliffs can be defined by their profile morphology into at least the six broad morphologies shown in Figure 4.1. Plunging cliffs are high cliffs (often 100–150 m or more high) developed in resistant rock; they have bases which extend into deep water (see Figure 4.1). Plunging cliffs occur on a range of lithologies including volcanic lava (e.g. Krakatau, Indonesia), resistant granite (e.g. Wilson's Promontory, Victoria, Australia), and Eocene–Miocene limestones (e.g. the Nullabor Plain in southern Australia). Very high cliffs are sometimes termed megacliffs (megafalaise). Examples include the 400-m-high cliffs of Kerry, Ireland, which are composed of resistant Old Red Sandstone and have been shaped by marine processes (Guilcher, 1966). Megacliffs up to 1000 m high, but with pronounced platforms at their base, occur on the uplifting coast of Chile, and are thought to have formed from a fault scarp (Paskoff, 1978). Vertical cliffs of more than 1000 m high are theoretically possible in homogeneous rock without discontinuities (Terzaghi, 1962). Elsewhere, there are plunging cliffs that can be related to Holocene faulting (e.g. Wellington fault, New Zealand; Cotton, 1952), rapid subsidence (e.g. Lyttelton Harbour, New Zealand; Bird, 1976), or glacial processes on the flanks of fjords in Norway and in southern New Zealand.

The spectacular plunging cliffs on Lord Howe Island (see Figure 4.4) presently reflect wave energy, and appear to have been cut in the past when sea level was lower. In some places they have a talus slope at their base, extending into deep water. Cliffs with similar talus cover occur in high latitudes where freeze–thaw generates scree slopes, for example on the northern coast of the Isles of Mull and Skye in Scotland (Flemming, 1965). Plunging cliffs can have a notch (or a nip) cut at, or close to, mean sea level. A horizontal incision is generally termed a notch when it is particularly deep (several metres) or asymmetrical, and a nip when it is shallow and symmetrical. Notches occur on less resistant rock types, such as tuff (Emery and Foster, 1956) or soluble lithologies, such as limestone. Seawater is generally supersaturated with respect to carbonate, and so there would appear to be no opportunity for solution to occur. Various studies have attempted to show (generally with little success) that undersaturation may occur in rock pools at night, perhaps biologically mediated (Trudgill, 1976).

Several factors influence the morphology of cliffs. Lithology and structure, exposure to wave action, and the relative balance between subaerial slope processes and marine processes are examined below.

Lithology and structure

Cliffs develop in a wide range of rock types, and lithology and structure clearly have a significant influence. Resistant rocks are more likely to form vertical cliffs than are less resistant rock types. For instance, Portland Stone, a Jurassic limestone occurring around the Isle of Purbeck in Dorset in southern England, is relatively resistant, but structural weaknesses such as joints, fractures and faults are exploited where they occur and influence rate of cliff retreat (Allison, 1989; Allison and Kimber, 1998).

There have been many descriptions of the striking cliffs in medium-resistance Cretaceous chalk along both sides of the English Channel in northwestern Europe, including the Seven Sisters of the Sussex coast, the White Cliffs of Dover, and similar cliffs along the coast of Normandy, France. The chalk is fine, white and friable, and contains flint (biogenic silica). The flint weathers out of the chalk supplying material to the shingle beaches of the shores of the Channel. This flint assists in the abrasion of the rock face as well as contributing to the beach at the cliff foot. Average retreat rates of 0.05–>1 m a^{-1} have been reported (May and Heeps, 1985). Retreat occurs by rockfall, mass movements and notching at the base, and is most rapid in winter months

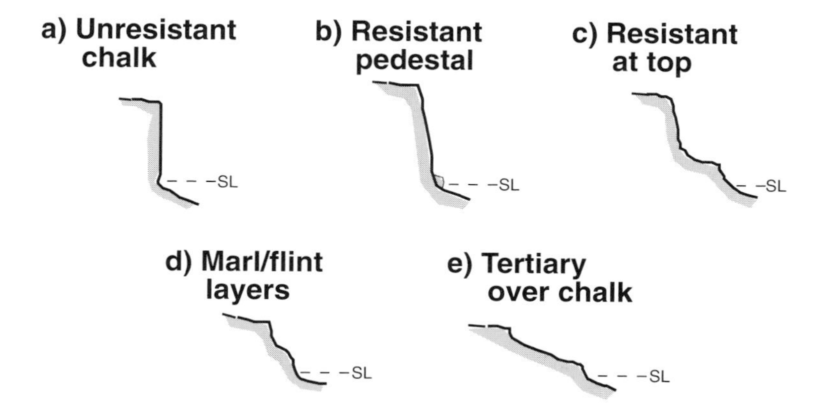

Figure 4.8. Cliff profile types in chalk characteristic of the coast of southern England and northwest France (based on Prêcheur, 1960; May and Heeps, 1985).

when flint nodules may be broken up by spalling and freeze–thaw (Robinson and Jerwood, 1987).

Five cliff profile types have been recognised in northern France (Prêcheur, 1960) and extended to the chalk cliffs of southern England (May and Heeps, 1985). These are illustrated in Figure 4.8. A near-vertical cliff with a basal ramp characterises those shorelines that are composed of homogeneous. weakly resistant chalk (Figure 4.8a). This type is characteristic of the cliffs between Brighton and Eastbourne in Sussex that reach a maximum height of 160 m at Beachy Head, with platforms of 100–150 m width at their base. Elsewhere, there are steep cliffs that have a resistant pedestal of harder material at their base resulting in a less vertical cliff (Figure 4.8b). A composite profile (Figure 4.8c) develops where the cliff consists of a prominent face that develops in an upper more resistant layer of chalk over a lower less resistant chalk characterised by landslides. Still more complex profiles occur where there is greater lithological variation, made of alternating bands of flint and chert of variable resistance (Figure 4.8d). In some places Tertiary sands and clays overlie the chalk (Figure 4.8e); mass movement occurs in these upper beds resulting in low-angle slopes and complex morphology (see Section 4.5). Landslides are common where gault clay overlies chalk (Hutchinson, 1983).

Wave exposure and climate

Wave action is an important factor in relation to cliff development. Waves maintain active cliffs, but there are many cases where equally impressive cliffs occur in areas of minimal fetch. For instance, there are steep cliffs on both coasts of the Isle of Skye in Scotland, although the

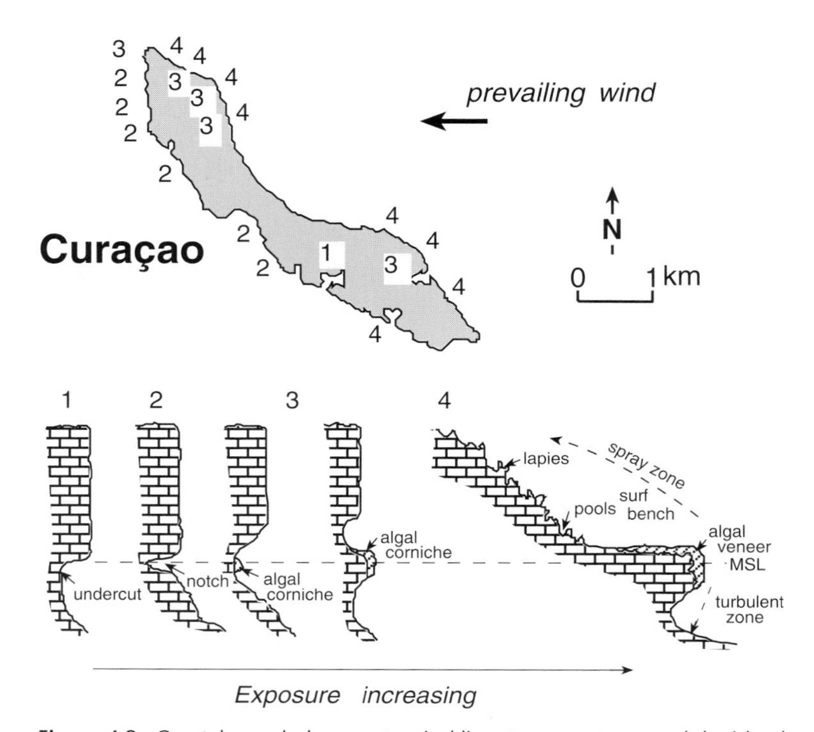

Figure 4.9. Coastal morphology on tropical limestone coasts around the island of Curaçao in relation to biodegradation and water turbulence which is a function of wave energy (based on Focke, 1978a,b). 1, sheltered; 2, leeward; 3, lateral (intermediate); and 4, windward.

west coast is exposed to high wave energy and the east coast is sheltered (Steers, 1973). The extent to which waves are necessary for rocky coast erosion is discussed in greater detail below in relation to shore platforms. First, an example of an apparent relationship between limestone cliff morphology and wave exposure is described.

Figure 4.9 shows limestone cliff morphology around the island of Curaçao in the West Indies (Focke, 1978a,b). The reef limestone is uniform in terms of lithology, structure and height and the cliff-profile is directly related to exposure to wave energy which affects water turbulence, duration of inundation and biological action. Associated with increase in energy is a broadening of the width of the zone affected. The role of biota is accentuated as energy increases, with a range of organisms both eroding the rock through grazing, and also having a major protective role in the intertidal zone. Organic corniches increase rock strength, with a double notch sometimes forming in mid-intertidal zone. In the most exposed settings a tidal terrace or surf bench forms, with organisms protecting the margin of the bench and a spray zone extending many metres above sea level. An example of a surf bench on the exposed shore of Atiu, in the southern Cook Islands, is shown in Figure 4.10. A surf bench is a characteristic feature of tropical reefal limestone shores that are highly exposed (Kaye, 1959), and it is often veneered by a range of organisms that both protect the rock, and serve to weaken it.

Similar cliff morphology occurs on other Pleistocene reefal limestone coasts in the West Indies (Woodroffe *et al.*, 1983b; Spencer, 1985), but has not been as convincingly related to wave energy. In the most protected areas, recession takes place subtidally primarily as a result of marine bioerosion (particularly borers). Undercutting, similar to that found on Curaçao, results where bioerosion exceeds subaerial processes of cliff retreat, for instance within Harrington Sound, a very sheltered interior water body on Bermuda (Neumann, 1966).

Limestone coastal morphology is also distinctive because of the rock strength of limestone. Figure 4.11 shows the coast of the island of Vatulele in Fiji which contains a series of prominent notches (Nunn, 1990, 1998). The continuity of height of the notch through sheltered sites away from direct wave action seems to imply that solution is important in addition to wave action. Resistant Palaeozoic limestone forms

Figure 4.10. A surf bench along a high-energy coast on the makatea island of Atiu in the southern Cook Islands, in the Pacific. The coast is composed of Pleistocene reef limestones. Waves break on the surf bench, which in places has a notch at its rear.

impressive tower karst with plunging cliffs on Langkawi Island in Malaysia, Phangnga Bay in Thailand and Ha Long Bay in Vietnam. At each of these locations there are notches at or around modern sea level and fossil notches indicating higher sea levels (Pirazzoli, 1986b; Tjia, 1996).

Limestone coasts also vary between different climatic and tidal settings (Guilcher, 1953). A series of characteristic morphologies are shown in Figure 4.12. Limestone shorelines in warm tropical or subtropical seas are often composed of Late Pleistocene (typically Last Interglacial) fossil reef limestone and exhibit a distinctive morphology. There is often a prominent notch that is generally asymmetric, floored by minor pools separated by constructional rims, termed vasques (Figure 4.12a). The upper roof to the notch has a pronounced visor (Hodgkin, 1970). The surface of the limestone is often tortuous, highly dissected into 'karst' limestone features, referred to as marine karren or lapies.

Limestone coasts in other climatic settings show different morphology. In the Mediterranean, a relatively warm but tideless sea, pools and lapies are typical of the supratidal (Figure 4.12b), whereas the narrow notch has a less pronounced visor than in the tropics. The lower lip, termed a corniche is encrusted by biota, often the coralline algae,

Figure 4.11. The limestone coast of Vatulele, Fiji. The resistant limestone is prominently notched indicating former stands of sea level (photograph S.G. Smithers).

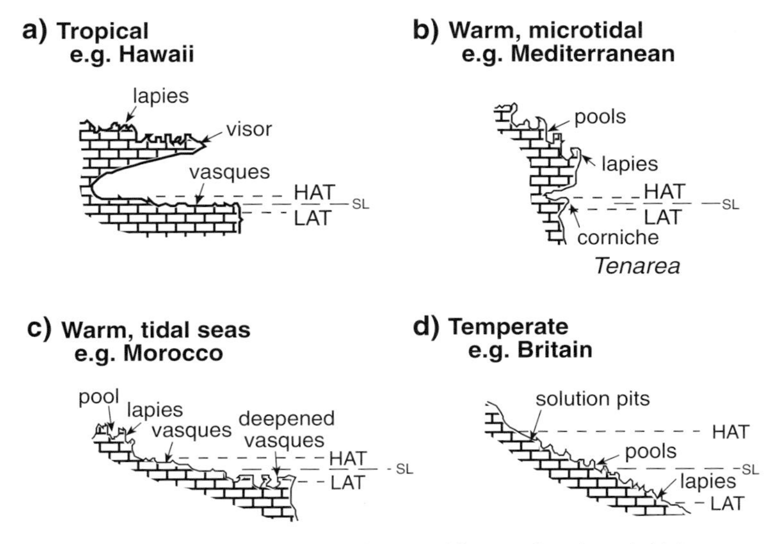

Figure 4.12. Limestone cliff morphology in different climatic and tidal range settings (based on Guilcher, 1953).

Tenarea tortuosa (Dalongeville, 1977; Laborel and Laborel-Deguen, 1996). In warm tidal seas, as for instance on the coast of Morocco, jagged lapies occur in the upper supratidal parts of the shore (Figure 4.12c). The lower part of the shore is characterised by small flat-bottomed pools with overhanging rims, and the pools are broken into a series of steps by vasques, terminating in a seaward scarp. Similar morphology has been described from calcareous eolianite (fossil dune) coastlines in Australia (Fairbridge and Johnson, 1978; Bird, 1993b). In cool temperate regions, relatively resistant limestone, such as the Carboniferous limestone exposed around the British Isles, contains zones that show the effects of differential dissolution and increased tidal range (Trudgill, 1987). The lowermost portion of the intertidal zone is heavily pitted with distinctive lapies. Within the mid-tide zone there may be pools, while in the upper intertidal and supratidal zones solution pits are characteristic (Figure 4.12d).

Subaerial versus marine processes

Subaerial slope processes play an important role in the morphological development of many cliffs. They are particularly important when considering the recession of soft-rock or unconsolidated cliffs, or cliffs no longer subject to wave action (Savigear, 1952, 1962). Even at the foot of cliffs which are still subject to marine processes there is evidence for the efficiency of subaerial processes, such as piping, in causing slope retreat (Nott, 1990).

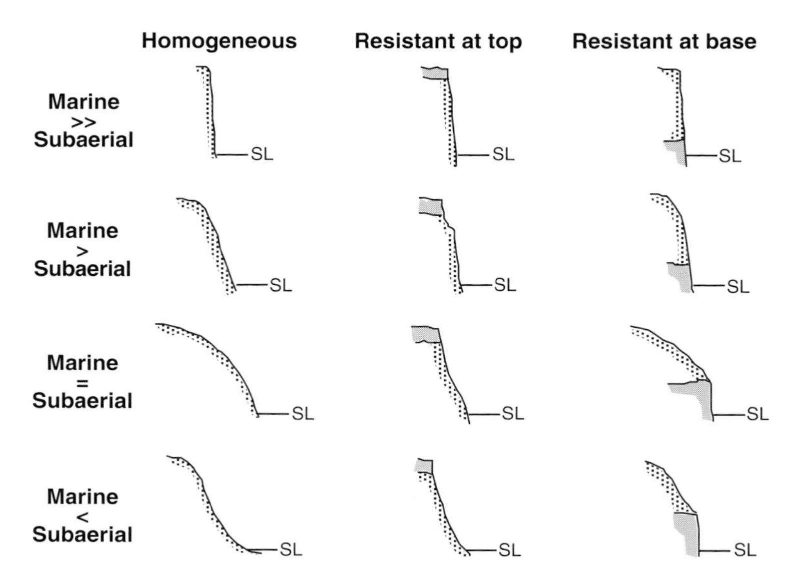

Figure 4.13. The morphology of cliffs of varying rock strength (more resistant rocks are shaded, less resistant are stippled) in relation to the relative effectiveness of marine and subaerial processes of erosion (based on Emery and Kuhn, 1982).

The response of the cliff to subaerial processes depends on material strength and rock mechanics. On high cliffs in resistant lithology, retreat is typically through block detachment and toppling. Igneous extrusive rocks develop steep cliffs, and more resistant dykes remain upstanding. In weaker rocks, however, weathering and degradation occurs by mass movement such as landslides (Figure 4.8e). The supply of material by mass movement from the slope above can overwhelm marine processes which are likely to depend on the rate of longshore movement to remove material from the base of cliffs (Carter *et al.*, 1990b).

The relative balance of marine and subaerial processes has been expressed schematically by Emery and Kuhn (1982), and is summarised in Figure 4.13. They suggested that the convexity of cliffs is a function of the relative balance of marine and subaerial processes of erosion. A steep cliff base indicates active marine truncation (as, for instance, on the exposed mid-oceanic island of Hawaii), whereas a gradual convex slope occurs where subaerial processes are dominant (e.g. St Bees Head in northwestern England and Las Palmas in the Canary Islands). The morphology of cliffs is further constrained by the relative stratigraphic position of resistant and less resistant strata. Where a resistant cap occurs and marine processes are dominant the cliff remains steep, but the profile becomes increasingly concave where subaerial processes are relatively significant. Cliffs are generally steep where resistant strata occur at their base, but the upper part of the cliff will be increasingly convex if subaerial processes are dominant. For instance, cliffs cut into laterite around Darwin in the Northern Territory of Australia have a different profile depending on whether water level coincides with the upper highly resistant ferricrete surface or the lower and softer pallid zone. Where the ferricrete outcrops at sea level, a broad shore platform occurs. However, where the indurated layer outcrops in headlands several metres above sea level, these are undercut by wave action acting directly on the less resistant pallid zone, and the ferricrete collapses (Young and Bryant, 1998). The importance of subaerial processes can be seen at Albany in Western Australia in Figure 4.14 where granite slopes extend into the sea.

4.3.3 Shore platforms

Shore platforms occur at the base of cliffs in many parts of the world. There is a wide range of platform morphologies, but it has become traditional to identify two forms, Type A, a gently sloping ramp (1–5°), and Type B, subhorizontal platforms terminating in a seaward or low-tide scarp (Figure 4.1b and c). Type B platforms can be further subdivided into high-tide and low-tide platforms, and low-tide platforms appear to

Figure 4.14. Granite cliffs at Albany, Western Australia, have been shaped by subaerial processes (photograph R.W. Young).

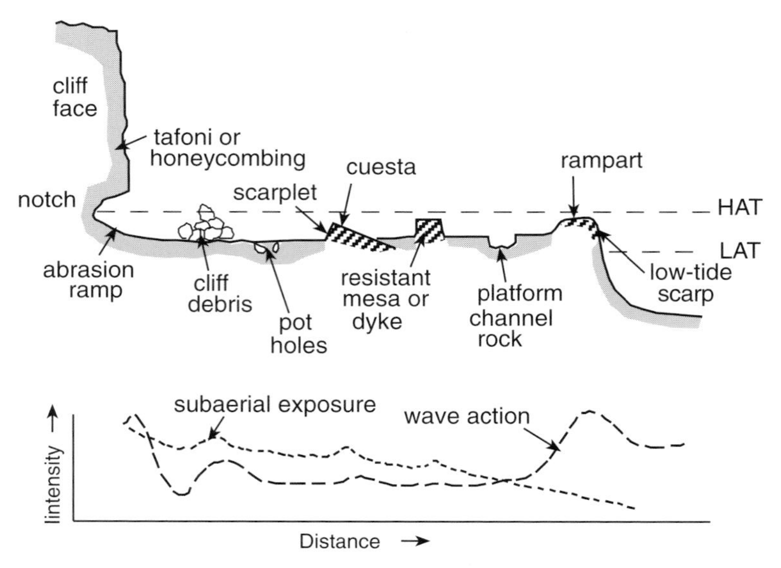

Figure 4.15. Schematic cross-section of a subhorizontal (Type B) shore platform.

occur on less resistant lithologies, such as eolianite (see Figure 6.32). Platforms truncated to low-tide level may represent a more advanced stage of planation of either Type A or high-tide Type B platforms (see Section 4.5); however, in many cases it seems a particular setting has platforms of only one morphology (Bird, 2000).

Subhorizontal platforms (Type B) are most characteristic of prominent headlands in mesotidal or microtidal settings. Sloping platforms (Type A) are found in some macrotidal settings (Davies, 1980); platform gradient seems to increase with tidal range (Trenhaile, 1987). Type A platforms are also characteristic where coarse sediment, such as shingle, leads to abrasion of the platform surface (Bird and Dent, 1966).

Figure 4.15 is a schematic profile across a subhorizontal shore platform, showing a range of features that can occur. The seaward margin of a platform is formed by a scarp (termed a low-tide scarp) that drops into the nearshore. On the seaward margin of some subhorizontal platforms there is a rampart, the form of which varies from site to site. Where there are lithological variations in the rock from which the platform is constructed, or distinct dipping beds, surface forms such as mesa, dyke, cuesta and scarplet can occur. Rock pools and potholes occur on almost all lithologies. Potholes are gouged by individual boulders that are often retained in the deepening pothole. The cliff face can be marked by tafoni (or honeycombing, see Section 4.4.1); there may be notches and cliff debris at its base. Figure 4.15 also suggests possible gradients in the intensity of wave action and subaerial exposure, both of which are distinct in the case of steep cliffs but can vary considerably across a shore platform.

There has been considerable debate about the origin of shore platforms and the relative roles of wave erosion and subaerial weathering processes (Trenhaile, 1980, 1987, 1999; Sunamura, 1992; Stephenson, 2000). In the older literature, the term wave-cut platform was used. Waves can abrade platforms, but other processes such as weathering also appear to be involved. Wave action may be the most important control on shore platform development in the storm-wave environments of the North Atlantic (Trenhaile, 1974, 1980). On the extensive shore platforms around Sydney and along the coast of New South Wales, such as those shown in Figure 4.16, there is also evidence of wave quarrying and differential wave erosion (Dana, 1849; Trenhaile, 1987). However, although the widest platforms seem to occur on the most exposed headlands, orientation of platforms is not related to direction of wave approach (Abrahams and Oak, 1975). The water that remains on the platforms at low tide is also likely to play a role (Figure 4.16).

Bartrum believed that in sheltered settings it was necessary for rocks to be weathered by subaerial processes before waves could remove material (Bartrum, 1935, 1938). Although the term subaerial processes is used, it is important to be clear that in this context it refers to a suite

Figure 4.16. Shore platforms on the southern side of Windang Island in the Illawarra, New South Wales, Australia. The ponding on the surface of the platform is characteristic of water layer weathering.

of physiochemical processes that operate on the shore platform surface, as distinct from subaerial slope processes which can be involved in the retreat of cliffs (see Section 4.4). Often termed 'water layer levelling' (Wentworth, 1938; Hills, 1949), physiochemical weathering has been emphasised in studies of the formation of shore platforms along the coast of New Zealand and southeastern Australia. Several researchers have suggested that weathering and wave action are needed in combination but that weathering is particularly important in sheltered locations where waves appear incapable of removing unweathered rock (Jutson, 1939; Hills, 1949, 1972; Kirk, 1977).

The debate about the relative roles of wave action and subaerial weathering has continued. The nature of the processes will be examined below (see Section 4.4). It is also important to recognise the significance of lithological variation. Rock structure appears to exert a control; for example, the height of the junction between the platform and the cliff behind it varies in relation to platform lithology in southern England (L.W. Wright, 1970). On the Otway coast of Victoria, platforms have been interpreted as resulting from differential erosion based on variations in lithology (Gill, 1972, 1973).

Although mechanical wave action has been emphasised in the storm-wave environment of the North Atlantic, chemical and salt weathering may be more important in Japan, Pacific islands and Australasia. It is not necessarily the case that the processes that erode cliffs are the same as the processes that planate shore platforms (Trenhaile, 1987). It is useful to distinguish between horizontal retreat of the major cliff face, vertical lowering of the shore platform, and horizontal retreat of the low-tide scarp, as different processes could be involved (Stephenson, 2000). It is also important to consider the time scale over which processes have been operating. The height at which platforms occur can vary considerably and in many circumstances it is likely that the platform is at least partially an inherited feature. In part, this results from variations in sea level over the time scales that are required for platforms to form. The role of higher sea level and increased storm activity in the past has been invoked in several explanations (Edwards, 1941; Cotton, 1963).

There can be little doubt that if subaerial weathering occurs on a platform surface, as implied by the pitting and ponded water on the platforms shown in Figure 4.16, it accelerates erosion by wave action. There are complicated interrelationships between mechanical, chemical and biological action (Sunamura, 1992); mechanical processes such as the abrasive action of wave-moved sediment, potholing, and physical rock disintegration through wetting/drying or freeze–thaw action are closely linked with chemical and biological processes. These processes will be examined in Section 4.4. Ramparts, where they occur on the seaward margin, could form from harder resistant rock, or they could result from being continually wet and therefore not experiencing the wetting and drying that weathers the adjacent platform (Gill, 1972; Gill and Lang, 1983; Sunamura, 1992).

4.3.4 Polygenetic rocky coasts and the role of inheritance

Cliffs and shore platforms generally erode at very slow rates and there is often little correspondence between the measured rates of process operation (see Section 4.4.3) and the morphological evidence of cliff or shore platform evolution (Rudberg, 1967). In many cases, this results because rocky coasts have not been eroding continually over geological time but have been punctuated by a series of changes in boundary conditions, particularly relative sea level. Many cliffs and shore platforms can be expected to have inherited forms that exert a control on their modern morphology. There seems little doubt that different processes operate on rocky coasts in different settings, and that many cliffs and shore platforms are polygenetic.

The clearest examples of cliffs that have undergone polycyclic evolution are those where marine processes now operate at the foot of slopes that have formed as a result of subaerial processes. The hogs back (or slope-over-wall) cliffs of Devon, for example, have been shown to result from re-excavation of interglacial cliffs buried by periglacial deposits during the last glaciation. Guilcher (1958) recognised that many European cliffs comprised steep subaerially formed slopes which have been trimmed at the toe by marine erosion, and he termed these 'false cliffs' (fausses falaises).

Inheritance is also a factor that is important in understanding plunging cliffs; there is increasing support for the view, proposed by Cotton (1963), that these are cliffs formed at lower sea level and drowned as a result of sea-level rise (or subsidence). Where the cliff face is bevelled with a series of fossil notches, as, for instance, on Vatulele (Figure 4.11), then it is clear that the sea has operated at different levels and that erosion has not been sufficient since that sea-level stand to remove the evidence. Notches may contain speleothems, such as stalactites and flowstones, indicating that time has passed since they were active, and providing an opportunity to derive minimum ages on past sea-level events.

Inheritance is also frequently invoked to explain elements of the morphology of shore platforms. Shore platforms along the coast of southeastern Australia, such as those shown in Figure 4.16, have been suggested to relate to inheritance from former sea-level stands (Bird and

Dent, 1966; Gill, 1973; Brooke *et al.*, 1994). Some could date back to Tertiary or Early Quaternary (Young and Bryant, 1993). Dating of ferricrete crusts on platforms, or former beach deposits, suggests that some features on shore platforms elsewhere have also been inherited (Woodroffe *et al.*, 1992; Trenhaile *et al.*, 1999).

Where boundary conditions are rapidly changing, for instance where sea level is non-stationary, efforts to link platform elevation with the operation of modern processes are likely to be frustrated. Computer modelling of shore platform response to wave erosion during sea-level variations over successive interglacials lends support to the suggestion that inheritance reinforces platform development (Trenhaile, 2001a,b,c).

4.4 Processes and rates of erosion

Processes on cliffs and shore platforms are difficult to study. They can rarely be measured directly because of poor accessibility to what are usually high-energy environments. There is generally a slow rate of change and infrequent occurrence of the events hinders observation or measurement (Griggs and Trenhaile, 1994). These difficulties mean that processes have more often been inferred from morphology than observed directly. However, morphology can be an ambiguous discriminator of process and process rate in these rocky coastal environments (Trenhaile, 1987; Spencer, 1988b; Stephenson, 2000).

4.4.1 Operative processes

A range of processes operating on rocky coasts is described in this section. Wave action is significant; it comprises hydraulic forces (associated with water movement) and mechanical forces (particularly associated with sediment that is carried by waves). The irregularity of the coast in planform concentrates wave energy on headlands (see Figure 3.7), and large waves which can move massive boulders would seem to play the dominant role in many cases. However, other processes are also important, particularly a suite of processes that operate on rock surfaces exposed above the level of the sea. The term subaerial has been used to describe this suite of processes, although it has been used in at least two different contexts, first to describe physiochemical processes which occur periodically on intermittently exposed rock, and second to refer to slope processes. In addition, there are biological processes. Often, more than one process is involved and there are significant interrelationships between the processes themselves. For instance, waves play an important role in removing the products of physicochemical weathering. Determining the relative roles between or within these types of processes is extremely difficult because they operate intermittently, can rarely be directly observed, and the processes operating today may not be the same as those that formed a particular coastal feature in the first place. The principal processes are examined below.

Hydraulic action

The hydraulic forces produced by waves include hydrostatic pressures, related to the mass of water, and dynamic pressures resulting from water movement (Trenhaile, 1987). Positive hydrostatic pressure is exerted on any rock surface that is submerged, and the amount of pressure increases with water depth. While most rock surfaces are constantly submerged, others are alternately covered by water and then exposed; these are subject to alternating hydrostatic pressure and its subsequent release, which can weaken the rock. Dynamic pressures include shock pressures and horizontal forces that can move material. When a wall of water within a non-breaking wave hits the rock surface it exerts an impact or shock pressure, also termed 'water hammer'. Lesser shock can also be transmitted by a breaking wave within its splash zone.

Hydraulic pressures exert compression, tension and shearing forces on the rock. These can abrade a rock surface and dislocate angular blocks from highly jointed surfaces by a plucking process known as 'wave quarrying'. Added to these purely hydraulic forces are pneumatic stresses associated with air that is trapped particularly by breaking waves in pockets between the wave and the rock surface. Significant pressures are exerted when air trapped in crevices in the rock is compressed. Cavitation (involving collapse of air bubbles at great pressure) can occur under extremely large waves, and a series of small-scale features such as flutes and grooves can be engraved into rock surfaces as a result of cavitation.

Mechanical action

Mechanical action, accomplished by waves that carry sediment, includes abrasion and attrition. Abrasion (also termed corrasion) is the wearing down of the rock surface as a result of grinding or scraping by sediment particles (Jones and Williams, 1991). Particularly clear examples of abrasion occur where potholes have been eroded into a shore platform by the concentrated action of a resistant boulder which, in some cases, can be seen at the bottom of the circular depression. Attrition is the breakdown associated with the impact stresses from particles that are carried by the waves (Williams and Roberts, 1995).

It is difficult to quantify the importance of these physical processes. However, the effectiveness of the entrainment of sediment, as opposed to the hydraulic action of waves alone, has been demonstrated by comparison of situations in which sediment is available and those where it is not. Dipping beds of resistant strata in a shore platform are likely to be worn smooth where pebbles and boulders are available to abrade them, for instance at Cape Doob in the Black Sea, but to remain as protruding ridges where there is no abrasive sediment (Zenkovich, 1967). Where beach sediment accumulates at the foot of a cliff, or on a shore platform, it can greatly increase the rate of cliff recession; for instance, on the Upper Lias shales around Whitby in Yorkshire northeastern England erosion rates are 15 times faster where there is sediment available than where no beach occurs (Robinson, 1977). It has been shown that sand abrasion increases the rate of erosion of tropical reef limestone in an intertidal notch from 1 mm a^{-1}, where this is achieved through bioerosion by grazing organisms alone, to 1.25 mm a^{-1}, where sand is available for abrasion (Trudgill, 1976). Beach sediment can also have more subtle effects such as 'wedging', whereby grains are forced into crevices, and 'beach etching' where the continual abrasion by beach material can result in a smooth surface developing on the underlying platform (Twidale and Campbell, 1999).

Physiochemical action

Physiochemical action involves physical and chemical weathering by seawater for which the term 'corrosion' is generally used. Corrosion includes solution, salt weathering and thermal stresses. The clearest indication of the corrosiveness of seawater is the pitting of rock surfaces by tafoni and honeycombing which occur above the water line. Tafoni, an Italian term, is often used as a generic term for cavernous weathering; tafoni pitting is attributed to halite crystal growth (Sunamura, 1996). Honeycombs, also called alveoles, imply cell-like structure but are not otherwise different from tafoni (Rodriguez-Navarro *et al.*, 1999). They develop in the spray zone above high water level on a range of lithologies, particularly sandstone. Blocks in seawalls can honeycomb in a few years.

A group of physiochemical processes are associated with water that lies on a horizontal rock surface during low tide. This ponded water is believed to undergo subtle chemical changes that can etch the surface, further weathering shore platforms. Initially termed 'water level weathering' by Wentworth (1938, 1939), these processes have subsequently been referred to as 'water layer weathering' or 'water layer levelling'. Wentworth considered that corrosion (solution benching and water level weathering) was more significant than ramp abrasion and wave quarrying in eroding a bench onto volcanic tuffs in Hanauma Bay, Oahu, Hawaii (see Figure 5.17). He considered that water that lay on the horizontal surface during low tide could undergo subtle chemical changes that would further etch the surface. This process is particularly effective in the tropics, where water layer weathering is considered a major factor in reducing the surface of shore platforms, especially on limestone. Solution pools, resulting from dissolution of the rock rather than mechanical action, occur predominantly on limestone; they tend to coalesce with time, in contrast to potholes which remain distinct (Emery and Kuhn, 1980).

Surface microtopography can also be modified by salt weathering or thermal stresses set up by wetting and drying (Guilcher, 1958). Such action is not limited to the tropics; basalt of the Giant's Causeway in Northern Ireland is similarly sculpted and corrosion has even been identified in high-latitude sites (Dionne, 1967). Rapid changes in conditions, wetting and drying, heating and cooling, but particularly freeze–thaw in high latitude settings, are important in weathering rock (Dawson *et al.*, 1987).

Biological action

Organisms influence the morphology of rocky coasts in several ways. In most cases, biological action is destructive and is termed bioerosion, which includes biomechanical or biochemical processes (Trudgill, 1987). However, in some cases organisms can protect the shoreline, coating rock surfaces and lessening the effects of other erosive forces. There are a series of grazing organisms, such as gastropods, chitons and echinoids. It has been realised for a long time that these rasp the rock surface as they graze over it, gradually eroding the rock (Hackshaw, 1878; Jehu, 1918). Several attempts have been made to quantify the rate at which these organisms erode the rock surface over which they graze, either by measuring the depth of incision by the radula as they graze, or by collecting and weighing faecal material (McLean, 1967; Andrews and Williams, 2000). Borers, such as sponges, molluscs and sea urchins, make holes in the rock surface; they play a significant role in the erosion of limestone coasts (Spencer, 1988b). However, there are organisms which coat and protect the rock surface, such as coralline algae, vermetid gastropods and serpulid worms, which can be prolific in high-energy environments (see Chapter 2).

Subaerial slope processes

Subaerial slope processes in operation on cliffs range from creep and slopewash, which occur on gentler slopes, to block fall, toppling and sliding, which involve mass movement on steeper slopes (Trenhaile, 1987). These can be triggered by oversteepening or undercutting of the slope, resulting from marine erosion, or groundwater-induced slides (Emery and Kuhn, 1982). This is particularly the case with soft-rock coastlines in which there is a low bearing strength, where mass movement occurs with large volumes of material slumping into the zone where it can then be mobilised by marine action. Slope failures tend to occur in advanced stages of regolith decomposition, occurring as structurally controlled translational slips along a predetermined surface, whereas rotational slumps occur in undifferentiated material (Trenhaile, 1987).

On poorly lithified cliffs processes range from avalanching of unconsolidated sands, to mud slides, spalling and slumping. Slopes may be periodically steady, or cyclic with retreat being parallel, or episodic recession of toe and top. Several examples of retreat in soft-rock cliffs are discussed in Section 4.5.1 (see Figure 4.22). Mass movement may provide material to shore platform surfaces which is then moved by marine processes, particularly longshore drift (McLean and Davidson, 1968). Landslide activity appears to be brought about primarily through groundwater rather than wave activity, and can be accentuated by human actions (Griggs and Trenhaile, 1994; Viles and Spencer, 1995).

4.4.2 Relation of processes to morphology

The distribution of processes in relation to wave characteristics on the major cliff and platform types is shown schematically in Figure 4.17, based upon Sanders (1968a; see also Trenhaile, 1997). Different processes are important in different situations. The erosive force of the waves is dependent on whether the waves break or are reflected off the vertical cliff face unbroken. Steep plunging cliffs, like vertical seawalls, are highly reflective and most of the wave energy associated with incoming waves bounces back to meet the next incoming wave, producing a standing wave or 'clapotis'. The standing wave will alternately rise and collapse as kinetic energy is converted into potential energy and back again. True clapotis is rare because the bottom is usually too shallow, or the cliff face too heterogeneous or sloping for there to be no interference, and small waves are likely to leak sideways (Trenhaile, 1987). The majority of the wave energy is reflected, however, and as the bottom is at considerable depth, there is likely to be little material available for abrasion, so wave action is relatively ineffective at mechanically eroding the cliff (Figure 4.17a). A notch can eventually form in a vertical cliff face as a result of hydraulic forces, such as water hammer, and pneumatic forces, such as air compression in joints, together with corrosion associated with wetting and drying (solution is particularly important in the case of limestone, see Figure 4.11). Over considerable time these processes can excavate sea caves on plunging cliffs (Cotton, 1963, 1969b). For

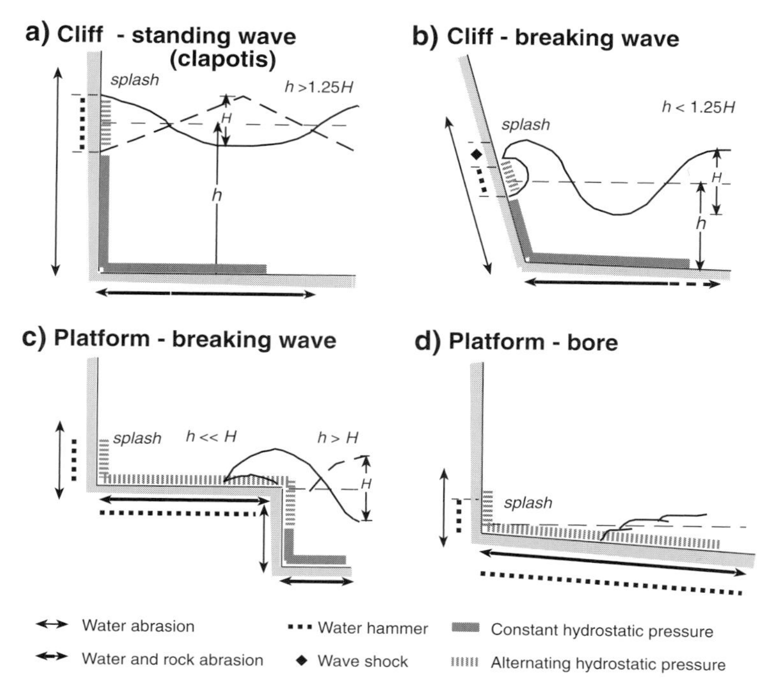

Figure 4.17. Relationship between wave form and cliff or platform morphology, and the processes that operate (based on Sanders, 1968a; Trenhaile, 1997).

instance, plunging cliffs more than 100 m high, descending 35–65 m below water on King Island, in Bass Strait, Australia, contain sea caves in which speleothems have been dated to Last Interglacial age (Goede *et al.*, 1979). These indicate that caves can form at water level in vertical cliffs but the ages imply negligible erosion over a long period despite high wave energy.

It has been shown experimentally that dynamic pressure is considerably greater under breaking waves than under standing waves or bores associated with broken waves (Sanders, 1968b; Sunamura, 1992). Waves will break at the base of a cliff if water depth in the nearshore is shallow relative to wave height (Figure 4.17b). In this case, not only are there increased hydraulic pressures, but there are also mechanical forces associated with entrainment of sediment.

Process operation across shore platforms is likely to be more complex and to vary across the platform. In those situations in which nearshore bathymetry is suitable, waves break on the seaward edge of the platform (Figure 4.17c). Waves breaking at the outer edge of the Type B platform will rush landwards across the platform as a bore, losing energy through turbulence and bottom friction, and decreasing

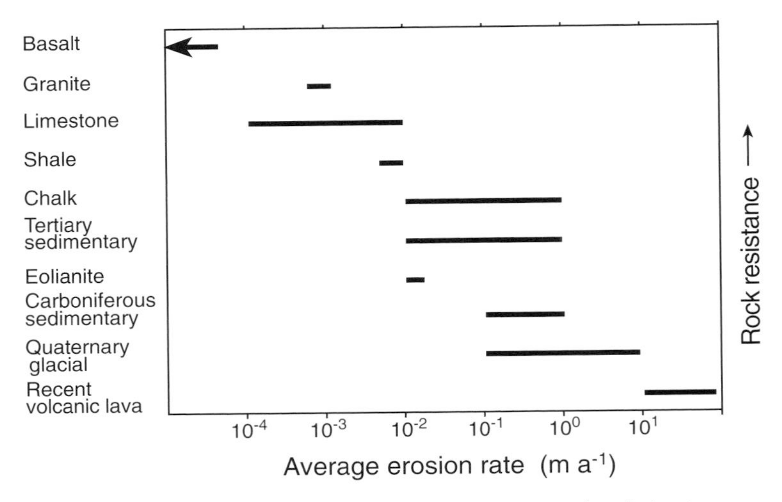

Figure 4.18. Schematic representation of cliff erosion rates in relation to lithology (based on Emery and Kuhn, 1980, using data summarised by Sunamura, 1992).

in height. On Type A platforms, waves are likely to spill and cross the platform as a bore (Figure 4.17d). As depth continually decreases across this type of platform waves will continue to shoal and to lose energy. Wave type depends on the gradient and width of the platform and computer simulation has indicated a tendency to progress from plunging to spilling waves with time, presumably maximising energy dissipation (Trenhaile, 2000). There has been very little study of propagation of waves across platforms in the field, and the distribution of dynamic pressures and mechanical forces is not altogether clear. Simple gradients are suggested in Figure 4.15, but they are conjectural. The operation of processes will depend on stage of the tide. For instance, on a shore platform the following sequence has been suggested by Carter *et al.* (1987). At low tide the incident waves are reflected off the low-tide scarp. As the tide rises, a threshold is reached, beyond which the wave breaks onto the platform and crosses it as a bore. At higher tidal stages the broken wave reforms into secondary waves, and at high tide waves could break at the back of the platform eroding the cliff. If physiochemical action and biological action are also implicated then patterns of alternate wetting and drying are likely to be at least as complex.

4.4.3 Cliff and shore platform erosion rates

The rate of erosion of cliffs and shore platforms, often termed recession, depends on a range of factors, particularly geological factors such as the lithology, structure and weathering properties of the rocks, and oceanographic factors such as wave climate, tidal range and degree of exposure to wave processes. The relative balance of these factors will vary from site to site. Rates of cliff retreat have been comprehensively reviewed by Sunamura (1983, 1992) and are summarised in Figure 4.18. Crystalline rocks are more resistant than sedimentary rocks which, in turn, are more resistant than unconsolidated rocks. Erosion on basalt is negligible. Limestone can erode at around 1 mm a^{-1}; other lithologies erode at up to 1 m a^{-1}, whereas young volcanic lava, discussed in Section 4.2.2, can exceed 10 m a^{-1}.

Particularly rapid rates of cliff retreat occur in the soft glacial deposits of eastern England. Rates of retreat for Holderness have averaged 1–2 m a^{-1} for the past 150 years (Valentin, 1954; Mason and Hansom, 1988), and rates of almost 1 m a^{-1} have been recorded for parts of the coast of Norfolk and Suffolk (Clayton, 1989). Similar rates can be extrapolated back at least until before 1600 AD for the sand cliffs at Dunwich, and the sites of former Roman ports are now well out to sea along this coast (Robinson, 1980). The retreat is episodic and extreme events can lead to excessive retreat; for instance, the 1953 storm surge accounted locally for retreat of 12 m on 12-m-high cliffs, and 27 m on 3-m-high cliffs near Covehithe in Suffolk (Williams, 1956). Figure 4.19 shows the cliffs at Easton Bavents in East Anglia indicating intermittent landslides generating talus along the cliff.

Figure 4.19. Rapidly eroding cliffs at Easton Bavents in East Anglia, eastern England. The erodible glacial deposits undergo collapse and intermittent talus can be seen protecting the foot of the cliffs (photograph D.A. Woodroffe).

Detailed descriptions of cliffs along the coast of California indicate a variable pattern of retreat (Kuhn and Shepard, 1984). Lithology appears more important than wave energy in determining rates of cliff retreat in California (Benum *et al.*, 2000). Cliff retreat in sandstone is episodic, with long periods of little change followed by large landslides causing several metres of retreat in a short time. More rapid retreat occurs in response to storms during El Niño years, as in March 1983, when the beach was stripped away and cobbles abraded the cliff foot directly (Storlazzi and Griggs, 2000).

Sustained rates of retreat of 0.20–0.25 m a^{-1} have been recorded for cliffs along the Lake Michigan shore (Jibson *et al.*, 1994). Recession rates vary from 1 mm a^{-1} recorded on eolianite around Perth by Hodgkin (1964), and around 2 mm a^{-1} in Barbados (Bird *et al.*, 1979), up to 5 mm a^{-1} on the basis of incision into a boulder emplaced by the 1883 Krakatau eruption (Verstappen, 1960).

4.5 Cliff morphodynamics

Short-term erosion rates can generally not be extrapolated to rates of lowering over millennia. This section examines some morphodynamic approaches that have been adopted to address issues of cliff and shore platform development. The challenge is to integrate process studies at an event time scale to enable effective comparison with longer-term patterns of change as determined from aerial photographs and other lines of evidence.

A recent resurgence of interest in rocky coast geomorphology has seen a wider range of techniques employed including accurate 3-dimensional photogrammetry to determine mass budgets of sediments mobilised in mass movements (Dixon *et al.*, 1998), application of the microerosion meter to determine long-term erosion rates (Stephenson and Kirk, 2000a,b), use of the Schmidt hammer to characterise rock hardness and degree of weathering (Haslett and Curr, 1998), cosmogenic dating studies to age terrace surfaces (Stone *et al.*, 1998), laboratory wave tank experiments (Sunamura, 1992), and computer simulation (Trenhaile, 2001a,b). Some of these techniques are discussed in this section.

4.5.1 Wave energy and rock resistance

The slow rate of change on rocky coasts means that field experimentation is less appropriate than on other coasts. Laboratory experimentation (Sanders, 1968b; Sunamura, 1992) has been able to demonstrate the development of basal notching and subsequent collapse, but there are often insuperable problems about scaling experiments realistically. Concern has been expressed as to whether notches that have been formed were the result of the waves, or whether they formed through corrosion of the materials used to simulate rock (see Sunamura, 1992). Conceptual and computer simulation models offer considerable potential (Cinque *et al.*, 1995; Anderson *et al.*, 1999; Trenhaile, 2001a,b). However, it remains difficult to test in the field the premises on which these models are based.

Wave energy

Attempts to relate shore platform morphology to wave characteristics alone have shown only poor correspondence. The wave energy actually received on a shore platform is not equivalent to open-water wave energy, and early modelling indicated that most waves break before they reach steep cliffs (Flemming, 1965). Offshore topography can be observed to attenuate wave energy in the field. Storm waves have been reported to break well offshore from shore platforms on the Kaikoura Peninsula in New Zealand, and wave energy is reduced by several orders of magnitude as a result of attenuation, never exceeding the compressive strength of the platforms (Stephenson and Kirk, 2000a). Modelling the wave component of platform evolution also implies that platform characteristics are unrelated to open-water wave height (Trenhaile, 2000).

If the erosive force of the waves on a cliff exceeds the resisting force of the rocks, then erosion occurs and a basal inflection forms which, with time, becomes a shore platform (Sunamura, 1983, 1991). A framework within which to consider both of these forces is shown in Figure 4.20. This approach is based on the studies of Tsujimoto (1987) who graphed measurements from sheltered and exposed sites around Japan in relation to wave pressure and the compressive strength of the rock (comprising resistant rocks like basalt as well as softer rocks). His studies implied that plunging cliffs occur on resistant rocks and shore platforms develop on weaker rocks, separated by a line at 45° on a logarithmic plot. Sunamura (1992) has augmented the data with several sites from elsewhere in the world. He simplified the representation of wave energy and rock strength and was able also to discriminate between Type A and Type B shore platforms (Figure 4.20). Sunamura (1992) demonstrated that extremely resistant rock types (i.e. basalts with compressive strength >3000 t m^{-2}) are unlikely to develop shore platforms, and usually persist as plunging cliffs. Weaker rocks maintain plunging cliffs if subject to little wave force, but will be more likely to develop shore platforms as wave forces increase. On intermediate lithology, shore platforms are more likely in the higher energy wave settings.

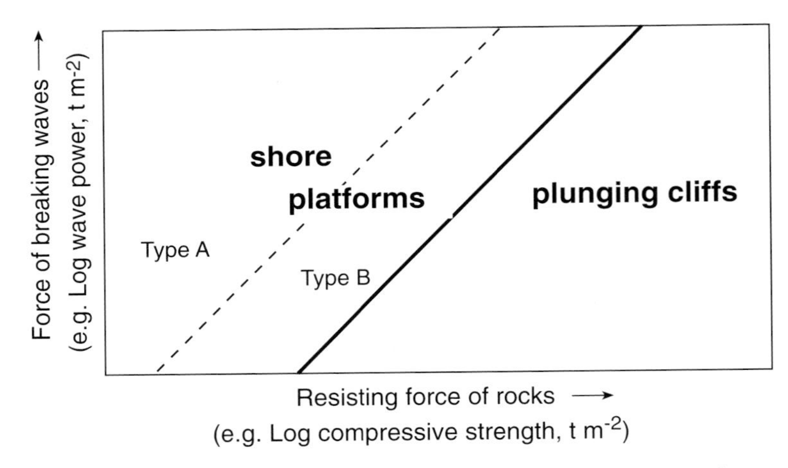

Figure 4.20. Generalised relationship between erosive force of waves and resisting force of rock, and the discrimination of plunging cliffs, and shore platforms (based on Sunamura, 1992). See text for discussion.

In Figure 4.20 the axes have been left unscaled, because it is not clear what parameters should be measured to represent wave energy and rock strength. As indicated above, it is difficult to determine wave energy reaching a cliff foot. Tsujimoto calculated maximum wave pressure at each site based on comparison of the open coast with representative sites for which there were wave data, allowing for some wave refraction into embayments (see Chapter 2).

Rock resistance

It is also difficult to assess rock resistance to erosion. Tsujimoto measured compressive strength of rock samples using a Schmidt hammer. Schmidt hammer tests (using a device that imparts an impact on the rock and measures compression) can indicate how weathered the rock has become, which has been used as an indication of platform age (Haslett and Curr, 1998). However, Schmidt hammer readings do not take account of structural factors. Whereas features such as tension cracking were accommodated in terms of sonic velocity measurements made by Tsujimoto (1987), they have not been incorporated by other researchers (e.g. Sunamura, 1992). Rock mass factors such as density and orientation of jointing and bedding planes or excavation along fault lines are significant in affecting resistance. They can be incorporated by using a Rock Mass Index that incorporates block volume (Budetta *et al.*, 2000); however, there remains much scope to extend this approach.

The comparison of erosive force of waves and resisting force of rocks indicated in Figure 4.20 offers great potential but it will be necessary to develop more rigorous parameters to measure appropriate wave energy and the resistance of the rock. Rock strength is likely to decrease with time because minor forces weaken the rock, and this approach could assist in evaluating the relative role of weathering of the platform as opposed to wave energy, which has become central to the debate about shore platform development (Stephenson, 2000). For instance, on shore platforms around the Kaikoura Peninsula rock strength, based on Schmidt hammer readings, is reduced as much as 50% by subaerial weathering (Stephenson and Kirk, 2000b). With time, continued subjection of even resistant rock to wave energy will lower its ability to resist. Furthermore, morphological changes may also result in adjustments to wave energy received, although as there is still no consensus on pathways of evolution, and little information on offshore bathymetry and shoaling wave attenuation, this will not be easy to resolve. At longer time scales, boundary conditions, such as sea level, do not remain stationary and changes influence cliff evolution.

Microerosion meter measurements of rock lowering have been undertaken on several lithologies, including limestone. Rates of lowering measured vary from 0.4 to 2 mm a^{-1} (Stephenson, 2000). Where short-term studies have been extended for one or two decades, similar rates are indicated to those based on only a couple of years (Stephenson and Kirk, 1996; Viles and Trudgill, 1984). In studies of the platforms around the Kaikoura Peninsula, long-term rates of lowering averaging 1–2 mm a^{-1} have been demonstrated over mudstone and limestone platforms (Stephenson and Kirk, 1996, 1998), and similar rates seem typical of chalk platforms in southern England (Andrews and Williams, 2000).

Equilibrium morphology

The similar cliff and shore platform morphology that results on different coasts (Figure 4.1) implies that the morphology adjusts towards equilibrium. If erosion of the low-tide scarp, retreat of the cliff face, and lowering of the shore platform surface are viewed separately, some interrelationships can be inferred. Various negative feedbacks can be envisaged; for instance, as wave energy is expended on the platform the forces acting on the cliff at the rear of the platform are decreased. Retreat of the cliff widens the shore platform which then decreases the rate of retreat of the cliff (So, 1965). Erosion of the low-tide scarp will narrow the platform and reactivate erosion of the cliff face at the rear of the platform, implying dynamic equilibrium (Challinor, 1949; Trenhaile, 1974).

There has been considerable debate as to whether the low-tide scarp retreats at all. Whereas some researchers have indicated that it does erode (Edwards, 1941; Hills, 1949; Trenhaile, 1987), others believe that the low-tide scarp marks the position of the old shoreline before erosion of the shore platform (Gill, 1950; Cotton, 1963). If there is no retreat of the low-tide scarp, a static equilibrium is implied (Sunamura, 1992). Important factors include nearshore topography and the availability of material to abrade the low-tide scarp.

Cliff recession is usually episodic. Recession of the base of the cliff, for instance by basal notching, leads to subsequent cliff failure, which produces talus that covers the foot of the cliff and reduces the direct impact of waves onto the cliff itself, thereby temporarily slowing erosion (Belov *et al.*, 1999). Figure 4.21 shows the cliffs at Hunstanton in eastern England. These cliffs are composed of sandstone overlain by red and white chalk. The relatively resistant chalk topples in angular blocks that can be seen in Figure 4.21 to persist temporarily at the cliff foot. The fine material is removed first and the larger clasts remain. Self-regulation occurs by negative feedback because recession produces sediments that protect or load the toe (Edwards, 1941). This pattern of collapse, toe trimming, further cliff steepening and subsequent collapse has been incorporated into models by Sunamura (1992), and is examined below (see Figure 4.22).

The relative rates of retreat of the front (low-tide scarp) and back (backing cliff) of a shore platform have been examined using computer simulation modelling (Trenhaile, 1989, 2000). Simulation modelling of

Figure 4.21. Cliffs at Hunstanton, eastern England. These cliffs demonstrate the role of variable lithology, composed of basal greensand, overlain by red chalk and capped by white chalk. The erosion of the cliffs leads to block fall which provides some protection to the foot of the cliff until the blocks disintegrate (photograph D.A. Woodroffe).

platform development shows divergence of platform morphology and indicates an insensitivity to wave parameters. The shape of the shore platform generated by this modelling attained a state of time-independent quasi-equilibrium, and many of the profiles did adopt either a ramp or a more horizontal platform form. The rate of platform width extension decreased with time, and platforms were developed which had widths of 10–325 m (Trenhaile, 2000, 2001c). However, it still remains very difficult to test any of the modelling outcomes. The role of extreme events is especially hard to determine.

4.5.2 Models of soft-rock retreat

Soft-rock cliffs retreat much more rapidly than other cliffs. Figure 4.22 is an attempt to synthesise the ways in which soft-rock cliff and poorly consolidated shorelines erode. Different modes of retreat are recognised in relation to sand, mud and soft rock lithologies, and for each a time perspective on the retreat of the top and toe of the cliff is hypothesised. The figure compares very poorly consolidated sand, mud and soft rock, identifying contrasting modes of retreat. Although cohesionless and porous sand is rarely a component of bold or rocky coasts (an exception being weakly lithified eolianite headlands, see Section 6.5.5), a consideration of these materials allows a more rapid analogy which can be extended to more resistant materials, such as sand-rock and mudstones.

Styles of retreat of sand substrates are based on studies in relation to foredune erosion (Carter and Stone, 1989), and although considered in detail in Chapter 6 are examined here because they illustrate how extremely loose sediment responds to marine erosion at the base. Dune sands retreat by avalanching, sliding or slumping. Avalanching is characteristic of dry unconsolidated sand and weakly lithified eolianite (see Chapter 6), involving cascades of sand, maintaining the slope at the angle of repose. This leads to parallel retreat in which top and toe are strongly coupled. In the case of foredunes, retreat is event-specific, recurring when basal wetting resumes during successive high tides.

Slides occur on more cohesive sandy slopes, such as vegetated foredunes or eolianite, either where there has been some initial cutting at the base, or where some internal tension cracking occurs, as for instance at the base of root penetration beneath vegetation (Carter and Stone, 1989). The material that slides accumulates at the toe and provides protection preventing further sliding, increasing the length of time between sliding events. Slumping occurs on materials that have greater shear strength and can maintain angles steeper than the angle of repose, such as well-vegetated dune sands.

In the case of mud deposits, several modes of retreat are possible.

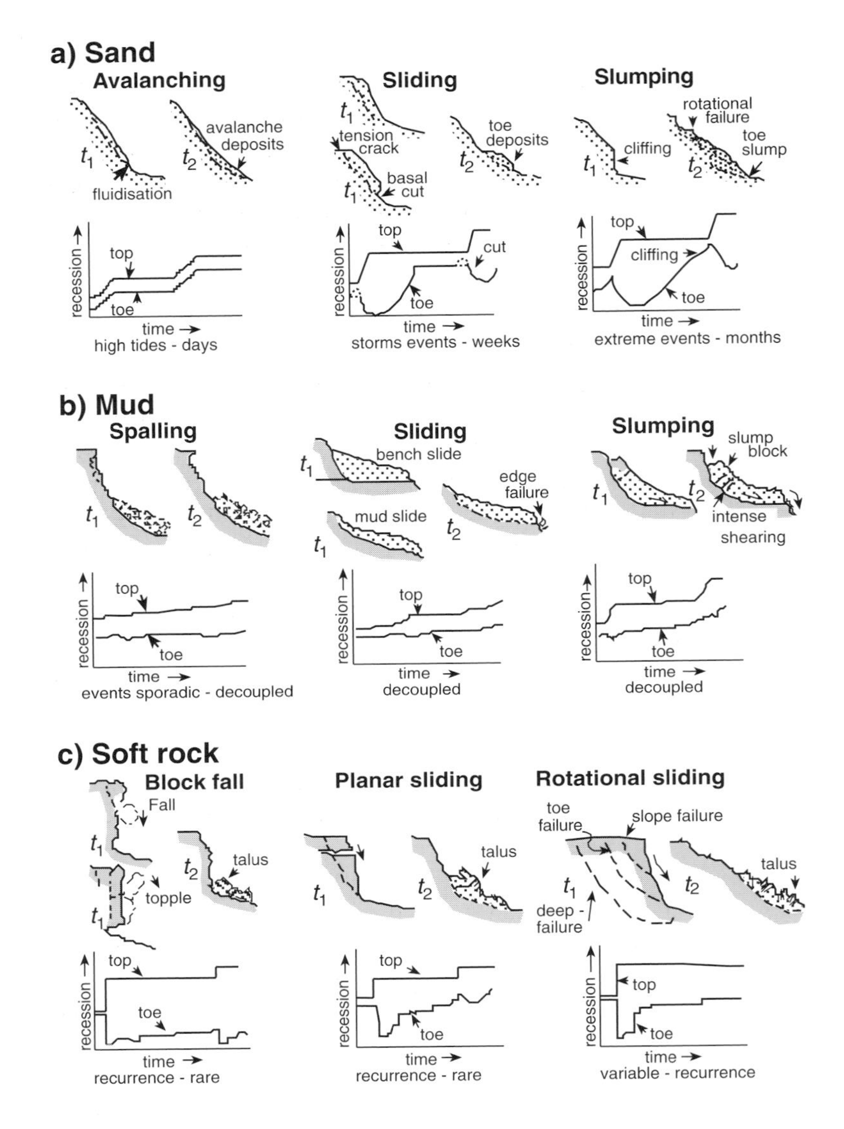

Figure 4.22. Recession of soft-rock cliffs. Several modes of retreat occur on each of (a) sand, (b) mud and (c) soft-rock lithologies, from least resistant at the top of the figure, and more rapid to the left. The rate of recession of the top and toe of the cliff is shown schematically; time scales are likely to vary considerably between examples and only very broad indications of recurrence intervals can be given. (The examples for sand cliffs are based on studies of foredune erosion by Carter and Stone, 1989; mud cliffs are based on Barton and Coles, 1984; soft-rock cliffs on Sunamura, 1992.)

The examples in Figure 4.22 are based on studies of the Tertiary clays of southern England, summarised by Barton and Coles (1984). These are also portrayed from free fall on the left to slumping on the right. Spalling is a process whereby a small mass of material is released from a cliff face to accumulate in a scree below. Spalling tends to occur as a result of subaerial processes, such as frost action, rainfall events or clay shrinkage. The retreat of the scree at the toe is decoupled from this cliff-face activity, and top and toe undergo recession at their own pace.

Sliding can take several forms. Particularly common in the Barton Clay cliffs of Hampshire are bench slides where the shear surface coincides with a bedding plane or stratigraphic discontinuity (Barton and Coles, 1984). Mud slides and debris slides also occur where movement is more localised and is related to groundwater outflow. The toe recession is generally independent of the sliding, with edge failure occurring. It does not seem that wave erosion processes are particularly significant in triggering mass movement in southern England (Hutchinson, 1983). Slumping involves rotational movement of material with intense shearing within the sediment and decoupling of top and toe.

On soft rock, such as glacial or alluvial deposits, further modes of retreat can occur as illustrated in Figure 4.22, following Sunamura (1992). Fall (vertical drop of a block) or toppling (rotation of a block) result in talus accumulation at the toe, persisting for a long time (centuries to millennia). Planar slides occur where a linear sliding surface occurs, but rotational slides are more common on soft-rock cliffs.

Figure 4.23 is a conceptual model of the pattern of toe erosion and its effect on the overall soft-rock cliff retreat, derived using aerial photographic evidence from the Great Lakes by Vallejo and Degroot (1988). The model implies that the rate of toe erosion can affect the overall mode of retreat of cliffs eroded into glacial material. Where toe retreat is rapid, subaerial processes translate recession up the slope, and parallel retreat of a uniform but relatively gentle slope occurs. With moderate toe erosion, defined as removal of debris that is transported down the slope, but no undercutting of the cliff itself into unweathered glacial deposits, slow parallel retreat occurs through slope processes alone. Where no toe erosion occurs, as when water level falls and the toe is no longer subject to wave processes, weathered material accumulates at the foot of the slope, resulting in a gentle, but stable, debris slope that does not recede.

Higher cliffs might be expected to retreat more rapidly because these can generate the higher shear stresses, but they also generate a larger quantity of material that may then require longer for wave action to remove, or break up, at the toe (Bray and Hooke, 1997). In glacial

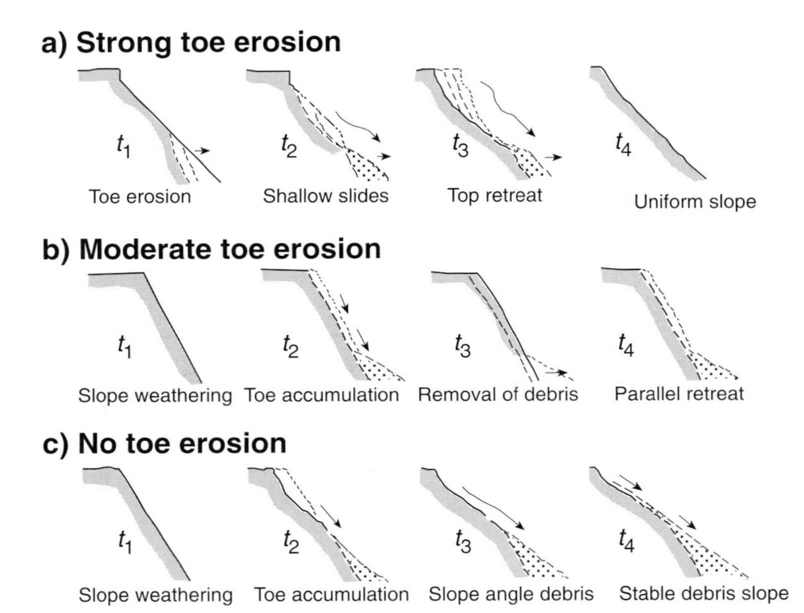

Figure 4.23. Model of the response of cliff retreat to rate of toe erosion (based on studies of glacial bluffs in the Great Lakes by Vallejo and Degroot, 1988).

deposits in Northern Ireland it appears to be infrequent events that actually lead to retreat of the cliff (McGreal, 1979). In these unconsolidated sediments the retreat rate depends on sand:mud ratio and shear strength (Jones *et al.*, 1993). Wave action appears more important on the shore of Lake Erie (Kamphuis, 1987); waves take silt away, but leave the coarser sand (Dalrymple *et al.*, 1986).

Coastal mudslide activity shows considerable short-term variability (Prior and Renwick, 1980). Mudslides remain inactive for long periods and then suddenly become active, as for example at Dowlands Chasm in southern England. This comprises Gault Clay over Triassic Lias, and it slipped at Christmas in 1839. A well-studied example is the Black Ven slide in Dorset. This appears to have been quiescent for more than 100 years until triggered into action in the 1950s, since which time it has maintained relatively constant form, slope angle and sediment budget in spite of rapid retreat (Brunsden and Jones, 1980; Chandler and Brunsden, 1995). Steepening and toe erosion appear cyclic, but maintain a longer-term dynamic equilibrium with characteristic scarp and bench morphology showing parallel retreat over time (Dixon *et al.*, 1998).

There is clearly a range of behaviour and modes of retreat for soft-rock cliffs. Figure 4.22 illustrates a variety of patterns of retreat from highly coupled parallel retreat of avalanching unconsolidated sands, to mudslides and slumping which may be periodically steady, or cyclic with varying degrees of decoupling between top and toe, to highly sporadic retreat in more lithified soft-rock cliffs. Generalised time scales of cyclic behaviour can be suggested, being 5–10 years in soft glacial sediments, 30–40 years for high cliffs in stiff clay (e.g. London Clay), 30–50 years in 100 m high cliffs in California, and 100–150 years in mudslides in interbedded clays of Dorset (Bray and Hooke, 1997).

The response of soft-rock cliffs to sequences of extreme events has been demonstrated by Phillips (1999) who showed that unconsolidated bluffs in North Carolina underwent accelerated retreat during the second of two major storms in 1996 because the first storm had already removed the protective toe. Similarly activation of mud slides in Hawke Bay in New Zealand followed the 1931 earthquake at Napier.

Cliffs appear relatively insensitive to the activities of humans, but erosion in California is accelerated by water added to the groundwater system as a result of people watering their gardens (Griggs and Trenhaile, 1994), and elsewhere building may also hasten erosion (Viles and Spencer, 1995). Where parts of cliffs are armoured or otherwise stabilised (and longshore sediment movement decreased), as along several sections of the south coast of England, adjacent sections are deprived of sediment input and erode. Even apparently harmless activities such as coastal road construction can interrupt transfers of sediment from mudslides or rivers and streams into the shoreline sediment compartments and lead to altered patterns of long-term change (McKenna *et al.*, 1992). These issues are examined further in Chapter 10.

4.6 Summary

Rocky coasts exhibit a rugged morphology that has developed over extended time scales. It is tempting to speculate on patterns of evolution based on processes more often inferred from morphology than observed to be operational. The literature is full of discussions of the relative significance of variations in wave action, weathering of the rock and the influence of lithology. However, it seems likely that at any particular location it is a subtle combination of material strength and operative processes, together with an element of geological inheritance, that determines morphology. Comparison of the relative eroding forces of waves and resisting forces of rocks as advocated by Sunamura (1992) offers a way of considering these factors in combination, but the challenge is to quantify these variables at an individual site, and to assess their variability over time.

Most of the attempts to model cliffs have been based on the premise that it is wave erosion that is dominant. Modelling of cliff retreat using both wave tank experiments and the relation between wave forces and rock resistance can continue to provide some rates to calibrate simulation models. Simulation modelling has indicated a decrease in erosion rate, and hence the development of cliffs or shore platforms, through time (Trenhaile, 1989). There can be little doubt that subtle feedbacks operate but it seems unlikely that boundary conditions will remain constant long enough for equilibrium to be established. Further developments in dating erosional landforms, particularly through the application of cosmogenic dating, will enable instances of geological inheritance to be determined. However, rates of operation of modern processes are likely to be inconsistent with the time-integrated landforms that evolve. Short-term and long-term studies will only be reconciled with a realistic representation of appropriate time scales and a clearer understanding of the role of extreme events.

Chapter 5
Reef coasts

> Yet these low, insignificant coral-islets stand and are victorious: for here another power, as an antagonist, takes part in the contest. The organic forces separate the atoms of carbonate of lime, one by one, from the foaming breakers, and unite them into a symmetrical structure. Let the hurricane tear up its thousand huge fragments; yet what will that tell against the accumulated labour of myriads of architects at work night and day, month after month. Thus do we see the soft and gelatinous body of the polypus, through the agency of the vital laws, conquering the greatest mechanical power of the waves of an ocean, which neither the art of man nor the inanimate works of nature could successfully resist. (Darwin, 1845)

This chapter is concerned with coral reefs and associated carbonate environments on tropical and subtropical coasts. Reefs are dynamic geomorphological systems demonstrating a complex interplay between physical and biological processes. They form solid limestone, simultaneously producing, breaking down and redistributing sediments of different sizes to construct a range of landforms. As a result of their ability to build rigid, wave-resistant structures, corals modify the environment in which they live, as expressed by Darwin in the quotation above. Reefs contain a variety of interacting subsystems operating over a broader range of time scales than generally seen on rocky coasts, comprising construction, destruction and various responses to extreme perturbations, such as storms.

The geomorphology of a reef, constructed over the relict topography of former surfaces, is an expression of complex biological interactions, including competition and grazing pressures, in response to physical factors such as wave action, nutrient levels, sedimentation, irradiance, desiccation and temperature. A reef contains a range of different energy settings; a near-continuous windward reef rim can experience relentless high-energy ocean swells, whereas lagoonal environments are more sheltered with scattered patch reefs of fragile corals. A coral-dominated reef is only one of a variety of carbonate environments; reefs may be dominated by other organisms or in some cases by inorganic precipitation of calcium carbonate (see Chapter 2), such as on carbonate banks (e.g. Great Bahama Bank). Reefs are particularly significant in terms of geomorphology because corals and, to a lesser extent, other components of the reef are amenable to radiometric dating, enabling reconstruction of former landforms to an extent that is generally not possible in other coastal environments.

The term 'reef' is difficult to define; it is derived from the Norse 'rif' meaning a rib of sand, rock or biological construction in shallow water, and was initially a nautical term used to describe a rocky outcrop at or close to the sea surface. In a geological context, it is generally used to imply some organic binding (Scoffin, 1987). A reef can be defined as 'a discrete carbonate structure formed by *in situ* or bound organic components that develop topographic relief upon the sea floor' (Wood, 1999, p. 5). In this chapter, coral reefs are the primary landform that is considered, although several of the concepts may also relate to other reefs, or reefs constructed by other biological agents (see Section 2.4.3, Table 2.5).

Coral reefs are geologically complex and ecologically diverse. They are best developed across the Indian Ocean and on the western margin of the Atlantic and Pacific Oceans, from the eastern margins of which they are generally excluded by upwelling of cold water. For instance, the luxuriance of coral development on the western Pacific margin contrasts with the poor development along the eastern margin. Corals occur in the Galapagos Islands in the eastern Pacific, but they are susceptible to major El Niño phases when sea-surface temperatures can exceed the limits of tolerance (Glynn and Ault, 2000).

The distribution of coral reefs is influenced by environmental factors such as light (symbiotic zooxanthellae in corals require light to photosynthesise), sea-surface temperature and carbonate saturation state (closely related to temperature). Corals are limited to waters where sea-surface temperatures rarely drop below 17–18 °C, or exceed 33–34 °C for prolonged periods. They prefer average seawater salinities. High salinities appear to limit corals, although there are prolific reefs in the Red Sea; low salinities, as experienced at the mouths of major rivers, inhibit reef development. Sedimentation is also a limitation on coral establishment and growth. Turbidity can result in smothering, shading and abrasion, as well as disruption to recruitment. However, some corals can survive better than others in muddy conditions and extensive reef growth is possible in turbid settings (Scoffin *et al.*, 1997; Edinger *et al.*, 2000). Reefs do not perform well where there is terrestrial input of clastic sediment, freshwater or nutrients. For example, in the Great Barrier Reef individual reefs are not well developed along the mainland continental coastlines where rainfall is high, whereas there are extensive fringing reefs within the Ningaloo Barrier Reef, along the drier coast of Western Australia.

Coral species diversity decreases with latitude in response to sea-surface temperature and currents of dispersal (Yonge, 1940). Reefs extend beyond the tropics where poleward-flowing warm ocean currents are favourable. The northernmost reefs occur in Bermuda where the Gulf Stream elevates sea-surface temperatures, and in Japan where there is a strong correlation between temperature and species diversity (Veron and Minchin, 1992; Yamano *et al.*, 2001). The southernmost reefs occur at Lord Howe Island in the southwestern Pacific as a result of the East Australian current (Harriott *et al.*, 1995). On higher latitude reefs, corals produce carbonate at a relatively high rate (Grigg, 1982; Harriott, 1992; Logan *et al.*, 1994), but corals can be weaker in structure and are less able to compete with macroalgal overgrowth (Crossland, 1984; Veron, 1995). In Florida, they are limited by cold water incursion (Roberts *et al.*, 1982). Beyond the latitudinal limit to reef development, corals form a community that merely veneers rocky substrates together with coralline algae; algal nodules (termed rhodoliths) dominate the shelves of eastern and western Australia beyond the limits to reef growth (Marshall *et al.*, 1998; James *et al.*, 1999).

5.1 Historical perspective

When Charles Darwin sailed around the world on the *Beagle* in the 1830s, the most widely held view on the origin of mid-ocean atolls was that coral growth veneered shallowly submerged volcanic craters (Lyell, 1832). Darwin came up with the remarkable deduction that reefs on different islands could be viewed as successive stages within a progression from fringing reefs through barrier reefs to atolls (see Chapter 1, Figure 1.3). This view of the evolution of atoll structure can be explained on the basis of 'prolonged subsidence of the foundations on which the atolls were primarily based, together with the upward growth of the reef-constructing corals . . . Fringing reefs are thus converted into barrier-reefs; and barrier-reefs, when encircling islands, are thus converted into atolls, the instant the last pinnacle of land sinks beneath the surface of the ocean' (Darwin, 1842: p. 109).

Darwin's subsidence theory was extended to other reefs initially by Darwin himself (1842) and subsequently by Dana (1872), who drew attention to the deep embayments around the coastline of high islands in the Pacific, representing valleys drowned by subsidence (see Chapter

4). The theory, however, was challenged by several scientists in favour of solution of limestone or on zoological grounds (Murray, 1889; Gardiner, 1931), on tectonically uplifted islands (Guppy, 1888), in relation to antecedent topography (Agassiz, 1899), and in terms of sedimentation (Wood-Jones, 1912).

Darwin's subsidence theory of reef development did not incorporate Quaternary fluctuations of sea level, and Reginald Daly proposed a glacial-control theory of reef development, envisaging that reefs had contracted from mid-latitude regions, termed 'marginal seas', during glacial periods (Daly, 1910, 1915, 1925). He proposed that reef platforms were planed off at lower sea level, mistakenly interpreting lagoon depths to be consistent, and that the rim of an atoll was constructed as a result of postglacial reef growth (Daly, 1934). W.M. Davis (1928) undertook an extensive review of the 'coral reef problem', discounting other theories and concluding that the Darwinian subsidence theory was the most credible theory, and Daly's glacial-control hypothesis was the only likely alternative.

Darwin's theory of the development of mid-oceanic reef structure was tested by deep drilling on atolls. The Royal Society sponsored a British–Australian expedition led by Sollas and Edgeworth David to the mid-Pacific atoll of Funafuti (Ellice Islands, now Tuvalu) in 1896–98. Shallow-water carbonates were encountered to a depth of 333 m, implying subsidence because these depths are below the modern depths at which the organisms grow, but no volcanic basement was encountered (David and Sweet, 1904).

Post-war deep-drilling on Bikini and Enewetak atolls in the Marshall Islands revealed more than 1000 m of shallow-water carbonates overlying a basalt core (Emery *et al.*, 1954). Subsequent drilling on Midway atoll in the Hawaiian Islands, and Mururoa and Fangataufa atolls in French Polynesia also reached the volcanic basement (see Table 5.1). These combined results support the subsidence theory of coral-reef development proposed by Darwin for many atolls 'with a certainty rarely obtained in geomorphology' (Guilcher, 1988, p. 70).

Several linear island chains in the Pacific (e.g. the Hawaiian and Society Islands) demonstrate successive stages in the Darwinian sequence, fringing reefs, barrier reefs, and coral atolls, along their length (Chubb, 1957). The sequence can be viewed in the context of plate-tectonic theory (see Section 5.3.1), with subsidence explained in association with the aging and contraction of the ocean floor as the plate moves (Parsons and Sclater, 1977; Scott and Rotondo, 1983).

Sea-level fluctuations were not envisaged by Darwin, but they have played an important role in determining the morphology and development of modern reefs and can be accommodated into his view of reef

Table 5.1. Results of deep drilling of atolls

Atoll	Location	Depth to basement (m)	Lithology	Reference
Funafuti, Tuvalu	Reef rim	>333	Terminated in Pliocene limestone	Judd, 1904
Kita Daito Zima, near Japan	Emergent atoll	>431	Terminated in Oligocene limestone	Hanazawa, 1940
Maratua, Indonesia	Emergent atoll rim	>550	Terminated in limestone	Kuenen, 1947
	Lagoon floor	>429		
Bikini, Marshall Islands	Reef rim	>777	Terminated in Miocene limestone	Ladd *et al.*, 1970
Enewetak, Marshall Islands	Reef rim	1283	Basalt underlying	Ladd *et al.*, 1948
	Reef rim	1408	Eocene limestone	Emery *et al.*, 1954
Mururoa, French Polynesia	Reef rim	415	Basalt underlying	Lalou *et al.*, 1966
	Reef rim	438	Miocene limestone	Labeyrie *et al.*, 1969
Midway, Hawaiian Islands	Lagoon	173	Basalt underlying	Ladd *et al.*, 1970
	Reef rim	503	Miocene limestone	
Fangataufa, French Polynesia	Lagoon	230	Basalt underlying	Guillou *et al.*, 1993
	Reef rim	360	limestone	

development (see Section 5.2.2). Stratigraphic studies of reefs have not supported truncation of the reef platform at low sea level with subsequent rapid postglacial reef growth as envisaged in the glacial-control theory by Daly (1934) and Wiens (1959, 1962). Instead modern reefs appear to be underlain by a sequence of previous highstand reef sequences marked by discontinuities, often termed solutional unconformities, recording episodic subaerial exposure and karst erosion as a result of Quaternary sea-level fluctuations (Schlanger, 1963). Erosion of reef limestone, as indicated by outcrops of Pleistocene reef limestone above sea level, is generally too slow for complete truncation to have occurred at lower sea level (Stoddart, 1969a). The idea that reef growth might accentuate karst erosional topography was proposed by MacNeil (1954) and gained particular support from extensive seismic studies by Purdy (1974a,b) who showed that modern reefs are established over topographic highs on the Belize shelf.

Coral reefs have played an important role in generating ideas about coastal evolution. Not only did observations of reefs stimulate Darwin, Dana, Davis and Daly to speculate on evolutionary sequences, but they also led to some of the first and some of the most successful examples of international and multidisciplinary expeditions. The Royal Society sponsored expeditions to Funafuti in 1896–98 led by Sollas and Edgeworth David and expeditions to the Great Barrier Reef in 1928 and

1973 led by Maurice Yonge and David Stoddart respectively. Subsequent intensive research on Pacific reefs by researchers from the United States in the late 1940s and early 1950s (Wiens, 1962), and in the 1970s and 1980s by French and Japanese researchers (Delesalle *et al.*, 1985; Yonekura, 1988, 1990), provided a firm basis for synthesis. These and other studies are reviewed in a comprehensive summary by Guilcher (1988).

5.2 Reef morphology and zonation

Coral reefs represent the largest biological constructions on earth. The Great Barrier Reef along the northeastern coast of Australia, for example, is more than 2000 km long. Reefs are composed primarily of framework constructed by corals, but with further skeletal sediment derived from a range of associated organisms and organic and inorganic cements. Although a reef is composed of the remains of organisms, the structure that is built is an aggregate of both the construction processes and the erosional or destructive processes.

Corals are colonial animals that secrete calcium carbonate forming an exoskeleton (or outer skeleton) in which they live. There is a wide range of coral growth forms, including massive hemispherical colonies, up to several metres in diameter, and branching corals. Polyps live in a calyx, a small depression on the surface of the coral, and contain symbiotic zooxanthellae (unicellular yellow-brown dinoflagellate algae) that limit coral growth to water depths of generally less than 30 m because they require light for photosynthesis.

The rate at which calcareous sediment is produced in a reef depends on the life history of the contributing organisms, and biological and chemical attrition. For instance, some sediment components, such as blades of the algae *Halimeda*, are produced rapidly, but break down quickly also. Physical processes on the reef redistribute and sort sediment, and mechanical breakdown is accelerated during transport making particles smaller and often more rounded (see Figure 2.15). The effects of organisms burrowing into and sorting sediments (bioturbation) are also significant in mixing sediments.

Coral reefs can be divided into an Atlantic province, centred on the West Indies, and an Indo-Pacific province, comprising much of the tropical Indian and Pacific Oceans. The Indo-Pacific province is much richer in species with about 500 species of corals; the dominant genus of branching corals, *Acropora*, contains over 150 species in the Indo-Pacific province, whereas only two species are prominent in this genus in the Atlantic province. Modern reefs are often prominently zoned, and there is generally a coincidence between geomorphological zones and their ecological composition. Where zones occur on reefs, they are most prominent to windward and generally less clearly defined on leeward reefs. Windward reefs are generally better developed and more productive than leeward reefs. They tend to be straight in planform, whereas the more sheltered leeward reefs are often irregular, without a prominent reef crest, and of lower slope (Stoddart, 1969a; Milliman, 1974).

5.2.1 Reef morphology

The cross-sectional morphology of a coral reef depends on the topography of the coast, the influence of geological inheritance, and the extent and pattern of reef growth. Sea-level change over the past few thousand years has been an important boundary condition, which has differed between and within the reef provinces resulting in morphological differences between reefs (Adey, 1978). Indo-Pacific reefs generally contain extensive reef flats that dry at low tide, often with supratidal conglomerate deposits, which have formed during a higher sea level in mid Holocene and have been exposed by emergence (late Holocene sea-level fall). West Indian reefs appear to have experienced a sea level rising up until present and do not have features attributed to emergence (Figure 5.1).

Figure 5.1 shows schematic profiles across a typical Indo-Pacific and a West Indian reef. Reefs that reach sea level are generally characterised by three zones: reef front, reef crest and backreef (Figure 5.2). The reef front is relatively steep and is usually covered with live corals. It merges into the forereef or reef slope that supports only sparse coral cover, but receives detrital material, termed talus, from the active reef above it. The reef crest is the high-energy margin, usually marked by a line of breaking waves. The backreef consists of more sheltered environments, comprising either a reef flat (a more-or-less horizontal surface in shallow water or exposed at lowest tides) or lagoonal environments. The backreef either abuts land, as in the case of the reef fringing Rarotonga (Figure 5.2), or grades into a lagoon (for instance on atolls, see Figure 1.8).

Isolated volcanic islands in the Pacific Ocean rise from deep water (often >4000 m) with flanks that have slope angles of 30–45° as a result of the extrusion of oceanic basalts. Reef limestone maintains steeper slopes. The forereef below depths of 30–40 m can be a near-vertical drop-off with overhangs and caverns. There are frequently a series of terraces, or forereef buttresses, identifiable on the seaward margin of Indo-Pacific reefs. A 10-fathom terrace or forereef buttress (about 15–21 m deep) described from several atolls in the Pacific (Emery *et al.*, 1954), often corresponds to the depth of contact between Pleistocene and Holocene limestone (Stoddart, 1969a). Pleistocene limestone, radiometrically dated to Last Interglacial (substage 5e), has been encountered beneath the reef rim of several atolls (Trichet *et al.*, 1984; Szabo *et al.*, 1985; Marshall and Jacobson, 1985). Similar terraces and forereef buttresses (termed bank barriers because they appear to have existed temporarily as barrier reefs) have been widely mapped in the West Indies (Macintyre, 1988). They seem to be short-lived postglacial reefs and those on Barbados have been drilled and dated to establish sea-level history (Fairbanks, 1989; see also Figure 2.10). The forereef buttresses are covered with coral that adopts an increasingly platy growth form with depth. They are separated by deep grooves or canyons that contain unconsolidated sediment that is swept down the reef front by currents (Roberts, 1983).

The slope of the reef front is more gradual than the deep forereef buttresses. The reef front in both West Indian and Indo-Pacific reefs is frequently dominated by a prominent spur and groove pattern (Figure 5.2). Spacing of grooves is of the order of 10–30 m, and length is often 30–80 m; they are most pronounced in areas that experience

Figure 5.1. Typical reef morphology and stratigraphy. (a) Schematic cross-section of an Indo-Pacific reef (based on Emery *et al.*, 1954; Stoddart, 1969a; McLean and Woodroffe, 1994) and (b) a West Indian reef (based on Shinn *et al.*, 1982; Macintyre, 1988).

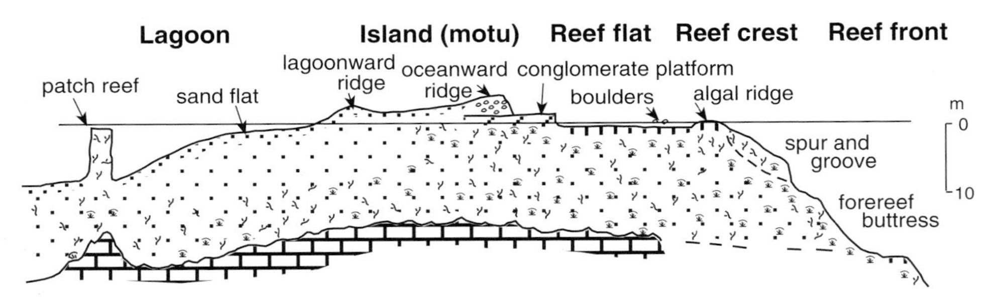

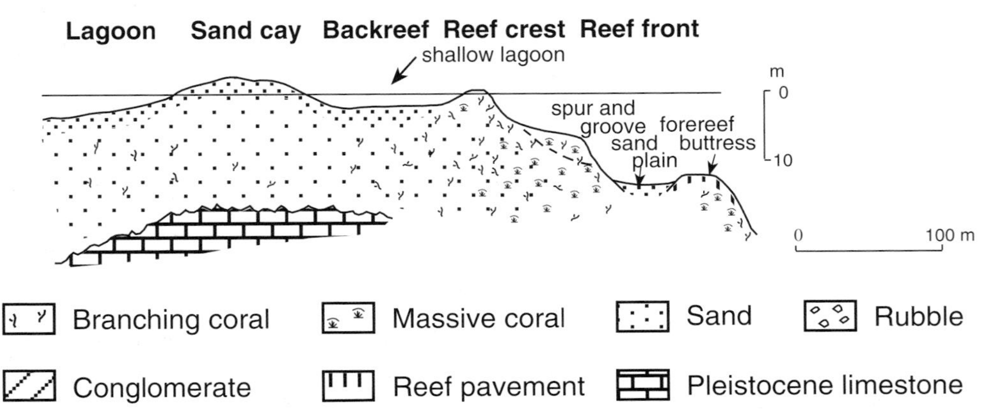

the greatest wave energy (Munk and Sargent, 1948). The origin of the spur and groove pattern has been debated. In some places it appears to result from erosion into a pre-Holocene surface (Newell *et al.*, 1951), but elsewhere spurs are built by coral growth (Shinn, 1963; Shinn *et al.*, 1981).

The reef crest is usually a sharply defined zone characterised by breaking waves. In the West Indies, this is often a zone of elkhorn coral, *Acropora palmata*, with branches oriented into the breaking waves (Figure 5.3). In higher energy settings, coral is replaced by coralline algae (Adey, 1975; Adey and Burke, 1976). An algal ridge (or rim), particularly of pink algae of the genus *Porolithon*, is a prominent feature on many Indo-Pacific reefs. It tends to be 5–15 m or more wide and rises at least 0.3 m above the reef surface, although locally, as on the southern shore of Tongatapu, Tonga, algal structures can build up to heights of a metre or more (Nunn, 1993). Algal ridges are much more common and are better developed on tradewind Indo-Pacific reefs that receive persistent swell, than on West Indian reefs.

There is considerable variation in the backreef zone. On simple fringing reefs, such as the fringing reef on Rarotonga shown in Figure 5.2, the backreef zone is narrow and the reef flat is contiguous with the shore. On many Indo-Pacific reefs, the reef flat is a pavement that can

Figure 5.2. Fringing reef on the young (Pleistocene) volcanic island of Rarotonga in the southern Cook Islands, Pacific. The forereef comprises spurs and grooves. The reef crest along which the waves are breaking consists of an algal ridge. The reef flat is covered by shallow water at low tide and has intermittent coral cover.

be more than a kilometre wide; it is cemented by coralline algae and dries at low tide. There can be a moat immediately landward of the algal ridge, as on several of the atolls of the Marshall Islands, or a boulder zone, in which rubble and reef blocks occur as a result of storms (see Section 5.4.3). Rubble ramparts occur on the reef flat where storms are frequent (Stoddart *et al.*, 1978). The extent to which live coral occurs over the reef flat depends on the degree to which water remains over the surface at low tide because corals cannot withstand exposure for long at low tide. On some reefs in the Great Barrier Reef the reef flat is colonised by mangrove forests, representing the final stages in reef-top development (Stoddart, 1980); elsewhere islands form (see Section 5.5). Where islands occur, those on Indo-Pacific reefs are often located on a conglomerate platform, whereas on West Indian reefs they are more often unconsolidated sand cays.

By contrast, in the West Indies the reef crest is often very narrow, and consists of a relatively sheltered shallow lagoon in the backreef, containing delicate branching corals, particularly the staghorn coral, *Acropora cervicornis*, and seagrass, *Thalassia testudinum*. In the case of barrier reefs and atolls, in either reef province (although both are more extensive in the Indo-Pacific), the backreef zone grades into a deeper lagoon, with isolated patch reefs.

Figure 5.3. Reef crest *Acropora palmata*, a shallow-water coral that dominates reef crests in the West Indies.

5.2.2 Reef zonation

A reef appears robust in geological terms being a massive structure with a high potential for preservation; however, reefs are also fragile ecosystems, and individual corals can be delicate. There are a series of environmental factors, such as light, sediment and wave energy, which vary across a reef and represent stresses to which corals have discrete tolerances. Corals show a variety of growth forms and the distribution of these is related to environmental gradients in these factors. These gradients and the range of growth forms are shown schematically in Figure 5.4 (Chappell, 1980).

Wave energy is one of the most important controls on reefs; reefs are best developed where the wave energy is high, but individual corals are limited by their susceptibility to damage under waves. Wave-induced stress is greatest beneath breaking waves, and decreases with depth down the reef front. Fragile branching corals are prone to breakage and can only survive where wave energy is low; they are replaced by ramose or digitate-plate corals where wave energy is moderate, and by globose or encrusting corals in higher wave-energy settings (Figure 5.4b). Corals are completely replaced by coralline algae on high-energy reef crests.

Corals also have to cope with a range of light conditions; for instance, the massive West Indian coral, *Montastrea annularis*, changes its form, becoming flatter and more platy with depth down a reef front. Platy corals are able to maximise light interception but are poorly adapted to tolerate high sediment loads. Ramose, branching and foliate corals are much better able to cope with high sediment loads and are more common where turbidity is high. Subaerial exposure during low tide is also a stress, to which encrusting forms, or discoid corals, termed microatolls (dead on top but live on the margins), are better adapted than those with a large surface area.

Figure 5.4a is a schematic model of coral growth forms in relation to environmental gradients proposed by Chappell (1980). On the basis of this model, growth forms are dominant at different points across the reef profile. The reef front is likely to have massive forms at depth. Towards the reef crest more sturdy branching forms of coral will occur, and the branches may be oriented into the dominant wave direction. In the highest wave-energy settings corals are either encrusting, or are replaced by encrusting coralline algae as occur on the algal ridges of Indo-Pacific reefs.

The broad scheme in Figure 5.4 can be illustrated with reference to West Indian reefs. Figure 5.5 is a schematic representation of reef zonation in relation to wave energy on West Indian reefs (based on Geister, 1977). The highest energy setting is a reef crest that is covered

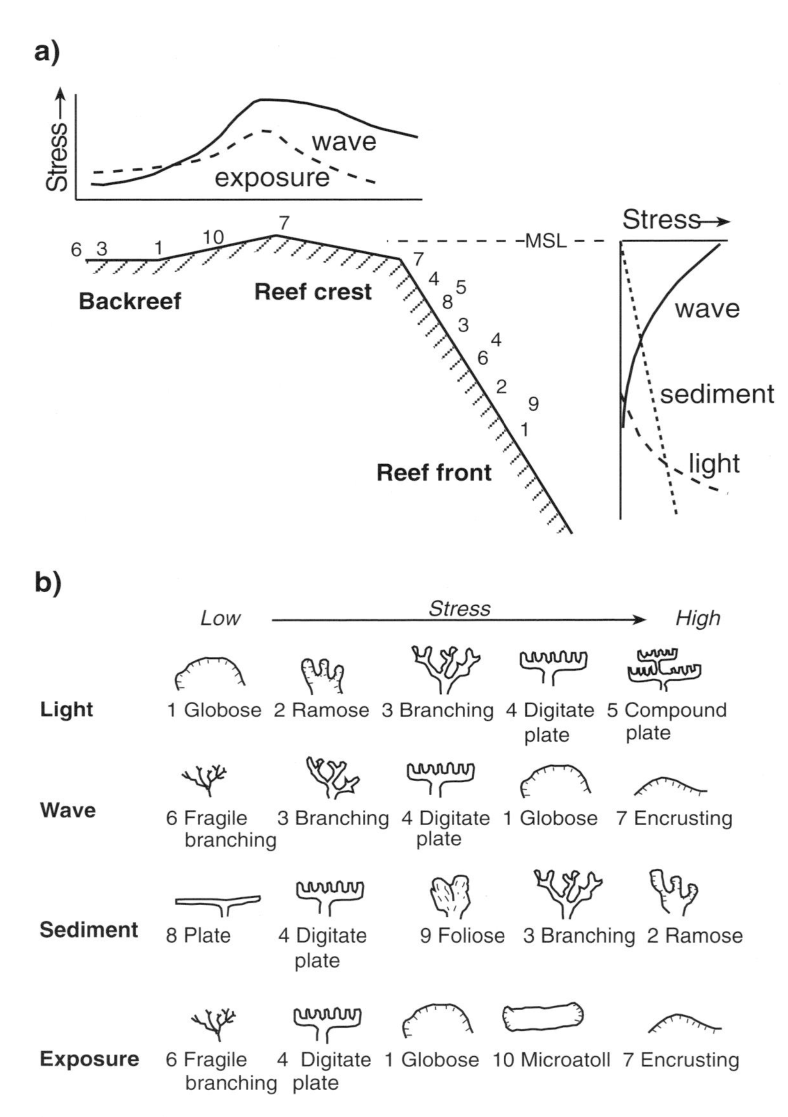

Figure 5.4. The response of coral growth form to physical stresses (environmental factors) in a forereef and reef crest (based on Chappell, 1980). (a) The distribution of those stresses and their influence on the distribution of coral. (b) Coral growth forms. The growth-form distribution on the reef flat and reef front in (a) is shown using the numbers in (b).

by coralline algae (Adey and Burke, 1977). The more delicate branching staghorn, *Acropora cervicornis*, occurs in the less energetic environments, whereas the elkhorn coral, *Acropora palmata*, often dominates the reef crest (see Figure 5.3). Wave energy is an important control; it determines the broad zonation that occurs across a reef in any particular wave-energy environment. The nature of the zonation can be seen to vary from low wave-energy to high wave-energy settings with the fire coral, *Millepora*, or coralline algae dominating in the higher energy settings.

5.3 Reefs in time and space

The geomorphology of reefs, the physical factors that influence the distribution of reefs, and the structure and functioning of reef systems can be viewed over a hierarchy of temporal and spatial scales. Reef structure, including the relationship between reef limestone and basement rock, is constrained by broad global plate-tectonic controls on the distribution of reefs at geological time scales. The surface morphology of reefs, comprising modern reef, lagoon and reef-island sediments, are the products of processes operating in response to morphodynamic adjustments at quite different scales (Stoddart, 1973a). Ecological processes also operate on a hierarchy of scales from evolutionary (or metapopulation) scales at which there have been mass extinctions, to population,

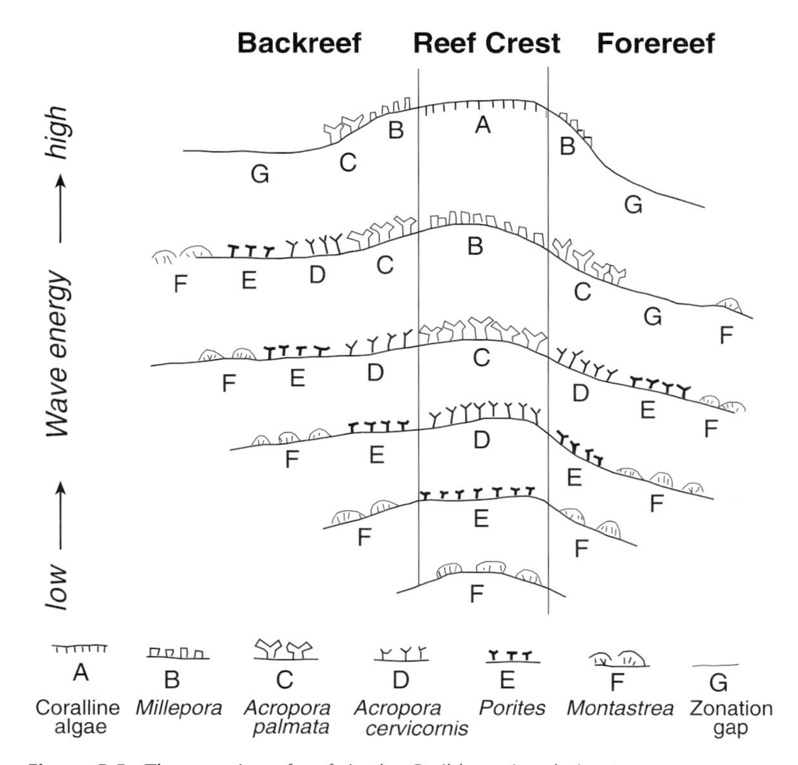

Figure 5.5. The zonation of reefs in the Caribbean in relation to wave energy (based on Geister, 1977).

community, and individual organism scales (Hatcher, 1997). There are considerable overlaps between the rates at which physical and biological processes operate (Figure 5.6); for instance, there has been a broad physical control on the pattern of evolution in corals, reflecting genetic connectivity and disruption by changing configurations of the continents and ocean basins (Veron, 1995).

Sea level has been a major control on reef geomorphology and studies of reefs and sea level have been of two types. First, there have been studies using reefs to determine sea-level history, as discussed in Chapter 2. Reefs are good indicators of the approximate position of sea level; many corals grow in water depths of several metres and can be used as directional indicators. Some, such as the West Indian reef crest coral, *Acropora palmata*, have been shown to be most abundant within 5 m of sea level (see Figure 5.3), and therefore provide an indication of the pattern of sea-level rise (Lighty *et al.*, 1982). Microatolls occur in the lower intertidal zone, constrained in upward growth by frequency of exposure, and can indicate sea level to within a few centimetres (Smithers and Woodroffe, 2000). In particular, reefs on mid-ocean islands serve as 'dipsticks' where data would otherwise be

Figure 5.6. Physical and biological processes in the reef system, and perturbations and response to them, occur at a series of different time scales.

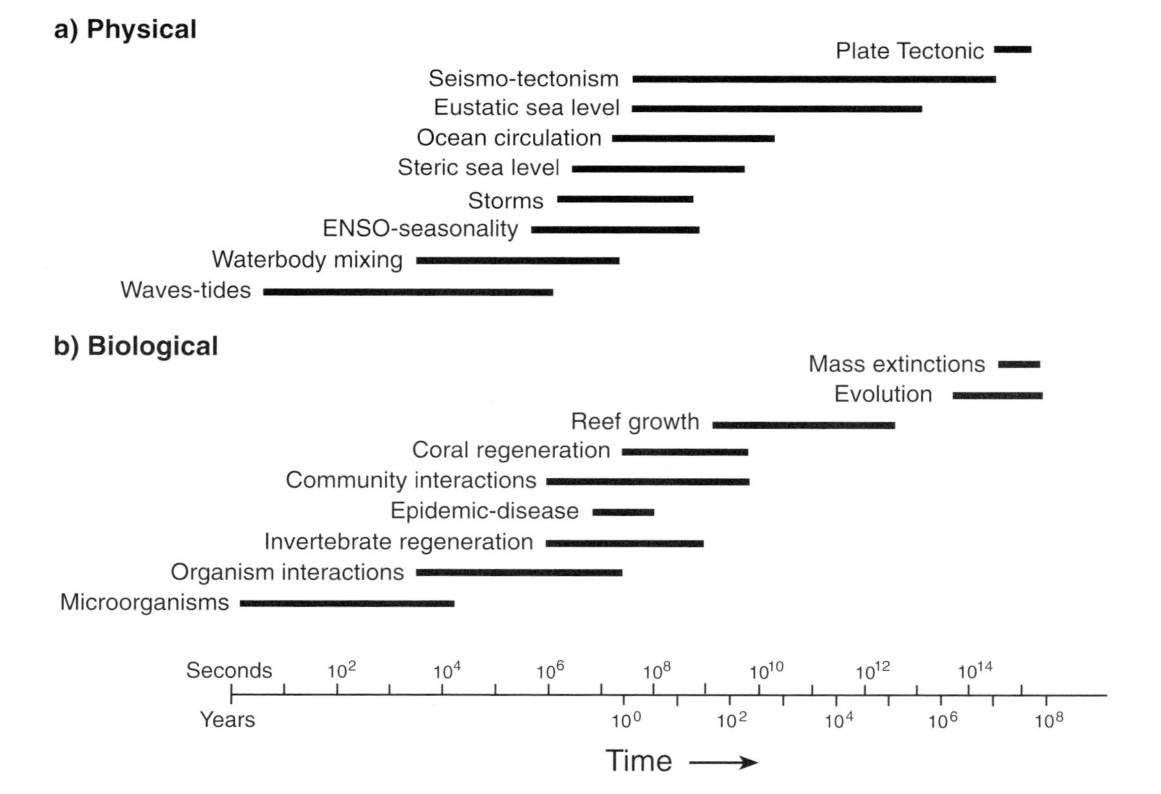

absent (Bloom, 1967). Fossil reefs have enabled reconstruction of the pattern of postglacial sea-level rise (Fairbanks, 1989; Chappell and Polach, 1991; Eisenhauser *et al.*, 1993; Montaggioni *et al.*, 1997), and have been used to test and constrain geophysical models of the response of a visco-elastic earth to changes in ice and water loads (Lambeck and Nakada, 1990; Peltier, 1999). Alternatively, geomorphological studies of reef-surface have viewed sea level as a boundary condition which has acted as a major control on reef development.

It is convenient to consider the geomorphology of reefs at four scales, although it is clear from Figure 5.6 that there are considerable overlaps in process operation. The four scales comprise (i) geological time which is relevant in terms of reef structure and distribution ($>10^6$ years), (ii) Quaternary ice ages during which sea level has fluctuated in response to ocean volume (10^5–10^4 years), (iii) mid–late Holocene, over which reef growth has varied in response to isostatic adjustments of relative sea level (10^3–10^2 years), and (iv) contemporary scale at which coral growth, sediment production and transport are active ($<10^2$ years). These provide a framework within which to examine the processes that influence reef morphology and morphodynamic adjustments between reefs, boundary condition changes and perturbations. They correspond with a similar framework used by Hubbard (1997), comprising macroscale, mesoscale and microscale, but he also recognises a nanoscale at which individual organisms function.

The biologically controlled processes that operate on a coral reef at the nanoscale, include ecological feedback mechanisms that imply a high degree of non-linearity in process interactions. These include larval settlement and establishment and interactions, such as competition. As the coral calcifies skeleton which contributes to the framework of the reef, it also sequesters a geochemical record of ambient conditions in the chemistry of the carbonate laid down. Sclerochronology, the determination of annual growth bands in corals, and geochemical analyses of corals provide a record of past conditions and insights into changing environmental conditions at a range of scales. For example, isotopic analysis of individual corals from a series of fossil reefs on Huon Peninsula, New Guinea, indicates timing and intensity of El Niño–Southern Oscillation variations across the Pacific over the past 120000 years (Tudhope *et al.*, 2001). These nanoscale investigations suggest a potential to examine fine-scale adjustments of reefs to individual events; they highlight the considerable potential of reefs to continue to refine our understanding of global scale and individual organism scale environmental variability (Gagan *et al.*, 2000; Grottoli, 2001).

5.3.1 Reef distribution and structure

The distribution of reefs at a global scale and across geological time is related to tectonic and climatic factors (Figure 5.7). Charles Darwin recognised that the distribution of major reef types correlated with broad seismo-tectonic factors, even though the theory of plate tectonics had not been formulated. He divided reefs into fringing reefs, barrier reefs and atolls (see Chapter 1, Figure 1.3). Fringing reefs are associated particularly with mainland coasts (for instance, there are localised fringing reefs along the mainland within the Great Barrier Reef, Queensland, and the Ningaloo Reef along the coast of Western Australia) and volcanic island arcs on active plate margins. They occur close to the shore, often being shore-attached; they are generally narrow and veneer bedrock. A barrier reef is a reef that is separated at some distance from

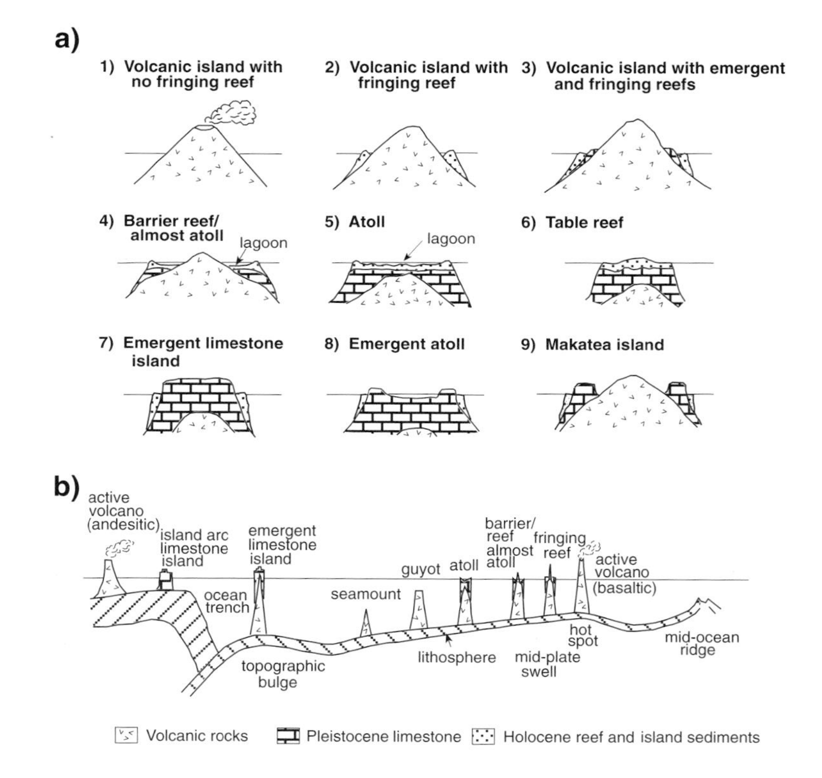

Figure 5.7. Island types and their occurrence in an ocean plate setting. (a) Types of island in mid Pacific, following Scott and Rotondo (1983), but showing relationships between Pleistocene and Holocene reefal limestones. (b) The distribution of island types along an ideal linear island chain.

land by a lagoon of 10 m or more water depth. Barrier reefs occur on mid-ocean islands (for instance Moorea, see Figure 5.8). They also occur along continental coasts, for example the Great Barrier Reef along the northeastern coast of Australia, and these are termed shelf reefs and described in Section 5.5.2. Atolls occur in mid-plate settings well away from zones of active tectonism and consist of an annular or irregular reef around a central lagoon (see Figure 1.8), often with a discontinuous ring of islands on the reef rim.

Fringing reefs, barrier reefs and atolls can be interpreted in the context of plate tectonics (Figure 5.7). Figure 5.7a shows schematically several types of islands from the Pacific Ocean and the relationships between volcanic basement and Pleistocene and Holocene reef limestones. A volcanic island, in reef-forming seas, is rapidly encircled by a fringing reef; for instance corals establish quickly on young lava flows on sections of the coast of the Big Island, Hawaii (Grigg, 1982). Young volcanic islands are generally surrounded by a fringing reef (for example, Rarotonga, see Figure 5.2). On some islands there may be an emergent (fossil) reef at a slightly higher level than the modern reef, often of Last Interglacial age (e.g. Oahu, Hawaiian Islands).

Figure 5.7 shows an island with a barrier reef and lagoon around it. Figure 5.8 shows cross-sections of several volcanic islands that have barrier reefs, with the former volcanic topography reconstructed as proposed by Menard (1983). The older volcanic islands have a greater inferred reef thickness, supporting the concept that barrier reefs occur around subsiding islands. An almost-atoll occurs where the volcanic peak has been eroded and has almost completely subsided beneath sea level. Truk in the western Pacific is an example of an almost-atoll, with several high islands rising in the middle of a symmetrical lagoon. The almost-atoll of Aitutaki resembles other atolls in the Cook Islands, but has a single volcanic remnant (Stoddart, 1975).

An atoll can result when the volcanic basement has subsided below sea level (Figure 5.7). A platform reef, or table reef, resembles an atoll but consists of a smaller reef platform without a lagoon, but often with one central reef island. The island types 1, 2, 4 and 5 in Figure 5.7 represent stages in the Darwinian sequence, several of which can be seen along some linear island chains in mid Pacific (Figure 5.7b). These chains have the youngest volcano at one end (generally the southeastern end), and atolls at the other end (Chubb, 1957). Horizontal movement of the plate into deeper water (as the lithosphere cools and contracts) provides a mechanism for subsidence (Scott and Rotondo, 1983). Islands can be carried beyond reef-forming seas, in which case they become submerged because there is no further reef growth to build reef

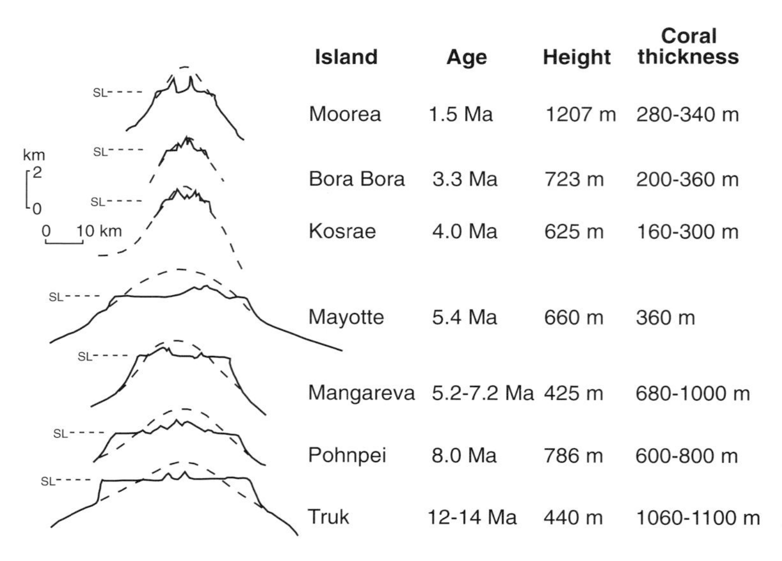

Island
Age
Height
Coral thickness
Moorea
1.5 Ma
1207 m
280-340 m
Bora Bora
3.3 Ma
723 m
200-360 m
Kosrae
4.0 Ma
625 m
160-300 m
Mayotte
5.4 Ma
660 m
360 m
Mangareva
5.2-7.2 Ma
425 m
680-1000 m
Pohnpei
8.0 Ma
786 m
600-800 m
Truk
12-14 Ma
440 m
1060-1100 m
SL
km
2
0
0
10 km

Figure 5.8. Volcanic island morphology, in cross-section, and its relationship to island age and barrier reef development. The former profile of the volcanic island is shown by the dashed lines and thickness of reef development is inferred from the break of slope associated with the reef morphology (based on Menard, 1983).

to sea level. The Hawaiian Island chain shows a sequence of guyots (the Emperor seamount chain) representing islands carried beyond the point where the rate of subsidence exceeds the rate at which corals near their limit can achieve vertical reef growth (Grigg and Epp, 1989). A similar process may have operated in the more remote geological past with northward migration of carbonate islands and their subsequent drowning as guyots in the Cretaceous (Wilson *et al.*, 1998).

The sequence of island types in mid plate is more complex than this because there are a range of islands on which Pleistocene reef limestone is exposed above sea level. It has already been shown that these are common along active plate margins where collision between plates leads to rapid uplift (see Chapter 2). Similar islands can occur in mid-plate settings, and three examples are shown in Figure 5.7 (7, 8 and 9). They can result from at least two causes. Emergent limestone islands occur on subducting plates flexed prior to subduction. The island of Nuie, just east of the Tonga trench, and Christmas Island in the Indian Ocean are islands that seem to have undergone uplift as the plate flexes prior to subducting into an ocean trench (Woodroffe, 1988).

Islands within a critical distance of relatively young volcanic islands have experienced uplift as a result of lithospheric flexure in response to loading of the plate by recent (Pleistocene) volcanism. The volcanic island represents a load on the plate and the lithosphere flexes forming a moat around the young island and an arch at a distance of several hundreds of kilometres from the volcano, depending on age and rigidity of the plate. Islands on the arch are uplifted (McNutt and Menard, 1978). The islands of Atiu, Mauke and Mitiaro in the southern Cook Islands lie on the arch around Rarotonga (Stoddart *et al.*, 1990). These are called makatea islands and comprise a degraded volcanic interior and a highly karstified Tertiary limestone periphery, named after the island of Makatea in French Polynesia (although Makatea is an emergent limestone island and not a makatea island). Uplift on the makatea islands in the southern Cook Islands appears to have continued, though at different rates on the different islands, as demonstrated by the different heights at which Pleistocene reefs are found (Woodroffe *et al.*, 1991a). Lithospheric flexure has also occurred in the Society Islands (Pirazzoli, 1994), and complex patterns of flexure are also apparent along the Hawaiian Island chain (Watts and Zhong, 2000).

5.3.2 Quaternary sea-level variation

Quaternary fluctuations in sea level and climate have had major impacts on the distribution and morphology of coral reefs. Deep drilling has supported Darwin's subsidence theory of coral-reef development but has revealed a series of solutional unconformities in the shallow-water carbonate sequences beneath atolls and other modern reefs (Schlanger, 1963). These mark periods of alternate exposure and drowning of reef platforms (Lincoln and Schlanger, 1991), and it is clear that at this scale reefs cannot be understood without an assessment of the role of sea-level change. A solutional unconformity, termed the 'Thurber discontinuity' (Thurber *et al.*, 1965), represents a hiatus between Holocene reefs, dated by radiocarbon, and underlying Pleistocene reef limestone, which has been dated by U-series dating and has been shown in many cases to be Last Interglacial in age. It records a period of subaerial exposure and karstification and is shown schematically in cross-section in many of the island types in Figure 5.7.

The glacial-control interpretation of atoll development proposed by Daly, and championed by Wiens (1962) implied that the reef platform was entirely truncated at the lower Pleistocene sea level. However, atolls are not underlain around their rim by thick enough sequences of Holocene limestone; drilling has revealed Pleistocene (Last Interglacial) limestone at shallow depth, and on emergent atolls exposed at the surface. For instance, Last Interglacial limestone has been encountered at a depth of 8–17 m below the reef rim of eastern Tarawa, the atoll shown in Figure 5.9 (Marshall and Jacobson, 1985). Similar 'Thurber discontinuities' occur at 7–14 m on Enewetak (Szabo *et al.*, 1985), at 6–11 m on Mururoa Atoll in the Tuamotus (Trichet *et al.*, 1984), at 6–16 m on the Cocos (Keeling) Islands in the eastern Indian Ocean (Woodroffe *et al.*, 1991b), and at 7–22 m beneath the lagoon on Aitutaki, Pukapuka and Rakahanga in the Cook Islands (Gray *et al.*, 1992). Some lowering of the previous atoll surface can have occurred through solution and a karstified surface occurs beneath the lagoon (Perrin, 1990; Purdy and Winterer, 2001), similar to that demonstrated by seismic profiling across the shelf reef of the Belize barrier reef (Purdy, 1974a). However, solution cannot account for the removal of thicknesses of 10 m or more of limestone since the Last Interglacial and subsidence is indicated.

Two possible models of atoll development during an interglacial–glacial–interglacial cycle are shown schematically in Figure 5.10, in comparison with sea level for the past 125 000 years. When sea level was high, during the Last Interglacial (marine oxygen-isotope substage 5e), an atoll similar to the modern reef presumably existed, as recorded in the stratigraphy of atoll rims (whether the sea was higher than present, or around present level remains a subject of discussion, see Chapter 2, and Lambeck and Nakada, 1992). During interstadials (such as oxygen-isotope substages 5c and 5a) shallow lagoonal sediments and peripheral

Figure 5.9. The eastern rim of Tarawa, an atoll in the Gilbert chain, Kiribati, central Pacific. Drilling beneath the reef islands has encountered Last Interglacial reef limestone at depths of 8–17 m.

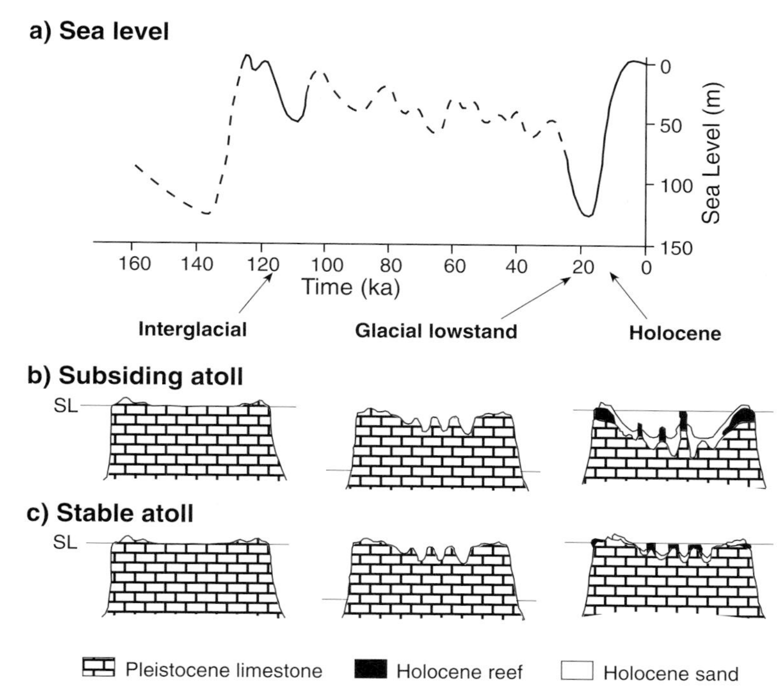

Figure 5.10. Evolution of atolls in the late Quaternary as a function of sea-level variations. Radiometric dating provides an insight into the environments of deposition and their change from the Last Interglacial (marine oxygen-isotope substage 5e) through the last glacial maximum to the modern Holocene reefs. Similar variations can be envisaged for previous glacial/interglacial cycles. (a) The sea-level curve of Chappell and Shackleton (1986). (b) The history of subsiding and stable atolls. The majority of atolls are subsiding and subsidence allows accommodation space in which 10–20 m of Holocene reef and lagoon sediments have accumulated. Atolls which are stable or even undergoing gradual uplift have Last Interglacial limestone exposed at or near the surface (based on Woodroffe and McLean, 1998).

reefs were presumably exposed, but sediments from this time have not yet been recognised in the subsurface stratigraphy of atolls.

Subaerial exposure and solutional weathering (karstification) during the glacial period can be inferred based on the widespread occurrence of solutional features, such as blue holes, on many reefs (Purdy, 1974a) and dating of speleothems from others (Lundberg *et al.*, 1990). As the sea rose, inundation of the Last Interglacial platform on subsiding atolls would have occurred before present sea level was reached. By contrast, on emergent atolls (see Figure 5.7), Last Interglacial limestone is exposed above sea level because the atoll has not experienced subsidence (Figure 5.10). Relatively restricted Holocene sedimentation has occurred on these reefs because there has not been a large horizontal area suitable for reef establishment (also described as limited accommodation space). Emergent atolls that would appear not to have undergone subsidence over the late Quaternary include several of the Phoenix Islands of eastern Kiribati (Tracey, 1972), Anaa atoll in the Tuamotu Archipelago (Veeh, 1966; Pirazzoli *et al.*, 1988), Aldabra Atoll in the western Indian Ocean (Braithwaite *et al.*, 1973), and Christmas Island in the southern Line Islands in the Pacific (Woodroffe and McLean, 1998). The gross morphology of subsiding and stable modern reef platforms reflects the topography of reefs during previous interglacials.

There is evidence that coral communities, during the peak of the Last Interglacial, extended and built reefs further south than the limits to modern reefs. For instance, corals have been dated to Last Interglacial at Rottnest Island on the west coast and Evans Head on the east coast of Australia (Marshall and Thom, 1976; Szabo, 1979). However, it is unclear to what extent the distribution of reef-building corals contracted equatorward during glaciations, into 'marginal seas' as conceived by Daly (1934). Tropical sea surface temperatures appeared cooler on the basis of isotopic analysis of foraminifera undertaken for the CLIMAP project (CLIMAP, 1976), but it now seems less likely that there were extreme temperature changes (Beck *et al.*, 1992; Tudhope *et al.*, 2001). Reefs at higher latitudes do not seem to have lagged behind tropical reefs in terms of Holocene colonisation of suitable substrates or growth to sea level; for example reefs in the Abrolhos Islands off the coast in Western Australia and Middleton and Elizabeth Reefs and Lord Howe Island in the southwestern Pacific, show a similar pattern of colonisation and vertical reef growth to more tropical reefs (Eisenhauer *et al.*, 1993; Collins *et al.*, 1997; Kennedy and Woodroffe, 2000).

5.3.3 Holocene reef growth and response to sea level

Modern reefs have evolved in response to Holocene sea-level changes. At least three patterns of reef response to changing sea level have been recognised in Holocene reefs: keep-up, catch-up and give-up (Neumann and Macintyre, 1985; Davies and Montaggioni, 1985). 'Keep-up' reefs are those that are able to track sea level as it rises; they maintain a rate of vertical growth that ensures that the reef crest remains at sea level. 'Catch-up' reefs are those where the vertical growth of the reef has lagged behind sea level, catching up with sea level either after it has stabilised, or slowed. They were established some time after the antecedent platform at depths of 10–20 m was inundated by sea-level rise. Nevertheless, reef growth, at a rate slower than that of sea-level rise, persisted and caught up with sea level later in the Holocene. 'Give-up' reefs are reefs which languish after an initial burst of growth, and in which

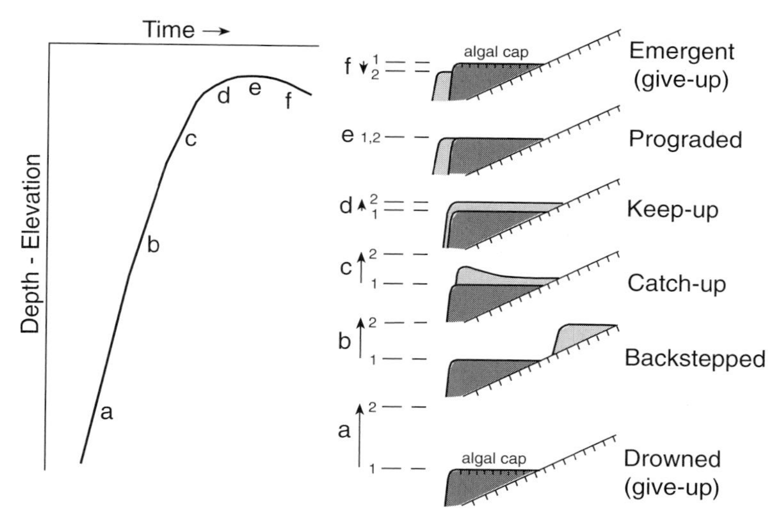

Figure 5.11. Modes of reef growth in response to relative sea-level change. The rate of sea-level rise determines whether or not reef growth can be achieved. The patterns of sea-level change most widely experienced in the Atlantic reef province and in the Indo West Pacific influenced the reef growth strategies (based on Neumann and Macintyre, 1985; Hubbard, 1997).

reef accretion slows, until the drowned reef persists only as a surface veneer at depths of 10 m or more (Figure 5.11).

Holocene sea-level change, as discussed in Chapter 2 (see Figure 2.11), has varied from place to place. The eustatic component of sea level was dominated by ice melt until around 6000 years BP; there has been little overall change in ocean volume since, but relative sea level has varied locally as the earth's crust and upper mantle have adjusted isostatically to the redistribution of mass. Local tectonic factors also mean that different localities have experienced their own individual relative sea-level history. Most coral reefs lie in the far field relative to former ice sheets and sea level reached a level close to present around 6000 years ago. This has been particularly true of the Indian and Pacific Oceans much of which has experienced a mid-Holocene sea level slightly above present, falling to present level in late Holocene. By contrast, the Caribbean Sea is in the intermediate field and experienced sea level rising at a decelerating rate up until present (Figure 2.11). These different sea-level histories explain some of the morphological differences between reefs in the Indo-Pacific and West Indian reef provinces (Adey, 1978; Davies and Montaggioni, 1985; Pirrazoli, 1991).

Figure 5.11 illustrates the way in which reef responses are related to sea-level change. The pattern is clearest for West Indian reefs experiencing sea-level rise at a decelerating rate. In the West Indies, the *Acropora palmata* reef crest has generally been able to 'keep up' with mid–late Holocene sea level. Reef accretion rates of up to 20 mm a^{-1} are interpreted, implying that reefs could keep up, or at least catch-up with sea-level rise of a similar magnitude; if the sea rises more quickly (particularly at rates of rise >45 mm a^{-1}), reefs are drowned and become give-up reefs (Blanchon and Shaw, 1995). Some reef areas dominated by massive corals or *Acropora cervicornis* lagged behind sea level and have adopted a 'catch-up' strategy (Neumann and Macintyre, 1985). 'Give-up' reefs occur throughout the Caribbean and west Atlantic Ocean expressed as terraces or forereef buttresses around islands, often in water depths of around 20 m. There has been some debate as to why these may have given up; reactivation of soil and runoff from the land appear to be factors in limiting their vertical growth (Lighty, 1977).

In the Indo-Pacific, sea level reached a level close to present around 6000 years BP. It is generally considered to have risen above that level, falling to its present level either gradually over the mid and late Holocene (Chappell, 1983a), or with a fall in late Holocene as seen in French Polynesia (Pirazzoli *et al.*, 1988). The nature of keep-up and catch-up reefs is not as well discriminated as in the Atlantic reef region because of the wider suite of corals and their greater environmental tolerances. Keep-up reefs are generally characterised by branching framework corals, whereas there are greater proportions of massive coral heads in catch-up reefs. Many reefs in the inner Great Barrier Reef appear to have kept up, whereas the outer reefs lagged behind, and caught up to sea level in the mid Holocene, reflecting hydro-isostatic flexure across the shelf (Chappell *et al.*, 1982). Give-up reefs in the Indo-Pacific, having accreted slowly if at all in late Holocene times, are represented by drowned banks at 20–30 m depth, such as the Chagos Bank in the central Indian Ocean (Stoddart, 1973b), and Saya de Malha (Guilcher, 1988).

The terms keep-up, catch-up and give-up have been largely applied to entire reef crests. However, it is also important to recognise that components of one reef can respond differently; for instance, whereas a reef crest might have kept up with sea level, a backreef area may have lagged behind and caught up later, and a forereef terrace on the same reef could have given up. In the Indo-Pacific province, a particularly significant type of give-up reef is the emerged reef flat which has grown up to a higher sea level, and subsequently given up as a result of a fall in sea level. On Tahiti, keep-up reefs occur on windward, catch-up on leeward, while reefs in muddy-patch-reef settings appear to have given up (Montaggioni, 1988). It is likely that different modes of reef growth, such as framework or detrital, have differing potential capacities to keep up. A sequence of reef responses is envisaged in Figure 5.11: (a) reefs are drowned at rapid rates of sea-level rise; (b) they back step, as appears to have occurred on West Indian reefs (Hubbard, 1997), when sea-level rise is slower; (c) they catch up as sea level slows and (d) keep up with moderate rates of sea-level rise that correspond to rates at which reefs can accumulate (see Section 5.4); (e) they prograde where there is no change in sea level; and (f) they give up when sea level falls. There can be other factors that lead to reefs giving up, but where there is a negative movement of sea level, reefs are left emergent, as has occurred throughout much of the Pacific (McLean and Woodroffe, 1994).

5.3.4 Reef growth and sedimentation

Processes of reef growth and sedimentation determine the surface morphology of the reef. In this section, reef growth will be described and then the production of sediment considered. It will then be shown that reefs are highly dependent on a series of environmental constraints. It is important to distinguish between the growth rates of individual corals and the rate at which reefs grow. Reef growth is an integrated process, involving consolidation of a framework constructed of the skeletal remains of individual corals infilled with sediment and bound together by many other organisms.

Reef growth

Reefs are almost entirely composed of skeletal carbonate sediments produced by marine plants and animals (biogenic), incorporated either as the rigid framework or as unconsolidated detrital sediments. The process of reef growth is complex, involving calcification by the organisms that contribute to the reef, various stages of breakdown, transport, redistribution and cementation. It is important to stress that reef growth is an aggregated (or integrated) consolidation of carbonate sediments, and is not the same as coral growth (the linear extension rate of individual corals).

The rate of coral calcification, and hence of coral growth, varies according to the growth form of the coral and environmental factors, such as water depth and temperature. Massive corals undergo skeletal extension at rates of 4–20 mm a^{-1}, whereas branching corals extend individual branches up to an order of magnitude faster, at 100–200 mm a^{-1} (Vaughan, 1916; Chave *et al.*, 1972). For instance, growth rates of the massive *Montastrea* reach a maximum of around 10 mm a^{-1} at a depth of about 10 m in St Croix, and then decrease to only about a quarter of that rate at 30–40 m (Hubbard and Scaturo, 1985). Corals at high latitude (25–30°) show lateral extension at similar, or only slightly

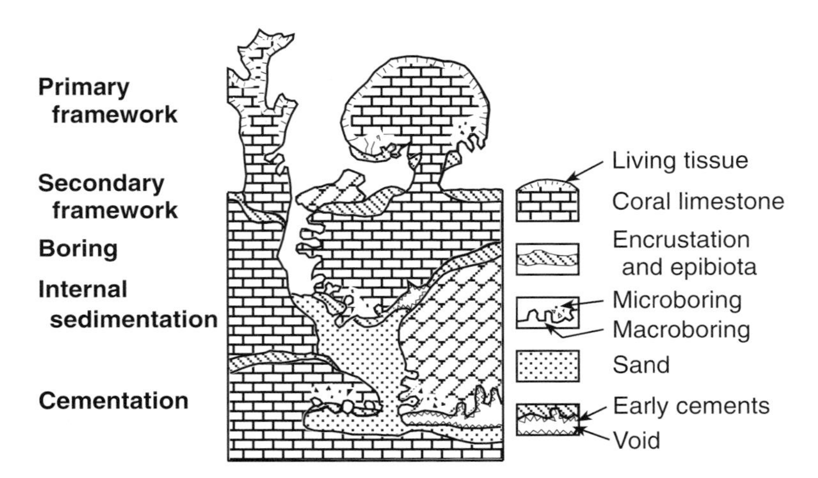

Figure 5.12. Reef growth consists of five processes which can be viewed in time, and also with increasing depth into the reef structure (based on Scoffin, 1987; Wood, 1999).

depressed, rates compared to their more tropical counterparts (Logan *et al.*, 1994; Harriott *et al.*, 1995). However, they may be stressed in terms of apportioning energy to calcification in these colder waters or because of a lower aragonite saturation level (Kleypas *et al.*, 1999).

Five processes are involved in reef growth and are shown schematically in Figure 5.12: primary reef framework growth, secondary reef growth, erosion by physical and biological processes, internal sedimentation and cementation (Scoffin, 1987; Wood, 1999). Primary reef framework is generally constructed by the scleractinian corals (hard corals of the family Scleractinia, see Chapter 2). Individual coral colonies are frequently modified by boring and breakage before being bound into a reef. A framework of coral growth, combining various generations of coral, forms the matrix within which other reef sediment accumulates. Secondary reef growth involves organisms that bind and encrust the primary framework, such as crustose coralline algae, and encrusting corals (such as *Agaricia* and *Millepora* in the Caribbean). In addition, bryozoans, bivalve molluscs, worms, foraminifera and sponges can play a role.

Erosion occurs by physical processes, including wave damage, and by biological processes, such as the actions of grazing and boring organisms. Bioerosion is a particularly significant process on reefs which serves to break down the primary and secondary framework and to provide the material for the next phase, that of internal sedimentation. Fine skeletal carbonate sediment is washed into the interstices of the reef, adding to the cohesiveness of the reef structure. Further sediment is provided by bioturbation and by the activities of cryptic organisms living within the reef. Finally, cementation is important in binding the reef together to preserve its structure. Aragonite or high magnesium calcite can be precipitated from seawater, filling cavities and cementing loose sediment. The more advanced stages of lithification and diagenesis (changing mineralogy) can occur over thousands of years, particularly after subaerial exposure of the reef (Moore, 1989). The cementation role is important in the long-term preservation of the reef.

The processes are shown schematically in Figure 5.12, based on studies of patch reef sedimentology (Scoffin *et al.*, 1980). The five processes may all be occurring contemporaneously within a section of reef, but are initiated in sequence. Primary framework is necessary for secondary framework builders to operate and, in turn, these stages must precede the erosion which provides material for internal sedimentation and subsequently cementation. It is also possible to view these processes occurring at progressively greater depths within the reef as illustrated in Figure 5.12. The primary framework might be the most prominent to the casual observer; the secondary encrusters and framework builders tend to veneer the surface, and often the undersurface, of the primary framework. The borers and grazers tend to operate at greater depth below the surface of the patch reef. Internal sedimentation occurs at a lower level again, with cementation only at the base of the reef (Scoffin, 1987).

There are several ways in which reef growth can be measured. The rates of productivity of corals and other individual reef organisms can be measured and a mass budget derived. However, these do not translate directly into reef accretion because of the complex sequence of coral breakdown, bioerosion and destruction by extreme events (especially storms), and the fact that the reef is the integrated end-product of the suite of processes described above (Scoffin *et al.*, 1980; Hubbard *et al.*, 1990).

Rates of reef growth have been determined on the basis of stratigraphic results and radiocarbon dating from shallow drilling through Holocene reefs. The dates (estimated age, generally time of death of the organisms that contributed to the reef) provide age control on primary framework. Coral material can also be incorporated into the reef in the form of detrital material, and some caution is needed in interpreting ages where there have been periods of transport and breakdown. Nevertherless, radiocarbon dating has enabled development of a series of morphodynamic models of reef geomorphology, some of which are examined in Section 5.5.

It is also possible to assess reef growth by measurement of the alkalinity of seawater and its change as a parcel of water passes over a reef. Termed the alkalinity anomaly or alkalinity depression method, it

Table 5.2. Calcification rates and potential for vertical growth determined from stratigraphic studies

	Calcification rate (kg $CaCO_3$ m^{-2} a^{-1})	Growth potential (mm a^{-1})
Outer slope	1.4	0.9
Algal rim	4.0	2.8
Reef flat coral zone	4.5	3.1
Sand flats	0.3	0.2
Lagoonal patch reef	1.5	1.0
Deep lagoon	0.5	0.3
Active reef (Kinsey)	10.0	7.0

Source: Based on Hopley, 1982.

involves the measurement of pH and total alkalinity upstream and downstream of a sampling site, with change represented by the removal of calcium from solution in the precipitation of calcium carbonate. This approach, which integrates photosynthesis and respiration and applies to the whole of a reef area, yields relatively uniform results for different reef habitats. Average values are up to 10 kg $CaCO_3$ m^{-2} a^{-1} for active reef, but more generally 4–5 kg $CaCO_3$ m^{-2} a^{-1} for reef flats, and less than 1 kg $CaCO_3$ m^{-2} a^{-1} for lagoonal areas (Smith and Kinsey, 1976).

The two approaches can be reconciled with consideration of specific gravity or density of reef habitats (aragonite 2.89 g cm^{-3}) and their porosity (40–60%) to convert one approach to the other, and results are summarised in Table 5.2. There is surprising agreement considering that the one measurement is based on water characteristics at very short (instantaneous) time scales, whereas the other is an integrated accumulation of material over centuries to millennia. Reef growth represents the total calcification of framebuilders, inorganic carbonate precipitation, local erosive regime, and a reflection of the accommodation space into which the reef is able to accrete. The significance of the latter point will be examined in more detail when specific reef types are examined in Section 5.5.

Reefs are not just framework structures built *in situ*, but contain a series of landforms that are constructed as a result of the erosion, transport and deposition of calcareous sediment. These processes of carbonate sedimentation are examined in the following section.

5.4 Processes on reefs

Coral reefs build large complex wave-resistant structures; the growth of reefs and the redistribution of the sediments produced on reefs modifies the rate at which processes operate (McManus, 2001). As indicated in Figure 5.4, reef morphology and ecology respond to a series of factors that vary across a reef, along a reef, or from one reef to another. In this section, the production, entrainment and deposition of sediment is examined in greater detail. The effects of wave and tide processes, and the transformation of these within and across reefs is considered. In some cases, as for instance in the Hawaiian Islands, large swell that has crossed a wide stretch of ocean impinges on the reef. In other cases, for example on the Great Barrier Reef, waves are derived locally, generated in the adjacent Coral Sea or on the shelf itself, and respond rapidly to shifts in wind direction. Many reefs are subject to the effects of extreme events, such as tropical cyclones, and the role of these high-magnitude events is also considered.

5.4.1 Sediment production on reefs

Sediment production, transport and deposition on reefs is closely related to the physical processes and the community of benthic organisms. An individual reef often shows complex zonation of landforms and organisms across its surface, as indicated in Section 5.2, responding to environmental factors. The biological component adds additional ecological complexity; reefs in apparently similar settings can be dominated by different benthic coral or algal communities. The relative dominance could relate to nutrient status and herbivorous grazing pressure. Corals dominate in low nutrient conditions (typically nitrate $<2.0\ \mu mol\ l^{-1}$, phosphate $<0.2\ \mu mol\ l^{-1}$), but require a high level of grazing by herbivores to control microalgae. Corals are replaced by turf algae where grazing pressure is low, by crustose coralline algae where nutrient concentrations are high, and by frondose macroalgae where nutrients are high and grazing pressure low (Littler *et al.*, 1991). Nutrients are unlikely to limit reefs that are adjacent to a landmass where nutrient runoff is large, or anthropogenically increased.

Waves and currents transport sediment from its point of origin to other parts of the reef, or remove it from the reef altogether. Calcification and production of sediment are greatest on the reef crest and reef front. These calcareous materials are transported by waves or wave-generated currents into backreef environments. Some material, particularly coarse rubble, is lost seaward down the reef front; for example, loss of detritus down the windward reef front on Lizard Island in the Great Barrier Reef occurs at an average annual rate of $2\ kg\ m^{-1}$ (Hughes, 1999).

The supply of sediments (and hence the rate of lagoonal accretion) increases as the area of productive reef rim increases (Tudhope, 1989).

Sediments are finer on the leeward margin and some material can be carried out of the reef setting over the leeward reef if hydrodynamic conditions are appropriate (Frith, 1983). Lagoon sediments can either be produced locally by organisms within the lagoonal habitats (*in situ* or autochthonous production) or transported in by currents (allochthonous). In fringing reefs where backreef environments comprise solid reef flats but no lagoon, sediment may be carried directly across the reef flat and accumulated on the shore. A sequence of ridges of coral boulders form in high-energy and storm-prone settings, whereas sand cays develop in lower-energy environments.

The type of biological components in the sediment influences sediment size (see Chapter 2). Although enormous boulders of reef limestone, termed reef blocks, can be deposited on the reef flat (see Figure 5.15), the coarsest material is generally coral boulders and shingle which can form ramparts and construct islands on the reef top (see Figure 5.27). Coral shingle (fine gravel) composed of sticks of *Acropora* is a significant component of reef sediments, although individual clasts break up through a series of discrete sediment size classes and contribute to sand and grit (see Chapter 2, Figure 2.15). Similarly, other organisms, such as the algae *Halimeda* and the tests of foraminifera, contribute directly to medium sand and may constitute more than 70% of reef sediments on some Pacific reefs (Weber and Woodhead, 1972).

Reef sediments are distinctive in terms of mineralogy (see Table 2.5), and the distinctive grain shapes and densities influence the way in which the grains respond to hydrodynamic forces, in contrast to clastic sediments which show less variability. Reef sediments are often texturally immature (not well-sorted), their grain-size characteristics being dependent on the contributing organisms. Fine-grained material is winnowed preferentially by 'currents of removal', and lag deposits of coarser material remain, as within boulder zones on the reef flat (Kench and McLean, 1996, 1997). By contrast, lagoonal areas which receive input of sediment from the reef front by 'currents of delivery' are more likely to be texturally mature with their size characteristics related to hydrodynamic conditions rather than the organisms from which the sediments are derived (Orme, 1973). Sediments often show polymodal distributions relating to transport by traction, saltation or suspension (Flood and Scoffin, 1978).

The rate at which sediment is produced depends on the life history of contributing organisms, and its persistence depends on biological and chemical attrition. *Halimeda* is highly productive and produces platelets rapidly. However, *Halimeda* grains are not durable, they rapidly break down into fine dust (10 phi) of aragonite microcrystals, in contrast to molluscan fragments and other robust components. Mud occurs on coral reefs as a result of mechanical attrition of coarser grains, biological breakdown of substrates, or physiochemical precipitation (Milliman, 1974; Adjas *et al.*, 1990). Bioerosion is particularly important. Sediment chips formed by the boring sponge *Cliona* are a major contributor to mud fractions on many reefs. Parrotfish (scarids) also have major impacts, causing direct erosion when they grind the surface of coral, breaking down material into smaller size groups, and also transporting it across the reef (Bellwood, 1995; Bruggeman *et al.*, 1996).

On a fringing reef in western Barbados, annual production of sediment from a range of sources has been shown to be about 10.7 kg m^2 a^{-1}, but bioerosion by sea urchins and parrotfish removes around 5.3 kg m^2 a^{-1} (Stearn *et al.*, 1977; Scoffin *et al.*, 1980). Similarly, on St Croix, about 60% of the reef that is formed is removed as a result of bioerosion, and much of the material that remains is later lost as a result of storms (Hubbard *et al.*, 1990). Organisms further modify the sediment by bioturbation. Borers, etchers, grazers and predators can have further effects on the sediment that is deposited; for example, holuthurians (sea cucumbers) continually ingest sediment and excrete faecal pellets. Callianassid shrimps are very significant in reworking sandy sediments, burying coarse fragments and resuspending fine material (Tudhope and Scoffin, 1984; Tudhope and Risk, 1985). Biota also play a role in stabilising sediments (Scoffin, 1970). Seagrass and algae form substrates on which other organisms live (epibionts) that produce further calcareous skeletal grains.

5.4.2 Wave and tide processes

Reefs grow best where wave energy is high. Reefs in tradewind settings receive persistent swell waves, and mid-ocean reefs can be affected by swell that has crossed the ocean. There are also reefs that are subject to local wind-generated waves on which wave periods are shorter and wave directions are more variable. In areas subject to monsoonal reversal, the direction of wind and wave occurrence is often seasonal.

Reefs are extremely effective at attenuating wave energy, and sheltered conditions are experienced in the backreef. It has been shown from measurements on Grand Cayman Island in the West Indies and the Great Barrier Reef that windward reefs reduce open-water wave energy by 70–90% (Roberts *et al.*, 1975, 1977; Hopley, 1982; Wolanski, 1994). Loss of wave energy occurs primarily as a result of waves breaking on the reef crest. Processes such as shoaling, refraction and wave scattering are relatively minor because most mid-ocean reefs rise steeply from deep water, resulting in little frictional attenuation until the spur and groove of the reef front. There appears to be virtually no reflection of wave energy from the reef crest, although the headward end of grooves on the reef front could reflect some energy.

Waves break on the reef crest and lose a significant proportion of their energy (Massel and Gourlay, 2000). The nature and shape of the reef edge effects the dissipation of wave energy. Wave set-up is significant (Munk and Sargent, 1954), and water levels can be raised by 20% of incident wave height as a result of breaking (Gourlay, 1996a,b). Waves propagate across shallow reef flats as bores, losing energy through turbulence (Gallagher, 1972), or reform into a range of shorter period secondary waves (Young, 1989). Relatively little research has examined the mode of energy dissipation across shallow lagoon or reef flats, or the pattern of sediment movement. Wave refraction is presumably important in determining sediment movement patterns in these backreef areas (Gourlay, 1994). Complex refraction patterns appear to result in concentration of sand cays on the leeward end of platform reefs (Davies *et al.*, 1976; Hopley, 1981; Flood and Heatwole, 1986).

Although it is convenient to envisage the 1-dimensional interaction of waves and reefs, this is not realistic for large reef regions where there are complex 2-dimensional patterns involving refraction and reformation of waves through the shallow-water lagoonal areas. On the Great Barrier Reef, individual reefs within the lagoon focus wave energy onto them, waves are refracted and break on the leeward sides of the reef, lowering wave height for significant distances in the between-reef areas (shadow) to their lee (Symonds *et al.*, 1995). Waves generated in the Coral Sea with frequencies of 7–10 s are filtered, and on the mainland locally generated waves are dominant and characteristically have frequencies of 8 s and 4 s (Wolanski, 1994).

Substantial differences in hydrodynamics, sediment movement and, consequently, geomorphological evolution are likely on different reefs depending on the degree of system closure. It is possible to distinguish unbounded reefs, land-bounded reefs and reef-bounded reef systems (Hatcher, 1997). Unbounded reefs are subject to almost unimpeded open-ocean water flow across them and are characterised by short residence time. For example, on the outer windward 'ribbon' reefs of the Great Barrier Reef, the eastern margin has a prominent spur and groove; there is a reef top of only around 400 m width, and leeward margin consisting of a detrital slope of sand. These ribbon reefs are dominated by flow across them and currents are unidirectional, with flow over the reef crests into the lagoon at all stages of the tide. Land-bounded reefs, such as fringing reefs, are influenced by the barrier formed by the landmass. Wave set-up can be significant in terms of piling water into the backreef or shore (Tait, 1972), resulting in retention of sediment in the backreef, unless there is a deeper channel through which water and sediment can escape. Reef-bounded systems, of which atolls with extensive reef-island development on their rims are the most extreme example, are characterised by longer water residence time leading to greater environmental extremes of high temperature, salinity, and the greatest degree of internal recycling and retention of sediment (Hatcher, 1997).

The pattern of wave energy typically received around the shoreline of Grand Cayman Island in the West Indies is shown in Figure 5.13. The reef morphology varies in relation to wave power. The higher-energy shore has closely spaced and well-developed coral spurs and deep grooves both on the upper and lower forereef terraces. The lower-energy coast has more widely spaced and poorly developed topography (Roberts, 1974). Wave processes are important on the shallow reef front, but currents are more significant on the forereef (Figure 5.13b). Currents of 0.071 m s^{-1} are experienced on high tide and 0.173 m s^{-1} on low tide (Roberts *et al.*, 1977). The wave-dominated zone is characterised by robust *Acropora palmata* and other bladed and encrusting corals, whereas the current-dominated zone is characterised by delicate *Acropora cervicornis* and plate-like *Montastrea*. A large percentage of wave energy is dissipated as the waves cross the fringing reef into a shallow lagoon, and circulation in the lagoon is also driven by influx over the reef crest, with net efflux from deeper passages (Roberts *et al.*, 1992).

Significant currents can result from tidal variations in backreef environments. This is illustrated in Figure 5.14 in the case of the Cocos (Keeling) Islands, an atoll in the eastern Indian Ocean. The atoll forms

Figure 5.13. Wave and current energy and morphology around Grand Cayman Island, West Indies. (a) The distribution of wave power around Grand Cayman. Arrow indicates field site. (b) The relative importance of wave and current processes in relation to fringing reefs. (c) Typical forereef morphology (based on Roberts *et al.*, 1992).

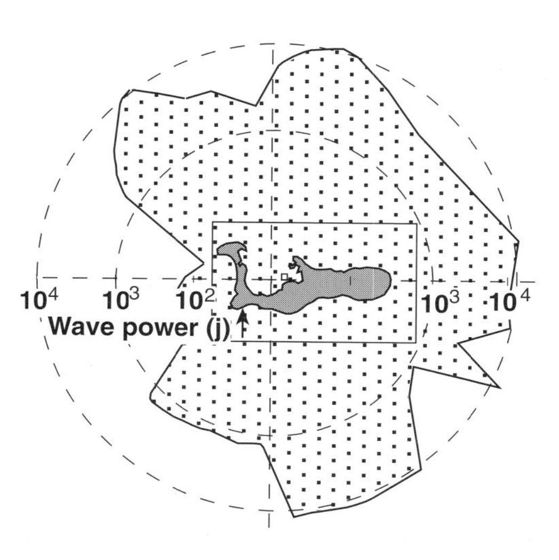

b) Current and wave energy

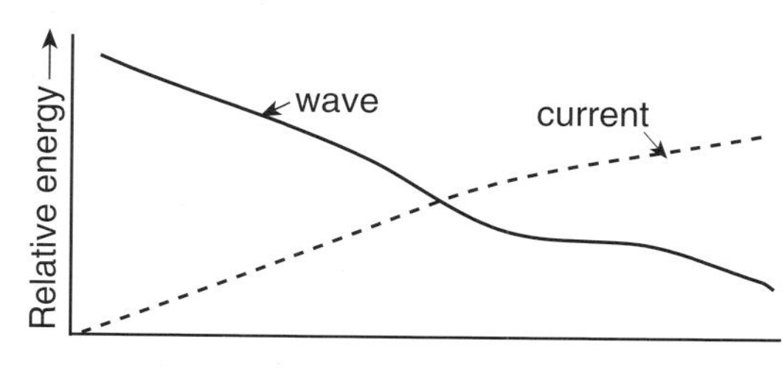

c) Forereef terrace

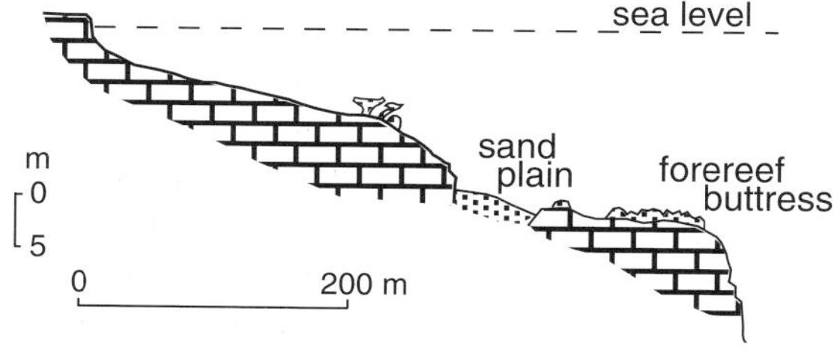

a) Cocos (Keeling) Islands

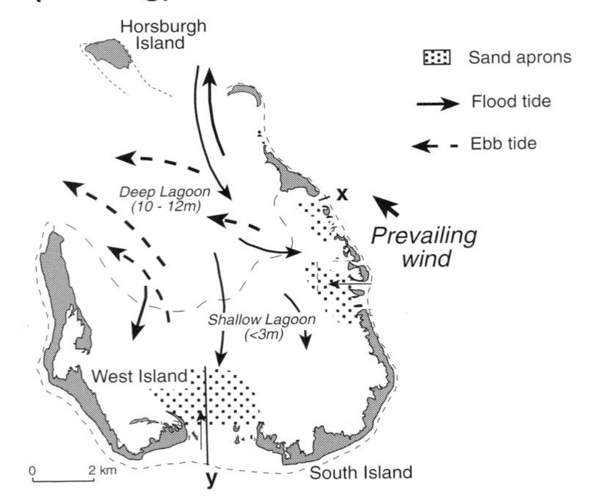

b) Transect x

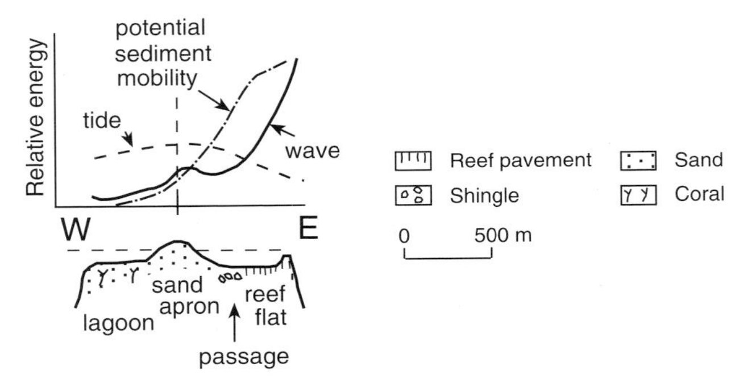

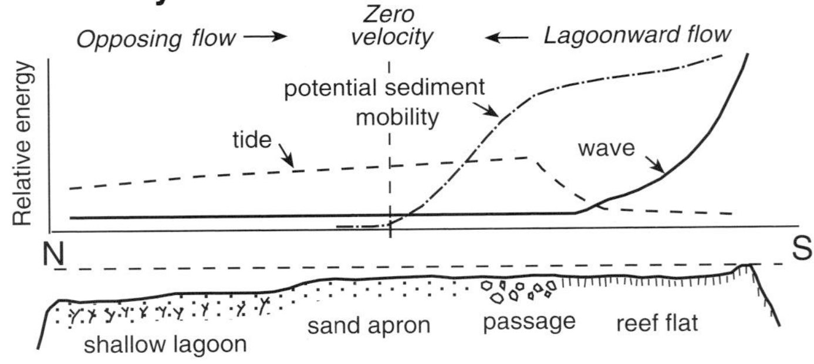

Figure 5.14. The distribution of wave and tidal forces across shallow-water environments of the Cocos (Keeling) Islands, Indian Ocean. (a) The atoll and the dominant flood and ebb tidal currents. (b) Transect x through an inter-island passage on the eastern margin, and (c) transect y on the southern margin of the atoll, and the relative importance of different processes and their potential to move sediment (based on Kench, 1998).

a horseshoe-shaped reef rim with deep passages to the northwest and northeast, but otherwise occupied by a series of 26 reef islands enclosing a lagoon of about 100 km^2. The lagoon comprises a shallower southern part in which there are reticulate reefs characterised by numerous deep 'blue-holes' to around 15 m depth, and a deeper northern section. Situated in the tradewind belt, the atoll receives persistent swell from the southeast for most of the year with modal wave height of 2 m. Currents in the lagoon are wave-induced, but are driven predominantly by tides. The inter-island passages at Cocos (see Figure 5.25) are flushed by tidally driven currents that are almost entirely unidirectional, from reef crest into the lagoon. The currents reverse only on the lowest spring tides through the eastern passages where the reef flat is not as high as to the south of the atoll (Kench, 1998).

Currents at Cocos reach speeds of up to 1.4 m s^{-1} on the east and 0.8 m s^{-1} on the south. These currents rapidly decelerate into the lagoon, and the lagoon itself is tidally flushed by circulation through the northern passages. It has been shown that the wave-induced currents through the passages are effective at moving sediment. However, this competency rapidly declines and sand is actively accreting in sand aprons on the eastern margin, and in a sand flat that progrades into the lagoon at the southern end of the atoll (Kench, 1997). As in the Great Barrier Reef, wave-induced frequencies of 8–9 s and 3–4 s are dominant. Energy is dissipated rapidly on the reef crest; of that energy which gets onto the reef flat, 76% is dissipated on the reef flat, and a further 68% is dissipated in the passages on the eastern margin, with values of 73% and 42% respectively on the southern reef flat (Kench, 1998). The balance between wave and tide energy is shown in Figure 5.14. The circulation of this relatively simple atoll is complex. The pattern of sediment build up cannot be stable in the long term as the evolving morphology alters the rates of operation of the main processes. Major storms may have a disproportionate impact in redistributing sediment, although neither Cyclone Frederick in 1988, nor Cyclone Graham in 1991 (resulting in a three to four times increase in sediment movement), had pronounced impacts on the atoll geomorphology (Kench, 1998).

Circulation patterns have been described from several other reef settings. Different factors have been identified as driving the circulation, with varying consequences for patterns of sediment movement. Circulation across open-ocean atoll lagoons can be driven primarily by wave set-up (von Arx, 1948). On fringing reefs around Kauai in the Hawaiian Islands, wave-driven currents lead to movement of sand shoreward in grooves in the reef flat surface. Accumulation of sediment on the reef top occurs because reef-flat morphology prevents significant lateral movement into the major channel through which water leaves the reef (Inman *et al.*, 1963). A similar pattern of circulation has been simulated for Great Pond Bay, St Croix, where currents drive water that comes across the reef out through a reef passage to the west of the lagoon (Prager, 1991). In contrast, circulation of a similar fringing reef on Ishigaki Island, Ryukyu Islands, Japan is driven primarily by winds (Yamano *et al.*, 1998). Much remains to be learnt about the nature of processes at the reef crest, and the consequences for water movement within a coral reef lagoon. There are similar circulation patterns on different reefs, but they can be variously driven by tides, wind or over-the-reef, wave-induced flow, or some combination of these.

5.4.3 Extreme events and disturbances

Reefs are extremely dynamic. Physical and ecological processes continually interact to influence reef morphology, and morphology induces feedback in terms of reef growth and sedimentation processes. Reefs appear to have a particular capacity to turn-on and turn-off, and it is not always easy to establish the cause (Buddemeier and Hopley, 1988). The reef system is subject to major disturbances, including physical stresses, such as hurricanes, floods, earthquakes, and low tides which expose reef organisms, and biological stresses such as predator outbreaks, pathogens and more subtle responses to factors which upset the ecosystem balance (Brown, 1997a).

There was a view that reefs, in common with other tropical ecosystems, were diverse because they remained stable for a long time, enabling the development of complex competitive and mutualistic species interactions. It is now recognised that many reefs undergo frequent disturbances, and some are in an almost continual state of disturbance or recovery from previous disturbances. Some reefs are adapted to such disturbances, and the greatest diversity can in some cases be associated with reef systems that undergo frequent disruption (Connell, 1978; Hughes and Connell, 1999).

Many reefs are impacted periodically by tropical cyclones (hurricanes, typhoons), which occur across the northwest and southwest Pacific, including the reefs of northern Australia, northern and southern parts of the Indian Ocean, and the Caribbean Sea (see Chapter 3, Figure 3.11). Impacts depend on the intensity and duration of the cyclone and the condition of the system when it strikes. Not all cyclones have the same effect on a reef. For instance, Hurricane Andrew (1992), although accompanied by very high wind speeds (up to 250 km h^{-1}), had relatively little effect on reef environments in northeastern Great Bahama Bank (Boss and Neumann, 1993). Where a reef is struck by two cyclones in succession, the second has different effects to the first. For example, comparison of Hurricanes Donna (1960) and Betsy (1965) that hit Florida Bay, or Hurricanes Gilbert (1988) and Hugo (1989) that hit the US Virgin Islands, demonstrated that although the paths were similar the impact of the second storm was different because the coast had not fully recovered from the first storm (Witman, 1992). In the Tuamotus in French Polynesia, the impact of a severe storm, after several decades that had been storm-free, was particularly devastating (Harmelin-Vivien and Laboute, 1986).

Large reef blocks, up to 20 m or more long, can be moved by these high magnitude events (Bourrouilh-Le Jan and Talandier, 1985). Figure 5.15 shows a reef block on the atoll of Suwarrow, where it constitutes the highest point on the atoll. The role of storms is hard to define, but storm deposits are known to persist in the reef environment (Scoffin, 1993). Many of the rubble ridges around Grand Cayman Island are composed of boulders up to 1 m in diameter that have been derived from 10–12 m water depths (Hernandez-Avila *et al.*, 1977). It has been suggested that the reef crests may also have been constructed by storm-derived material (Blanchon *et al.*, 1997, see also Section 5.5.1).

Tropical cyclone events deliver sediment into the system, ripping much of it from the reef front (Blumenstock, 1961). Often the coarser material is deposited on the reef top, not far from the reef crest. A good example was Cyclone Bebe which struck Funafuti (Tuvalu) in 1972, and

Figure 5.15. A reef block on the rim of Suwarrow atoll in the northern Cook Islands, Pacific. This block of reef limestone was ripped from the reef rim presumably by a tropical cyclone. This atoll was struck by a severe cyclone in 1942, which may have been responsible.

deposited a ridge or rampart of rubble along the southeastern part of the reef flat (see Figure 5.27), adding the equivalent of 10% to the area of land on the atoll (Maragos *et al.*, 1973). This material was redistributed over the decades that followed (see Section 5.5.3).

The intensity of destruction to reef environments is related to the distance away from the eye of the storm (Stoddart, 1971). Storms play a role in shaping coral communities, with protracted periods between storms required for establishment and growth of large coral colonies (Massel and Done, 1993). Rubble ridges and ramparts are common on the windward-facing margins of low wooded islands of the Great Barrier Reef. These ridges have steep inner faces and intermittent shingle tongues at right angles to the reef crest on the windward margin and at a steeper angle around the periphery (Stoddart *et al.*, 1978).

Tropical cyclones with wind speeds of 120–150 km h^{-1} cause effects that can be undetectable after only a decade or so. Tropical cyclones of the intensity of Hurricane Hattie, with wind speeds of 200 km h^{-1} take much longer to heal, perhaps around 50 years (Stoddart, 1985; Scoffin, 1993). Recovery from the impact of tropical cyclones is likely to depend on the severity of the damage and the intensity and duration of the storm; for instance, where shingle islands (motu) are hit by storms, the removal of vegetation may render the islands more susceptible to mobilisation under moderate energy events which would otherwise have had no impact (Nunn, 1994).

Reefs are subject to other physical disturbances, such as earthquakes (Stoddart, 1972). Movement of large reef blocks on the outer reef crest of the Great Barrier Reef, and imbrication of rubble inshore has been attributed to tsunami (Nott, 1997). Disease, such as the white band disease and predators can also impact reefs. The crown-of-thorns starfish, *Acanthaster planci*, has been widespread throughout Indo-Pacific reefs since the 1970s. It feeds on live coral by smothering and ingesting living tissue. Initially, infestation by outbreaks of *Acanthaster* was explained as due to human over-collection of the triton shell, *Charonia*, which preys on the starfish. Alternatively, infestations have also been considered as natural outbreaks, with evidence of past outbreaks preserved in the sediments around reefs (Frankel, 1977). Although the causes of recent outbreaks remain contentious, there seems to have been some recovery of infested reefs (Done, 1999). Mass mortality of a herbivorous Caribbean echinoderm, *Diadema antillarum*, in 1982–84 also had alarming impacts on coral reefs, leading to a protracted algal bloom. Impacts were especially noticeable in hurricane-affected areas (Hughes, 1994). Whereas the reef stratigraphy can preserve evidence of past storm impacts, other extreme events, such as outbreaks of *Diadema*, are generally not preserved in the sedimentary record (Greenstein, 1989). In some cases, it appears that multiple stable states are possible and that perturbations serve to flip the system from one state to another – termed a phase shift (Done, 1999). In other cases, it may be multiple stresses that exert an impact on the reef and cause it to switch from one state to another (Hughes, 1994; Hatcher, 1997; Hughes and Connell, 1999). It will become increasingly important in geomorphological studies of reef systems to attempt to decipher the causes of phase shifts when considering human impacts and their management implications.

Corals respond to some stresses and natural disturbances by expelling their symbiotic algae, which is called 'bleaching'. A series of coral bleaching events were reported in the 1980s (Glynn, 1984). Global warming has led to increased sea-surface temperatures and more frequent and persistent coral bleaching which appears to be threatening the survival of coral reefs in some equatorial locations (Hoegh-Guldberg, 1999; Lough, 2000). Recovery from stress seems to be interlinked with response to further stresses (Brown, 1997a,b). Coral from an area of the Caribbean little impacted by human effects has been observed to have been devastated by bleaching but was then not heavily impacted when hit by a subsequent hurricane (Ostrander *et al.*, 2000). The impact that human-induced global warming might be having on the bleaching of coral reefs is also of major concern, although it appears that the physiological responses of coral are more complex than suggested by visual surveys (Fitt *et al.*, 2001).

5.5 Morphodynamic models of reefs

Coral reefs are especially amenable to geomorphological reconstruction because their past form is generally preserved in the evolving reef structure and can be examined using techniques such as bathymetry, seismic profiling and coring. Much of the carbonate skeletal material can be aged using radiometric dating techniques. Consequently, reefs have been investigated to derive a record of past environmental changes, including sea level and sea-surface temperature history at a range of time scales. This has also enabled the development of morphodynamic interpretations of the way in which reefs and associated landforms have evolved. Several morphodynamic models are outlined in this section, starting with fringing reefs, extending the concepts to barrier and shelf reefs, and finally examining the surface morphology of atolls. Most of the studies involve the integration of processes that are observed at event scales to develop models over time scales of decades to millennia.

5.5.1 Fringing reef development

Fringing reefs are simple reefs being close to shore or shore-attached. The contribution of sediment from the land is generally small in comparison with the rapid production of calcareous reefal material. However, terrestrial influences, particularly the runoff of freshwater, nutrients and sediment, can limit reef development. A fundamental distinction can be made between fringing reefs in which the backreef contains a reef flat, and those which comprise shallow lagoons. The prime control on this distinction appears to be the different sea-level history that has been experienced (Adey, 1978). Reef flats are widespread in the Indo-Pacific region as a result of Holocene emergence, whereas shallow lagoons are more common in the Caribbean (see Figure 5.1), but shallow lagoons can also occur where reef growth has adopted a different mode of development as will be shown below.

Modern fringing reefs have adopted their present form in the Holocene and preserve a record of past morphology within the calcareous sediments that form a relatively thin, seaward-thickening wedge over non-reefal topography. Whereas antecedent Pleistocene topography, particularly Last Interglacial reef limestone, frequently forms the base on which atolls, barrier reefs, and shelf reefs are established (see Figure 5.10), it is less commonly a substrate for the development of fringing reefs.

Detailed studies of fringing reefs at Galeta Point, Panama, in the Caribbean (Macintyre and Glynn, 1976) and Hanauma Bay, Oahu, Hawaii, in the Pacific (Easton and Olson, 1976) produced remarkably similar chronologies of reef growth despite different sea-level histories (Figure 5.16). Further morphodynamic studies indicate that at least six models of fringing reef development can be identified, based on stratigraphy and chronology (Kennedy and Woodroffe, 2002). These are described below: first the classic Panamanian and Hawaiian studies are described, then the evidence, particularly from fringing reefs within the Great Barrier Reef, but also elsewhere, is synthesised into the six models.

Hanauma Bay and Galeta Point

The classic studies of fringing reefs at Galeta Point, Panama (Macintyre and Glynn, 1976), and Hanauma Bay, Oahu, in the Hawaiian Islands (Easton and Olson, 1976) involved detailed stratigraphy and dating, enabling reefal age structure to be effectively reconstructed and isochrons drawn (Figure 5.16).

The Galeta Point fringing reef is 200 m wide. Macroalgae and seagrass presently dominate the surface of the reef, with mangroves to landward, and the main carbonate secreting organisms are coralline

a) Galeta Point, Panama

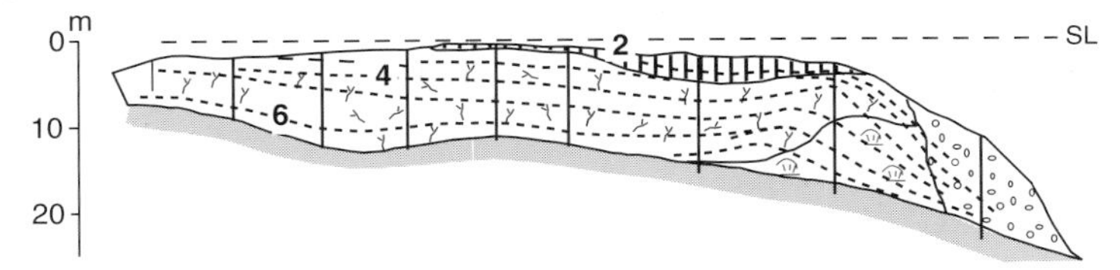

b) Hanauma Bay, Oahu, Hawaii

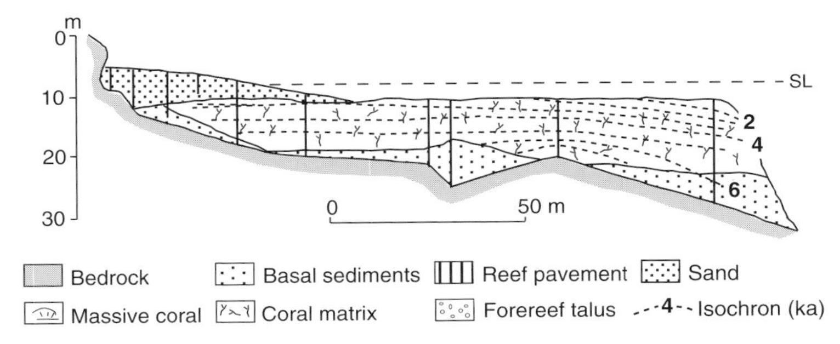

Figure 5.16. The stratigraphy of fringing reefs. (a) Galeta Point, Caribbean coast of Panama (based on Macintyre and Glynn, 1976), and (b) Hanauma Bay, Oahu, Hawaiian Islands, and isochrons of reef development (based on Easton and Olson, 1976).

algae and *Halimeda*. The stratigraphy of the outer part of the reef, examined in 13 drillholes, comprises Holocene reef up to 14.3 m thick established over a Miocene calcareous siltstone (Macintyre and Glynn, 1976). The subsurface facies of the reef are dominated by the staghorn coral, *Acropora palmata*, and massive corals. Reef growth at Galeta Point commenced around 7000 years BP, dominated by coral heads with a rate of accretion of around 3.9 mm a^{-1}. From 6000 to 3000 years BP, *Acropora palmata* dominated the reef, accreting at a rate of between 1.3 mm and 10.8 mm a^{-1}, masking the antecedent topography (Figure 5.16a). Coral heads and detritus replaced the *Acropora* after 3000 years BP with the modern reef-flat facies developing at around 2000 years BP. Little accretion appears to have occurred after 2000 years BP.

The fringing reef in Hanauma Bay, a sheltered breached volcanic crater, subject to low wave energy, is only about 90 m wide (Figure 5.17). It is presently dominated by algae, especially *Porolithon*, although growth of the coral *Porites* appears to have dominated the deeper portions of the reef. The reef also established about 7000 years BP. The early stages of reef growth between 7000 and 5700 years BP showed most rapid vertical accretion on the fringing reef crest. The main phase of growth occurred between 5700 and 3500 years BP, vertically accreting at a rate of 2.9 mm a^{-1}, masking the antecedent topography. From 3500 years BP to present only 1 m has been added vertically; however, the reef has prograded seaward (Figure 5.16b).

The pattern of reef accretion appears very similar between these two fringing reefs despite the areas having experienced a different sea-level history. Sea-level history in eastern Panama, as elsewhere in the Caribbean (Lighty *et al.*, 1982), was characterised by rising, but decelerating, sea level (see Figure 2.11). The Galeta Point reef adopted a keep-up growth strategy. A similar interpretation was proposed for the Hanauma Bay reef by Easton and Olson (1976) who rejected a suggestion by Stearns (1935) that sea level had been higher than present in mid Holocene on Oahu. Easton and Olson compiled a sea-level curve based on the dated coral from Hanauma Bay reef, but their curve recorded reef growth rather than sea level (Montaggioni, 1988). More recently, evidence for mid-Holocene sea level higher than present throughout the larger Hawaiian Islands has been reaffirmed (Fletcher and Jones, 1996; Grossman and Fletcher, 1998; Grossman *et al.*, 1998). Reef growth at Hanauma Bay has therefore been reinterpreted to have adopted a catch-up reef-growth history (Grigg, 1998).

Other fringing reef models

Studies of other fringing reefs, particularly on the Great Barrier Reef (see Figure 5.22), show that several different models of fringing reef evo-

Figure 5.17. Hanauma Bay, Oahu, Hawaii. The reef is about 90 m wide. The bay is formed in tuff. A bench occurs just above sea level along the tuff to the left of the picture.

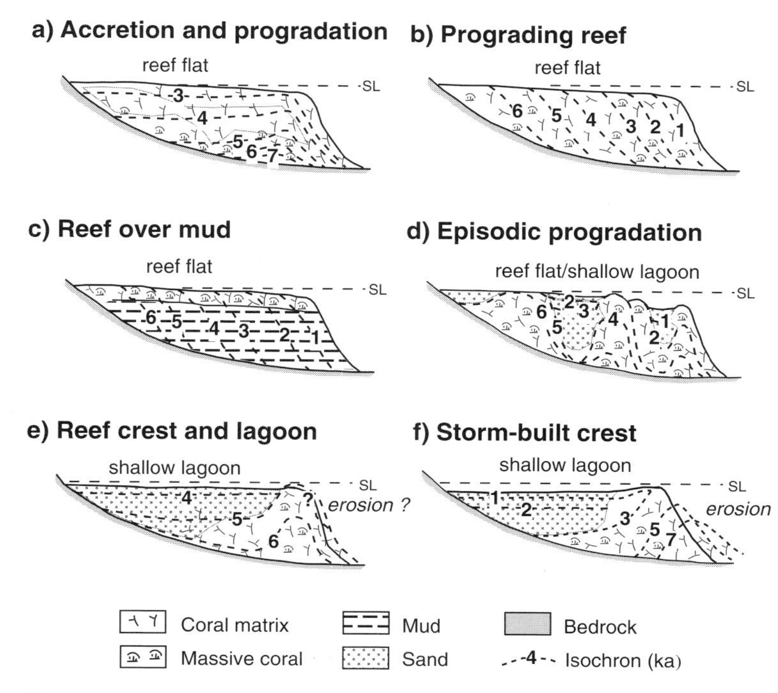

Figure 5.18. Six models of fringing reef development, showing the pattern of isochrons (in ka) and reef composition (based on Kennedy and Woodroffe, 2002). See text for details.

lution are possible. Figure 5.18 shows six models that characterise fringing reef development based on isochrons that are drawn with respect to radiocarbon dating of reef material (Kennedy and Woodroffe, 2002). The first model, based on Hanauma Bay and Galeta Point, indicates initiation of reef growth around 7000 years BP towards the modern reef margin, and more rapid vertical growth at the reef crest. In this model, a shallow lagoon formed but became infilled as the backreef caught up with sea level. Of the fringing reefs studied on the Great Barrier Reef, only the reef on Hayman Island seems to have adopted this pattern of growth, with more rapid vertical accretion at the reef crest, reaching sea level by 4000 years BP (Hopley *et al.*, 1978).

The second model of reef growth is one in which the reef builds out parallel to the reef front (Figure 5.18b), and can be seen in several reefs on the Great Barrier Reef. For example, fringing reef growth on Orpheus Island began around 7000 years BP over a pre-Holocene surface at a depth of about 6 m. The inner reef flat had reached sea level by 6250 years BP and by 5500 years BP the reef had commenced progradation parallel with the reef front (Hopley and Barnes, 1985). Rates of seaward progradation have averaged 50–100 mm a^{-1} on the leeward side and 500 mm a^{-1} on the windward side of the island (Hopley *et al.*, 1983). The reef on Fantome Island has prograded in a similar manner, with older sediments being exposed at the surface of the reef as a result of mid-Holocene sea-level fall (Johnson and Risk, 1987).

Fringing reefs around many rocky islands on the inner shelf of the Great Barrier Reef and along the mainland coast have prograded over muddy substrates, adopting the third model (Figure 5.18c). At Cape Tribulation, north of Cairns, reefs comprise a shallow reefal veneer over detrital sand and rubble (Hopley, 1994). Progradation of similar reefs over terrigenous mud as sea level has fallen has been demonstrated by radiocarbon dating of sequences of microatolls along much of the inner part of the northern Great Barrier Reef (Chappell, 1983a; Chappell *et al.*, 1983). On many reefs around islands on the inner Great Barrier Reef and in Torres Strait, the upper 1–2 m of reef has prograded over terrigenous mud (Kleypas and Hopley, 1993; Woodroffe *et al.*, 2000). Many inshore reefs throughout southeast Asia have undergone similar lateral fringing reef progradation over mud; for example, fringing reefs at Phuket, Thailand are composed of a framework of *Porites* corals, which have prograded seaward over mud through a process of block toppling as sea level has fallen during the mid–late Holocene (Tudhope and Scoffin, 1994; Scoffin and Le Tissier, 1998).

A fourth model involves episodic reef growth and occurs where prior topography exerts a control or some other factor causes the reef periodically to build seaward (Figure 5.18d). The modern reef is characterised by a sequence of linear shore-parallel reefs becoming younger seaward, with the reef flat between them infilled with unconsolidated reef-derived sediment. This pattern of development was inferred for the tide-dominated Torres Reefs on the basis of seismic reflection profiling (Jones, 1995) and is demonstrated by radiocarbon dating of microatolls and subsurface corals on Yam Island, also in Torres Strait (Woodroffe *et al.*, 2000). On Mangaia in the Cook Islands, emergent reef flat contains an inner fossil reef crest and an outer younger reef crest, each with its own backreef area infilled with unconsolidated sediment (Yonekura *et al.*, 1988). A similar morphology and mode of development is reported in the Mariana Islands (Kayanne *et al.*, 1988).

The fifth model is characterised by a shallow lagoon in the backreef area, rather than an emergent reef flat as in many of the other models, and the lagoon has filled in with sand after the reef crest has caught up with sea level (Figure 5.18e). A narrow reef crest and shallow lagoon is typical of the backreef on West Indian reefs. It also occurs on other reefs where there is either no evidence of the sea having been higher in the Holocene (such as Mauritius), or where massive corals are absent, as in some high-latitude reefs (such as Lord Howe Island). On Mauritius and

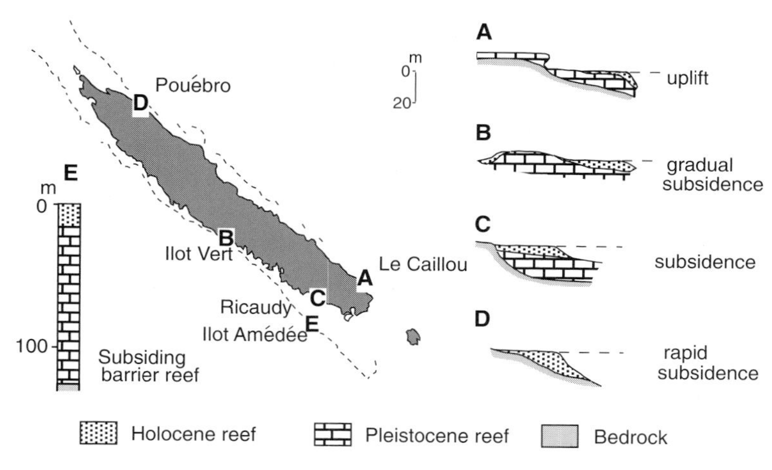

Figure 5.19. Fringing and barrier reef morphology and stratigraphy in New Caledonia as related to rate of uplift or subsidence (based on Cabioch *et al.*, 1999a).

in the Seychelles, there has been a history of lagoon infill behind the reef crest (Montaggioni and Faure, 1997; Braithwaite *et al.*, 2000). The reef on Lord Howe Island (see Figure 4.4) appears to have caught up with sea level around 5000 years ago, and the lagoon has infilled subsequently adopting this reef growth pattern (Kennedy and Woodroffe, 2000).

The sixth model shows a similar morphology but has developed as a result of different processes (Figure 5.18f). Storm processes detach and rework reef framework, which can be stabilised during subsequent quiet periods by the next generation of reef-top biota. Storms have been interpreted to be the major agents that have built reefs around Grand Cayman Island in the West Indies (Blanchon and Jones, 1997; Blanchon *et al.* 1997). In this model the reef front undergoes erosion and reworking during storms (see Section 5.4.3); these extreme events have been important in building some reef structures.

The models shown in Figure 5.18 apply to stable sites in which sea level has been the prime control on fringing reef development. If the land is undergoing uplift or subsidence, the pattern of fringing reef development can differ from that shown. This can be particularly effectively demonstrated in Figure 5.19 where the stratigraphy and morphology of reefs at different sites around New Caledonia differ according to uplift or subsidence history (Cabioch *et al.*, 1999a). Holocene reefs form only a thin veneer over Pleistocene reefs where the coast is uplifted, whereas much thicker Holocene reefs have developed in locations where subsidence is rapid. In this case, it appears that the accommodation space available for reefs to fill has been a constraint on the extent of reef development. Fringing reef growth has been interrupted by uplift on

Huon Peninsula in New Guinea (Chappell, 1980). Fringing reefs have also prograded seawards at accelerated rates in the Ryukyu Islands, south of Japan, as a result of tectonic uplift superimposed on a gradual fall in sea level (Kan *et al.*, 1997; Webster *et al.*, 1998).

5.5.2 Barrier reefs and shelf reefs

A barrier reef differs from a fringing reef in that the reef is separated from the land by a lagoon of at least 10 m depth. Classic barrier reefs occur around several gradually subsiding mid-oceanic islands, such as Moorea in the Society Islands, Mayotte in the western Indian Ocean, and Pohnpei and Truk in Micronesia (see Figure 5.8). Barrier reefs also occur on larger islands; for example there is a barrier reef along the western and southern shores of New Caledonia and, as indicated in Figure 5.19, this appears best developed where the coast is subsiding. Much more complex barrier reefs occur along continental margins, termed shelf reefs. The Great Barrier Reef of Australia and the Belize Barrier Reef in Central America have been studied in detail. In this section, morphological and stratigraphical characteristics of selected island barrier reefs are described, and morphodynamic development of individual reef platforms is examined.

Barrier reefs on mid-oceanic islands

A barrier reef forms a more-or-less continuous barrier that protects the land from the open-ocean swell. The reef crest can contain one or more major passages through it, and sand cays develop where wave refraction concentrates sediments, such as on the ribbon reefs of Mayotte Island or on barrier reefs around several islands in the Society Islands chain. The inner or lagoon slope of the barrier reef is generally composed of sediments washed over the reef, entering the lagoon as protruding fans or aprons. In some cases there are multiple barrier reefs, for example along parts of Mayotte Island or in northern New Caledonia (see Figure 5.19). An elevated double barrier reef at Marovo on New Georgia appears to be related to faulting and tectonic uplift (Stoddart, 1969b).

The sediments that infill the lagoon behind the barrier reef reflect the terrestrial hinterland only if it is a large area and contributes significant runoff. Generally, calcareous reef sediments dominate; for example, carbonate comprises >95% of sediments within about 5 km of the reef in the Mayotte lagoon, with non-reefal sediment up to 50% in embayments directly fed by the island catchments (Guilcher, 1988).

Barrier reefs are typical of subsiding islands and, as indicated for the case of New Caledonia in Figure 5.19, the 'Thurber discontinuity'

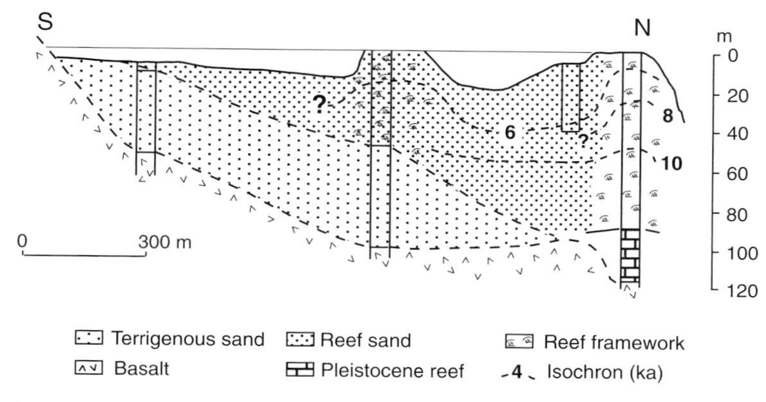

Figure 5.20. Morphostratigraphy and chronology of development of barrier reef on the northern shore of Tahiti (based on Cabioch *et al.*, 1999b).

marking the contact between Holocene reef framework and underlying Pleistocene limestone will be below sea level. Figure 5.20 shows a morphostratigraphic cross-section on the north coast of Tahiti in the Society Islands. Beneath the reef crest, Holocene limestone is more than 80 m thick overlying Pleistocene limestone. Drilling and extensive dating have enabled reconstruction of isochrons of reef growth, and demonstrate that lagoonal sedimentation has lagged behind reef growth on the crest that has kept up with sea level (Montaggioni *et al.*, 1997; Cabioch *et al.*, 1999b).

Shelf reefs

The Belize Barrier Reef is a continental-shelf reef system that demonstrates several geomorphological relationships that can be extended to the larger and much more complex Great Barrier Reef. Seismic reflection profiling has indicated that Holocene reefs are established on high points in the underlying Pleistocene topography and that antecedent topography has been a significant control on modern reef morphology on the Belize Barrier Reef (Purdy, 1974a,b). Radiometric dating has shown that some of the high points in the underlying topography are Last Interglacial in age (Gischler *et al.*, 2000).

The distribution of broad sedimentary environments across the Belize Barrier Reef is shown in Figure 5.21. The reef gets deeper, and terrigenous sediments are more apparent to the south. Fine muds, formed of faecal pellets (peloidal), accumulate in the northern part of the reef where it is sheltered behind Ambergris Cay. Mangrove forests were probably widespread on the Belize shelf during mid Holocene; their remains are preserved as mangrove peat. Some mangrove islands, termed mangrove ranges have persisted and are underlain by continuous

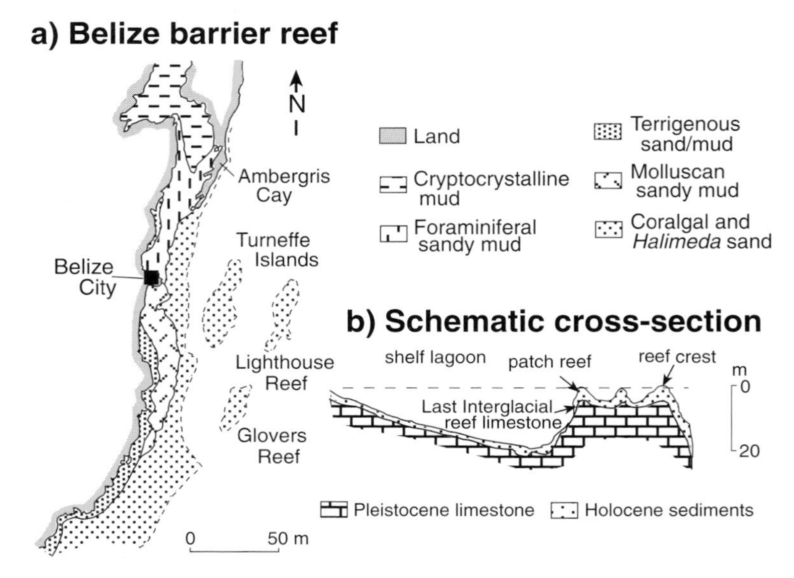

Figure 5.21. The Belize Barrier Reef showing the distribution of environments of deposition (based on Purdy, 1974a,b) and typical morphostratigraphic cross-section (based on Halley *et al.*, 1977; Shinn *et al.*, 1982; Gischler *et al.*, 2000).

peat (Macintyre *et al.*, 1995) whereas other mangrove islands have grown over the top of carbonate banks (Macintyre *et al.*, 2000). Coralgal (coral and algae) and *Halimeda* sand accumulates in the outer part of the lagoon. Lime mud occurs in the south, where reefs are established on deltaic sediments (Choi and Ginsburg, 1982).

The Holocene evolution of the Belize Barrier Reef involved drowning by postglacial sea-level rise followed by establishment of reef on high points in the Pleistocene topography (Halley *et al.*, 1977). The development of platform reefs and islands within the lagoon reflects a combination of karst control and differential reef growth, although the honeycomb reef mosaic may also be controlled by fault patterns (Macintyre *et al.*, 2000).

The Great Barrier Reef is much larger than the Belize Barrier Reef; it extends for 2300 km, over 14° of latitude, and comprises more than 2500 reefs (Hopley, 1982). The northern part of the Great Barrier Reef is a narrow (<50 km) shelf with distinctive reefs, termed ribbon reefs, that run parallel to the shelf edge (Figure 5.22). Ribbon reefs form continuous barriers up to 25 km long; they are cut by deep passages that appear to mark prior channels. The middle shelf has platform reefs, up to 25 km long, which on the inner shelf contain distinct low wooded islands. These are reef platforms on which a complex of geomorphological features can be seen, including rubble ramparts to the windward, a

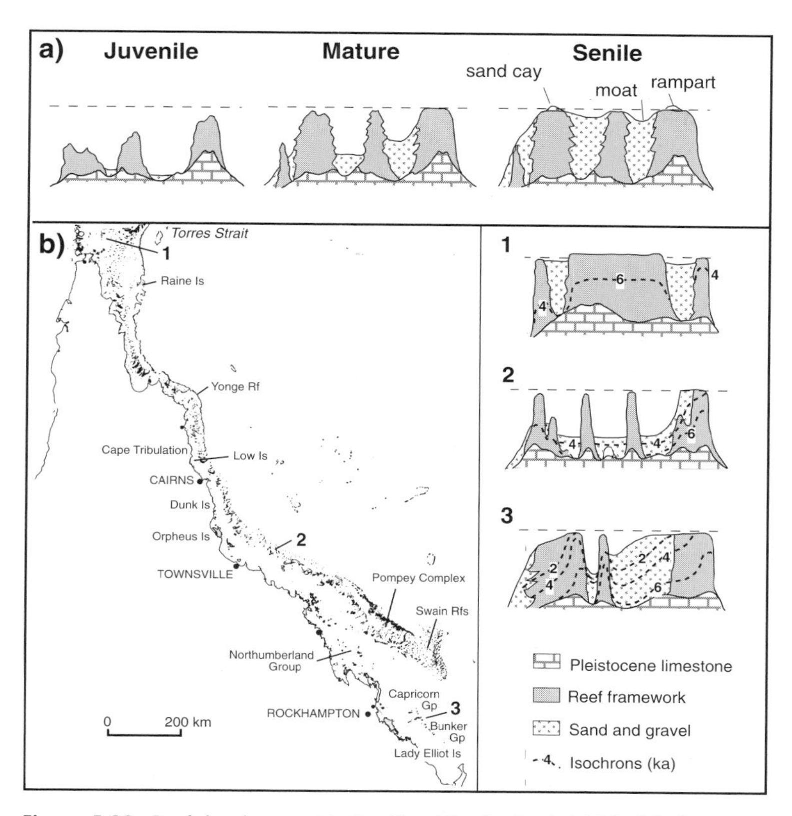

Figure 5.22. Reef development in the Great Barrier Reef. (a) Model of stages of development of a platform reef, showing juvenile, mature and senile stages (based on Hopley, 1982). (b) The Great Barrier Reef, and examples of platform reefs which have accreted in different ways: (1) Warraber Reef (based on Woodroffe *et al.*, 2000); (2) Davies Reef (based on Tudhope, 1989); and (3) One Tree Reef (based on Marshall and Davies, 1982).

leeward sand cay and stands of mangrove woodland. Three Isles is an example (Figure 5.23). South of Cairns the shelf widens but reefs, which are irregular patches or crescentic in outline, tend to be limited to the outer margin of the shelf. Further south the shelf becomes wider (300 km) and tidal range increases (to more than 4 m). The Pompey complex on the outer margin comprises a network of reefs intersected by well-defined channels. The Swain Reefs are further south, comprising smaller flat-topped reefs with numerous sand cays. The southernmost part of the reef is less than 100 km wide and comprises the Bunker–Capricorn group of 22 platform reefs and vegetated sand cays. Fringing reefs are well developed around high islands on mid to inner shelf, and are intermittent along the mainland.

The Great Barrier Reef has evolved as the Australian mainland has moved north on the Indo-Australian plate. The oldest part is to the north, but much is younger than 500000 years old (Alexander *et al.*, 2001), and the reef sequence is less than 300 m thick in places. Sea-level oscillations have repeatedly affected the Great Barrier Reef. When the sea was lower, deltaic environments appear to have characterised much of what is now carbonate province. Holocene reefs are generally founded on former Pleistocene reefal features (Thom *et al.*, 1978a), and the control of antecedent topography is strong (Hopley, 1982; Walbran, 1994). Karstic development during low sea level is convincingly demonstrated by the antecedent topography around blue-hole type solution features (Backshall *et al.*, 1979). Reef growth on the windward margins and Pleistocene highs has enhanced the original relief.

There were extensive *Halimeda* banks around the margin of the shelf about 10000 years BP, preceding reef re-establishment (Marshall and Davies, 1988). Sea-level rise across the Great Barrier Reef shelf 8500–7500 years BP triggered a number of significant geomorphological changes (Davies *et al.*, 1976; Woolfe *et al.*, 1998b). There are likely, however, to have been significant changes in water quality that accompanied this inundation, and water quality probably still limits reef development in some places (Hopley, 1994). A mud wedge appears to have delayed initiation of reef growth, particularly on the inner shelf (Woolfe

Figure 5.23. Three Isles, a low wooded island in the central Great Barrier Reef viewed from the southeast. The windward margin (to right of photo) contains rubble ramparts behind which there are extensive mangrove forests. The sand cay (centre background) has accumulated on the leeward margin of the reef platform (photograph S.G. Smithers).

and Larcombe, 1999; Larcombe and Woolfe, 1999). Turbidity is also likely to have been especially high where tidal range is large, as in the Cumberland and Northumberland Groups (Kleypas, 1996). The reefs that did establish in the early Holocene grew rapidly; about 80% of the reef framework in the Great Barrier Reef was laid down in the period 8000–6500 years BP. Nevertheless, many of the reefs on the outer shelf lagged behind sea level (which reached its present level by around 6000 years BP, and has been slightly above since then; Chappell, 1983a), and 'caught up' in mid Holocene (Hopley, 1982). Various alternative interpretations had been proposed to explain this phenomenon. One of the possible explanations was the concept of a mid-Holocene high-energy window suggested by Neumann (1972), such that reef growth has acted as a filter for high-energy swell and storm waves (Hopley, 1984). Much higher energy waves were considered able to cross the shelf in mid Holocene before reefs grew to modern sea level and attenuated wave energy, an example of an internal or intrinsic control on coastal geomorphology. This different pattern of reef growth appears better explained by hydro-isostatic flexure of the shelf in response to flooding by the sea (Chappell *et al.*, 1982; Hopley, 1987). Flexure of the shelf has resulted in evidence for a sea-level high around 5500 years BP being elevated along the mainland coast and on the inner shelf, whereas sea level appears to have been below modern level on outer reefs at that time.

Sea-level stabilisation allowed reefs to catch up with sea level, and commence lateral development. A sequence of stages in the development of individual platform reefs within the Great Barrier Reef has been recognised (Maxwell, 1968; Hopley, 1982). Much of the carbonate production of the period 6500–3000 years BP has gone into detrital accumulation as a result of storm reworking, instead of framework growth. This was the period during which reef flats developed, providing a substrate for and yielding the sediment to build cays on the reef top.

Three stages of platform reef development are shown in Figure 5.22a. The juvenile stage involves initial colonisation on an antecedent foundation and upward growth. The mature stage is one in which reefs reach modern sea level and laterally accrete, infilling lagoons and widening reef flats. The senile stage is characterised by complete infilling and bypassing of sediment from windward reefs to leeward extension. Different reefs within the Great Barrier Reef can be used to illustrate these successive stages. Juvenile stages incorporate unmodified antecedent platform reefs, submerged shoal reefs, and irregular patch reefs. Mature reefs include crescentic reefs and lagoonal reefs, and senile reefs are planar reefs that are infilled by lagoon sedimentation (for example, Three Isles; see Figure 5.23), and which show coalescence of patch reefs, moating, microatoll development and shingle ridge extension (Hopley, 1982). Once reef flats developed on the upper surfaces of these reef platforms, a series of processes of island formation and reef-flat colonisation by mangroves could begin (Stoddart, 1980).

The model shown in Figure 5.22a provides a generalisation as to how platform reefs evolve. The actual pattern of development differs according to size and orientation of antecedent topography and the processes of sedimentation. Also shown in Figure 5.22 are schematic cross-sections of several platform reefs for which stratigraphy, chronology and mode of development are relatively well studied.

The model is a particularly appropriate generalisation of the development of One Tree Reef. Reef growth began around 8000 years BP with high-energy coral head facies on the windward margin and lower energy branching facies on leeward patch reefs. Vertical growth occurred at up to 8 mm a^{-1}, until the reef reached sea level, after which the platform filled in and leeward progradation occurred (Marshall and Davies, 1982).

A model of lagoonal infill has been proposed for Davies Reef by Tudhope (1989). A lagoon persists on Davies Reef and there is less extensive reef framework (Figure 5.24). It is continuing to infill with material produced particularly on the windward margin. A different

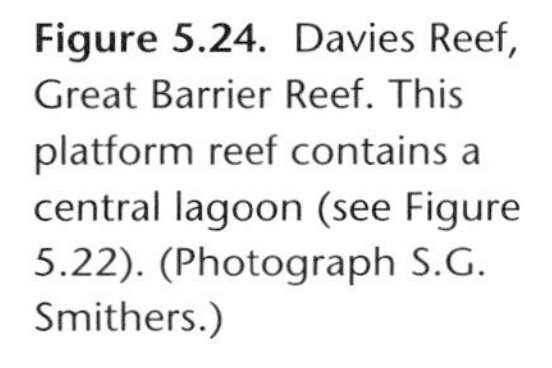

Figure 5.24. Davies Reef, Great Barrier Reef. This platform reef contains a central lagoon (see Figure 5.22). (Photograph S.G. Smithers.)

pattern of development is shown on Warraber Island in Torres Strait. The central core of this platform reef is composed of Holocene reef framework that established over a Pleistocene topographic high. The core of the platform is composed of fossil reef that had caught up with sea level by 5000 years BP, and which has become emergent as a result of sea-level fall (Woodroffe *et al.*, 2000). Holocene reef has built out on all sides of the platform by the episodic progradation of reef crests (as shown in Figure 5.18d).

The morphodynamic concepts developed on barrier and shelf reefs can be refined by examining their application to atolls which are generally simpler in structure and offer the opportunity to extend ideas of reef sediment production and reef-island formation.

5.5.3 Atolls and atoll reef islands

Atolls are composed entirely of calcareous reef-derived sediments and show similar morphology in different parts of different oceans (Stoddart, 1965). They are most common in the Indian and Pacific Oceans where they frequently occur in archipelagoes. The term atoll derives from the Maldives where there is a double chain of 22 atolls. There are 76 atolls in the Tuamotus in French Polynesia. There are relatively few atolls in the Caribbean, although the reefs immediately offshore from the Belize Barrier Reef (see Figure 5.21) comprise the atolls of Glovers Reef, Turneffe Islands and Lighthouse Reef.

Holocene atoll development

The development of the surface morphology of atolls has been constrained by the pattern of sea-level change during the Holocene. The late Quaternary history of atolls is summarised in Figure 5.10. Holocene reef sediments forming the atoll rim are underlain by lithified Pleistocene limestone (often of Last Interglacial age), generally at a depth of around 10–20 m below sea level, except on those few atolls that are not subsiding where Last Interglacial limestone is exposed at the surface. The poorly consolidated Holocene limestones have formed as a result of reef growth and the transportation of reef sediments, as sea level has risen in the final stages of the postglacial marine transgression. The reef rim shows better lithification towards the reef crest; lagoonward, the sediments are the least consolidated and merge imperceptibly into the sands of the lagoon floor.

Holocene reef growth occurred in three phases on atolls in the eastern Indian Ocean, such as the Cocos (Keeling) Islands (Woodroffe and Falkland, 1997) and the Maldives (Woodroffe, 1993b), and atolls in the central Pacific, such as the Marshall Islands (Buddemeier *et al.*,

1975) and Kiribati (Marshall and Jacobson, 1985; Falkland and Woodroffe, 1997). The three phases include an early to mid-Holocene phase of catch-up reef growth, a mid-Holocene phase of reef-flat formation and a late Holocene phase of reef-island formation and lagoonal infill (McLean and Woodroffe, 1994). Atoll platforms were initially inundated about 8000 years ago as the sea rose over them, as interpreted from locations where sea-level evidence is preserved, and from geophysical modelling. Reef establishment on atolls appears to have lagged slightly behind sea level and the time at which the reef rim of atolls reached present sea level differs between atolls, and in some cases around the margin of an individual atoll, with leeward reefs generally lagging windward reefs.

In the case of the Cocos (Keeling) Islands, reefs all around the atoll rim reached sea level about 4000 years ago when the sea was around 0.8 m above the present level (Woodroffe *et al.*, 1990b). In parts of French Polynesia reefs reached sea level around 5500 years BP (Pirazzoli and Montaggioni, 1988). On atolls elsewhere in the Pacific this occurred around 4500 years BP (Hopley, 1987; Grossman *et al.*, 1998; Dickinson, 1999), and in southern Tuvalu and northern Kiribati it occurred only in the past 2500 years (McLean and Woodroffe, 1994; Woodroffe and Morrison, 2001).

On many atolls the second phase of Holocene development is marked by development of a conglomerate platform that records extensive reef-flat development as vigorous reef growth ceased and lateral reef consolidation occurred. Boulder deposits, cemented into a conglomerate platform, have been interpreted either as evidence of a sea level higher than present (Daly, 1934; Newell, 1961), or alternatively as a result of storms (Shepard *et al.*, 1967; Newell and Bloom, 1970). In the Cocos (Keeling) Islands, the platform has been dated to 4000–3000 years ago (Figure 5.25), and the occurrence of microatolls that were constrained in upward growth by sea level indicates that the sea was slightly higher than present at that time (Woodroffe *et al.*, 1990a). Cemented coral conglomerates have been radiometrically dated to about 4000–2000 years BP in the Marshall Islands (Buddemeier *et al.*, 1975), Kiribati (Woodroffe and McLean, 1998) and the Cook Islands (Yonekura *et al.*, 1988; Woodroffe *et al.*, 1990b). In the Tuamotu Archipelago similar deposits yield ages between 4000 and 1500 years BP (Montaggioni and Pirazzoli, 1984; Pirazzoli *et al.*, 1987; Pirazzoli and Montaggioni, 1988).

During the third phase, covering the past 3000 years, Cocos experienced gradual fall of sea level to its present position and lagoon sedimentation (Smithers *et al.*, 1993). The conglomerate platform provided the basement on which reef islands could form (Figure 5.25) and on

Cocos gradual incremental growth of reef islands has also occurred in the past 3000 years (Woodroffe *et al.*, 1999).

Reef islands

Figure 5.26 illustrates schematically the morphology of an atoll rim in areas of different storm frequency and intensity. Where storms are frequent, reef flats generally contain rubble deposits on the more exposed (windward) side of the atoll. There is often a well-developed spur and groove on the reef front, an algal ridge on high-energy windward reef crests, and a conglomerate platform can cover much of the reef flat and underlie islands (Figure 5.27). Where storms are not as frequent or as severe, there are less extensive rubble deposits, the algal ridge is generally less prominent, and the conglomerate platform less extensive (e.g. Cocos (Keeling) Islands; some islands in Kiribati). In storm-free areas, rubble is not a major component in island sediments; instead sand cays are found even on the outer atoll rim (e.g. many islands in Kiribati, Maldives).

Reef islands are naturally dynamic; sediment production in adjacent reef environments, and erosion, deposition and cementation of sediment on islands can occur concurrently. Former beach positions around reef islands are often marked by beachrock, lithified beach sand

Figure 5.25. Reef islands on the eastern rim of the Cocos (Keeling) Islands, Indian Ocean. This atoll contains a horseshoe-shaped reef rim (see Figure 5.14). The islands are founded on a conglomerate platform that can be seen protruding from the oceanward shore.

cemented with calcareous cement and dipping parallel to the beach. There have been relatively few studies of the processes and chronology of development of reef islands (Stoddart, 1969a). Reef islands can be divided into 'motu' composed of sand and shingle, and generally characteristic of high-energy environments, and 'cays' composed of sand and typical of lower energy environments (Stoddart and Steers, 1977).

Figure 5.26 also shows possible responses of reef-island morphology to variations in process operation (Bayliss-Smith, 1988; McLean and Woodroffe, 1994). Storm events play an important constructive and destructive role (Stoddart, 1971; Bourrouilh-Le Jan and Talandier, 1985); they result in an input of rubble to motu but cause erosion of sand from cays. Reef blocks and rubble ramparts occur initially as lag deposits; with medium-term breakdown and redistribution of some of this material the islands readjust towards equilibrium. Cays lose sand during storms, but are rebuilt by beach recovery through normal processes. Rubble and shingle are significant components of island sediments in the storm belt.

The impact of Hurricane Bebe on the atoll of Funafuti, Tuvalu has been studied in particular detail (Maragos *et al.*, 1973). A rampart, com-

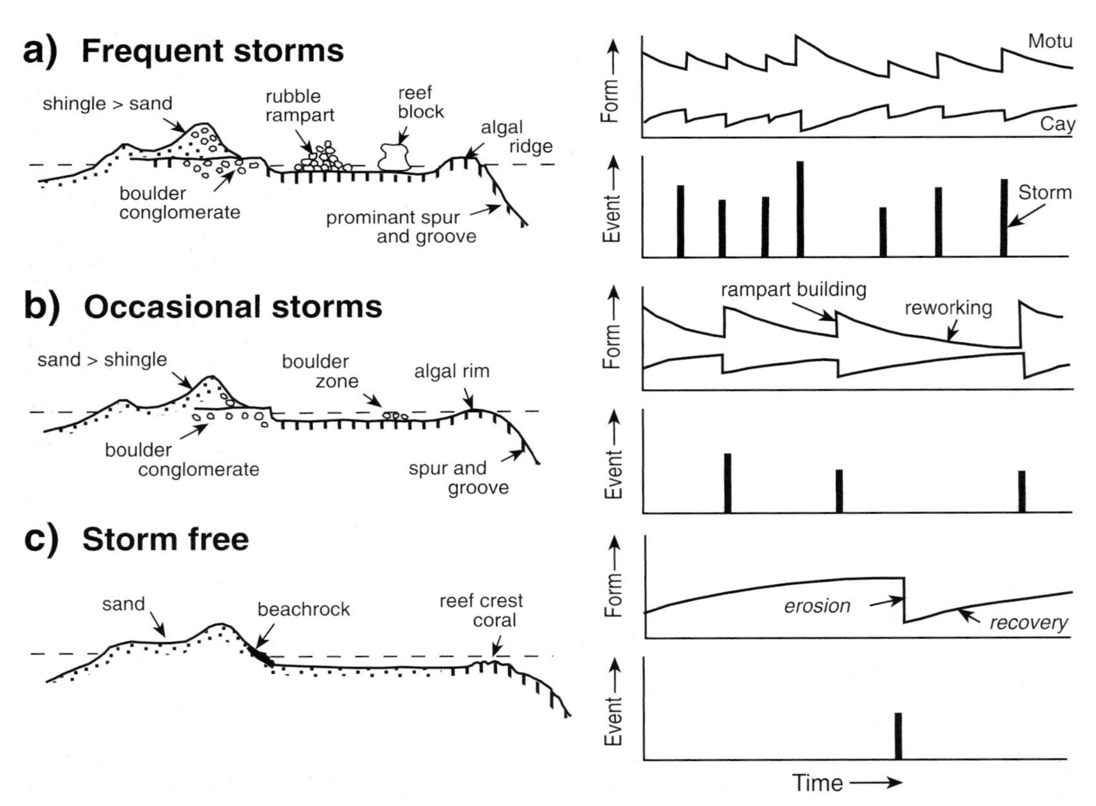

Figure 5.26. Model of reef-top morphology and response to storm perturbation, and the concept of equilibrium (after McLean and Woodroffe, 1994). See text for details.

posed of coral rubble from the reef front, was thrown onto the reef flat or onto the elongate reef islands, adding about 10% to the total land area of the atoll (Figure 5.27). Regular less severe storms have broken down and redistributed the storm rubble (Baines *et al.*, 1974; Baines and McLean, 1976a,b). Reef-island adjustment occurs over a range of other time scales. Seasonal changes characterise islands in the Maldives in response to reversal of the prevailing monsoon. Accretionary and erosional phases on Tarawa, in Kiribati, can be related to interannual El Niño–Southern Oscillation variations in windfield and sea level (Solomon and Forbes, 1999).

When storms are very frequent (or very severe) motu and cays can move their location frequently. When storms are occasional, complete recovery is possible between events. Islands can be gradually increasing in size because of the production and supply of new sediment, perhaps as a result of storm devastation of adjacent reefs, or through the more normal processes of wave and current transport for sand-sized material. In other circumstances islands may be undergoing reduction in size, as implied for islands on Ontong Java through periodic storm cut (Bayliss-Smith, 1988).

Figure 5.27. Funafuti, Tuvalu. The reef island of Fongafale is fronted by a rubble ridge that was formed by tropical cyclone Bebe in 1972. The rubble has been reworked across the reef flat onto the island. Similar deposits of rubble conglomerate occur on the island presumably deposited as a result of previous storms. The prominent spur and groove can be seen on the reef front.

It is likely that reef islands have varied in the rate at which they accrete over time in response to boundary condition change, such as sea level, or intrinsic thresholds within the reef system. For instance, as a result of gradual sea-level fall on the Great Barrier Reef, reef-flat habitats became suitable for foraminiferal production around Green Island, contributing sediment to the island (Yamano *et al.*, 2000). It has been suggested that a gradual fall of sea level may be necessary for reef islands to form on the rim of atolls (Schofield, 1977). However, although the examples described above are from areas in which sea level has fallen, similar reef islands occur in areas such as the West Indies where the sea has been rising and is continuing to rise.

Reef islands can accumulate in several different ways. Some build up by vertical accumulation or by oceanward or lagoonward accretion. Shingle ridges on Lady Elliott Island in the southern Great Barrier Reef and in the Palm Islands north of Townsville have accreted uniformly over the past few thousand years (Chivas *et al.*, 1986; Hayne and Chappell, 2001). Alternatively, rollover or washover processes can be significant. In some cases islands build in one episode of island accretion. For instance, radiocarbon dates on reef islands in the northern Great Barrier Reef indicate major phases of shingle island and sand cay building 3500–3000 and around 1500 years BP (McLean and Stoddart, 1978). In other cases, islands can be primarily erosional. Different models of island building probably apply to different islands. Sediment accretion on atoll reef islands outside the major storm belts appears to have been relatively continuous. For example, on West Island on the Cocos (Keeling) Islands, islands have built over the past 3500 years (Woodroffe *et al.*, 1999), and on Makin in northern Kiribati, they have accreted over the past 2500 years (Woodroffe and Morrison, 2001). Radiometric dating offers an insight into timing of geomorphological events. In addition, reef islands may become cemented by the precipitation of cements that form beachrock or cay rock and give the beach greater stability. The morphodynamics of beaches and associated nearshore sediments is examined in the next chapter.

5.6 Summary

In this chapter it has been shown that reef structure is a function of long-term geological processes, with broad global patterns reflecting plate-tectonic processes. Large-scale sea-level changes (100–150 m amplitude) have been superimposed on this pattern of reef structure throughout the Quaternary. Modern reef morphology is primarily the product of inherited foundations and the pattern of Holocene sea-level rise; 'like ancient cities, many reefs are built upon the ruins of their predecessors' (Hubbard, 1997, p. 58). It is the good state of preservation of former reefs, and the fact that they are amenable to dating which has provided insights into their morphodynamic behaviour over a wide range of time scales.

Surface morphology of reefs, lagoons and reef islands varies considerably within and between reefs, and reflects finer-scale variations in boundary conditions, such as wave-energy levels, sea-level trends and other environmental factors. Geomorphological variations on the top of reef platforms respond to wave and tidal currents that redistribute the sedimentary products of skeletal calcification. Reefs seek a balance between geomorphological form and contemporary processes. However, they are subject to a series of perturbations, both physical, such as storms, and biological, such as epidemics and predator–prey fluctuations. In some cases, these extreme events maintain reef diversity; in other instances they lead to major phase shifts.

Human factors represent a further set of impacts on reefs and this issue is examined in more detail in Chapter 10. A challenge for geomorphologists, reef scientists in general, and coastal managers in particular will be to disentangle the response of reef systems to natural and anthropogenic perturbations.

Chapter 6

Beach and barrier coasts

> The piling up of the ridge goes on, its height and steepness increasing, until the wash can reach no higher, and the steepness of the ridge at each point is such that the assistance which gravity gives to the down-flowing surface stream counterbalances the loss of transporting power due to percolation at that level. This is the equilibrium profile or regimen. . . . The greater the volume of water flung forward by the breaker, the greater is the depth of the back-flowing surface stream, and thus for the same size of beach material the carrying power of the back-wash is more nearly equal to that of the on-wash. Consequently, in a given locality, the regimen slope of beach proper to a rough sea is not so steep as that for a quiet sea. . . . Practically, as heavy seas on our coasts do not continue for long, the equilibrium form due to such seas is comparatively seldom seen, and the initial stage of cutting into a steep beach-ridge is often mistaken for a real change from on-shore to off-shore action. (Cornish, 1898)

Beaches represent some of the most dynamic coasts; they are attractive, not only from aesthetic and recreational points of view, but also as field areas for geomorphological research. The term beach describes wave-deposited sediment. As sediments continue to accrete on beaches they build a larger feature much of which is no longer actively reworked by waves, termed a barrier. Barriers are large shore-parallel accumulations of sediment that form over a longer time scale than beaches (of which a beach composes the seawardmost part); they are often topped by dunes. This chapter is concerned with beaches and barriers, and the morphodynamic interrelationships within and between them. First, the form and processes typical of beaches in planform and in profile are described, and integrated into a 3-dimensional view of beach morphodynamics. This is then expanded across spatial and temporal scales in a discussion of the geomorphology of barriers.

Most beaches are composed of predominantly sand-sized material.

However, the term beach derives from the Anglo Saxon word for shingle; and gravel (pebbles and cobbles) is the dominant component of many mid to high-latitude northern hemisphere beaches. Gravel on many of these beaches has been derived from coarse glacial deposits which mantle the landscape, such as moraines, or has been reworked by rivers or weathered from cliffs by processes such as freeze–thaw. The beach that accumulates is a lag deposit of material too large to be winnowed away from the shore; silt and clay are generally washed away from such relatively high-energy environments.

Beaches and the associated landforms have been studied by coastal geomorphologists who have recognised sequences of landforms that appear related to processes operating on the coast. Beaches have also been studied in the geological context, examining deposition and preservation of sediments and their relationships in terms of associated barrier deposits, across geological formations. These deposits are relatively poorly represented in the geological record, and appear more likely to move and be reshaped with a landward migrating shoreline than to be preserved, except when sea level is relatively stable or falling. Engineering approaches to beach systems have been concerned primarily with predicting beach change at scales that are relevant to coastal management and assessing the role of human interference. In recent years, there has been greater integration between process-based studies undertaken in the field by geomorphologists and modelling and wave tank experimentation concerned with short time scales, and longer-term assessments of environmental change over Holocene time scales (Komar, 1998; Short, 1999a). This has seen the emergence of beach morphodynamic studies that bridge event time scales ($<10^1$ years) and geological time scales ($>10^3$ years), and the development of models that are relevant to 'large-scale coastal behaviour' at decadal to century scales.

The prime requirements for a beach to develop are sufficient wave energy and available sediment. Figure 6.1 shows satellite imagery of a part of the coast of southern New South Wales and illustrates a series of beaches that have accumulated between the bedrock control exerted by several headlands. On this wave-dominated coast, there is a sequence of curved beaches contained between rocky headlands. Rivers and creeks that drain into coastal embayments become impounded behind sand barriers, and lagoons or barrier estuaries are formed which fill in progressively with time. There are a series of beaches of different exposures, and barriers of different sizes that will be discussed in the sections that follow. First, the historical perspective for beach and barrier studies is examined.

Figure 6.1. Satellite image of the coast of southern New South Wales (Landsat 7 ETM + Panchromatic band, November 2000), showing a series of different beaches along an embayed coast. The southernmost beach, Bhewherre Beach is swash-aligned and occurs on a broad dune barrier that separates St George's Basin from the sea. Rocky headlands surround Jervis Bay, within which a series of small curved beaches have accumulated. Seven Mile Beach to the north of the area is a drift-aligned beach that is the active part of a prograded beach-ridge plain. This beach appears to have received river sediment from the Shoalhaven River in the middle of the figure (see Chapter 7).

6.1 Historical perspective

There has been a long interest in the behaviour of beaches from the perspectives of natural curiosity, scientific enquiry and as a result of applied needs to manage beaches for recreational use and within the context of the development of coastal towns and cities. It has been realised that there is some geometrical consistency to beach planform at least since Leonardo da Vinci drew plans to drain the Pontine marshes and depicted smooth arcuate beach shorelines. There have also been various attempts to explain changes to beaches and the barriers on which they form. For instance, Bremontier (1833) recognised that beaches and dunes on the French coast reacted to storms by adjusting their profile; foredunes were eroded during storms and sediment was moved to the nearshore causing waves to break further from the shore. Similarly, on English beaches, storm waves transport beach material offshore, whereas during post-storm calmer conditions sediment is gradually returned back onto the beach (Cornish, 1897, 1898; see quotation at beginning of this chapter).

6.1.1 Beach studies

The concept of a cross-shore equilibrium profile is generally attributed to an Italian, Paolo Cornaglia (1889), who considered that the shoreface developed a concave-up profile in equilibrium with wave conditions. He suggested a neutral line (subsequently called a null point), at which sediment of particular grain size would be at equilibrium between the effect of wave asymmetry tending to move the grains onshore and gravity tending to move the sediment offshore. The null point hypothesis of Cornaglia was refined by Cornish (1898).

The idea of an offshore equilibrium profile in response to wave action was enthusiastically adopted by geomorphologists (Fenneman, 1902), and appeared to be supported by bathymetric surveys (Johnson, 1919). Observed grain-size distributions, fining seawards, suggested that the offshore profile was graded (Swift, 1970), a concept borrowed from fluvial geomorphology by Johnson (1919). The null point concept was extended by Ippen and Eagleson (1955) and Eagleson and Dean (1961), although their studies implied that fine material would be carried onshore and coarse offshore, the reverse of that observed (Bowen, 1980).

An alternative explanation of the offshore profile is that it adopts an entropy-maximising equilibrium in which there is uniform dissipation of energy across the shoreface, with sediment movement to eliminate sharp gradients (Keulegan and Krumbein, 1949). This is demonstrated by fitting exponential curves through the concave-up shore profile. For instance, Bruun (1954, 1962) demonstrated an exponential pattern to profiles at Mission Bay in California and in the North Sea off the west coast of Denmark, and Dean (1977) demonstrated a strong relationship for over 500 profiles along the southern and eastern coasts of the United States.

The preliminary studies of beaches were largely descriptive

(Andrews, 1916). They generated important observations, such as the fact that flat, wide beaches are often associated with large waves, but they provided little explanation and had little predictive value. Beach research with a strong process emphasis began in the middle of the 20th century, particularly as a result of complications during World War II in terms of landings on beaches (Shepard, 1948; Williams, 1960; Bascom, 1964). Changes in offshore profiles were demonstrated for Californian beaches by monitoring along the pier at the Scripps Oceanographic Institution (Shepard and Lafond, 1940). These beaches, and many other northern hemisphere mid-latitudinal beaches, were shown to be eroded during winter storms, but restored to a steep beach during summer. Studies in North America and in Europe indicated similar trends and gave rise to the concept of erosional storm (winter) and constructional swell (summer) profiles (King, 1972).

Field experimentation, for instance using tracers to track sediment movement (Zenkovich, 1946; Ingle, 1966), and laboratory studies, for example using wave tanks (Bagnold, 1940; Johnson, 1949) showed association between factors such as beach gradient, wave height, sediment size, and wave steepness (McLean and Kirk, 1969; King, 1972). Research supported by the Geography Program of the US Office of Naval Research and engineering organisations in North America and Europe (see Section 1.3.3) was particularly important. Empirically based relationships provide better predictive capability than the descriptive studies, but little explanation of the processes involved. Theoretical approaches, based on physical laws and quantifiable equations, were also successful. For instance, Inman and Bagnold (1963) predicted a relationship between beach gradient and the rate of percolation based on fluid dynamics, which they then tested and found to hold.

The planform shape of beaches suggested that they adjust their longshore morphology to the dominant or prevailing wave energy (Ward, 1922; Lewis, 1931, 1938; Jennings, 1955). Davies (1958a), in particular, distinguished wave climates based on the prevalence of swell or storm waves (see Chapter 3). He emphasised the important role of refraction of swell, reasserting the significance of longshore transport, noting the way that wave crests approaching the beaches in embayments refract to foreshadow the shape of the shoreline. There was also considered to be an equilibrium between the curvature of the beach in planform and the energy required to transport sediments (Tanner, 1958). Geometrical descriptions of beach planform were adopted by engineers within the context of project design and planning (Silvester, 1960, 1974).

During the 1970s, an impressive body of descriptive, empirical, theoretical and mathematical data on the coastal geomorphology of North American and European beaches existed (Meyer, 1972; King

1972; Davis and Fox, 1972; Komar, 1976). This was extended by studies of beaches from elsewhere (Zenkovich, 1967; Davies, 1980). Detailed study of the 3-dimensional morphology of Japanese beaches showed a pattern of rhythmic topography alongshore (Homma and Sonu, 1962). Extending these studies, Sonu demonstrated a sequence of change at Nags Head, North Carolina, and developed a model of change on the subaerial beach, using a 'Markov-chain' approach to determine the probability that one morphology would be followed by another (Sonu and van Beek, 1971; Sonu and James, 1973). Particularly important ideas on the morphodynamic relationships between different beach types were reinforced by time-series of Japanese beaches (Sasaki, 1983; Horikawa, 1988) and studies of beach types and associated hydrodynamic conditions in southeastern Australia (Short, 1979; Wright *et al.*, 1979).

The concept of coastal morphodynamics, expounded by Wright and Thom (1977), has provided a framework in which energy sources (dynamics) and sedimentary response (morphology) can be related. It has become clear that wave energy can be either reflected from the shoreline, or dissipated through frictional attenuation, wave breaking, and entrainment of sediment. These concepts, introduced in Chapter 3, became the focus of research on the interaction between waves and beaches, and the importance of infragravity frequency nearshore interactions, also discussed in Chapter 3, has become increasingly recognised. Morphodynamic concepts have also provided an opportunity to link short-term changes on beaches with the longer-term dynamics of barriers. These ideas are examined in greater detail in this chapter.

6.1.2 Barrier studies

The origins of sand and gravel barriers, including both sequences of beach ridges, and detached forms such as spits and barrier islands, have been the subject of considerable debate amongst geologists. The issues can be succinctly summarised in relation to barrier islands, which are elongate sandy (or sand and gravel) islands parallel to the shore but separated from it by a back-barrier lagoon. Three hypotheses have been suggested regarding barrier-island formation: that they have been built vertically from submarine bars; that they are the tops of drowned bars; or that they formed by longshore movement of sediment.

The view that barrier islands evolved from submarine bars or platforms was proposed by de Beaumont (1845) and has been supported by some researchers (Johnson, 1919; Otvos, 1970). This was challenged by Evans (1942), who indicated that waves breaking on any bar or platform would erode it rather than build it above high water level. The evolution of barrier islands as a result of drowning of coastal ridges was proposed by McGee (1890), and further supported by Hoyt (1967). A similar interpretation has been used to explain the formation of beach ridges as the result of emergence of nearshore bars under low-energy conditions (Bigarella, 1965; Curray *et al.*, 1969). An argument against either of these origins is that there do not seem to be submarine bars that are presently evolving into barrier islands.

The term barrier was first used by Gilbert who favoured the view that sandy shore-parallel ridges on the shores of Lake Bonneville resulted from spit elongation by longshore drift (Gilbert 1885). Similarly, Penck (1894) considered that the Frisian barrier islands, off the coast of the Netherlands and Germany, were remnants of a sand spit broken by storms, and he used the term 'nehrungsinsel' (spit islands) to describe them. Evans also considered that longshore transport was important in spit construction (Evans, 1939, 1942).

Various types of sandy barrier deposit were described by Johnson (1919, 1925) in relation to the dissection of drumlins (see Chapter 1, and also Figure 6.21). Zenkovich (1967), following Johnson, classified the forms as attached forms, free forms (including spits), barriers, looped forms and detached barriers. Cuspate forelands, which appear to have grown at the point of convergence of longshore drift from two directions, received particular prominence in the literature (Gulliver, 1896; Lewis, 1932). Spits across the mouths of rivers also attracted special attention (Kidson, 1963; Carr, 1965), and there has been particular concern with rates of spit extension (King, 1969) and attempts to simulate spit growth (King and McCullach, 1971).

The extent to which storms have been important in building barriers has also been a subject of debate. The term 'beach ridge' was applied to shingle ridges on the south coast of Britain by Redman (1852), who associated formation of these ridges with storm events. Gilbert (1885) believed that individual beach ridges were constructed by discrete storm events. However, Johnson (1919) stressed the significance of lower-energy waves in beach-ridge formation, considering that storms were important in adding material to the front of beach ridges, but that there were insufficient ridges for one to have been formed during each storm. Psuty (1965) also considered that beach ridges in Mexico were built as a result of high-energy storm waves supplying material to the coast, but that lower-energy waves appeared to build them into ridges. Tanner and Stapor (1971), however, attributed similar ridges to swash construction, and did not regard storms as important. In Australia, Davies (1957, 1958b) favoured the view that a berm was built by waves during calm weather and that it provided a ridge on which a foredune would begin to accrete and vegetation could become established. Erosion and scarping of foredunes was considered to result in the separation of successive ridges (Bird, 1960).

The recognition of the extent of shoreline migration as a result of sea-level fluctuations, and the greater availability of dating techniques has enabled a clearer time frame for the development of barriers (Thom, 1978). It has become clear that barriers have formed episodically; in many cases their location is subtly constrained by bedrock topography or previous barrier deposits, and in others they are responding to patterns of sea-level change (Kraft, 1971; Swift, 1975). Nevertheless, it remains the case that the significance of longshore transport versus *in situ* build up by reworking and transgression, and the relative significance of extreme events are unresolved issues in barrier geomorphology, which might vary between particular cases. These and other issues concerning barrier evolution are examined in Section 6.5.

6.2 Beach morphology

Beaches are sedimentary deposits that remain at the shore despite the action of waves. Although these unconsolidated sediments would appear fragile, their persistence in the face of variable, and often highly energetic, wave regimes indicates that they are in some balance with the processes that shape them. Their success lies in their malleability, their ability to be reshaped into a form that either reflects or dissipates wave energy, and ensures their longer-term survival. Beaches, and adjacent subaqueous nearshore environments, can adopt a relatively small range of longshore and cross-shore morphologies in response to wave conditions.

Beaches vary in size and shape. They can be long and straight, running for tens or hundreds of kilometres, where sediment supply has been sufficient to completely veneer the underlying topography, or they can comprise a series of pocket beaches separated by headlands on embayed coasts with considerable inheritance from underlying bedrock. Figure 6.1 shows a section of the New South Wales coast and demonstrates the geometrical patterns that beaches can adopt along this embayed coastline. Bhewherre Beach, at the southern end of this segment of coast, is a wide sandy beach that has accumulated on a broad barrier which is topped by dunes. The barrier has cut St George's Basin off from the sea, with limited tidal exchange through a sinuous creek (Sussex Inlet). Seven Mile Beach at the northern end has developed a sequence of beach ridges that front the coast and which have impounded several swampy wetlands, but have completely veneered the bedrock topography. This beach extends north of the mouth of the Shoalhaven River and is fed by sand that is transported from south to north unimpeded. Jervis Bay is a large sediment-starved embayment, where rocky headlands, including impressive cliffs, exert control on the shape of the overall embayment, around which there are several small beaches.

The characteristics of deep-water waves, and the transformations that occur as the waves move into shallower water are outlined in Chapter 3. As the wave shoals, its speed decreases and it interacts with the seafloor, losing energy as a result of sediment mobilisation and through friction with the bed. Considerable wave energy is expended on breaking. Waves need to balance energy and momentum; radiation stress is the excess flow of momentum, and it results in set up of waves in the surf zone. Within the surf zone, there are complex water-level variations at infragravity frequencies, discussed in Chapter 3, resulting in undertow, rip currents and longshore currents (Section 3.3). Their interactions with topography in the beach and shoreface, both in planform and profile, are described in the following sections.

6.2.1 Beach planform

The planform shape of beaches is related to the direction of wave approach, which is transformed as a result of refraction (Davies, 1958a). Refraction occurs because the wave velocity in shallow water is slower than in deeper water, and as waves encounter a shallower seafloor they are slowed with respect to other parts of the crest that are in deeper water. When waves approach at an angle to a shore which shelves gradually and uniformly, the wave orthogonals are turned towards the shore, and the wave crests become increasingly parallel to the shore (see Figure 3.7). The wave will break when it reaches water depths that are similar to wave height (generally $1.2 \times H_b$).

In Chapter 3 the consequences of wave refraction in terms of sediment transport were discussed, and it was indicated that when waves approach the shore as parallel wave crests they do not move sediment alongshore. Waves approaching the shore at an angle of around 45° cause the maximum rate of longshore transport. There is a tendency for the shoreline to be either swash-aligned or drift-aligned (Stapor, 1971; Davies, 1980). Swash-aligned shorelines are those on which waves arrive parallel to the shore, and they tend to be curved and are likely to have negligible longshore transport. Drift-aligned shorelines are those on which waves break at a sharp angle to the shoreline; they are usually longer and straighter, and are characterised by longshore transport (Figure 6.2). There is also a significant contrast between beaches dominated by long-period swell, and those that are dominated by more variable, shorter-period storm waves. Swell dominates on those shores

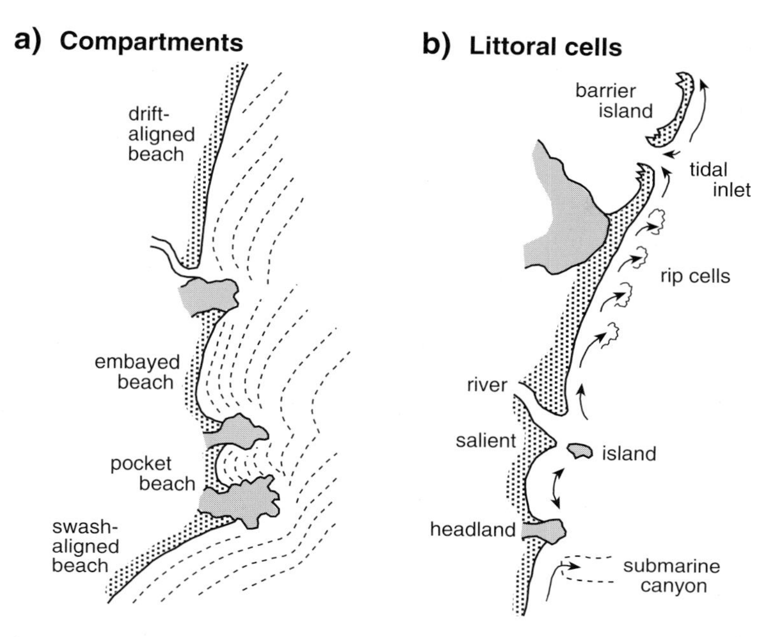

Figure 6.2. Sediment compartments and littoral cells along a coast. (a) Several sediment compartments similar to these examples can be seen in Figure 6.1. (b) Littoral cells can occur bounded by 'fixed' natural topographic features, or they may be 'free', and liable to change.

facing wide oceans, whereas many mid- to high-latitude northern hemisphere shorelines are dominated by storm waves, particularly by winter storms (see Figure 3.11).

Littoral cells and compartments

Figure 6.2 examines the concepts of sediment compartments and littoral cells. Compartmentalisation of the shore occurs where there are major obstacles, particularly headlands on deeply embayed coasts, to the longshore transport of sediment. The most enclosed beaches are impeded beaches, often called pocket beaches, on which the sediment volume remains constant; they are closed compartments (Davies, 1974). In these very sheltered beaches the waves that reach the beach are usually refracted and each small beach within an embayment is 'swash-aligned', with no net longshore movement of sediment (Davies, 1980).

It is also important to what extent there is a supply of sediment. Where there is a river that supplies new sediment to the shore, the beach is often drift-aligned beach. Waves are likely to have the energy to move sediment because if sediment protrudes from the shore, the waves will approach at oblique angles (Figure 6.2a). An embayed beach is one that

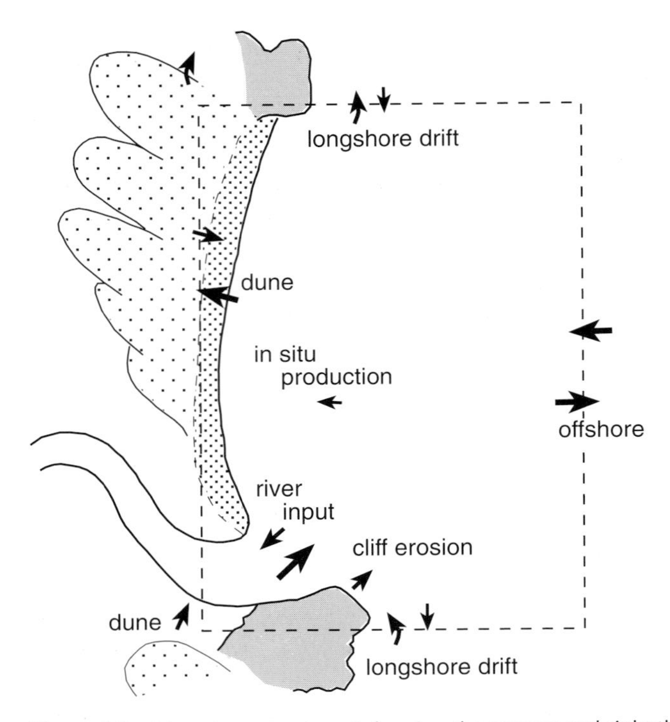

Figure 6.3. A beach compartment showing the sources and sinks that can occur. Sediment can be moved along the coast by longshore drift, which can occur in either direction, but is likely to be predominantly in one direction. The extent to which transport is possible around headlands affects the extent to which a compartment is closed; sediment bypassing around the headland can occur during storms. The compartment may receive new sediment from cliff erosion, from a river, or *in situ* production, for example by shell production. A river or estuary can also be a temporary or long-term sink for sediment from the nearshore. Sediment can be lost from the beach into dunes; periodic erosion of dunes or bypassing from embayments upwind can result in return of sands from the dune system into the beach system. There can also be loss of sand to offshore, or an input of sand from the seafloor.

is controlled in its shape by a headland at the updrift end; refraction into the embayment results in a shadow zone, and a longer, straighter beach downdrift. These beaches are examined below.

The concept of littoral cells is similar to that of compartments (Figure 6.2b). Cells are parts of the coast within which sediment is circulated. Although compartment and cell have been used to mean the same thing, they are here differentiated in that compartment is appropriate for coastlines on which there are intermittent headlands that interrupt longshore transport. Cells can occupy a hierarchy of scales. They are usually of a smaller dimension than compartments. Individual rip cells can act as sediment cells. Elsewhere cells can comprise entire compartments. Cells need not be closed; they can have leaky budgets. The boundaries of cells can be marked by one of several features. Headlands can mark divisions between cells; for example, littoral cells have been identified along the south coast of Britain, with prominent headlands such as Start Point and Lizard Point being points of divergence (Bray *et al.*, 1995). Submarine canyons mark the southern boundary of a series of littoral cells along the narrow continental shelf off the Californian coast (Bowen and Inman, 1966; Inman and Frautschy, 1966). For example, Scripps canyon appears to lose seawards around 200 000 m^3 of sediment which corresponds to the amount carried alongshore within the cell (Komar, 1996).

Cell boundaries can also occur as a result of river mouths. Rivers tend to supply material to the coast; for example, the coast of the Bight of Benin, West Africa, is drift-aligned with longshore drift of material east from the delta of the River Volta (Anthony, 1995). However, estuarine embayments can be sinks for sediment carried alongshore. Along the eastern coast of the United States, tidal inlets represent boundaries to cells (Pierce, 1969). It is useful to discriminate fixed cell boundaries and free cell boundaries (Lowry and Carter, 1982). Topographic features such as headlands and canyons represent fixed boundaries to compartments. Free cell boundaries occur where divergences can change their location because of wave refraction patterns. Cell boundaries can also migrate where they are related to rip cells, or river mouths, which can move over longer time scales.

The concept of a sediment budget is often a useful one in the management of coastal compartments and/or cells (Komar, 1996). There are a number of sediment sources and sinks within a beach system, shown schematically in Figure 6.3. Input of sediment can occur from a river, from cliff erosion, from offshore, or alongshore from adjacent cells or compartments. Measuring these sources is generally difficult; rivers are often gauged which enables a first approximation of discharge, but calculating sediment loads (dissolved, suspended and bedload) is rarely easy. Although rivers can bring large quantities of sediment to the coast, it is also possible that this is deposited in the estuary before reaching the littoral cell (see Chapter 7). Cliff erosion contributes little where there are intermittent rocky headlands, but can be a significant source where cliffs are eroding more rapidly (e.g. along the coast of California, or in eastern England, see Chapter 4). Offshore sources are difficult to measure. Further inputs can occur as a result of *in situ* production by calcareous organisms such as shells, and as a result of erosion of foredunes (including when offshore winds blow dune sand into the embayment, or winds blow sand across headlands and into neighbouring compartments).

Losses of sediment include sand blown into dunes, offshore losses and longshore transport to adjacent cells. If the rate of supply from sources and the rate of loss can be determined it may be possible to gain an insight into the sediment budget. Changes to the pattern of sediment movement can occur naturally; for instance, in 1825 the Los Angeles River diverted from the Santa Monica cell to the San Pedro cell, altering the sediment budget of both cells (Wiegel, 1994). Human actions can also alter the sediment budget; damming of rivers and sand mining represent unnatural negative interruptions to the budget, whereas beach nourishment can represent a positive alteration (see Chapter 10).

Compartments or cells are usually leaky to some extent. In many cases, headlands serve to divide a stretch of coast into compartments, although the extent to which headlands prevent longshore transport depends on whether or not bypassing of sediment can occur around the headland. Often, some leakage occurs around headlands under storm conditions or over headlands as a result of wind-blown dune sand. Similarly exchange in either direction may occur with the offshore zone – with sediment transported into greater depths on the inner shelf during storm conditions or remobilised during a subsequent storm phase.

The topography of the coast influences the extent to which there are likely to be sediment compartments, or changing littoral cells. A rocky embayed coastline, such as that of southeastern Australia, has compartments that are influenced primarily by headlands. It appears that longshore sediment transport along this coast would have been achieved more easily during times of lower sea level when the bedrock topography was not as embayed, but is now interrupted by the more numerous headlands that occur at the modern shoreline (Roy *et al.*, 1980, 1994). The occurrence of distinctive minerals within compartmentalised embayments along the coast of Oregon implies that bypassing of headlands along that coast also occurred with less impediment during lower sea level (Clemens and Komar, 1988).

The size of an embayment, its topography and the materials available influence the pattern of circulation (Figure 6.4). The sediment budget for Kujukuri Beach near Tokyo, Japan, was assessed by Sunamura and Horikawa (1977), who demonstrated that it is a closed system (pocket beach). They estimated that input from two rivers amounted to 469 000 $m^3 a^{-1}$, leading to progradation of the beach by 2–3 $m a^{-1}$. In small embayments controlled by the headlands, a strong circulation can be set up, with a large rip current (megarip) adjacent to one or both headlands (Short and Masselink, 1999). On some coasts, the headlands themselves can change over time. For instance, eroding drumlins, such as those along the coast of New England, form temporary

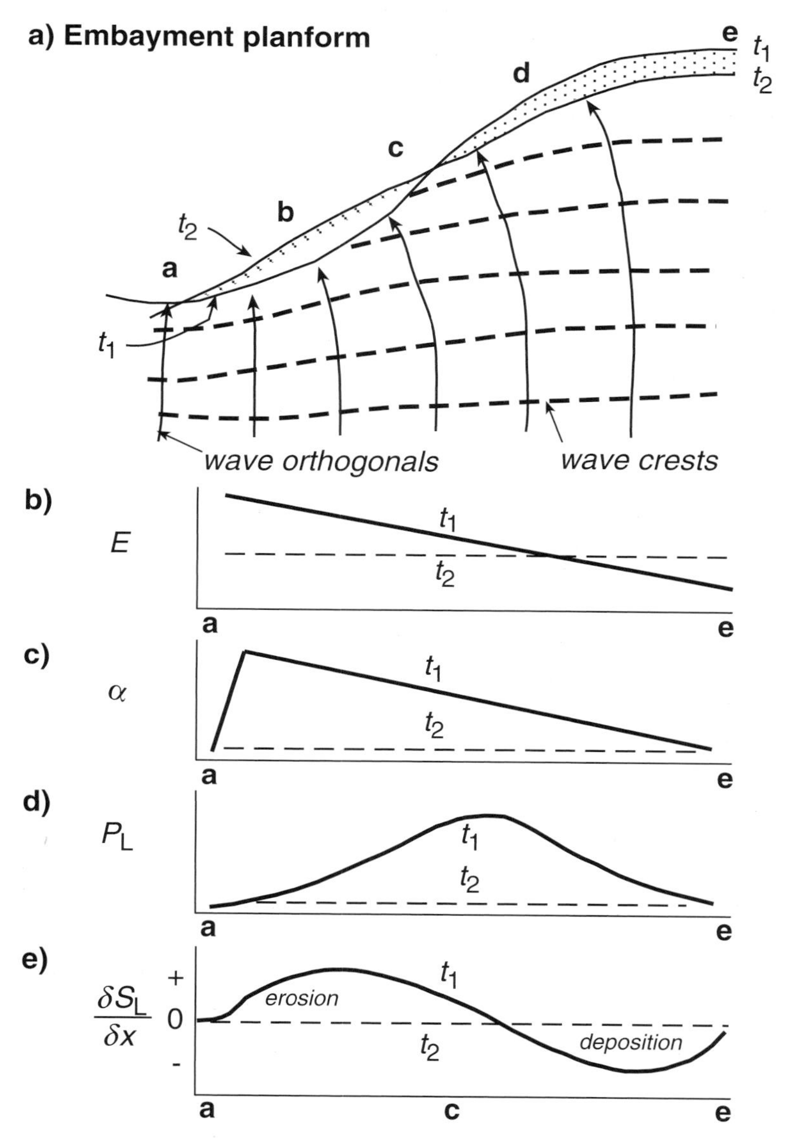

Figure 6.4. The concept of littoral power gradient (after May and Tanner, 1973). (a) Embayment planform, (b) wave energy, (c) angle of wave approach, (d) longshore component of wave power, and (e) sediment transport rate. Wave energy decreases into an embayment whereas longshore wave power is a function of the angle that waves make with the shoreline. Where the waves reach the shoreline parallel, there is no angle of incidence and hence no net movement of sediment. Points (a)–(e) define the cell. See text for discussion.

headlands (see Figure 1.4), that are active points of divergence supplying littoral cells that evolve as the coast changes. The cells that are associated with such systems have been described by Carter *et al.* (1990b), using the concept of a littoral power gradient, and are described below (Section 6.5.4; Figure 6.28).

Littoral power gradient

The concept of littoral power gradient was developed by Stapor (1971) and extended by May and Tanner (1973) to explain movement of sand in embayments. Gradients in wave energy exist as a result of wave convergence onto headlands and divergence into embayments (see Figure 6.4). Longshore wave power is a function of the angle of incidence between waves and the shoreline (see Chapter 3). Where wave approach is parallel, there is no angle of incidence and hence no longshore movement of sediment. The convergence of wave rays (orthogonals) onto a headland increases wave power (because wave height increases), and hence the competence to undertake work. In the schematic example shown in Figure 6.4, waves reach the headland (a) and the back of the embayment (e) parallel to the shore (swash-aligned) and therefore longshore wave power is low and insufficient to move sediment. However, between these points (a and e), the waves are at an angle to the shore (drift-aligned), and have the capacity to carry sediment.

An embayment can be envisaged that is initially (t_1) not in equilibrium. Concentration of wave energy on the headland results in movement of sediment where there is a large angle of approach, reshaping the shoreline, and in turn shifting towards an equilibrium (t_2) in which longshore wave power and sediment transport rate is zero (Figure 6.4). The concept of a littoral power gradient (a–e) is a useful way to identify cells within which sediment movement is initiated (a) and ceases (e), particularly along shorelines that are mobile, such as eroding drumlin headlands. Spits that develop downdrift of a drumlin headland can be divided into sections that are drift-aligned and experience sediment movement and those that are swash-aligned and are stable (see Section 6.5.4, and Figure 6.28).

The concept is extended in Figure 6.5, based on modelling by Rea and Komar (1975). If wave crests are assumed to have a parabolic form refracted over nearshore bathymetry, an initially straight beach will adjust by longshore drift to reduce the angle of wave approach at all points to become zero (as indicated in Figure 6.4). This equilibrium will be stable under unchanging wave conditions (Figure 6.5a). Figure 6.5b shows the case where wave trains come alternately from two different directions, the beach planform is likely to be similar, but its position will oscillate; this is called a metastable equilibrium (see Chapter 9).

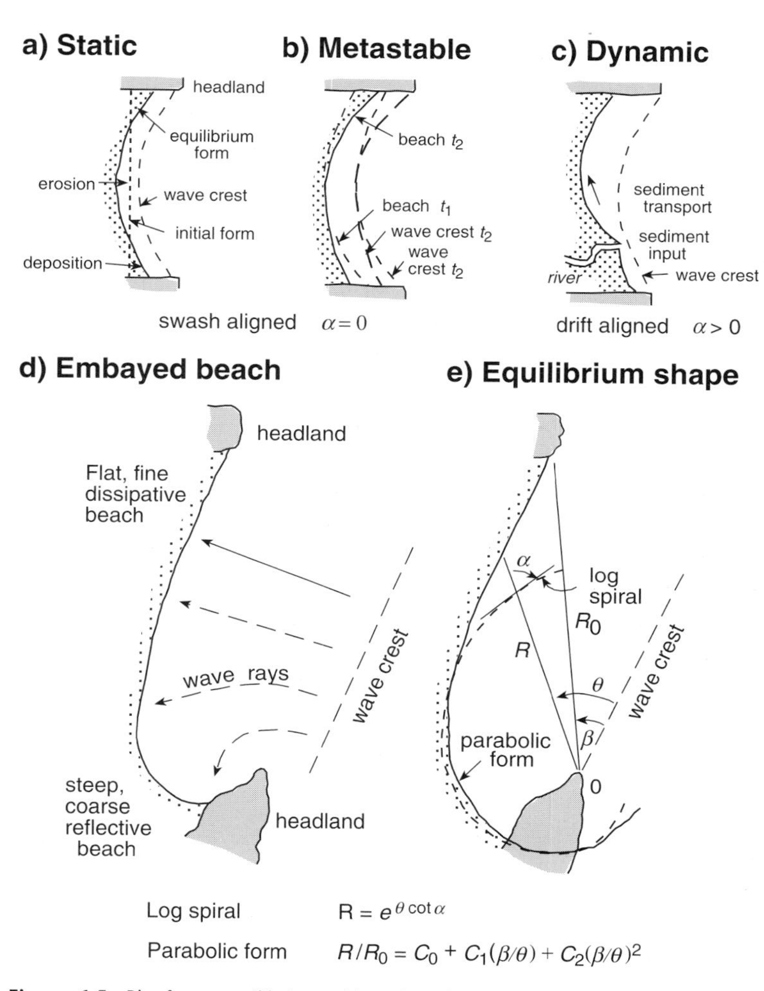

Figure 6.5. Planform equilibrium of beaches. (a) An embayed beach that has reached equilibrium (based on Rea and Komar, 1975; Komar, 1998). An initially straight beach will adjust to reduce the angle of wave approach at all points to become zero. (b) Where wave trains come from different directions, the beach can oscillate or rotate, an example of metastable equilibria. (c) If there is a source of sediment, in this case a river, the beach will continue to accrete and the angle that wave crests make to the sand added to the beach at the river mouth ensures its distribution along the beach. The beach is drift-aligned, and in dynamic equilibrium. (d) An embayed, or zeta-form, beach, showing the divergence of wave orthogonals behind a controlling headland. (e) Contrasting characterisation of an embayed beach by log-spiral or parabolic geometry (based on Hsu *et al.*, 1989a).

Oscillation of beaches in embayments, also termed rotation, can contribute to occasional leakiness of sediment compartments as a result of bypassing around headlands (Short and Masselink, 1999).

If a river brings sediment into the embayment, the shoreline is likely to be 'drift-aligned', with waves reaching the shore at an angle enabling longshore transport away from the sediment source. In this case (Figure 6.5c), the sediment budget is not fixed; a dynamic equilibrium exists, wave crests do not reach all the beach at the same time, but the angle of wave approach to the shore ensures that sediment is moved along the beach (Komar, 1998).

Embayed beaches

Beaches along an embayed coast, such as those along the New South Wales coast shown in Figure 6.1, demonstrate regular geometrical planforms. Particularly common is an embayed beach called a 'zeta-form' beach (also termed a half-heart shaped, a headland-bay beach, crenulate-shaped bay, spiral or log-spiral beach, or curved or hooked beach). This is characterised by an asymmetric planform, with a strongly curved shadow zone behind the headland that fixes the upcoast end. Waves refract and diffract into the relatively sheltered hook or shadow zone (Figure 6.2a). Embayed beaches occur on swell-dominated coasts where the dominant wave approach is at an angle to the shore. The centre of the embayment is mildly curved and the further end is relatively straight to the downcoast headland. Figure 6.5d shows a generalised wave refraction pattern with waves refracted into the shadow zone. Beaches in the shadow zone are likely to be steep and reflective. Beaches at the more exposed far end of the embayment tend to be flatter, finer and more dissipative, and they are more parallel with wave crests from the open ocean. Several beaches in Figure 6.1 show this form.

Systematic variation of grain size along a beach, called 'grading', is a common feature of many of these beaches, but can result from more than one cause. In some cases, a beach gets finer downdrift, which can occur because fine-grained sediments are carried further (Yasso, 1965), or because gravel clasts decrease in size as a result of attrition (Bird, 2000). In other cases, shingle beaches become coarser downdrift, which can be related to variations in wave energy, or has been inherited from previous sets of conditions.

The regular geometric characteristics of embayed beaches imply that there is an equilibrium shape (Silvester, 1974; Wong, 1981). At the simplest level, a relationship seems to exist between the ratio of the embayment length (defined as the control line, R_0 in Figure 6.5e) and embayment depth; the greater the angle of wave approach, the deeper the embayment. The hooked part of the beach can be approximated by a logarithmic spiral (Yasso, 1965). However, the spiral does not fit the straighter segment at the distal end of the beach, and the centre of the spiral does not match the point from which diffraction occurs. A parabolic curve relates the curvature to the angle of wave approach (Hsu *et al.*, 1989a). The point of origin is also difficult to relate to the headland that fixes the beach at the updrift end (Figure 6.5e). However, several engineering studies have examined beach shapes behind artificial obstructions, termed headland control (Hsu *et al.*, 1989b; González and Medina, 2001). Such generalisations can be useful design tools, but a beach adjusts to variations in the open-ocean swell, and it is unrealistic always to expect a constant or consistent geometric relationship between form and process (Phillips, 1985).

6.2.2 Beach profile

A beach system is shown schematically in profile in Figure 6.6 (extending concepts introduced in Figure 1.1). The usage of terms to describe morphology or process zones has varied between coastal scientists, depending on the discipline or country of origin. On a typical microtidal wave-dominated beach, there are four zones defined on the basis of wave characteristics: a zone of shoaling waves, a breaker zone (which can extend across a considerable width if there are waves of different sizes arriving at the shore), a surf zone, and a swash zone (on the beachface). Not all zones need be present on any one beach, and they vary in width and through time. The surf zone represents a complex mix of wave and current motions, operating at a range of frequencies. As indicated above, the width of the surf zone is time-variant, as not all waves break at the same point. Indeed in the extreme case small waves reach a steep beach (reflective) and surge up it as swash, and the surf zone can be absent. The swash zone migrates across the beach as the tide varies; on macrotidal beaches its rate of migration can be as rapid as 5 mm s^{-1}. Figure 6.6b shows a generalised distribution of the relative dominance of swash, surf and shoaling activity averaged across a schematic macrotidal beach. Waves shoal at a different position across the wide beach depending on the stage of the tide (Masselink and Hegge, 1995; Masselink and Turner, 1999).

The zone above the high-tide level is termed the backshore; it comprises the subaerial beach and aeolian landforms such as dunes. There is often a distinctive 'berm', a relatively narrow wedge-shaped, high-tide bench marked by a subtle steepening of the beach, or by a prominent berm crest (Otvos, 2000). The beach between high and low tide is referred to as the beachface, and it grades into the shoreface. Although the term shoreface was introduced by Barrell (1912) and adopted by

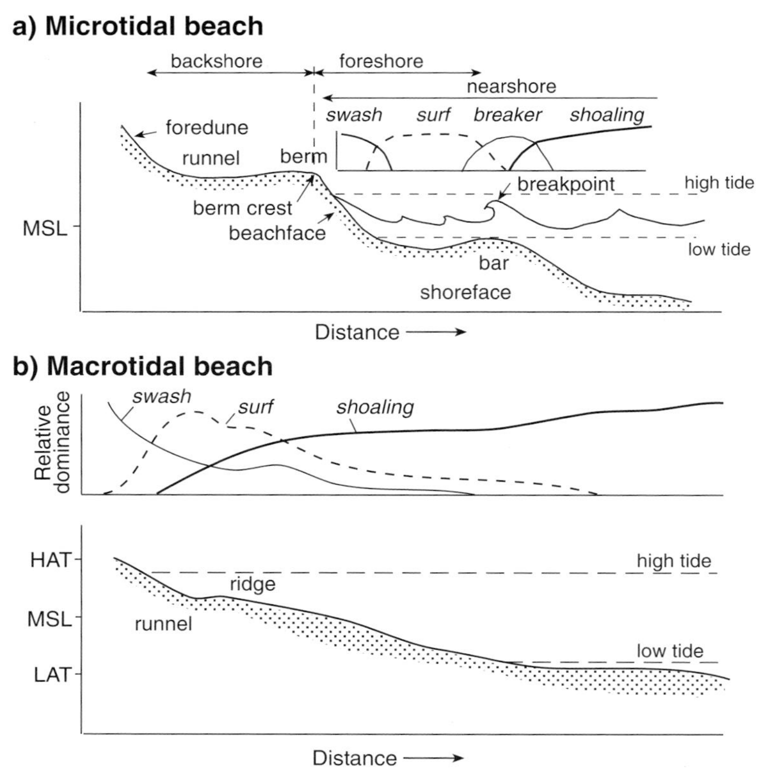

Figure 6.6. Beach in profile. (a) Profile of a microtidal wave-dominated beach. (b) Typical morphology of a mega- or macrotidal beach. Generalised distribution of the relative dominance of swash, surf and shoaling activity is shown averaged over time across the beach (based on Masselink and Turner, 1999).

Johnson (1919), it has not been used consistently. It can be considered to extend from the limit of wave run-up (hence including the beachface) onto the continental shelf to a poorly defined seaward limit, generally called closure depth. Closure depth is the depth at which the shoreface profile ceases to be eroded and redeposited by wave action. In practical terms, there is a limit to the amount of change that can be measured (presently about ±30 cm), and therefore to defining closure (Hallemeier, 1981). Closure becomes important where volumes of sand are of interest (as discussed below).

Shoreface equilibrium

The shoreface adjusts to a time-averaged profile that is relatively constant over decades to millennia. The history and basis for the concept of a shoreface equilibrium is examined and approaches to modelling it are discussed in this section. The concept of a concave-up equilibrium profile, suggested by Paolo Cornaglia (1889), involved what he termed a neutral line (subsequently called a null point). The neutral line represents equilibrium for any particular grain size between the effect of wave asymmetry, which tends to move sediment onshore, and gravity which tends to move the sediment offshore. Figure 6.7 examines the concept in more detail, contrasting an unstable linear slope with a stable concave

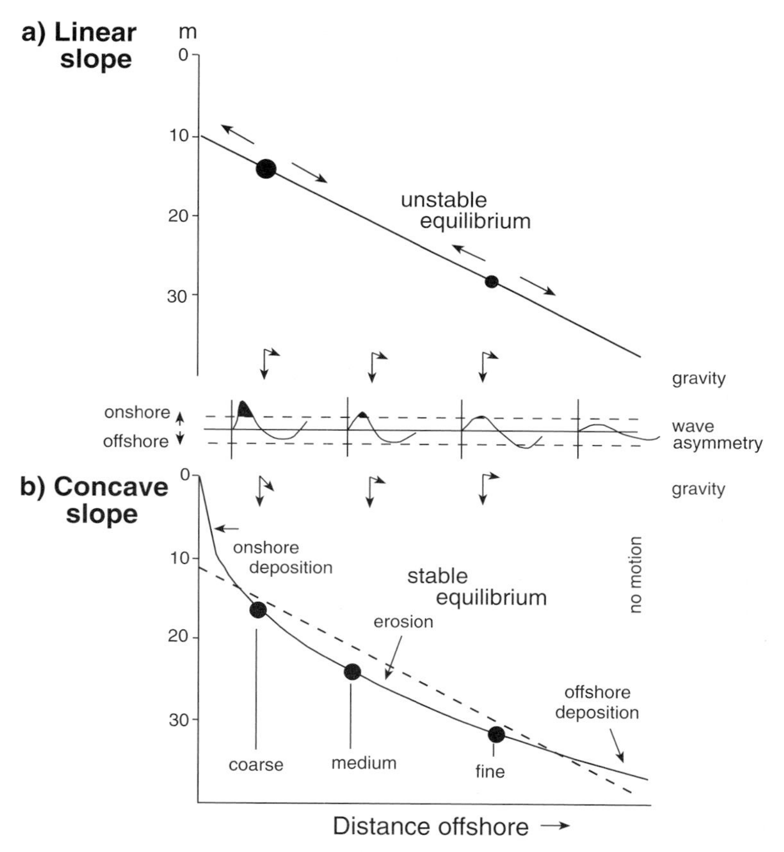

Figure 6.7. The concept of an equilibrium profile as envisaged by Cornaglia (1889). (a) The case of a linear slope, and (b) the case of a concave slope. Wave asymmetry increases closer to the shore. A neutral line (null point) exists at which sediment of particular grain size reaches an equilibrium between the effect of wave asymmetry tending to move the grains onshore and gravity tending to move the sediment offshore. On a linear slope the null point is an unstable equilibrium because if the grain is displaced from the null point it will be moved shoreward by wave velocities or seaward by gravity. On a concave profile the effect of gravitational force increases shoreward as slope increases, and a stable neutral line can be envisaged where gravitational force and wave force are equal.

slope. Orbital motion under waves at the seafloor becomes more asymmetric in shallow water, leading to an increased capacity to transport sediment as the shore is approached. The increase in asymmetry means that increasingly coarser sediment can be moved closer to shore. Cornaglia suggested that the effect of gravity on a linear slope would be the same everywhere. He envisaged an equilibrium point (neutral line) where landward movement because of orbital-velocity asymmetry and gravitational movement downslope (offshore) balance. This neutral line is closer to shore for coarse grains than it is for fine grains (Figure 6.7a). However the equilibrium is unstable; if a sand grain is agitated from that point it will either be carried offshore by gravity or onshore by wave action. A sand grain, once displaced shoreward, will become entrained and carried even further shoreward by the asymmetric orbital movements beneath waves, or, if displaced seaward, will be carried still further seaward by gravity, until it reaches the depth at which velocities are too small to move particles of that size.

Cornaglia envisaged that a concave slope, by contrast, would be stable, because the gravitational force would increase shoreward as slope increased (rather than being constant as on a linear slope). The neutral line for each grain size occurs where gravitational force and wave force are equal (Figure 6.7b). The shoreward deposition of coarse grains by waves steepens the upper part of the profile, whereas offshore transport of fine grains by gravity flattens the profile with deposition of fine sediments at or near wave base. The stable equilibrium implied in Figure 6.7b involves a grading of sediment size from coarse onshore to fine offshore.

These concepts were reinforced by Cornish (1898). Johnson (1919) presented a series of offshore profile data to support this concept of an equilibrium shore profile, and Bagnold (1940) showed that grading occurred in wave-tank experiments. However, there have been mixed results in attempts to test the ideas in the field (Miller and Zeigler, 1958; Eagleson and Dean, 1961). There are many assumptions that appear simplistic. Attempts to refine the approach to characterising beachface and shoreface profiles have involved three approaches, (i) mathematical formulations of rates of cross-shore sediment transport, (ii) larger-scale, curve-fitting procedures that make assumptions about the dissipation of energy on the shoreface, and (iii) computer simulations. These are examined briefly below, and are discussed in greater detail in Komar (1998).

Cross-shore sediment transport

Attempts to determine cross-shore sediment transport have been based on the energetics and wave power approach of Bagnold (Bagnold, 1963, 1966). This assumes that orbital velocities above the seafloor can be used to determine the quantities of sediment moved onshore and offshore as bedload and suspended load. It extends the assumption that equilibrium exists between the relative onshore and offshore components of bedload and suspended sediment load in relation to grain size and slope. Various attempts have been made to build on these concepts (Bowen, 1980; Bailard and Inman, 1981; Nairn and Southgate, 1993), with transport rate related to the velocity at a selected reference level above the seafloor.

The various models make a series of assumptions about simple wave–bed interactions and linear relationships; for instance concentration–velocity models derive the transport rate by integrating the product of the predicted mean sediment concentration and velocity profile through the water column (Hedegaard *et al.*, 1991). Sediment transport models based on energetics are still not sufficient to predict morphological change or even reliably forecast the direction of transport (Foote and Huntley, 1994). There are many reasons why most of the models perform poorly in terms of determining the amount of sediment moved across the shoreface as a whole. For example, irregularities of bedforms and lags between instantaneous velocity and the sediment entrained violate the assumptions relating velocities and sediment transport rates (Nielsen, 1992; Cowell *et al.*, 1999).

The mathematical and theoretical models that examine cross-shore sediment transport are focused on the instantaneous velocities beneath waves and, in view of the highly variable nature of water movements in time, these offer few insights into the longer-term geomorphological change on beaches. Schoonees and Theron (1995) evaluated ten approaches, each based on a different theoretical basis, and indicated that each involved uncertainties, particularly relating to initiation of motion and closure depth, but that different models may be applicable under different conditions.

Entropy-maximising curves

An alternative concept to the null-point hypothesis of Cornaglia and Cornish has been the concept of an entropy-maximising equilibrium, involving uniform dissipation of wave energy across the shoreface, with sediment movement to eliminate sharp gradients (Keulegan and Krumbein, 1949). This is a macroscopic morphological approach, deriving a concave-up shore profile with an exponential increase in depth with distance from shore, that sidesteps the need to scale up the sediment dynamics from the instantaneous time scale (Larson and Kraus, 1995). The nature of the relationship is shown in Figure 6.8.

To test the entropy-maximising equilibrium concept, curves are fitted analytically through bathymetric data to demonstrate regularity

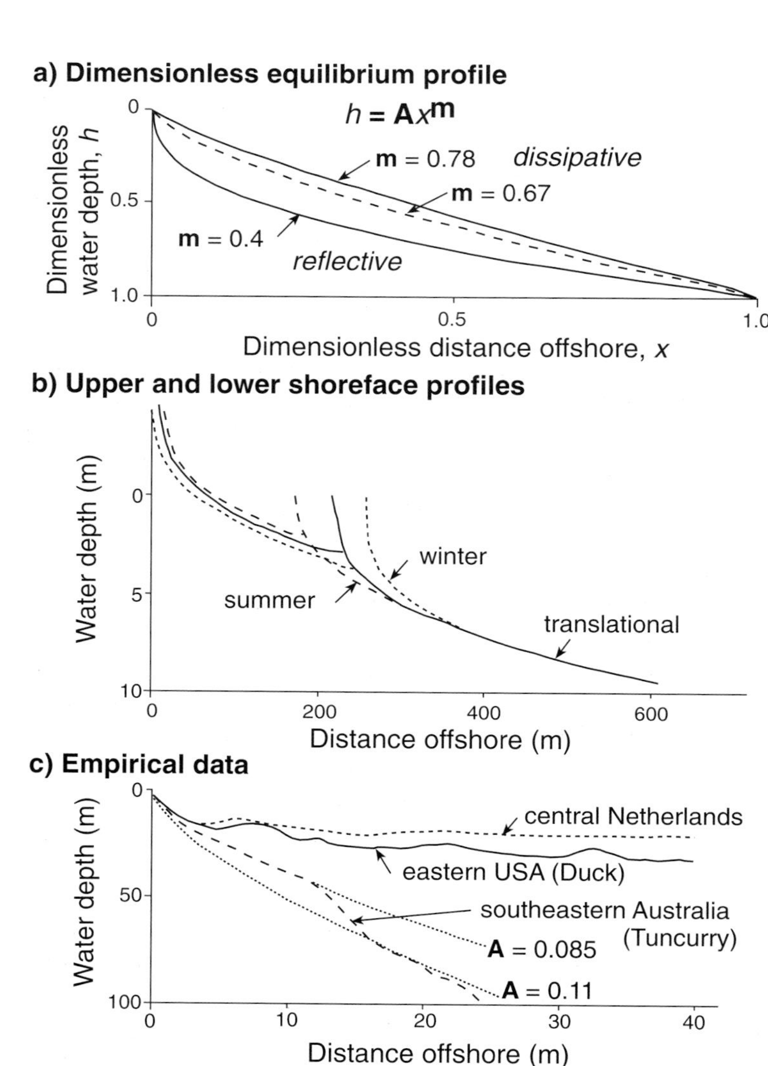

Figure 6.8. Equilibrium shore profile following the concepts proposed by Dean (1977, 1991). (a) Profile shape expressed on dimensionless scale (after Cowell *et al.*, 1999). The constant A expresses the shape factor. (b) Variation of winter and summer profiles recognising an upper bar-berm and a lower shorerise (after Inman *et al.*, 1993). (c) Comparison of offshore profiles from central Netherlands, eastern USA and southeastern Australia (after Cowell *et al.*, 1999).

in offshore shape. For instance, curves have been fitted through shoreline bathymetric data from Mission Bay in California and the North Sea off the western coast of Denmark by Bruun (1954, 1962). A strong relationship was also demonstrated for over 500 profiles to a depth of around 6 m on the Atlantic and Gulf of Mexico coasts of the United

States by Dean (1977, 1991). It is the nature of the assumed energy dissipation that influences the morphology (Figure 6.8a), and water depth (h) is shown to be a function of distance from shore (x), in terms of a constant (A) and exponent (m). Bruun assumed uniform energy dissipation per seafloor area (uniform bed stress), and derived an 'm' exponent of 0.67. Dean examined the relationship in terms of energy dissipation per volume of water that indicated an 'm' exponent of 0.4. The value of 0.67 is generally used, especially in the shoaling zone; but a value of 0.4 may apply in the surf zone (Kotvojs and Cowell, 1991; Inman *et al.*, 1993; Cowell *et al.*, 1999). In some cases it is more appropriate to fit different curves to the shoaling and surf zones or to summer and winter profiles (Hardisty, 1990). For example, Inman *et al.* (1993) identified an upper shoreface (bar-berm) and a lower shoreface (shore-rise) profile (Figure 6.8b), meeting at the breakpoint-bar. Their translational profile, shown in Figure 6.8b represents an average of the summer and winter profiles.

There does appear to be some regularity to the shoreface profile of a beach, but curves are fitted to individual sites on the basis of site-specific factors; for instance, the nature of the 'A' constant, termed the shape factor, is also variable and relates to sediment characteristics along the profile (Dean, 1997). Figure 6.8c shows that there is a wide divergence of profiles from different coasts and the degree of fit is dependent on site-specific empirical input. There are other uncertainties; for example, a parabolic curve would be vertical at its origin, which is impracticable (the beachface cannot maintain a vertical cliff form), and in practice curves are determined by iteration to get the best fit. The equilibrium profile has been incorporated into the 'Bruun rule' which is used to predict a new equilibrium profile that moves landward under conditions of sea-level rise. Its appropriateness in this context is discussed in Chapter 9. It has been particularly criticised in terms of application to continental shelves, where inheritance from previous morphology and bedrock topography is an important factor (Riggs *et al.*, 1995; Thieler *et al.*, 2000), and an ongoing supply of sediment appears necessary (Rodriguez *et al.*, 2001).

Modelling beach profiles

Considerable progress has been made using quasi-equilibrium models that determine rate of shoreface profile change by simple or bulk parameters. Formulation assumes existence of equilibrium profile shape based on sediment size, under incident wave conditions, and translates that to a new state. For instance, SBEACH is an empirical model, based on wave-tank and natural-beach equilibrium profiles and predicts change as a function of the degree to which surf-zone dissipation is out of phase with the equilibrium rate (Larson and Kraus, 1989). There are a number of other mathematical and computer simulation models available to calculate cross-shore sediment transport. One-dimensional models include GENESIS (Hanson and Kraus, 1989), SLOPES (Hardisty, 1990) and ONELINE (Kamphius, 2000). Similarly, a shoreline translation model (STM), based on a generalised Bruun rule, has been used to simulate translation of the shoreface across substrates at low angles of 0.1–0.8° on the New South Wales coast (Cowell and Thom, 1994; Cowell *et al.*, 1995). Modelling has indicated that at steeper angles the accumulation of sediment in a beach would not be stable but would be displaced into deep water (Roy *et al.*, 1994). These models are examined in more detail in Chapter 9 (see Section 9.4).

It is extremely useful to examine the shoreface profile and to monitor the change in the active, or 'sweep', zone. Nevertheless, it needs to be recognised that 'the complex phenomenon of beach-profile development defies all attempts at a perfect solution. This complexity is due to the difficulty in describing breaking waves, and the fact that once sediment is entrained in the fluid the hydrodynamic equations that describe a fluid flow are no longer applicable because both sediment and fluid are now moving with respect to each other' (Swain, 1989, p. 215). Other unresolved issues include the importance of bedforms, turbidity flows, the improbability that gravity exerts a perceptible effect on such low gradients (Cowell *et al.*, 1999), and the fact that wave conditions are continually changing, and are non-stationary. In summary, 'models that consider equilibrium solely in terms of waves offer insights into how waves mould slopes, but they cannot explain real profiles' (Wright, 1995, p. 24).

A quasi-equilibrium form may be reached in a matter of days on the upper shoreface, but changes in the lower shoreface probably occur over much longer time scales (Wright, 1995; Cowell *et al.*, 1999). The lower shoreface exerts control on horizontal migration of the upper shoreface (Stive and de Vriend, 1995), but it seems that the link is weaker now than it has been during periods of more rapid sea-level change, such as the postglacial transgression (Wright, 1995). Extension of modelling to longer time scales, described as 'large-scale coastal behaviour', involves still further generalisations about parameters such as closure depth. A value for annual closure is often adopted, corresponding to around 20–30 m on high-energy shelves in southeastern Australia. Sand movement can occur seaward of that, which may actually amount to a substantial volume. However, closure varies; during large storms sediment can be mobilised from considerable depths on the shelf (Wright, 1995; Cowell *et al.*, 1999). At geological time scales, no clear-cut seaward limit is evident for long-term shoreface change

6.3 Beach morphodynamics

The previous section has examined the concepts of 2-dimensional beach morphology. Beaches appear to adjust towards equilibrium in terms of both their planform and their profile. However, 2-dimensional studies are limited because they cannot account for the movement of sand in three dimensions. It is clear that beach adjustment involves both cross-shore and alongshore transport of sand, and the development of rhythmic topography along a beach is an indication of the significance of 3-dimensional movements of sand. In this section the 3-dimensional topography of beaches and their morphodynamic adjustments are examined.

North American and European studies have recognised erosional storm (winter) and constructional swell (summer) profiles (King, 1972). Relationships have been observed between beach slope (although no consensus exists on how this should be measured), wave height and period, and grain size (Sunamura, 1984). Net offshore sediment transport in the surf zone builds a nearshore bar, whereas prolonged onshore sand movement constructs a berm. Bars act not only as a temporary sediment store but also as mobile, flexible obstacles to waves and wave-induced currents (Horikawa, 1988); thus morphology can be seen to have a clear feedback in terms of process.

Incident waves undergo a series of transformations as they approach the shore; where they break depends on beach gradient in relation to wave steepness (see Chapter 3). Spilling waves are associated with flat beaches and fine sediments. Surging waves are associated with steep beaches and coarse sediments. The former involve dissipation of wave energy, and are associated with dissipative beaches, whereas the latter are characteristic of beaches on which much of the wave energy is reflected. Intermediate beach morphology occurs between these two extremes, with both dissipative and reflective components, and often characterised by plunging waves. The concept of a continuum of beaches from dissipative to reflective, defined in terms of indices such as the surf scaling parameter, the surf similarity index or dimensionless fall velocity (Table 6.1), has become central to understanding 3-dimensional changes that occur on beaches.

6.3.1 Beach types

There have been a series of 3-dimensional models that have been developed to address the response of beaches to wave-energy conditions. Based on the studies by Sonu, Guza and others (e.g. Sonu and van Beek, 1971; Guza and Inman, 1975), observation of a series of different beach types on the moderate to high wave-energy coast of southeastern Australia has led to development of concepts of beach–surf-zone morphodynamics (Short, 2001). Three beach types, dissipative, intermediate and reflective, have been recognised (Figure 6.9). Dissipative beaches are flat and resemble northern-hemisphere mid-latitude winter or storm profile beaches, whereas summer or berm profile beaches are reflective. The intermediate type has been further subdivided into four beach types in southeastern Australia, and similar morphologies have also been described from elsewhere (Wright and Short, 1984; Short and Wright, 1984). The occurrence of different wave motions, oscillatory flows related to incident waves, oscillatory and quasi-oscillatory standing waves, and standing edge waves, net circulations and rip currents and non-wave currents can be related to the different beach types (Wright *et al.*, 1979), and is examined below.

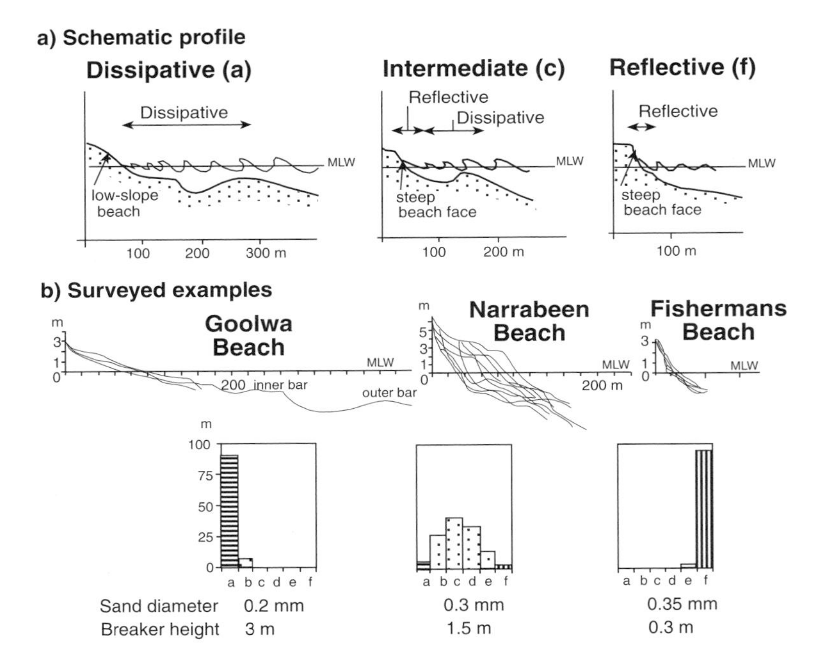

Figure 6.9. Dissipative–reflective beach types. (a) Characteristics of dissipative, intermediate and reflective beaches in profile. (b) Examples from southeastern Australia showing a range of surveyed profiles (above) and a histogram of states experienced by each beach from which modal beach state can be determined (based on Short, 1999b). The types (a)–(f) correspond with the planform beach states shown in Figure 6.15.

Table 6.1. The dissipative reflective continuum and parameters used to discriminate between these in terms of wave energy

Parameter	Expression	Dissipative (Spilling waves)	Intermediate (Plunging–collapsing waves)	Reflective (Surging waves)	Source
Breaker coefficient	$B_b = \frac{H_b}{gmT^2}$	>0.068	0.068–0.003	<0.003	Galvin, 1968
Surf scaling factor	$\varepsilon = \frac{2\pi a}{gT \tan^2 \beta}$	>20	20–2.5	<2.5	Guza and Inman, 1975
Phase difference	$P = \frac{t}{T}$	>1.0	1.0–0.5	<0.5	Kemp, 1975
Surf similarity index (Iribarren Number)	$\xi_b = \frac{\tan \beta}{(H_b/L_o)^{0.5}}$	<0.64	0.64–5.0	>5	Battjes, 1994
Dimensionless fall velocity	$\Omega = \frac{H_b}{T w_s}$	>6	5–2	<1	Dean, 1973; Wright and Short, 1984

Notes:
H_b, wave height at break point; T, wave period; L_o, open water wave length; a, wave amplitude; β, beach slope; t, time that elapses between wave breaking and reaching beach; m, breaker position relative to beach, corresponds to surf-zone gradient; w_s, sediment fall velocity; g, gravitational constant = 9.8; $\pi = 3.1429$.

The degree of wave reflection or dissipation can be defined using a series of parameters which are outlined in Table 6.1. The parameters are based on the principles discussed in Chapter 3. For instance, the surf scaling factor is a function of incident wave amplitude, wave period and beach slope (Table 6.1); it is characteristically <2.5 where reflection dominates, but if it is >20 where the beach is primarily dissipative. Beach types can also be differentiated using a parameter that Dean (1973) called the dimensionless fall velocity based on breaker height, sediment size (expressed as fall velocity) and wave period. Where dimensionless fall velocity is <1 a beach is reflective and where it is >6 a beach is dissipative. The parameters are useful indices of beach behaviour, although it is often appropriate to use more than one of the parameters (Bauer and Greenwood, 1988).

A dissipative beach is a low flat beach, typically composed of fine sand. It has a relatively small subaerial volume, compared with the volume of sand stored in the shoreface, and is dominated by large high-energy waves. The energy within these waves is dissipated across the broad surf zone; the waves are spilling waves, and initially break

200–500 m offshore (Figure 6.9b). The broad surf zone parallel to the shore is generally flat and secondary waves can reform within it (Figure 6.10). There is little longshore variation in morphology, and wave set-up and set-down are dominated by shore-parallel standing waves at infragravity (surf beat) periodicity (Short, 1999b).

A reflective beach is a steep beach, dominated by an accretionary berm and low breakers (generally <1 m high) which travel right to the beach, surge up it, and rapidly return. Reflective beaches are associated with longer-period waves, and coarser-grained sediment (Bryant, 1979, 1982); for example, gravel beaches are generally steep and reflect wave energy. Sediment storage in the subaerial beach is large in proportion to the subaqueous beach. Steep beaches are often dominated by subharmonic edge waves, which can lead to the development of beach cusps (discussed below). There is often a step (low-tide terrace) at the base of the beach where coarse material, such as shells, accumulates. Reflective beaches occur in sheltered areas, as for instance where waves have been extensively refracted, for example within an embayment or sheltered behind a reef (Sanderson *et al.*, 2000). These beaches tend to be stable because the low wave energies are not capable of rapid geomorphological change. Pearl Beach in Broken Bay, New South Wales

Figure 6.10. A dissipative beach, Middleton Beach, Port Eliot, South Australia (photograph P.J. Cowell).

Figure 6.11. Pearl Beach in Broken Bay, New South Wales, Australia. This is a reflective beach with characteristic beach cusps (photograph P.J. Cowell).

(Figure 6.11) is an example; it shows a sequence of beach cusps. Waves arrive at this beach parallel to the shore having been refracted within Broken Bay irrespective of offshore wave conditions. Berms represent the landward extent of wave action, beyond which aeolian activity can be important. Berms on reflective beaches can erode as a result of swash uprush, or where the water table is close to the surface (Chappell *et al.*, 1979).

Gravel beaches, generally associated with a discrete source of supply of material, for example volcanic or coral gravel, flint eroded from chalk, or gravel from glacial deposits, are ususally reflective. Run-up can be amplified during big waves overtopping the beach crest and building it up; washover and landward migration are typical (Carter and Orford, 1984). Swash tends to infiltrate into the very porous beachface and backwash is considerably reduced in its effectiveness. Gravel barriers are discussed in Section 6.5.4. Mixed beaches on which sand and gravel occur together can behave differently both under ambient conditions and storm conditions (Kirk, 1980; Carter *et al.*, 1984; Shulmeister and Kirk, 1993).

The beachface is a notoriously difficult place to study because the land–water boundary is very shallow and moves with the tide. Additionally, the fluid layer and the sediment layer become indistinguishable (it may be difficult to distinguish overlying granular fluid layer from underlying stationary bed), with the mobilisation of a sediment layer and saturation or partial saturation of the beach. The uprush of swash is the most observable of phenomena, whereas the backwash, partly effected by gravity and with infiltration into the beach, is less clear and may interfere with the subsequent uprush. The concept of equilibrium, expressed in the quotation by Cornish at the beginning of this chapter, implies a dynamic equilibrium that exists when net transport is averaged over several swash cycles (Hughes and Turner, 1999).

There are a series of intermediate beach types, which result from the complex current patterns that develop in the nearshore. In the case of southeastern Australia, four types have been described: longshore bar and trough, rhythmic bar and beach, transverse bar and beach, and ridge-runnel (or low-tide terrace). Similar morphologies, although recognising two further types, have been proposed from studies of a sequence at Duck, in North Carolina (Lippman and Holman, 1990). A transitional beach can show an upper reflective portion and lower dissipative portion (Figure 6.9a), tending to be more reflective at high tide than at low tide. For instance, longshore bar and trough, which occurs on exposed southeastern Australian beaches where wave height is typically around 2 m, has a wide beachface that is steeper than on a dissipative beach, but an upper beachface that may even have beach cusps. A nearshore bar is separated from the beachface by a prominent trough. The bar is usually straight, but there are instances where it may be crescentic with weak rip development.

Other intermediate beach types also show components of both reflection and dissipation of wave energy, together with different patterns of surf-zone currents and circulation. In the case of the rhythmic bar and beach, typically experiencing waves of 1.5–2 m height, bars alternate with rip current channels, at which the beach erodes. The broad embayment has been termed a megacusp and beach cusps may occur on the horns between megacusps. The transverse bar and rip beach state, typically occurring in association with waves of 1–1.5 m high, consists of shore-attached bars which have been welded onto the shoreface, separated by very strong rips and pronounced scarping of the embayment between bars. The ridge-runnel type of beach consists of a low-tide terrace around mean low water level; minor rips may occur where runnels drain at low tide. The relationships between these beach types are examined below after a description of the nature of circulation patterns and nearshore topography.

6.3.2 Three-dimensional beach morphology

As discussed in Chapter 3, infragravity frequency waves can be important in establishing nearshore rip currents, longshore currents and undertow. Figure 6.12 reinforces these concepts. Standing or progressive edge waves (see Figure 3.9) can be trapped at the shoreline as a result of oblique wave approach. At the shore, their crests are normal to the beachface and to the approaching wave crests. Edge waves are of greatest amplitude (only a few centimetres) at the beachface decreasing exponentially into the surf zone (Holman, 2001).

Beach cusps

Beach cusps are rhythmic crescentic features formed by swash action. They are particularly common on reflective beaches exposed to relatively low wave energies with waves breaking or surging directly onto the beach (Masselink *et al.*, 1997). They are typical of coarse-grained beaches, occurring on coarse sand and gravel (Russell and McIntire, 1965). There are actually a series of crescentic features that can form along a coast, varying from swash cusps (8–25 m long), storm cusps (70–120 m long), giant cusps (700–1500 m long) to headland-dominated undulations, such as those typical of the barrier island coast of the eastern United States (Dolan, 1971). Cusps at still closer spacing (from only a few centimetres to 2–3 m) can occur on lake shores (Johnson, 1919).

Swash cusps consist of an arcuate berm, called an embayment, with horns (Longuet-Higgins and Parkin, 1962; Williams, 1973). Sediments

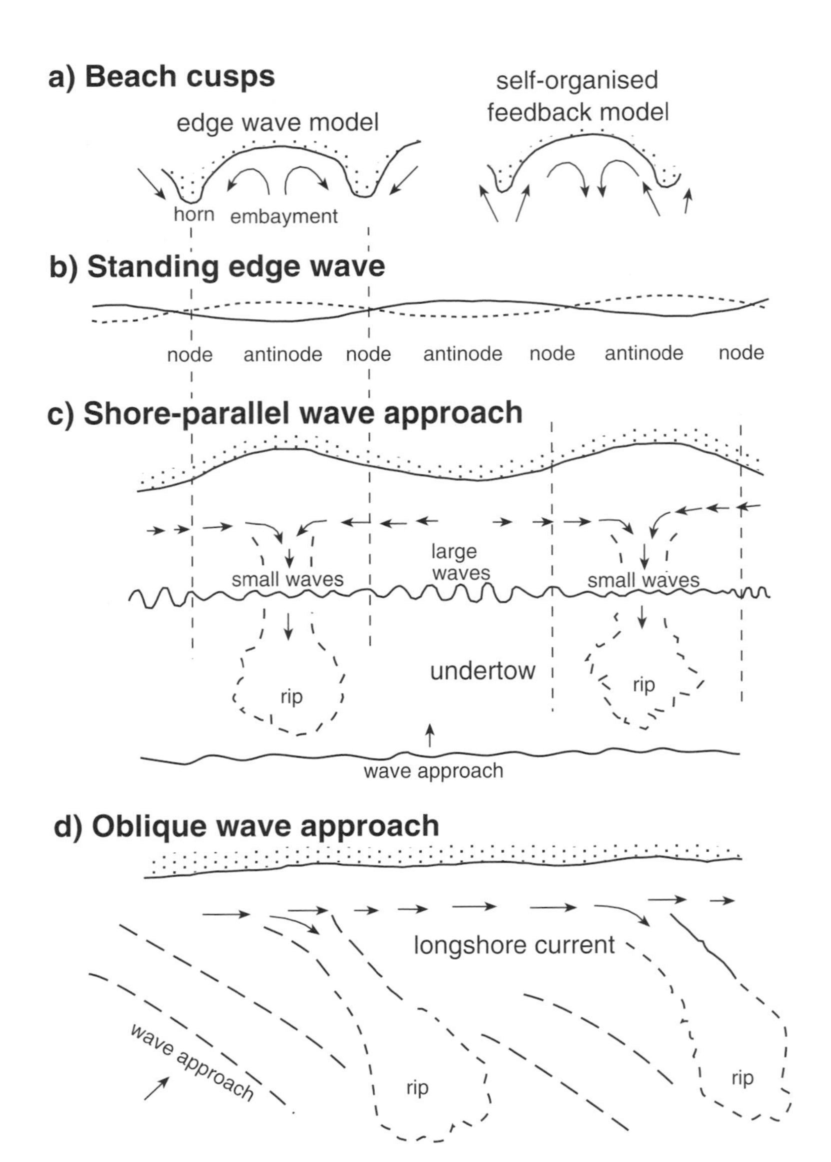

Figure 6.12. Rhythmic cusps and the occurrence of rip currents, undertow and longshore currents in the surf zone. (a) Beach cusps may form either as a result of edge waves (here shown in phase with edge waves), or through self-organisation and feedback (based on Holland and Holman, 1996). (b) Development of a standing edge wave along the shore; no variation occurs in water level at nodes. (c) Relation of rip currents to edge waves (based on Komar, 1998). (d) Longshore current where waves approach the shore obliquely.

on the horns tend to be coarser than those in the intervening embayments (Figure 6.13). There has been considerable disagreement as to how cusps form, and it is likely that different processes are responsible in different settings (Komar, 1998; Hughes and Turner, 1999). Cusps tend to form as a result of swash uprush being deflected from the horns into the embayment (horn-divergent), but they may also be sculpted by circulation where swash entering the embayment overtops the horn (horn-convergent) and causes erosion (Masselink, 1998).

There is much evidence that implies that cusps are related to edge waves (see Figure 6.12a). Edge waves that are subharmonic (with a period twice that of incident waves) are particularly implicated, though cusps could also form on synchronous edge waves (Bowen and Inman, 1969; Guza and Inman, 1975). The edge waves accentuate the elevation of swash (Holman, 1983). However, there is disagreement whether horns coincide with nodes where run-up is a minimum, or antinodes where run-up is a maximum (Hughes and Turner, 1999). Positive feedback is likely to accentuate the cusp morphology (Inman and Guza, 1982).

An edge-wave origin for cusps has proved difficult to confirm on the basis of statistical analysis of spacing (Holland and Holman, 1996; Nolan *et al.*, 1999). Cusp spacing should vary in relation to beach gradient if the edge-wave theory applied (Figure 6.12b), but no systematic variation in cusp spacing has been observed at Palm Beach in Sydney (Masselink, 1999). Moreover, observations of cusp re-establishment following a storm suggest that cusp morphology is largely controlled by the occurrence of former cusps higher on the shoreline (Masselink *et al.*, 1997).

An alternative view is that beach cusps are self-organised features that develop by positive feedback between swash flow and morphology, accentuating random morphological irregularities (Johnson, 1919). Computer simulation of cusp development by Werner and Fink (1993) showed that incipient topographic depressions are amplified. This occurs because the depressions attract and accelerate water flow which enhances local erosion, whereas topographic highs repel and decelerate water, encouraging deposition (Coco *et al.*, 1999).

Bars

Infragravity variations in water-level in the surf zone become increasingly apparent because these occur at too long a wavelength to be dissipated. Longshore variations in wave height occur, both as a result of interaction between incoming waves and edge waves, and in response to variations in topography in the surf zone. It is the development of this

Figure 6.13. Beach cusps on a beach in South Australia. The horns of the cusps are coarse, and the surging waves can be seen (photograph S.G. Smithers).

topography that is the subject of this section, and that relates to the generation of rips, nearshore currents and rhythmic topography alongshore (Figure 6.12c). Although wave height varies alongshore, wave periods are too short for the associated water-level variations to drive nearshore circulations. These result instead from the longer period infragravity variations in water level that are associated with set-up in the surf zone (Chapter 3).

Bar morphology can vary from single to multiple, shore-parallel or

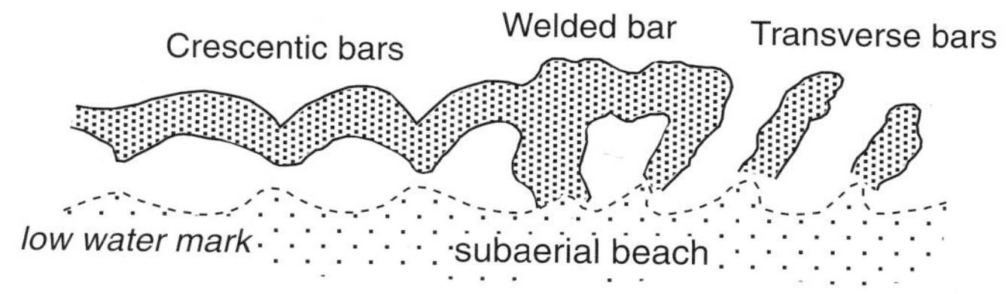

Figure 6.14. Bars found on beaches. (a) Beach planform, showing types of shore-parallel bars. (b) Types of rhythmic bars.

oblique, continuous or compartmented, and linear, sinuous or crescentic (Figure 6.14). Broad flat beaches, across which there are a series of up to four to five bars, often with the intervening channels draining through breaches in these ridges on the ebbing tide, have been called ridge and runnel beaches (King and Williams, 1949). On the high-energy beaches of southeastern Australia, there is generally only one nearshore bar present. On other beaches, there can be a sequence of multiple nearshore bars (Greenwood and Davidson-Arnott, 1979). In tideless seas, a sequence of nearshore bars can result from waves of different heights (King, 1972).

Rhythmic bars form with variation alongshore. A crescentic bar is generally closer to the beach than the nearshore bar, and shows variations in morphology. Bars can become welded to the shore (Figure 6.14). In other situations, transverse bars develop where the bar is oblique or shore-normal (Konicki and Holman, 2000). Rhythmic bars are often associated with the presence of rip currents and edge waves. Circulation cells establish in association with this rhythmic topography, as shown in Figure 6.12. The extent to which bars are modified by edge waves, or are responsible for the generation of edge waves, remains difficult to establish, and it seems highly likely that there are tight feedbacks between form and process (Kirby *et al.*, 1981). Bars often represent disequilibrium features which migrate (Gallagher *et al.*, 1998), and they can be continually readjusting their configuration and location (Aagaard and Masselink, 1999; Ruessink *et al.*, 2000). There may be a transition from outer more dissipative bars to inner more rhythmic bars (see Figure 6.15).

6.3.3 Beach variation over time

Dissipative and reflective beach types are endpoints in a continuum, and an individual beach, or any part along a single beach, can vary within the continuum, and change between types (then termed beach states). In the case of beaches dominated by pronounced seasonal variations in energy conditions, as typical of the northern hemisphere exposed beaches, there will be a seasonal adjustment between states. However, southeast Australian beaches show less prominent seasonal change, but respond most conspicuously to major storm events that tend to erode the subaerial beach.

The morphology that a beach shows at any point in time is a function of its sediment characteristics, incident and antecedent wave conditions, and the antecedent beach state. Tide and wind conditions can also have an effect. Although there is a strong relationship with incident wave conditions, it is important to recognise that a beach does not respond immediately, but takes a finite time to adjust from one state to another, and can therefore be out of equilibrium with the wave regime that is operating at any time. Exposed beaches can change quite rapidly between states, eroding more rapidly than accreting (Short, 1993).

Some beaches remain within one state; sheltered reflective beaches deep within embayments, for instance, are often highly reflective and may show little morphological variation over time because waves always refract and reach them parallel irrespective of wave energy. Open-ocean beaches, subject to a larger range of wave conditions, can vary between several states (Wright *et al.*, 1979). It is not necessarily the case that any one beach or section of a beach will adopt all of these states; many beaches vary over only a couple of states. Frequently, there is one state which the beach adopts more often than others, referred to as the modal beach state (Wright and Short, 1983, 1984). There is a greater likelihood that dissipative or reflective endmember beaches remain in, or close to, these endpoints (Short, 1999b). Intermediate beaches are more likely to vary through a wider range of states. Figure 6.9b, for instance, shows a series of surveys of Narrabeen Beach in Sydney (Short, 1979, 1999b), defining the active 'sweep' zone, and showing that it can vary throughout the range of beach types. It demonstrates the entire sequence of intermediate states, from the more dissipative forms, such as longshore bar and trough and rhythmic bar and beach, to the more reflective transverse bar and rip, and low-tide terrace.

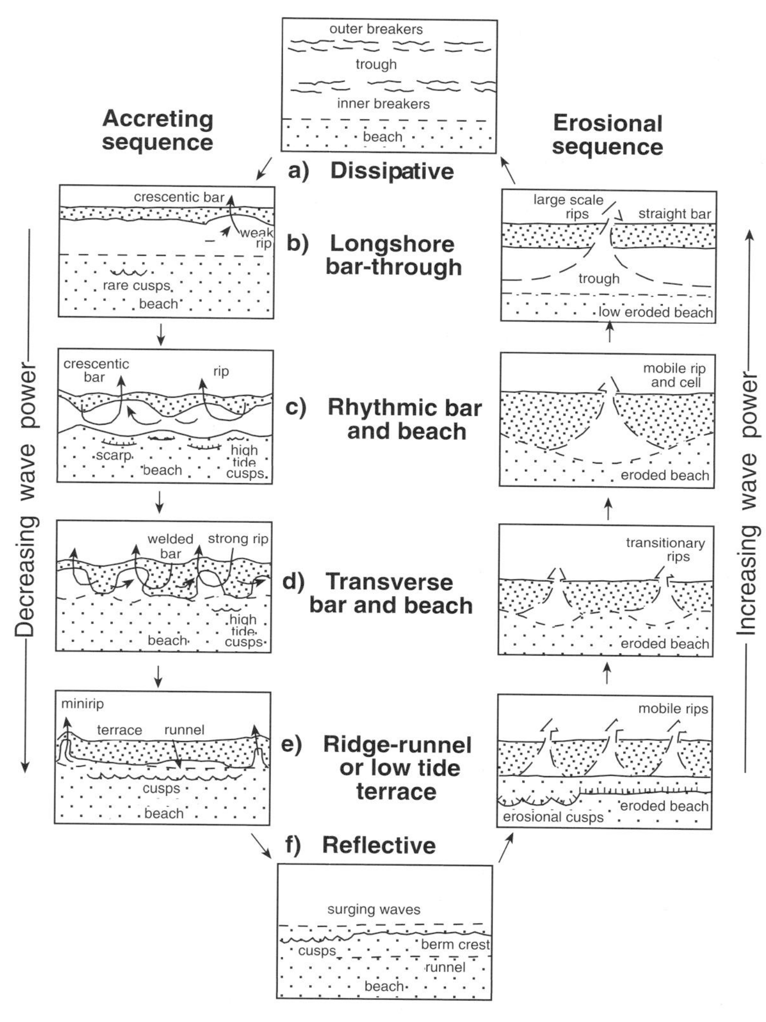

Figure 6.15. Planform evolution of beach states (based on Short, 1979, 1999b; Wright and Short, 1984; Sunamura, 1989; Lippman and Holman, 1990b). Beaches may vary over a part of the range through erosion or accretion (see text for discussion).

It is also the case that differences exist between beaches that are accreting and those that are eroding. This is illustrated in Figure 6.15, based upon Sunamura (1989) and Short (1999b). In the accretionary mode, the nearshore bar migrates towards the shore and welds on to it. In the eroding mode the beachface is eroded and rips intensify. In the accretionary sequence a dissipative beach becomes increasingly rhythmic (see Figure 6.15). As the bar migrates landward, standing shore-parallel

Figure 6.16. An intermediate beach, showing the development of rhythmic topography and rip currents, Kurnell Peninsula, Sydney (photograph D.M. Kennedy). The bar in the foreground has welded to the shore, whereas further along the beach there are prominent rhythmic bars and rip currents.

waves in the surf zone are replaced by standing edge waves, and cell circulation is initiated with rip currents. It may take several months for a beach to recover after it has been eroded by a storm.

The morphological changes reinforce the dynamics by positive feedback; for example rip currents tend to scour sediment from the beach, enlarging the channels between bars (Figure 6.16). It is likely to take a beach weeks to change to the transverse bar and rip state, and there may be a tendency for the beach to become arrested in a more dissipative state than incident conditions would suggest or for variation along the beach to become apparent. The beach is able to rearrange itself to accommodate bigger waves in the erosional sequence more quickly as larger waves, by their very nature, provide more energy to do this geomorphological work (Wright *et al.*, 1995; Benavente *et al.*, 2000). Large waves increase the set-up, so the water table is raised and the beachface slumps. The larger waves and rips move sand seawards, eroding the low-tide terrace and extending the bar seawards (Short, 1999b).

The Wright–Short model has been modified to accommodate multiple bars (Short and Aagaard, 1993; Short, 1999b). There have been several recent attempts to model multiple bar formation, migration offshore and decay over months to years (Lee *et al.*, 1999; Ruessink and Terwindt, 2000). Bars are generally considered either to form at the breakpoint, at the point at which there is sediment convergence (Roelvink and Stive, 1989), or to form as a result of standing infragravity waves (Bauer and Greenwood, 1990; Holman and Sallenger, 1993).

Ridge and runnel beaches are characteristic of low-energy dissipative macrotidal settings, especially in large embayments or straits, such as at Blackpool in England or Normandy in France (Orford and Wright, 1978). The bars can represent multiple phases of formation under spilling waves but that have been preserved because the tide rises and falls across the beach more rapidly than the beach can readjust (Carter, 1988). Although the bars tend to persist, they are destroyed or more subdued after storms (Orme and Orme, 1988; Mulrennan, 1992). However, there appear to be other processes that can operate to form ridge-runnel patterns, and it is likely that similar morphology results from different causes (Michel and Howa, 1999).

6.3.4 Beaches in other settings

Studies of beaches on microtidal wave-dominated coasts have enabled the development of morphodynamic models that provide powerful insights into the way that beach morphology and process are interrelated. Recently these ideas have been extended to examine distinctive beaches in other settings. Three examples are discussed in this section: estuarine beaches, beaches on solid rock or reef platforms, and megatidal and macrotidal beaches.

Beaches that occur within estuaries have been the subject of several studies (Nordstrom, 1989; Nordstrom and Roman, 1996). These often demonstrate pronounced variations in the relative balance of wave and tide processes that are related to the opportunity for wind waves to be generated within the embayment and attenuation or distortion of the tide (Nordstrom and Jackson, 1992). In coastal lagoons, there can be negligible tide and wave action is often entirely dependent on fetch (See Section 7.5.4).

Beaches that form over solid substrates, such as reefs, are not able to adjust their profile to adopt a shoreface equilibrium (see Chapter 5). Rock outcrops, or reef configuration, have significant effects on wave energy and the beach is likely to differ significantly from open-ocean, wave-dominated beaches (Smith, 2001). Cross-shore modelling of shoreface adjustment has generally been based on the assumption that there is free exchange of sediment between shoreface and beachface. This is not the case where beaches are perched on an underlying non-erodible platform, as is the case for reef islands or along rocky coasts where sediment is limited. Several modelling approaches have been extended to take account of reef-protected beaches (Muñóz-Pérez *et al.*, 1999; Larson and Kraus, 2000).

Tides have a direct effect on surf-zone water level as the tide migrates across the shore (Strahler, 1966; Thornton and Kim, 1993). Elements of the beach profile change from shoaling to surf to swash zone and even to backshore during the course of a tidal cycle (see Figure 6.6b), with the indirect effect of influencing the water table within the beach (Masselink and Turner, 1999). Bars migrate more slowly under high tidal range, for instance in the case of megatidal areas such as Carmarthen Bay in Wales where tidal range exceeds 10 m (Jago and Hardisty, 1984) and Contentin, near Mont St Michel, Normandy, in France (Levoy *et al.*, 2000).

Megatidal and macrotidal beaches tend to be long and flat, often with a prominent change of gradient in mid beach or near low tide as with low-tide terrace beaches (Figure 6.6b). During low tide there can be considerable dissipation of wave energy across a broad surf zone, rips may be stronger (Wright *et al.*, 1982; Jago and Hardisty, 1984). At high tide, much of the foreshore that was surf and swash zone, or subaerially exposed, is submerged, and becomes shoaling zone with relatively little dissipation of wave energy. The steeper beachface can be very reflective at high tide. Shoaling waves have a more pronounced role on macro-tidal beaches, as opposed to swash and surf-zone processes. Offshore-directed bottom flow can be important on the falling tide, and shore-parallel tidal currents play an increasingly significant role on lower intertidal and subtidal beaches.

The concept of relative tidal range (the ratio of tidal range to breaker height) has been proposed by Masselink and Short (1993), extending the relative scaling of these factors proposed by Davis and Hayes (1984). Masselink and Short (1993) proposed a classification of beaches based on both the dimensionless fall velocity and the relative tidal range; wave conditions remain more important than tidal processes. Mudflats are typical of areas that have a high tidal range but low wave energy, and these have been termed ultradissipative by Masselink and Short (1993). These features are examined in Chapter 8.

6.4 Beach and backshore change over decadal–century time scales

In terms of coastal morphodynamics, the changes that can occur in the form of the beach are perhaps better understood than on any other coastal type. There is an instantaneous response by the fluid to a physical change on the sediment surface. As wave energy changes, the beach moulds to adjust to wave energy by the movement of sediment (Reeve *et al.*, 2001). This response however is lagged, with morphology lagging behind process adjustments because of the finite time required for movement of sediment. Incident conditions are extremely changeable across a wide range of energy settings. A beach adjusts towards equilibrium in terms of shape, but non-stationarity is likely at a variety of time scales, and equilibrium is a moving target (Horikawa, 1988; Wright, 1995). This is discussed in more detail in Chapter 9.

Beaches are changing continually at the event time scale; they alter in response to tidal, diurnal, seasonal and longer-term (ENSO) cycles, as well as reacting to events (e.g. storms). They can also be viewed at geological time scales. Sediment on a beach is mobile under many conditions, and a beach is often a highly changeable landform. As with other coastal landforms, it is often the case that an enormous amount of work is done to achieve a small result. A single grain on the beach can travel kilometres, only to end up a few metres from where it started. Short-term monitoring programmes may not be representative of longer-term trends; for instance tracking of individual grains for only a few tidal cycles can show a pathway of transport that is contrary to the long-term average (Kidson and Carr, 1959). Other beaches respond to sea-breeze characteristics which compound their response to swell (Masselink and Pattiaratchi, 1998). Under these circumstances, with the complicated reversals of flow which typify the beach, it is not surprising that sediment transport equations, often linearising complex non-linear relationships, are poor predictors of long-term patterns of change. Under these circumstances, morphodynamic approaches offer valuable alternatives to the scaling up of fluid dynamics.

6.4.1 Recession, accretion and stable shorelines

Erosion of beaches can be indicated by cliffing of the backshore, truncation of the dune vegetation, or exposure of remnant beachrock (lithified beach sands, generally with calcareous cement). A broad global survey suggested that as much as 90% of the world's beaches were eroding (Bird, 1985a, 1996); however, erosion may be part of longer-term erosion/accretion cycles, and longer-term studies are necessary to establish the underlying cause and significance.

Beach erosion can occur for many reasons. Relative sea-level rise can result in retreat of a beach that can be either washed into backshore environments or eroded with sediment transport offshore (see Chapter 9). Beach erosion may also occur when a littoral cell no longer receives the sediment to which its budget is adjusted. A decrease in river sediment input can result in erosion (discussed in Chapter 7). Where sediment is supplied from cliffs, a reduction in that rate of supply, as when cliffs have been stabilised, or where groynes prevent longshore transport of material, as along much of the south coast of Britain, can lead to erosion in the area now deprived. Similarly, if a beach had adjusted to sediment supply from dunes or from the seafloor and this flux of sediment is interrupted, erosion is likely to result. Increase in wave energy at Pakiri in northern New Zealand following dredging resulted in erosion (Hesp and Hilton, 1996).

Many beaches undergo short-term erosion (storm cut) and subsequent accretion as part of a cycle with no overall sediment loss or gain. The variability of beaches can be shown where a series of beach profiles taken at different times are superimposed (see Figure 6.9b) and the envelope that contains these profiles is referred to as the 'sweep zone'. The difference between recession (long-term retreat of the coast), stability, and long-term accretion is shown schematically in Figure 6.17; it requires repeated surveys to determine the sweep zone, and to distinguish these trends. Boundary conditions, such as sea level, are presumed to be stable; if sea level changes, then further influences are likely on whether or not the coast recedes, as discussed below and in Chapter 9.

There are few studies of sufficient duration to assess the relative significance of major storms and determine the long-term pattern of change (Dolan and Hayden, 1983). The best data come from those beaches that have been surveyed at regular intervals; a less clear picture is available by using aerial photographs to monitor beach position or volume (Winnant *et al.*, 1975; Dolan *et al.*, 1991, 1992). For small isolated, clearly defined beaches, comparison of oblique aerial photographs may extend the record. A comparison of the utility of these different approaches can be made from the New South Wales coast.

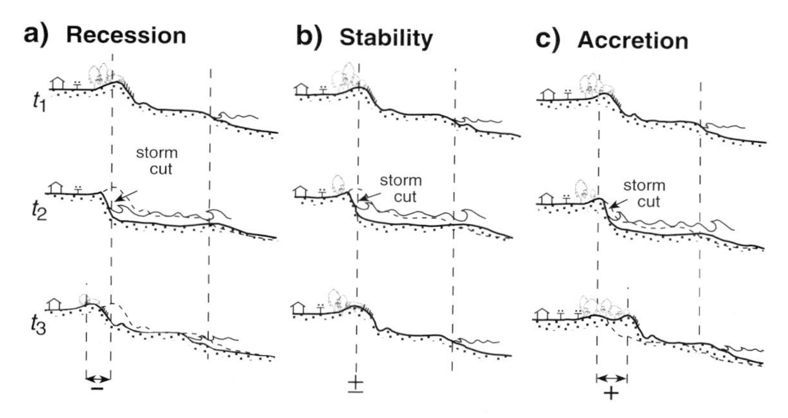

Figure 6.17. A beach undergoes erosion and subsequent recovery, but the extent to which the long-term profile shows (a) recession, (b) stability and (c) accretion can vary.

Empirical orthogonal function analysis of fortnightly surveys at a series of transects along Warilla Beach in New South Wales over the period 1975–85 demonstrated low-frequency changes in sediment volume, changing from berm to bar, at annual to biennial scales, but long-term oscillation around a mean position (Clarke and Eliot, 1983, 1988a,b). A longer-term study, comparing oblique photographs over several decades for Stanwell Park Beach (further north on the same coast) indicated the significance of beach erosion by major storms (Bryant, 1988).

Moruya Beach is a prograding bay-barrier on the New South Wales south coast (Figure 6.18a) on which regular surveys have been maintained for about 28 years. In this case, it is possible to compare this period of record with a geomorphological reconstruction of the behaviour of the embayment (Figure 6.18b) which shows that the coast has been prograding over the mid–late Holocene (Thom *et al.*, 1981; see discussion below). The modern beach has been dominated by a period of erosion by storms in the 1970s, with particularly extensive sediment loss from the subaerial beach in 1974 (Thom and Hall, 1991). Since that erosional phase there has been a period of accretion, followed by a relatively stable period of fluctuations (Figure 6.18c). The amount of erosion (above mean sea level) by the 1974 storms was of the order of 100 m^3 m^{-1} beachface at Moruya, whereas the maximum range within which the beach has fluctuated is over 200 m^3 m^{-1} beachface (see Figure 6.18).

Long-term monitoring, such as that at Moruya, indicates the degree of natural variability of one individual beach. The volume of the sweep zone (the volume of sand that can be considered active on the subaerial beach or shoreface) is likely to vary from site to site. If either recession or accretion dominate, then definition of the sweep zone will change with time. It is significant to note that engineering design volumes for cut–fill are often of the order of 25–30 m^3 m^{-1} beachface in North America. This compares with observed variations of 150–200 m^3 m^{-1} beachface on measured profiles such as Moruya in Australia, and up to 320 m^3 m^{-1} beachface for volume variation at Byron Bay in northern New South Wales. At the scale of several decades substantial volumes of sand accumulate in the backshore and are subsequently reactivated during storms. An appreciation of these dynamics is an important element of the coastal management of these systems.

6.4.2 Beach ridges

Construction of a series of beach ridges can occur on accreting coasts over decades–centuries, as sediment is transferred to the backshore and eventually becomes inactive as a result of progradation of the shoreline. Sand and gravel can form wave-built beach ridges, and in some cases the sand is blown further inland to form a dune (as discussed in the next section). If there is too little sediment available to construct a beach-ridge crest the beach will be overtopped, whereas if sediment is abundant the beach-ridge crest is built up (Anthony, 1995; Thompson and Baedke, 1995). Storm surges also result in deposition of material

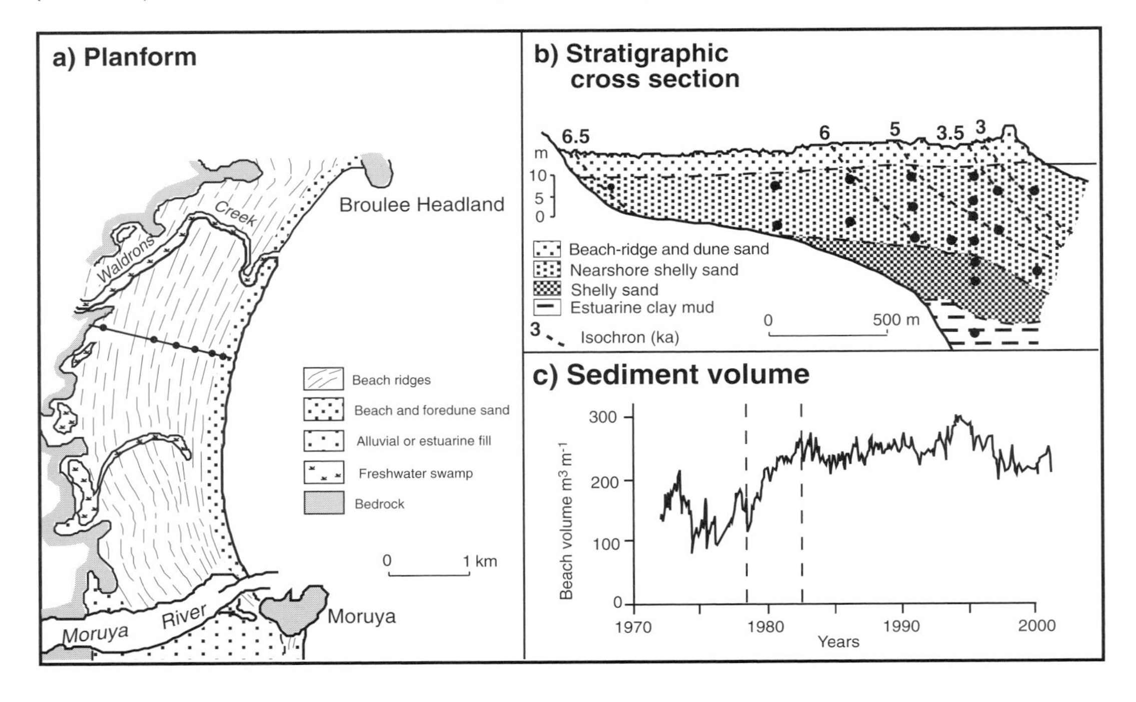

Figure 6.18. A prograded beach-ridge plain at Moruya in southern NSW. (a) Generalised distribution of beach ridges; (b) stratigraphy and chronology based upon drilling and radiocarbon dating (after Thom *et al.*, 1981), and (c) pattern of beach volume change during period 1972–2001 (based on Thom and Hall, 1991, with additional unpublished data from R.F. McLean). The beach has shown an early phase of erosion (1974–78), a phase of accretion (1978–83), and a phase of relative stability (since 1983).

beyond the beach; for instance a record of past surges has been derived from the deposition of shelly material within marsh and dunes in Holland (Jurgen, 1993; Jelgersma *et al.*, 1995).

Johnson (1919) first used the term beach ridge to describe wave-built features. Beach ridges are defined as 'wave- and wind-built landforms that originated in the inter- and supratidal zones' (Otvos, 2000, p. 83). Beach-ridge plains are also called strandplains. Otvos (2000) has suggested that the term beach ridge is also used to describe ridges that have a foredune on top of them, called relict foredune ridges by some authors, on the grounds that these represent minor eolian veneer over predominantly swash-built landforms.

The debate as to whether ridges are built by storms or accretion by calmer swell on top of a berm has been contentious. Abundant sediment and a shallow offshore gradient seem necessary for beach-ridge formation. It has been demonstrated that dunes are best developed behind dissipative beaches, whereas berms are a feature of reflective beaches (Short and Hesp, 1982). Hesp (1984a,b) considered that beach ridges are incipient foredunes formed on dissipative beaches, but this interpretation has been disputed (Bird and Jones, 1988). Although general sea-level trends may be indicated by a sequence of beach ridges, which are especially common where sea level has been falling, individual ridges are rarely related to distinct oscillations of sea level (Taylor and Stone, 1996). Although Tanner (1993, 1995) has suggested that it is sometimes possible to reconstruct sea-level oscillations on the basis of detailed grain-size characteristics, this procedure needs further investigation before it is extended to other ridge sequences, in view of the likelihood that different factors are implicated in ridge construction in various geographical settings.

A simple model that links four geomorphological types of beach-ridge plain to processes involved in their development has been proposed by Shepherd (1987), expanding on observations by Davies (1980). On rapidly prograding coasts, a succession of beach ridges occurs with the building of each new beach ridge starving the former ridge of sediment and resulting in a series of low ridges. On a slowly prograding coast where there is less sand supply, the new ridge does not form so rapidly and the existing beach ridge is topped by a higher foredune, resulting in a series of higher foredune ridges. In the case where a formerly rapidly prograding beach-ridge plain is now stable, a sequence of low beach ridges would terminate at the shore with a single foredune that had built higher than the earlier beach ridges. This is the case in much of southeastern Australia and has been suggested to result from depletion of the seaward sand supply (see below). If a previously rapidly prograding coastal plain is now eroding because the sand budget is in deficit, then old beach-ridge deposits are eroded and former soil horizons may be exposed on the shore and covered by an erosional lag.

The model described above suggests that morphology could, under some circumstances, provide an indication of changed patterns of sediment supply, and could be a powerful coastal management tool. However, relatively few tests of the model have been undertaken. Radiocarbon dating of ridges around Perth, Western Australia, has demonstrated local rapidly prograded ridges deposited 6000–5000 years BP, but ridge accretion has been complicated by attachment to, and incorporation of, small islands along the coast (Shepherd, 1987, 1990). It can be shown that phases of rapid build out occur where islands have temporarily blocked longshore supply of sediment.

It appears that sand can be transported landward from the beach by wind during either long-term accretion or erosion, depending on sediment supply, beach morphology and and the processes that are operating (Hesp, 1999). Similar beach ridges in different geographical settings may form as a result of different processes (Carter, 1988; Taylor and Stone, 1996). The several situations in which dunes form are described in the next section.

6.4.3 Beach–dune interactions

Coastal dunes only develop on certain coasts; they are most extensive where there are suitably strong onshore winds, sufficient sediment supply of medium to fine well-sorted sand suitable for entrainment, and vegetation to assist in the trapping of that sand (Goldsmith, 1989). Foredunes (the first dune ridge forming at the back of the beach) play an important role storing sediment and protecting the land from extreme wave and tide conditions. The dune system acts independently of the beach. Dunes represent a sand reserve from which material can be borrowed (by erosion) under extreme conditions to reshape the nearshore (as described above), with return of those sand volumes during ambient conditions. Erosion and scarping of dunes by waves begin a cycle (see Figure 4.22) from which the dune recovers when it has achieved a stable slope and becomes revegetated (Carter *et al.*, 1990a). The links between beach and dune can be broken temporarily, a process known as decoupling. For instance, the growth of vegetation on dunes slows the rate of sand movement, maintaining sand on the dune rather than enabling offshore winds to mobilise it. Another example is armouring of the beach surface that can occur when fine sand is blown from the beach, leaving coarse shell or gravel as a surface lag or pavement temporarily protecting the beach surface from further winnowing (Carter, 1988).

Dunes are particularly extensive in mid latitudes in the tradewind belts, behind high to moderate energy beaches (Jennings, 1964). They also occur in other settings. For instance, the dunes of Landes on the southwest coast of France reach 80–90 m high and 8–9 km wide. There was a view, now no longer accepted, that dunes were rare in the tropics (Jennings, 1965; Bird, 1965). Extensive dunefields have been described from Sri Lanka (Swan, 1979), northern Queensland (Pye and Rhodes, 1985), and occur in Fiji and Hawaii and on other tropical islands. Elsewhere, such as in Namibia, Oregon, Oman, Peru and parts of Australia, coastal dunes grade into inland dunes.

The backshore represents a critical boundary between the highly dynamic wave- and tide-dominated surf and swash zones, and coastal dune environments built by wind and involving a role for sand-binding vegetation. Beach sand dries when exposed subaerially and even when slightly wet can be entrained by wind. There appears to be a link between beach type and backshore processes, depending on whether the beach is stable eroding or accreting (Hesp and Short, 1999). The broader, flatter beaches of generally fine sand associated with the dissipative end of the beach continuum are more conducive to supply sand for entrainment by wind allowing relatively unimpeded transport. Larger foredunes occur at the rear of dissipative beaches (Short and Hesp, 1982; Hardisty and Whitehouse, 1988; Sherman and Lyons, 1994). Tidal range is also important, because broader source areas are likely to be exposed on the upper intertidal zone in macrotidal areas. Sand movement is usually by saltation with movement of larger grains by creep.

Wind velocity is slowed around any obstacle, such as a bottle on the beach, or a clump of vegetation, and an impeded dune can form in the shadow zone around the object. The extent of vegetation cover on the foredune influences its effectiveness in trapping sand (Hesp, 1988, 1999). The vegetation on dunes usually shows a zonation of species, with sand-binding grasses and herbs on the foredune and shrubs and trees at greater distance from the beach. Whether sand dune vegetation shows an orderly successional change needs to be viewed in the context of the nature of the dune system. One single dune ridge is more likely to show a stable zonation reflecting environmental gradients in factors that vary with distance from the sea, such as salt content in the sand, exposure to wind and humic content of the soil. Where there are a series of foredune ridges that have prograded, then distance may also be equivalent to time, and a temporal sequence of vegetation is possible (Carter, 1988).

Dune succession is likely to be episodic because geomorphological change is episodic. Some dunes, such as the grass-covered machair dunes of Scotland become arrested in their development due to repeated disturbance (Ritchie, 1979). Human impact on dunes is often considerable. Uses range from extraction, for example sand mining, to trampling, offroad vehicle use, deforestation and introduction of exotic species. There have been various attempts to stabilise dunes using fencing and other approaches, and even to construct completely artificial dunes (Nordstrom, 2000); some of these issues are examined in Chapter 10.

There have been various theoretical, empirical and experimental attempts to quantify the volume of sand transported by wind (see Table 3.4), but there are deficiencies to existing models (Berg, 1983; Pye and Tsoar, 1990). Observed transport rates differ from predicted because of a range of factors to do with the surface crusts or armouring, moisture, vegetation or microtopography (Sarre, 1989; Sherman and Hotta, 1990; Bauer *et al*, 1990). Discrepancies can also result because there is insufficient sediment available for transport (Bauer, 1991).

A range of dune types can form behind a beach in relation to beach and dune budgets (Sherman and Bauer, 1993). Beaches experiencing a positive sediment budget can supply sand to a dune, but dunes also form on eroding beaches that have a slight negative budget (Psuty, 1992). Figure 6.19 illustrates a sequence of dune morphologies and the factors which appear to lead to their occurrence (based on Pye, 1990; Psuty, 1992; Trenhaile, 1997). Where sand supply to the beach is abundant but winds are insufficient for dune building, a beach-ridge plain of swash-built beach ridges can form. In a similar setting with positive beach budget but higher energy, and efficient vegetation, a series of foredune ridges can be constructed (as discussed above), comprising beach ridges topped with eolian accumulations (Eliot *et al.*, 1998). In situations with a positive beach budget, moderate wind energy but ineffective vegetation cover, hummocky dunes can form. Transverse dunes, with ridge crests roughly perpendicular to the prevailing wind (and hence often shore-parallel), and steep lee slopes but gentle slopes towards the wind may occur as a variant of this form. They form where there is a good supply of sand, but an absence of anchoring vegetation, as on the margin of deserts.

If vegetation is very dense and efficient at trapping sand, a single vertically accreting foredune ridge is formed (Hesp, 1988; Arens, 1996). The foredune accumulation can show mesoscale variations which comprise alternating phases of build up during sediment excess and winnowing when sediment is scarce (Cooper *et al.*, 1999a). Parabolic dunes are U-shaped dunes with a low-gradient but poorly vegetated slope towards the shore, and a steep inland face on the distal part of the dune. They form when the sand budget supplied to the dunes is slightly greater than that to the beach, and the beach can be undergoing net loss of sed-

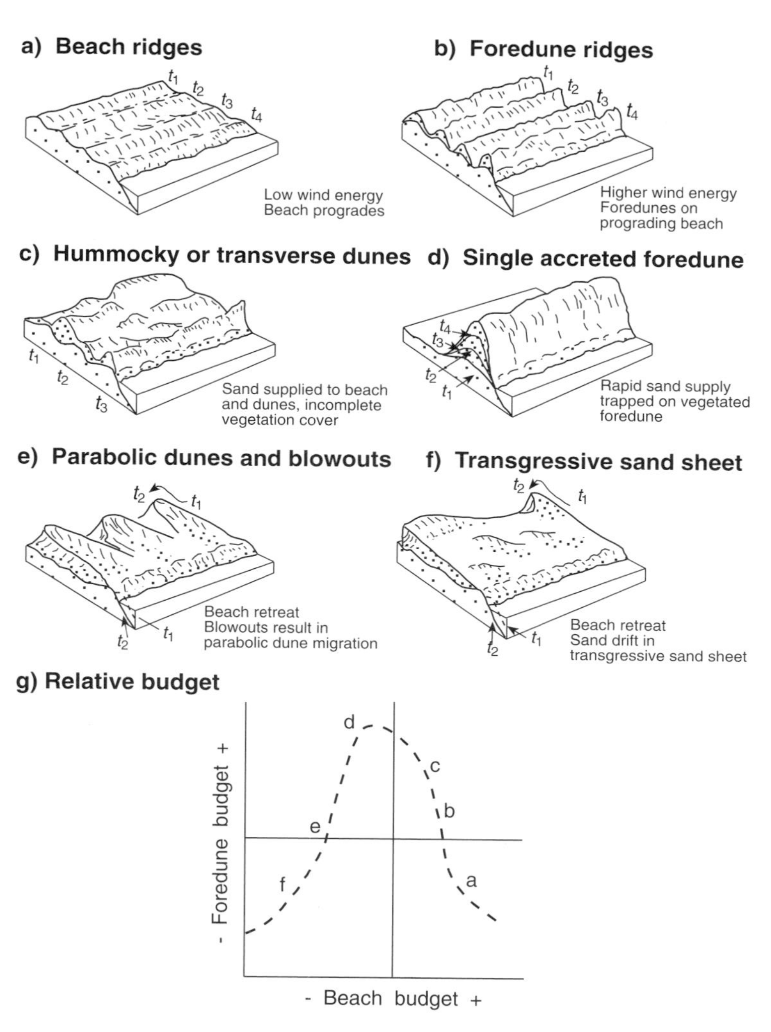

Figure 6.19. Dune formation and occurrence in relation to beach and foredune sand budgets. The morphology of (a) beach ridges, (b) foredune ridges, (c) hummocky or transverse dunes, (d) single accreted foredunes, (e) parabolic dunes and blowouts and (f) transgressive sand sheets (based on Pye, 1990; Trenhaile, 1997). (g) Occurrence of dune types in relation to whether beach and foredune sand budgets are positive or negative (based on Psuty, 1992).

iment. They tend to migrate inland as a result of the development of blowouts on the seaward slope (Jungerius and van der Muelen, 1989; Ritchie, 1992; Gares, 1992; Plius, 1992). A blowout shows a positive feedback, because the loss of vegetation not only permits deflation, but also means there is nothing to bind the sand (Calderoni *et al.*, 1999).

A transgressive sand sheet represents the largest scale of dune development; the entire sand sheet is poorly vegetated, and advances landwards with a steep wall of sand (Hesp and Thom, 1990). These large dune fields have also been called long-walled transgressive dunes or sand sheets, and 'precipitation ridges' (Cooper, 1958, 1967). They are actively moving inland (transgressing); the term 'transgressive' has been used to describe both vegetated and unvegetated sand sheets. These much larger dunes are a major component of some barrier systems, and they are examined further in that context in Section 6.5.5.

Other dune types include wrap-around dunes formed around particular obstacles, and cliff-top dunes, which result from deposition during times of high sea level or have formed by dunes climbing up a cliff face (Jennings, 1967; Short, 1988; Roy *et al.*, 1994). Longitudinal dunes are less frequent in a coastal setting than inland. It has been suggested that they might result when parabolic dunes break down through over-extension (Thom *et al.*, 1992). Oblique dunes can form at an angle to the coast where wind direction is not perpendicular to the coast. These can grade into barchans (Pye and Tsoar, 1990).

The various dune types shown in Figure 6.19 can also occur with distance away from a source of sediment such as a river. In these cases, there is likely to be a sequence of dune types indicating the variable balance between sand supply to the beach and that to the foredune (Psuty, 1992). The foredune height is likely to increase with distance away from the sediment source (Sherman and Bauer, 1993), and temporal patterns can be seen where rivers are cut off and cease sediment supply (Carter *et al.*, 1992). Dune types can also show a progression where there is a gradient in wind energy, as along the beach at Hawks Nest in New South Wales (Thom *et al.*, 1992). Figure 6.20 shows the dune sequence at Magilligan in Northern Ireland, where dune reworking contributes to barrier extension (Carter, 1986). The pattern of formation of the sand barrier on which beach and dune systems form is examined in the following section.

6.5 Barriers and barrier islands

Many contemporary beaches are the seawardmost active deposit on more extensive sand or gravel sedimentary 'barriers' which have accumulated over a longer time scale. A barrier is an elongate accumulation of sand and/or gravel formed by waves, tides and wind, parallel to the shoreline, rising above present sea level, often impounding terrestrial drainage or blocking off a lagoon. The example at Moruya, seen in Figure 6.18, shows that the modern beach occurs at the eastern extreme of a sequence of relict beach ridges, forming a prograded barrier. There have been various attempts to produce a morphological classification of sand and gravel coastal landforms. For instance, Zenkovich (1967)

Figure 6.20. The sequence of coastal foredunes at Magilligan Point, Northern Ireland. The foredunes have extended because of reworking of Holocene beach ridges (photograph R.W.G. Carter).

recognised (i) attached forms, including beaches and forelands, (ii) free forms such as spits, (iii) barriers such as looped and multiple forms, (iv) looped features including tombolos, and (v) detached forms which included barrier islands. Attached barriers comprise three components: the sand barrier complex itself with beach ridges or dunes; an enclosed lagoon or barrier estuary (see Chapter 7); and inlets or channels which cross the barrier and connect impounded waters with the sea.

Barriers represent a time-integrated depositional record of past beach, backshore and dune environments. They undergo extensive wave reworking during periods of activity, interspersed with periods of inactivity that are characterised by drowning, stranding above sea level, or burial by other sediment. Formerly active barriers can be preserved as relict shorelines indicating previous sea levels. Individual barriers offer only a brief sedimentary record of periods during which sea level has been relatively stable (stillstands). Although barriers occur on a large percentage of the world's coasts at present, they are not prominent in the overall sedimentary record because they can be reworked by rising sea level, or left stranded by falling sea level and reworked by subaerial processes.

The morphology and stratigraphy of barriers vary and can be related to several factors, particularly the pattern of sea-level change. In many cases, the modern barrier lies adjacent to an older Pleistocene feature formed during a previous stand of the sea. Along the east coast of the United States there are ridges and beach-ridge plains forming remnant barriers of Quaternary and perhaps late Tertiary age (Oaks and DuBar, 1974). Similar barriers, some formed of calcareous sands, occur at rising elevations in South Australia behind the modern barrier of the Coorong (Sprigg, 1979; see also Chapter 2, and Figure 2.6b).

Geological inheritance plays a significant role in sand-barrier construction, both in terms of present morphology (Pilkey *et al.*, 1993) and in terms of stratigraphy and evolution (Roy *et al.*, 1994). For instance, barrier islands of the eastern United States are frequently based on Pleistocene deposits (i.e. Sapelo Island, Georgia) or abut the Silver Bluff Pleistocene shoreline (Oertel, 1979). The barrier islands of the Wadden Sea are similarly located over a Pleistocene high (Streif, 1989), and the barriers of northern New South Wales, Australia, parallel Pleistocene inner barriers (Roy and Thom, 1981).

The evolution of barriers on wave-dominated coasts has been reviewed by several researchers (Swift and Thorne, 1991; Thorne and Swift, 1991a,b; Roy *et al.*, 1994). This section considers first the morphology of barriers (Figure 6.21), then examines the stratigraphy and evolution of barriers in relation to sea-level change (Figure 6.22). The range of landform sequences that can develop in association with stillstand barriers is examined. Transgressive barriers, and particularly barrier islands are described. Gravel barriers are discussed in Section 6.5.4, and finally dune-building phases and the development of transgressive dunes are examined in the context of the longer-term development of barriers.

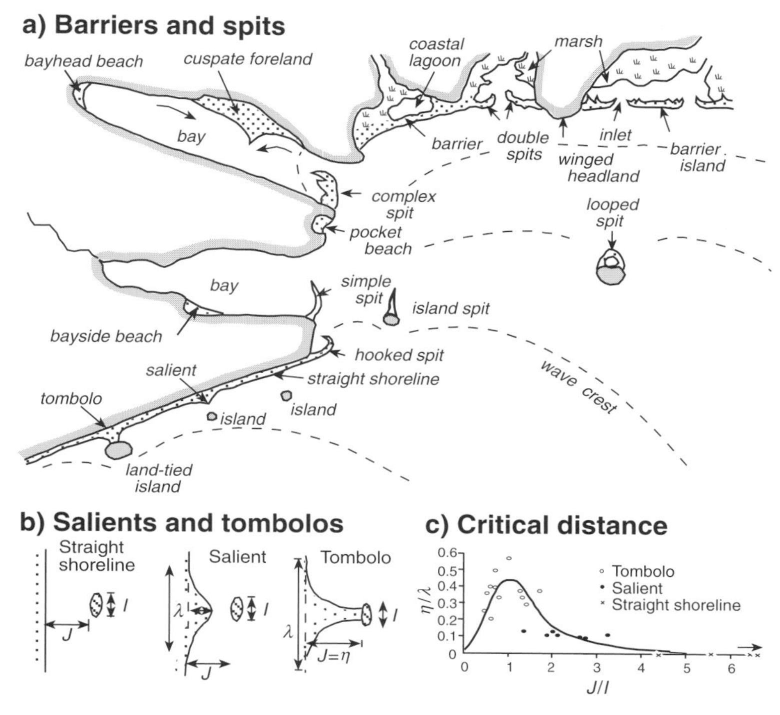

Figure 6.21. Various types of spits, tombolos and cuspate forelands. (a) The schematic distribution of features in relation to wave approach (based on Horikawa, 1988). (b) Defintion of parameters and the occurrence of salients and tombolos in relation to island diameter and (c) critical distance from shore (based on Horikawa, 1988).

6.5.1 Barrier morphology

The term 'barrier' was first used by G.K. Gilbert (1885) in relation to shoreline ridges of former Lake Bonneville. Lake shoreline barriers occur on the shores of Lake Erie (Bray and Carter, 1992) and Lake Michigan (Petty *et al.*, 1996), but adoption of the term by Johnson (1919) led to recognition of a much wider suite of coastal barrier landforms. Barriers occur across the mouths of embayments or within bedrock embayments as attached features, and examples of these have been extensively studied in southeastern Australia (see Section 6.5.2). In some cases, sandy (or gravelly) sediments accumulate as a detached beach; for instance, barrier islands are particularly extensive in southern and eastern United States (see Section 6.5.3).

Figure 6.21 shows a series of different barrier types, based on the range of features recognised by Johnson (1919). Spits represent subaerial projections of sediment generally accumulating as a result of alongshore sand or gravel transport. Spits commonly occur where there is an abrupt change in the direction of the shoreline, extending along the trajectory of the shore (Figure 6.21a). For example, Farewell Spit at the northwestern corner of the South Island of New Zealand extends more than 30 km north beyond the bedrock limit to the coast (McLean, 1978). Spits require both a supply of sediment and wave energy to move that sediment; they are rare in areas of large tidal range. Spits can be simple, straight, curved or hooked. Compound spits include several ridges, and recurved ridge crests usually mark former termini of the spit (Hine, 1979). Spits often build out across the mouths of estuaries (Kidson, 1963); for instance, Orfordness in eastern England is a shingle spit which has built more than 18 km long diverting the River Alde south behind it. Spits, or other barriers, at the mouth or within bedrock embayments are often called bay barriers.

Cuspate forelands (also called cuspate barriers) are pointed protrusions from the general alignment of the coast, usually containing a sequence of ridges (McNinch and Luettich, 2000), often enclosing lagoonal areas or swampy ground (Figure 6.21a). Dungeness in southern England is a cuspate foreland comprising ridges that have silted up Romney Marsh (Lewis, 1932; Long and Hughes, 1995). Cuspate forelands indicate swell approach from more than one direction. They have attracted particular attention on paraglacial coasts (those subject to glacial activity in the late Pleistocene, see Section 6.5.4), such as along the coast of Maine in the eastern United States (Duffy *et al.*, 1989) or in Patagonia (González-Bonorino *et al.*, 1999). However, they can also occur in other situations where refraction causes regular wave approach from two or more directions, for instance behind rocky outcrops or individual reefs in Western Australia (Semeniuk *et al.*, 1988; Sanderson and Eliot, 1996).

A tombolo is a similar spit-like projection from the shore that links an island to the mainland (Figure 6.21). Wave refraction in the lee of the island shapes a tombolo, although where the water is deep tombolos can also be built by diffraction (Flinn, 1997). Sediment to build a tombolo can be derived from the island, from the mainland beach, from the seafloor, or from a combination of these sources. Where wave refraction processes in the lee of an island are insufficient to connect with the island, a cuspate feature, called a salient, is formed, that can grow into a tombolo if further progradation occurs. Whether tombolos or salients form appears to depend on configuration of the coast. A tombolo forms where the ratio of the island's offshore distance to its length is equal or less than 1.5 (J/I; see Figure 6.21b); a salient forms where it is 1.5–3.5, and no protrusion of the coast occurs where it is greater than 3.5 (Sunamura and Mizuno, 1987).

Several explanations can account for the development of spits and other barriers. Although longshore sediment transport seems to be important, as proposed by Evans (1942), landward (transgressive) reworking of a seaward source of sediment can result in the development of a barrier that can adopt a spit or tombolo morphology. For instance, Chesil Beach forms a tombolo that joins the Isle of Portland to the Dorest coast in southern England. It contains coarse gravel to the east, fining to the west, which has been difficult to explain in the context of longshore transport, and the barrier appears to have formed offshore and been worked landward to its present position (Bray, 1997). A similar explanation is likely to explain other spit-like features such as Blakeney Point in north Norfolk, England (Funnell and Person, 1989), or the gravel beach at Nelson, South Island, New Zealand (Dickinson and Woolfe, 1997; see Figure 6.30). Spits that have not been built primarily by longshore drift are often formed on top of a spit platform (Novak and Pederson, 2000).

6.5.2 Stillstand barriers

Four factors appear to be critical to the development of barriers: substrate gradient, wave energy versus tide energy, sediment supply in relation to accommodation space, and rates of sea-level change. Figure 6.22 indicates the principal morphological features and stratigraphy of barriers on transgressive, stillstand and regressive coasts. Coasts fall into one of three categories in this respect, low-lying coastal plain coasts, more irregular embayed coasts, and relatively steep, cliffed or protruding sectors of coast (Roy *et al.*, 1994).

The range of landforms observed along a wave-dominated coast on which sea level is stable (stillstand) are shown in Figure 6.22. These are drawn primarily from extensive morphostratigraphic studies of sand barriers in southeastern Australia (Bird, 1973; Roy and Thom, 1981; Thom, 1978, 1984a; Thom *et al.*, 1978, 1981, 1992; Chapman *et al.*, 1982), but with evidence for similar patterns elsewhere (Dickinson *et al.*, 1972). In the case of central and southern New South Wales, headlands and submerged bedrock ridges restrict sediment exchange between compartments and the contrast between different morphologies resembles that shown in Figure 6.22. Where pre-existing bedrock control is less embayed, as in the Gippsland Lakes areas of Victoria (Thom, 1984b), barriers arise from reworking of former (Pleistocene) barriers and longshore sand movement (Bryant and Price, 1997). In this generalised diagram, different barrier types are shown with increasing volumes of sand stored in the subaerial portion of the barrier to north; several of the barrier types are apparent in Figure 6.1.

Holocene barriers first occupied their present positions along the New South Wales coast around 6500 years BP at the time that sea level stabilised (Thom and Roy, 1985). In northern New South Wales (as also in Gippsland, Victoria), the Holocene outer barrier abuts an older Pleistocene (often Last Interglacial) inner barrier. In southern New South Wales, this has been eroded away, either because the steeper shelf allowed higher wave energy at the shore (Chapman *et al.*, 1982), or because of stripping by high-energy events (Young and Bryant, 1992). It has been shown that where seafloor gradients are too steep for sand to accumulate on the shore, sand bodies form on the shelf instead (Roy *et al.*, 1994).

On steep coasts, or where there is a deficiency of sediment, a receded barrier generally forms. This is narrow and has retreated over back-barrier or estuarine sediments that can be exposed periodically on the foreshore. Receded barriers show stratigraphic relationships that are common in transgressive settings (see below), and both sandy barrier islands of the east coast of the United States, and gravel barriers, as in Nova Scotia, can retreat in this fashion (Orford *et al.*, 1991).

The balance between sand supply and the ability of wave energy to move sand onshore since onset of the stillstand influences the nature of the sand barrier. For instance, where sand was rapidly moved onto the barrier and the offshore supply then depleted as in isolated compartments, a stationary barrier is formed, usually with a complex foredune

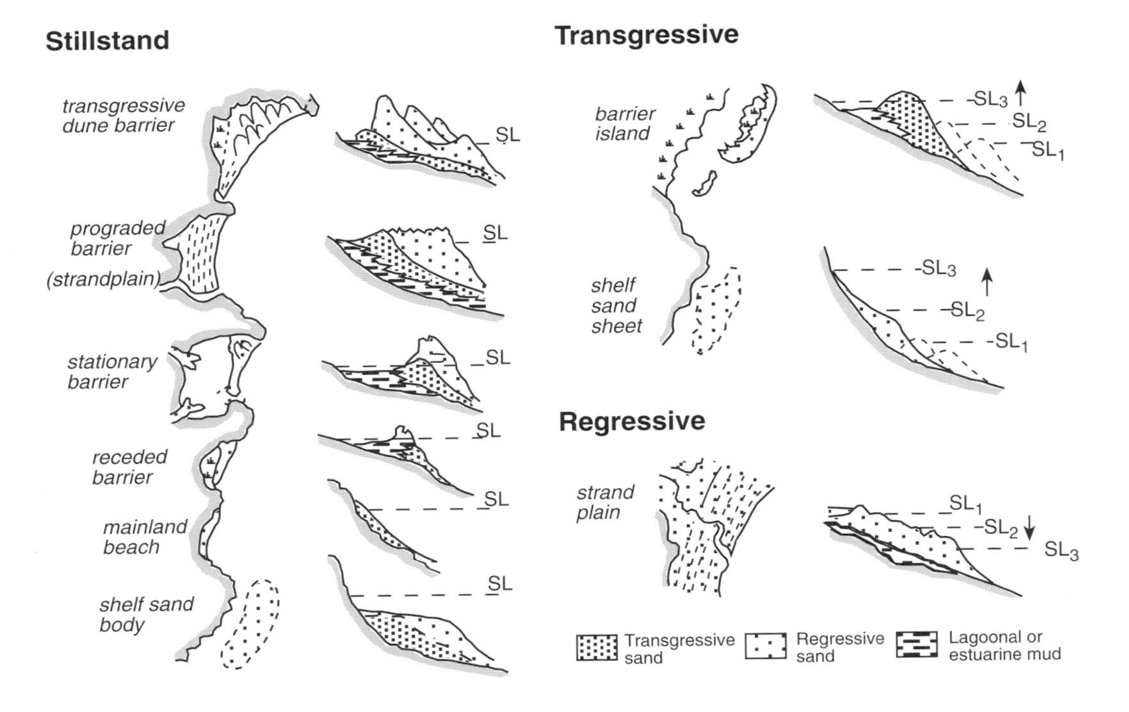

Figure 6.22 Barrier types in relation to sea-level change. Generalised morphology and stratigraphy of stillstand barriers, transgressive barriers, and regressive barriers (based on Roy *et al.*, 1994).

structure. By contrast, the ongoing supply of sand leads to a prograded barrier, blocking off lakes and swamps in drowned valleys. Prograded barriers are equivalent to the strandplain or beach-ridge plains of other writers (Boyd *et al.*, 1992). Rates of progradation for six prograded barriers in southeastern Australia fall in the range 0.2–0.6 m a^{-1} (Roy *et al.*, 1994). Many show rapid build out in the millennia following sea-level stabilisation, and then decelerating increase in volume, perhaps resulting from exhaustion of the offshore sand supply. Barrier progradation can be maintained where supply of sand from a fluvial source occurs, as, for instance, at Seven Mile Beach north of the Shoalhaven River mouth (L.D. Wright, 1970; see Figure 6.1).

Figure 6.18 shows a cross-section of the Moruya prograded barrier based on several radiocarbon-dated drillholes (Thom *et al.*, 1981). A sequence of shelly sand with gravel overlies basal estuarine clay (which was presumably deposited behind a barrier) and records the landward migration of the shoreline as sea level rose during the postglacial transgression. This is overlain by a prograded (regressive) beach-ridge sequence. The embayment does not experience sufficient winds to have developed transgressive dunes. Radiocarbon dates on shells, which have been transported from the position in which they lived, record the gradual extension of the coastal plain seawards. The volume of sediment, and also the rate of progradation were greatest in the period 6500–6000 years BP with gradual decline thereafter, and with negligible progradation since 3000 years BP.

There are prograded plains, generally dominated by sequences of foredune or beach ridges, elsewhere in the world. Indeed, it is clear that if there is sufficient sediment supply, then even under conditions of gradually rising sea level it is possible for the shoreline to prograde a sequence of ridges. Figure 6.23 shows the sequence of ridges and tentative isochrons on their progradation at Nayarit in Mexico (Curray *et al.*, 1969), on Galveston Island (Bernard *et al.*, 1962; Morton, 1994), and Kiawah Island in South Carolina (Moslow and Heron, 1979; Hayes, 1994). In each of these cases, as in the case of Moruya, there has been a deceleration of the rate at which the coast has prograded. This appears not only to be a function of building into deeper water, but also to indicate a smaller volume of sand accumulating over time. Other progradational beach-ridge plains which have been examined in detail include Padre Island (Fisk, 1959), Tabasco in Mexico (Psuty, 1965), eastern Malaysia (Nossin, 1965), northern and southwestern Florida (Stapor, 1975; Stapor *et al.*, 1991), and Sherbro Island in Sierra Leone (Anthony, 1991).

Sequences of ridges can also be formed on a regressive shoreline that has experienced sea-level fall (Figure 6.22). This occurs in high latitude situations where there has been a high rate of isostatic adjustment,

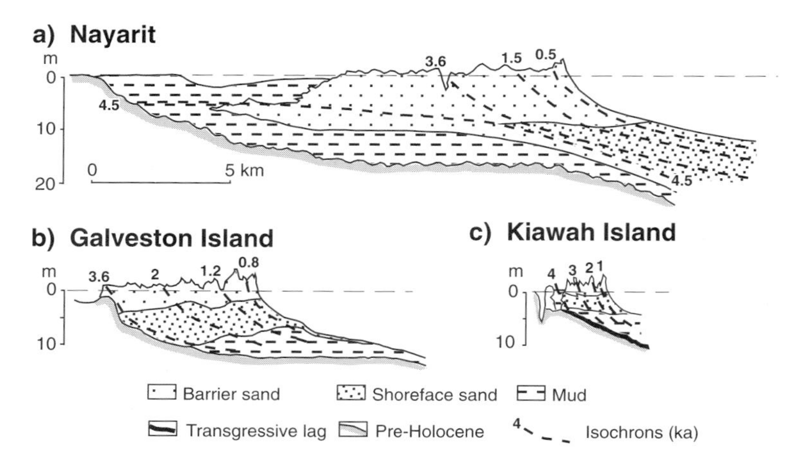

Figure 6.23. Morphostratigraphy of prograded barriers from Nayarit, Mexico (based on Curray *et al.*, 1969), Galveston Island (based on Bernard *et al.*, 1962), and Kiawah Island, South Carolina (based on Hayes, 1994).

and in situations such as the western and northern Black Sea where water level has fallen (Zenkovich, 1967). Regressive shorelines which receive an abundant supply of sediment can prograde as a sequence of parallel foredune ridges despite sea-level fall, as in eastern Hudson Bay, Canada (Ruz and Allard, 1994).

Transgressive dune barriers are transitional between barriers with complex foredunes and non-barrier desert coastlines where eolian dune sheets are influenced by shoreline processes (Hesp and Thom, 1990). The use of the term transgressive in this case refers to the movement inland (transgression) of the dune sand, and should not be confused with transgression of the sea during sea-level rise (see Figure 6.22). Transgressive dune barriers can form during stillstand or regression, or even during rising sea level; radiocarbon dating of soil horizons in southeastern Australia indicates that some transgressive dune barriers had begun accumulating as the sea was rising (see Section 6.5.5). If there is an increase in wave energy, sediment is supplied to the beach more rapidly, leading to initial progradation and then exhaustion giving a period of shoreline stability. These barriers are also called episodic transgressive barriers, and evidence for phases of activity is discussed below.

6.5.3 Barrier islands

Barrier islands are elongate, shore-parallel islands, composed of sand or gravel, separated from the mainland by a back-barrier lagoon. They have become popular sites of residential and tourist development.

Barrier islands are found on much of the world's coastline, but are most prominent on trailing-edge coasts experiencing relative sea-level rise (Chapter 2), although there are barrier islands on collision coasts, for example, on the tropical coast of Colombia, South America (Martinez *et al.*, 1995). They require a supply of sediment (Davis, 1994). Barrier islands are wave-dominated features, with waves shaping their characteristically long straight or gently curving seaward shore, causing longshore drift, and leading to episodic washover that forms washover fans on the inner margin (Figure 6.24). The relative significance of waves and tides has been debated. Hayes (1979) indicated that barrier islands are common on microtidal coasts. It is convenient to consider the relative influence of wave and tide processes, and to recognise mixed-energy barrier islands (Davis and Hayes, 1984). Three geomorphologically distinguishable types of barrier islands are shown in Figure 6.24: wave-dominated, mixed-energy and gravel barriers.

Barrier islands are often found where sea level is rising (Figure 6.22), but it seems that there are several situations in which analogous landforms can develop (Schwartz, 1971). Many barrier islands appear to form as a result of vertical accumulation and landward transport of

Figure 6.24. Major types of barrier island. Wave-dominated and mixed-energy types are distinguished based on Davis and Hayes (1984). Gravel barriers are examined in greater detail in Figure 6.28.

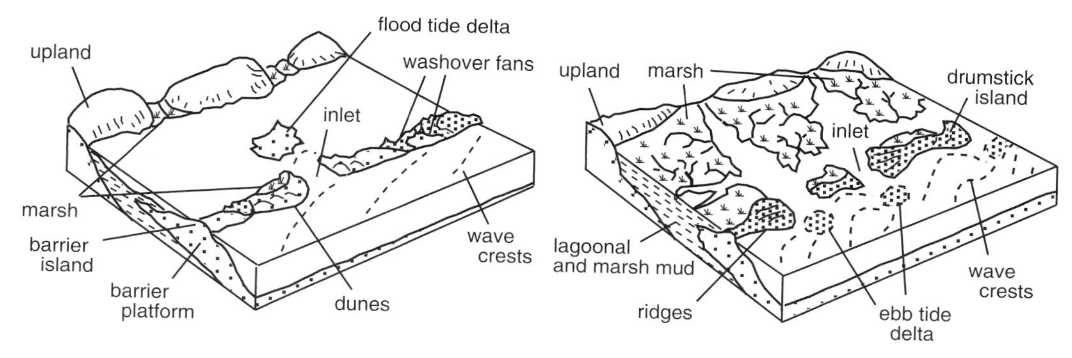

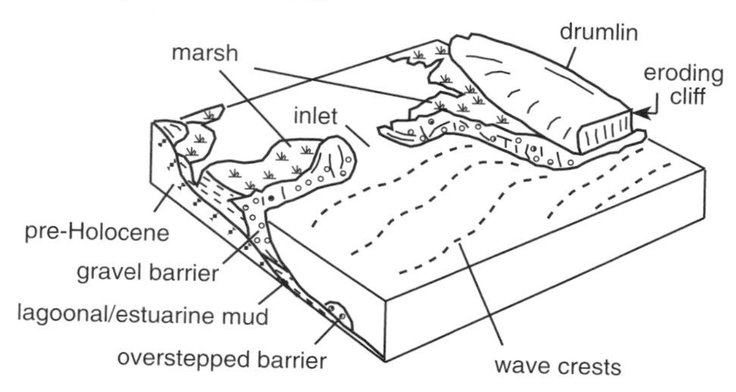

sand (Davis, 1994). Figure 6.25 shows Grand Isle, a barrier island adjacent to the Mississippi delta in southern Louisiana, where there are a range of relative sea-level changes as a result of subsidence of the delta (see Chapter 7). Barrier islands can also form where the sea is stable. These regressive barrier islands show progradational stratigraphy, or in rarer instances may build up *in situ* through aggradation (Davis, 1994). Aggradational barrier islands have also been described from Mississippi Sound (Otvos, 1981, 1985).

An example of distinctive landforms associated with mixed-energy conditions is a drumstick island (Figure 6.24b). Barrier geometry, inlet spacing and the nature of tidal deltas are related to tidal processes (Nummedal *et al.*, 1977). Tidal influence is greater where there is a large tidal prism in the back-barrier lagoon, resulting in more or larger inlets to facilitate exchange (see Chapter 8 for a discussion of inlet equilibrium). Inlets can have flood tide deltas that extend into the back-barrier lagoon, or ebb tidal deltas that are better developed seaward of the inlet where there is a larger tidal range. Wave refraction alters the direction of longshore drift in the vicinity of the inlet, building the drumstick nature of the end of the island which becomes a sediment trap, resulting in the distinctive 'drumstick' shape (Hayes, 1979; Davis, 1994). The morphology can also allow sediment to bypass inlets (FitzGerald *et al.*, 1992).

Barrier islands and spits are virtually absent from macrotidal coasts (Hayes, 1975), although gravelly barriers, forming incipient cheniers

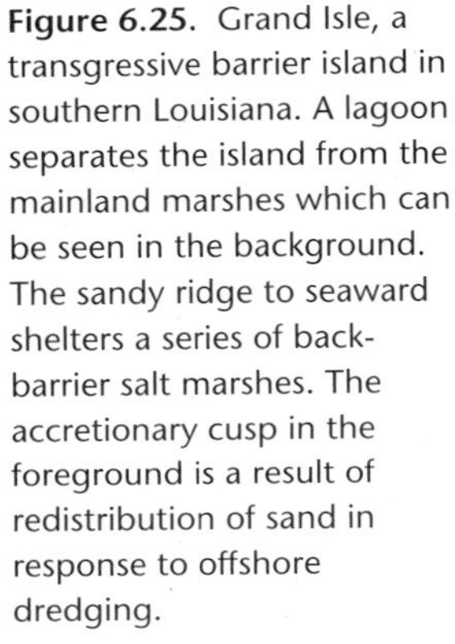

Figure 6.25. Grand Isle, a transgressive barrier island in southern Louisiana. A lagoon separates the island from the mainland marshes which can be seen in the background. The sandy ridge to seaward shelters a series of back-barrier salt marshes. The accretionary cusp in the foreground is a result of redistribution of sand in response to offshore dredging.

have been described from King Sound, Western Australia (Jennings and Coventry, 1973). The Dutch and German barrier islands illustrate the effect of tidal range; the range increasing from 1.2 m at Den Helder in the Netherlands to more than 4 m in the German Bight, where the mouths of the Ems, Weser and Elbe are funnel-shaped in response to tide (see Chapter 3). At this extreme of tidal range in the apex of the German Bight, barrier islands do not form.

Barrier islands do occur, especially as isolated gravel barriers on the tide-dominated coasts of northern Maine (FitzGerald and van Heteren, 1999). FitzGerald *et al.* (1994) and FitzGerald and van Heteren (1999) have expanded both the early classification of Hayes (1979) based on tidal range, and the classification of Davis and Hayes (1984), in which the interaction of wave and tide energy was considered, to include compartmentalisation of the coast by headlands or inlets (Figure 6.26). The New England barrier islands are complex depending on the degree of shelter and the sediment available (Kochel *et al.*, 1985). They can be divided into distinct geographical groupings (FitzGerald and van Heteren, 1999) related to tidal range and wave energy (Figure 6.26).

Barrier island migration is highly dependent on inlet characteristics (Aubrey and Gaines, 1982; Davis *et al.*, 1989). Instability of inlets is demonstrated by inlet migration, particularly where there is longshore movement of sediment, or by inlet closure. At least eight different responses are possible. These include: lateral movement (erosion on one end and spit elongation of the other); advance through seaward progradation; dynamic equilibrium with long-term movement but maintaining

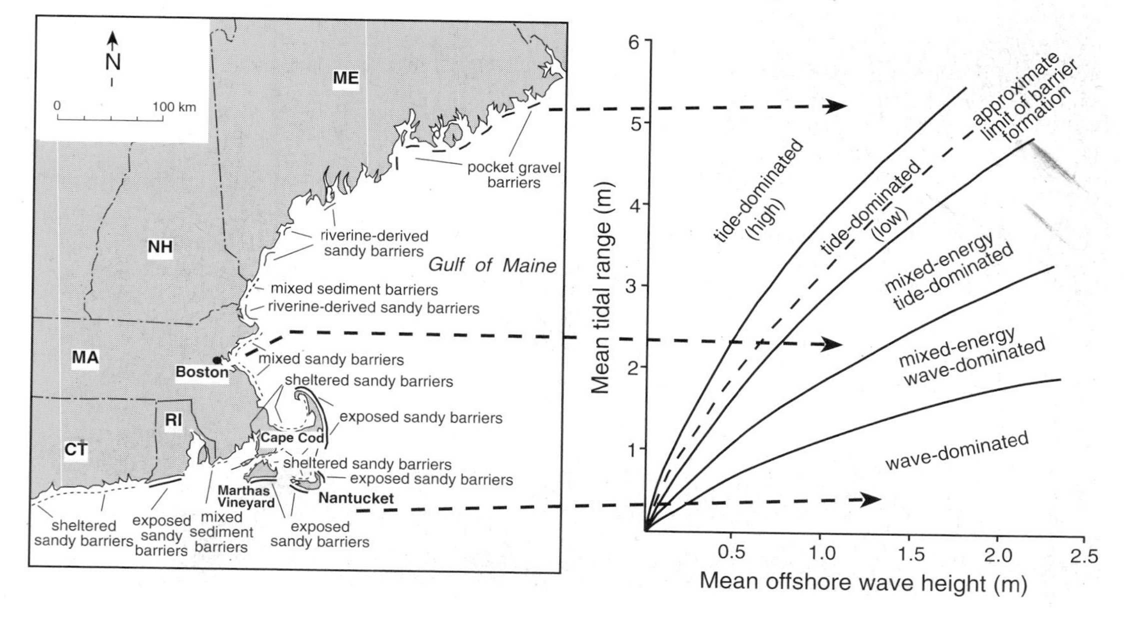

Figure 6.26. Barrier islands along the coast of New England (based on FitzGerald *et al.*, 1994), and a comparison of the wave- versus tide-dominated classification of barrier island coasts (after Davis and Hayes, 1984). The occurrence of selected parts of the New England coast on the classificatory diagram is shown (based on FitzGerald and van Heteren, 1999).

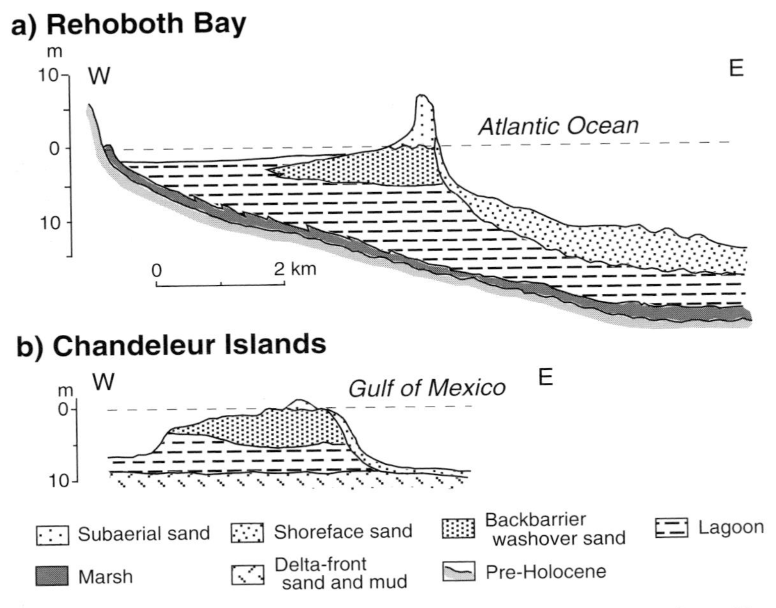

Figure 6.27. Morphostratigraphy of transgressive barrier islands as indicated by Rehoboth Bay, Delaware (based on Belknap and Kraft, 1981), and the northern Chandeleur Islands, Mississippi Delta (based on Otvos, 1986).

a constant shape; retreat through shoreline erosion; narrowing; landward rollover as a result of overwash during storms; break up through the development of new inlets; and rotational instability (McBride *et al.*, 1995; Kana *et al.*, 1999).

The barrier islands along the east coast of the United States have been studied in detail (Leatherman, 1979, 1983; Nummedal, 1983; Tanner, 1990; Davis, 1994). Over short time scales barrier islands appear stable despite rising sea level, with aggradation and landward widening resulting from overwash and eolian deposition if sand supply is sufficient. Where sand supply is limited, barrier islands migrate in a landward direction (Niederoda *et al.*, 1985). However, it is clear that many barrier islands are undergoing a longer-term landward migration related to relative sea-level rise.

It is possible to recognise transgressive barrier islands and regressive barrier islands each with distinctive stratigraphy (Kraft and John, 1979). The lagoonal muds form a continuous sequence beneath transgressive barrier islands, as shown at Rehoboth Bay, Delaware, eastern United States (Figure 6.27). Barrier islands off the Mississippi Delta (Chandeleur Islands, see Chapter 7) are transgressive because subsidence of the abandoned delta plain means that they are subject to rapid rates of relative sea-level rise (Figure 6.27). An evolutionary sequence by which an abandoned distributary is reworked forming an erosional headland with flanking barriers, which through progressive subsidence becomes a chain of barrier islands and subsequently a shelf sand shoal has been described by Penland *et al.* (1985, 1988) and is examined in the next chapter (see Figure 7.13). A regressive barrier island is one that has prograded seawards; Galveston Island (Figure 6.23b) is a good example, showing the wedge of shoreface sediment over which the island has built.

It has been shown by Kraft (1971) that the sediments deposited beneath a transgressive barrier island are reworked when sea-level rise is relatively slow, but are bypassed quickly, and remain in stratigraphic succession if sea-level rise is rapid (Fischer, 1961; Field and Duane, 1976). There has been some debate about whether barrier islands are able to overstep, that is retreat landward by more than the island's width, under rapid sea-level rise. Rampino and Sanders (1980) described evidence of former barriers on the shelf off Long Island, New York indicating that overstepping would occur if sea level rose too rapidly. This was challenged by Leatherman (1983), who believed that barrier islands would migrate continuously landward. It is certainly the case that gravel barriers, described below, can overstep when sea level rises rapidly (Forbes *et al.*, 1991; Ruz *et al.*, 1992, see Figure 6.28). The gradient and topography of the seafloor appears to control whether a sand barrier keeps pace with a rise in sea level; modelling of sand barriers on the New South Wales coast implies that gradients of $>0.8°$ are too steep and the sand remains entirely subaqueous forming a shelf sand body (Roy *et al.*, 1994). Further modelling also simulates a stage at which a barrier can overstep (see Figure 9.9c).

Barrier island formation, rates of evolution, and stratigraphy vary both locally and regionally in response to several factors, such as wave and tide energy, sediment supply and particularly relative sea-level change (Hesp and Short, 1999). In the case of many of the Dutch barriers, the rate of sediment supply in mid–late Holocene has been sufficient that they have prograded despite a gradual rise in sea level. Other examples of regressive (prograded) barrier islands were described in the previous section (see Figure 6.23), including the Nayarit Coast of Mexico, Galveston Island (Texas), and Kiawah Island (South Carolina). These regressive sequences, therefore, emphasise the point that barriers can prograde despite gradual sea-level rise if there is sufficient sediment supply (Nummedal, 1983), but most of the east and south coast of North America is dominated by transgressive stratigraphies.

6.5.4 Gravel barriers

Gravel barriers are common on paraglacial coasts, which are coasts that have been conditioned by glaciation or are built from glacial deposits, although they are not presently subject to glacial processes (Forbes and Syvitski, 1994). Shorelines associated with drumlin fields have received particular attention since the detailed geomorphological observations of Johnson along the coast of New England and Nova Scotia (Johnson, 1925). Drumlins are consumed by wave action, and provide an anchor from which attached barriers form, and many of the geomorphological processes originally proposed by Johnson (1919, 1925; see Figure 1.5) have been reinforced by more recent studies.

Erosion of drumlins on the eastern shore of Nova Scotia provides a temporary source of poorly sorted material which becomes increasingly sorted with time and with distance as it is carried alongshore (Carter, 1988). An individual drumlin continues to provide a mixture of coarse and fine material to adjacent beaches over a period of about 2000 years (Carter *et al.*, 1989, 1990b). Waves erode drumlins and then transport sediment grading the beaches according to size on sandy barriers (Boyd *et al.*, 1987; Nichol and Boyd, 1993) and on gravel barriers (Carter and Orford, 1993). In time, and with distance from the headland (drumlin source), the barrier becomes more organised in terms of form and sediment size.

A morphodynamic model of the way that these coasts adjust is shown in Figure 6.28; there are significant contrasts between drift- and swash-aligned sections of the coast (Orford *et al.*, 1991). Pronounced gradients in wave height, angle of wave approach, longshore currents and wave power account for sediment distribution. Drift-aligned barriers, along which sediment is rapidly transported, flank eroding drumlins in their early stages, and are later replaced by swash-aligned barriers (Orford and Carter, 1995). Swash-aligned beaches form parallel to wave crests of incoming waves; sediment transport along them is minimal and they are dominated more by transgression (Forbes and Syvitski, 1994; Forbes *et al.*, 1995). The sandy spit platform appears to be decoupled from the overlying gravel barrier (Shaw and Forbes, 1992). Spits form and elongate during periods when the sea rises but they become sediment starved, thin and breached under stable sea-level conditions (Carter, 1988). The barriers can then overstep, with penetration of wave energy into formerly sheltered back-barrier environments. Alternatively, crestal build up may continue with a steep aggraded barrier forming. A third possibility is that ridges prograde forming a prograded gravel beach-ridge plain, that has been termed a concatenated sequence

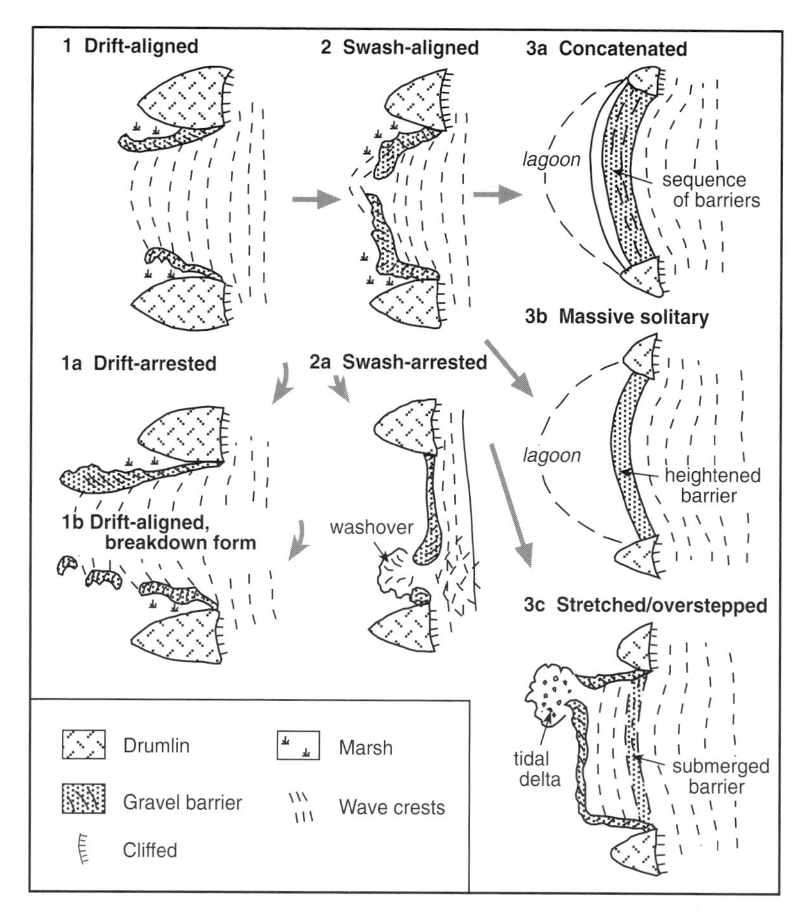

Figure 6.28. Evolutionary stages of gravel barriers in Nova Scotia (based on Orford *et al.*, 1991, and Carter and Orford, 1993). Individual eroding drumlins break down with removal of material following the sequence of stages from 1 to 3.

(Figure 6.28). Catastrophic switch from one state to another occurs when a heightened solitary barrier oversteps (Forbes *et al.*, 1991; Carter and Orford, 1993).

Gravel barriers show a tendency to prograde if sediment supply is abundant, but they move landward with reworking of the ridges, termed cannibalisation, even under stable sea-level conditions, if there is decreased sediment supply (Carter and Orford, 1984; Forbes *et al.*, 1994). Figure 6.29 shows the eroding Grand Desert drumlin, flanked by a mixed gravel and sand spit. Phases of cannibalisation (where sediment is eroded to maintain longshore growth) have occurred in other settings, for instance in Tierra del Fuego (Isla and Bujalesky, 2000). Figure 6.30 shows a gravel barrier along the coast at Nelson in the South Island of New Zealand. Stratigraphy and dating of this barrier imply that it has moved landward as a transgressive barrier as sea level has risen (Dickinson and Woolfe, 1997).

It was particularly to coarse reflective beaches, such as gravel barriers, that Cornish was referring in the quotation at the beginning of this chapter. He was aware of the subtle changing equilibrium forms of these beaches and how they responded to storms. He was also aware that natural adjustments could be misinterpreted to indicate serious erosion. Nevertheless, human impact exerts additional pressures on these beaches and understanding of the natural morphodynamics is an essential component of their management. The example of gravel barriers at Hallsands in Devon southwestern England demonstrates management problems. A seawall was built at Hallsands in 1841, but gravel was subsequently dredged from Start Bay between 1896 and 1902 for harbour works at Plymouth. Extraction of gravel from Start Bay had disastrous consequences for the village of Hallsands which was washed away in 1917 (Ward, 1922; Robinson, 1961). There are other gravel beaches around the coast of Britain, such as those at Porlock in Somerset and on the coast of Sussex, which have undergone erosion in recent years and where human intervention is implicated (Orford *et al.*, 2001).

Figure 6.29. Grand Desert drumlin, an eroding drumlin on the eastern shore of Nova Scotia, looking east across the eroding drumlin face, showing the mixed sand and gravel that is fed to the drift-aligned spit (photograph R.W.G. Carter).

6.5.5 Dune-building phases

Coastal dunes that form on barriers have undergone episodes of sand accumulation interspersed with stabilisation and soil formation (Orme, 1990; Muckersie and Shepherd, 1995). Transgressive dune fields are defined as those that move inland over older surfaces, marked by buried soil horizons or other signs of periodic activity. The development of podsol soils on dunes would appear to take thousands of years (Thompson, 1981; Bowman, 1989). The age of the dunes can be established using several dating techniques, including radiocarbon dating of material from paleosols, or luminescence dating of quartz sand grains. Although siliceous dunes can be more extensive, calcareous cementation of carbonate dunes (eolianite) preserves a more detailed history of dune activity.

Transgressive dunes in eastern Australia contain large volumes of sand and occur in the most exposed locations, often downdrift in terms of longshore transport. Long-walled transgressive dunes are typical, and occur on several large islands such as Fraser and Stradbroke Islands. Paleosols within the dunes indicate several phases of activity, and dating has indicated that extensive Late to Middle Pleistocene deposits have been reworked during the last glaciation (Cook, 1986;

Figure 6.30. Gravel barrier at Nelson, South Island, New Zealand. This barrier appears to have migrated landward as the sea has risen.

Tejan-Kella *et al.*, 1990; Bryant *et al.*, 1994, 1997). In northern Australia dune building during the last glaciation, 24 000–17 000 years BP, in early Holocene 9000–6500 years BP, mid Holocene (around 4000 years BP), and at least two phases in the late Holocene 2700–1800 years BP is indicated (Lees *et al.*, 1990, 1995). The dunes at Cape Flattery in Queensland have yielded dates that were beyond radiocarbon dating (>48 000 years BP) and early Holocene ages (9000–7000 years BP; Pye and Rhodes, 1985). Several explanations have been offered for variations in dune activity, including fluctuations in sand supply, wind regime and vegetation vigour.

Phases of dune activity have been related to changes in sea level which might influence the availability of a sand supply (Figure 6.31). One view is that sand is available for incorporation into dunes immediately following a highstand of sea level, when subtidal sediments are exposed to wind action (Wright, 1963). An alternative view is that dunes formed at times of low sea level. This view was advanced by Ward (1977) who believed that dunes were formed from sand exposed on the Australian continental shelf during glacial stages. In this case, coastal dunes might have coincided with dunefields that were active within central Australia (Bowler *et al.*, 1976; Wasson and Clark, 1987). Although this is clearly not the only time of dune activity, thermoluminescence (TL) dating indicates this to have been a period of extensive reworking of dunes (Thom *et al.*, 1994).

Early Holocene dune activity implies that sea-level transgression supplies sand for dunes as the coast is eroded. Active parabolic dunes in Queensland were associated with shoreline erosion, together with destruction of vegetation (Pye and Bowman, 1984). Phases of dune activity in the late Holocene, during which time sea level has been relatively stable around Australia, imply that mechanisms other than sea-level change must also be significant triggers for dune instability. Climatological causes, such as periods of storminess have been inferred (Thom, 1978). Human causes, such as grazing or aboriginal burning may also have been responsible (Bird, 1974). Dune activity appears to have shown some cyclicity; for instance, dunes appear to have been reactivated every 1500 years, in Jutland, Denmark (Clemmensen *et al.*, 2001).

Eolianite

Eolianite is fossil dune limestone, generally of bioclastic sand (calcarenite). Although when the term was first used (Sayles, 1931) it was intended to cover any lithified wind-blown deposit, it has tended to be used for coastal dune calcarenites of Quaternary age (Fairbridge and

Johnson, 1978). Most often the sands are calcareous, although they can be siliceous, but cemented by calcareous cement. The cement preserves the bedding of the dune deposits, frequently at angles near the angle of repose for sand of around 34° (Figure 6.31a). In some cases the dune morphology, with transverse ridges parallel to the shore, is also preserved. Eolianite often contains prominent paleosol horizons marking periods of soil development during a hiatus in dune activity (Figure 6.31b). There may also be lesser protosols marking shorter intervals between depositional events. Eolianite occurs on isolated mid-ocean islands, on carbonate banks, and on continental shelves. The degree of cementation reflects the abundance and composition of carbonate

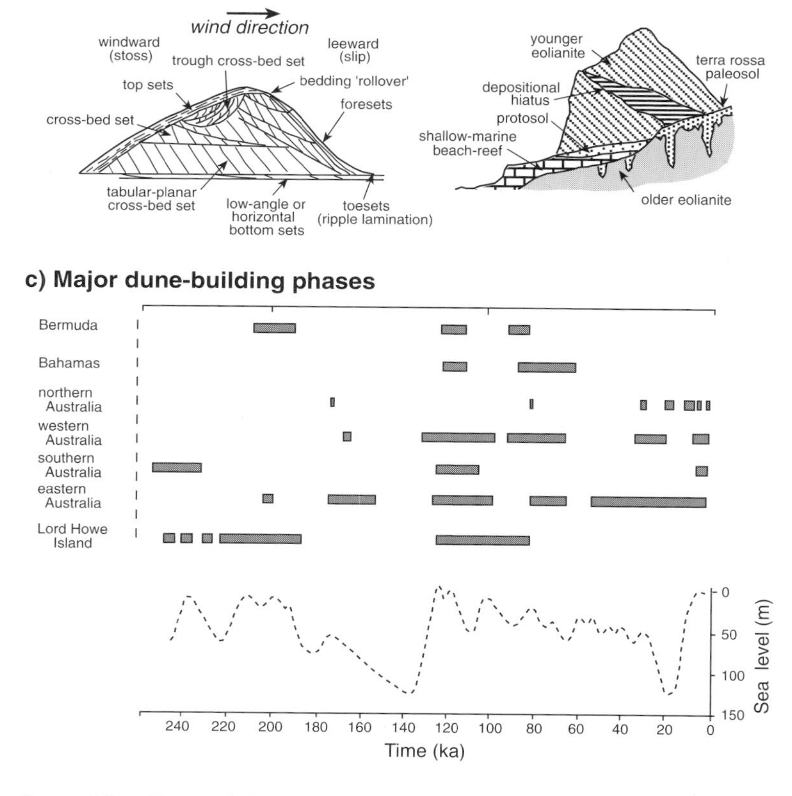

Figure 6.31. Phases of dune activity. (a) Bedding in modern dunes (based on Pye and Tsoar, 1990). (b) Eolianite stratigraphy as typical on Bermuda (based on Hearty *et al.*, 1992). (c) Dune activity as indicated by siliceous dunes and eolianite in the north Atlantic and the Australian region (based on Brooke, 2001).

sediment, the topographic setting and hydrodynamics, the effects of vegetation and soil moisture, and the dominant wind regime.

It has been realised that eolianite sequences preserve a record of changing environments since Darwin examined outcrops on St Helena and Ascension Islands and on the south coast of Australia (Darwin, 1851). He interpreted that these had been deposited as dunes during times of lower sea level. Particularly impressive sequences of dune units occur on Bermuda, separated by prominent *terra rossa* soils (paleosols). Sayles (1931) interpreted these dunes to have formed when the sea was low during glacial periods exposing the shallow platform around the islands, and the soils to have formed in interglacials. In fact the dune units can be seen to grade into interbedded shallow-marine beach and reef units, such as the Rocky Bay unit which contains Last Interglacial corals (see Figure 6.31b). It has subsequently been shown that the dune units were deposited during highstands of the sea (Bretz, 1960; Land *et al.*, 1967; Vacher and Rowe, 1997). Extensive lithostratigraphic and chronostratigraphic studies (using uranium-series and amino-acid racemisation dating) support the differentiation from Bermuda of dune units at oxygen-isotope stages 11, 9, 7 and substage 5e, with the youngest sequence of dunes belonging to the Southampton Formation deposited during substage 5a (Hearty *et al.*, 1992; Vacher *et al.*, 1995).

Eolianite has formed on many islands flanking major carbonate shelves with a warm climate favourable for carbonate production and in the tradewind belt so that onshore winds mobilise beach sand (McKee and Ward, 1983). The extensive eolianite deposits of the Bahamas have formed at similar times to those in Bermuda (Figure 6.31c), though they may have been active for longer periods after each interglacial (Hearty, 1998; Hearty and Kaufman, 2000). Dunes appear to have also been formed in and immediately following interglacials on islands within the Hawaiian chain and on Lord Howe Island (Figure 6.32) at the northern and southern limits to Pacific reef growth respectively (Stearns, 1970; Woodroffe *et al.*, 1995; Sherman *et al.*, 1999; Brooke, 2001). On these island settings, high sea level has resulted in highly productive shallow banks, similar to the Great Bahama Bank. Sediments formed on these banks have become available for incorporation into dunes when a slight fall in sea level has exposed them. The dune units may be horizontally or vertically stacked (Vacher *et al.*, 1995; Hearty and Kindler, 1993).

There are also eolianites on the continental shores of the Mediterranean (Hearty *et al.*, 1986; El-Asmar, 1994), southern and eastern South Africa (Marker, 1976) and Yucatan, eastern Mexico (Szabo *et al.*, 1978). Cool-water carbonates form off the coast of South Australia, and have built up an extensive heterozoan carbonate province (Gostin *et al.*, 1988). A lateral succession of eolianite ridges occurs

Figure 6.32. Eolianites on Lord Howe Island, showing a prominent shore platform that has eroded into the bedded units most of which are of marine oxygen isotope stage 5 in age (see Figure 6.31).

across the Coorong coastal plain (see Figure 2.5); the Woakwine is Last Interglacial (Belperio *et al.*, 1996; Murray-Wallace *et al.*, 1998) and other interglacial and interstadial ridges have also been identified (Huntley *et al.*, 1993; Murray-Wallace *et al.*, 2001). There are extensive calcarenites forming a sequence that extends along 800 km of the west coast of Australia, including offshore to Rottnest Island (Fairbridge,

1948; Playford, 1997). Termed the Tamala Limestone, this also overlies a reef unit of Last Interglacial age exposed on Rottnest Island (Veeh, 1966; Stirling *et al.*, 1998). Calcarenite extends to more than 70 m below sea level, and dune activity continued over much of the glacial (Price *et al.*, 2001). The better preservation of eolianite extends the chronology that can be determined for phases of dune activity, but implies that dunes were active at different times on different parts of the coast.

6.6 Summary

Sandy or gravelly beaches and barriers are some of the most obvious, some of the most intensively studied, and some of the best understood of coastal landforms. They occur on a large proportion of the world's coastline. Wave energy is particularly important in determining sediment transport patterns and differential movement of material of different sizes. Other factors such as wind and tide regimes, sediment supply, underlying topography and geological inheritance are also significant in influencing which of a series of landforms develop.

Beaches adjust to the wave energy that is incident on the shore, and they are malleable and can be reshaped when wave energy changes. There have been geometrical patterns recognised in planform and in profile which imply adjustments towards an equilibrium. These approaches have been considerably advanced by the concept of 3-dimensional beach morphodynamics. Beaches adopt a series of morphological states ranging from those that reflect wave energy to those that dissipate it. Intermediate beach types contain various bars, cusps and circulation patterns. Beaches undergo variations over time, adjusting between states as energy conditions alter. Beach response to extreme events and longer-term trends towards accretion or erosion are harder to decipher, particularly if boundary conditions such as climate or sea level do not remain constant.

Beach ridges, including gravel barriers, build up where there is an ongoing supply of sediment. There can be episodes of erosion and phases of dune activity. Over Holocene time scales barriers and barrier islands respond to alterations in sea level and sediment supply. Many can also be seen to have preserved a legacy of earlier events, particularly previous interglacials. Although the beaches and barriers that are found today are adjusting to incident wave energies, the larger sediment bodies of which they are a part represent the integration of sediment accumulation over much of the Holocene, and contain evidence of previous phases of erosion or deposition.

Chapter 7
Deltas and estuaries

> The process of delta formation depends almost wholly on the following law: the capacity and competence of a stream for the transportation of detritus are increased and diminished by the increase and diminution of the velocity. (Gilbert, 1885)

Deltas and estuaries are dynamic systems associated with the mouths of rivers. Deltas are accumulations of river-derived sediment whereas estuaries are the tide-influenced lower parts of rivers and their valleys. The distinction between them is sometimes difficult to discern, and it is useful to consider a continuum of deltaic–estuarine landforms. Deltaic–estuarine morphology is influenced by geological setting and topography, and landforms are shaped by hydrodynamic processes. Riverine and coastal sediments are affected by both alluvial and marine influences, together with minor local processes, such as direct input of colluvium from hillslopes, cliff retreat, wind redistribution of sediment, and chemical and biological action.

River discharge and the rate of delivery of sediment to the ocean or embayment vary in relation to catchment size, lithology and climate (Milliman, 2001). The rivers that drain from the continental area of southern and eastern Asia, with highly tectonic hinterlands and prominent monsoon climates, for instance, deliver large volumes of sediment to the oceans (Milliman and Meade, 1983). However, steep, tectonically active island catchments, such as those throughout the Indonesian island arc, also contribute disproportionately large sediment volumes to the ocean (Milliman and Syvitski, 1992). For instance, rates of sediment delivery by the Solo and Brantas Rivers in Indonesia (1200–1600 t km^{-2} a^{-1}; Hoekstra, 1993), or the Sepik River in New Guinea (>1000 t km^{-2} a^{-1}; Chappell, 1993) are amongst the highest anywhere, and are directly comparable with that delivered down the Ganges–Brahmaputra–Meghna system. A delta forms where the supply of sediment down a river exceeds the rate of removal by nearshore processes.

The geomorphology of deltas and estuaries is strongly influenced by sea-level change. Estuaries adopted their present morphology on most coasts as a result of postglacial sea-level rise drowning river valleys (Roy *et al.*, 1994). Worldwide expansion of deltas has occurred in the mid Holocene as a result of deceleration of postglacial sea-level rise, and the coincidence of sea level with extensive low-gradient topography (Stanley and Warne, 1994).

The geomorphological history and the future dynamics of deltas and estuaries are of considerable significance to the people who live adjacent to them. The emergence and growth of societies based on cultivation occurred after the mid-Holocene expansion of deltaic–estuarine plains, particularly in semiarid areas such as the mouth of the Nile, the Indus valley and the Fertile Crescent of Mesopotamia where flat, irrigable land enabled agricultural production that could support urbanisation (Cooke, 1981). In contrast, densely vegetated deltas in humid tropical regions remained inaccessible because of disease and uncontrollable flooding (Büdel, 1966). These alluvial deltaic plains have also subsequently become rich agricultural land, suitable for wet-rice cultivation on a large scale, favourable for human settlement and supporting many of the world's megacities (Nicholls *et al.*, 1999). Associated sedimentary basins often contain oil and gas resources. In many cases, it is now becoming clear that anthropogenic intervention, such as land clearance, damming and diverting of rivers, and groundwater extraction, has had irreversible impacts on these dynamic geomorphological systems (see Chapter 10).

The study of deltas and estuaries has involved two approaches: short-term process research and a longer-term stratigraphic and geochronological perspective. Process studies have concentrated on short time scales, examining circulation patterns and sediment concentrations and fluxes. Process studies have indicated that morphology adjusts in response to the forces acting on the system. The longer-term perspective on the stratigraphy and evolution of deltas has been a particular focus of the petroleum industry. The development of facies models that can be applied to the geological record has involved integration of stratigraphy, particularly sequence stratigraphy, with contemporary process operation. In this chapter these two approaches are combined, examining Holocene deltaic and estuarine morphodynamics. After a brief historical perspective, the processes that operate in deltas and estuaries are examined and the morphology of associated landforms is outlined. It is shown that there is a continuum of river, tide and wave processes, not only between different deltas or estuaries but also within an individual system. The evolution of several individual deltas and estuaries is examined to illustrate the longer-term morphodynamics of deltaic–estuarine systems.

7.1 Historical perspective

Many deltas and estuaries are associated with a point source of fluvial discharge and sediment input. Rivers play an important role in their geomorphology, but the significance of tide and wave processes has long been recognised. Figure 7.1 shows a classification of both deltas and estuaries that places them within a continuum in terms of the relative dominance of river, wave and tide parameters. This section examines the historical perspective behind these classifications.

7.1.1 Deltas

The term delta, used by Herodotus in 450 BC, is derived from the triangular-shaped Greek letter 'delta' as applied to the wedge-shaped sediment accumulation at the mouth of the Nile River. A delta has been defined as 'a deposit partly subaerial built by a river into or against a body of permanent water' (Barrell, 1912, p. 381), or 'a deposit built by jet flow into or within a permanent body of water' (Bates, 1953, p. 2119).

Gilbert (1885), in his classic description of lake shorelines, described the sedimentary structure of deltas that had formerly discharged into Lake Bonneville. He identified the importance of capacity and competence of water flow, meaning the amount of sediment and the size of the sediment that can be carried (see quotation at beginning of this chapter). Gilbert recognised that velocity had rapidly decreased as

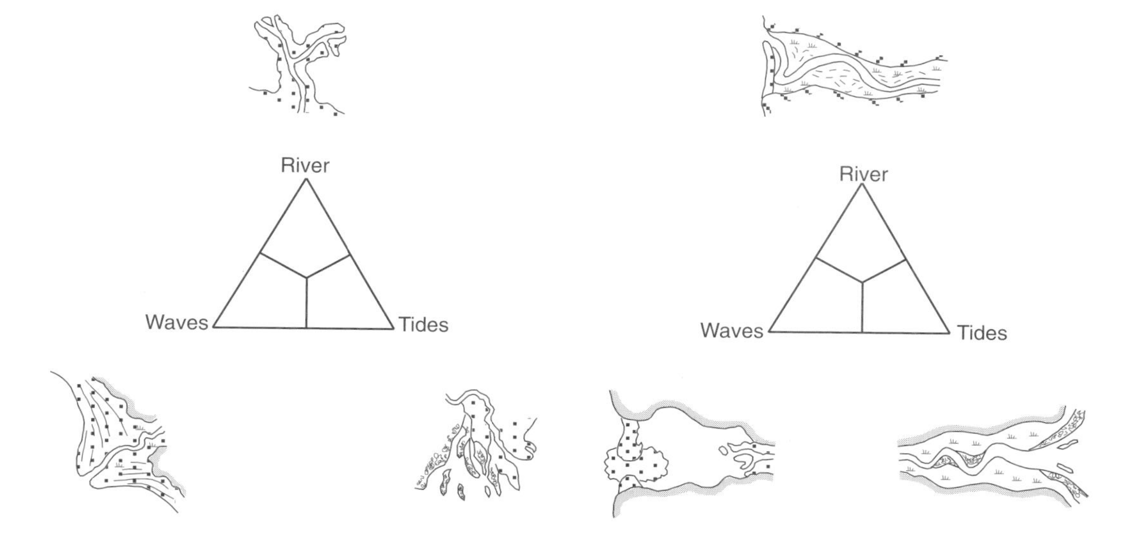

Figure 7.1. Relative domination of river, wave and tide processes and its relation to the broad morphology of (a) deltas (based on Galloway, 1975; Wright and Coleman, 1973), and (b) estuaries (based on Dalrymple *et al.*, 1992; Cooper, 1993).

river flow had emptied into the lake. The coarser component of the load was deposited first, forming bars at or just beyond the river mouth. Sandy sediment was deposited on the delta front as the flow continued to decelerate, forming dipping beds, termed foreset beds. The finer sediment was deposited later, and formed a low-gradient prodelta. This sedimentary pattern, involving dipping foreset beds, is now called a Gilbert-type delta (see Figure 7.2a), and is characteristic of the mouths of rivers that are of similar density to the basin into which they discharge.

Where a river discharges into an embayment, a sequence of embayment filling, from youthful bayhead delta, through adolescent bay-delta, to mature delta protruding into the sea was envisaged by Gulliver (1899). This model fits into the cycle of erosion adopted by Davis (1909) and Johnson (1919), but it is simplistic and does not recognise various other factors such as the substantial inputs of sediment into estuarine embayments from seaward, or the importance of relative rate and direction of sea-level change. Nevertheless, progressive infill of barrier estuaries (see Figure 1.7) does occur if sea level remains constant for long enough, and models which have built on these ideas are described in this chapter.

There are a variety of delta forms recognised in the literature. Lobate deltas have prominent lobes of deltaic sediments, for instance the delta of the Niger River in West Africa (Allen, 1975). Lobate deltas may be unilobate, such as the River Ebro in Italy, or multilobate such as the River Volga. Cuspate deltas comprise concave delta margins shaped by waves, meeting at the cusp at which the river channel discharges, such as the Tiber and the Vistula–Gdansk Rivers. Digitate or birdsfoot deltas occur where the river extends with levée growth into deeper water, as with the modern Mississippi River.

These descriptive classifications (e.g. Samoilov, 1956; Volker, 1966) suffer from the disadvantage that terms are not used consistently (lobate deltas are also referred to as arcuate), and their utility is hampered by the fact that 'every river mouth offers the investigator a new problem' (Zenkovich, 1967, p. 585). For instance, the Danube, Rhône and Nile are intermediate between lobate and cuspate (Russell, 1942), while other deltas have been termed blunt (the São Francisco in Brazil) or blocked (the Senegal in West Africa). It was already clear that there were complex compound deltas, such as the Ganges–Brahmaputra–Meghna, where tidal processes are dominant after abandonment of distributaries (Strickland, 1940; Morgan and McIntire, 1959), and the Irrawaddy, where monsoonal flooding occurs (Stamp, 1940).

Deltas also change over time. The contemporary Mississippi River delta, having now reached the edge of the continental shelf, is no longer lobate and the modern Balize delta is birdsfoot (digitate). The sedimentary history of the lower Mississippi River was examined in great detail by Russell and Fisk in the 1930s to 1950s (Russell, 1936, 1940; Fisk, 1944; Fisk *et al.*, 1954). They recognised the significance of sea-level fluctuations, with incision and entrenchment of the river channel during low sea level. The chronology of lobe occupation and abandonment was reconstructed for laterally stacked delta lobes on the basis of pottery sherds, Indian mounds and shell middens (Scruton, 1960; Kolb and van Lopik, 1966). Frazier, in a classic study, established a detailed chronology based on radiocarbon dating, indicating the complexity of overlapping delta lobes (Frazier, 1967), and his work remains the basis for subsequent geomorphological studies described below.

The significance of density differences was recognised by Bates (1953), and has been used as a basis for classifying deltas in terms of process. The Gilbert-type delta formed where there are no density contrasts between the inflowing river water and the receiving water body, and is termed homopycnal. If river flow is denser than the receiving basin water it is termed hyperpycnal, and if river flow is less dense it is termed hypopycnal (Bates, 1953). The sedimentary processes in the Mississippi delta result in the coarser sediment from the plume being deposited closer to the river mouth, and progressive fining of the sediments from foresets to bottom sets (Scruton, 1960). The interrelationship between river supply and reworking of sediment, and wave power, itself a function of subaqueous slope, was demonstrated by studying offshore profiles and morphological patterns of several deltas (Wright and Coleman, 1971, 1972). These studies led to recognition by Wright and Coleman, and others, of the continuum of deltaic forms associated with the relative dominance of river, tide and wave processes as shown in Figure 7.1a (Wright and Coleman, 1973; Wright *et al.*, 1974; Galloway, 1975; Orton and Reading, 1993). The scheme is examined in more detail in Section 7.2.

7.1.2 Estuaries

The term estuary, derived from the Latin 'aestus' meaning tide, refers to a tongue of the sea reaching inland. A definition that is widely used describes an estuary as 'a semi-enclosed coastal body of water having a free connection with the open sea and within which sea water is measurably diluted with freshwater derived from land drainage' (Cameron and Pritchard, 1963, p. 306). The upstream limit of an estuary can be defined in terms of the limit of saline water penetration, or the uppermost point at which tidal oscillation of water level is discernible. In this chapter, bedrock embayments and coastal lagoons are also considered because they show many similarities to estuaries. The term 'ria', which is Spanish for estuary, was used by von Richthofen (1886) to refer to mountainous-sided estuaries which have not been glaciated and which cut across the geological strike. It has been used more broadly to refer to an estuary in which there is control by bedrock (Cotton, 1956). Fjords are steep-sided valleys that have been glaciated, but in which present riverine processes are generally limited; their origin was examined by Gregory (1913).

Coastal lagoons, some of which may be periodically open to the sea, are considered estuarine environments because they show many similarities to other estuaries, such as barrier estuaries, in morphology and development. They were interpreted as infilling as a result of wave processes transporting sediment through inlets in a classic model developed on the eastern North American coast by Lucke (1934). Large elongate coastal lagoons often develop segmentation (initially described as septation; Price, 1947), which has been the subject of debate in terms of the relative role of wind, waves and vegetation in forming spits that separate discrete lagoons (Zenkovich, 1959; Bird, 1967).

Embayments were viewed within the context of a cycle of coastal development by Johnson (1919, 1925), and the barriers and associated landforms that he recognised (see Sections 1.3.2 and 6.5.3) have been widely used. The geomorphology of estuaries was recognised as an important constraint on their ecology and functioning in several compilations of estuarine research in the 1960s (Fosberg, 1966; Lauff, 1967; Russell, 1967). There have been further refinements by Fairbridge (1980), and recognition that some estuaries are only periodically, rather than permanently, open to the sea (Day, 1980). Fairbridge's classification of estuaries, with some minor amendments (Perillo, 1995; Kench, 1999), recognises: (i) fjords (deep U-shaped valleys in areas which have had ice cover); (ii) rias (V-shaped river valleys drowned by sea-level rise, and generally with bedrock margins); (iii) coastal plain estuaries on sedimentary coasts; (iv) delta-front estuaries (delta distributaries); (v) bar-built estuaries; (vi) blind estuaries (blocked or periodically closed coastal lagoons); and (vii) structural or tectonic estuaries (which includes inlets that owe their origin to a range of tectonic factors).

There are a continuum of river, tide and wave processes that operate within estuaries and Figure 7.1b indicates distinct estuarine morphology that occurs at the extremes of this continuum. Recognition of process as a basis for estuarine classification has been more recent than for deltas; wave-dominated and tide-dominated estuaries were discriminated by Dalrymple *et al.* (1992), whereas river-dominated estuaries have been described by Cooper (1993). Such a morphological approach refers to the relativities of the operation of these processes, not necessarily to any absolute measure of operation (Davis and Hayes, 1984).

The classification scheme shown in Figure 7.1 is useful because it provides a framework within which overall deltaic–estuarine morphology can be placed. However, most deltas and estuaries are influenced by river, tide and wave processes to different extents, and the influence varies geographically from one part of the system to another, and through time, as the system evolves. A broader classification of coastal depositional environments has been proposed by Boyd *et al.* (1992) in the context of sequence stratigraphy, and the relative balance of river, wave and tide processes (see Figure 7.5). They distinguish between transgressive shorelines (in which the shoreline is retreating landward), and prograding (regressive) shorelines in which it is building seaward. Transgressive sequences are associated with relative sea-level rise but regressive sequences are not always associated with sea-level fall. Progradation of the coast can occur where the rate of sediment supply is sufficient to allow the shoreline to build seaward even if there is gradual rise of sea level (Curray, 1964). The variability of processes and morphology within modern deltas and estuaries is examined in the next section, and these ideas are explored in the context of longer-term deltaic–estuarine evolution and morphodynamics later in the chapter.

7.2 Deltaic and estuarine processes

The processes operating in a delta or estuary are a function of the characteristics of the catchment and of the receiving basin into which the river discharges. Although the relative balance of wave, tide and river processes will be shown to exert considerable control over morphology, deltas and estuaries are the product of longer-term evolution in time and space over which boundary conditions change, and such a steady-state, time-independent examination of deltas is generally unrealistic (Postma, 1995). Boundary conditions include the broad climatic, geophysical and oceanographic setting, and are a function of antecedent conditions and relative sea-level history. In the short-term, this is related to Holocene sea-level change and isostatic factors, and in the longer term it expresses the balance between tectonism and subsidence (see Chapter 2).

7.2.1 Delta-mouth processes

The processes that operate at the mouth of a delta depend on the relative density of river flow and the water body into which it discharges (Bates, 1953). Processes and resulting morphology have been described by Wright (1985) and are examined schematically in Figure 7.2. A delta is termed homopycnal where the density of river water and the basin into which it discharges is similar, as, for instance, where a river discharges into a freshwater lake. The morphology of the delta that forms depends on the nature of the flow, the geometry of river mouth, relative depths of river and the basin into which it empties, sediment load and grain size and water-level variations (Axelsson, 1967). The river water enters the basin as a jet in which turbulent mixing occurs across a zone of flow establishment (see Figure 7.2a). Gradual deceleration of flow, as envisaged by Gilbert (1885), decreases its competence to carry sediment in suspension and the coarsest material is deposited first in a lunate bar. In cross-section the delta consists of near-horizontal topsets, dipping foresets and a muddy prodelta.

Homopycnal deltas are rare where rivers discharge to the sea because density differences occur as a result of salinity or sediment load. Freshwater flows over the more dense saltwater (hypopycnal); however, if the river carries a large sediment load, it may be denser than seawater and sink beneath it (hyperpycnal). The effect that density differences have can be measured using the densimetric Froude number to assess the shear stress between water bodies (flow depth (h) corresponds to the depth to the density interface; cf. Figure 3.2). Wright (1985) has

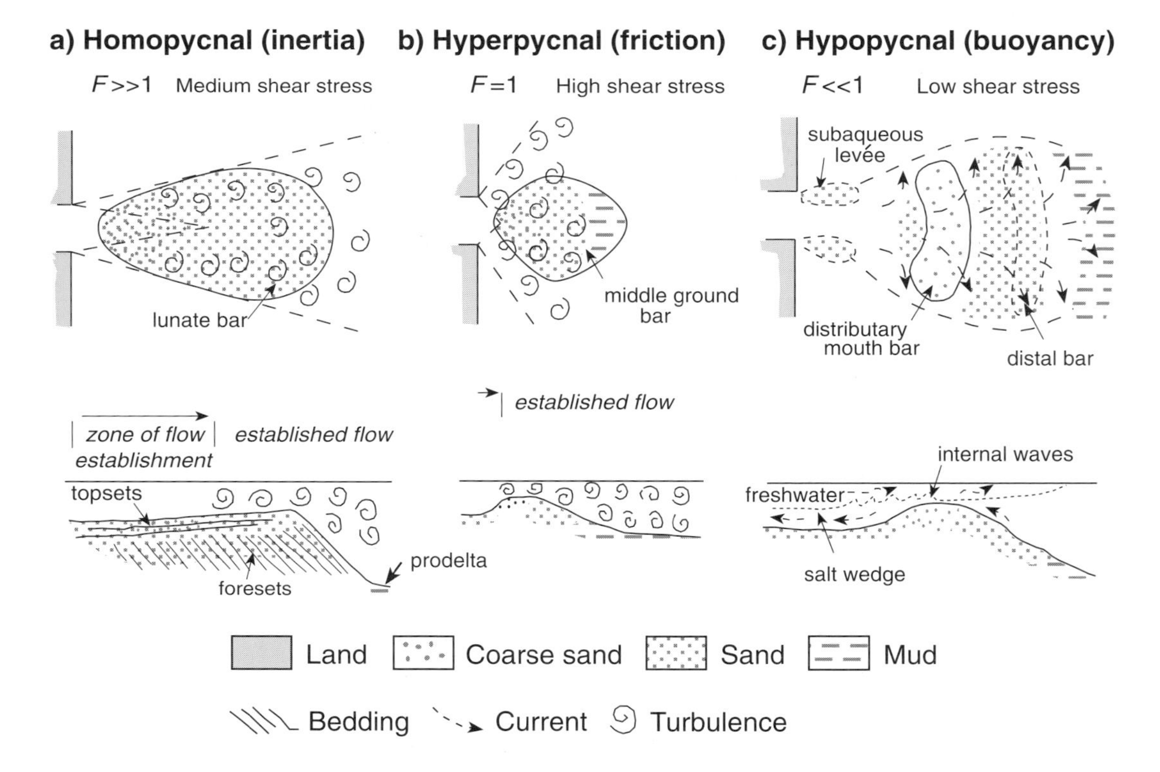

Figure 7.2. Delta-mouth processes, and the morphology (above) and profile and stratigraphy (below) characterising inertia- (homopycnal), friction- (hyperpycnal), and buoyancy-dominated (hypopycnal) river mouths (based on Wright, 1985). There is considerable variation in densimetric Froude number (F, see Figure 3.2 for the definition of the Froude number), reflecting density differences, and shear stress between flows of different density (Carter, 1988).

expressed these relationships in terms of domination by inertia (homopycnal), friction (hyperpycnal) and buoyancy (hypopycnal).

In hyperpycnal delta mouths the inflowing water is in contact with the floor of the basin. The Huanghe (Yellow) River is hyperpycnal, for example, and river waters are extremely sediment-laden and disperse down the delta front as low-density flows, or as high-density channelised underflows extending beyond the mouths of active distributaries (Wright *et al.*, 1986). Friction with the seafloor, either bedrock or more often with sediment bodies deposited as bars at the river mouth, slows flow and leads to a far greater lateral spread of the sediment-laden waters. Hyperpycnal flows are characteristic of shallow basins; sediments are deposited in a mid-channel (middle-ground) bar within the mouth of a channel, and the flow bifurcates around the bar (Figure 7.2b). The flow in each of the bifurcated channels, or distributaries, is also slowed by friction with the seafloor, promoting further mid-channel bar deposition and channel bifurcation. Middle-ground bars and regular branching of channels can be seen in the subdeltas of the Mississippi River (discussed in more detail in Section 7.4.1), or the rapidly prograding Atchafalaya River delta, the most recent lobe of the Mississippi (van Heerden and Roberts, 1980).

Hypopycnal deltas are characterised by buoyant flow, resulting from density differences between the inflowing water and the receiving basin. Fresh-water discharging into the sea is buoyed up by the salt water, and experiences little if any friction with the seafloor. This is termed a salt wedge (Figure 7.2c) and enables the river discharge to flow seaward relatively unimpeded. Where a salt wedge occurs, as for instance at the mouth of the Mississippi River, the distributary mouth bar (distal bar) is deposited further into the receiving basin than where friction or inertia are dominant (Wright, 1985).

These simple delta-mouth responses to inertial, frictional and buoyancy forces, shown schematically in Figure 7.2, represent endpoints of a continuum. There are likely to be a series of additional factors that complicate the riverine input of sediment. However, a further range of variability is associated with the receiving basin. The effects of tides and wave conditions in particular, but also of longshore currents, density contrasts, and basin-floor topography, are significant in modifying deltaic processes and form. In almost all examples, flow is variable; high magnitude flood and storm events are likely to operate very differently from more regular flow conditions.

Delta form is a function of the size and steepness of the drainage basin and the coarseness of sediment load (Postma, 1990, 1995). Sediment can be carried in rivers as dissolved load (washload), suspended sediment, or bedload, depending on grain size and flow velocity. Suspended or dissolved load channels (bedload <3%) are typical of the largest catchments. Their banks are muddy and cohesive and channels are usually meandering or anastomosing, for example the Ganges–Brahmaputra–Meghna system (Coleman, 1969). River channels can change their position as a result of lateral migration or through avulsion, switching from one course to another and not migrating across the intervening plain.

Bedload channels (bedload 11–50% of total load) occur on glacial outwash plains and sandurplains, such as those on Van Mijenfjorden, central Spitsbergen (Nemec, 1995). In the Arctic, for instance, processes such as frost action which produces ice-wedge polygons and ice-heaved conical hills, tend to result in coarse-grained deltas including the Ob, Yenisey and Lena Rivers in Russia (Are and Reimnitz, 2000), and MacKenzie, Yukon and Colville Rivers in North America (Walker, 1998). Sand channel bank stability is enhanced by permafrost in winter, and subject to wind action in summer, and the landscape is composed of peaty swamps, although with restricted vegetation (Hill *et al.*, 1994). Coarse deltas, dominated by bedload channels (in which bedload comprises >50% of the load) can form steep fan deltas (McPherson *et al.*, 1987; Nemec and Steel, 1988). For instance, coarse fan deltas formed as scree slopes on the margins of fjords have slopes of 10–25° and debris avalanches disturb the surface frequently enough to ensure that plants cannot establish (Prior and Bornhold, 1990). Coarse deltas are generally associated with small steep catchments, such as the Copper River in Alaska (Galloway, 1976) or in the Zagros Mountains of Iran in the northern Persian Gulf (Baltzer and Purser, 1990). Coarse-grained deltas usually demonstrate steeply dipping delta-front foreset bedding, such as in the Yallahs fan delta in Jamaica, and little wave energy is dissipated, resulting in steep reflective seaward beaches that are subject to erosion during storms (Westcott and Etheridge, 1980).

7.2.2 Estuarine hydrodynamics

Estuaries are distinguished from deltas by the fact that the basin water penetrates inland into the embayment; however, many aspects of estuarine circulation are also appropriate in consideration of the flows of water at the mouths of deltas. Estuarine circulation is generally a function of the interaction of marine processes, particularly tides, with fluvial discharge and water already in the system. In order to understand estuaries, it is necessary to consider the nature of flows, and the consequent flux of material, in some detail. Many of the principles of river hydrology apply to estuaries, although a significant difference is that tidal flows in estuaries are bi-directional.

The way in which the tide propagates up an estuary is described in Section 3.4.2. There are also residual currents that result from density differences. Velocities of flow vary as the tide propagates up a channel, with both progressive and standing wave components. In small embayments, tidal flushing ressembles a standing wave and fluxes can be approximated by considering the tidal prism. Longer estuarine channels are flooded by a tide that progresses up them, and tidal flows show increasing asymmetry upstream (see Figure 3.14). Tidal range tends to be amplified in those estuaries and embayments that get narrower upstream, and to decrease in those that widen. Where channel banks are unconsolidated or poorly consolidated, there is a tendency for a negative exponential decrease in width with distance upstream (see Chapter 3). This tapering of tidal channels can be seen in the macrotidal creek system shown in Figure 7.3.

Velocity asymmetry has consequences for sediment movement. If velocity of flow is faster on the flood, the greater bedload transport can be seen in flood-formed bedforms, and the system is flood-dominated. Bedforms in the Ord River in northwestern Australia support the concept of bedload transport upstream during the dry season (Coleman

Figure 7.3. A mangrove-lined tide-dominated estuary in the Northern Territory of Australia. This macrotidal system shows a clear tapering in width with distance upstream.

and Wright, 1978). Vegetation can modify estuarine morphodynamics through its effects on sediment trapping and through its influence on the shear strength of channel banks. In Figure 7.3 the banks of the tidal estuary are lined by mangrove forests. Landward-directed tidal transport of sediment would choke estuarine systems in macrotidal seasonal tropical settings, such as the Ord River or the Normanby River in north-eastern Australia, except where fluvial discharge during river flood events carries sediment downstream (Bryce *et al.*, 1998). Some macrotidal systems may be flood-dominated on neap tides, but ebb-dominated on spring tides (Lessa, 2000).

In some estuaries, flow takes separate pathways on flood and ebb tides. This appears to occur in the delta of the Fly River in Papua New Guinea (Baker *et al.*, 1995). It may explain distinct estuarine meanders, also called cuspate meanders, that have a cuspate inner bank to the meander (Ahnert, 1960), which may result from separate flood and ebb channels with ebb flow confined to deep parts of the channel where it can achieve transport downstream.

Estuarine circulation has generally been classified into salt-wedge, partially mixed and well-mixed (Pritchard, 1952). Salt-wedge estuaries typically occur where the tidal prism is small and tidal currents too slow to mix the inflowing and receiving water bodies (Figure 7.4). Salt-wedge estuaries and salt-wedge delta mouths (see Figure 7.2c) remain stratified with a near-horizontal halocline (a sharp transition in salinity); fresh water is buoyed up over the salt water and the river flow is buffered from interaction with the seafloor. The influx of river discharge drives the seaward movement of fresh water, and the salt water is moved landwards by a process of residual flow (Figure 7.4a).

Well-mixed estuaries occur at the other end of the continuum of circulation characteristics (Dyer, 2001). Estuaries with a large tidal range, where tidal volumes (prisms) are considerably greater than freshwater river discharge and most of the water in the estuary is moved up and down the estuary by tidal rise and fall, are well mixed. Flow is highly turbulent and mixes the river water effectively with the salt water. Isohalines (lines of equal salinity) are nearly vertical in well-mixed estuaries and grade from high salinity at the mouth to low salinity upstream (Figure 7.4c).

Partially mixed estuaries lie between these extremes (Figure 7.4b). They are typically found in mesotidal areas; much of the volume of the estuary is moved to and fro with tidal oscillations, but some salinity stratification occurs with deeper waters more saline than surface waters.

Figure 7.4 indicates the generalised pattern of isohalines that characterise salt-wedge, partially mixed and well-mixed estuaries. It also shows the patterns of flow typical of flood and ebb tides. A salt-wedge

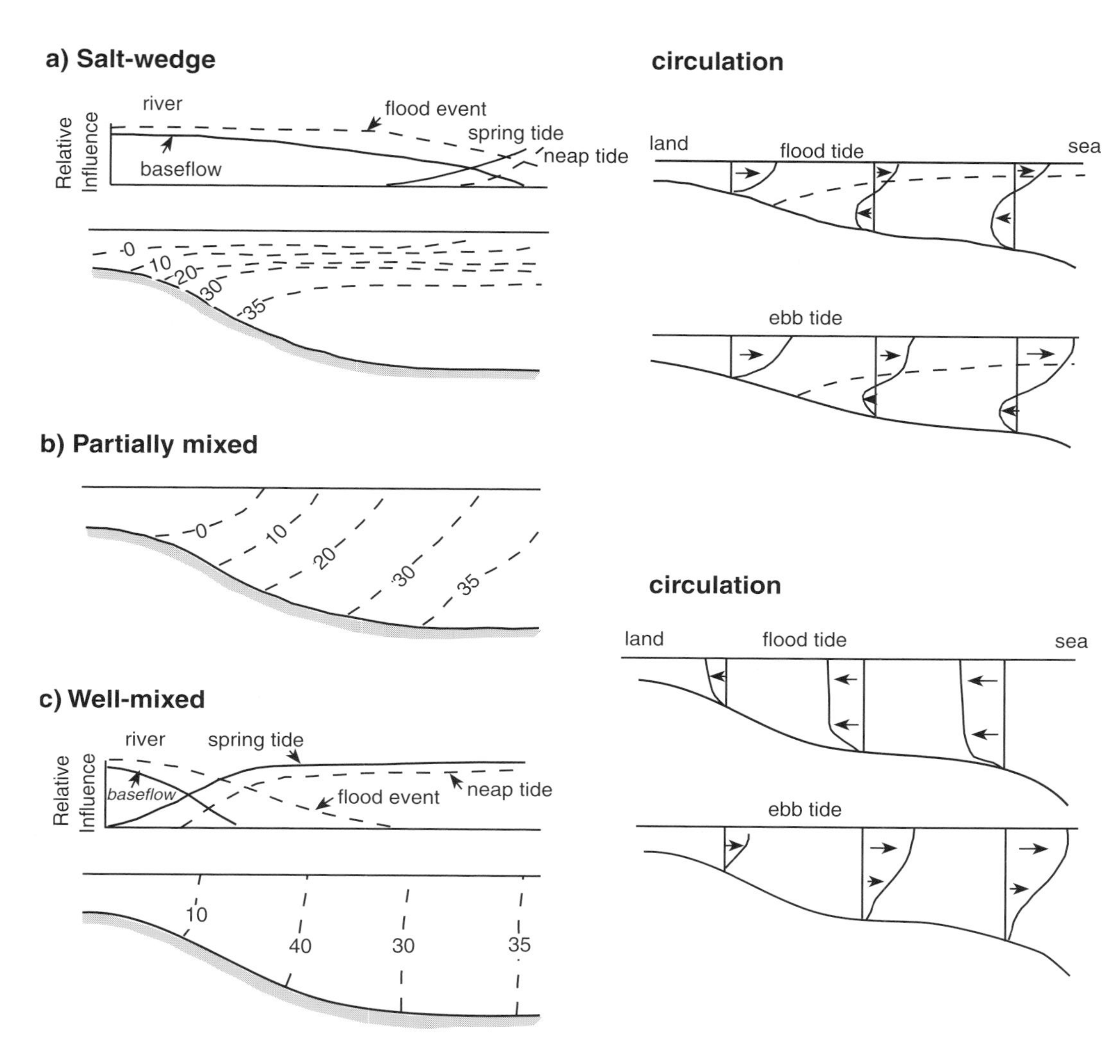

Figure 7.4. Circulation within estuaries (based on Pritchard, 1952), showing (a) salt-wedge, (b) partially mixed and (c) well-mixed. Isohalines (lines of equal salinity) are shown, as is a schematic indication of the time-variant relative influence of river and tidal processes along salt-wedge and well-mixed estuaries, for which the pattern of circulation and flow velocity on flood and ebb tide is also shown schematically (based on McDowell and O'Connor, 1977).

estuary can maintain upstream residual flows within the salt wedge even during ebb tides. Flood-tide flows in well-mixed estuaries often show relatively uniform flow velocity with depth, whereas ebb flows are more variable with depth. The relative influence of river and tide processes is also shown schematically, indicating that this can be very variable according to neap and spring tides and river discharge. A salt-wedge can be swept out of an estuary during a river flood event and can take weeks to months to re-establish.

In hot arid climates where there is intense evaporation, an inverse or negative estuary can occur, in which the system becomes hypersaline (El-Sabh *et al.*, 1997). For example, water diffuses into Spencer Gulf in South Australia by gravity currents to replace that lost by evaporation (Lennon *et al.*, 1987). Seasonal hypersaline conditions at the mouth of tropical rivers can serve to trap waters, and hence sediments, within an estuary (Wolanski, 1986). Evaporation-driven systems in monsoonal northern Australia can result in concentration of sediment upstream as a result of net upstream flows (Wolanski, 1994).

7.2.3 Deltaic–estuarine sedimentation

The transport and deposition of sediments in deltas and estuaries is usually complex and is a function of the hydrodynamic circulation. A turbidity maximum is present in many estuaries. This occurs generally towards the landward end of the estuary where freshwater encounters salt water of low salinities (1–5‰). Vigorous tidal action maintains fine sediment in suspension, with maximum turbidity of around 200 mg l^{-1}, except in the highest energy estuaries with large tidal ranges, where turbidity reaches as much as 1–10 g l^{-1} (Dyer, 1995). The position of the turbidity maximum can move with the dominant flows in the estuary, being pushed downstream during river floods (Uncles and Stephens, 1989). Fine sediment can be effectively trapped in an estuary, remaining in suspension for a long time where turbidity maxima occur.

Mud-sized particles in estuaries can experience electrolytic flocculation or organic flocculation which increases settling by reducing the resistance of the flocs as a result of the decrease in surface area to volume ratio (rather than the increase in mass). Electrolytic flocculation refers to the total surface ionic charge on particles. As clay particles move down river and into an estuary the negative charges on the clay particles are overcome by the positive charged ions in seawater, activating strong attractive forces between the clay particles. This depends on pH and organic matter content, but as washload is carried into saline waters the surface charge is intensified and clay particles aggregate into macroflocs of up to several millimetres diameter (Hunter and Liss, 1982; Wolanski *et al.*, 1997).

Organic flocculation occurs where particles are organically bound. Faecal pellets, for instance, can form a significant component of sediment in which mucus binds particles together (Ginsburg, 1975). Surface sediments may be stabilised by micro-organisms or their by-products, including carbohydrate polymers known as extracellular polymeric substances. Biologically mediated removal of material from suspension may also play other important roles. For example, phytoplankton remove suspended material from the water column of the Fly River outflow limiting spread of the sediment plume seaward (Ayukai and Wolanski, 1997).

Lateral trapping of suspended sediment in estuaries can occur where sediment-laden water enters tributaries and other secondary water bodies on an incoming tide, but is bypassed, or only gradually mixed on the ebb tide (Wolanski, 1994). Trapping can also occur in deltaic–estuarine mangrove forests where onshore wind, wave or current activity restricts the seaward spread of a sediment plume (Francis, 1992).

Suspended sediment concentration can vary substantially both spatially and temporally in estuaries, with a pronounced transition in concentration, called a lutocline. Muds that are deposited during periods of reduced water movement (i.e. at slack tide), often settling as mud drapes over rippled sandier bedforms, can be resuspended after tidal reversal or during later larger tides. Resuspension of fine sediments can occur as individual flocs are re-entrained or through mass failure (Dyer, 1995). The dynamics of muddy shorelines, discussed in Chapter 8, is often influenced by micro-organisms, such as diatoms and algae, that coat the surface or play a role within the sediment.

Estuarine sediments can be derived from the catchment or from off the continental shelf. They can come from the atmosphere; they may be eroded from the shoreline (i.e. from cliffs or the estuarine margin) or they may be produced *in situ*, as biogenic material. Salt-wedge estuaries are more likely to retain fine sediment in suspension because they are stratified and exchange between the underlying saline water and the upper freshwater only occurs as residual flows. Sediment is more likely to be carried from landward out to sea, as seen in the case of buoyant delta mouths, such as that of the Mississippi River (see Figure 7.2c).

In partially mixed estuaries, such as Chesapeake Bay in the eastern United States, there can be net sediment influx from seaward. Fine sediment can be carried into the turbidity maximum where it is likely to be deposited. Fully mixed estuaries occur where tidal energy is large in comparison to fluvial input, for instance in macrotidal settings such as the Bay of Fundy in eastern Canada. The tidal currents break down any stratification, and keep the finest sediment in almost continual motion, and the estuarine bed is usually dominated by sand.

Vertical profiles of suspended sediment concentration in the macrotidal South Alligator River in northern Australia vary through the tide cycle. High velocities and shallow water depths early in the flood tide generate turbulent mixing through the water column. Sediment settling at high-tide slack water, followed by lower ebb velocities, leads to formation of a lutocline with lesser velocities in the underlying, high suspended-sediment concentration water than in the upper, low suspended-sediment concentration water (Wolanski *et al.*, 1988). Vertically integrated velocity–density profiles differ between flood and ebb flows (see Figure 7.4c), which can result in upstream drift of suspended sediment. Salt-wedge structures, which develop when there is freshwater outflow, also generate upstream sediment transport by residual flows in the underlying saltwater.

Despite generalisations about sediment inputs, and the relative movement of sediment within estuaries, it is important to recognise considerable temporal variability. In some cases, this may be seasonal. For instance, at the mouth of the Mekong River during the wet season a salt wedge forms in the distributaries, but during the dry season they are well mixed with a turbidity maximum, and flood tides can move fine sediment upstream (Wolanski and Ridd, 1986). In this case, fine silt deposited in the nearshore in the wet season is reworked up the channels in the dry season (Wolanski *et al.*, 1998). Deltaic–estuarine circulation can vary over longer time scales, in relation to river floods or storms. Extreme events represent anomalies in terms of the normal sediment movement patterns, and it may be during these events that significant geomorphological modification of the system can be achieved as a result of individual or repeated events.

7.3 Deltaic–estuarine morphology

Broad geomorphological descriptions of deltas and estuaries indicate that there are some morphological features that are more likely to be related to wave or tide processes but they do not always adequately describe the range of variability that can occur within any one delta or estuary. In this section, a classification of landforms in terms of river, wave and tide process operation and sea-level trend is outlined, and the components of individual deltas and estuaries are examined. The concept of domination can be applied at a series of scales. Entire deltas or estuaries can be differentiated in terms of river, wave and tide domination, as indicated schematically in Figure 7.1a and 7.1b, but more often individual sectors of one particular delta or estuary can be river-, wave- or tide-dominated to different extents.

7.3.1 The influence of tide and wave processes

Figure 7.5 shows a schematic representation of the relative dominance of tide, river, and wave processes in terms of coastal depositional

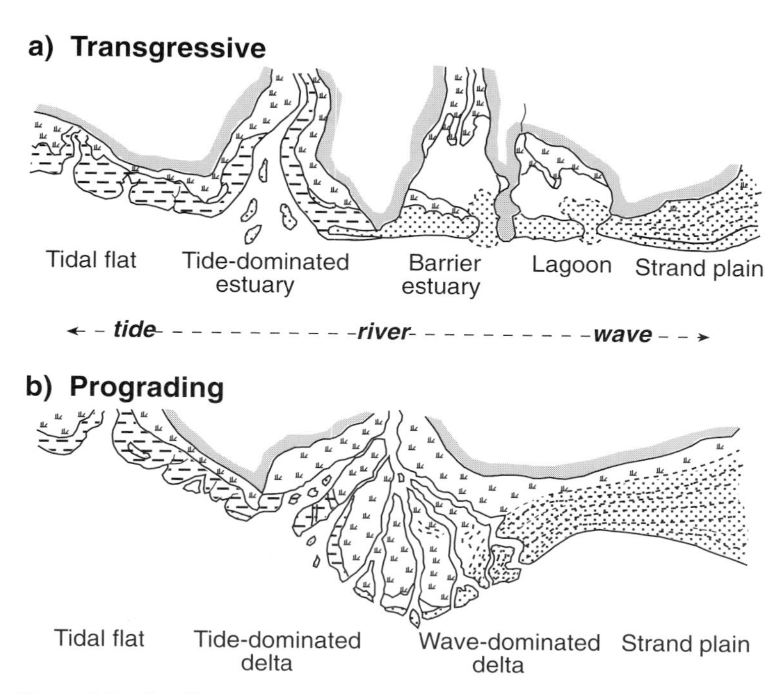

Figure 7.5. Classification of coastal depositional landforms within the context of river, wave and tide domination, and relative sea-level tendency, (a) transgressive, and (b) prograding (based on Boyd *et al.*, 1992).

environments based on Boyd *et al.* (1992). In an evolutionary context, in which either transgression or progradation can be occurring, these characteristic morphological environments are linked through time. River-dominated environments contain deltas with distributaries, flanked by levées, building out into a receiving basin. Where tides dominate, there is a progression from tide-dominated deltas and tide-dominated estuaries, often with extensive tidal flats. On wave-dominated coasts, there are barriers (elongate shore-parallel sand bodies, see Chapter 6) comprising beach, dunes, tidal deltas, washover deposits and spits. Coastal lagoons occur where waves build barriers across embayments. Although waves do not overwhelm the effect of river or tide processes, they often build wave-dominated coastal landforms across river mouths. A continuum of wave-influenced deltas and wave-influenced estuaries can be envisaged.

An individual deltaic–estuarine system might occupy several different positions within the continuum. This is particularly true of highly seasonal, monsoonal deltas, such as the mouth of the Mekong River described above, which are river-dominated for only a short period of the year. In the case of the Mississippi River, although the modern birdsfoot

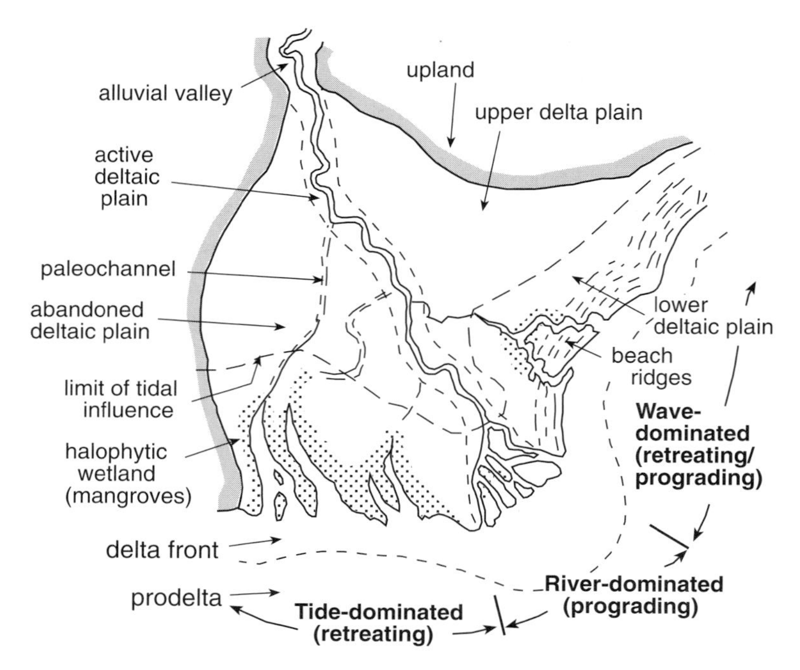

Figure 7.6. Components of a typical delta and the dominance of river, wave and tide processes (based on Wright *et al.*, 1974).

Balize complex is river-dominated, wave processes erode sandy headlands flanking abandoned distributaries and shape barrier islands (see Section 7.4.4). By contrast, the Ganges–Brahmaputra–Meghna is river-dominated in the east (Meghna), but tide-dominated throughout the Sundarbans. In the following section, the landforms within an individual delta or estuary are described and the extent to which they are tide or wave influenced is considered.

7.3.2 Delta and estuary components

Modern deltas can be divided into several major geomorphological components (Figure 7.6). The subaqueous part of the delta consists of a shallow delta front, which is relatively steep (though gradients of less than 1° characterise the Mississippi delta), and composed of sandy foreset beds and a prodelta, which occurs in deeper water and comprises muddy bottom set beds. The subaerial delta generally consists of an upper deltaic plain influenced by fluvial processes, and a lower deltaic plain dominated by estuarine processes. These are generally separated at the innermost extent of tidal influence which often corresponds with the limit to the distribution of mangrove or salt-marsh vegetation (see Chapter 8). Both upper and lower deltaic plain can be divided into an active section through which the river channel reaches the sea and an abandoned section containing paleochannels marking former river courses (Wright *et al.*, 1974).

Figure 7.6 indicates the major components of a typical delta, and illustrates how the shoreline shape can indicate the relative domination of river, wave or tide processes. For example, the Red River delta in northern Vietnam has three distinct sections, the upstream portion that is river-dominated whereas the downstream portions are wave- and tide-dominated respectively (Mathers and Zalasiewicz, 1999). The river-dominated section of a delta is characterised by numerous scroll bars and former river courses. The tide-dominated section has a sequence of tapering channels and numerous tidal creeks, whereas the wave-dominated section is composed of series of shore-parallel ridges.

River domination is indicated by natural levées flanking the river which are often augmented by human actions to reduce flooding. Levées constrain river flow within the confines of the bank. On the deltaic plain, levées can be breached by crevasses, and sediment-laden water spills through crevasse splays into interdistributary basins, comprising shallow bays, floodplains, lakes or tidal flats. The flood basins and backwaters behind the levées receive fine sediment that settles out from seasonal flooding maintaining soil fertility and providing the critical enrichment of areas of intensive agriculture in many deltaic areas around the world. This sedimentation also counters compaction and subsidence that tends to lower the deltaic plain surface.

The influence of tide and wave processes is most clearly seen on the abandoned deltaic plain. Tide-dominated sections of the delta front contain characteristically tapering tidal channels. Former distributaries on the abandoned portion of the delta become increasingly tide-dominated, adjusting their dimensions to taper upstream, with highly sinuous tidal creeks along their flanks. Where waves dominate the abandoned delta front, shore-parallel sandy ridges occur. These can form part of a sequence of beach ridges (see Chapter 6) or cheniers (see Chapter 8) where sediment supply has been sufficient and wave domination has persisted.

Rivers can continue across a delta as a single channel or, more commonly, bifurcate into several channels of different sizes (Knighton, 1998). The active, river-dominated portion of a delta is often characterised by lateral migration of meandering channels. Distributaries lengthen as a result of progradation of the delta, lessening channel gradient and resulting in switching of channel courses (avulsion) as alternative routes are found with greater hydraulic efficiency (Suter, 1994). Distributary channels avulse, adopting an entirely different course, as a result of a large flood event that bursts the bank. Most deltas, and many estuarine plains, contain conspicuous paleochannels or anabranches, as distributaries are abandoned but do not dry entirely. The abandoned delta front becomes increasingly influenced by tide or wave processes, as both assume a greater importance relative to river processes. Several examples are considered in the following section.

7.4 Morphodynamic development of deltas

Rivers are the principal sources of sediment for most deltas, but the extent to which river processes dominate the geomorphology of a delta depends on the relative influence of waves and tides. The Mississippi River has been studied in the greatest detail, and the geomorphology of delta lobes and subdelta cycles is examined as an example of the dynamic processes that can occur at the mouths of a river-dominated delta. Even in the case of the Mississippi, however, wave processes are important in shaping the shoreline of the abandoned parts of the delta. The significance of wave processes is discussed in relation to several other large rivers, and it is shown that waves and tides tend to have some effects, even where river discharge is very large. It is shown that, where wave processes are significant, the shoreline is generally composed of shore-parallel ridges and, where tide processes are significant, there are usually rapidly tapering tidal channels. The examples that are used in this section include deltas of big rivers; however, it is important to stress that much smaller rivers can also build deltas that show many of the same geomorphological characteristics on a smaller scale.

7.4.1 River-dominated deltas

The Mississippi River is often regarded as the typical river-dominated delta, and is described in this section because it has been studied in detail. The Mississippi River drains a large catchment (over 3×10^6 km^2) and empties into the Gulf of Mexico, in which wave energy is relatively low and tidal energy (tides are diurnal) is almost completely overwhelmed by meteorological variations in water level. Although it has a large catchment, the annual sediment load of about 3.5×10^8 t a^{-1} is not as large as that of several Asian rivers (Table 7.1).

The Mississippi River flows through an alluvial valley, and the morphostratigraphy of the meandering section of the river was the subject of intensive study by Russell (1936) and Fisk (1944). By about 9000 years ago, a significant part of the alluvial valley had been filled, and the present deltaic plain began to build out. Evolution of the Mississippi River delta involved the successive occupation of a series of delta lobes (Figure 7.7); each of the abandoned delta lobes covers an

Table 7.1. River mouths of the major rivers

	Drainage area (10^3 km^2)	Mean annual discharge (m^3 s^{-1})	Sediment load (10^6 t a^{-1})	Wave energy	Tidal range	Fate of most sediment
Amazon	6300	199634	900	Moderate	Macro	Longshore movement of mud
Ganges–Brahmaputra–Meghna	1650	30769	1670	Low (surges)	Macro/Meso	By-pass to Bengal Fan
Mississippi	3300	15631	349	Very low	Micro (diurnal)	Beyond shelf and longshore
Nile	2700	1480	111	Moderate	Micro	Presently deposited in Lake Nasser
Changjiang (Yangtze)	1800	28519	478	Moderate	Macro/Meso	Deep water and longshore
Huanghe (Yellow)	860	1552	1080	Moderate	Micro	Hyperpycnal plumes and longshore
Zaire (Congo)	3800	1250	43	Moderate	Micro	Most to Angola Basin down canyon

Source: Data based on Orton and Reading, 1993; Eisma, 1998; Milliman, 2001.

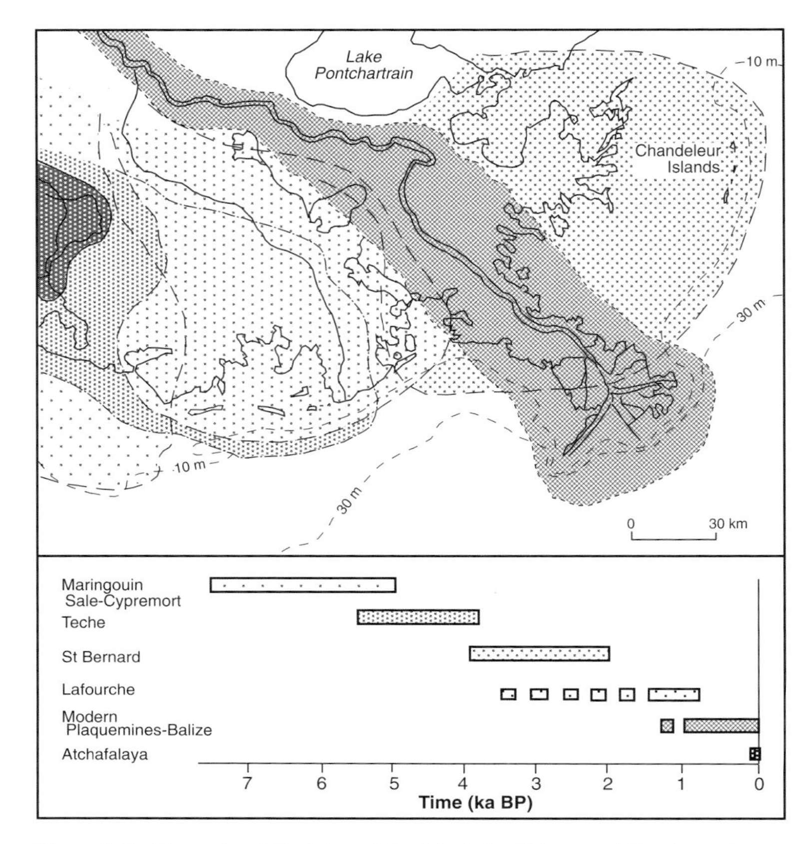

Figure 7.7. The major delta lobes of the Mississippi River, and the chronology of their occupation (based on Coleman, 1988; Törnqvist *et al.*, 1996). See text for details.

area of approximately 30 000 km^2, and has a sediment thickness of about 35 m. The delta has been the site of rapid subsidence, with deformation of the shelf continuing to play a substantial role in the geomorphological development of the landforms associated with the deltaic plain (Coleman *et al.*, 1998).

Each delta lobe has undergone a cycle, consisting of initial progradation, enlargement by further progradation, distributary abandonment and transgression (dominated by subsidence). After abandonment, there is often a hiatus as a delta lobe progrades elsewhere, followed by a further stage of distributary progradation over the abandoned sequence of older deltaic sediments (Coleman, 1988). The chronology of individual delta lobes has been refined by successive studies (Scruton, 1960; Kolb and van Lopik, 1966; Frazier, 1967; Penland *et al.*, 1981; Coleman, 1988; Törnqvist *et al.*, 1996; Roberts, 1997); it appears that an average life cycle takes about 1500 years as summarised in Figure

7.7. The oldest delta lobes are the Maringouin and Sale Cypremort (7500–5000 years BP) and the Teche lobes (5500–3800 years BP) which occur on the western part of the deltaic plain. A major lobe to the east, the St Bernard, is considered to have been active between 4000 and 2000 years BP. The Lafourche lobe is younger, but it is not clear when sedimentation began within this delta lobe. It may have become active around 3500 years BP in the Bayou Terrebonne region (Frazier, 1967); but elsewhere activity is estimated to have begun about 2500 years BP (Coleman, 1988), or as recently as 1500 years BP (566–608 AD) (Törnqvist *et al.*, 1996). Recent evidence suggests that the Plaquemines–Balize modern lobe began progradation 1000–800 years ago (radiocarbon dates 1300 years BP, calibrated to 664–744 AD, Törnqvist *et al.*, 1996), only shortly after the Lafourche, and the two operated simultaneously for a period of time.

In its most recent phase of progradation, the river has reached the edge of the continental shelf and the modern Balize delta is now building into deep water and consists of a number of levée-flanked distributaries (Figure 7.8). This modern delta is a digitate, or birdsfoot, delta in contrast to the previous lobes. The Mississippi River delta distributaries deliver sediment-laden waters to the Gulf of Mexico because buoyant freshwater is discharged over a salt wedge. Sand is deposited in discrete distributary-mouth bars (see Figure 7.2c) as distributaries build

Figure 7.8. An active distributary of the Mississippi River extending out into the Gulf of Mexico. The distributary is flanked by levées. The canals have been excavated to gain access for hydrocarbon extraction.

seawards, forming bar-finger sands that can represent important hydrocarbon traps in ancient analogue delta sequences. The delta front is also extremely dynamic with complex subaqueous delta-front processes, including rotational slides, slumps and other mass movements carrying a substantial proportion of the sediment into deep water (Coleman, 1988). The current in the Gulf of Mexico (as a result of the Coriolis force) carries mud to the west and builds a chenier plain (see Chapter 8).

The Mississippi River distributaries overflow their banks through crevasses during major floods, supplying sediment to the intervening interdistributary bays. These crevasse splays, termed subdeltas, also go through a cycle of build up and destruction. Subdelta cycles give rise to stacked bay-fill sequences, and coalescence of numerous overbank or crevasse splay deposits builds a natural levée. The history of bay fill has been reconstructed for four subdeltas (Coleman and Gagliano, 1964). Each initiated as a crevasse splay and then went through a cycle of growth and deterioration, covering 300–400 km^2 in area and composed of sediments 10–15 m thick, with life spans of around 115–175 years (Wells and Coleman, 1987). The example of Cubits Gap is particularly well known; the initial breach into Bay Rondo that formed the Cubits Gap subdelta was made by a local fisherman in 1862. The breach had become 740 m wide by 1868, an indication of the rapid processes that can occur. The subdelta extended as a result of deposition of middle-ground bars (see Figure 7.2b) in the bifurcating channels, with marsh growth on sediments when the bars became subaerial. However, the interior of the subdelta became increasingly isolated and the marshes covering the surface disintegrated into a series of pools because sedimentation was no longer occurring rapidly enough to keep pace with subsidence. Wetland disintegration has become widespread in this part of the delta because sediment no longer spills over into the crevasse splays and there is much concern about the marsh loss that is occurring (Baumann *et al.*, 1984).

The Mississippi River is presently attempting to adopt a new, much shorter course, along a distributary called the Atchafalaya River. Engineering structures on the Mississippi River divert about 30% of flow down the course of the Atchafalaya River (Kolb, 1980; Davis and Detro, 1980). The Atchafalaya River has built a delta into Atchafalaya Bay since the 1950s. Its growth has involved deposition of a series of coarse-grained, middle-ground bars that have resulted in channel bifurcation (van Heerden and Roberts, 1980). This initial phase of channel extension and bifurcation, with development of sinuous overbank channels, has been followed by a mature phase of upstream lobe growth and consolidation (Vaughn *et al.*, 1996; Roberts, 1998).

The extent to which other deltas worldwide are river-dominated and the landforms that result vary considerably. Where discharge is large, or where tide and wave influences are minimal, river processes tend to dominate; for instance, deltas are river-dominated where they drain into inland seas, such as the delta of the Volga River, emptying into the Caspian Sea (Kroonenberg *et al.*, 1997). The Zaire River in western Africa has a structurally controlled 450-m deep canyon at its mouth. Much of the fluvial sediment load is carried down the canyon and directly into the Angola Basin offshore (Eisma and van Bennekom, 1978). The Amazon River drains a large catchment and experiences a macrotidal range, but it delivers a large mud load to the continental shelf where it is reworked by strong currents (Nittrouer *et al.*, 1986; Allison *et al.*, 1996). It has developed neither a subaerial delta, nor an estuary in the usual sense, but supplies one of the most remarkable mudflat coasts to its northwest (see Chapter 8). River-dominated deltas also occur where steep rivers drain from tectonically active islands. For instance, along the north coast of Java in Indonesia rivers, such as the Citarum and Cimanuk Rivers, have undergone rapid progradation at rates of >1 $km^2\ a^{-1}$ over the last 100 years (Hollerwöger, 1966; Bird, 1985b).

Each of these deltas is river-dominated, but parts of each delta are influenced by other processes. For instance, wave processes, combined with ongoing subsidence, rework the abandoned distributary sediments of the Mississippi River delta and form barrier islands (Penland *et al.*, 1981). Barrier islands occur at various stages of development around the margin of the Mississippi River delta. The Chandeleur Islands form a narrow chain of transgressive barrier islands that are separated from the disintegrating shoreline of the St Bernard delta lobe by a lagoon more than 30 km wide (Figure 7.7, see also Figure 6.27b). Broader barrier islands occur closer to shore, adjacent to more recent delta lobes (see Figure 6.25). The wave reworking of these barrier island shorelines in the context of the delta–lobe cycle, and in relation to subsidence, is examined in Section 7.4.4.

7.4.2 Wave influence on deltas

Shore-parallel sandy ridges are characteristic of deltas that are subject to strong wave action. Where waves approach obliquely they move sand alongshore and form spits; for example the Danube River delta in the Black Sea and the Ebro River delta in Spain have prominent spits, and Cape Bowling Green is a recurved spit on the Burdekin River delta in Queensland. Wave action is the dominant process shaping cuspate deltas, such as occur at the mouth of the São Francisco River in Brazil (Suter, 1994), or the Tiber and Ombrone Rivers in Italy (Innocenti and Pranzini, 1993; Bellotti *et al.*, 1995).

Wave processes can shape deltas where fluvial domination is reduced, either because of human actions, or as part of a delta cycle where distributaries are abandoned. The Niger River has a lobate delta with a barrier along its shore. The river bifurcates into a series of distributaries that are influenced by fluvial and tidal processes (Oomkens, 1974; Allen, 1975), but the shore-parallel sand barriers are shaped by wave processes and are experiencing erosion away from the active distributaries that supply new sediment to the coast (Ebisemiju, 1987). Deltas of the Danube and Rhône Rivers are wave-influenced in part and comprise sequences of shore-parallel ridges which record the sequential evolution of delta lobes (Giosan *et al.*, 1999; Hensel *et al.*, 1999). In other deltas, wave-influenced regions are recorded by isolated individual ridges, for instance in the Mekong River delta (Nguyen *et al.*, 2000) or by drowning barriers, for instance in the Mississippi River delta (Penland *et al.*, 1988).

The Nile River flows across the Sahara in a narrow alluvial valley and then empties through a classic wedge-shaped delta into the almost tideless, low-wave-energy Mediterranean Sea (Figure 7.9a). The river used to carry an annual sediment load of the order of $>100\times10^6$ t prior to closure of the Aswan high dam in 1964, and deposited about 9.5×10^6 t a^{-1} over the delta region, an average annual sediment build up of around 1 mm thickness. The majority of this sediment (estimated at 98%) is now intercepted and deposited in Lake Nasser behind the dam, depriving the delta of sediment. The gross stratigraphy and geochronology of the Nile River delta have been established, based on over 100 radiocarbon-dated cores to depths of 10–50 m (Stanley and Warne, 1993b). The basal stiff alluvial mud is overlain by shallow marine deposits, followed by coastal transgressive sands, above which are coarsening-upwards deltaic sand, silt and mud, and beach-ridge sands (Figure 7.9b). The sequence records decelerating sea-level rise, and it is thickest to the east where subsidence in excess of 4 mm a^{-1} appears to have occurred (Stanley and Warne, 1998). The Nile delta has experienced millennial-scale geomorphological adjustments or climatic change, recorded in the sediments of former distributaries such as the Canopic, Sebennitic and Tanitic, and indicated by archaeological remains. At present, the Rosetta and Damietta distributaries are artificially maintained and although these have promontories at their mouths indicating some progradation of sediment at the river mouths, the reduction in sediment load has also led to wave erosion around their outlets (Figure 7.9).

The Huanghe (Yellow) and Changjiang (Yangtse) Rivers are the largest rivers in East Asia and carry substantial sediment loads to the coast, in this case enhanced by land-use change in the catchment (Ren

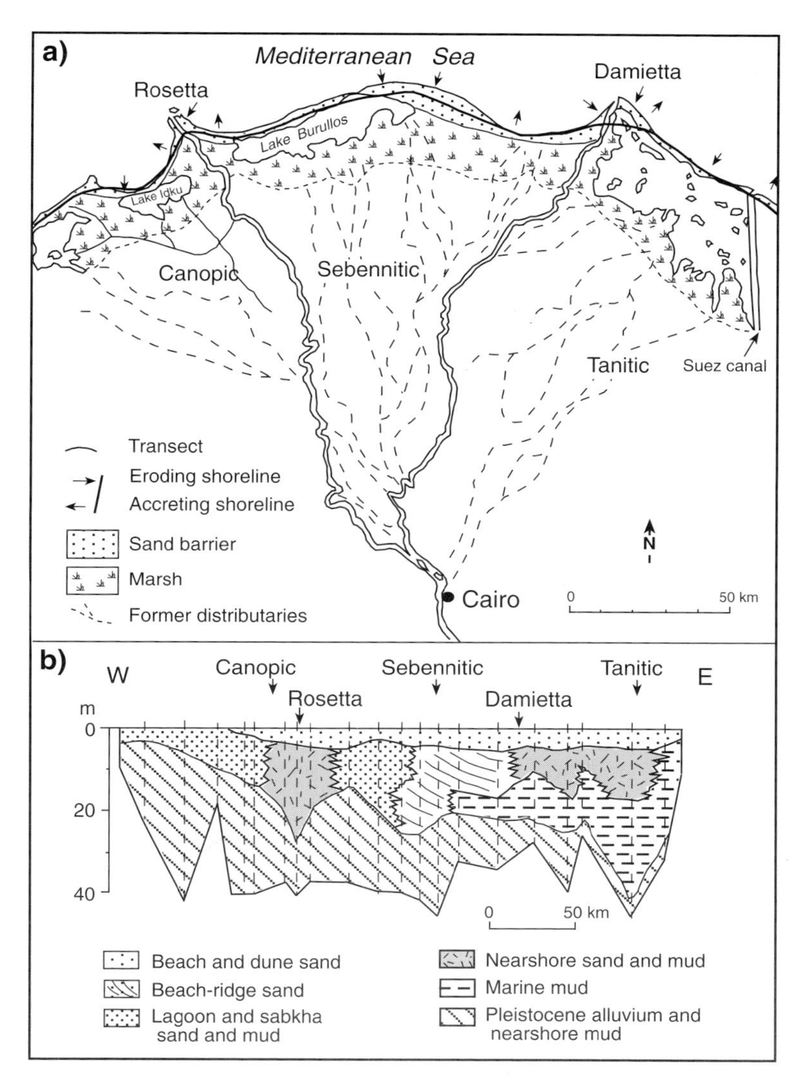

Figure 7.9. (a) The delta of the Nile River, and (b) a schematic morphostratigraphic section (based on Stanley and Warne, 1993b).

and Shi, 1986). The deltas of these rivers are characterised by mudflats that are prograding at rates of up to 100 and 60 m a^{-1}, respectively (Milliman *et al.*, 1989; Yan *et al.*, 1989). The Huanghe River switches between several subdeltas on a regular basis, with rapid progradation (Figure 7.10), adding 20–25 km^2 a^{-1} before avulsing to a new active sub-delta with a periodicty of around 10 years (Wang and Liang, 2000). Although the Huanghe River used to empty into Bohai Bay, it was diverted as a result of man-made structures to occupy a southern

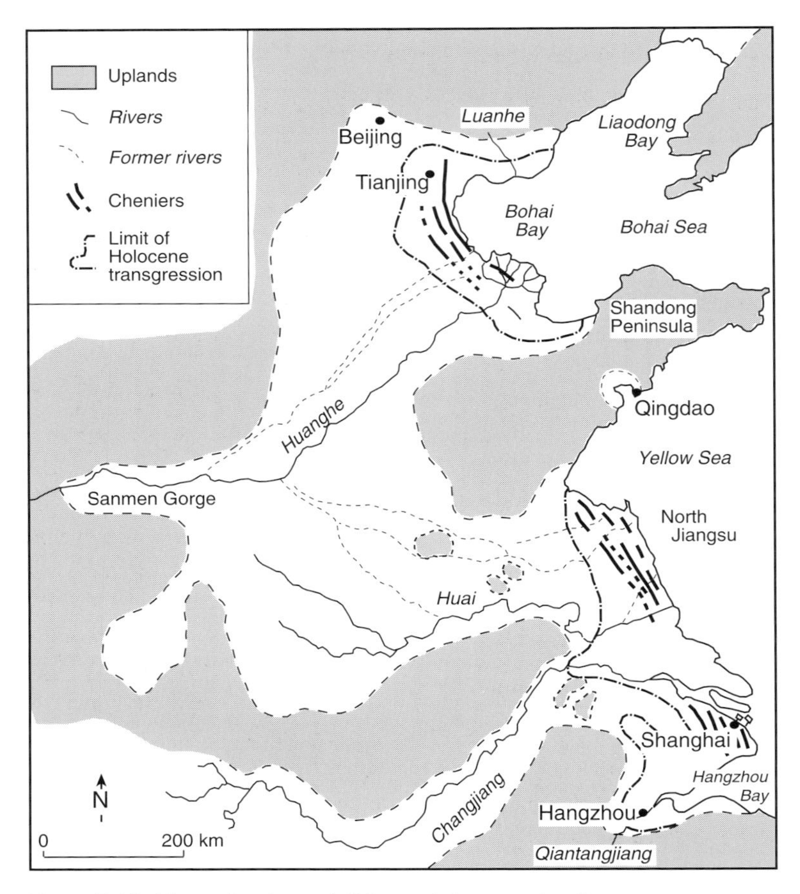

Figure 7.10. The major rivers of China and the coastal sedimentary environments, including chenier ridges, associated with their mouths (based on Yan *et al.*, 1989).

course, discharging into the Yellow Sea on the Jiangsu coast in 1128. For several periods it drained both to the south and the north, and then drained entirely to the Yellow Sea until 1855 when it underwent a natural switch to reoccupy its present delta on the southern shore of Bohai Bay (Saito *et al.*, 2000). Although rivers are very important suppliers of sediment to the coast, wave processes rework the sediment, and the tidal flats of Bohai Bay and North Jiangsu contain prominent shore-parallel chenier ridges. The Luanhe River in northwestern Bohai Bay provides a smaller example of the relative dominance of processes. Since it was diverted in 1915, the river has brought sediment to the coast and built a fan-shaped delta, but a decrease in sediment load since 1980 has resulted in wave reworking of sandy barriers around the margin of the delta (Feng and Zhang, 1998).

7.4.3 Tide influence on deltas

The most prominent characteristic of a delta that is influenced by tide processes is the pronounced tapering of channels (see Chapter 3). A regular decrease in width is shown with distance on channels that occur in macrotidal settings. It also becomes increasingly prominent on former distributaries and other shoreline indentations after abandonment or where river input has been decreased for other reasons. The macrotidal delta of the Ganges–Brahmaputra–Meghna Rivers is examined as an example, and then other deltas where a strong tidal influence is apparent are considered.

The Ganges–Brahmaputra–Meghna delta is one of the largest in the world, and exhibits many features that are typical of deltas in Southeast Asia. The Ganges River flows from the Himalayas to meet with the Brahmaputra River, and the broad channel south of their confluence is called the Meghna. The rivers are very flood prone as they flow through the Bengal Basin, and most of the low-lying areas of Bangladesh are under water in the wet season, with the exception of the Barind Tract and Madhupur Terrace which are Pleistocene terraces (Morgan and McIntire, 1959). The large tidal range in the Bay of Bengal means that much of the delta is tide-dominated. The shoreline of eastern India and western Bangladesh comprises the Ganges Tidal Plain, which is the abandoned deltaic plain of the Ganges. This area is composed of sediments derived from seaward (Allison and Kepple, 2001), and is now dominated by tidal influence and covered by extensive mangrove forests, termed the Sundarbans (Figure 7.11). The active deltaic plain in the east, called the Meghna Deltaic Plain, experiences a large fluvial discharge (96% of the discharge of the rivers; 4% is carried by the former distributaries of the Ganges, such as the Gorai). The Bay of Bengal has a substantial tidal range reaching around 6 m at Calcutta, and adjacent to Sandwip Island, but around 3 m for much of the rest of the delta (Barua, 1991). In addition, the mean water level varies over an amplitude of 85 cm as a result of wind set-up, barometric changes, freshwater input and density differences. Cyclones occur in the May to October period and can raise water levels by 2–3 m, locally increasing them by up to 6 m (Umitsu, 1996).

The tide affects currents in the various channels of the Meghna; the flood tide is of shorter duration than ebb flows, and landward transport of fine sediment is indicated in some areas around Sandwip Island (Barua, 1990). Active coastline change is especially prominent in the river-dominated eastern sections, with deposition of sandbanks, termed chars, that are stabilised after floods by mangroves (especially *Sonneratia apetala* and *Avicennia officinalis*). Sandwip Island has been

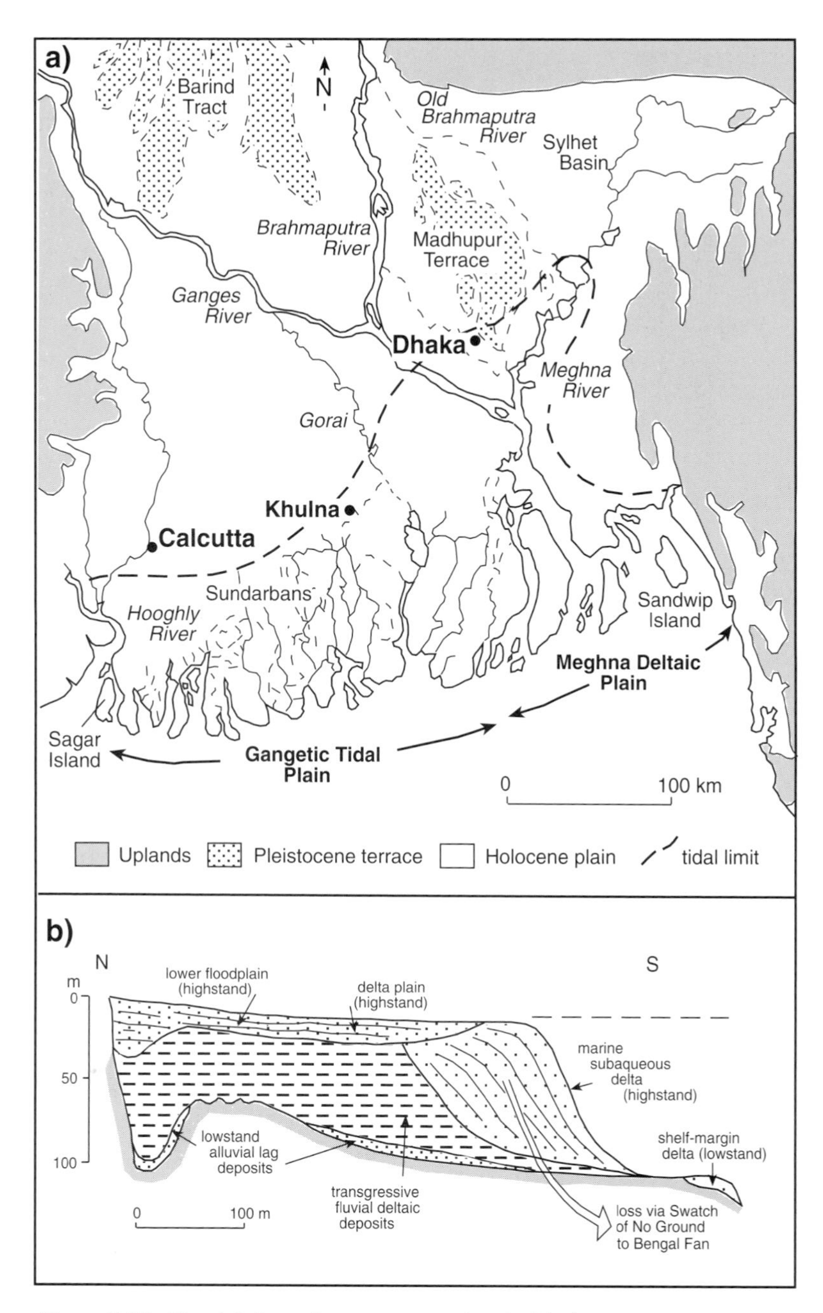

Figure 7.11. The deltaic environments associated with the Ganges–Brahmaputra– Meghna, and shore-normal morphostratigraphic cross-section (based on Goodbred and Kuehl, 2000a).

eroding on the northern side at rates of up to 250 m a^{-1}, but has undergone at least 50 km of progradation southwards in the last 200 years, though still subject to erosion during individual storm surge events (Umitsu, 1996).

Tides dominate water movement in the Gangetic Tidal Plain, and the coastline of the Sundarbans is dissected by numerous, tapering, tidal creeks and is gradually retreating (Allison, 1998; Stanley and Hait, 2000). The Hooghly River, which used to be an active distributary of the Ganges, is characterised by erosion on the southern side, and Sagar Island, as well as many of the mudflats of the Sundarbans are being eroded (Paul *et al.*, 1987; Saenger and Siddiqi, 1993).

Sea level has been a major constraint on the Late Quaternary evolution of the Ganges–Brahmaputra–Meghna delta. Drilling has identified an oxidised gravel layer at depths of 50 m or more in the central plain, which radiocarbon dating indicates to be related to subaerial exposure during the last glacial maximum (Goodbred and Kuehl, 2000a). The postglacial transgression, 11 000–7500 years BP, is recorded by the presence of marine shelly silts and sands. Radiocarbon dating of 7000–6500 years BP on shell and wood at depths of 11–18 m indicates transgression in the Khulna area (Umitsu, 1993). The Ganges–Brahmaputra–Meghna delta appears to have been prograding during sea-level rise because of the extremely high sediment discharge of the rivers amplified by a strengthened monsoon, in contrast to most deltas worldwide that were undergoing landward retreat at the time (Goodbred and Kuehl, 2000b). Mid–late Holocene progradation of the delta is recorded in the uppermost muds and peats radiocarbon dated around 6000–2600 years BP at Calcutta, and 4000–3000 years BP near Khulna (Vishnu-Mittre and Gupta, 1972; Barui and Chanda, 1992; Umitsu, 1993). Although the delta front is undergoing progradation, sediment is also deposited over the active floodplain, and some is lost down the Swatch of No Ground, a submarine canyon, to contribute to the Bengal Fan (Kuehl *et al.*, 1997).

Tidal influence is also apparent by the more tapering channels in the abandoned parts of other deltas. For instance, the Mahakam River delta in eastern Borneo, with spring tidal range of up to 3 m, is often cited as an example of a tide-dominated delta (Figure 7.12a). The stratigraphy of the delta consists of prodelta clays which have been compacted as the delta prograded over them. These are overlain by delta-front shelly sands, which are in turn overlain by tidal-flat muds, and delta-plain organic-rich clays (Allen *et al.*, 1979). The delta has prograded south along the distributary shown in long-section in Figure 7.12b. Much of the delta-front sand and mud was deposited around 6000–5000 years BP, whereas near the modern shoreline dates of around 2000 years BP have been obtained. Active distributaries undergo lateral

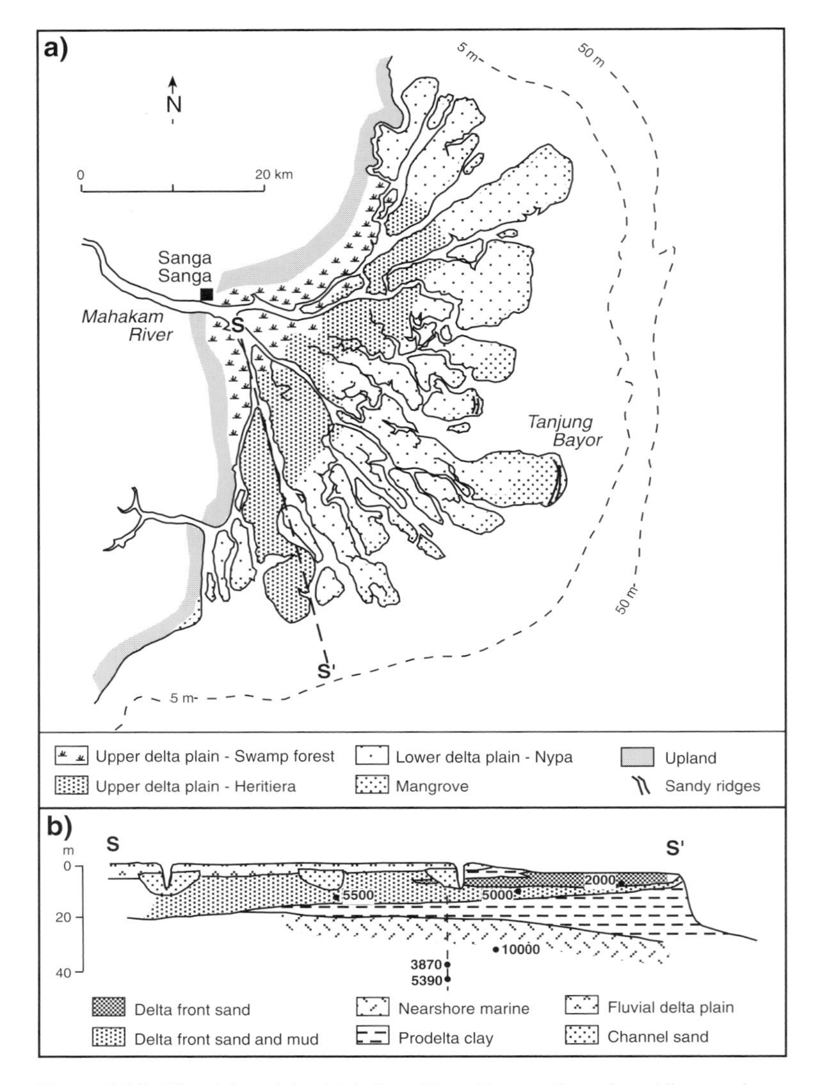

Figure 7.12. The delta of the Mahakam River, Borneo (based on Allen *et al.*, 1979), (a) the major environments of the delta, (b) a morphostratigraphic cross-section of the southern distributary showing radiocarbon dates indicating time of sediment deposition.

accretion of sandy point–bar deposits, suggesting episodic distributary build out and abandonment (like that shown by the Mississippi River delta). By contrast, the eastern distributaries of the Mahakam River delta receive little river discharge, form part of the abandoned delta, and are tide-dominated. Tidal channels meander and taper, with numerous shore-perpendicular shoals, and creeks are infilling with mud (Allen *et al.*, 1979). Erosion is typical of the delta shoreline in the areas that are no longer actively receiving fluvial sediment (Figure 7.12). Sand, shell and organic fragments, particularly of lignite, are concentrated by wave erosion into small shore-parallel ridges, 1–2 m high and 2–3 m thick, on the eastern margin of the delta (Gastaldo and Huc, 1992).

Deltas that occur in areas of large tidal range, such as the Purari River and the Fly River deltas in southern Papua New Guinea have broad funnel-shaped mouths (Thom and Wright, 1983; Baker *et al.*, 1995). Tidal influence can also be important on deltas affected by a smaller tidal range. For instance, the Solo River is a low-energy, mud-dominated, rapidly extending 'single finger' delta, for which sediment load is 95% mud carried in a buoyant plume over a salt wedge that can extend up to 100 km upstream in the dry season (Terwindt *et al.*, 1987). Although the delta is predominantly river-dominated, tidal processes do play a role in shaping the levées flanking the distributaries (Hoekstra, 1993).

The channels that occur on some deltaic plains behave like estuaries, and the distinction between tide-dominated deltaic and estuarine landforms is blurred (Bhattacharya and Walker, 1992; Walker, 1992). For instance, there is a large tidal range at the mouth of the Changjiang River (Yangtse), and the river has prograded across and infilled an incised prior valley (Li *et al.*, 2000). The mudflats in the mouth are very dynamic (Yang *et al.*, 2001). Neighbouring Hangzhou Bay shows prominent tapering (Figure 7.10); it is macrotidal, but very little sediment is supplied to the head of this embayment by the Qiantangjiang River. Most of the sediment in this embayment is derived from the Changjiang River to the north and is moved upstream by the strong tidal currents (Eisma, 1998). It is an example of a tide-dominated estuary.

7.4.4 Patterns of delta distributary change

Deltas are shaped by river, wave and tide processes, and the balance between these can alter with time. They also respond to external factors, particularly the relative rate of sea-level change. The role of external boundary conditions can be seen by considering overall patterns of delta channels. For instance, the predominant distributary morphology of the Rhine–Meuse deltaic system between Rotterdam and Utrecht underwent a transition from meandering to anastomosing around 7000 years BP, with ribbon-like channel belts, deposited in channels subject to frequent avulsion, preserved in lacustrine muds and organic sediments (Törnqvist, 1993). Since around 4000 years BP, the system has reverted to meandering single-channel distributaries with more homogenous overbank deposits and less frequent avulsion (Törnqvist, 1994; Berendsen and Stouthamer, 2000). The changes appear to reflect changing rates of sea-level rise (Beets and van der Spek, 2000).

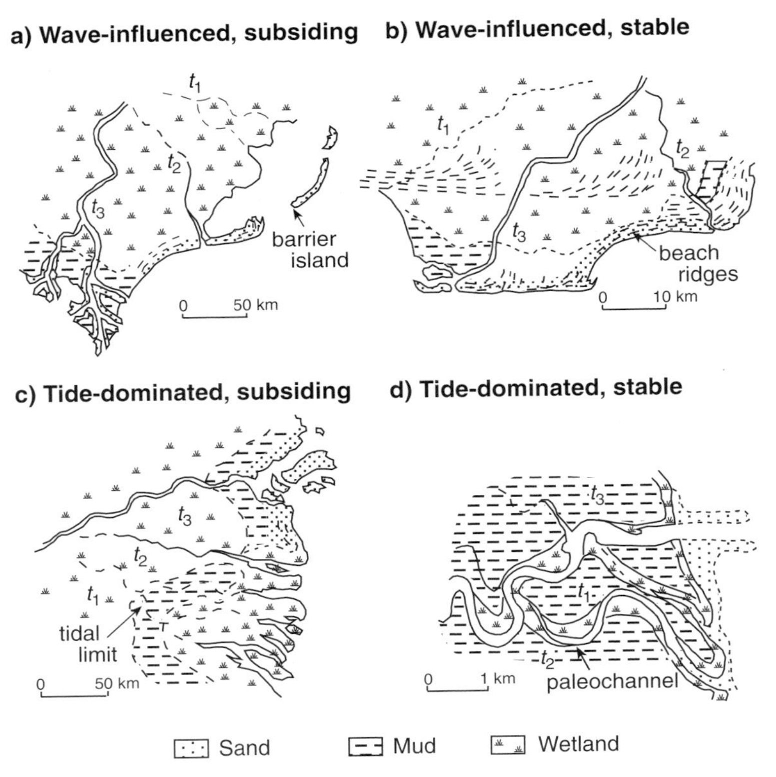

Figure 7.13. The shoreline of abandoned deltaic plains as influenced by wave and tide processes and the rate of subsidence. (a) Subsiding coast where wave action exceeds tide processes, e.g. Mississippi delta (Penland *et al.*, 1988), (b) stable coast where wave action exceeds tide processes (e.g. Rhône River), (c) subsiding coast where tide factors exceed waves, e.g. Ganges–Brahmaputra (Morgan and McIntire, 1959), and (d) stable coast where tide action results in infilling of former distributaries, e.g. McArthur River (Woodroffe and Chappell, 1993).

The relative balance between processes is discernible from patterns of distributary change before and after abandonment. Distributaries switch (or avulse), abandoning one route in favour of a hydrodynamically shorter route. The cycle of distributary extension and abandonment described for the Mississippi River delta by Frazier (1967) can also be seen in the sandy barrier and barrier islands that result when abandoned delta lobes are reworked by wave processes (Penland *et al.*, 1988). Figure 7.13a shows a schematic example for a subsiding delta where wave influence is more pronounced than tide influence, based largely on the Mississippi River. When a distributary ceases to be active distributary-mouth deposits are reworked by wave action, forming a sandy erosional headland with flanking barriers. Isle Dernieres is a sandy barrier island related to abandonment of the Lafourche delta lobe.

The delta lobe disintegrates because, as the coast retreats, the volume of sediment in the eroding barriers is too little to keep pace with subsidence. The flanking barriers become partially submerged, forming a barrier-island arc. The Chandeleur Islands (see Figure 7.7), forming a crescentic chain of barrier islands, dominated in the north by inlets and to the south by overwash, relate to the St Bernard delta lobe and its disintegration (see Figure 6.27b). At a later stage, the barrier islands themselves become submerged and form a submerged shoal; Ship and Trinity shoals relate to the Marigouin and Teche lobes to the west of the deltaic plain (Stone and McBride, 1998). The cycle is rejuvenated at the stage when this subsided shoreline offers a shorter route to the sea than the extended path of the active distributary, and a new distributary occupies it (Penland *et al.*, 1988).

The pattern of change as seen in the Mississippi River delta can be considered cyclic, driven by rapid subsidence beneath this large delta and by switching in delta-lobe position. On smaller stable deltas, sand ridges form at eroding distributary mouths but are not submerged and persist as a series of ridges or as a beach-ridge plain within the delta (Figure 7.13b). Deltas such as those of the Rhône and the Danube Rivers record periods of active distributaries, and subsequent wave action forming the series of ridges across its surface (Zenkovich, 1967). There are similar ridges on many other deltas, for example the Acheloos River delta in Greece (Piper and Panagos, 1981) and the Mekong River delta in Vietnam (Nguyen *et al.*, 2000).

In tide-dominated deltas, a similar cycle of distributary extension and abandonment is common. Former distributaries are reshaped by tidal currents into tapering, funnel-shaped channels and well-developed tidal-creek networks, rather than being reworked by wave action (Figure 7.13c). A classic example is the Ganges tidal plain of the Ganges–Brahmaputra–Meghna delta (see Figure 7.11). A similar contrast is evident between the active and abandoned deltaic plains of many Southeast Asian deltas, including the Mahakam Delta (see Figure 7.12). The Purari River in southern Papua New Guinea has a tide-dominated abandoned delta (tidal range around 3 m), which is covered in mangrove forests (Thom and Wright, 1983). In these large deltas, subsidence of the abandoned deltaic plain accentuates the erosion of these tide-dominated shorelines. In smaller stable deltas, there can also be tidal domination of abandoned distributaries. For instance, in the case of the McArthur Delta on the southern shore of the Gulf of Carpentaria,

Australia, periodic switching of distributaries has occurred and abandoned distributary courses have been infilled with sands so that previously parallel-bank river-dominated channels are now tapering (Woodroffe and Chappell, 1993). The Gulf of Carpentaria has a predominantly diurnal tide with a small range (<1.5 m) and the infill and reshaping of distributaries is clear evidence of tide-domination of these abandoned channels (Figure 7.13d).

7.5 Morphodynamic development of estuaries

Most estuaries and major embayments in bedrock have been drowned by the postglacial rise in sea level. Rivers tend to deposit their sediment load when their flow decelerates on entering an estuary, often building fluvial deltas. Tide processes also influence estuaries and can import sediment from seaward, in some cases forming a tidal delta at the seaward end of the estuary. Wave processes can also shape an estuary. Waves construct a sand barrier; this can partially enclose an estuary, forming a barrier estuary, and it can also result in total closure of the embayment, encapsulating a lagoon.

An evolutionary classification of estuaries in southeastern Australia into drowned river valleys, barrier estuaries and saline coastal lakes (which correspond to coastal lagoons) has been proposed by Roy (1984), and has application around and beyond the Australian coast (Kench, 1999). The type of estuary that occurs depends on inherited geological factors, such as bedrock topography, and the relative balance between river discharge and wave energy. Deep, bedrock-controlled embayments, such as Port Jackson (Sydney Harbour) and the Hawkesbury River (Broken Bay), have relatively wide entrances, with a sandy tidal delta that forms a sill, over which full tidal exchange can take place. They tend to show little sediment infill relative to the size of the embayment and are termed drowned river valleys. Barrier estuaries are separated from the sea by a sand barrier, with an entrance through which tidal exchange is attenuated. They occur where bedrock is relatively shallow (<30 m), in contrast to drowned river valleys, forming a basement on which barrier accumulation has focused. Strong wave conditions maintain the barrier at the mouth of an incompletely filled embayment, but river discharge is large enough to keep an inlet open at least intermittently. Coastal lagoons, termed saline coastal lakes by Roy (1984), tend to be saline to brackish (but not tidal) waterbodies, which remain closed for long periods, although they can open temporarily during major floods and large seas. Progressive infill of each of these can be seen as part of a sequence; different estuaries along the New South Wales coast are at different stages (Roy *et al.*, 1980, 2001). This classification is used as a basis for the discussion of estuarine morphodynamics that follows, but is extended by addition of tide-dominated estuaries, of which seasonal macrotidal rivers in northern Australia provide an example.

This section examines the morphodynamics and evolution of estuaries. The simplest estuaries are drowned river valleys, including fjords, rias and incised valleys. Tide-dominated estuaries are found in areas that have a large tidal range, or where other processes exert only a minor influence. Barrier estuaries are influenced by wave processes and are common on wave-dominated coasts, and coastal lagoons are often found in similar wave-dominated settings to barrier estuaries. It should be emphasised again that these forms exist across a continuum, and that any deltaic–estuarine system may be influenced by river, tide and wave processes to different extents varying throughout an individual system, or over time.

7.5.1 Drowned valleys

The majority of estuaries are drowned valleys that owe their origin and, in particular, their modern morphology to Holocene sea-level rise. The term is especially appropriate for those systems in which the prior bedrock configuration exerts a primary control on morphology, such as fjords and rias. Incised valleys, where incision has been into older valley fill sediments, are also considered in this section (Dalrymple *et al.*, 1994).

Fjords

Fjords are glaciated troughs with mountainous shores and occur poleward of 45° N or S, occurring in western Canada, Norway, southern Chile, Greenland, Labrador, Iceland, Antarctica, southwestern New Zealand, and in Scotland where they are called 'lochs'. The term fjord is Nordic for inlet and fjords are steep-sided rocky basins, excavated by land-based ice, with flat floors as a result of sediment deposition. The steepness of the walls of fjords depends on lithology. Fjords can contain a series of landforms such as tributary fjords and fjord lakes, cirques (ice-formed depressions) and hanging valleys, often occupied by waterfalls (Syvitski *et al.*, 1987). The mouth of a fjord often contains a sill (or threshold), and the deep basin behind the sill can be 1000 m or more deep, and is typically 2000 m deep in Antarctica (Syvitski and Shaw, 1995). The reason that a distinct sill develops is not always clear; it can result from the deposition of moraine, but in some cases it is formed by over-deepening of the rock basin. Circulation within fjords is often stratified, and the deep basin can be stagnant and anoxic because the sill hinders exchange.

Four different types of fjord can be recognised which can be inferred to represent different stages of formation. Fjords that are ice-filled represent the earliest stage of development, for instance several fjords in Greenland. Subsequently the ice melts and the basin becomes filled with seawater with floating glaciers, as on Baffin Island in Canada. Retreat of glaciers until they remain only on land represents the third stage, as for instance in many fjords in Alaska and Norway. Finally ice melts completely leaving U-shaped basins, such as the sea lochs in Scotland or Milford Sound in New Zealand.

Fjords are major sediment traps and account for a significant proportion of global transfer of sediment from land to sea (Syvitski and Shaw, 1995). Extrapolation of modern rates of sediment supply to fjords explains only a small proportion of the sediment (<10%) in fjord basins, and implies that most have experienced an earlier period of periglacial activity, characterised by more rapid sediment supply (Forbes and Syvitski, 1994). Fjords fill primarily as a result of glacial or fluvioglacial input, generally receiving little, if any, sediment from seaward. They do undergo submarine sediment redistribution by turbidity flows, slides and slumps (Syvitski and Shaw, 1995). Coarse debris can accumulate at the head of fjords in fan deltas, sandur plains and outwash deltas (Forbes and Syvitski, 1994). Despite rapid rates of sediment input, fjords are incompletely filled and generally contain only a small proportion of the sediment that they could accommodate.

Rias

Rias are drowned river valleys where the bedrock topography plays a prominent role in estuarine morphology, as opposed to those systems where estuarine or alluvial sediments line the channel (Castaing and Guilcher, 1995). The term 'ria' is particularly used for high-relief estuaries of the Atlantic coast of the Iberian Peninsula and Brittany in France (where the term 'aber' is also used). Rias also occur in Devon and Cornwall in southwestern Britain, for example the mouths of the Teign, Dart and Tamar Rivers. Rias also occur in western Ireland and Wales (aber) in areas in which some glaciation has been experienced, but where fjords have not formed.

European rias usually consist of deep entrances (up to 30 m), cut during lower sea levels, and contain sediment sequences reflecting Quaternary fluctuations in sea level. For instance, interglacial beaches from oxygen-isotope stage 5 or 7 are found on the valley sides of some rias in Devon (Castaing and Guilcher, 1995). The drowned river valleys are incompletely filled, particularly where they have received a relatively small fluvial load, or where late Holocene sea-level rise and postglacial rebound have continued to increase the volume that can be filled with sediment (accommodation space).

Examples of stratigraphic sequences found in rias include Pleistocene basal sands of fluvial origin and fan deltas along the valley sides of the Ria de Muros y Noya, northwestern Spain (Somoza and Rey, 1991). Continuing sea-level rise increases the accommodation space and leads to upward and outward accretion, as in the Sado Estuary, Portugal (Psuty and Moreira, 2000). Mud is often derived primarily from erosion of periglacial sediments (Guilcher and Berthois, 1957; L'Yavanc and Bassoullet, 1991). Sand appears to be reworked landwards into rias by waves and flood tide processes as sea level rises, and many rias have complexes of spits and bars, or sandy tidal deltas dominated by marine processes at their mouths (Castaing and Guilcher, 1995).

Drowned river valleys are one of the principal types of estuary on the coast of New South Wales around Sydney, and have been described by Roy (1984). These also show sandy sediments at their landward and seaward ends, with mud accumulating in between. The rate at which the fluvial delta progrades into the embayment at the landward end depends on the size and lithology of the catchment. Towards the mouth, a sandy tidal delta is formed; in the case of Port Hacking this reaches to water level and is intermittently exposed at low water; although in other drowned river valleys the tidal delta is permanently submerged. The central part of the embayment is dominated by a mud basin, in which fine sediments accrete vertically (Roy, 1994).

Many estuaries are valleys incised into sedimentary sequences rather than into bedrock, and there are varying degrees to which previous sequences of valley fill have been removed during successive lowstands. For instance, Chesapeake Bay occurs within sediments that have accumulated on a trailing-edge coast over a series of glacial cycles or longer. Rias in Europe and along the eastern coast of North America are generally still at a relatively early transgressive stage of infill, as a result of continued relative sea-level rise. There appear to be some broad similarities that can be recognised in the pattern of infill whether or not the drowned river valley can be strictly described as a ria. A sequence of incised valley fill for the Hawkesbury River valley in New South Wales records the progressive extension of fluvial sediments seawards and tidal-delta sediments landwards over the vertically accreting central mud basin (Nichol *et al.*, 1997). Eventually, the entire valley will be infilled and the river then adopts a channelised course directly to the seaward end of the embayment (Roy, 1984, 1994). A similar sequence is described for barrier estuaries in Section 7.5.3.

The prominent drowned river valleys along the coast of New South Wales, described by Roy and coworkers (Roy *et al.*, 1980, 2001; Roy, 1984, 1994), demonstrate a sequence of stages of infill. The earliest comprises a transgressive phase that terminated around 6000 years BP followed by a regressive phase during which individual embayments have infilled at rates that reflect the relative fluvial inputs and wave-energy characteristics. In their infilled stages they evolve like wave-dominated estuaries (described below). Similarly bedrock-controlled embayments along the northern and northwestern coast of Australia are also drowned river valleys or rias, and have also undergone a transgressive phase and subsequent progradational phase (Semeniuk, 1985a). However, estuaries that have been largely infilled with sediments have been dominated by tidal processes, and evolve like tide-dominated estuaries which are described in the next section.

7.5.2 Tide-dominated estuaries

Estuaries that occur in macrotidal settings are dominated by tidal currents and, where they occur on prograded coastal plains, they are typically funnel shaped with wide entrances tapering upstream (Chappell and Woodroffe, 1994; Wells, 1995). Sandy shoals, elongated parallel with the flow, are often prominent features within the mouth. These estuaries are well mixed as a result of strong bi-directional tidal currents. Rapid flows resuspend and transport sediments. Three zones with distinct channel morphology can be recognised in many macrotidal estuaries: an upstream river-dominated zone, a central mixed-energy zone and a seaward marine-dominated zone (Dalrymple *et al.*, 1992). The upstream zone is often relatively straight with net seaward sediment movement by fluvial processes. The central zone is a zone of sediment convergence influenced by both river and marine processes, often with a highly sinuous channel. The seaward zone is relatively straight, its banks are funnel shaped, and sediment moves in a net landward direction (Figure 7.14).

There are many macrotidal estuaries in northern Australia influenced by strong tidal flows throughout the year but dominated by fluvial floods during the monsoon season (see Figure 7.3). Tides become progressively asymmetrical with distance upstream from the mouth (see Chapter 3), and flood tides have shorter duration and higher flow velocities than ebb tides (Woodroffe *et al.*, 1989). Where highest mean tidal flows are directed upstream the net bedload sediment movement is also upstream, until tidal asymmetry is offset by downstream fluvial discharges during the wet season (Bryce *et al.*, 1998). The smaller the fluvial influence, the weaker the seaward transport of sediment during wet

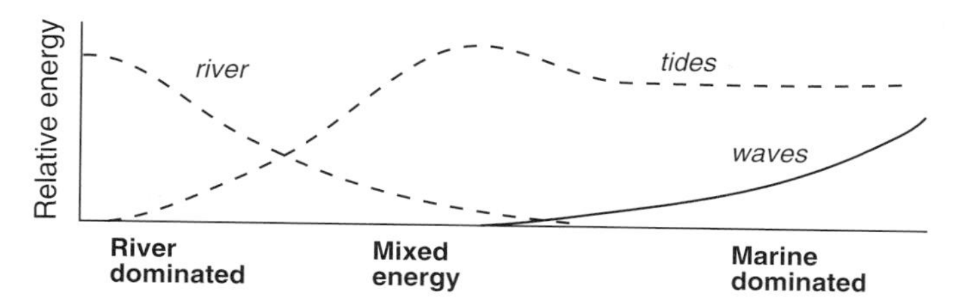

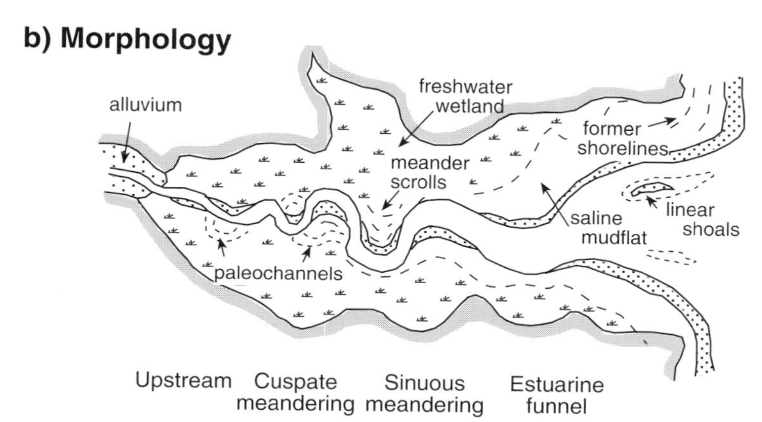

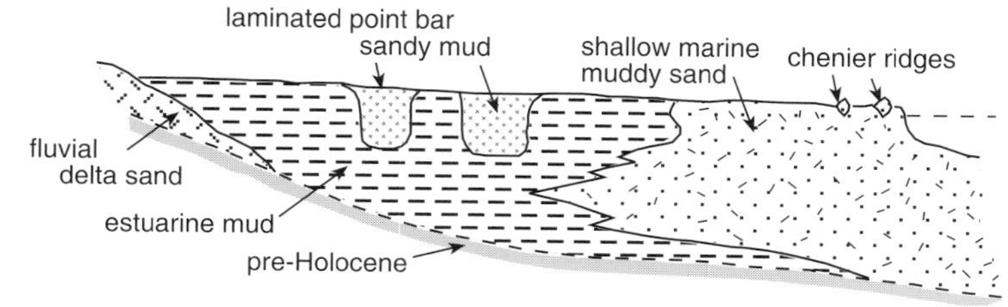

Figure 7.14. Morphology and stratigraphy of a typical macrotidal estuary. (a) The relative influence of river and marine domination; (b) typical morphology based on estuaries in northern Australia, and (c) schematic morphostratigraphy (based on South Alligator River in northern Australia, Woodroffe *et al.*, 1989).

season floods and the stronger the tendency for upstream-moving sediment to remain within the estuary.

Antecedent topography controls the gross form of contemporary macrotidal estuaries in northern Australia. Some valleys have been drowned by sea-level rise and form rias; for instance, Darwin Harbour comprises a dendritic network of drowned valleys (Semeniuk, 1985a,b). Larger river systems have infilled their lower valleys with sediment, and traverse extensive Holocene deltaic–estuarine plains. For instance, the

Figure 7.15. Sinuous estuarine meanders on the macrotidal Adelaide River estuary in the Northern Territory, Australia. These meanders appear to be inherited from a prior fluvial channel which was drowned by rising sea level in the early Holocene. There are prominently zoned mangrove forests on the insides of the meanders, whereas the deltaic–estuarine plains are covered by grasses and sedges.

Ord, South Alligator and Daly Rivers which have a tidal range of around 6 m and are tidal for more than 100 km upstream (Thom *et al.*, 1975; Woodroffe *et al.*, 1989; Chappell, 1993). Tidal range is up to 12 m in King Sound in Western Australia and extensive saline mudflats flank the broad estuary of the Fitzroy River (Jennings, 1975; Semeniuk, 1981). The highly sinuous meanders of the central zone of the Adelaide River, shown in Figure 7.15, are unusual because they appear to have been inherited from a prior fluvial river that traversed the area when sea level was lower (Woodroffe *et al.*, 1993).

The evolution of the deltaic–estuarine plains that flank these rivers has been studied in detail for the South Alligator River, draining into the macrotidal van Diemen Gulf in the Northern Territory. Schematic cross-sections of the plains are shown in Figure 7.16. The floor of the prior valley, comprising pre-Holocene oxidised sands, gravels or a stiff mottled clay, was inundated by the sea at about 8000 years BP and

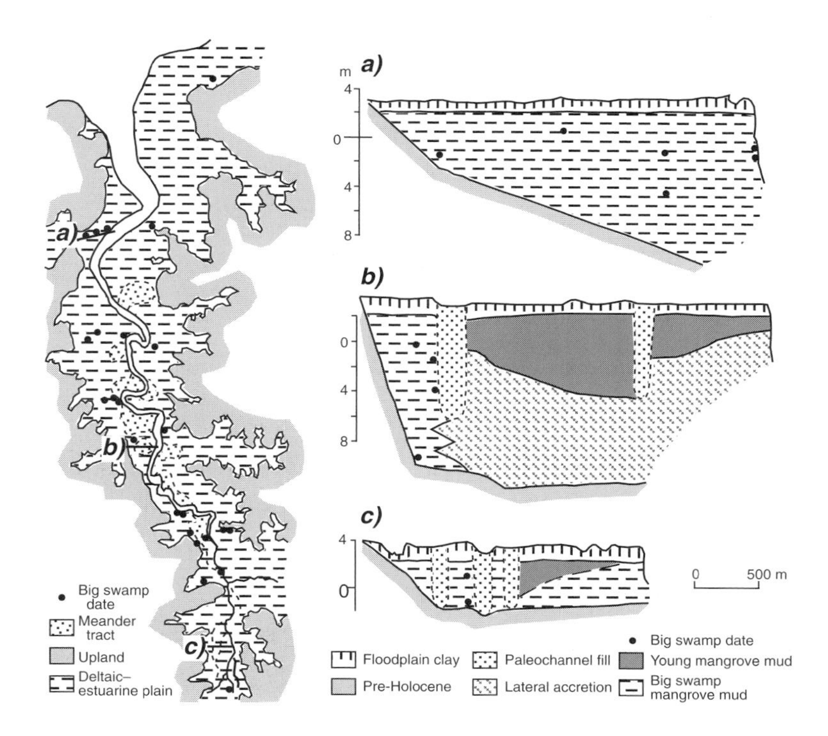

Figure 7.16. Schematic morphology and stratigraphy of the South Alligator River plains in the Northern Territory, Australia. The sinuous section contains lateral accretion sands and muds deposited as the river migrated. The upstream section contains channels that have avulsed, and which have not reworked the older Holocene muds between them (based on Woodroffe *et al.*, 1989).

transgressive muds were deposited as sea level rose. Around 6800 years BP when sea level was within 5 m of present, extensive mangrove forests, termed 'big swamp', covered much of the plains, indicated by an organic-rich layer of mangrove mud 2–5 m below the plains surface. This layer has been radiocarbon-dated at 6800–5300 years BP (Woodroffe *et al.*, 1985).

The 'big swamp' phase has been widely recognised throughout estuarine systems in northern Australia (Woodroffe *et al.*, 1993; Crowley, 1996). Pollen analysis has indicated successional change from mangrove forest dominated by seaward *Rhizophora* to more landward stands of *Avicennia*. This was subsequently replaced by freshwater clays supporting freshwater wetlands dominated by grasses and sedges (Woodroffe *et al.*, 1985; Clark and Guppy, 1988; Grindrod, 1988). After the 'big swamp' phase channel migration has reworked part of the plains, and abandonment and infill of paleochannels has added to the complexity of habitats. In the central zone, meanders have migrated across the plains forming a meander tract by lateral accretion and meander cutoff. In the upstream section, the channels have switched by avulsion as shown by the presence of mangrove muds of big swamp age between the younger paleochannels (Chappell and Thom, 1986).

The three phases of estuarine infill identified on the plains of the South Alligator River appear to have occurred on many of the adjacent rivers (Woodroffe *et al.*, 1985, 1993) and a schematic model of evolution is shown in Figure 7.17. The three phases are: (t_1) a transgressive phase (8000–6800 years BP) of marine incursion into the prior valley, (t_2) a big swamp phase (6800–5300 years BP) of widespread mangrove forest development, and (t_3) a phase of alluvial floodplain development since 5300 years BP. The extent to which each macrotidal estuarine channel meanders, the rate at which lateral accretion occurs, and the proportion of floodplain reworked depend on the relative magnitude of tidal flows in the dry season, in comparison with river-dominated floods in the wet season (Chappell and Woodroffe, 1994). The Daly River has a large catchment and fluvial floods have reworked much of its floodplain with meanders migrating at up to 40 m a^{-1}; extensive scroll bars across the plains mark former meander locations of this actively migrating river channel (Chappell, 1993). The South Alligator River on which tidal flows and maximum river floods may be of similar discharge, has changed from a sinuous river, and is now characterised by relatively stable, cuspate estuarine meanders. On the adjacent, but smaller, Mary River, coastal sedimentation appears to have impounded the estuary which has infilled with marine sediment, and much of the system has become inactive (Mulrennan and Woodroffe, 1998).

Many macrotidal systems are dominated by landward movement of sediment. In the case of the South Alligator River, it appears that much

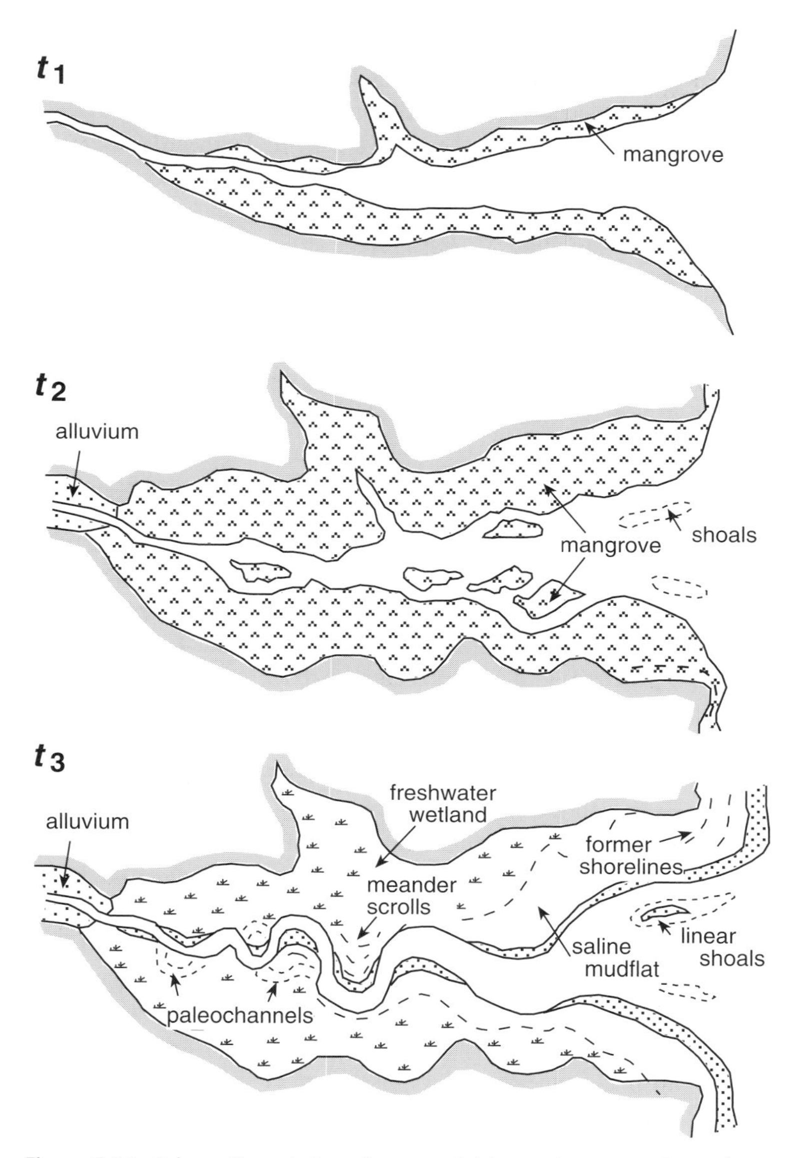

Figure 7.17. Schematic evolution of a macrotidal estuarine system in northern Australia, involving (t_1) transgressive phase as the sea level rises, (t_2) big swamp phase at the time of sea-level stabilisation, and (t_3) stillstand phase involving extension of floodplain clays over mid-Holocene muds (based on Woodroffe *et al.*, 1993).

of the large volume of mud that has infilled the deltaic–estuarine plains has been derived from seaward. Other macrotidal systems demonstrate similar morphology and processes. Cobequid Bay in the Bay of Fundy, Canada, with the largest tidal range in the world, contains prominent sandy bedforms exposed at low tide, indicating landward sediment

estuary, from the transgressive stage, during which the sea level was still rising towards present level, and progressive infill during the subsequent stillstand. Rivers that drain a large catchment composed of weathered rocks appear to infill most rapidly (Jennings and Bird, 1967). Infill involves extension of the fluvial bayhead deltas across the mud basin, and the estuarine area decreases with flow through the barrier estuary becoming increasingly concentrated into channels and, consequently, with tidal influence extending further upstream. As a barrier estuary progresses through the evolutionary sequence, the complexity of its shoreline can increase, until the river course becomes channelised in the final stages of infill. Tidal energy, as indicated by tidal range, is rapidly attenuated through narrow or shoaled entrances, and there tends to be water-level fluctuations of only a few centimetres in barrier estuaries, such as Burrill Lake. In the final stage of infill, when flow is completely channelised, fluvial influence is experienced right to the coast, but with tidal variations in water level, determined by the relationship between tidal prism and fluvial discharge, experienced further upstream than would have occurred in earlier stages of infill (Roy, 1984). Examples of rivers in several stages of infill have been described by Roy (1984, 1994).

The Shoalhaven River in southern New South Wales has extensive deltaic–estuarine plains at its mouth, and represents the ultimate stage

Figure 7.19. Burrill Lake, a barrier estuary on the southern New South Wales coast. The flood tidal delta has built several kilometres into the embayment (photograph R.J. Morrison).

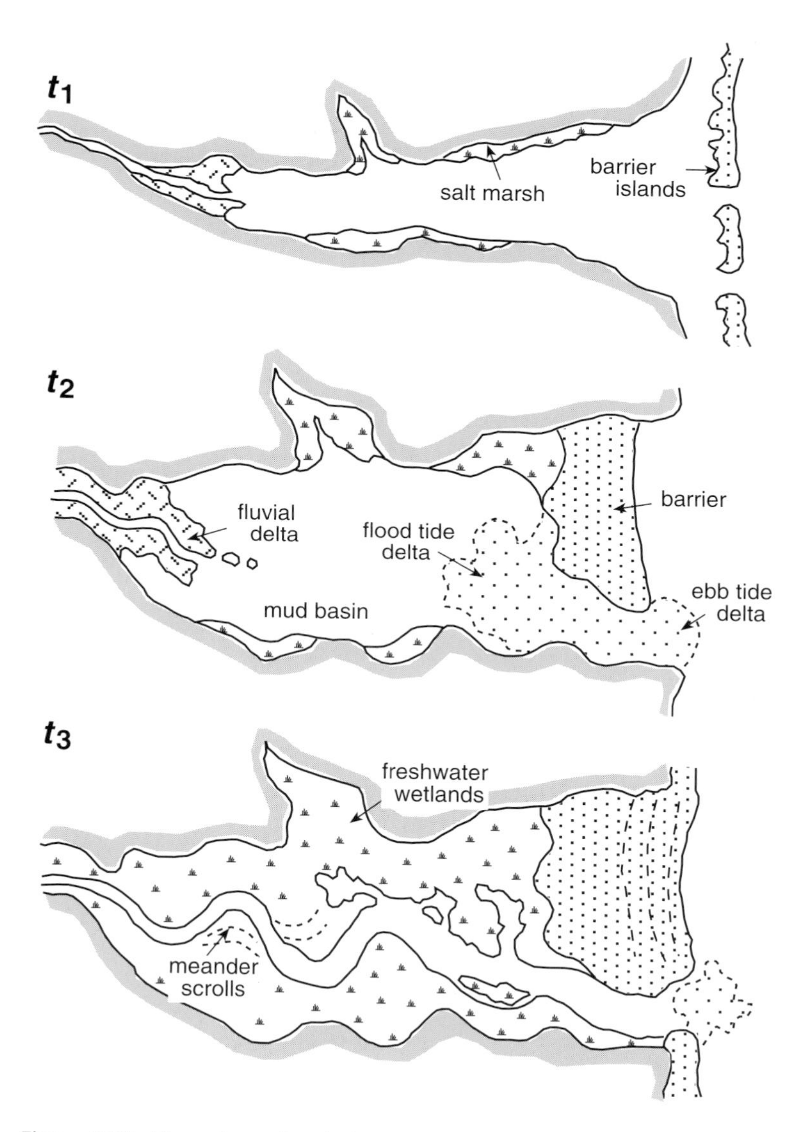

Figure 7.20. The pattern of evolution of a barrier estuary, involving (t_1) a transgressive phase prior to (t_2) sea-level stabilisation, and (t_3) during stillstand (based on Roy, 1984).

in the infill of a barrier estuary (see Figure 6.1). It is one of the largest rivers in southern New South Wales and its sediment load has completely infilled the estuarine embayment (Figure 7.21). The Holocene deltaic–estuarine plains of the Shoalhaven River are underlain at depths of 4–5 m, but at 20–25 m beneath the former course of the river, by oxidised alluvial gravelly sands which have been shown by TL dating to be

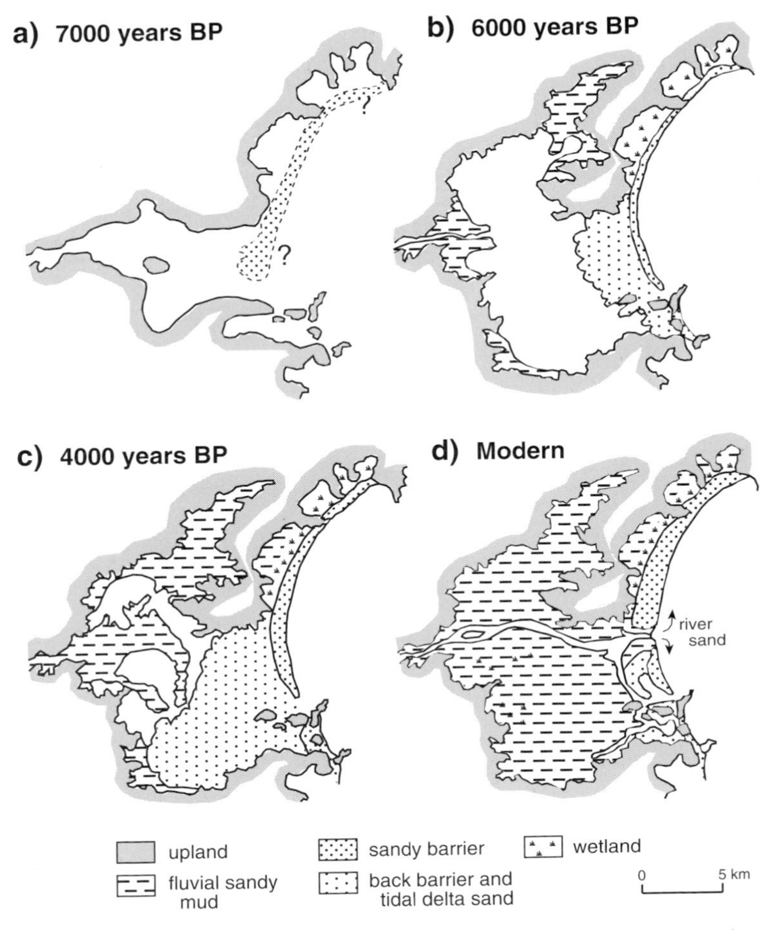

Figure 7.21. The evolution of the Shoalhaven deltaic–estuarine plains, southern New South Wales. (a) Transgression during the final stages of postglacial sea-level rise, (b) around 6000 years ago, (c) around 4000 years ago, and (d) the modern plains (based on Roy, 1984, 1994; Umitsu *et al.*, 2001). The mouth of the Shoalhaven River can be seen in the satellite image in Figure 6.1.

late Pleistocene in age (Young *et al.*, 1996b). The topography of these older sediments, incised by the river at lower sea levels, has been a major constraint on the morphology and pattern of infill of the Holocene estuarine environments. In the final stages of the postglacial transgression, the sea inundated the valley and the shoreline migrated landwards until the sea reached its present level around 6000 years ago. Most of the area of the present plains is underlain by mid-Holocene muds containing marine and brackish assemblages of molluscs and deposited as part of a large mud basin interconnected with the open sea (Umitsu *et al.*, 2001). The fluvial delta of the Shoalhaven built across the embayment, impounding a series of flood basins between the distributaries, and during the late Holocene has delivered sediment directly to the coast. This is in contrast to most estuaries along the southern New South Wales coast which are in less mature stages of infill; sand from the Shoalhaven River has contributed to the progradation of the Seven Mile Beach barrier as a beach-ridge plain (L.D. Wright, 1970; Roy, 1984). The surface of the plains is composed of alluvial sediments deposited under freshwater conditions. Levées indicate the course of several fluvial channels and the river has adopted more than one course to its mouth. Wave energy is sufficient to close the mouth of the river which opens during large river flood events. Diversion of the river through an artificial canal to join the adjacent Crookhaven system now ensures that this river is continually open to the sea through this southern entrance (Figure 7.21).

7.5.4 Coastal lagoons

Coastal lagoons are water bodies impounded by a sand barrier; they represent an extreme form of barrier estuary (Cooper, 1994; Isla, 1995). There have been several attempts to classify coastal lagoons in terms of their hydrodynamics. Coastal lagoons can be connected intermittently to the ocean by one or more restricted inlets through the barrier. Lagoons are balanced if inflows and outflows of water are equal. Where river flows dominate, lagoons have lower salinity and resemble estuaries (Carter, 1988). Hypersaline lagoons occur where evaporation exceeds freshwater input, for instance parts of the Coorong in South Australia or Laguna Madre in Texas.

The relative balance of tide, river and wave energy varies, influencing the hydrodynamics and salinity of the lagoon (Barnes, 2001). Waves are important outside the lagoon and influence the entrance, if there is one. The effect of tides is usually restricted. Flood tidal deltas appear more common in microtidal settings and ebb tidal deltas are better developed in meso–macrotidal settings (Hubbard *et al.*, 1979; Hicks and Hume, 1996). The influence of rivers varies over time; floods can be important in breaching or enlarging an inlet that has sealed, or partially filled. Wind-waves often influence circulation and sediment movement in the central parts of large lagoons (Figure 7.22). Some sediment can be autochthonous, produced within the lagoon by organisms, or chemically precipitated (Bird, 1994, 2000). In view of their role as a net sediment sink, lagoons are prone to accumulation of pollutants (Barnes, 1980) and are susceptible to changes in rate of sediment supply as a result of human factors, such as land-use change in the catchment (Martin *et al.*, 1981).

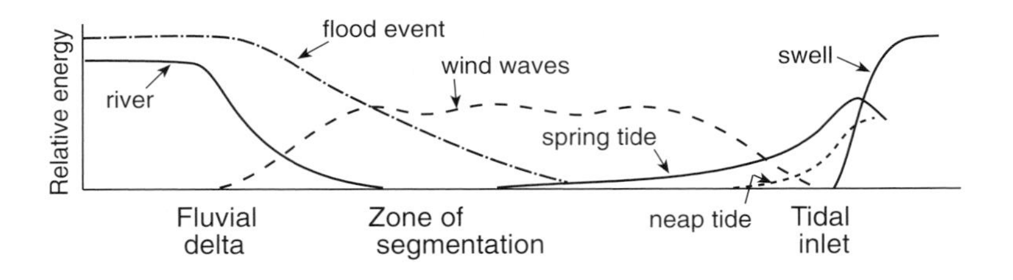

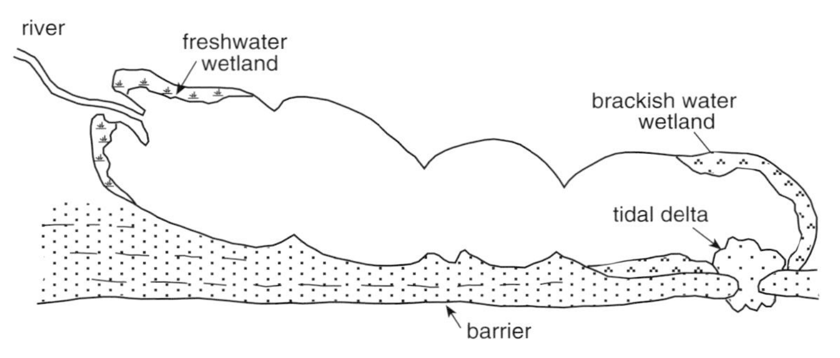

Figure 7.22. Schematic coastal lagoon, showing the relative influence of different factors (based on Bird, 1967, 2000).

On a geological time scale, coastal lagoons are short-lived features, controlled by sea level, climate and tectonic setting. They are most numerous on microtidal coasts (Barnes, 1980). On rugged or embayed coasts, lagoons are generally formed in undulating topography, in which case they can be deep and irregular in outline. On sediment-rich, trailing-edge coasts, particularly behind barrier islands along the eastern coast of the United States which is undergoing gradual relative sea-level rise, lagoons are often shallow and elongated parallel to the coast (Kjerfve and Magill, 1989; Nichols, 1989).

Nichols (1989) has suggested a classification of lagoons on the basis of whether they are filling in or increasing in size. He has defined equilibrium lagoons as those where increase in volume as a result of sea-level rise equals the rate of sediment supply. Deficit lagoons occur where sea-level rise is greater than sediment supply, as occurs along parts of the eastern coast of the United States. For instance, detailed assessment of all sources of sediment input (including eolian transport, tidal exchange, storm washover, runoff, longshore and autogenic *in situ* production, which total to <1 mm a^{-1}) into Laguna Madre in Texas suggests an increase of accommodation space as a result of relative sea-level rise (c. 4 mm a^{-1}) exceeding the rate of sediment infill (Morton *et al.*, 2000b). Transgressive settings, such as the trailing-edge barrier island coast of the eastern United States, contrast with regressive settings, such as Australia and South Africa, where sea level has been relatively stable, or even fallen slightly (Cooper, 1994). The former appears to be infilling with sediment derived from seaward, due to tidal-delta deposition as initially suggested by Lucke (1934). In eastern Australia different coastal lagoons, also called saline coastal lakes, are at different stages of infill primarily because of the rate of supply of sediment from fluvial sources as indicated in the model of barrier-estuary evolution outlined above (Roy, 1994). Coastal lagoons in eastern Australia are surplus lagoons in the classification by Nichols (1989), becoming smaller because of sediment infill. Figure 7.23 shows Coila Lake, a coastal lagoon in southern New South Wales that is still largely unfilled; it is at an early stage in the progression of infill similar to that described for barrier estuaries.

The infill of some coastal lagoons, particularly those that are elongated shore-parallel, involves the development of cuspate divisions along the lagoon shore dividing the lagoon into a series of segments (see Figure 7.22). Segmentation (or septation) has been attributed to winds blowing along the length of the lagoon producing waves which build spits isolating the lagoon into separate basins (Zenkovich, 1959). Shoreline vegetation, particularly reeds and rushes such as species of *Phragmites*, *Scirpus*, *Juncus* and *Typha*, can also be important in modifying the lagoon shore in the Gippsland Lakes in Victoria (Bird, 1967, 1994). Similar segmentation has occurred in Laguna dos Patos in southeastern Brazil (Isla, 1995; Toldo *et al.*, 2000), and Lake Songkhla in southeastern Thailand (Pitman, 1985). In South Africa, sandy Kosi

Figure 7.23. Coila Lake, a coastal lagoon in southern New South Wales, at an early stage of infill. The sand barrier can be seen to have closed the entrance in the foreground (photograph R.J. Morrison).

lagoon is segmented, whereas the muddier Lake St Lucia is less so (Cooper, 1994). Figure 7.24 shows the Coorong, a hypersaline lagoon in South Australia impounded behind the Younghusband Peninsula which forms a Holocene coastal barrier on which transgressive dunes are migrating inland. The irregular patten of the back-barrier shoreline in this lagoon is a function of local dune activity.

7.6 Summary

The gross form of a delta or estuary is related to geological, climatic and oceanographic setting which determine the gradient of the river and the coarseness of its load, in addition to wave and tide influences. The processes that operate at an individual delta mouth influence the mixing of river water with the receiving basin waters and determine the pattern of sedimentation that occurs, for example the location and shape of distributary-mouth bars. Similarly, the hydrodynamics of an estuary depends on the relative balance of river discharge and tidal prism and affects sediment accumulation, for example the location of a turbidity maximum. These processes vary over time, at several scales, such as spring–neap tides, or seasonally, or in response to extreme river flood or storm events.

The relative significance of river, wave and tide processes varies not

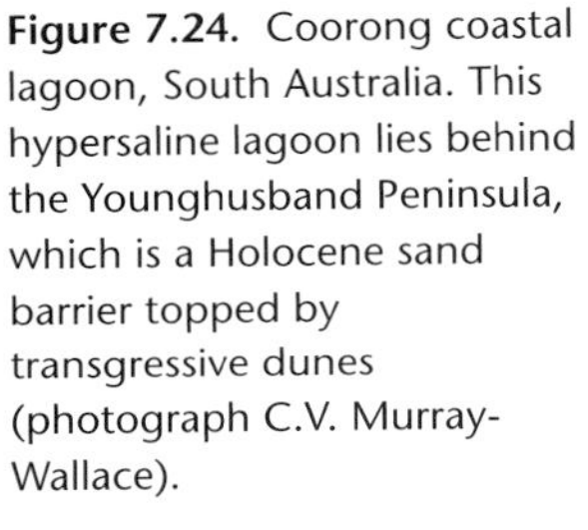

Figure 7.24. Coorong coastal lagoon, South Australia. This hypersaline lagoon lies behind the Younghusband Peninsula, which is a Holocene sand barrier topped by transgressive dunes (photograph C.V. Murray-Wallace).

only in the short term, as shown by delta-mouth or estuary circulation, but also spatially across a delta or estuary, and in response to longer-term changes associated with boundary condition or intrinsic thresholds. Those parts of deltas or estuaries that are river-dominated are characterised by active channels that extend rapidly and bifurcate, often flanked by levées. Tidal domination is characterised by channels that taper upstream and dendritic tidal creek networks, and the influence of waves is most clearly demonstrated by the occurrence of shore-parallel sand barriers and ridges. Whereas active distributaries are river-dominated, abandoned distributaries become reshaped by tide and wave processes. In large deltas, as shown for the Mississippi River delta, there can be major cycles of distributary progradation and abandonment that are linked with cycles of wave reworking of delta lobes as they disintegrate because of subsidence.

Estuaries are sinks for sediment. Consequently, their development over time involves gradual infilling of drowned valleys or embayments. Estuaries fill as a result of sediment input from landward and reworking of estuary-mouth sediments by wave and tide processes. The relative roles of each of these differ; sediment from landward is the dominant input into fjords, but wave or tidal input is more important in the case of coastal lagoons on arid coastlines that are experiencing relative sea-level rise, such as Laguna Madre.

Sea level has been an important boundary condition and relative changes in sea level have exerted a major control on deltaic–estuarine geomorphology. Worldwide expansion of deltas in the early to mid Holocene occurred as a result of deceleration and stabilisation of sea level following the postglacial transgression, and most estuaries also resulted from inundation of embayments during this period. Significant geomorphological changes can occur in response to changes other than in the relative rate of sea-level change. Abrupt changes can result from external factors, or from intrinsic thresholds within the system. The Mississippi River delta has undergone a sequence of distributary cycles triggered by an intrinsic threshold, the overextension of a delta lobe and its abandonment when a more efficient route is adopted during a major river flood. The change from delta lobes to a digitate (birdsfoot) morphology has resulted from the infill of the shallow continental shelf and the extension of the modern delta into deep water, a factor external to the system.

Australia is relatively stable and adjacent estuaries on the wave-dominated coast of southeastern Australia and tide-dominated estuaries in northern Australia have undergone similar Holocene sea-level histories. These systems changed from transgressive to stillstand at a similar time as a result of a change in the sea-level boundary condition (around 6000 years BP, see Figures 7.17 and 7.20). Each estuarine system has subsequently undergone a more individual pattern of infill based on the relative significance of fluvial input in relation to tide or wave influence, and other factors, such as prior or inherited topography, or the type and rate of sediment supply. Intrinsic thresholds can be identified in the infill of these estuarine systems; for instance, when a barrier estuary has infilled the embayment behind a sand barrier, it contributes sediment to the nearshore which can lead to renewed barrier progradation. The relative balance between landward movement of sediment into macrotidal estuaries and fluvial sediment loads in the wet season appears to have triggered a change in channel morphology in the South Alligator River, from sinuous meandering to cuspate meandering. Before 4000 years BP paleochannels appear sinuous, but the river has adopted a cuspate meandering form since 4000 years ago, in response to an intrinsic threshold, such as upstream movement of channel shoals as a result of asymmetry of macrotidal flows (Chappell and Thom, 1986). In areas of rapid uplift or glacio-isostatic response, such as northwestern Europe and eastern North America, estuaries experienced highly individual relative sea-level histories, and only by careful comparison between individual estuaries can evidence for regional changes in boundary conditions be discriminated from intrinsic changes (Long *et al.*, 2000).

Determining the past response of deltaic–estuarine environments to changes in boundary conditions or intrinsic thresholds provides a basis from which to assess the probable direction of response to future changes, such as are anticipated under conditions of enhanced greenhouse gas concentrations (see Chapter 10). However, many deltas have been modified by human actions which can increase or decrease sediment input. Where rivers have been dammed, the amount of sediment supplied to the coast has decreased dramatically with erosion resulting. The example of the Nile River delta is particularly pronounced (see Figure 7.9), with retreat near the mouth of the Rosetta distributary of up to 120 m a^{-1} since closure of the Aswan High Dam (Sestini, 1992). Similar effects have occurred on the Chao Phraya in Thailand (Wolanski *et al.*, 1998), and the Volta in West Africa (Anthony and Blivi, 1999). Because these are fertile areas with extensive near-horizontal land they have been preferentially used for a range of anthropogenic activities, and it is often difficult to resolve to what extent changes that can now be observed are the result of human factors (see Chapter 10).

It is important to recognise that deltas and estuaries are subject to a series of natural morphodynamic adjustments and that there are likely to be areas where deposition is occurring and other areas that experience erosion. Response to a regional change in boundary conditions is likely where similar changes occurred on several adjacent systems. However, changes in more than one boundary condition can occur concurrently, for instance, changes in sea level, climate and sediment supply rate may be interrelated. Adjustments that are intrinsic to a system are more likely to occur at different times on different systems. Anticipating these natural morphodynamic adjustments, or discriminating them from the impact of human activities, will continue to be a challenge, but comparison of different systems and further development of morphodynamic modelling offers many exciting avenues of research.

Chapter 8
Muddy coasts

> The appearance of the marsh soil, indicates a gradual formation from the grasses, aided by the fine rich sediment which the high tides occasionally deposit. The saline grasses grow only above high water mark, and as the roots in the lowest part of the soil, even eight or more feet below the surface, are in their natural position, showing no distortion, we must conclude that their *situs* was above the high water line, and that subsidence has been so gradual that the growth of the plants has never been interrupted. (Mudge, 1858)

Coasts composed predominantly of mud, comprising silt and clay-sized sediment, occur in low-energy settings, generally sheltered from wave action. Muddy coasts can be part of, or adjacent to, deltaic–estuarine coasts as described in the previous chapter, and are usually dominated by tides. Fine sediments are transported considerable distances in suspension but if subject to flocculation into low-density deformable aggregates can behave like larger particles. After deposition, muddy sediments tend to be cohesive, making them more resistant to resuspension, and highly organic, supporting a diverse biota.

On muddy coasts, sediments become finer onshore in contrast to beach and barrier coasts which become coarser onshore. Mudflats characterise much of the intertidal zone, but the lower intertidal and subtidal zones tend to contain extensive sandy deposits. The upper intertidal and supratidal zones often support halophytic (salt-tolerant) vegetation, comprising mangroves on tropical shorelines but dominated by salt marshes in mid and high latitudes.

Mudflats, salt marshes and mangroves are most extensive on macrotidal coasts (Hayes, 1975). They are especially associated with large embayments where tidal currents are strong and wave action limited. However, muddy landforms can occur wherever there is a large supply of fine sediment irrespective of tidal range. For instance, extensive mud banks develop downdrift of major deltas and, over time, these prograde to form a muddy coastal plain. Coarse sediment is either deposited within deltaic landforms (see Chapter 7), or concentrated into narrow shore-parallel ridges, termed 'cheniers'. A prograded muddy coastal plain is called a chenier plain where it is composed of mud with intermittent coarse chenier ridges marking former shorelines.

Mudflats, salt marshes and mangroves are significant for two reasons. First, they are productive ecosystems which contribute to the broader nearshore food web, a process termed 'outwelling'. They provide habitats and perform a nursery role for a wide range of species that have a high rate of primary productivity. Second, they are areas of fine sediment build up, and preserve a record of changing environments within their sediments. Muddy shorelines have been used to reconstruct paleoecological and paleoenvironmental changes. Their depositional stratigraphy contains sensitive records that, where bioturbation has been limited, offer potential for detailed reconstructions of the history of sea level and coastal evolution. Muddy coasts also assist in flood and erosion control, filter nutrients and pollutants from the water, are of considerable recreational value, and, as in the case of associated deltaic and estuarine plains (see Chapter 7), have been heavily utilised because of the extensive low-lying plains that are formed (Vernberg, 1993).

The apparent paradox, by which muddy coasts act as a productive biological source but a geological sink, involves an understanding of the processes through which these coastal systems are flooded and drained via tidal creeks, over different spatial and temporal scales. The bi-directional flux of the tide results in movement of large volumes of sediment, nutrients and organic material landward with the flood tide and seaward with the ebb tide, and the net budget (deposition or export) represents a small proportion of the total volume. In the short term (days to years), a range of tidal and meteorological factors influence deposition of the fine-grained sediments, but it is becoming increasingly clear that muddy coasts also experience phases of erosion. Over several years to decades, mud tends to fill the available accommodation space but erosion and accretion cycles occur and sediment stored in the upper intertidal salt-marsh or mangrove systems can be mobilised during major high-energy events. Over centuries, the system is likely to adjust to changes in external boundary conditions, including sea level or climatic factors such as cycles of storminess or direction of wave approach. Over millennia, a fining-upward sedimentary sequence develops that is generally interpreted to indicate progradation of the shoreline with salt marsh or mangrove extension over mudflats, which have in turn prograded over sand flats, but which is often composed of many phases of erosion and redeposition at a range of scales (Allen, 2000).

Dutch tidal flats provide an appropriate example of processes at a

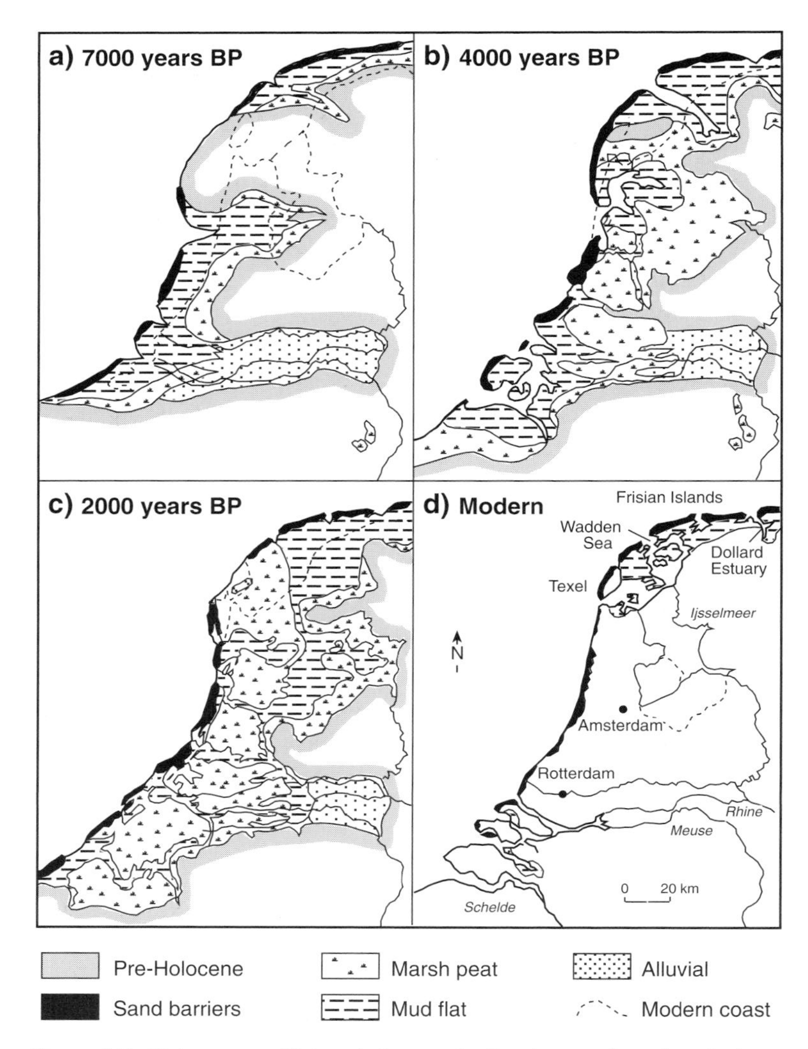

Figure 8.1. Holocene mudflat evolution on the Dutch coast (based on De Jong, 1977; Beets and van der Spek, 2000).

range of scales. Much of the coast of the Netherlands is built from muddy tidal flat deposits, and there has been considerable interest in the processes which operate in Dutch tidal environments (van Straaten and Kuenen, 1957; Postma, 1961). Figure 8.1 shows the pattern of Holocene development interpreted from extensive coring throughout the Netherlands. Particularly extensive tidal flats occur associated with the Wadden Sea in the north (extending along the German coast to Denmark), and to a lesser extent around the mouths of the Rhine and Meuse Rivers. Mudflats and marsh peat underlie most of the

Netherlands, formed in lagoons landward of coastal sand barriers (De Jong, 1977; Beets and van der Spek, 2000). The northern plains and the Wadden Sea have evolved behind a sequence of barrier islands, and the area of tidal flats decreased as the space between the barrier islands and hinterland decreased as a result of sea-level rise (Vos and van Kesteren, 2000). There remain significant areas of mudflat, and the processes of sedimentation are of concern in relation to extensive reclamation, as for example in the Ijsselmeer.

Sedimentation rates in the intertidal zone can be rapid initially, but decrease as the muddy substrate builds towards a surface morphology that is adjusted to the processes operating. There has been relatively little research on the effect of high-magnitude but low-frequency events on muddy coasts, making extrapolation of short-term process measurements to longer-term evolutionary trends inappropriate. In this chapter, the morphodynamic responses of mudflat, salt-marsh and mangrove environments are examined, primarily in relation to tides, although the influence of rivers and wave processes can also be significant. After a brief historical perspective, the processes and morphology of tidal flats are examined in this chapter. The role of tidal inlets and tidal creeks is described, and the geomorphological role of salt-marsh and mangrove vegetation in terms of landform evolution assessed. Morphodynamic feedbacks are examined and the potential for modelling muddy coasts is discussed.

8.1 Historical perspective

The study of muddy coasts has comprised long-term history as determined from stratigraphy, and shorter-term examination of processes operating on these coasts. Ecological and sedimentological aspects of muddy environments have been extensively studied, but only recently from geomorphological perspectives.

8.1.1 Stratigraphy

Submerged forests associated with estuaries in southern Britain provided evidence for relative sea-level changes (Huxley, 1878). It was realised in the 19th century that there were sedimentary accumulations that exceeded the present tidal range in thickness beneath salt marshes. Mudge (1858), for instance, in the quotation at the beginning of this chapter, recognised that the roots of salt-marsh plants in sediments below the depth to which they presently rooted indicated that subsidence had occurred. This interpretation was challenged by Shaler (1896), who believed that marsh colonised and built seaward over the top of intertidal flats. The correctness of Mudge's interpretation was demonstrated for the eastern coast of the United States by Davis (1910), who recovered cores from several marshes showing that the roots continued below the present rooting zone. This evidence was used by Johnson (1919, 1925) in recognition of what he termed submergence along this coast. Johnson distinguished New England coastal plain marshes (primarily underlain by peat) and Bay of Fundy marshes (underlain by inorganic sediments), from inorganic coastal plain marshes further south on the eastern United States coast. Further studies of salt-marsh stratigraphy confirmed the interpretation of submergence (Redfield and Rubin, 1962), and reinforced differences between New England sites and other West Atlantic locations, indicating that they had undergone different relative sea-level histories (Redfield, 1967, 1972).

In the case of mangrove shorelines, the prominent prop root system of *Rhizophora* (see Figure 8.10) together with the fact that it produces viviparous propagules (the fruits already have a well-developed root before they fall from the tree) supported the suggestion that mangroves are actively building seaward over calcareous mudflats in Florida Bay (Vaughan, 1909; Bowman, 1917; Davis, 1940). This view was similar to that proposed by Shaler for salt marshes, but it gathered further support from ecologists because zonation of mangrove species was interpreted as evidence of a succession as suggested for terrestrial plant communities by the ecologist Clements (1936). Zonation of mangrove species was regarded as indicative of a sequence through time, by which seaward zones are replaced by more landward zones, with eventual replacement by non-mangrove 'climax' vegetation (Davis, 1940; Chapman, 1944).

A broad zonation related to the frequency of inundation was recognised on the more diverse mangrove shorelines in the Indo-Pacific region (Watson, 1928; Macnae, 1966, 1968) and elsewhere (Chapman, 1976). It does not follow from the recognition that shoreline vegetation is zoned, that zonation represents succession though time; on shorelines which are not undergoing progradation, the patterning may be a static equilibrium in relation to environmental gradients in habitat factors, such as salinity or water-logging (Sauer, 1961). A series of studies emphasised that the patterning of mangrove species was a response to an ever-changing series of geomorphological habitats (West, 1956; Vann, 1959; Carter, 1959; Fisk, 1960; Fosberg, 1966). This was shown particularly convincingly for mangrove habitats on the coast of Tabasco, Mexico, by Thom (1967), who demonstrated that mangrove species are segregated in relation to landform types and respond opportunistically to habitat change induced by geomorphological processes, such as point bar migration and distributary avulsion.

Reassessment of the successional view of mangroves building

seawards across intertidal mudflats can be effectively demonstrated based on more detailed stratigraphy and dating of sediments in the Florida Bay and Everglades region. Mangrove peat reaches several metres thickness, exceeding the tidal range (and the range within which mangrove peat is presently formed), and must have been deposited as mangroves advanced landwards into the Everglades freshwater wetland system, as a result of a relative sea-level rise (Egler, 1952). Detailed stratigraphic study and radiocarbon dating of mangrove peats documented the pattern of sea-level rise (Scholl, 1964; Scholl *et al.*, 1969). Subsequently, the calcareous mud beneath the mangrove peat was shown on the basis of petrology and pollen analysis to have been deposited under freshwater conditions, indicating a transgressive freshwater–intertidal–marine sequence, recording gradual sea-level rise at a decelerating rate up to present (Spackman *et al.*, 1966). In parts of southwestern Florida, a lower transgressive sequence has been recognised, with the shoreline moving landwards as the sea rose at a rate greater than 3 mm a^{-1}, and an upper regressive sequence of mangrove progradation with oyster and vermetid reefs, as sea-level rise slowed below 3 mm a^{-1} (Parkinson, 1989).

Several lines of evidence indicated that muddy coasts have been subject to alternating phases of sedimentation and erosion. On many muddy coastal plains there is a stratigraphy in which peat is interbedded with inorganic sediment; for example, the fenlands around the Wash in eastern England (Godwin and Clifford, 1938). Chenier plains, such as that to the west of the Mississippi River delta, in western Louisiana and eastern Texas, appear to be linked to episodes of mud supply from the Mississippi River (Russell and Howe, 1935). Particularly detailed sedimentological and micropaleontological reconstructions, involving some of the first radiocarbon dating, were undertaken to demonstrate the alternate phases of mud progradation and reworking indicated by coarse chenier ridge deposition (Leblanc and Bernard, 1954; Gould and McFarlan, 1959). The concept that there are periods of erosion and periods of deposition has become significant in terms of several scales of adjustment on muddy coasts and will be examined in more detail in this chapter.

8.1.2 Process studies

Process studies on muddy coasts are complicated because the processes occur slowly and are difficult to measure. Geomorphological processes are also closely interrelated with ecological processes. For instance, the primary production of the salt marsh has been shown to be considerable and marshes have been attributed an important role in terms of outwelling of nutrients and organic material (Teal, 1962; Woodwell *et al.*, 1979). An ecological classification of mangrove forests which partly recognised geomorphological habitats was proposed by Lugo and Snedaker (1974). It distinguished riverine (tall mangrove stands that flank river channels), fringe (mangroves that occur along the seaward fringe of mangrove forests and are tidally flushed), and basin mangroves (interior of forests, often low in stature, further from either type of flushing). It also differentiated scrub (dwarf) mangrove, overwash mangrove (which comprise isolated islands of mangroves that are perpetually tidally flushed) and hammock mangroves (outposts of mangrove within the Everglades wetland environment).

It has long been recognised that flooding and sedimentation of muddy coasts occurs through a network of tidal channels, called tidal creeks where these traverse salt-marsh or mangrove wetlands. Unlike river channels where flow is draining a catchment and is runoff-dependent, flows in these tidal systems are bi-directional and are driven primarily by the stage of the tide. In the case of salt marshes and mangrove systems, tidal creek networks are routes through which much of the upper intertidal is flooded and drained (Jakobsen, 1954). Attempts to gauge the flows in these systems showed that there are complex distortions as the flow leaves the creek and inundates the marsh surface during tides that reach sufficient heights (Langbein, 1963; Pestrong, 1965).

As discussed in Chapter 3, the discharge through the cross-sectional area of a tidal inlet is closely related to the volume of water between high and low tides, termed the tidal prism. A relationship was first proposed by O'Brien (1931) who demonstrated a good linear correlation between the cross-sectional area of the inlet mouth (below mean sea level) and the tidal prism for inlets on the Pacific coast of the United States. A stable equilibrium relationship in terms of the velocity of flow through the inlet has been demonstrated more generally for tidal inlets around the United States (O'Brien, 1969; Johnson, 1973) and in Japan (Shigemura, 1980). If other conditions are suitable, a critical flow velocity of 1 m s^{-1} has been suggested, well in excess of the velocity at which fine sand is entrained (around 0.3 m s^{-1}). If inlet size increases, flow velocity decreases and sedimentation will occur, returning the inlet to the equilibrium size. If the inlet becomes smaller as a result of sediment deposition, velocity will increase, and hence sediment will be eroded, enlarging the inlet, unless it shoals to a point of instability (the velocity drops below 1 m s^{-1}) and the inlet closes (Escoffier, 1940). While some tidal embayments and lagoons appeared to conform to these relationships, intertidal mudflats appear more complex.

On Dutch mudflats, long-term mud accumulation can be inferred (see Figure 8.1) but, paradoxically, creeks are ebb-dominated which would appear to imply net export of material (van Straaten, 1954). In order to explain this paradox, Postma (1954, 1961) proposed tidally driven sedimentation as a result of 'settling and scour lag'. Sediment deposition occurs at slack high tide, but during the time between velocity dropping below the critical threshold and the time that the sediment actually settles it is carried further landward with the slowing flood tide (settling lag). On the ebb tide, a velocity considerably higher than that under which it settled is required to entrain the sediment again (scour lag). Material is therefore moved further landward than it can be returned on the ebb tide, resulting in a net landward flux (van Straaten and Kuenen, 1957).

Sedimentation rate is likely to be spatially variable on muddy coasts, and vegetation, such as seagrass, can cause more rapid deposition as a result of slowing the velocity of water flow. Rates of sedimentation beneath salt-marsh vegetation have been the subject of numerous studies (Yapp *et al.*, 1917). Particularly detailed studies were undertaken on Scolt Head Island on the north Norfolk coast of England using marker layers of sand, and indicating sedimentation rates of around 3–8 mm a^{-1} (Steers, 1960). The higher the marsh accretes within the upper intertidal zone the less frequently it is inundated and the slower the sedimentation rate (Richards, 1934; Randerson, 1979; Letzsch and Frey, 1980). Early measurements of sedimentation rates beneath mangroves indicated rates of around 1 mm a^{-1} using marker layers in New Zealand (Chapman and Ronaldson, 1958), but locally in Australia rates of up to 8 mm a^{-1} were recorded (Bird, 1971; Spenceley, 1977).

A series of detailed studies of the modern geomorphological processes and past geological accumulation history of muddy shorelines were undertaken in the 1970s, including the Wash in eastern England (Evans, 1965, 1975; Kestner, 1975) and Morecambe Bay in northwestern England (Kestner, 1970), the Bay of Fundy in Canada (Grant, 1970), the Bay of Mont Saint Michel in Brittany (Larsonneur, 1975), Shark Bay in Western Australia (Logan *et al.*, 1970, 1974) and Broad Sound in Queensland (Cook and Mayo, 1978). Many of these studies involved both process studies and longer-term geological evolution of the landscape (Reineck, 1972; Klein, 1985). Already, it was clear that it was not possible to extrapolate rates from one to the other because of complex feedbacks, for instance the autocompaction of organic material within sediments as the mass of material accumulated (Kaye and Barghoorn, 1964). These morphodynamic feedbacks are examined in more detail in this chapter, initially in relation to tidal flats, and then in terms of salt marsh and mangrove systems.

8.2 Tidal flats

The term 'tidal flat' is very broad, and covers a range of generally muddy low-gradient intertidal or supratidal surfaces (Amos, 1995; Dyer, 1998; Eisma, 1998). Intertidal flats are defined as sandy to muddy flats emerging during low tide and submerging during the highest tides. These are often colonised by halophytic vegetation (salt marshes and mangroves) at their upper margins; the geomorphological role of these vegetated wetlands is considered in more detail in Section 8.4. Supratidal (supralittoral) flats are near-horizontal flats that occur beyond the regular reach of the tides and are rarely inundated except under exceptional storm-surge conditions. Subtidal or sublittoral flats also occur seaward of low tide level but have received little study and are only discussed briefly below. Although tidal flats are particularly associated with areas with a large tidal range, extensive plains can develop that are flooded rarely, or by seasonal rainfall rather than tidal processes. For instance the mangrove areas of the coast of Tabasco on the Gulf of Mexico (Thom, 1967) and those fringing the Gulf of Carpentaria in northern Australia (Rhodes, 1982) are inundated seasonally.

There have been various attempts to classify tidal flats, based on their sedimentary characteristics, hydrography, tidal range or geomorphological setting (Amos, 1995; Dyer, 1998; Dyer *et al.*, 2000). Carbonate sediments can be discriminated from clastic (siliciclastic) sediments (see Chapter 2). There are distinctive calcareous tidal flats; muddy subtidal environments occur in the centre of carbonate banks such as the Great Bahama Bank (see Section 2.4.3, and Table 2.6). Intertidal flats are also common on carbonate coasts (see Figure 2.16); those in Shark Bay in Western Australia contain living stromatolites. Muddy tidal flats can be differentiated from sandy tidal flats primarily on hydrodynamic conditions, although both sand and mud occur in varying proportion in most flats. In other cases, the proportion of mud depends on source factors; for instance an increase in the extent of sand flats in comparison to mudflats occurred in the Gulf of California after damming of the Colorado River which decreased mud supply (Thompson, 1968).

Figure 8.2 shows a very generalised planform and profile of a muddy coast. Each muddy coast shows a progression from sandy channels in the lowest part of the tidal range, to sand flats, replaced by mudflats in the upper intertidal with salt-marsh or mangrove wetlands in the highest part of the tidal range extending supratidally (Figure 8.2a). In some cases the intertidal zone merges into extensive supratidal plains, elsewhere it is flanked by upland. Where supratidal plains are extensive, these are usually underlain by a fining-upward sedimentary sequence that records the vertical accretion of the subtidal, intertidal and supra-

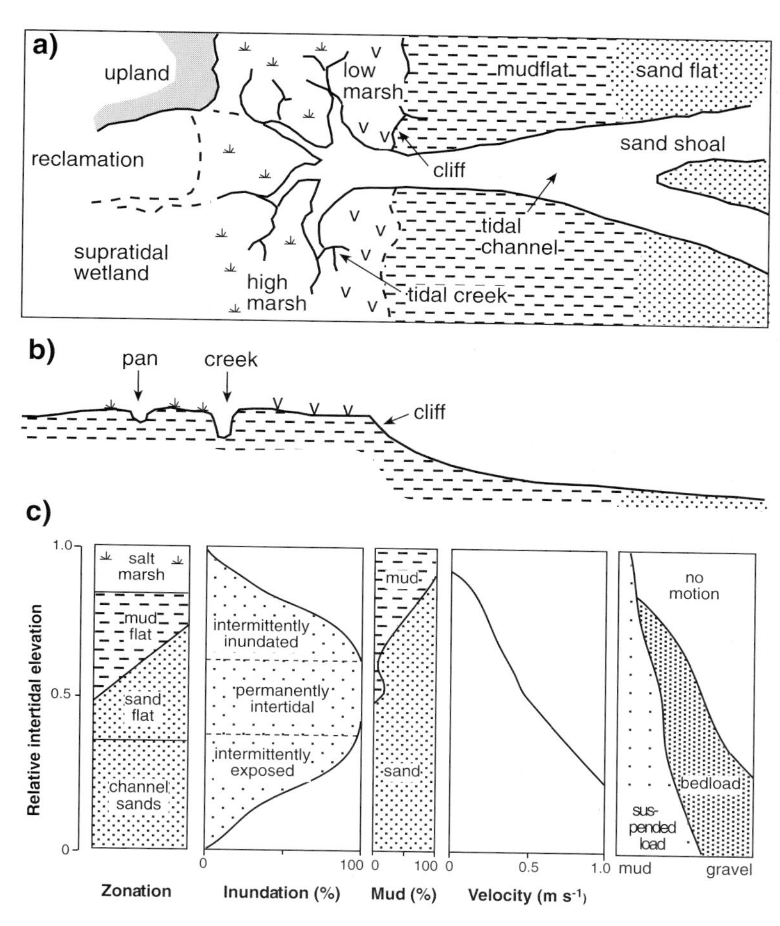

Figure 8.2. Variation of sedimentary characteristics across a typical muddy coast. (a) Schematic planform of muddy coast showing low-tide sand flat, intertidal mudflat and upper intertidal halophyte (salt-marsh or mangrove) area. (b) Schematic morphostratigraphic profile of mudflat. (c) Variation in sediment characteristics, inundation, mud content, peak tidal velocity and behaviour of sediment of different size fractions (based on data from the Wash and Bay of Fundy after Amos, 1995).

tidal habitats (Figure 8.2b). Vertical accretion results in horizontal progradation of the shoreline if there has been a period of relative sea-level stability (Curray, 1964). In many cases, the supratidal plains have been 'reclaimed' with embankments that exclude tidal inundation for agricultural, industrial or residential use. Networks of tidal creeks (examined in Section 8.3) and halophytic (salt-marsh and mangrove) vegetation (examined in Section 8.4) are important components of the muddy coast, and morphodynamic adjustments of these systems are examined in Section 8.5.

8.2.1 Sedimentary characteristics

Intertidal flats show site-specific shore-parallel zonation of sedimentary characteristics, bedforms and ecological communities, varying in relation to tidal range. Figure 8.2c generalises some of the trends in terms of relative position within the tidal range (elevation normalised between LAT and HAT, see Table 3.3) based on individual systems from northern Europe, North America, Australia and China (Amos, 1995). A decrease in grain size from coarse sediments at low water to mud at high water occurs over most intertidal flats in response to decrease in tidal current speed.

The velocity of tidal currents is important, and muddy coasts tend to show an asymmetry which is a function of the morphology of the system. Flow is often defined in terms of peak velocity and hence the capacity to move sediment, although flow duration varies and can also determine net fluxes. It was shown in the previous chapter that tidal flows can be either flood-dominated, in which case there is likely to be a net influx of sediment, or ebb-dominated with a tendency for net export (see Chapter 7). Sedimentation on the upper parts of tidal flats implies landward movement of sediment, although ripple patterns after an ebb tide may misleadingly indicate flow direction only during that most recent ebb tide (Collins, 1981).

Sediment source

The source of the fine sediment can be from the catchment, from cliff erosion or from a seaward source (cf. Figure 6.3) and varies between different systems. Muddy coasts associated with major river deltas often contain mud banks composed of sediment carried as suspended load from the river catchment. For instance, almost 1 Gt of sediment is brought down the Amazon River annually and extensive mud banks occur along the 1200 km of the northeastern coast of South America downdrift of the Amazon and Orinoco Rivers (Wells and Coleman, 1981; Froidefond *et al.*, 1988; Augustinus *et al.*, 1989). These mud banks, which are shore-attached, 10–60 km long, and average 5 m thick, undergo episodic movement (Eisma *et al.*, 1991). The mud banks occur in an exposed situation, and the mud that is deposited on the shoreface has a high water content. It is termed ‘fluid mud’ and dissipates the incoming wave energy. Open-coast mud bank development and progradation also occurs along the Gulf of Mexico west of the Mississippi River (Wells, 1983; Alexander *et al.*, 1991), and around the Bohai Sea downdrift of the Huanghe (Yellow) River, extending to the western shore of Korea (Wang, 1983). Mud is also extensive along the coast from the Changjiang (Yangtse) River into Hangzhou Bay (Wang and

Eisma, 1988; see Figure 7.10), and on the coast of Kerala, India (Jiang and Mehta, 1996).

The processes associated with fluid mud mobilisation have been studied in several settings (Wells, 1983). Models indicate that breaking waves effectively erode mud which is then transported alongshore primarily in the surf zone. However, non-breaking waves cause streaming of fluid mud, as part of a viscous continuum (Rodriguez and Mehta, 1998, 2000). Mud banks migrate progressively, perhaps triggered by shifts in the trade winds, and may take 1000 years to cover the distance that mud in suspension otherwise travels in about a month (Allison *et al.*, 2000).

On other muddy coasts, there appears to be more mud than has been supplied by rivers. For instance, around the North Sea, input from rivers is insufficient to account for the volume of mud (McCave, 1979). Rapid erosion of the glacial boulder clay cliffs on the east coast of Britain is presumably the source for the majority of the mud input into the Wash (Ke *et al.*, 1996). In the case of the Netherlands, only about 10% of the Holocene sediment budget has been derived from river input, and the majority appears to be from offshore, including erosion of Pleistocene deposits (Beets and van der Spek, 2000). Similarly erosion of glacial boulder clay cliffs provides a local source of material to the Bay of Fundy, and a similar local Pleistocene periglacial source for estuarine sediments has been identified in northwestern France (Guilcher and Berthois, 1957). In the Severn River estuary, the erosion of cliffs contributes 30% of the current annual sediment input (Davies and Williams, 1991). Local production of carbonate sediment, particularly shell material, plays a significant role in the Bay of Mont Saint Michel in northwestern France (Larsonneur, 1994). A seaward source for muddy sediment is inferred in northern Australian macrotidal rivers, where the mid-Holocene 'big swamp' phase of widespread mangrove forests (see Figure 7.17) indicates a disproportionate amount of mud compared with that apparently derived from the catchments (Woodroffe *et al.*, 1993).

Sediment redistribution

The 'settling and scour lag' hypothesis proposed by Postma (1954) provides a mechanism by which sediment could be moved landward in large tidal embayments (Section 8.1.2). However, it has proved difficult to measure the movement of fine sediment in the field because of factors such as variability in floc size and cohesiveness. Studies in the Forth estuary, Scotland, provide some support for a scour lag, indicating deposition at velocities of 0.3 m s^{-1}, whereas resuspension occurs at velocities above 0.6 m s^{-1} (Lindsay *et al.*, 1996). The behaviour of individual particles may be less important than diffusion along concentration gradients (Groen, 1967), and a lag between currents and suspended sediment concentration (Dronkers, 1986a). Median settling velocity increases as suspended sediment concentration increases, as a result of increased collision of grains and flocculation so that a greater proportion of the material settles at slack water (Pejrup, 1988; Mikkelsen and Pejrup, 1998). Time-averaged transport of sediment is possible even if no net residual flux of water occurs, or ebb and flood flows are equal, based on tidal and turbulent mixing which result in a seaward exponential decline in suspended sediment concentration. This down-gradient exchange appears to indicate an equilibrium capacity of a water mass to hold sediment independent of local variations in sediment supply and water flow (Amos, 1995).

Mudflat processes have been studied in detail in the Dollard Estuary in the Netherlands. This system, which has developed in the past millennium as a result of human modification of surrounding areas (Vos and de Kesteren, 2000), is ebb-dominated in terms of currents, but is actively accreting muddy sediments (Dankers *et al.*, 1984). It appears that organically mediated flocculation of sediment, increasing settling velocity to around 1 mm s^{-1}, leads to deposition at high tide. Suspended sediment concentrations reach a maximum at the beginning of the flood tide and the end of the ebb (Christie *et al.*, 1999). However, during windy days (wind speeds >6 m s^{-1}) residual ebb fluxes of sediment occur, except when diatom production is high (Dyer *et al.*, 2000). The settling and scour lag hypothesis of Postma (1961, 1967) seems broadly substantiated by these studies, but the factors involved are much more complex than previously envisaged.

The Dollard study indicates that any simple assessment based on asymmetry of tidal currents will be inappropriate. The threshold erosion of mud depends on biological as well as sedimentological criteria, and these are seasonally variable (Widdows *et al.*, 1998), resulting in a seasonal cycle of erosion and deposition. The benthic organisms are characteristically zoned in relation to elevation and give the surface greater resistance to erosion (biostabilisation) (Dittmann, 2000). The role of diatoms and other microbial biofilms is being increasingly appreciated; poisoning of the mudflat surface has been shown to reduce the strength and lead to rapid erosion (Coles, 1979; Decho, 2000). For instance, mud can adhere to organic films (biodeposition) at relatively low current velocities (<0.05 m s^{-1}), whereas erosion occurs at greater velocities (0.1–0.5 m s^{-1}). Sediment movement patterns in the Dollard Estuary are a function of local wind direction and intensity, and are mediated by diatom population dynamics (Dyer *et al.*, 2000). Appreciation of the significance of local factors implies that it may be difficult to extrapolate results to other estuaries where the range of available sediment types, biotic activity and local wind patterns are likely to be considerably different. For instance, on similar mudflats in Denmark, the role of biological films is also significant, but these can be grazed by the gastropod *Hydrobia*, reducing the erosion threshold (Andersen, 2001).

8.2.2 Tidal flat morphology

This section examines the shape of mudflats. At a general level, correlation between the tidal prism of a basin and the cross-sectional area through which it is flooded and drained (as suggested by O'Brien, 1931) implies an equilibrium. This concept is examined further in Section 8.3. Although the concept of shoreface equilibrium profile has had a long history in the context of sandy coasts, it has only been applied to muddy coasts relatively recently. It is possible to hypothesise a dynamic balance between landward transport due to tidal asymmetry and seaward diffusion as a result of the concentration gradient (Amos, 1995). An equilibrium profile on muddy shores can also occur if wave and current stresses are uniformly dissipated across the entire width of the shore. 'Schematically, tidal asymmetry induces onshore sediment transport, then generates accretion on the upper flat leading to a convex bottom profile, which in turn favours an ebb dominance that enhances the seawards sediment transport . . . Finally a tidal equilibrium is likely to occur, with a resulting slope and convexity. Such an equilibrium can be upset by wavy episodes that erode the flat, prevent deposition on the upper flat and favour offshore transport' (Le Hir *et al.*, 2000, p. 1456).

Profile shape

Kirby (1992) first suggested that the mudflat profile might be convex-up where it is dominated by accretion, and concave-up where it is erosional, based on profiles from the Severn River estuary, western Britain. Much of the Severn River estuary is concave-up, eroded into Holocene muds (discussed below; see Figure 8.4). However, the construction of a barrier within Cardiff Bay which attenuated wave energy, has resulted in a change of profile to an accreted, convex-up profile (Kirby, 2000).

Mud behaves differently from sand. Whereas sand is moved across the shoreface primarily as saltation or bedload, mud is moved in suspension and can continue to be transported even when flows reduce below threshold velocities (the concept of settling lag suggested by Postma). Mud is also more difficult than sand to erode once deposited (scour lag), because stability is increased as a result of subaerial exposure, drying and changed sediment properties (including biofilms). The greater cohesion of muddy sediments is shown by the greater entrainment velocities required to erode it, as suggested by Hjulström (1935; see Figure 3.1). Whereas maximum wave dissipation occurs in the breaker zone on sand beaches, fluid mud with a dense near-bottom layer (lutocline) into which wave orbital motion is transmitted dissipates wave energy through increases in viscosity (Azam and Mokhtar, 2000).

Although the tide appears to be the prime force shaping mudflat/wetland form, waves have been shown to play an important role in several muddy settings (Maa and Mehta, 1987). Waves, even small-amplitude waves, can exert a higher bed shear stress than tides because of the backward and forward motion, and because they can be accentuated under stormy conditions. Wind-waves of small amplitude resuspend sediment, especially as the tide floods over mudflats, and significantly enhanced suspended sediment concentrations occur on windy days (Anderson, 1972; Dyer *et al.*, 2000). Similarly, windy conditions have been shown to resuspend mud in and around nearshore reefs in turbid areas of the inner Great Barrier Reef, Australia (Larcombe *et al.*, 2001).

There is increasing evidence that on many muddy coasts wave processes shape the erosional concave-up profile, and tide processes shape the more convex-up accretionary profile (Roberts *et al.*, 2000). On open coasts, muddy shores can receive substantial wave energy, and mudflats represent an end-member of the beach continuum (Masselink and Turner, 1999). 'Intertidal mud beaches thus experience temporally periodic processes of erosion and deposition – tidal energy drives the morphology-enabling depositional processes while subsequent morphology adjustments are carried out in response to wind wave energy' (Pethick, 1996, p. 186).

Figure 8.3 shows the stages of entrainment of a fluid mud layer under waves which lose height exponentially in response to the wave attenuation coefficient, and the shore profile that existed under calm conditions. Suspended sediment concentration is likely to increase towards the shore because there is less water volume in the shallower water. Sediment can be transported inshore or offshore as a result of gravitational movement, and the new profile which establishes under calmer conditions is shown in Figure 8.3a (Lee and Mehta, 1997). Modelling by Lee and Mehta (1997) has indicated that a more concave profile results if the wave attenuation coefficient is high.

A convex-up profile is indicated, particularly in macrotidal settings, where uniform bottom shear stresses related to tidal currents are modelled across intertidal surfaces (Friedrichs and Aubrey, 1994). The end-member shapes have been expressed mathematically (see Figure 8.3). They can be expressed in dimensionless form in terms of relative depth

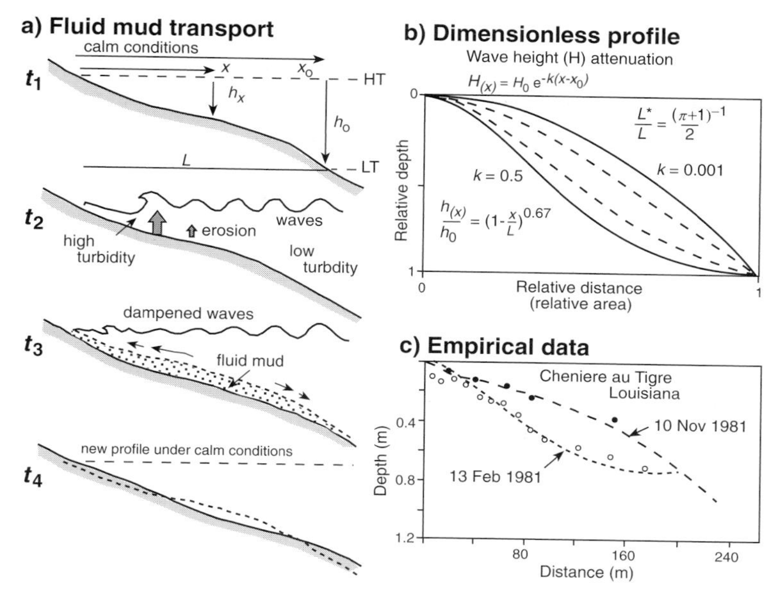

Figure 8.3. Mud behaviour and the evolution of mudflat shape. (a) Erosion and transport of fluid mud. Under calm conditions (t_1), the bed is not agitated (parameters are defined). Under high wave energy maximum erosion occurs in the breaker zone (t_2), and there is a gradient in turbidity. After a short time waves are dampened through attenuation in a fluid mud layer (t_3), and net sediment is either moved shoreward by advection or offshore by gravitational movement. The new profile under subsequent calm conditions (t_4) indicates the legacy of wave action on the profile. (b) Mudflat shape can be modelled under different conditions and degree of concavity relates to the wave attenuation coefficient (k). The formulae express endpoints in the accretionary or erosional profile spectrum. (c) Data on offshore profiles in southwestern Louisiana indicate that observed depths (dots) correspond well with modelled (dashed) profiles for both erosional (February 1981) and accretionary (November 1981) profiles (based on Lee and Mehta, 1997).

and relative distance (Figure 8.3b), and can also be plotted effectively as 'hypsometric' curves (indicating accommodation space as a proportion of total volume of the system) where the 3-dimensional bathymetry is known with sufficient precision (Kirby, 2000).

The equilibrium shape of mudflats (expressed in dimensionless form in Figure 8.3b) is approximated by empirical data from the nearshore off Cheniere au Tigre, Louisiana. A more concave profile is shown after a storm in February 1981 than during calm conditions in November 1981 (Figure 8.3c). These data are subject to similar shortcomings to those outlined for equilibrium profiles on sandy shores. For instance, the modelled profile is unrealistic at the origin, and it is difficult to define the depth to seaward beyond which there is no adjustment (termed closure depth). It is especially difficult to model response under circumstances where bedforms develop (Dalrymple *et al.*, 1978, 1990), or to simulate the behaviour of mixed sand and mud under shallow water waves in depths of 2–3 m or less (Lee and Mehta, 1997).

Erosional episodes

Muddy coasts are often assumed to be accretionary, but there is increasing evidence that phases of erosion occur at a variety of time scales and response of the mudflat depends on factors such as tidal range and sediment supply. Morphological response of the intertidal profile to high-energy events is buffered by sediment storage in the marsh. For instance, in the open-water marshes of Dengie Peninsula, Essex, eastern England, a storm event resulted in erosion of the seaward edge of the marsh, with sediment redistribution seawards onto the mudflat. Subsequently sediment was reworked back onto the marsh, re-establishing the pre-storm profile, with marsh vegetation playing a role trapping and retaining the reworked mud (Pethick, 1996). If there is an abundant sediment supply, landward transport of sediment is limited primarily by accommodation space; but more often the volume of sediment supplied is not great enough to fill the accommodation space and muddy coasts are supply-limited (Amos, 1995).

Episodes of accretion and subsequent erosion at time scales that are tidal, seasonal (eroding in autumn and winter and accreting in summer and spring) and decadal have been described from the Severn River estuary in western England (Whitehouse and Mitchener, 1998; O'Brien *et al.*, 2000). Evidence for longer-term patterns of alternate accretion and erosion is recorded in the stratigraphy beneath the mudflats (Figure 8.4). Figure 8.4a shows the distribution of erosion along the modern shoreline of the Severn River estuary (Allen, 1992). The stratigraphy, shown in schematic section in Figure 8.4b, indicates episodes of organic (peat) and inorganic sediment deposition. Marsh sediment units, successively eroded into each other, have been given geological formation status by Allen (1987a, 1993), comprising the oldest Rumney Formation which began accreting in the 16th century, an 18th century Awre Formation, and the youngest Northwick Formation dating from the 20th century. Subtle textural patterns, and silt–peat couplets accompanied by a set of infilled paleochannels, enable discrimination of units and determination of the position of the shoreline (Allen, 1997). The mid- to late-Holocene sediments underlying these have been called the Wentlooge Formation, and contain a sequence of peat beds within the Middle Wentlooge, indicating alternating conditions over recent millennia (Figure 8.4b). The deposition of mudflat and peat sedimentary units

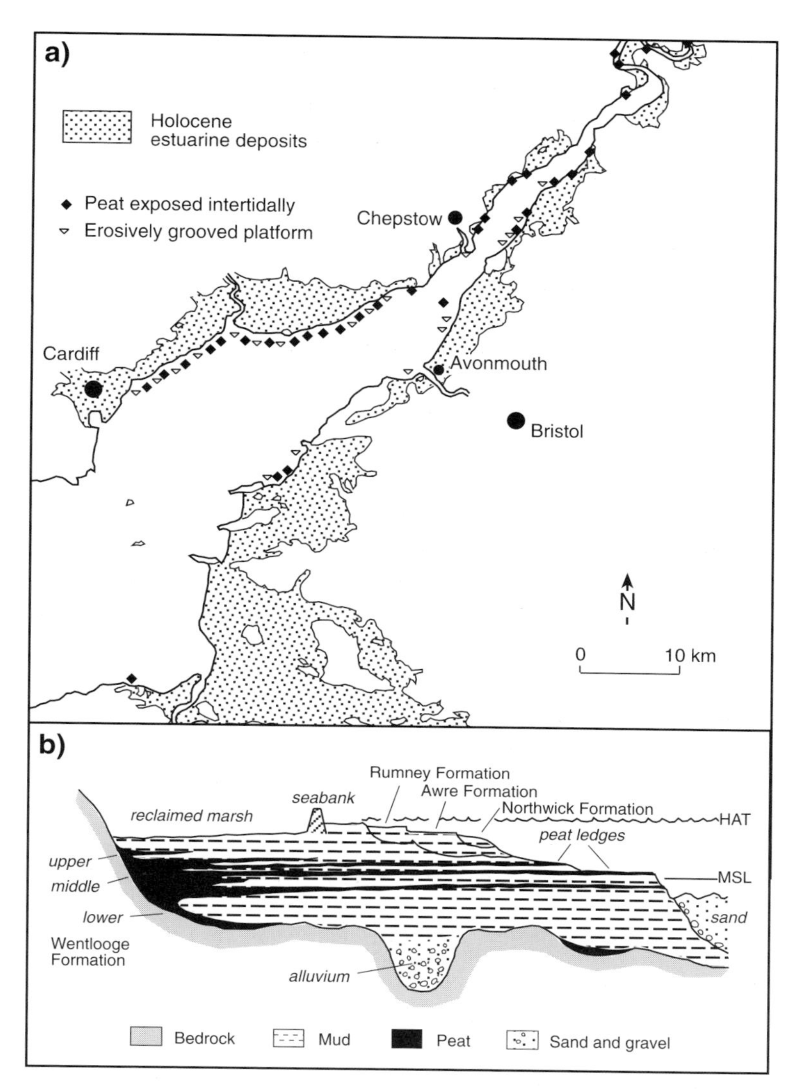

Figure 8.4. Severn River estuary and schematic stratigraphy. (a) Estuary showing alluvial areas that have accreted as Holocene coastal/estuarine plains, and (b) schematic cross-section showing morphostratigraphic units (based on Allen, 1992).

indicated in the Severn River estuary has been part of the evolution of the estuary and conforms with the broad pattern of macrotidal estuarine infill indicated in Chapter 7. The significance of the peat layers is discussed in the following section and in Section 8.5.2.

There are extensive mudflats fringing the macrotidal King Sound in Western Australia. The strong tidal currents promote mud transport and deposition, but the mud flats also show prominent patterns of erosion (Semeniuk, 1980a). Erosion has truncated earlier sequences of tidal flat sedimentation. In this case, these mudflats are underlain by sediments that were deposited beneath mangrove forests, and former mangrove muds, shown by the remains of mangrove stumps in growth position, are revealed by the erosion (Semeniuk, 1981).

8.2.3 Longer-term coastal plain development

Peats interfingered within muds, giving a stratigraphic pattern as shown by the extensive peats in the Middle Wentlooge Formation of the Severn River estuary (Figure 8.4b, Allen, 1987b; Allen and Fulford, 1996), are found throughout the muddy coasts of Britain and northern Europe (e.g. Godwin and Clifford, 1938). The typical sequence begins with a basal peat overlying weathered and pedologically altered bedrock or Pleistocene deposits. The peat is overlain by mud within which several wedges of peat, spanning the interval between 6000 and 2500 years BP, can occur. The peat is generally attributed to an organic wetland during a local regression of sea level, as will be shown below (see Figures 8.16 and 8.20) and records sea-level tendency (Shennan, 1986a,b). Silt over peat marks a transgression, and widespread transgression typifies the period 3250–2250 years BP (Spencer *et al.*, 1998). Increasingly the detail of the story as revealed by peat layers is being expanded by analysis of microfossils, especially foraminifera, diatoms, ostracods and testate amoebae (Murray and Hawkins, 1976; Devoy, 1979; Horton *et al.*, 1999).

A similar stratigraphic sequence is common along the German, Dutch and Belgian coasts (Streif, 1989; Baeteman *et al.*, 1999; Vos and van Kesteren, 2000), and results from coastal evolution as the sea has risen (see Figure 8.1). Most of the peat intercalations date from 3500–2800 years BP, implying regressive–transgressive cycles of vertical amplitude of around 0.8 m, and of 300–600 years duration (Vos and van Kesteren, 2000). During the past 2000 years there are fewer peat layers. In the eastern United States there is also a similar stratigraphy. Basal peat in New England has been dated around 3670 years BP (Orson *et al.*, 1987). Accelerations of sea level can be determined from microfossil analysis (Gehrels *et al.*, 1996; van de Plassche *et al.*, 1998; Gehrels, 1999; Shaw and Ceman, 1999). Deceleration of sea level led to widespread salt-marsh peat development which occurred between 3000 and 4000 years BP in Massachusetts, New Hampshire and Connecticut (Orson *et al.*, 1987).

There can be problems correlating on the basis of lithostratigraphy despite the apparent simplicity of the sequence of sediments. There can be several peat layers or peat may be absent, and dating of the peat indicates that it was formed at different times in different places, termed diachronous (Wheeler and Waller, 1995). Tidal range itself can have changed as the geomorphology has evolved. Modelling has shown that the tidal range in the vicinity of the Bay of Fundy has amplified as sea level has risen in the late Holocene (Grant, 1970; Scott and Greenberg, 1983; Amos *et al.*, 1991). Local modifications to tidal range have occurred as the Wash has infilled (Hinton, 1992, 2000; Austin, 1991) and can be inferred to have been a feature of the evolution of the Dutch coast (Roep and van Regteren Altena, 1988; van der Molen and de Swart, 2001). The significance of autocompaction means that distortion of the true depositional positions of distinctive strata may be considerable (Streif, 1972; Allen, 2000).

There have also been extensive muddy coastal plains develop along tropical coasts dominated in this case by mangroves. Mangrove forests accrete an organic peat, composed of the roots of *Rhizophora* where the tidal range is small and the input of inorganic sediment limited (as in Florida and the West Indies). The stratigraphic record preserved on low-energy muddy tropical coastlines, indicates significant switches in the rate of sea-level rise; for instance, when sea level slowed in northern Australia, an organic-rich facies, called the 'big swamp' was deposited (see Chapter 7). In northeastern Australia the sedimentary record has been interpreted to contain fluctuations within it (Gagan *et al.*, 1994). Mangrove pollen has been used to detect a transition from mangroves to supratidal environments (Woodroffe *et al.*, 1985; Clark and Guppy, 1988; Crowley and Gagan, 1995). In semiarid areas of Western Australia the change was to bare supratidal flats (Thom *et al.*, 1975; Jennings, 1975); in the monsoonal Northern Territory it was to convex alluvial grass–sedge-covered floodplains (Woodroffe *et al.*, 1989, 1993); and in the wet equatorial regions of Southeast Asia it was to peat swamp forests (Anderson, 1964; Staub and Esterle, 1994). The transition from mangrove muds to freshwater peat swamp environments, as part of a progradational sequence resulting from gradual build out of the coast, has also been demonstrated by microfossil studies in Peninsular Malaysia (Coleman *et al.*, 1970; Haseldonckx, 1977; Kamaludin, 1993). Extensive mangrove forests persisted throughout many of the deltaic–estuarine areas of Southeast Asia until around 4000 years ago, shortly after which they were replaced by the initial stages of peat swamp development in each of the Baram, Rajang and Klang Rivers (Woodroffe, 2000).

8.2.4 Cheniers and chenier plains

Cheniers are sandy, gravelly or shelly elongate ridges which are perched on finer-grained, nearshore sediments. They form shore-parallel high ground on broad, prograded, supra/intertidal chenier plains (Otvos and Price, 1979). The term 'chenier' derives from a broad plain in Louisiana to the west of the Mississippi River delta, where there are ridges on which the live oak (la chêne) grows (Russell and Howe, 1935). Single or bifurcating ridges indicate episodic coarse sediment deposition by waves in what are otherwise mud-dominated environments. Figure 8.5 shows a chenier plain in northern Australia where the modern shoreline is lined by mangrove forests. Sandy ridges are preserved on the muddy coastal plain.

Chenier plains occur on low-gradient, low to intermediate wave-energy, microtidal to macrotidal settings in all latitudes, but are especially characteristic of muddy tropical coasts (Augustinus, 1989). They have been classified into bight coast settings such as Louisiana west of the Mississippi River delta and the coast of Suriname downdrift of the Amazon, and bayhead settings, such as the Gulf of California (Thompson, 1968; Otvos and Price, 1979). They can also occur in bayside settings, such as Broad Sound in Queensland, Australia (Cook and Mayo, 1978) and the Firth of Thames in New Zealand (Woodroffe *et al.*, 1983a). There are numerous chenier plains along the north Australian coast (Chappell and Grindrod, 1984; Short, 1989), including much of the southern shore of the Gulf of Carpentaria (Rhodes, 1982).

Figure 8.5. View across a northern Australian chenier plain, north of the Burdekin River, Queensland. The sandy ridges are perched on muddy sediments. The modern shoreline (foreground) is fringed with mangroves.

Cheniers are most common where there is an abundant supply of mud, longshore transport of sand-sized (or coarser) sediments, and a mechanism which can lead to episodic alternation between these conditions. One such mechanism is the switching between delta distributaries or delta lobes (Otvos and Price, 1979). In the case of the Mississippi River, it has been suggested that rapid mud progradation on the chenier coast occurs when the river discharges through distributaries to the west of the broad deltaic plain, and that erosional processes dominate when the river discharges to the east of the plain (Gould and McFarlan, 1959). Cheniers appear to form as a result of winnowing of the coarse sand fraction from the eroding foreshore when it does not receive large quantities of mud (Todd, 1968; Hoyt, 1969). More recent studies have significantly revised the chronology of delta switching in the Mississippi delta, and indicate that more than one delta lobe may have been active at any time over much of the mid to late Holocene (see Figure 7.7). Nevertheless, the chronology of the chenier plain appears linked to the alternation between delta lobes (Kaczorowski, 1980; Penland and Suter, 1989).

Perhaps the clearest evidence that delta progradation and chenier building are linked comes from the Huanghe (Yellow) River in China. As shown in Chapter 7, this has switched between outlets in Bohai Bay and Jiangsu (see Figure 7.10). Cheniers mark erosional interruptions to the supply of mud (Xitao, 1989; Wang and Ke, 1989; Saito *et al.*, 2000). Radiocarbon ages of shells within chenier ridges do not relate directly to the date of ridge deposition, but to the time of shell production before ridge formation. Incorporation of shells from fossil shellbeds is possible where these are excavated by chenier ridge retreat (Gould and McFarlan, 1959).

Other mechanisms have been suggested by which chenier ridge formation and mudflat progradation alternate. Sea-level changes were thought to have triggered ridge formation in the Firth of Thames, New Zealand, and oscillations were deciphered from ridge height variation (Schofield, 1960). However, although the internal structure, particularly the base of cheniers, has been used for sea-level reconstruction elsewhere (Rhodes, 1982), other evidence from the Firth of Thames does not support such sea-level variations (Woodroffe *et al.*, 1983a). Along the coast of Suriname and Guyana, lateral migration of mud banks is considered to trigger ridge-forming episodes (Augustinus *et al.*, 1989; Daniel, 1989). In northern Australia, regional climate change has been inferred to lead to ridge formation during long dry periods of lowered river discharge and mudflat progradation is considered to respond to wetter periods (Rhodes, 1982; Lees and Clements, 1987; Lees, 1992). Chenier ridges appear to have formed intermittently along the southern shore of van Diemen Gulf, northern Australia. However, the muddy sediments over which these have formed have been derived from seaward (Woodroffe *et al.*, 1993), and it seems unlikely that their chronology relates to supply of mud from the catchments. Chenier plain evolution in Princess Charlotte Bay (Figure 8.6) appears to have alternated between a 'cut and recover' mode of relatively slow build out and a 'rapidly prograding' mode of mudflat formation as a result of pulses of mud input which reduce shellfish production (Chappell and Grindrod, 1984). This intrinsic threshold is unrelated to changes in regional boundary conditions.

Storms are considered to be an important influence on chenier formation and chenier-plain morphodynamics. Ridges can be moved inland by storms and storm surges (Anthony, 1989; Qinshang *et al.*, 1989). In Australia, storms have devastating impacts on mangroves fringing chenier plains, enabling storm waves to drive sand and shell landwards and deposit a ridge behind the mangroves (Cook and Polach, 1973). However, Cyclone Tracy, a very destructive cyclone in 1974, did not appear to have a major effect on the movement of individual cheniers, which migrate landward as a result of regular events of considerably smaller magnitude (Woodroffe and Grime, 1999).

Cheniers provide clear evidence of the episodic nature of sedimentation on muddy coasts, and support recent studies that suggest that erosion can occur frequently in these environments. Several different mechanisms have been proposed to explain phases of mud progradation and episodes of ridge building; in some cases this represents a response to external factors, while in other cases it appears to be triggered by internal factors.

8.3 Tidal inlets and tidal creeks

Muddy coasts are generally dominated by tidal processes and tidal inlets and tidal creeks are particularly important because they are the route by which the system is flooded and drained. Discharge is determined by tidal prism, and is related to cross-sectional area of the channel through which the tide flows, as outlined in Chapter 3 and discussed in greater detail in Chapter 7 in relation to estuarine channels. This section examines tidal inlets which are associated with tidally flooded lagoons and back-barrier environments, and tidal creeks that develop across mudflats and dissect salt-marsh and mangrove wetlands.

8.3.1 Tidal inlets

Estuaries generally adjust towards an equilibrium form in which tidal energy is expended uniformly per unit area and nowhere exceeds the

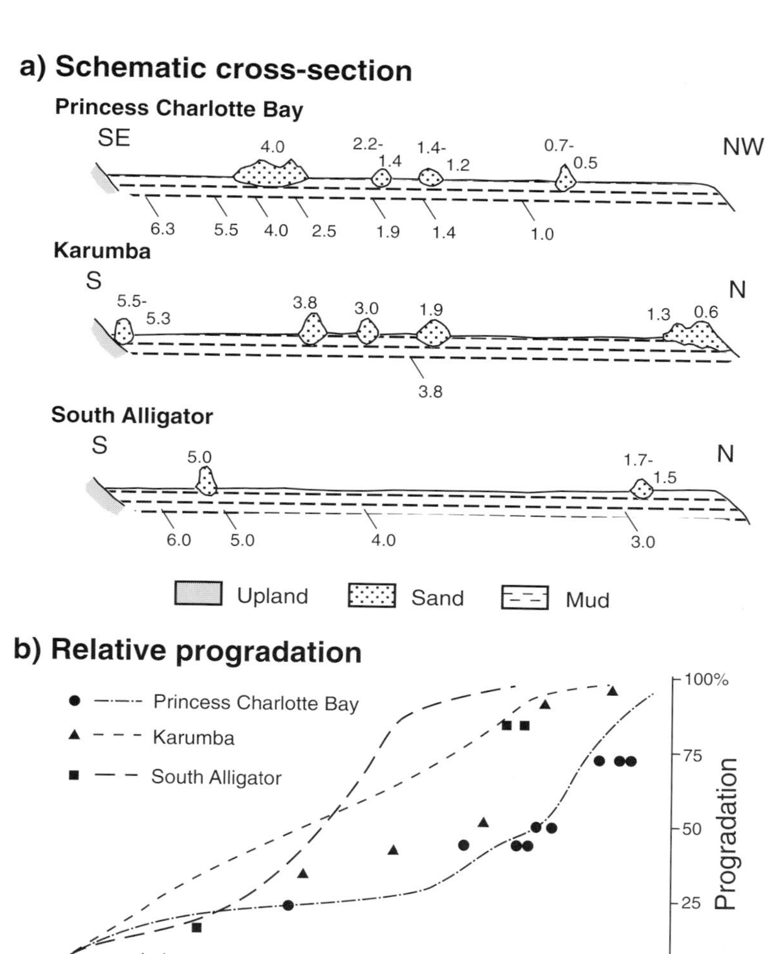

Figure 8.6. Stratigraphy and chronology of several chenier plains in northern Australia. (a) Schematic representation of the spacing and age (ka) of chenier ridges for parts of the Princess Charlotte Bay chenier plain (based on Chappell and Grindrod, 1984), a chenier plain at Karumba in the Gulf of Carpentaria (based on Rhodes, 1982), and a chenier plain at the mouth of the South Alligator River in the Northern Territory (based on Woodroffe *et al.*, 1989). (b) The relative timing of plain progradation and ridge formation, showing some disparity between chronology determined from shell ages (symbols), and that determined from the underlying stratigraphy (lines).

strength of material comprising the estuarine banks (see Chapter 7). A tidal inlet is a restricted, relatively narrow channel developed across a barrier where tidal currents are accelerated in a jet-like fashion (Isla, 1995). Figure 8.7 examines the relationship proposed by O'Brien (1931, 1969) between tidal prism and cross-sectional area of inlets for a series

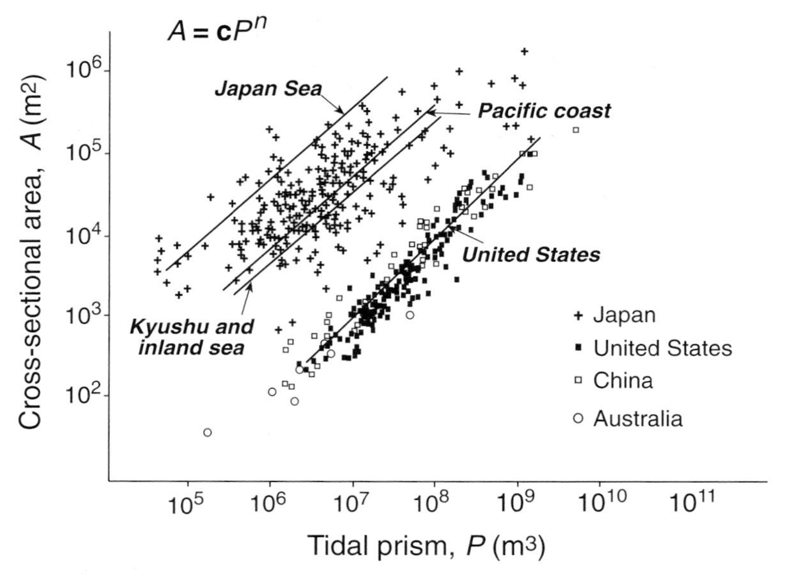

Figure 8.7. Relationship between cross-sectional area of tidal inlet (*A*) and tidal prism (*P*), showing data from Japan (Shigemura, 1980) and Australia (Williams, 1983) plotted against data for China and United States (based on Gao and Collins, 1994).

of inlets along the coast of the United States, Japan (Shigemura, 1980), China (Gao and Collins, 1994) and Australia. It is clear that there is considerable variability in the data. Inlets of the United States and those of Japan are offset largely because the latter are controlled by physical structures, either jetties and training walls, or are bedrock-controlled inlets that cannot be reshaped by hydrodynamic processes, as in uncohesive sediments.

Many coastal lagoons (which are also tidal basins in this context) are not in equilibrium (see Chapter 7) because they are subject to barrier transgression and relative sea-level change, and input of sediment from seaward as well as landward sources (Nichols, 1989). The relative duration, as well as asymmetry of tidal velocity on flood and ebb tides, freshwater input into the system, and the supply and transport of sediment and its characteristics, are also important (Gao and Collins, 1994). Inlets can be tide- or wave-dominated, which in turn is related to the extent of flood/ebb tidal delta development, and wave shoaling at the mouth (Hubbard *et al.*, 1979; McBride, 1987). The relative balance of tidal prism and the longshore component of wave power appears to explain inlet stability (Bruun, 1978), and it is significant whether the inlet is dominated by flood or ebb tides (Boon and Byrne, 1981).

Inlets to barrier estuaries and saline coastal lagoons (or lakes) along the wave-dominated coast of New South Wales, whose Holocene development is described by models (Roy, 1984) discussed in Chapters 6 and 7, do not fit the relationship well (see Figure 8.7). The degree to which these inlets are open varies; several barrier estuaries close completely for several months until a large rainfall or storm event re-opens the mouth (Williams, 1983). Under these circumstances, tidal prism does not seem to determine cross-sectional area as implied by O'Brien and it appears more likely that inlet dimensions affect tidal characteristics (Nielsen and Gordon, 1981).

The O'Brien relationship applies best to large embayments or tidal basins. Nevertheless, it is clear that the size and shape of many tidal systems adjusts to the volume of flood and ebb tides. This can be approximated by simple 1-dimensional models relating inlet dimensions to the area of the basin, as developed for inlets in the Netherlands (van Dongeren and de Vriend, 1994). More detailed models, termed hypsometric models, can be developed where the 3-dimensional topography of the tidal system is known in detail (Strahler, 1952b; Boon and Byrne, 1981; Dronkers, 1986b; Oertel, 2001). The topography of intertidal surfaces is difficult to measure directly, but can be derived from remote sensing imagery. For instance, the water-line approach using radar imagery taken at different stages of the tide enables interpolation of a digital elevation model from water-covered surfaces at different tidal stages (Mason *et al.*, 1998). Where tidal systems, such as Morecambe Bay in northwestern England, have been monitored over time it has been found that the intertidal topography varies at a range of time scales (Mason and Garg, 2001).

8.3.2 Tidal creeks and creek dynamics

Tidal creeks perform the role of flooding the marsh and returning overmarsh flows to the sea; they may also serve to dissipate tidal energy across the mudflats and halophytic wetlands (Pethick, 1996). They have been studied particularly in relation to salt marshes, and although similar channel systems occur across intertidal flats and through mangrove forests, those on salt marshes will be described in the following section. Creeks can be more or less branching and sinuous and can show a range of forms: linear, dendritic, meandering, reticulate or superimposed (Eisma, 1998).

Although tidal creeks bear a superficial resemblance to dendritic river channel networks, flow is bi-directional (Pestrong, 1965), creeks are distinctively tapering (Fagherazzi, 1999; Fagherazzi and Furbish, 2001), and discharge is determined by the tidal prism (Langbein, 1963). The tidal prism can be divided into the volume within the creek system and the volume over the marsh surface. Hydroperiod (the time for which the marsh is inundated) and hydraulic duty (the depth of flooding) differ between tides (Allen, 2000).

Figure 8.8 illustrates the functioning of marsh creek systems. Tides can be divided into undermarsh tides (neap tides) and overmarsh tides (particularly spring tides); the threshold is at bankfull tidal stage (Bayliss-Smith *et al.*, 1979). Flooding of the marsh surface occurs primarily by tidal water that enters the creek mouth and fills the creek as the tide rises, flooding overbank onto the marsh surface. On neap tides, flow is generally moderate (often 0.1–0.2 m s^{-1}). On spring tides, water fills the creek system but continues to rise, flooding over the marsh surface until slack high water, after which the tide ebbs and the water falls, draining from the marsh surface through the creek system.

Flow characteristics can be broadly simulated by continuity models based on the hypsometry (the 3D topography) of marsh surface and the volumes of overmarsh and creek prism (Pethick, 1981; Boon and Byrne, 1981). Empirical studies of the nature of flow through creek cross-sections, however, indicate that there are complex non-linear responses especially in terms of velocity surges and water-surface gradients (Figure 8.8b). Typical responses are illustrated in Figure 8.8c, which shows flood and ebb velocities plotted in relation to tidal stage (the height of water during the tidal cycle). While flows are contained within the creek (t_1), they are of moderate velocity. When the tide does rise above the banks there is a pulse of rapid inflow, as the prism increases dramatically with marsh surface inundation (t_2). The water surface is unlikely to be completely horizontal as the wetting front spreads more slowly through the vegetation of the marsh surface. At high tide the water surface is presumed to be near horizontal (t_3), and flow slackens. In fact, there may be some inertial influx after the water has ceased to rise, and it is also possible that the height to which the water reaches is less at the rear of the marsh (Pye, 1992).

On the falling tide, the flow reaches maximum velocity shortly after water level has dropped below bankfull, as a large volume of water drains off the marsh surface (t_4). The water surface is not horizontal, but adopts a gradient into the creek as water is impeded by vegetation and levées (Healey *et al.*, 1981). The banks of tidal creeks on European salt marshes are often characterised by dense vegetation as shown in Figure 8.9. During the later part of the ebb tide (t_5) water continues to drain through the creek beyond low-tide slack water, and through the early part of the flood tide, until the next flood tide enters the creek.

Geomorphological processes associated with creek development include deposition of sediment on the marsh surface at high water, and

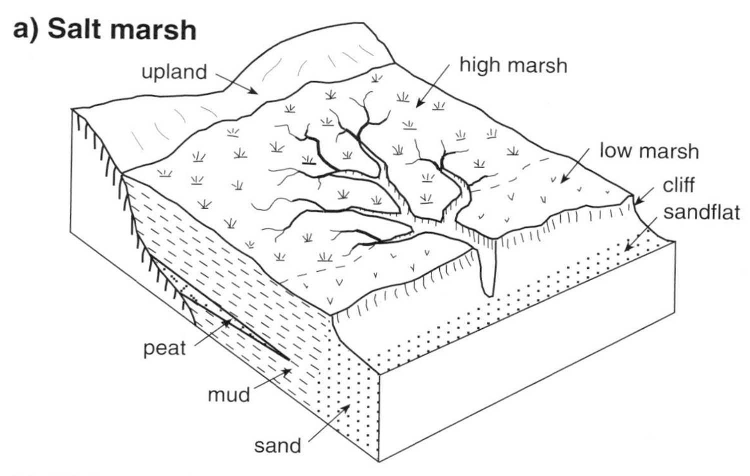

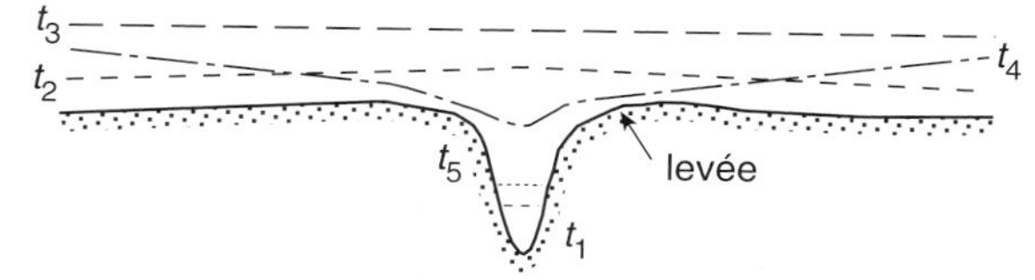

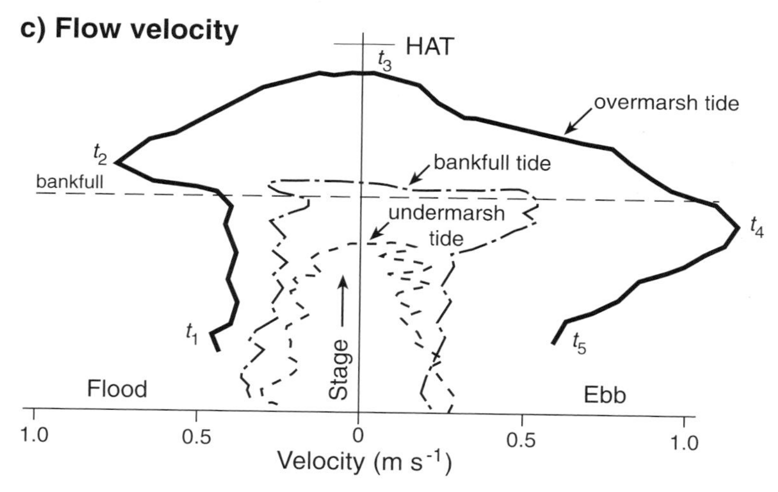

Figure 8.8. Schematic illustration of the significance of tidal creeks in the processes of marsh sediment flux. (a) A creek system dissects the mature marsh. (b) The water surface relative to a cross-section across the creek system and flanking marsh is unlikely to be horizontal as the tide floods over marsh or as it drains the marsh surface. (c) The velocity of flow, plotted as velocity against tidal stage (i.e. water height). Note the considerably greater flow velocities on overmarsh tides, compared to bankfull or undermarsh tides. The tidal stages at t_1–t_5 correspond to the water surface slopes in cross-section in (b).

erosion of the creeks by high velocity flows around bankfull stage on the ebb. Creeks are actively changing and can migrate across the surface of the marsh (Ashley, 1988; Zeff, 1988); they are often dominated by bank slumping, undercutting and rotational slips, and in cold climates can also be affected by freeze–thaw. Whether or not there is pronounced tidal asymmetry and whether it is the flood or ebb tide that is dominant are likely to influence net import or export of material. It is difficult to calculate a net flux between a large volume of material that is carried in, and a large volume that is carried out (Zarillo, 1985; Woodroffe, 1985). For example, in a budget for South Carolina salt marshes there appeared to be a net export of organic suspended sediment but a balance or import of inorganic suspended sediment (Settlemyre and Gardner, 1977; Ward, 1981). Ecological functioning of a marsh, and the extent to which organic products are exchanged with, or outwelled to, adjacent waters, is related to geomorphological setting, particularly shape and hydrodynamic processes (Mann, 1982; Odum, 1984). Tidal creeks are the routes by which organic production is exchanged between marsh and open water, and the conduits by which sediment is imported (Wolaver *et al.*, 1988; Pillay *et al.*, 1992; Carpenter, 1997).

There are subtle morphodynamic feedbacks. Frictional variation occurs over the marsh surface, primarily resulting from vegetation, but also microtopographic features such as levées which are likely to be the sites of enhanced deposition as overbank flows decelerate (French and

Figure 8.9. A tidal creek bisecting the salt marsh at Wareham, North Norfolk, England. The creek has slumped on the right bank, whereas the left bank appears to be undercut. The creek banks are covered with the salt-marsh plant *Halimione portulacoides*.

Clifford, 1992). Sediment redistribution via creeks in turn alters the hydrodynamics of flows (Bayliss-Smith *et al.*, 1979; French and Stoddart, 1992). Over time 'these events balance in their effects, so that channels neither silt up nor enlarge erosively but remain essentially constant in cross-section. Some creeks may shift laterally, but they experience no appreciable change in cross-sectional area. The tidal hydraulic forces drive the processes of erosion, transport and deposition over such intervals and at such rates as to maintain in sum over the very short term a neutral condition, or dynamic equilibrium' (Allen, 2000, p. 1177).

There are many reasons why it is not possible to assess the tidal prism accurately. Some muddy coastal systems occur at the downstream end of a river or stream and receive a direct freshwater input. For many tidal creek systems, water also floods over the seaward margin of the marsh during high spring tides (Miller and Gardner, 1981); this can account for as much as 60% of the spring-tide prism on north Norfolk marshes (Pethick, 1981). There can be other direct inputs of water, such as groundwater inputs and overmarsh flow received from adjacent tidal creek systems. Water losses occur across tidal catchment boundaries and as a result of evaporation, accentuated by ecological processes including bioturbation (Carr and Blackley, 1986).

The role of mangrove creeks is at least as complex as that associated with salt-marsh creeks. Trapping of sediment beneath mangrove forests has been described by Wolanski and Ridd (1986). The creeks may be either flood-dominated or ebb-dominated (Wolanski, 1992; Furukawa *et al.*, 1997). The form of the creeks and their relationship to adjacent supratidal flats can determine the balance of import or export through mangrove creek systems (Boto and Bunt, 1982; Aucan and Ridd, 2000). Many systems in the short term may be in a dynamic equilibrium as implied for salt-marsh creek systems. Over longer time scales they may reach a threshold and switch from flood-dominated systems that import sediment to ebb-dominated ones that export sediment.

8.4 Salt-marsh and mangrove shorelines

Salt marshes and mangrove forests are upper-intertidal environments in which a generally muddy substrate supports varied and normally dense stands of halophytic (salt-tolerant) plants. Salt marsh and mangrove ecosystems are important, both in terms of the aquatic food web they support and nutrients they export to coastal waters, and terrestrial wildlife such as migratory birds. They also act to protect coasts, dampening wave activity and helping to disperse and filter water flows from inland. There has been a long human interaction, particularly with marshes, exploiting the natural resources and also embanking marshes for land claim (Allen, 2000). Mangroves are utilised for timber and charcoal derived directly from the forest, fish and other aquatic resources, including aquaculture, and also provide less-tangible services, such as shoreline protection and sediment and nutrient trapping (Ewel *et al.*, 1998).

Salt marsh is particularly characteristic of temperate shores (Teal, 2001), although there are salt marshes that occur in association with mangroves in the tropics, such as in South and Central America (West, 1977), India (Blasco, 1975), and Australia (Adam, 1981), which has broad areas in which the two intergrade (Clarke and Hannon, 1967). Salt marshes are largely absent in polar latitudes, even where these are macrotidal, as at the head of the Sea of Okhotsk (Eisma, 1998).

8.4.1 Salt-marsh and mangrove vegetation

Halophytes are plants that can survive in saline substrates. They show a series of adaptations to the stresses of inundation and exposure, heating, desiccation and salinity that are experienced in the intertidal zone. Salt-marsh vegetation is azonal and is dominated by grasses and herbs throughout the world. The salt marshes of the eastern United States are dominated by the smooth cordgrass, *Spartina alterniflora*, whereas other temperate marshes tend to have a greater diversity of species (Chapman, 1974; Adam, 1990).

Arctic marshes are species-poor, characterised particularly by the grass *Puccinellia phryganodes*, and subject to ice action which gouges the surface. Boreal marshes, such as those of Hudson Bay, James Bay, southern Alaska and the Baltic, and Hokkaido in Japan, are slightly more diverse with sedges such as *Eleocharis* spp. and *Carex* spp. Physical disturbance during the break-up of ice lifts blocks of vegetation from the surface of the marsh (Dionne, 1967, 1999; Guilcher, 1979).

Temperate marshes are often divided into low marsh and high marsh on the basis of elevation and the frequency of inundation. The high marsh tends to show the highest salinities and a greater degree of oxygenation, with the low marsh being around seawater salinities and subject to frequent inundation. Low marsh is characterised by pioneer vegetation such as samphire, *Salicornia*, and the grass *Spartina* (Long and Mason, 1983). High marsh contains perennial grasses such as *Puccinellia maritima*, and herbs such as *Halimione portulacoides*, *Suaeda maritima*, *Limonium vulgare* (sea lavender), *Plantago maritima* (sea plantain) and *Aster tripolium* (sea aster). Glycophytes (non-halophytes) can establish on the upper high marsh unless some extreme flooding event prevents them. In Britain and much of Europe, a division can be recognised between the marshes which are predominantly muddy (east coast of Britain), and those which are much sandier (west

Figure 8.10. The prop root systems of a stand of *Rhizophora stylosa* in northern Australia.

coast of Britain) with grasses such as *Puccinellia* which are suitable for grazing.

Other temperate salt marshes are broadly similar, although with the addition of different species regionally; for example *Distichlis spicata*, *Sporobolus virginicus* and *Spergularia* spp. occur on marshes along much of the western coast of North America, with *Spartina foliosa* in California. *Zoysia sinica* var. *nipponica* occurs in Japan and *Sarcocornia quinqueflora* and *Juncus kraussii* occur on marshes in Australia, New Zealand and southern Africa (Adam, 1990).

Mangrove forests include more than 50 species (from 20 genera in 16 families) of trees and shrubs, distinguished on the basis of their ability to live in adverse intertidal conditions rather than any taxonomic affiliations (Spalding, 2001). They are most diverse in the Indo-Pacific region, with a second centre of diversity centred on the West Indies (Chapman, 1976; Hogarth, 1999). Mangroves appear to be limited by the 20 °C winter isotherm and their sensitivity to frost (Woodroffe and Grindrod, 1991). They show a range of adaptations to waterlogged soil, particularly above-ground roots which include pneumatophores in the genera *Avicennia* and *Sonneratia*, prop roots in *Rhizophora* (Figure 8.10), knee roots in *Bruguiera*, and sinuous buttresses in *Xylocarpus*.

Most mangrove species grow best in relatively low salinity waters (<30‰); *Avicennia* appears to be the most tolerant genus, coping with salinities in excess of 90‰. They appear to require a large root mass in order to meet the demand for water, and the root: shoot ratio increases with increasing salinity (Ball, 1998). Water-use efficiency is measured by the ratio of carbon assimilated to water used, a subtle balance between minimising water expenditure, holding down leaf temperature and maximising carbon dioxide acquisition and growth.

In the West Indies, the three species of mangrove tend to occur in discrete locations, *Rhizophora mangle* to seaward, *Avicennia germinans* in more landward locations, and *Laguncularia racemosa* mixed with the other species, or in areas that have been disturbed. In the Indo-Pacific region there are more species of mangroves, with up to 30 species in the most diverse locations. On open prograding shorelines, there is often zonation of species, though this is by no means universal (Watson, 1928). The seaward zone is often dominated by *Avicennia marina* and *Sonneratia alba*, occasionally with *Camptostemon*. *Rhizophora* spp. generally dominate a zone in mid shore with one or more species of *Bruguiera*. This is replaced to landward by *Ceriops*, and the landward-most fringe is generally dominated by *Avicennia*, often with samphire and less salt-tolerant mangrove species such as *Excoecaria* and *Lumnitzera*, and with non-mangrove species (Macnae, 1966, 1968). An example of a coast that has prograded over the past 6000 years and that has a pattern of zonation similar to this is shown in Figure 8.11.

Figure 8.11. A prograding coast, van Diemen Gulf, northern Australia. The mangrove fringe is prominently zoned. It comprises a seaward zone of *Avicennia* and *Sonneratia*, then a zone of *Rhizophora*, then *Ceriops*, and a landward zone of stunted *Avicennia*. The plain behind the mangroves is supratidal and seasonally flooded.

However, in many cases the distribution of mangroves is complex in response to geomorphologically defined habitats, and there are no clearly definable zones. Mangroves are replaced in areas that are only slightly brackish, or may be freshwater-flooded, by a range of other wetland species such as *Nipa fruticans*, *Heritiera* (*H. fomes* and *H. littoralis*), *Oncosperma* and *Acrostichum*. In northern Australia, there is often bare saline mudflat with a fringe of succulent herbs and paperbark trees (*Melaleuca* spp.).

8.4.2 Species distribution

Salt-marsh and mangrove vegetation have often been described as showing a shore-parallel zonation of species which has been interpreted to indicate a temporal succession whereby one species prepares the way for another. On salt marshes, the lower limit to which 'pioneer' plants grow on low marsh appears limited by tidal inundation. *Spartina* can grow as low as mean sea level on the eastern coast of the United States (Frey and Basan, 1985; McKee and Patrick, 1988). Elsewhere, marshes are confined to levels within the upper quarter of the intertidal zone, being generally above mean high neap water in Britain (Randerson, 1979; Allen, 2000). On high marsh, distribution appears to be controlled more by competition between species, but relatively few manipulative experiments have been undertaken, and the patterning is variable over short distances (Scholten and Rozema, 1990). Subtle microtopographical differences can determine duration of inundation and anaerobic conditions (Adam, 1990). Species can have overlapping distributions and 'it must not be assumed that zonation along an environmental gradient (a spatial factor) has been wholly, or even partly, generated by succession (a temporal phenomenon)' (Gray, 1992, p. 71). For instance, in Poole Harbour in southern England, salt marsh on one side of a small island was found to be accreting, while on the other side the salt marsh was undergoing erosion, with redeposition of sediment elsewhere on the marsh (Gray, 1992).

The rapid invasion of the hybrid *Spartina anglica* into the seaward-most zone of many salt marshes is an example of a temporal disruption to salt-marsh systems, implying that there is not an ecological equilibrium, and illustrates the flawed logic of an obligate successional view. *Spartina anglica* was first collected from Hythe on Southampton Water in 1870, and has since spread dramatically. Rapid sediment accretion and its ability to tolerate submergence meant that it was considered valuable as an aid to reclamation and it was initially planted not only in Britain but elsewhere, such as Australia, New Zealand and China. Its history in Poole Harbour, southern England, is typical. It first appeared there in the 1890s, attaining its maximum areal extent in the 1920s, with rapid sediment build up through trapping and binding. It declined in the late 1920s, partly because of a loss of vigour, but also because of horizontal erosion of swards and replacement by other species where it had built up to sufficient heights (Raybould, 1997).

On open coasts, the elevation at which individual mangrove species grow within the upper intertidal range appears to relate primarily to frequency of inundation (Watson, 1928; Woodroffe, 1995a). However, there is variation both in the mangrove species that are found, and the elevation at which they grow, up estuaries (Bunt *et al.*, 1982; Bunt, 1999). In complex mangrove communities and at local scales, species commonly display different sequences from transect to transect, and one species can recur at more than one elevation, termed bimodality (Bunt, 1996; Bunt and Bunt, 1999).

In many cases, the distribution of mangrove species is opportunistic and mangroves are not undergoing succession. In the Palisadoes of Jamaica where the concept of mangrove succession was proposed by Chapman (1944), long-term studies of the extent of mangroves have not detected any substantial build out of the shore (Alleng, 1998). Recent studies of gap dynamics indicate growth of the same species as already growing at a site, rather than a successional sequence (Clarke and Kerrigan, 2000).

Mangrove zonation is often a static response of species to conditions along an environmental gradient (Lugo, 1980). There are numerous exceptions to the classic zones, for example *Avicennia* is found bimodally at the landward and seaward margins on mangrove shorelines throughout the Indo-Pacific. Zonation of mangrove species has been reported for shorelines in northwestern Australia that are eroding (Semeniuk, 1980b). Such landward retreat of a zoned mangrove shoreline might also be expected to have occurred at times in the Holocene when sea-level rise was rapid and the shoreline was migrating landwards.

In many cases, mangroves respond opportunistically to the geomorphological habitats that are available, adjusting to morphodynamic changes in the landforms (Vann, 1959; Thom, 1967). In the case of deltas, for instance, it has been shown in Chapter 7 that there are active parts of the deltaic plains in which mangrove vegetation is limited, but that the abandoned deltaic plain contains a range of geomorphological environments in which mangroves can establish. The distribution of mangroves in estuarine environments in northwestern Australia has also been related to available habitats (Thom *et al.*, 1975; Semeniuk, 1986). It is useful to distinguish broad geomorphological settings, and to subdivide these into landforms with which there are particular associations of salt marsh and mangrove species, and this approach is examined in the next section.

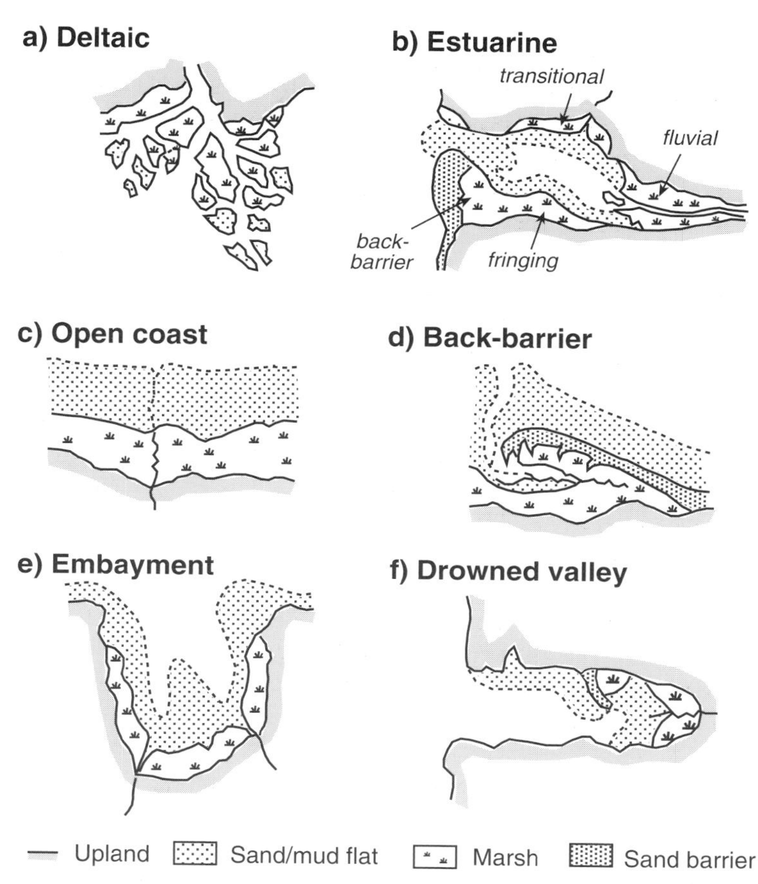

Figure 8.12. Geomorphological settings for salt-marsh and mangrove systems (based on Dijkema, 1987; Allen and Pye, 1992; French, 1997; Allen, 2000, and incorporating classes from Thom, 1982; Woodroffe, 1992). See text for details.

8.4.3 Geomorphological settings

This section defines a series of geomorphological settings (Figure 8.12) within which salt marsh or mangroves occur. Typical salt-marsh settings have been described from northwestern Europe (Dijkema, 1987; Allen and Pye, 1992; Pye and French, 1993; French, 1997; Allen, 2000). Similar types of salt marsh occur in North America; for instance, back-barrier, transitional, fluvial and bluff-toe have been identified in Maine, New England (Kelley *et al.*, 1988). A series of geomorphologically defined settings can be recognised within which mangrove forests occur (Thom, 1982; Woodroffe, 1992); these are broadly similar to those recognised for salt marshes and build on the recognition of different deltaic and estuarine landforms examined in Chapter 7. These broad categories are not exclusive, but represent a continuum, with overlap between categories, for example between estuarine and embayment, back-barrier and deltaic. The extent to which they are river-dominated or tide-dominated is examined in Figure 8.14.

Deltaic marshes occur in southern Europe (Camargue in the Rhône delta, Ebro delta, Rio Guadalquivir), but are especially extensive in association with the Mississippi River delta. Deltaic mangroves occur in river-dominated settings in association with large tropical rivers. Within a delta, mangroves occur in a variety of geomorphological habitats in response to the relative distribution of fluvial, tidal and wave energy, and microtopography associated with landforms such as levées, point bars and abandoned distributaries (Thom, 1967). Mangroves are generally not well-developed at the freshwater mouth of the river, but tend to be much more extensive in the abandoned deltaic plain that is generally extensively tide-dominated, for instance the Purari River delta in Papua New Guinea (Thom and Wright, 1983). Large subsiding deltas with tide-dominated abandoned distributaries provide ideal habitat for mangrove establishment (see Figure 7.13); they are flooded largely by tidal flows and shoreline retreat is typical, as in the western Sunderbans (Blasco *et al.*, 1998; Ellison *et al.*, 2000).

Estuarine settings also contain a suite of landforms that enable the development of estuarine marshes. Those in northwestern Europe, such as those in the Oosterschelde (Netherlands), Dyfi (Wales), Gironde (France), Severn and Thames River estuaries (England), have been subdivided into restricted entrance embayment, estuarine fringing, and estuarine back-barrier marsh types by Pye and French (1993) and Allen (2000). Variations in the replenishment of marsh surfaces by riverine sediment input occur along estuaries (Craft *et al.*, 1993). For example, Chesapeake Bay salt marsh has been divided on the basis of distance upstream into coastal, submerged upland, estuarine meander and tidal freshwater marshes, and rates of sedimentation vary between types (Stevenson and Kearney, 1996). Estuarine mangroves are extensive where there is a large tidal range, but also where there is substantial clastic sediment and it is reworked primarily by tidal action, even though the range may not be large.

Open-coast salt marsh occurs in some exposed locations, as for instance on Schleswig-Holstein, Wadden Sea (see Figure 8.1), north Norfolk (see Figure 8.19), Essex, or in northwestern Florida (Hine *et al.*, 1988). These marshes are generally sandy, and are open to direct wave action and storms which can be important either in the addition of sediment directly to the marsh surface (Goodbred and Hine, 1995), or can contribute to sediment removal. Dengie Peninsula in Essex, eastern England, has received particular attention because the marshes are eroded by wave action and serve an important role as coastal defence

(Harmsworth and Long, 1986; Reed, 1988; Leggett, 1994). Storm waves in the winter of 1989 led to vertical and horizontal retreat; the surface was lowered by 6 mm and recovery to the pre-storm profile took several years (Pethick, 1992). Mangroves can also occur in open-coast settings, particularly along the mudbank and chenier plain environments that are discussed above (Section 8.2.4). In these cases, the wave energy is either attenuated by the extensive mudflats that accrete to seaward, or wave action builds a chenier ridge of coarser material. Figure 8.13 shows a part of the coast of southern van Diemen Gulf in northern Australia. The mangrove fringe is irregular and a sandy upper beachface is active. Elsewhere along this coast, there are zoned mangroves that gradually prograde (see Figure 8.11). However, wave action works sand through mangroves. The process has been described in detail for shelly ridges in New Zealand (Ward, 1967), but is also recorded on other mangrove shorelines exposed to wave action (Woodroffe and Grime, 1999).

Salt marsh and mangrove are not extensive on wave-dominated coasts, preferring sheltered locations, such as lagoon or back-barrier settings. Lagoonal marshes occur in Europe on the Adriatic Sea coast, and in other areas where wave-dominated coasts lead to the development of coastal lagoons, such as southern France and Slapton Ley in southwestern Britain. Back-barrier salt marsh is common along much of the

Figure 8.13. An eroding part of the chenier plain coast of southern van Diemen Gulf, northern Australia. The sandy deposits to the left are being worked through the mangrove fringe which remains as isolated individuals of *Avicennia marina*. This is just to the east of the mangrove forest shown in Figure 8.11.

eastern coast of the United States, the Frisian Islands (northern Netherlands and northwestern Germany), and has been extensively studied at Scolt Head Island and other sites in northern Norfolk (Steers, 1960). These marshes tend to be subject to overwash processes, and are sensitive to the reshaping of the barriers in the lee of which they have formed. This is more important in those areas where relative sea-level rise is rapid. Marsh areas on the rapidly changing barrier islands (such as the Chandeleur Islands (see Chapter 6) associated with the Mississippi River delta) are prone to more rapid change than barrier islands on the more stable coast of Norfolk. Mangroves also occur in lagoonal and back-barrier settings, where they are protected from direct wave action.

Embayment marshes are found in relatively large, shallow coastal embayments, which can have a restricted entrance and receive large freshwater inputs, including the Wash and Morecambe Bay, and smaller embayments such as Portsmouth, Poole and Langstone Harbours in Britain and the Bay of Fundy in North America. These grade into drowned valley settings which include loch or fjord-head marshes, as their name indicates found on the predominantly rocky coasts of Brittany, western Ireland and northwest Scotland (e.g. Loch Crinan in Scotland). Mangroves are also found within bedrock embayments (or rias), such as the drowned river valleys of the New South Wales coast (Thom, 1982), and the mangroves of Darwin Harbour in the Northern Territory (described below; see Figure 8.15).

Salt marsh, but particularly mangroves, can be extensive in carbonate settings where there is not a large clastic sediment supply, but in which sediments are calcareous. These include reef settings, where mangroves can develop in the lee of reefs, or on sheltered reef flats (Stoddart, 1980); carbonate banks, such as the Great Bahama Bank or Florida Bay (see Figure 2.16), and the mangrove forests developed behind beach ridges on emergent limestone islands (Woodroffe, 1992).

8.4.4 Geomorphologically defined habitats

Salt marshes and mangroves function differently between these different settings, but also in geomorphologically defined habitats within any one setting, in terms of the relative dominance of river and tide processes, which are the principal sources of flooding. Wave action is generally limited, but several mangrove areas are far from any flooding, and are termed interior on the continuum shown in Figure 8.14, which is also an indication of sedimentary processes. Where there is a substantial river flow, sediments are deposited from the river, and organic production, enhanced by the freshwater, can be exported with the flow, supporting the concept of outwelling (Fleming *et al.*, 1990; Furakawa

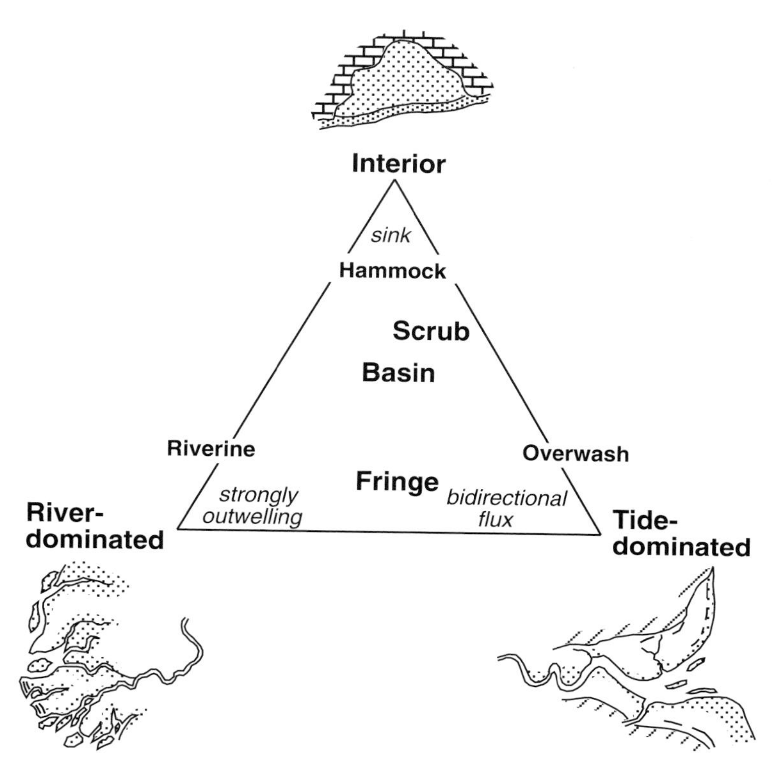

Figure 8.14. Continuum of processes affecting mangrove shorelines and settings within which mangroves are found, and representation of mangrove habitats in terms of the relative dominance of river, tidal or interior processes. The functional, ecological classification of Lugo and Snedaker (1974) is shown in relation to these gradients. The extent to which there is pronounced outwelling is also shown (based on Woodroffe, 1992).

et al., 1997). Mangroves which are flooded by the tides (fringe and overwash mangroves in the classification by Lugo and Snedaker, 1974) experience bi-directional flux between productive areas of mangrove and open coastal waters, and more limited outwelling can occur (Jimenez and Sauter, 1991; Lee, 1995, 1999). Other types, such as scrub or basin mangrove, are interior mangrove relatively remote from direct flushing; they are sinks for organic matter, nutrients and sediment, and are unlikely to export sediment or organic production (Twilley, 1985).

The ecological classification proposed by Lugo and Snedaker (1974) is essentially a static view of mangrove shorelines. There has been broader recognition of the wide range of geomorphological habitats, building particularly on the description of deltaic mangrove habitats in Tabasco, Mexico, by Thom (1967). The approach of dividing the intertidal habitats into geomorphologically based mangrove associations is a useful way to characterise geomorphological and ecological variability and has been extensively adopted in northern Australia (Thom *et al.*, 1975; Semeniuk, 1985a).

The mangroves flanking Darwin Harbour in the Northern Territory of Australia cover more than 25 000 ha. A broad pattern of geomorphologically defined mangrove communities has been described by Semeniuk (1985b), who has recognised a tidal creek habitat, a tidal flat habitat and a hinterland margin habitat. A typical tidal creek draining into the harbour is shown schematically in Figure 8.15. A tidal creek zone, dominated by *Rhizophora*, and tidal flat zone, dominated by low stature *Ceriops* and *Bruguiera exaristata* can be recognised, resembling fringe and basin mangroves respectively. A pronounced bare salt flat, associated with elevated salinities in the dry season, marks the transition between hinterland influence and the relatively flatter areas of tidal flat mangrove, and appears to be a feature of mangroves in areas which experience high evaporation (Spenceley, 1976). Coarse colluvium is carried into the hinterland margin by slopewash during the wet season and results in distinctive mangrove associations. Alluvial fan and chenier ridge mangrove types are also distinctive landform-related habitats. Mangrove forests are very productive; litter production is one measure of this important role as primary producers (Snedaker and Snedaker, 1984; Twilley *et al.*, 1992; Saenger and Snedaker, 1993). The geomorphologically defined habitats have coherence in terms of factors such as sedimentological characteristics of the substrate, geochemical characteristics of the soil, as well as functional and ecological factors, such as litterfall beneath the mangroves (Woodroffe *et al.*, 1988), indicating that primary production is also a function of geomorphological setting.

8.5 Salt-marsh and mangrove morphodynamics

Figure 8.16 is a schematic diagram showing three stages in development of marsh or mangrove in the upper intertidal zone. In the initial stage, an intertidal sand or mudflat, which may already have a cover of diatoms or seagrasses such as *Zostera*, is colonised by pioneer plants. The second stage is a mature stage in which there is a seaward low marsh and a more extensive high marsh, dissected by a well-developed tidal creek network. Salt marshes or mangrove sediments comprise a wedge-shaped accumulation of inorganic or organic sediments that progressively fill the volume (accommodation space) between the lowest level at which halophytic vegetation can establish and an upper level, close to highest astronomical tide. A further stage is envisaged, in which the marsh has accreted to a level at which its upper surface is rarely inundated, and wetland vegetation of non-halophytes occurs. This wetland is frequently

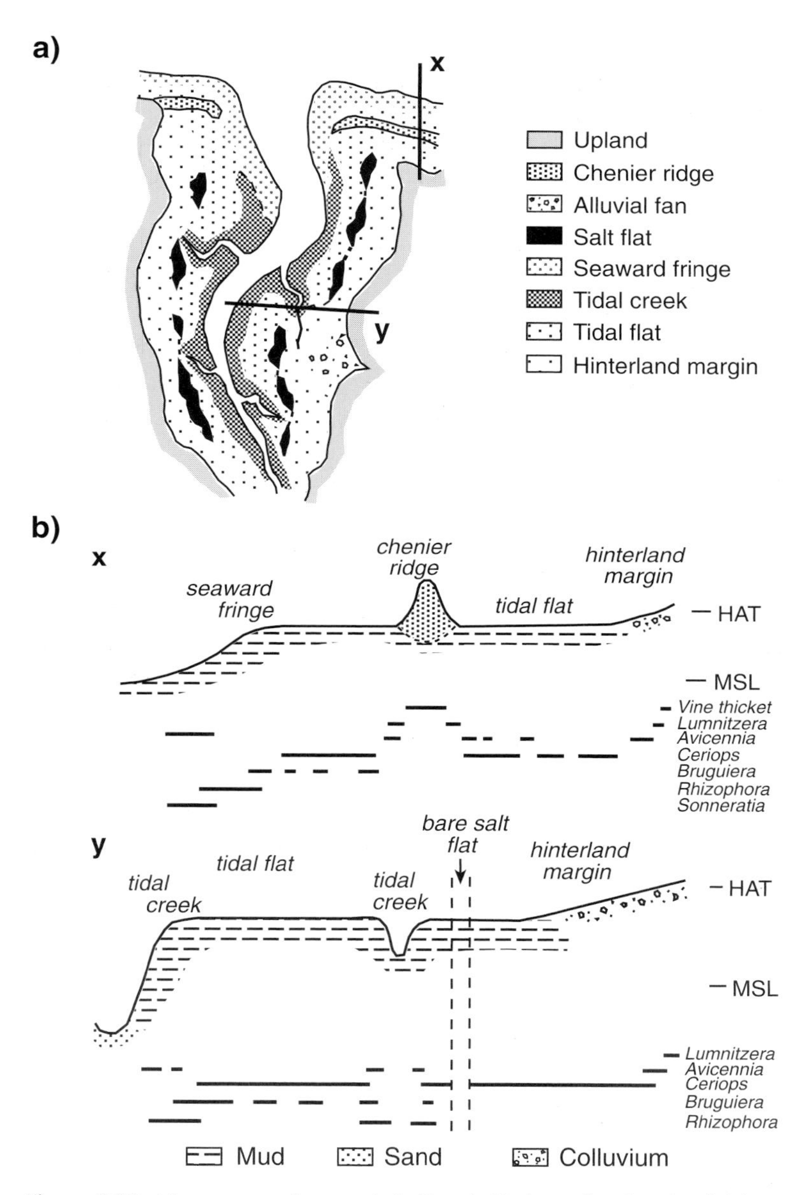

Figure 8.15. Mangrove environments in Darwin Harbour showing the division of geomorphologically based habitats within which mangrove associations can be recognised (habitat classification based on Semeniuk, 1985b; mapping based on Woodroffe *et al.*, 1988).

waterlogged and aerobic decomposition is inhibited, leading to peat accumulation. In wet, rain-fed areas of Britain, a domed bog characterised by *Eriophorum*, *Calluna* and *Sphagnum* can be inferred from pollen analysis, or 'fen' wetlands of birch and alder; these no longer exist, presumably as a result of human activity, including reclamation and

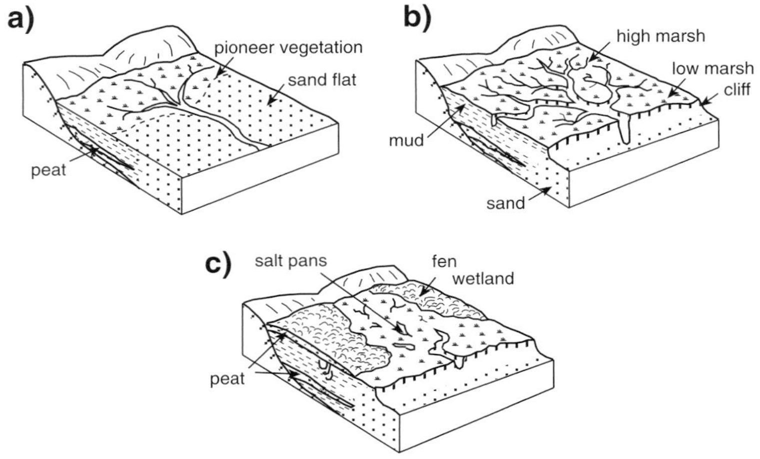

Figure 8.16. Schematic block diagram of (a) initial stages of salt-marsh formation; (b) mature salt marsh, and (c) salt marsh in which the upper surface has reached its maximum level and peat-forming non-halophytic environments are forming.

burning. In the wet tropics, peat swamp forest develops over a woody peat, replacing mangrove forests (Anderson, 1964; Woodroffe, 1993a).

8.5.1 Landforms and sedimentation

There are a series of subtle landforms that develop on the marsh surface; tidal creeks can be dynamic, eroding one bank through slumping, shifting position and infilling (see Figure 8.9). Nested sequences of creeks have been described on the basis of detailed morphostratigraphic mapping of muddy sediments in the Severn River estuary (Allen, 2000). Other subtle landforms include levées that develop along the margins of the creeks, and a small cliff that is a common feature of the seaward margin of the marsh (Figure 8.17). Slumping of the cliff is an active process that mobilises sediment that is generally entrained and deposited elsewhere on the marsh surface (Stumpf, 1983).

The marsh surface acts like a tidal floodplain and, during overbank tides, the vegetation slows the flow of turbid waters, and sediments settle as flocs that are not resuspended by the ebb tide; larger tides bring in and deposit more sediment (Christiansen *et al.*, 2000). Flow direction as water flows over the marsh and back into the creek during ebb can be highly variable (Eiser and Kjerfve, 1986; Wang *et al.*, 1993). Overmarsh flows depend on viscosity and friction, which differs below the canopy, within the canopy and above the canopy (Pethick *et al.*, 1990; Allen, 2000). The role of vegetation is likely to vary seasonally with many

Figure 8.17. Cliff eroded into the seaward margin of salt marsh, Eskmeals, northwestern England.

annual species and with seasonal variations in biomass, and on some marshes vegetation plays a minor role (Brown, 1998; Yang, 1999). The role of mangroves is likely to differ from that of salt marsh because most of the biomass is above the maximum flood level (Wolanski *et al.*, 1980).

Salt pans are significant features of the surface of mature marsh (Figure 8.18), for which various alternative origins have been proposed.

Uneven vegetation colonisation of the prior surface may lead to the development of residual areas without vegetation, called primary pans (Yapp *et al.*, 1917; Steers, 1969). Such areas persist because they become unfavourable for drainage, and saline water as a result of evaporation can make vegetation colonisation of these areas impossible. Elsewhere it appears that disintegration of the tidal creek system leads to pan formation (Guilcher and Berthois, 1957). In other cases, rafts of vegetation can lead to dieback. Below-ground piping has been suggested as a mechanism for pan formation (Kesel and Smith, 1978). It has even been suggested that tidal creeks can have developed through wind erosion and extension of salt pans (Perillo *et al.*, 1996). Bare areas also occur in mangrove forests and can arise for several different reasons (see Figures 8.15 and 8.21).

Vegetation appears significant in dampening waves on open-coast marshes (Knutson, 1988). It has been suggested that along the northern bank of the Severn Estuary the high-tide level reaches a metre lower where there are wetlands than where these have been replaced by a seawall (Brampton, 1992). If this is the case in other places also, then destruction of such wetlands should be a major concern; extensive areas of mangroves throughout the tropics have been cleared. Waves have only 20% of their energy left 180 m into open-coast marshes in north Norfolk (Moeller *et al.*, 1996; Shi *et al.*, 2000). Wave velocities on these

Figure 8.18. Salt pan on the surface of high marsh, Eskmeals, northwestern England.

marshes are of a similar order to tidal velocities, and can be much larger in times of storms, so dominate in their ability to achieve work over time (Möller *et al.*, 1999). Storms have been observed to be a major means of erosion in lower marsh, supplying sediment to the upper marsh (Ranwell, 1964; Houwing, 2000).

Several techniques have been used to measure sedimentation rates on the marsh surface. Accretion has been measured against stakes, marker layers (Steers, 1960; Stoddart *et al.*, 1989), or chance markers (such as a storm sand layer), silica flour, and filter paper sediment traps (French, 1993; Brown, 1998). It can also be measured using short-lived isotopes such as ^{210}Pb (Oertel *et al.*, 1989), or the isotope ^{137}Cs which has become detectable since bomb testing, indicating sediments deposited after 1954 (DeLaune *et al.*, 1987; Oenema and DeLaune, 1988). The significance of accretion in relation to subsidence or compaction can be measured with respect to fixed structures (Cahoon *et al.*, 2000). Long-term sedimentation studies tend to support a decreasing sedimentation rate with elevation (Pethick, 1980), whereas repeated measurements at short intervals indicate greater spatial and temporal variability, with phases of deposition interspersed with phases of erosion, particularly on the lower marsh (French and Stoddart, 1992; Reed *et al.*, 1999).

On Scolt Head Island sedimentation rates of 8 mm a^{-1} were observed on the lower marsh in the 1930s, declining to 5 mm a^{-1} in the 1980s; whereas rates of 2 mm a^{-1} were observed on the upper marsh and these appear to have remained relatively constant (Stoddart *et al.*, 1989; French *et al.*, 1995). Sediment concentration and sedimentation rate have also been observed to decline with distance onto the marsh elsewhere (Richards, 1934; Thom, 1992; Woolnough *et al.*, 1995). On North Norfolk marshes, sedimentation rate appears to decrease with distance from creeks over single tidal cycles but, at annual or longer scales, elevation is the key variable (French and Spencer, 1993). Average rates of 20–45 mm a^{-1} on mudflats to seaward, 9–37 mm a^{-1} on pioneer *Spartina*, and 2–4 mm a^{-1} on high marsh have been observed at Flax Pond on Long Island, New York (Richard, 1978). Sedimentation rates are likely to vary with relative stage of development that the marsh has reached, hydrological factors and below-ground organic production (Reed, 1988, 1990). For instance, roots, especially of high marsh plants such as *Puccinellia* and *Limonium*, serve to bind sediment and prevent retreat of the cliff found to seaward on European marshes (van Eerdt, 1985). Ratios of below ground to above ground biomass can vary from 4:1 to 1:3 (Good *et al.*, 1982; Hemminga *et al.*, 1996).

The salt-marsh seaward boundary can adopt one of several forms. It can be cliffed and retreating as a result of erosion (Figure 8.17). Elsewhere the marsh can have a ramped seaward shore, or a smoothly sloping transition, often interpreted as accretionary. In some circumstances a spur and runnel topography develops, which may serve to attenuate wave energy on open marsh shores (Allen and Pye, 1992). For instance, 82% of energy was observed by Möller *et al.* (1999) to be dissipated by friction on the marsh surface, in contrast to 29% over the adjoining sand flat surface.

Sedimentary processes also vary between mangrove forests of different settings. Organic production is a major element of substrate development beneath many mangrove forests, sometimes forming mangrove peat. There has been discussion of whether mangroves accelerate sedimentation, or whether they merely grow in areas that are already accreting (Carlton, 1974). The topography beneath mangroves is often relatively steep in contrast to the more gradual slope of the mudflat offshore which would appear to provide morphological support for more rapid sedimentation beneath mangroves (Bird, 1986). Sedimentation rates beneath mangroves are difficult to measure, and the few attempts that have been made have yielded a wide range of estimates. Maximum rates have been reported up to 4.6 mm a^{-1} in northern Australia and up to 8 mm a^{-1} in southern Australia using stakes (Bird, 1971; Spenceley, 1977, 1982), and around 3 mm a^{-1} in Mexico using ^{210}Pb and ^{137}Cs dating (Lynch *et al.*, 1989). Longer-term average rates of up to 6 mm a^{-1} during the early Holocene have been determined in northern Australia from radiocarbon dating of stratigraphic sequences (Woodroffe *et al.*, 1989). The methods are very different, and the results are not directly comparable. Still higher rates have been indicated by recent studies using buried plates in eastern Malaysia, with some indication of highest rates near the seaward margin, declining into the mangrove forest as the substrate increases in height (Saad *et al.*, 1999).

Studies from north Norfolk appear to indicate that sedimentation decelerates on the marsh surface as a marsh matures (Stoddart *et al.*, 1989). This would be anticipated on the basis that the frequency of inundation reduces as the intertidal surface accretes, and so increases its elevation. Pethick (1980, 1981) has examined a sequence of marshes of different ages since inception (Figure 8.19). The data have been reviewed by French (1993) who has revised the topographic height of several of the marshes, and suggested that marshes adjust to an equilibrium surface level that is 0.8 m below HAT. There appears to be a clear relationship that the older marshes have reached a higher elevation and are no longer accreting as rapidly.

8.5.2 Modelling marsh accretion

A number of time-step models have been developed in order to simulate plant dynamics (Randerson, 1979), and the response of the marsh

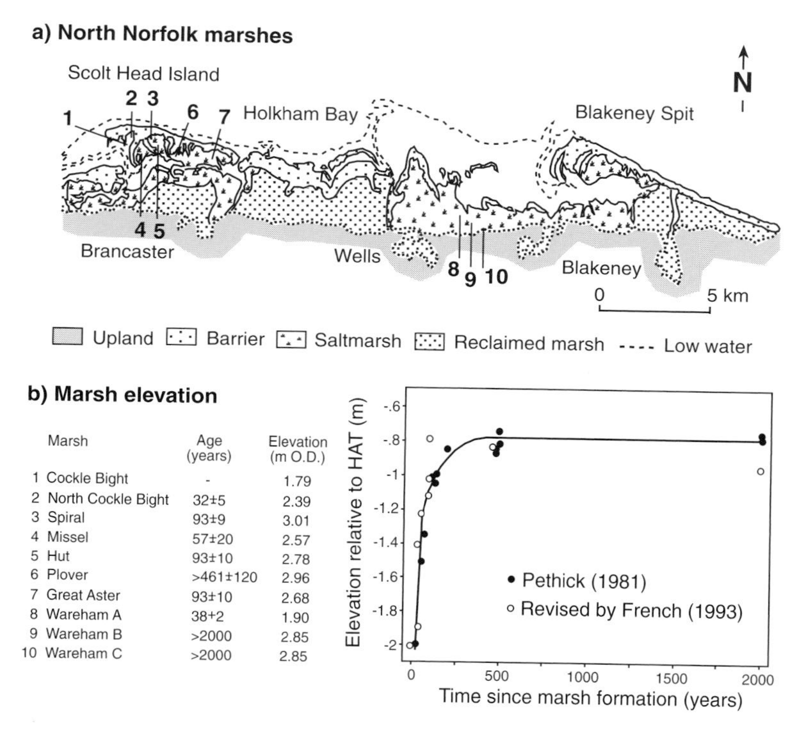

Figure 8.19. North Norfolk salt-marsh systems and the degree to which they have accreted and height of salt-marsh surface (based on Pethick, 1980; Stoddart *et al.*, 1989; French, 1993); (a) location of marshes, and (b) marsh age (relative to 1990) and elevation of highest vegetated surface (based on Pethick, 1981; French, 1993).

systems to sea-level changes (Krone, 1987; Allen, 1992; French, 1993; Schwimmer and Pizzuto, 2000). Models based on the mass budget and change in elevation involve both inorganic sediment accretion and organic accumulation, and variations in the accommodation space resulting from autocompaction or relative sea-level change.

In salt-marsh models developed by Allen and French, the rate of inorganic sediment input decreases exponentially as elevation increases. Mangrove sediment is also likely to fill the accommodation space between mean sea level and highest tide level, at an exponentially decreasing rate, as tidal inundation declines with surface accretion. Organic input, from *in situ* plant production, is considered constant; this is clearly an oversimplification, but it recognises that the accumulation of plant biomass contributes to the increase in sediment mass. Under these circumstances, it is possible for a system to accrete above the highest tide level through organic accumulation, with peat development as implied in Figure 8.16. This also provides a framework within which to explain peat layers found within the stratigraphy of many temperate muddy coasts (see Section 8.2.3). Allen (2000) has modelled situations where alternations between organic and inorganic sedimentation occur, and a summary of his conclusions is shown in Figure 8.20.

Figure 8.20 is a schematic representation of marsh morphodynamics; Figure 8.20a illustrates the situation where external boundary conditions are held constant (steady-state) and a marsh accretes with increase in overall marsh elevation, at a decelerating rate. The marsh approaches the maximum upper threshold, defined for convenience by HAT (although in practice modal upper spring tide level may be a more appropriate limit). Within the models developed by Allen and French, sedimentation by organic matter is independent of marsh elevation, so a marsh will switch from predominantly inorganic (intertidal) to predominantly organic (supratidal, peat) when it reaches this upper threshold (equivalent to the commencement of peat-forming fen wetland shown in Figure 8.16c). Autocompaction, the lowering of the surface through the oxidation and compression of organic material, can initially keep the marsh surface from exceeding this threshold, but compaction decreases through time and eventually organic accumulation will exceed the threshold level if external factors remain constant.

Figure 8.20b is a situation more typical in northwestern Europe where sea level is gradually rising. In this case, the youthful marsh is followed by a mature marsh which keeps pace with sea-level rise (Allen, 2000). If there is a rapid rate of relative sea-level rise, as for instance where marshes are subsiding, the marsh sediments may never fill the accommodation space. Average sedimentation rates have been estimated to be around 0.8 mm a^{-1} in Chesapeake Bay in comparison with sea-level rise of 3 mm a^{-1} (Stevenson *et al.*, 1986). There is a still more pronounced sediment deficit in the subsiding marshes of Louisiana. As indicated in Chapter 7, the Mississippi River delta region is rapidly subsiding, which provides the opportunity to examine whether rates of sedimentation can keep up. Interior marsh builds slowly and suffers nitrogen deficiency and sulphide toxicity (Mendelssohn and McKee, 1980), leading to die-back and loss of marsh through waterlogging. Sedimentation rates are highly variable, but generally occur at 13 mm a^{-1} on levées adjacent to creeks and 7 mm a^{-1} in backmarshes away from creeks (Hatton *et al.*, 1983). These marshes are generally not flooded by tides, the area being microtidal and diurnal, but storms play an important role both in water circulation patterns and sedimentation (Hopkinson *et al.*, 1985; Reed, 1989). Sedimentation rates generally reach around 8 mm a^{-1}, but subsidence occurs at around 12 mm a^{-1} (DeLaune *et al.*, 1987). Similarly in the subsiding Atchafalaya Delta, sedimentation rates lag behind the relative sea-level rise rate (DeLaune *et al.*, 1990).

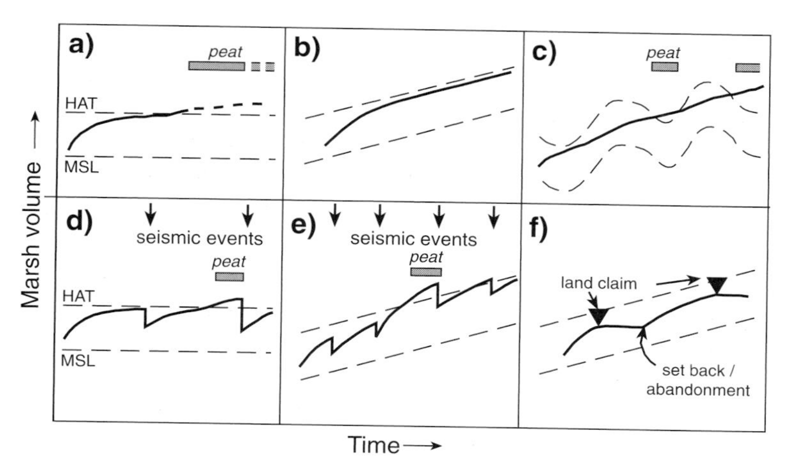

Figure 8.20. Schematic representation of the morphodynamics of salt marshes (based on Allen, 2000). (a) Steady state, sea level and other external boundary conditions held constant, the salt-marsh surface (volume) accretes asymptotically until organic accumulation takes it out of the tidal range (above HAT), after which peat accumulates. (b) Scenario in which sea level rises at a constant rate and marsh reaches maturity and then continues to accrete at a similar rate to sea-level rise. (c) Scenario in which sea level varies through time, rising irregularly. The marsh accretes but at times is above HAT; at these times peat accumulates. (d) Stable sea level, but episodic seismic subsidence rejuvenates marsh. (e) Episodic subsidence (seismic events) and rising sea level lead to complex salt-marsh morphodynamic response. (f) Situation where embankment and land claim lead to period of no change in embanked sedimentary volume (except subtle autocompaction) but, in face of sea-level rise, periodic set back or abandonment required.

It has been inferred that in some situations mangroves keep up with sea level (Cahoon and Lynch, 1997). In carbonate environments where inorganic sediment input is low it has been suggested that mangrove substrates may collapse if sea-level rise exceeds a few millimetres per year (Ellison and Stoddart, 1991). It is clear that mangrove sedimentation cannot keep up with the fastest rates of sea-level rise experienced during the postglacial transgression. Pollen, which may be recovered in ocean-floor cores, indicates that mangroves must have been extensive, especially during past highstands (Grindrod *et al.*, 1999), but there is little information directly available on Pleistocene mangrove shorelines. Mangrove peat is found at various places across the broad continental shelves, as for instance in the Sunda and Sahul shelves (Hanebuth *et al.*, 2000), and the mangrove shoreline has backstepped as the sea has risen rapidly. Modern coastal sedimentary wedges are rarely more than 15–20 m thick, recording continuous mangrove sedimentation only during the mid to late Holocene, as sea-level rise began to slow.

There are a number of marshes or other muddy environments where the rate of sedimentation appears roughly equivalent to the rate of sea-level rise (Dijkema *et al.*, 1990; Suk *et al.*, 1999; Dyer *et al.*, 2000). It is important to establish whether this covariance is coincidental or the result of morphodynamic adjustment by feedback mechanisms between mudflat or marsh morphology and active processes. In fact, sea level is not changing constantly, and marsh response to unsteady sea-level change is more realistic (Figure 8.20c). During periods when sea level undergoes a negative tendency, the marsh can exceed the upper threshold of marsh accretion and peat accumulation is likely. Alternations of inorganic and organic accumulation, such as those described typically in the stratigraphy of northwestern Europe (see Section 8.2.3) would occur.

Various other perturbations can be simulated. The role of storms can be either erosional or can result in deposition of a sedimentary layer (Ehlers *et al.*, 1993; Goodbred *et al.*, 1998). In Figure 8.20d episodic subsidence as a result of seismic events is portrayed. This is particularly the case in plate-margin settings, such as the marshes of the Pacific coast of North America. Subsidence of 1.0–1.5 m, followed by gradual uplift of 0.5–1.0 m associated with tectonism is reported with recurrence of 300–1000 years from Washington and Oregon coasts (Darienzo and Peterson, 1990; Long and Shennan, 1994; Cundy *et al.*, 2000). Subsidence of marshes, in some cases accompanied by deposition of sediment by tsunami, results in drowning, and sets the marsh back in the evolutionary sequence. Earthquake activity on the coast of Chile can lead to catastrophic drowning of forests and establishment of marshes (Reed *et al.*, 1988). Similarly, on rapidly uplifting coasts this means the episodic seaward movement of marsh communities, although the transition is punctuated by other scales of ecological change and is not a progressive succession (Cramer and Hytteborn, 1987). A complex pattern can be anticipated where marsh evolution accompanies episodic seismic events and changing sea level (Figure 8.20e). The models indicate marsh drowning if the rate of local relative sea-level rise is greater than the rate of marsh aggradation, marsh maintenance where the two rates are equal, and marsh expansion if marsh accretion exceeds relative sea-level rise (Orson *et al.*, 1985). Human intervention, particularly through land claim, can also be portrayed as a perturbation and simulated (Figure 8.20f). If the sea level is rising, land claim is followed by post-embanking stabilisation, pedogenesis and enhanced compaction which terminate when the embankment is either set back or abandoned (Allen, 2000).

Similar modelling could be envisaged for mangrove shorelines. A number of perturbations can occur to disrupt mangrove forests. The most obvious is the impact of storms (Lugo *et al.*, 1976). Storms can have far-reaching impacts on mangrove shorelines, defoliating the trees, but leaving many still standing for 20–30 years without recovery (Stoddart, 1962; Craighead, 1964; Glynn *et al.*, 1964; Cintron *et al.*, 1978). Cyclone Tracy devastated mangroves around Darwin in northern Australia on Christmas Day 1974. There has been negligible re-establishment, leaving substrate that is now unfavourable for recolonisation (Woodroffe and Grime, 1999). In contrast to salt marshes, where storm events form a temporary hiatus in sedimentation, tropical storms can have a persistent legacy. Figure 8.21 shows Twin Cays, an island, termed a mangrove range, on the Belize Barrier Reef (see Section 5.5.2). The bare areas contain the remains of mangrove stumps, and it seems likely that these areas are the result of devastation by Hurricane Hattie (Stoddart, 1962). The hurricane may have caused clearing which other factors, such as sulphide toxicity prevent from becoming revegetated (Woodroffe, 1995b; McKee and Faulkner, 2000).

Mangroves can be subject to other disturbances. The role of lightning strikes and gap dynamics has received particular attention (Chen and Twilley, 1998). In order to understand these extreme events and their role in the long-term development of mangrove shorelines, it is necessary to consider the time scales over which biological and physical

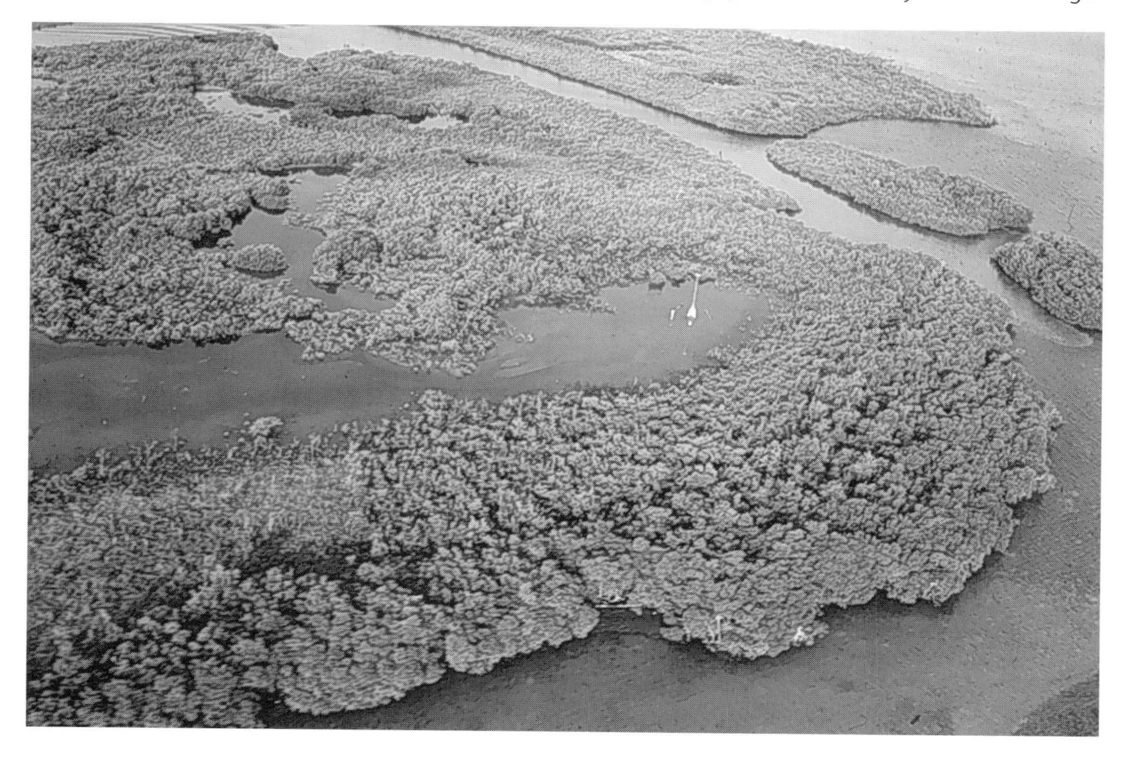

Figure 8.21. Twin Cays mangrove range, Belize Barrier Reef. The bare areas in the mangrove forests contain stumps of former mangroves and appear to have been cleared by hurricane damage.

processes operate in mangrove environments. Figure 8.22 examines the scales appropriate for each. It is clear that the life history of individual plants in the mangrove forest may be greater than the recurrence of disturbances and, as in the case of reefs (cf. Figure 5.6), an element of resilience in the long-term is implied.

8.5.3 Muddy coast morphodynamics and sea-level change

The simulation of muddy coastal environments suggests that they adjust to changes in external boundary conditions (see Figure 8.20) particularly sea level. There are likely to be subtle equilibria and thresholds. 'The morphology of salt marshes is the attainment of an equilibrium between stress and strength: the physical expression of the critical erosion threshold. Depositional processes may act up to, but not beyond, such critical thresholds and it is the definition of these that allows explanation of the distinctive salt marsh morphology' (Pethick, 1992, p. 41).

In the early stages of marsh growth, flows are likely to be flood-dominated because the shorter duration of the flood-tide limb is likely to be associated with greater velocities, as shown for estuaries (see

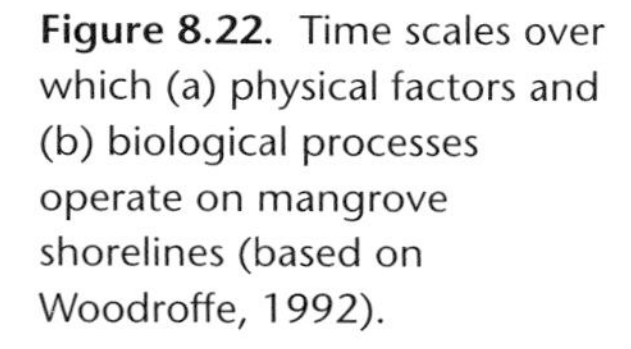

Figure 8.22. Time scales over which (a) physical factors and (b) biological processes operate on mangrove shorelines (based on Woodroffe, 1992).

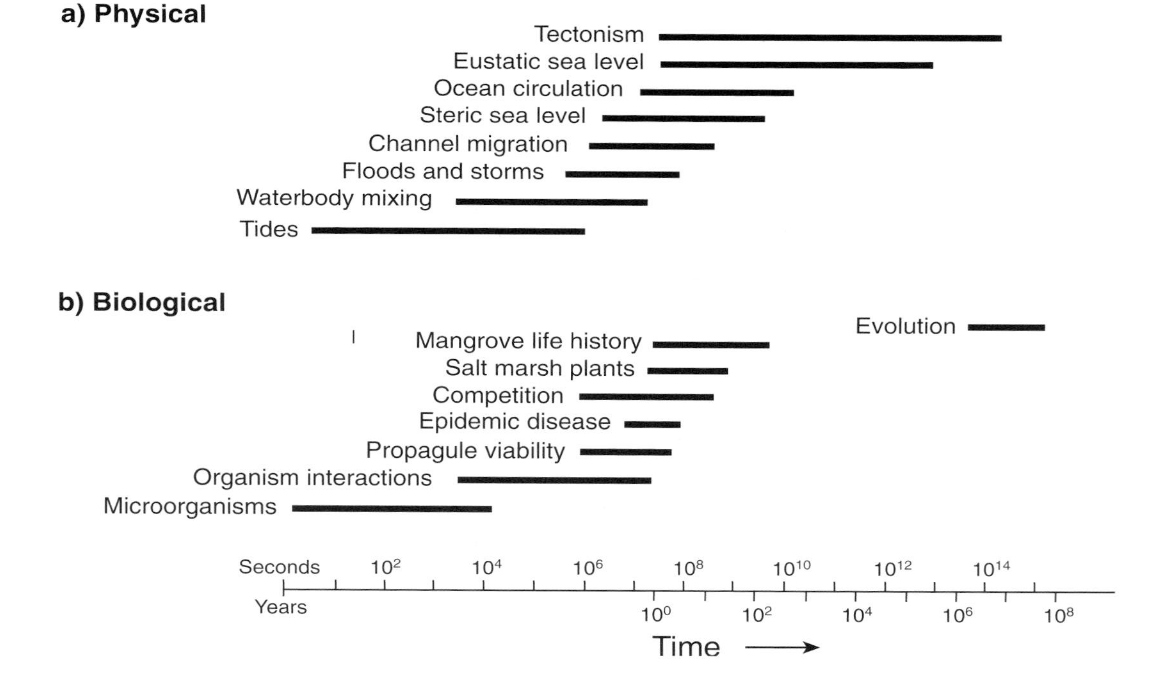

Chapter 7). On the mature marsh, where ebb flows are accentuated by substantial overbank storage, flows become ebb-dominated and ebb flows maintain channel dimensions (see Figure 8.8c). Despite the equilibrium between flows and hypsometry, there still remains accommodation space and sedimentation will continue, at least in terms of organic accumulation from the plants themselves. The rate of inorganic sedimentation is strongly controlled by frequency of tidal inundation, decreasing as the surface reaches higher elevations within the tidal range. Autocompaction is an important feedback, which also decreases asymptotically through time. It occurs as a loss of sediment volume through both inorganic processes, such as rearrangement of the mineral architecture, silica solution, clay dehydration and other diagenetic processes, and through decomposition and compression of organic matter (Kaye and Barghoorn, 1964; Cahoon *et al.*, 1995; Pizzuto and Schwendt, 1997; Allen, 2000).

There appears to be a cycle of creek network development which goes through initiation, elongation, elaboration (extension), ending with abstraction, although views differ as to the details of how the networks develop (Allen, 2000). One view is that channels become deeper over time as the marsh surface accretes around them (Jakobsen, 1954; Gardner and Bohn, 1980; Beeftink and Rozema, 1988). An alternative view is that the creek network develops with deepening of the creeks themselves (French and Stoddart, 1992) and elaboration of creek systems as marsh matures (Pethick, 1992). None of the models explains how 'during upward marsh growth, progressive channel enlargement and densification square with the implied decrease in the overmarsh tidal prism' (Allen, 2000, p. 1193). Allen (2000) has hypothesised a situation in which overmarsh tidal prism decreases as marsh height increases over time, but drainage density and bankfull prism reach a maximum and then subsequently decrease.

Spatial heterogeneity occurs in terms of autocompaction which responds to antecedent topography and to the sedimentary composition of the accumulated sediments. Microtopographic differences in such features as levées, former channel courses and their persistence as salt pans, as well as ecological variations and the effect of extreme events (such as storms or wave impact) also make the concept of equilibrium applicable only at a broad scale.

Mangrove sediments provide a relatively good record of sea-level rise, particularly where the sediments are of high bulk density (inorganic), low compactability, and sea-level rise has been rapid (Woodroffe, 1990). Along the northern Australian coast, mangrove muds record the pattern of sea level from around 8000 years BP to around 6500 years BP

(Grindrod and Rhodes, 1984; Woodroffe *et al.*, 1989). As sea level decelerated, the accuracy with which sea-level position can be inferred from dates on mangrove peat declines as a result of compaction and root-contamination issues. Mangrove sediments are relatively poor indicators of emergence; indeed organic accumulation under mangroves can continue after the substrate has reached the highest tidal level, as the accretion of peaty substrates above tidal levels beneath mangroves behind beach ridges on islands in the West Indies indicate (Woodroffe, 1982).

Such studies enable an interpretation of how mangroves have developed in previous situations, and also give a perspective on how mangroves may respond to several different sea-level change scenarios in the future and variations in the supply of sediment (Wolanski and Chappell, 1996). In view of the possible role of salt marshes as a temporary store of sediment, with interaction between salt marsh and mudflat, described above, it will be important to develop a better understanding of the morphodynamics of mangrove shorelines. As the sea rises, it will flood across low-gradient Holocene plains (see Chapter 10). Generally, tidal waters may find low-lying points and exploit these to cut back into the plains. Rapid incision can occur through easily erodible sediments, and expansion of tidal creeks into the plains represents a means by which salinisation can progress. An example of such tidal creek expansion, and salinisation of coastal plains has been described from the lower Mary River in the Northern Territory (Knighton *et al.*, 1991). In this case there is no evidence that this incursion has occurred as a result of sea-level rise. Indeed there appears to have been negligible sea-level rise and some other cause needs to be invoked. Nevertheless, evidence is emerging that muddy coasts are subject to frequent depositional and erosional episodes, and the rapid morphodynamic change that can occur needs to be incorporated into strategies for coastal management of these systems.

8.6 Summary

Muddy coasts have received far less intensive study than sandy coasts, but their complexity is becoming increasingly apparent. Muddy coasts generally comprise an intertidal zone that can contain significant quantities of sand, and upper intertidal salt-marsh or mangrove areas, fed by a tidal inlet or dissected by tidal creek networks. The long-term accumulation of mud leads to a sedimentary record within which there is evidence of cycles of erosion and accretion that are likely to have occurred at a range of scales. Physical factors, and biological response occur at several overlapping time scales (Figure 8.22).

Mudflats, salt marshes and mangroves, can also be considered at several different spatial scales. There are regional or latitudinal variations, for instance between salt marsh and mangrove, or admixtures of the two. It is also convenient to recognise broad geomorphological settings and, at a smaller scale, there are geomorphologically defined habitats. The stratigraphy of prograded coasts indicates that, although halophytic plant communities can prograde seawards over intertidal mudflats in some open-coast settings, the long-term pattern of development is subject to external controls, such as sea level and sediment supply. Studies of individual muddy coasts show that there is generally a more complex relationship between vegetation and landforms with ecological response to microtopographical variation.

The main external controls on salt-marsh or mangrove development are sediment type and supply, tidal range and regime, wind–wave climate and the relative movement of sea level. Muddy coasts undergo landward migration under conditions of relative sea-level rise. Intrinsic controls include sediment autocompaction (which together with sea-level rise defines the accommodation space for sequestering further sediment) and halophytic vegetation that traps and binds sediment. The morphodynamics of salt marshes can be divided into the study of the marsh surface (platform) where there appear to be feedbacks between vegetation and sediments, and the study of the tidal creek systems.

Muddy coasts appear to undergo periodic alternations between accretion and erosion. This can be most clearly seen in the case of chenier plains, but occurs at other spatial and temporal scales. Morphology is sometimes an indicator of process. Eroding mudflats appear to be concave-up, and accreting mudflats convex-up (Dyer, 1998). Association of eroding mudflats with cliffed and retreating salt marsh, or eroding mangrove shorelines also suggests that there may be significant interrelationships between mudflat and adjacent wetland (Kirby, 2000). Response of muddy landforms to seasonal and extreme events needs to be further assessed. Detailed 3D hypsometry may offer a rapid way to assess mudflat dynamics (Mason and Garg, 2001).

There is likely to be considerable spatial variability in the sedimentation rate at different locations within salt marshes or mangrove forests, depending on factors such as the degree of tidal inundation, inorganic accretion rate, organic below-ground production and autocompaction. Characteristics of the setting, and the individual landform habitats within any one setting, will be important in understanding muddy shoreline evolution. Modelling offers insights into how these processes might operate, but it will be important to undertake field studies to understand the complexity of individual systems.

Recognition of equilibria between inlet dimensions and prism appears simplistic. The morphodynamic adjustments of creek systems, by contrast, are inadequately studied and are likely to be complex at a range of scales. It is useful to view salt marshes and mangroves as adjusting towards a morphology that resists long-term morphological change from tidal processes, but perhaps also in terms of wave impacts where these are significant 'If mudflat morphology does experience internally induced periodic change, then it is of crucial importance for coastal managers and conservationists who may otherwise mistake erosion for long-term deterioration rather than natural homeostasis and take inappropriate mitigating action' (Pethick, 1996, p. 192).

Chapter 9
Morphodynamics of coastal systems

> In order that a particular portion of shore shall be the scene of littoral transportation, it is essential, first, that there be a supply of shore drift; second, that there be shore action by waves and currents; and in order that the local process be transportation simply, and involve neither erosion nor deposition, a certain equilibrium must exist between the quantity of the shore drift on the one hand and the power of the waves and currents on the other. On the whole this equilibrium is a delicate one, but within certain narrow limits it is stable. That is to say, there are certain slight variations of the individual conditions of equilibrium, which disturb the equilibrium only in a manner tending to its immediate readjustment. For example, if the shore drift receives locally a small increment from stream drift, this increment, by adding to the shore contour, encroaches on the margin of the littoral current and produces a local acceleration, which acceleration leads to the removal of the obstruction. Similarly, if from some temporary cause there is a local defect of shore drift, the resulting indentation of the shore contour slackens the littoral current and causes deposition, whereby the equilibrium is restored. (Gilbert, 1885)

Coastal landforms adjust towards, but rarely find, delicate and dynamic balances with the processes that operate. This is expressed very clearly in the above quotation from G.K. Gilbert, and has been a recurrent theme in the preceding chapters. Coastal systems adjust to shore-parallel and shore-normal gradients in wave, tide or other forces. This chapter examines these gradients in process operation, linking them to associated variation in landforms. The principles introduced in Chapter 1 are placed into a broader coastal morphodynamic framework (Wright and Thom, 1977; Cowell and Thom, 1994). The value of models and the role of hypothesis generation and testing, and the significance of different time and space scales are investigated.

9.1 Models in coastal geomorphology

Description of coastal landforms and the interrelationship between form and process provides the foundation for theories about how the coast changes. A theory is often expressed in terms of a model of how a coastal system behaves under controlled conditions. Modelling provides a powerful tool for describing how coasts have changed in the past and how they could change in the future. Models include physical models (such as wave tanks), conceptual models (such as Darwin's theory of atoll development) and computer models (such as SBEACH). They are not predictors of future conditions, but are frameworks within which various scenarios can be simulated and hypotheses tested, often based on statistics or probabilities.

9.1.1 Models and hypothesis generation

The best models are simple models that ignore much of the detail of individual systems, but embody most of the variation that is observed. For instance, a model of wave conditions simplifies variables into a small number of representative conditions, such as particular wave periods and approach directions. A model is particularly useful if it suggests hypotheses that can then be tested. It is described as robust if it accounts for the range of landforms that occur and does not contradict specific sets of observations at particular scales of interest.

A model is a simplification based on assumptions, particularly the premise that the response of a variable to a given process is known (Fox, 1985). Nevertheless, within the constraints set by the assumptions, models enable rapid testing of hypotheses, analysis of sensitivity to perturbations and changes in boundary conditions, and extrapolation to the potential range of past or future system responses. Sophisticated computer models are not a true representation of reality, but are simplified tools for experimentation, explanation, prediction and hypothesis generation (Lakhan and Trenhaile, 1989a). It is often the case that models can be manipulated to produce almost any behaviour, and the validity of the assumptions and hypotheses on which a model is based must be continually and rigorously reassessed.

Coastal research should ideally be directed towards addressing a particular question. Figure 9.1 indicates the successive steps in the generation of hypotheses and the refinement of models. The scientific method involves five discrete steps that are generally undertaken in the following order: (i) recognition (formulation) of a question, (ii) collection of data, (iii) induction of hypotheses, (iv) deduction and testing of original hypotheses, and (v) revision and interpretation of hypotheses

(Beveridge, 1980). In some instances, theories are deduced from existing knowledge and can be tested with an appropriate research design. In other cases, field data are collected and analysed to derive hypotheses through induction (Rhoads and Thorn, 1996).

Hypotheses are particularly important because they can guide data collection. It is generally not possible to verify or validate earth-science hypotheses about the course or cause of landform evolution because the events occurred in the past and the evidence is often not available (Oreskes *et al.*, 1994). Scientific knowledge expands by identifying and dismissing incorrect hypotheses (falsification), narrowing the field of probable correct answers to a question (Popper, 1968; Kuhn, 1970). However, if scientific problems are approached with only one preconceived model or 'ruling hypothesis' there is a danger that bias may lead to collection only of data that seem to confirm that hypothesis (Gilbert, 1886). In the earth sciences, the concept of multiple working hypotheses has long been advocated to enable more objective evaluation of the way that geological processes shape landforms (Chamberlin, 1890). Nevertheless, there is often a series of conventions amongst scientists in specialised fields constraining the extent to which studies outside a generally accepted set of theories, or paradigm, will be accepted (Dott, 1998; Miall and Miall, 2001).

The origin of a landform is usually considered partially or temporarily resolved if weak pieces of evidence are consistent with only one of several competing hypotheses (Rhoads and Thorn, 1996). Multiple working hypotheses can be tested, in parallel or in sequence, and either rejected or retained and incorporated into more detailed models (Schumm, 1991). In practice, major controversies in the earth sciences have rarely been resolved in favour of one particular model. More often a compromise has been reached with components of several theories

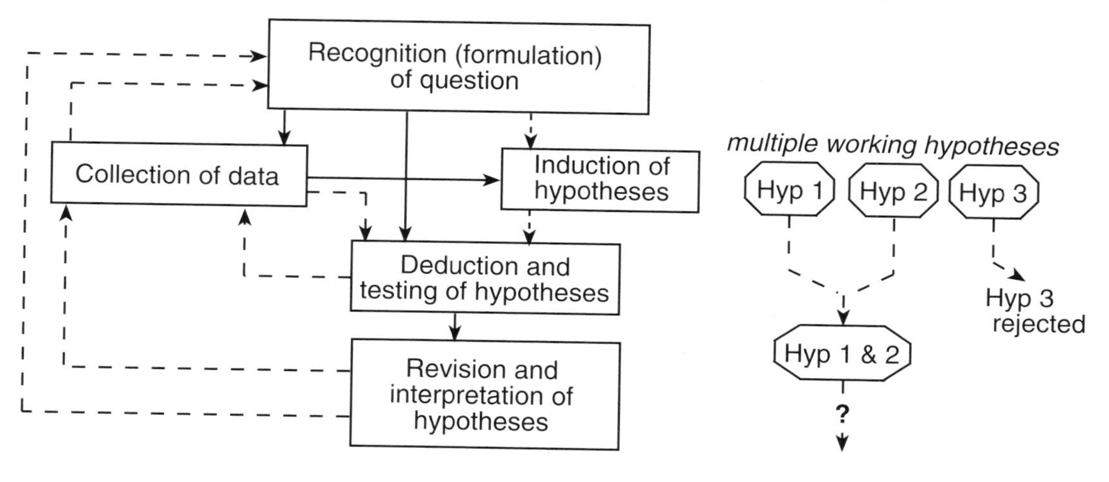

Figure 9.1. Generation and testing of hypotheses and refinement of multiple working hypotheses.

amalgamated into a composite model. One hypothesis may be valid in one instance, but an alternative may seem more appropriate in other cases (see Figure 9.1).

The history of sea-level studies, discussed in Chapter 2, provides examples of stages in the refinement of hypotheses. Early in the 20th century, it was recognised that global sea level had fluctuated with the ice ages. In the mid 20th century there were attempts to construct a model of global postglacial sea level (a sea-level curve) based on dating of fossil shorelines. A major controversy arose as to whether Holocene sea level had risen to present level at a decelerating rate, or had oscillated (including periods both above and below present), or had been above present level in the mid Holocene and fallen to present. Attempts to develop a single postglacial eustatic sea-level curve were unsuccessful because data from different parts of the globe appeared to indicate spatially divergent sea-level histories (e.g. Fairbridge, 1961). No single eustatic model was accepted, but a more complex model, incorporating elements of earlier models together with hypotheses about the geophysical deformation of the earth in response to changing ice and water loads, has been found to explain better much of the field data. Holocene sea level appears to have behaved differently in different places because of the interplay of eustatic, isostatic and local tectonic factors (see Chapter 2). Relative sea-level curves, in turn, provide constraints on the evolving geophysical theories of isostasy, as well as setting the geomorphological boundary conditions for an individual section of coast.

The controversy about Holocene sea level also provides examples of interpretations that were constrained by preconceptions. Although a highstand of sea level had been proposed on Pacific islands by several geologists (David and Sweet, 1904; Wentworth, 1931; Daly, 1934), acceptance of a Pacific mid-Holocene highstand was slow amongst those conditioned by studies on coasts where the sea has not exceeded its present level (Russell, 1967). For instance, the CARMARSEL expedition to the Caroline and Marshall Islands in the central Pacific in 1967 failed to resolve the question of the pattern of sea-level change. Different researchers interpreted field evidence to support each of the alternative models: (i) that the sea had been higher, (ii) that there had been a period of greater storminess, and (iii) that there had been sea-level rise decelerating towards present (Shepard *et al.*, 1967; Bloom, 1970; Newell and Bloom, 1970). A mid-Holocene highstand is now recognised to have occurred widely through the Australasian and Pacific Ocean regions (Hopley, 1987; Pirazzoli, 1996; Grossman *et al.*, 1998; Dickinson, 2001).

A similar controversy developed in Australia in the 1970s. In New South Wales, sea level appeared to have remained close to its present level for the past 6000 years, whereas in Queensland there was evidence for a highstand (Thom *et al.*, 1972; Gill and Hopley, 1972). In either case, the sea-level curve relative to Australia was substantially different from that generally accepted in Europe or North America (Thom and Chappell, 1975). As a result of an international, multidisciplinary expedition in 1973 it became widely recognised that the sea had been higher than present on the Great Barrier Reef (Stoddart *et al.*, 1978; McLean, 1984). The pattern of relative sea-level change varies around the coast of Queensland (Chappell *et al.*, 1982), and this variation has provided constraints on geophysical modelling (Lambeck and Nakada, 1990). The details of sea-level variation along the New South Wales coast are less completely preserved; evidence for a highstand has been reported from several locations and the details of Holocene sea-level change remain contentious (Young *et al.*, 1993; Baker and Haworth, 2000).

9.1.2 Types of models

A model is a framework within which relationships between variables can be represented. Models vary from simple conceptualisations to complex mathematical formulations. Several levels of refinement and explanation can be envisaged, including descriptive, empirical and theoretical models (Wright and Thom, 1977; Hardisty, 1990). Descriptive models are based on association of observations. For example, the recognition that flat beaches are associated with large waves, or the relationship of beach planform to swell direction (Davies, 1958a). The description usually identifies a valid relationship, but the explanatory and predictive power of such an observation or model is generally poor. Empirical models are based on field or laboratory observations; for instance, measurements of several beaches demonstrated a relationship between sediment size and incident wave energy (Bascom, 1951; King, 1972). In this case, predictive capability is increased but explanatory power remains poor (Harrison, 1970). Theoretical models are developed on the basis of theory. For instance, Inman and Bagnold (1963) predicted that there should be a relationship between beach gradient and the rate of percolation of water into the beachface on theoretical grounds, and then set out to demonstrate or test this relationship. These models have greater explanation and predictive capability (Hardisty, 1990).

Models have been classified on the basis of various criteria. In the context of coastal geomorphology, it is useful to distinguish physical models that have played an important role in enabling experimentation, conceptual models that have provided a framework within which further research can be designed and executed, and computational models that have been developed from conceptual models using mathematics, statistics or computer simulation. Each of these will be examined below.

Physical models

Physical models, often called analogue or hardware models, are scaled representations of a system. There has been a tradition of study of coastal processes using wave tanks or other physical systems (Bagnold, 1946; Schwartz, 1965). These models offer several advantages. Some processes that are difficult to measure in the field can be represented in such a system. For instance, the effect of wave impact on cliffs can be monitored under laboratory conditions (Sunamura, 1992). It is possible to study the effects of waves on beaches using a wave tank, overcoming field complications such as the inability to see below the sediment surface, the natural irregularity of wave conditions and the simultaneous variation of the tide (Bagnold, 1940). However, scaling issues are a major limitation on these physical models (Kamphuis, 2000), because relationships between sediment size and the size of waves that can be generated in the laboratory are non-linear (see Chapter 3, Figure 3.1).

Conceptual models

Many of the models of the way in which coastal systems evolve that are described in this book are conceptual. In the case of the structure and development of coral reefs, Darwin deduced his subsidence theory of reef evolution (see Figure 1.3). The theory has been criticised (e.g. Ross, 1855; Murray, 1889); in particular, recognition of the significance of sea-level fluctuations has meant refinement of Darwin's ideas (see Chapter 5). However, Darwin's model still provides a framework for understanding the structure of mid-oceanic reefs (Guilcher, 1988).

The geographical cycle, proposed by W.M. Davis (1909), and extended to coasts by Johnson (1919), provided a conceptual framework within which youthful and mature stages of coastal development could be recognised. These cyclical, or life-history, models of coastal landform development continue to provide a simple but powerful template for understanding coasts, and underpin the concept of space–time substitution (see Chapter 1). The refinement of radiometric dating techniques has assisted the testing of these evolutionary models, and enabled interpolation between the geological time scale and the scale of process studies to address issues relevant to decadal to millennial time scales. Models of infill of estuaries (for instance, the model of Roy, 1984; see Figures 1.7 and 7.20) and distributary progradation and switching in deltas (see Figure 7.8) are examples developed on the basis of a long-term overview of landforms on a geological time scale, but incorporating understanding of the rate at which modern processes operate. It has been possible in some cases to incorporate these conceptual models into computer-based simulation models, such as the shoreline translation model of barrier response to sea-level change which is discussed below (Cowell and Thom, 1994; Cowell *et al.*, 1995).

Computational models

Computational models include mathematical models, based on equations, logical models, based on heuristic rules, and computer simulations. A simulation model is a codification of a system, enabling manipulative experiments to identify the sensitivity of particular variables to changes in other variables in the system. Computer simulation models offer a particularly powerful tool to examine the function of coastal systems that are too complex to be evaluated empirically (Lakhan, 1989). They should be based on clear specification of the question and the objectives of modelling. Often theoretical probability distributions based on limited data are used to generate outcomes with which further field data can be compared. The sensitivity of individual variables (or combinations of variables) can be examined (sensitivity analysis) by holding other variables constant and then undertaking a series of computer iterations under different boundary conditions, such as wave height or rate of sea-level rise. Such experiments can rarely be done effectively in the field, or in physical models such as wave tanks, and provide insight into which of a series of factors may be most significant in relation to observed landform variability (Lakhan and Trenhaile, 1989b).

An early example of computer simulation involved the evolution of Hurst Castle Spit in southern England, demonstrating sensitivity to boundary conditions such as wave conditions (King and McCullach, 1971). Subsequently, many other models have addressed the issue of coastal shape and evolution (e.g. Niedoroda *et al.*, 1995). Inverse modelling or 'reverse engineering' involves running a behaviour-oriented model through a period of Quaternary time for which morphological, stratigraphical or geochronological control provides data against which to evaluate model results (de Vriend *et al.*, 1993a; Cowell and Thom, 1994). Holocene coastal stratigraphy provides independent outcomes that can be compared with model simulations (Cowell *et al.*, 1995; Raper, 2000).

It is important to emphasise that modelling can only produce probable scenarios, it cannot predict future outcomes; stochastic factors and non-linearities within coastal systems prevent such exact extrapolation (Oreskes *et al.*, 1994). Despite considerable advances in computing it is still the case that models cannot be as complex as necessary to represent reality effectively. They must be simplified to reduce computational time (using time steps instead of allowing parameters to vary continuously through time) or volume (aggregating data into classes instead of using the total range of variables measured in the field). Static models do not incorporate time. For instance, short-term sediment transport models can be based on hydrodynamic characteristics of coastal systems; however, longer-term dynamic morphological models need to incorporate morphodynamic feedbacks (van Rijn, 1993). Dynamic models treat time as continuous or, more often, iterate outcomes using discrete time steps (de Vriend *et al.*, 1993b), which means that long-term model predictions may be unrealistic, despite the fact that they are based on valid relationships at shorter hydrodynamic time scales (Haff, 1996). Lumped parameter models aggregate data and generalise spatial variability, whereas distributed parameter models allow variables to be assigned different values in space. Deterministic models contain a progression of states that is clearly defined, whereas stochastic models are based on probabilities (Lakhan and Trenhaile, 1989b; Lee *et al.*, 2001).

Nevertheless, computer simulation of the range of potential behaviour provides a powerful means of examining coastal systems and understanding sequences of landforms (Ashton *et al.*, 2001). The nature of a coastal system, and the interrelationships between processes and landforms are examined in further detail in the following section, building on concepts introduced in Chapter 1 and using examples from the subsequent individual chapters.

9.2 Coastal systems

The complexity of coastal systems makes complete descriptions difficult and accurate predictions improbable. As emphasised in Chapter 1 and illustrated with examples throughout this book, the coast varies in both time and space. This provides a constraint on what it is possible to study, but also offers the opportunity to compare and contrast one coast with other examples drawn from another time and another place.

The behaviour of a coastal system in relation to boundary conditions can be examined in terms of interrelationships between subsystems, portrayed as a 'black box', 'grey box' or 'white box' depending on the degree to which it is broken down into its component variables (see Figure 1.11). Reducing a system down to a series of components to study the interrelationships between them is called reductionism. It is increasingly realised that although this can provide much explanation, it rarely allows total understanding. For many systems, the 'whole is more than the sum of its parts' and the system needs to be viewed 'holistically', as a whole.

The coast is a complex non-linear system. Non-linear responses

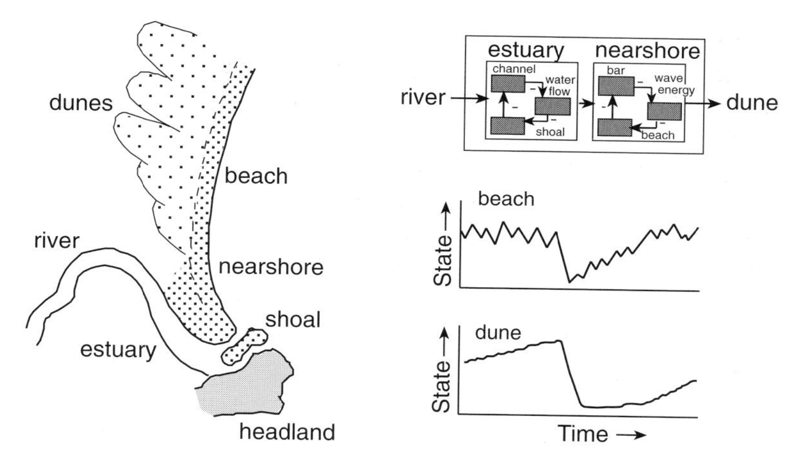

Figure 9.2. Subsystems within a coastal system can be interrelated but operate at different time scales in response to forcing events. In this example a schematic coastal embayment, like that shown in Figure 1.11, is shown. The beach and dune are likely to respond differently. They may be closely coupled in response to an extreme event, such as a storm that erodes both the beach and the dune. However, at other times they appear to be decoupled and operate in response to different factors; the beach responds to incident wave energy, whereas the dune responds to wind winnowing of the subaerial beach.

result from complex interactions and feedbacks, and the existence of thresholds in the system. Reductionist approaches to science often represent linkages between variables as simplified linear relationships, and linear thinking is especially entrenched in the approach that policy makers adopt towards coastal management.

The concept of feedbacks was introduced in Chapter 1 (see Figure 1.11), and it is explored further in relation to the beach morphology in Figure 9.2. A beach changes subtly with the passage of each wave, but remains more or less constant in gross form over time adopting an equilibrium adjusted to 'average' process conditions (see Chapter 6, and the quotation by Gilbert (1885) at the beginning of this chapter). An associated dune also changes gradually, but at a different rate and in response primarily to wind rather than wave conditions. However, coupling and interdependence between the morphological responses of each can become apparent when a major storm erodes both (Ritchie and Penland, 1990; McLachlan, 1990). Constancy or recurrence of form often implies equilibrium; the concept of equilibrium, based on thresholds and feedbacks, is examined below.

9.2.1 Feedback and thresholds

In Chapter 1, the processes of feedback were discussed, and there have been numerous examples in the descriptions of individual coastal systems where negative feedback has been shown to limit the extent to which coastal morphology changes. In other instances, positive feedback operates to accelerate change from one state to another. Distinctions between particular states, or coastal landforms, are often marked by thresholds in processes operation.

Thresholds mark major changes in system response to a variable reaching a critical level. For instance, when a critical shear stress is exceeded sediment is eroded from the floor of a tidal channel, and when waves are large enough to exceed a threshold velocity sediment of a particular size is entrained and carried in suspension (Bagnold, 1941). The nature of the sediment response to the acceleration or deceleration of process operation can differ. The velocities at which sediment will settle to the bed, and the velocities at which it will be eroded from the bed, generally differ (see Chapter 3, Figure 3.1). For instance, in a tidal channel, tidal current accelerates during the flood tide, ceases altogether at high (slack) tide, and accelerates again on the outgoing ebb tide (see Chapter 8, Figure 8.8). Such asymmetries are termed hysteresis and result in complex net sediment fluxes. The role of flocculation of fine grains and the effect of micro-organisms in binding the sediment further complicate sediment behaviour. Similarly, on a rocky coast there can be abrupt thresholds in behaviour as the tide rises. For instance, the low-tide scarp of a shore platform causes wave energy to be reflected at low tide, but at a higher stage of the tide waves may break over the platform, rushing across it as a bore and dissipating energy (Carter *et al.*, 1987). Under these circumstances there are abrupt thresholds (jumps) in process operation, and relationships between variables are inherently non-linear (Phillips, 1992).

Whereas a coastal system can change behaviour in response to alteration in some boundary conditions external to the system, changes are also triggered by intrinsic factors responding to thresholds that are internal to the system (Chappell, 1983b). These intrinsic or self-organised thresholds, also termed self-organised criticality, cannot be predicted on the basis of external stimuli. Self-organisation through positive feedback can be illustrated by the accumulation of a pile of sand grains. If new sand grains are continually added to the top of the pile, the pile will build up, at an angle at, or close to, the angle of repose, until internally determined thresholds are reached at which small avalanches of grains cascade down the face (Bak *et al.*, 1988). Intrinsic changes in coastal systems are considered further in Section 9.3.2.

Figure 9.2 illustrates changes to the exposed beach from accreted to eroded morphology as a result of a storm event (an external factor, exceeding a threshold); however, its return from eroded to accreted form occurs as internal thresholds are exceeded. For instance, it may be necessary after a major erosional event for the beach to accrete to a particular extent before sand becomes available for wind transport from the beach into the dune, demonstrating how, in this case, the dune is dependent on a threshold condition in beach form. Such internal thresholds will be particularly important to recognise, because a system that has changed state as a part of a natural morphodynamic adjustment may be very hard to differentiate from one that has changed as a result of an external factor, such as sea-level change or human impact.

9.2.2 Equilibrium

Coastal landforms tend to adjust towards, or oscillate around, equilibrium, particularly by negative feedback between variables. The concept of morphological equilibrium is a very useful generalisation, because it provides insight into future changes, but rarely is it possible to identify exactly how equilibrium is achieved or what variables to measure. Particularly significant variables can be determined through inductive generalisation from field-based observations, and these are termed emergent variables (Baker, 1996; Rhoads and Thorn, 1996). Apparent equilibrium morphology has been discussed in Chapter 4 in relation to rocky coasts, in Chapter 6 in relation to sandy coasts and in Chapter 8 in relation to muddy coasts. In some cases there appears to be a simple equilibrium that changes little over time; in other cases the coast appears to adopt a dynamic equilibrium (Scheidegger, 1994). These concepts are examined in Figure 9.3 and discussed below.

Simple equilibrium

The terms static, stable and steady-state have all been used to describe a simple equilibrium. They imply subtle differences, although this has not always been apparent from the way that the terms have been used. A static equilibrium is one that does not change over time; a stable equilibrium is one where processes are balanced such that there is no net effect on the landform, and a steady-state equilibrium is one in which there are no changes in boundary conditions.

Figure 9.3a examines the concept of equilibrium in relation to a simple coastal example, pebbles in rock pools on a shore platform, extending the analogy of balls on a pinball table used by Hardisty (1990). A pebble comes to rest (reaches equilibrium) at the series of points indicated in Figure 9.3a. Pebbles in rock pools are in stable equilibrium; a minor external stimulus (a small wave – low excitation, equivalent to jerking the pinball table) will not disturb them from equilibrium (a point attractor in the language of chaos theory). An unstable equilibrium (a point repellor) exists for the individual pebble that has come to

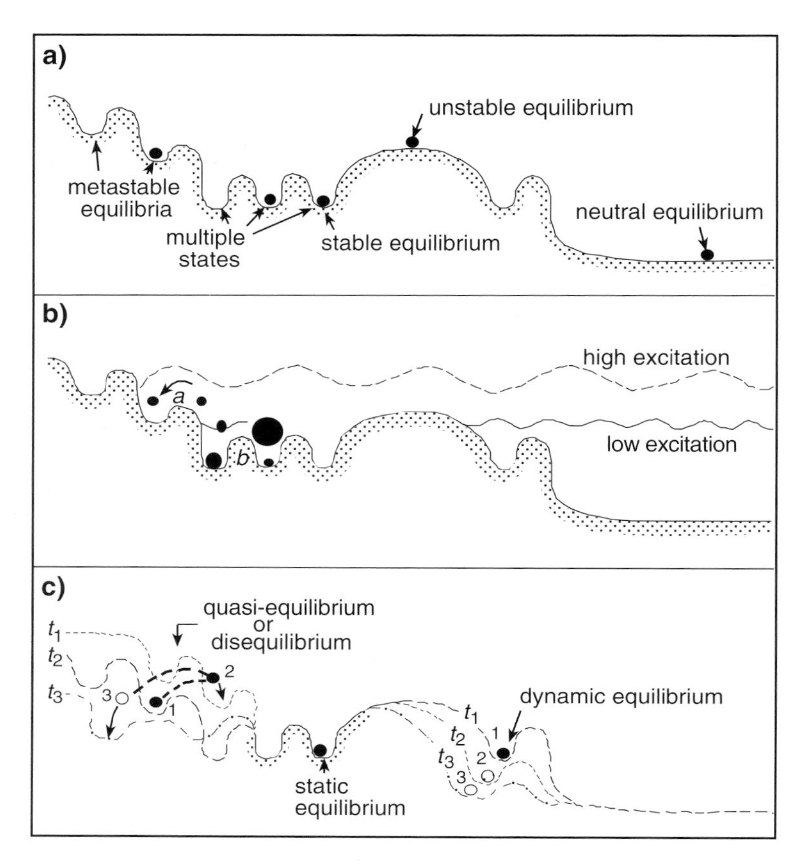

Figure 9.3. Schematic representation of the concept of equilibrium. The system may be thought of as a series of rock pools within which pebbles adjust to equilibrium, or as a pinball table (see text for discussion).

rest on the crest between pools. It is in equilibrium, but only minor agitation (a small wave, jerking the table) will cause it to move from the equilibrium point, and to accelerate as a result of positive feedback (the further it moves, the steeper the slope and the more it accelerates). An example of an unstable equilibrium was discussed in relation to the null-point hypothesis of Cornaglia (see Chapter 6, Figure 6.7a), where sand grains on a linear shoreface might reach a point at which the wave force moving them landwards equals the gravitational force moving them seawards. However, if displaced from the equilibrium point they would accelerate away from it. A neutral equilibrium exists for those pebbles removed seawards from the system (the bottom of the pinball table which represents out of the game).

Figure 9.3b takes the analogy further. In this case, the system is subject to disturbance by the passage of a wave across the shore platform. A small wave represents low excitation, a larger wave represents high excitation. Metastable equilibria are stable states but at different levels of energy input. In Figure 9.3b pebbles in rock pools at point b are in stable equilibrium, and the different rock pools represent multiple stable states each of which has a similar probability of adoption. Low excitation can shift pebbles from one of the multiple stable states to another; however, it requires a higher level of energy input to nudge the pebble up to a metastable equilibrium state. One might envisage, as shown at point b, that different pebbles will have different responses, if any, to particular levels of disturbance. One wave may agitate some pebbles, but it might have no effect on others because they are too big (insensitive) or trapped in the rock pool in which they were last deposited by larger pebbles above them.

A steady-state equilibrium is possible where the boundary conditions (independent variables) which operate external to the particular system do not change over the period of study. A period of unchanging boundary conditions is termed stationarity. It is often useful to presume a steady-state in which boundary conditions remain unchanged to enable modelling in which interactions between components of the system can be simulated. In the case of the river–estuary–beach–dune system shown in Figure 9.2, the input of sediment from the river might be considered statistically stationary through time for modelling purposes. In practice, this stationary long-term pattern will contain variability on the scale of individual flood events, but average conditions might fluctuate around a long-term mean, as long as the climate, and hence sediment supply, is not changing. If human factors influence the catchment, for instance changing land use, then the riverine input boundary condition is no longer stationary, but will eventually demonstrate a trend to which the system is likely to respond.

Boundary conditions are generally not unchanging and systems often show non-stationarity, as demonstrated from examples in preceding chapters. For instance, it is clear that climate has changed over past millennia, and could presently be changing still more rapidly as a result of anthropogenic factors. Sea level is a major controlling factor that has changed at a range of time scales, and that seems likely to change in largely unpredictable ways in the future (Pirazzoli, 1996). However, it appears that sea level has remained relatively constant close to present level in the far field (away from major ice sheets) during the past 6000 years (see Figure 2.11). This has enabled evaluation of the significance of other factors, for example in the infill of estuarine systems in both eastern and northern Australia. The sea-level boundary condition has been relatively constant (stationary), in contrast to other coastlines that have experienced a range of relative sea-level changes and on which sea level may have been the primary boundary condition driving coastal geomorphology. However, stationarity is generally unlikely over long time scales.

Dynamic equilibrium

Dynamic equilibrium is more complex and involves an equilibrium morphology that changes over time. This occurs either because there are interactions within the system that constantly rework it, or because there is a long-term change in boundary conditions to which the system adjusts. The analogy in Figure 9.3c consists of a mobile substrate such as rippled sand in which the troughs (in which a coarse grain or pebble can reach equilibrium, equivalent to the rock pools in the earlier analogy) adjust through time. It is still possible for the pebble to occupy a dynamic equilibrium, adjusting with the lowest point in the ripple field. In Figure 9.3c a static equilibrium, equivalent to the stable equilibrium previously defined, remains where the system is not changing.

Coastal morphodynamics involves co-adjustment of form and process. Whereas process adjusts instantly to form (the dynamics of fluid motion adjusts immediately to any changes in substrate configuration), it takes a finite time for the form of the substrate to adjust to the processes operating. This is because adjustment of substrate requires the movement of sediment. Further complexity results from the variability of process operation in nature (for instance the irregularities of incident wave conditions) and the non-stationarity in boundary conditions. It is more realistic, therefore, to view equilibrium as a state that the morphology strives to attain. For instance, the pebbles shown to the left of Figure 9.3c are moving towards an equilibrium but morphology may change before they get there. Equilibrium requires time to be attained; it is a moving target to which the coast is continually striving to adjust, often approached but never quite reached (Wright and Thom, 1977). Coastal systems are, therefore, often in a state of quasi-equilibrium, a condition close to equilibrium, and the extent to which they fall short of true equilibrium can be described as disequilibrium stress (Wright, 1995).

Whether or not equilibrium is considered static or dynamic is partly a function of the time scale at which the system is examined (Schumm, 1991). This can be illustrated with reference to the beachface. The shape of a beach may appear static because each day at low tide it seems to have the same shape. However, beneath the swash and backwash sediment is moved up and down the beach. If the net volume of sand moved up by swash processes is equivalent to the net amount that returns down the beach through a combination of backwash and gravity, then the system is in dynamic equilibrium at that scale of enquiry. It is possible

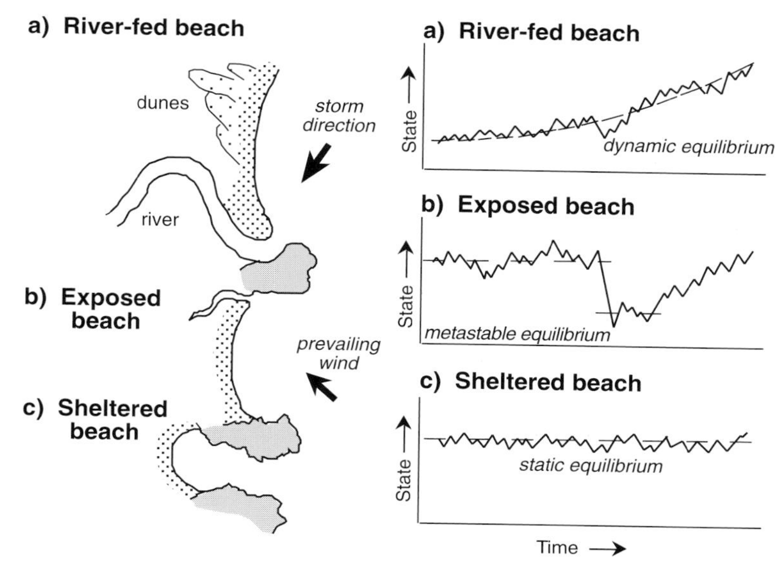

Figure 9.4. Three different beaches exhibiting three different types of equilibrium. (a) The river-fed beach receives new sediment from the river and it shows a long-term dynamic equilibrium with increase in volume. (b) The exposed beach varies from accreted to eroded, as the result of storm cut; these alternative states represent a metastable equilibrium. (c) The sheltered beach is not disturbed by major storms and does not receive new sediment; it is in a static equilibrium.

to address the question of whether or not there is an equilibrium at each of the scales from individual swash–backwash events, tidal cycles, lunar cycles, seasonal or annual scales, or longer-term scales of enquiry. It is important to consider different coastal landforms, or landform elements, at a range of time scales, and to distinguish modes of behaviour in response to perturbations and changes of boundary conditions.

Equilibrium also depends on the scale at which the system is examined, and energy regimes relative to thresholds of sediment entrainment. The different types of equilibrium are shown in relation to different beaches in Figure 9.4. The simplest is a sheltered pocket beach that appears to be in static equilibrium because the beach, measured by a suitable variable such as beach volume, varies little over time (Figure 9.4). It is too sheltered to change state in response to variations in wave energy or storms, and is not receiving any new sediment. An exposed beach shows a series of fluctuations in response to wave conditions, being eroded by major storms (an increase in energy input), as shown schematically in mid sequence with change to an eroded metastable equilibrium (Figure 9.4). The river-fed beach undergoes fluctuations, but its overall volume is subject to long-term increase as a result of the supply of sediment from the river. While storms may temporarily erode the beach, over time there is a dynamic equilibrium around which the beach volume fluctuates (Figure 9.4). In this situation, the increasing volume of sand might also support the development of dunes.

9.3 Geomorphological change over time

Geomorphology in the early 20th century adopted a view that time was progressive, orderly and irreversible, and coastal landforms were interpreted to undergo sequences from youthful to mature (Johnson, 1919). However, it is clear that time has no beginning and no end, and that many coastal landforms experience cycles or other trajectories of change (e.g. beach morphodynamic adjustments, delta distributary switching, coastal lagoon and barrier estuary inlet opening and closing). Sequences of landform development are also responsive to adjusting boundary conditions and internal adjustments or self-organisation. In this section, the role of convergence or divergence of landforms, and the significance of extreme events and other causes of disruption are considered.

Morphodynamic studies examine coasts at a range of temporal and spatial scales. At the 'instantaneous' scale, a series of deterministic linear relationships are generally used to describe what are commonly non-linear interactions (see Figure 1.9). The 'event' scale, typically of months to years, includes variability, for instance the seasonal patterns of change on a beach, or the response of a coast to, and recovery from, a storm event. This is the scale at which process studies have been concentrated, but non-linearities in coastal behaviour mean that it is not possible to reach a truly predictive capability (Cowell and Thom, 1994). The 'geological' time scale, at the other extreme, is the scale over which Quaternary sea-level and climate fluctuations influence coastal development, an incomplete record of which may be preserved in sedimentary sequences or erosional landforms.

A decadal to century time scale has been variously referred to as the 'historic', 'engineering', 'societal' or 'planning' time scale because it is the scale over which we know from historic records that there have been changes of special significance in terms of planning or engineering projects. It is difficult to study because it falls between the scope of conventional disciplines (Hanson *et al.*, 2002a). This is the intermediate or 'mesoscale' time scale over which human society affects and is affected by the pattern of global environmental change (Sherman, 1995). It involves significant global oscillations in climate and oceanographic factors, such as the El Niño–Southern Oscillation and the Pacific Decadal Oscillation. Such oscillations can influence coastal processes and behaviour; for example, causing erosion and subsequent recovery of beaches in eastern Australia (Bryant, 1983; Thom and Hall, 1991; Short *et al.*, 1995), the recession of cliffs in California (Storlazzi and Griggs, 2000), and the build up and reshaping of islands in the central Pacific (Solomon and Forbes, 1999). It is becoming clear that human society has had, and is having, an impact on global climate that is apparent over this time scale which is discussed in the next chapter.

Although it is convenient to discriminate these four time scales (Figure 1.9), the range which any one of these spans can vary considerably for different coastal types, as examined in the preceding chapters. For instance, the event scale covers storms and erosional events that recur at frequencies of days to decades in the case of beaches, but which can extend to centuries or longer in the case of cliffs. The degree of explanation or prediction in morphodynamic systems becomes less reliable as time and space scales expand (Wright and Thom, 1977). It is usually only practicable to make observations of coastal systems over a short period of time; individual studies rarely exceed 3–5 years ('thesis' time), and this is often inadequate to understand the longer-term behaviour of the system or to appreciate the significance of low-frequency, high-magnitude events (Bartholdy and Aagaard, 2001). Extrapolation of these short-term studies to longer time scales depends on adequate historical data where surveys or sequences of aerial photography are available or, in a few instances, on suitable surrogate or proxy paleoecological data. Longer-term extrapolation has been particularly successful where data sets based on monitoring for several decades are available, for example, Moruya or Narrabeen beaches in New South Wales (see Figure 6.18), and Nags Head and La Jolla beaches in the United States (Larson *et al.*, 2001).

Understanding geographical variation in coastal landforms provides considerable insights into what processes are most significant in terms of coastal evolution (Davies, 1980). Location (space) is a unique attribute, but particular sections of the coast are often at different stages of evolution. The concept of time–space substitution has been an especially valuable tool in coastal geomorphology, and was demonstrated well by Johnson's study of the New England–Acadian coastline (Johnson, 1925). More detailed examination of the wave-dominated coast of Nova Scotia indicates that coastal systems become more organised through time as mixed sediments eroded from drumlins are sorted by transport processes and gravel barriers adjust to wave climate and sea level (Orford *et al.*, 1991; Forbes *et al.*, 1995).

Plants and animals can play an important geomorphological role in coastal morphodynamic systems, and the mutual interaction of geomorphological and ecological dynamics is called biogeomorphology

(Viles, 1988; Spencer, 1988a, 1992). Plants play roles in binding and stabilising sediments; the root systems of dune grasses increase the threshold velocity required for transport of sand on dunes (Godfrey *et al.*, 1979) and seagrasses and algal or diatom mats stabilise subtidal sediments (Scoffin, 1970; Decho, 2000). Biogeomorphological links between geomorphology and biota are even clearer on reefs (Chapter 5), which can adopt one of several multiple states dominated by corals or coralline algae (Knowlton, 1992). Complex non-linear biotic responses can affect geomorphological change (Phillips, 1992); for example, the foraging activity of lesser snow geese in La Pérouse Bay, Manitoba, Canada, has had an unpredictable impact on the salt marsh (Packham and Willis, 1997).

Species zonation is a prominent feature of some intertidal salt marshes and mangrove shorelines. This can represent a temporal succession of communities where the coast is undergoing active geomorphological change. Actively prograding dunes or beach ridges often consist of sequences of ridges of progressively increasing age. In these cases, the more landward communities contain mature vegetation communities, reflecting not only the more benign environmental conditions further from the shoreline but also the fact that these substrates are older and the community has had longer to adjust. Similarly, halophytic communities may be undergoing continual sediment accretion and contain a succession of communities through time, but in rare instances plant communities can maintain dynamic equilibrium with an eroding substrate (Semeniuk, 1980b). In other cases, the zonation may be static, indicating variable species response along an environmental gradient (Lugo, 1980). Species associations can be an indicator of subtle changes in environmental or microtopographic factors, and can be related to geomorphological units (Semeniuk, 1985a,b). The patterning of species can result from several different causes and is an example of convergence, which is discussed below.

9.3.1 Inheritance, convergence and polygenetic landforms

Coastal landforms often bear the imprint of past events and inheritance from the past constrains future outcomes. In numerous examples through the preceding chapters it has been shown that coastal evolution is time-bound and has followed a sequence of antecedent states and changes in boundary conditions. A change in any step in the sequence alters the final outcome. The modern coast is therefore dependent, or contingent, on the 'unerasable and determining signature of history' (Gould, 1989, p. 283). Time-bound systems are unpredictable and are not as well-behaved as the timeless laboratory experiments of chemists or physicists (Schumm, 1991). Whereas process studies, based on physical principles, tend to be highly repeatable, the geomorphological sequence of past coastal landforms demonstrates 'state dependence'; the morphology that they adopt is at least partly contingent on apparently unique past events.

At geological time scales many modern coastlines are influenced by inheritance, particularly as a result of repeated interglacial shorelines at levels close to those of the present sea level (Roy *et al.*, 1994). Coastal evolution at shorter time scales is also a cumulative process, and is a function of antecedent states in which the morphological outcomes of one phase of coastal evolution are amongst the physical constraints for the ensuing stage. The form of a beach, for instance, shaped as it would appear by the incident waves arriving at the shoreline, is partly a function of the form that it assumed under wave conditions during preceding periods. This state-dependence (expressed in the language of non-linear dynamical systems as sensitive dependence on initial conditions) together with the stochastic variability of forcing factors, means that the pattern of coastal evolution is essentially an historical accident, and is largely unpredictable, unrepeatable and irreversible (Cowell and Thom, 1994).

Convergence and divergence

When similar coastal landforms result from operation of different sets of processes or sequences of events, it is termed convergence, equifinality or multicausality. For example, a horizontal peat bed within muddy sediments in the stratigraphy of estuarine shores can result from at least two different depositional histories. This mud–peat–mud stratigraphy has been related to periodic seismic events, termed the earthquake deformation cycle, on the coast of Washington and Oregon in the United States (see Chapter 8). Similar stratigraphy relates to phases of marine and wetland conditions caused by sea-level fluctuations (see Figure 8.20) in Britain and northwestern Europe (Long and Shennan, 1994). A consequence of equifinality is uncertainty relating to any inferences that are made about antecedent conditions (Phillips, 1996).

Different evolutionary sequences could explain similar coastal landforms. For instance, beach cusps could be generated by edge waves in some circumstances, but might originate through self-organisation in others; beach ridges might represent relict foredunes in one case, but be formed by wave action in others (see Chapter 6). A sequence of two emergent shorelines could result from one of several scenarios, such as stable sea level but periodic coseismic uplift of the coast, punctuated sea-level fall, or sea-level fluctuation superimposed on a gradual uplifting trend (Pirazzoli, 1996). Figure 9.5 shows an example from Cayman Brac in the West Indies. The prominent notch indicates a sea level 6–7 m above present, whereas the reef limestone comprising the Ironshore Formation implies a sea level around 2–4 m above present, and U-series dates indicate that this was associated with the peak of the Last Interglacial (Woodroffe *et al.*, 1983). There appears to have been a sequence of sea-level highstands that have been recorded in the Cayman Islands, and it remains the case that sea-level indicators could be related to several combinations of highstands (Vézina *et al.*, 1999), or to fluctuations within a stand such as the Last Interglacial which have been identified elsewhere (Neumann and Hearty, 1996; Blanchon and Eisenhauer, 2001; Johnson, 2001).

In other circumstances, landforms can diverge and coastal systems that have undergone similar events can end up in different states. For example, two coasts might have experienced similar eustatic sea-level histories, but the relative sea-level curve could have been different because of isostatic or tectonic factors. Examples in the preceding chapters indicate that the response of the coast to a change in sea level is not simple, but reflects a series of parameters such as the rate of sea-level rise, the volume of sediment available, and the accommodation space available. Modelling is a powerful way to approach the issue where observational data are incomplete. However, it is often the case that

Figure 9.5. Fossil shorelines on Cayman Brac, Cayman Islands, West Indies. The prominent notch in the foreground is cut into Bluff Limestone of Tertiary age. The fossil reef in the background comprises the Ironshore Formation from which corals have been dated to Last Interglacial age. The evidence could be consistent with several different sequences of events (convergence) and would require further age control on the time of formation of the notch to discriminate (photograph D.R. Stoddart).

several model formulations are possible, and these do not yield unique results. This is true of complex geophysical models; for instance parameters in either ice- or earth-model can be altered to match simulations and field evidence of past sea levels in northeastern Britain (Shennan *et al.*, 2000a). Models lose their power when the solutions they produce are not unique. Non-uniqueness often precludes the coastal scientist from reconstructing or forecasting a systems response with certainty.

9.3.2 Response to changes over time

Coastal landforms respond to a series of changes, including regular or cyclical adjustments to swash–backwash, tidal and seasonal factors (Aubrey, 1979; Clarke and Eliot, 1988); responses to short-term disturbances, termed perturbations; changes in boundary conditions; and internal, or intrinsic, changes. Responses to several forcing factors are likely to be complex and can occur simultaneously with the stimuli, or lag behind. These various forcing factors and response times (comprising reaction and relaxation time) in relation to recurrence interval are examined in Figure 9.6, and are discussed below.

Perturbations

A perturbation in some independent factor, such as an extreme climatic event, can have an effect on a coastal system (Figure 9.6a). The response often varies between systems and can be an immediate, instantaneous reaction, or can be delayed (lagged) because a finite time is required for sediment entrainment, movement and deposition. There is also likely to be a longer-term relaxation time (also called recovery time) back towards any pre-existing equilibrium (or quasi-equilibrium) condition.

Where a major storm affects a beach it can be quickly cut back into its erosional form during the storm when wave energy is great enough to move large volumes of sand offshore (see Figure 9.2). It will take considerably longer for the beach to regain its former volume, with the gradual return of sediment from nearshore onto the beach during normal conditions. Relaxation represents the progressive decay of geomorphological evidence of an event; in some cases, more moderate events are not competent to shape landforms or to move coarse material, such as storm blocks, and these do not decay, but are preserved. For example, the fact that the larger waves associated with a storm have a deeper wave base may mean that sediment is mobilised in deeper water during the storm than can be reactivated during normal conditions. There may be a time lag in reaction; for instance coral reefs take time to respond to a rise in sea level, and reef growth lags behind or catches up with sea level (see Chapter 5).

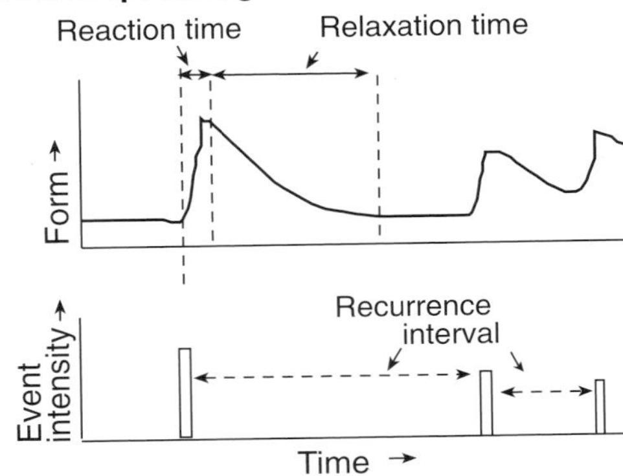

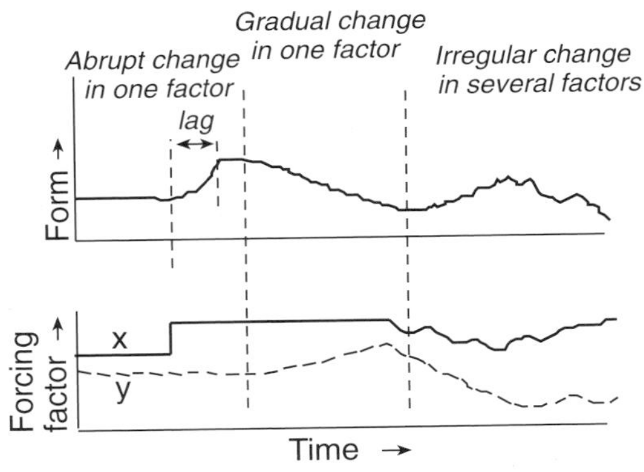

Figure 9.6. Change in coastal systems. (a) Response to a perturbation. The sequencing of events is shown in terms of event intensity. Response of coastal form to each event is a function of reaction and relaxation time, and the relationship between this response and recurrence interval. (b) Response to a change of boundary conditions. Response may be lagged or gradual, and where several forcing factors are involved change is likely to be irregular.

Whether the system recovers completely from a short-term disturbance depends on the temporal scale of the perturbation and the system's response in relation to changes in boundary conditions. If external (boundary-condition) changes occur more rapidly than the response time (particularly relaxation/recovery time), then re-equilibration may not be achieved (Phillips, 1995). The system is described as intransitive (or pulsed) where the landform returns to its pre-disturbance state, or transitive (ramped) where there is a change in boundary conditions and it does not return to its previous state (Brunsden and Thornes, 1979; Chappell, 1983b). The sequence in which perturbations occur can be critical. Where two storms occur in succession, the second one can have quite different impacts from the first. This is shown schematically in Figure 9.6a where a landform responds to the first perturbation and has time to return to its pre-disturbance state, whereas the second perturbation is followed by a third perturbation before the system has regained its equilibrium state (see Section 5.4.3 for examples of the impact of successive storms on reefs). Resilience describes the degree to which a system recovers to its initial (pre-disturbance) state (Brunsden, 2001).

Boundary-condition changes

A change in boundary conditions can also cause a change in the state of a system, as long as the system can adjust before the next change in boundary conditions. Figure 9.6b shows schematic response of a landform to two boundary conditions; a lagged response occurs to a change in one boundary condition (or forcing factor), but a gradual change occurs in response to the other. If both boundary conditions change in an irregular manner, then the landform could follow a highly irregular, and largely unpredictable, trajectory. For instance, sea level has been shown to be a very important boundary condition for most coasts and the nature of the response to changes in sea level differs as shown for different types of coast in the preceding chapters. Gradual climate change can also lead to changes in wave climate or the rate of sediment supply. Complex changes to boundary conditions are likely which will operate at different time scales; sea-level adjustment is likely to be felt more rapidly than the altered sediment supply, and the combined effect may involve further feedbacks as internal thresholds are exceeded.

Intrinsic changes

Coastal systems are usually more complex than the simple scenarios in Figure 9.6. Different landforms adjust differently (or show no response) to particular perturbations or boundary conditions. Most systems are influenced by several boundary conditions and several types of irregular disturbances. Moreover, there may be intrinsic adjustments within the system itself. In some cases, a system is relatively resilient as a result of negative feedback. In other cases, changes persist and accumulate through positive feedback until the system crosses a threshold and adopts a new state.

Intrinsic changes are abrupt erosional or depositional adjustments that come about as a result of accumulated change without specific external stimuli (Schumm, 1979). For instance, continued sedimentation on intertidal mudflats can build up the substrate to the level at which it is no longer subject to inundation and, consequently, the dominant processes change independently of any external threshold. A feature of non-linear dynamical systems is that perturbations can lead to switches in mode of behaviour and processes without relaxation back to the original state.

When a change is observed in a coastal landform, it could be a response to a short-term perturbation, a change in boundary conditions, or an intrinsic change in the system. Dissociating the responses to different stimuli is complex. Coastal systems undergo changes on a range of scales, some of which are short term, others of which are lagged responses or reactions to intrinsic thresholds. It is important to determine whether a change in a coastal landform has been caused by external factors (such as a change in sea level, sediment supply or human impact) or intrinsic factors. It may be possible to discriminate responses to climatic change from the consequences of extreme events with due attention to the interaction between variables and their response to thresholds (Chappell, 1983a).

Understanding beaches and their recovery after disturbances such as major storms has improved as a result of several long-term monitoring studies (see Figure 6.18), on the basis of which patterns of behaviour can be determined (Thom and Hall, 1991; Morton *et al.*, 1994). Similarly, on reefs, the impacts of storms and recovery after a storm have been observed (see Chapter 5). It is harder to distinguish whether past changes have resulted from perturbations, boundary-condition changes, or are an intrinsic response of the system where evidence is based only on stratigraphical or paleoecological evidence. In these cases changes often have to be inferred rather than reconstructed from direct evidence, and interpretations can rarely be verified.

Discrimination between intrinsic thresholds and adjustments due to boundary conditions may be possible through a comparison of adjacent systems. A boundary condition change is likely to have occurred regionally and adjacent systems are likely to have reacted simultaneously. This is particularly the case for changes in sea level. An intrinsic threshold is more likely to be exceeded locally and at different times. For instance, it is frequently the case that coastlines have changed from transgressive systems to prograding systems during the Holocene as sea-level rise slowed. This can be seen clearly in the chronology of sedimentation in estuaries on the wave-dominated coast of southeastern Australia (see Figure 1.7). Estuaries changed from transgressive to mud basin systems simultaneously as sea level reached a level close to its present around 6000 years BP (Roy, 1984), a regional boundary-condition change. Each estuary has subsequently infilled progressively at an individual rate (see Chapter 7, Figure 7.20). At the final stage of infill, the river becomes fully channelised, as indicated in Figure 9.7. This represents an intrinsic threshold that will occur at different times for each system. It has

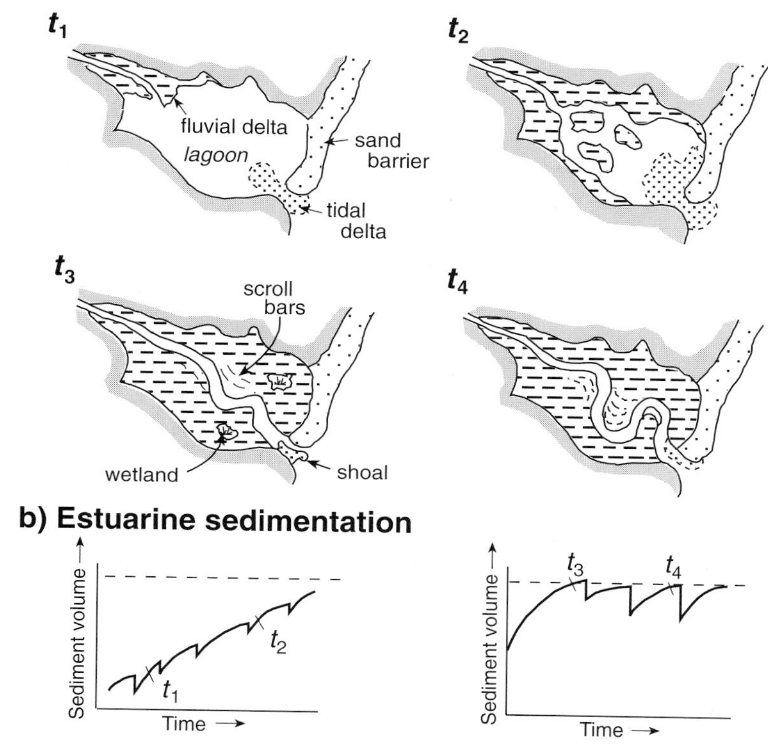

Figure 9.7. Conceptual model of stages of infill of estuarine system on a wave-dominated coast such as southern Australia or southern Africa. During the early phases (t_1 and t_2) the estuary infills progressively in response to fluvial delta progradation and barrier formation and tidal inlet dynamics. The system changes to river domination when the estuarine basin has totally infilled, which is an intrinsic threshold. In the later river-dominated phases (t_3 and t_4), the system ceases to accumulate more sediment, but is cyclical, characterised by channel mobility with sediment bypassing, and phases of erosion associated with large flood events (based on Cooper, 1993; Cooper *et al.*, 1999b).

occurred only on the largest systems, such as the Shoalhaven River (Umitsu *et al.*, 2001, see Figure 7.21), where sufficient sedimentation has occurred. Similar systems in southern Africa that have infilled completely are now river-dominated estuaries. These undergo channel meandering and migration, but sediment bypasses the estuary and is discharged into the sea (Cooper, 1993, 2001). Patterns of cyclical change may be associated with these infilled systems (Figure 9.7), with flood events eroding the channel followed by subsequent infill (Cooper *et al.*, 1999b; Roy *et al.*, 2001).

The change from transgression to stable sea level occurred simultaneously throughout Australia, representing an external boundary-condition threshold marking the beginning of a regressive phase during which the barriers prograded and estuarine systems infilled. In northern Australia, the transgressive phase was replaced by the 'big swamp' phase of widespread mangrove forest establishment as a result of stabilisation of sea level, a similar external boundary condition (Woodroffe *et al.*, 1993). However, the replacement of mangrove forests by freshwater wetlands when the intertidal accommodation space was filled occurred at different times on each system (see Figure 7.17), and the degree of river reworking of the floodplain responded to intrinsic thresholds (Chappell and Thom, 1986; Woodroffe *et al.*, 1993).

There are similar examples from the coast of Britain but the differentiation of the relative significance of boundary conditions and intrinsic factors is more difficult because the relative sea-level history is so variable around the coast (see Figure 2.12). For instance, comparison of stratigraphic evidence from estuaries in southern England suggests that they may have each undergone adjustments at similar times in response to regional patterns of sea-level and climate-change boundary conditions (Long *et al.*, 2000). Similarly, geomorphological studies in the Culbin Sands region of northeastern Scotland indicate that the rate of sea-level adjustment has been a prominent factor to which landforms are still adjusting with greater compartmentalisation since sea-level change has slowed in recent millennia, now affecting sediment supply to gravel spits (Hansom, 2001).

Studies of other estuarine systems have also suggested changes, for instance from ebb-dominated to flood-dominated or from tide-dominated to wave-dominated, which imply intrinsic thresholds in particular system behaviour (Pethick, 1996; Pendón *et al.*, 1998).

Hypotheses about intrinsic thresholds within the system require careful testing to establish their validity. An example in relation to changing conditions on reefs in the Holocene can be used to illustrate this. The presence of mid-Holocene coral rubble deposits and conglomerates on many reefs and tropical islands led to an innovative suggestion that the coast had been subject to more storms because the reefs had lagged behind sea-level rise during the final stages of the postglacial transgression (Neumann, 1972). The presence of storm deposits of mid-Holocene age could indicate (i) a greater frequency and intensity of storms at that time, (ii) sea level had been higher, or (iii) the reefs had not grown into shallow enough water to attenuate wave energy to the extent that they do now (high-energy window). Each of the hypotheses has received widespread support; for instance, in the case of the Great Barrier Reef of Australia, each has had its supporters (see Section

5.5.2). The concept of an intrinsic threshold associated with a 'high-energy window' is consistent with evidence that reefs on the outer Great Barrier Reef lagged behind those on the mainland (Hopley, 1984). There is widespread evidence that the sea has been higher in this region (Chappell, 1983a), and that isostatic flexure has elevated evidence along the mainland coast (Chappell *et al.*, 1982). Radiometric dating of storm deposits offers an insight into the frequency of these extreme events, suggesting that this has not changed during mid–late Holocene (Hayne and Chappell, 2001). The role of extreme events is examined in the next section.

9.3.3 The role of extreme events

One of the significant features of coastal morphostratigraphic studies is that they can offer an insight into the magnitude and frequency of events that have occurred in the past (Zong and Tooley, 1999; Liu and Fearn, 1993, 2000). Storms play an important role in shaping many coasts (Morton *et al.*, 1995; Lee *et al.*, 1998). Deposition of gravel ridges by storms can leave a record of successive events that can be dated, for example radiocarbon dating of coral within ridges on Curacao Spit in the northern Great Barrier Reef implies that storm frequency has remained essentially unchanged over mid and late Holocene (Hayne and Chappell, 2001; Nott and Hayne, 2001). On muddy coasts, burial of coarse sediments deposited during high-energy events preserves a record of past storm surges or hurricanes (Ehlers *et al.*, 1993; Shinn *et al.*, 1993; Jelgersma *et al.*, 1995; Donnelly *et al.*, 2001a,b).

Other high-energy events, such as tsunami can also be recorded within the coastal stratigraphy. The sedimentary signatures of tsunami are known from those coasts that are susceptible to seismic activity, for example tsunami occur around much of the Mediterranean Sea whereas other extreme events are rare (Pirazzoli *et al.*, 1999; Mastronuzzi and Sansò, 2000; Dominey-Howes *et al.*, 2000). On higher-energy coasts, deposits have been examined from known tsunami, for instance in Japan (Minoura *et al.*, 1994), Flores in Indonesia (Minoura *et al.*, 1997), and at Sissano Lagoon on the north coast of Papua New Guinea where a localised tsunami occurred in 1998 (McSaveney *et al.*, 2000). Along the western coast of North America, earthquakes generate tsunami and distinctive sand sheets are deposited in estuarine marsh sediments (Clague *et al.*, 1994; Benson *et al.*, 1997). Stratigraphy and dating of sedimentary sequences enables the chronology of past events to be established on this coast (Atwater and Moore, 1992; Clague and Bobrowsky, 1994a,b; Williams and Hutchinson, 2000).

Recognition of anomalous sedimentary evidence can indicate that events have been experienced on coasts on which they are presently unknown, or can be used to infer 'mega-events', extreme events of a magnitude for which there appears to be no contemporary, or historical, analogues. A well-documented example relates to a distinctive micaceous sandy layer that occurs widely in the muddy sediments (termed carselands) of several estuaries in eastern Scotland. This was initially interpreted as a storm deposit (Smith *et al.*, 1985). However, its prominence within several estuarine systems exposed to the northeast, but its absence in western Scotland, has supported its interpretation as a deposit from a tsunami generated by the Storegga slides off the northwest coast of Norway around 7200 years BP (Dawson *et al.*, 1988; Long *et al.*, 1989a,b).

It is particularly important to establish signatures for high-magnitude events. Tsunami-deposited sediments appear to have several distinctive features, in some cases recording the swash and backwash of up to three waves in one event (Dawson, 1994, 2000). Careful comparisons suggest that it is possible to discriminate between sandy material deposited by the tsunami associated with the Storegga slide and material deposited at about the same time as a result of deceleration in the rate of transgression (Bondevik *et al.*, 1998). It also appears likely that tsunami deposits can be differentiated from storm surge sediments (Foster *et al.*, 1993; Nanayama *et al.*, 2000; Witter *et al.*, 2001). Earthquake-related sedimentation changes (distinguished on the basis of grain size) are different from anthropogenic-related sedimentation changes (which are chemically distinct) in New Zealand (Chagué-Goff *et al.*, 2000; Goff *et al.*, 2001).

Much of the impetus for recognising a role for mega-tsunami has come from the study of deposits inferred to have had a tsunami origin in the Hawaiian Islands (Moore and Moore, 1984). Coral-bearing gravels reported at elevations of up to 375 m on Lana'i (Moore and Moore, 1988), and similar deposits on neighbouring Molokai (Moore *et al.*, 1994a) have been considered to have been laid down as a result of these giant waves. However, detailed survey, stratigraphy, and U-series dating of these has indicated that they are highly bimodal in age, and it appears that they were formed by more normal processes operating during the penultimate and last interglacials (Rubin *et al.*, 2000).

Giant waves have been inferred by the presence of very large boulders on Pleistocene shorelines around the world (Paskoff, 1991; Jones and Hunter, 1992; Hearty, 1997; Perry, 2001). In southeastern Australia tsunami waves are considered responsible for movement of large boulders and bedrock sculpting (Young and Bryant, 1992; Bryant and Young, 1996). They are inferred to have been important landscape-shaping processes along much of this aseismic coastline (Young *et al.*, 1996; Bryant *et al.*, 1996; Bryant, 2001).

There will continue to be recognition of anomalous sediments and large boulders, and the suggestion that these have been deposited during unusually high-magnitude events deserves rigorous testing, although independent confirmation of the impact of tsunami is difficult to achieve where events have not been experienced during the historical period. One of the major justifications for undertaking paleoenvironmental reconstructions is that they enable insight into the past and hence an indication of the magnitude of events that might be possible in the future. It remains unclear to what extent large reef blocks on modern reefs may have been moved by hurricanes, or tsunami (Talandier and Bourrouilh-Le Jan, 1988). Large boulders scattered across reefs at Agari-Hen'na Cape on the eastern side of Miyako Island in the southern Ryukyu Islands (Figure 9.8) resulted from tsunami associated with seismic activity along this active plate margin and the emplacement of some can be related to specific events (Kawana and Nakata, 1994). However, cyclones can produce waves in excess of 20 m high (Earle, 1975), and they seem to have achieved more geomorphological work on isolated islands in deep water than tsunami which are generally of small

Figure 9.8. Boulders scattered across reefs at Agari-Hen'na Cape on the eastern side of Miyako Island, southern Ryukyu Islands. These have been attributed to tsunami (photograph T. Kawana).

amplitude in mid ocean. Calibrating the size of event that this geomorphological evidence indicates remains a challenge; hydrodynamic comparisons indicate that tsunami waves, as a result of their long wavelength, are more powerful (Nott, 1997). Nevertheless, large reef blocks, like that shown in Figure 5.15, occur on reefs that are associated with the passage of hurricanes, and are not found on low-energy reefs outside the cyclone belt. It seems likely that similar morphology can result from several causes (equifinality), and it will require further comparison to discriminate the cause of particular deposits.

9.4 Modelling coastal morphodynamics

Many of the basic issues that face coastal geomorphologists in the 21st century have much in common with those that faced geographers and geologists at the beginning of the 20th century. Cyclic concepts underlay the elegant conceptual models of coastal evolution proposed by Johnson (1919) but later developed and modified by other coastal geomorphologists such as Steers (1953), Guilcher (1958), Zenkovich (1967) and Russell (1967). The concept of equilibrium was clearly outlined by Gilbert (1885) in his interpretation of lake shoreline processes, summarised in the quotation at the beginning of this chapter. Coastal morphodynamics provides a framework that bridges the dichotomy between historical or evolutionary geomorphology and process geomorphology.

The analytical tools available to assist the coastal geomorphologist are increasingly sophisticated. Data loggers and telemetry enable time-series measurements of hydrodynamic forces, including current velocity fields and wave parameters (Wright and Thom, 1977). Flow meters have been improved and include impellors, acoustic doppler, acoustic and optical backscatterance sensors, heat-transfer and electromagnetic current meters (Wright, 1995). Quartz pressure sensors for water level, and digital sonar altimeters record seafloor erosion and accretion. This has proceeded more rapidly than measuring sediment transport and morphological change. Suspended sediment concentrations can be assessed using acoustic backscatterance or nephelometers. High-frequency and high-resolution time-series of flow structure and sediment concentration are possible using short-range acoustic doppler current profilers and laser doppler velocimeters, with laser diffraction systems giving an insight into sediment size (van Rijn, 1993; Wright, 1995).

Increasing sophistication in geophysical techniques such as seismic reflection profiling, digital side-scan sonography and sub-bottom profiling is now available while, on land, ground-penetrating radar offers further extension of drilling and seismic refraction (Bristow *et al.*, 2000;

Neal and Roberts, 2000). Amongst the major methodological advances during the 20th century the development of absolute dating techniques has provided a firmer insight into past rates of deposition (see Chapter 2). Cosmogenic isotope techniques offer the prospect of a similar perspective on patterns of erosion. Although this provides age control on models of coastal evolution, many of the models had been partially developed prior to the availability of suitable dating methods.

It is still relatively difficult to acquire both nearshore and hinterland topography for a coastal area at appropriate scales for local coastal issues (Wright and Bartlett, 1999). Depth data are often sparse in the shallow areas, which are generally of the greatest interest to the coastal scientist, but least accessible to the hydrographer (Raper, 2000). New spatial data techniques, such as computer-aided photogrammetry, enable a greater precision to the 3D morphological definition of coastal systems. For instance, the water-line method based on radar imagery acquired at different stages of the tide can be used to derive a digital terrain model (DTM) of intertidal areas (Mason *et al.*, 1999) and can form a basis for subsequent morphodynamic models (Wang *et al.*, 1995; Mason and Garg, 2001).

Geographical Information Systems (GIS), computer-based systems for inputting, storing, manipulating and outputting data that are geographically referenced, are being increasingly used as an advanced database system, and an effective means of displaying and visualising coastal information (Ricketts, 1992; O'Regan, 1996). Global positioning systems (GPS) offer a rapid way to map the gross morphology of coastal systems (Morton *et al.*, 1993; Dail *et al.*, 2000).

Remote sensing offers the advantage of rapid and economic coverage of large areas synoptically. It enables geomorphological mapping in shallow coastal areas, such as reefs, where traditional ship-based survey and research is hampered by poor access. For instance, Landsat MSS satellite imagery has been used to undertake shallow-water mapping over the Great Barrier Reef (Jupp *et al.*, 1985), and laser airborne depth sounder (LADS) enables depth determination and mapping in clear water around reef areas for a fraction of the cost of ship-based surveys (Abbot, 1997). Wave height data are increasingly available from satellite altimetry (Lawson *et al.*, 1994; Krogstad *et al.*, 1999). Airborne scanners, such as CASI, allow mapping of geomorphological reef features (e.g. Maritorena, 1996; Clark *et al.*, 1997; Borstad *et al.*, 1997; Mumby *et al.*, 1998). Mapping of salt marsh (e.g. Donoghue *et al.*, 1994; Ramsey and Laine, 1997) and mangrove (Everitt *et al.*, 1996; Long and Skewes, 1996) can be undertaken from suitable imagery. However, the subtle differentiation of those still poorly understood ecological and microtopographic factors which influence process operation and the subtleties of sediment movement in such systems are still beyond the scope of most remote sensing techniques.

Comparison of time-series of photographs or images enables detection of change (Thieler and Danforth, 1994). Large-scale geomorphological changes can be effectively determined in estuaries (Riddell and Fuller, 1995) and deltas (Stringer *et al.*, 1988; Vaughn *et al.*, 1996; Guo and Psuty, 1997) using remote sensing, depending on grid cell resolution. Changes to intertidal communities such as the fragmentation of salt marsh and other wetlands (Larson, 1995; Kastler and Wiberg, 1996; Miller *et al.*, 1996) can be very effectively determined.

Often, broad databases at very coarse resolution are available which allow generalisations about the extent of low-lying areas that may be subject to inundation by sea-level rise or other hazards. Attempts to extrapolate simple models of vulnerability using such data in combination with sparse societal data for areas of the coast remain simplistic and unrepresentative (Jelgersma *et al.*, 1993; Kay *et al.*, 1996). Nevertheless, incorporation of geomorphologically based models of process, form and materials, within more flexible GIS (e.g. dynamic modelling, geostatistics, fuzzy classifications, object-oriented data structures) offer great prospects for future development of more powerful coastal-zone GIS applications (Raper and Livingstone, 1995; Wilson and Burrough, 1999; Raper, 2000). Similarly, videography (Eleveld *et al.*, 2000) and animation techniques (G.M. Smith *et al.*, 2000) are becoming more widely used. Two examples that are examined in more detail below are the modelling of equilibrium shore profiles and multi-dimensional morphological models.

9.4.1 Equilibrium shore profiles

The concept of an 'equilibrium profile' on sandy shorefaces was examined in Chapter 6. It has been proposed by Dean (1977, 1991), but is still controversial because site-specific application of equilibrium morphology requires empirical tuning. There also appear to be many exceptions where the shoreface does not adopt an equilibrium shape because, although processes adjust instantaneously to form, the movement of sediment takes a finite time to adjust morphology (Riggs *et al.*, 1995). Although a 2D oversimplification of a 3D system, such equilibrium concepts are useful, particularly where engineers are required to reconstruct or nourish beaches. The concept of an equilibrium profile has been extended in relation to the 'Bruun rule' (discussed below) to infer beach shape under conditions of higher sea level.

Scientific and engineering approaches to the coast differ because whereas science strives for an ultimate truth, engineering design operates within the practicality of a particular level of safety. Baker (1994) has described these approaches in relation to flooding, but the ideas can be readily translated to the coast. 'Science uses prediction as a tool to test explanations built upon its temporary state of understanding. Engineering assumes that temporary state of understanding in proceeding directly to a useful prediction. The engineer must always be concerned that the basis in understanding is inadequate, while the scientist must always strive to elucidate the inadequacy of that understanding' (Baker, 1994, p. 147). Whereas engineering may adopt these generalisations as the most appropriate working hypotheses, science should continually seek to reassess the assumptions on which the principles are based.

Cause and effect relationships and the nature of any morphological equilibrium are dependent on the time scale of investigation or frequency of monitoring. For instance, short-term studies of beach erosion and recovery might indicate oscillation around a more-or-less constant average over months, but show longer-term trends. The pattern of cut and recovery on Moruya beach (see Chapter 6, Figure 6.18) provides an example (Thom and Hall, 1991). Surveys over several decades indicate changes, but it is not possible to determine whether beach volume remains constant and in a steady-state. However, viewed over millennia, there has been overall net build up, and any recurrent morphology would appear to represent a dynamic equilibrium. Elsewhere, other factors, such as human influence may have altered beach volumes and it may be harder to consider any kind of equilibrium (Morton, 1979).

The Bruun rule

A simple model, termed the Bruun rule, has been widely adopted to predict sandy beach response to sea-level rise. Bruun proposed a rule by which shorelines would re-equilibrate to changes in sea level based on conservation of mass, such that sea-level rise causes an upward and landward translation of the equilibrium profile (Figure 9.9), involving erosion of the shoreline and redeposition of sand in the nearshore (Bruun, 1962, 1988). The Bruun rule has been shown to apply in very controlled wave-tank experiments (Schwartz, 1965, 1968). It also appears to be appropriate for some shorelines (Dean, 1990, 1991; Daniels, 1992). However, application of the model to shorelines on Lake Michigan indicated a lag, implying that it might be most appropriate to represent average equilibration over a decadal scale (Hands, 1983). There have been many attempts to predict the amount of erosion under conditions of rising sea level using the Bruun rule, but the simplicity of the concept implies that the models do not represent effectively what

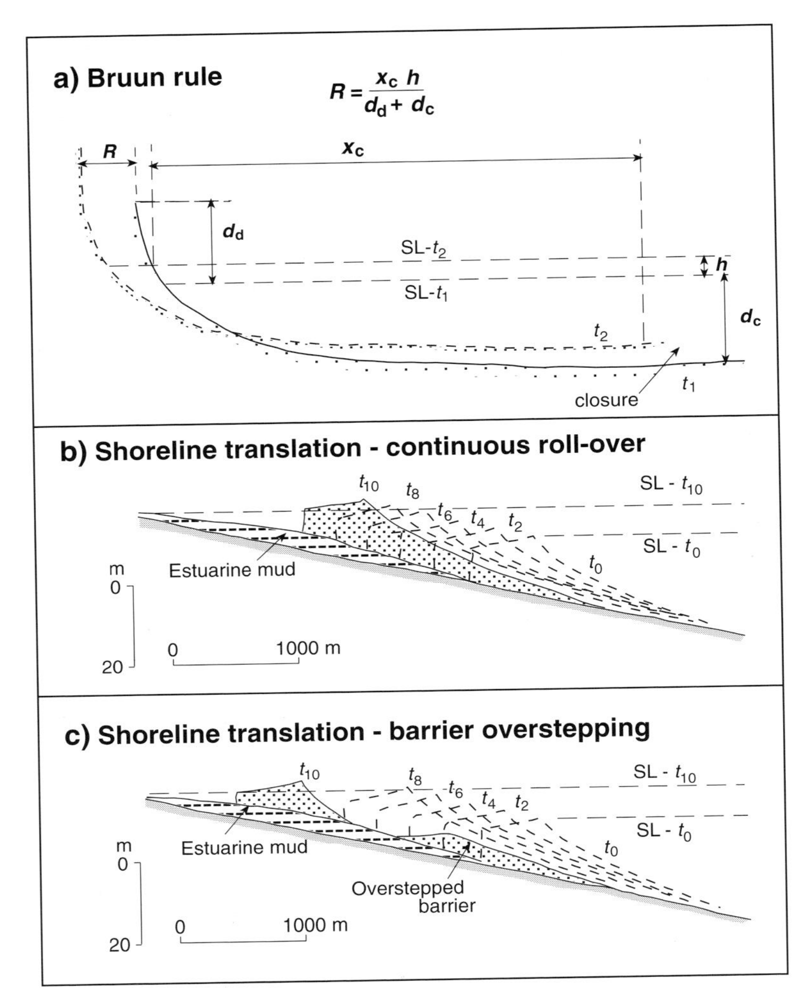

Figure 9.9. Simulation modelling of sandy shoreface response to rising sea-level. (a) The Bruun rule, showing the definition of variables and the response to a rise in sea level. (b) The shoreline translation model in continuous roll-over mode, and (c) the shoreline translation model in barrier overstepping mode (based on Cowell and Thom, 1994).

happens and cannot be used in a predictive capacity (Pilkey *et al.*, 1993; Thieler *et al.*, 2000). For instance, the gradual landwards migration of barrier islands during relative sea-level rise involves landward movement of sediment whereas the Bruun relationship implies net seaward movement (Dean, 1987; Eitner, 1996). Studies of the behaviour of sandy coasts indicate that during the postglacial transgression sand barriers have undergone net translation shoreward rather than seaward as the Bruun rule predicts (Thom and Roy, 1988).

The Bruun rule provides a rule of thumb, which appears to indicate correctly the direction of change, but which appears a poor predictor of the amount of erosion (Komar, 1998). There are limitations in terms of longshore transport and the time frame within which the rule should be applied (Larson and Kraus, 1995). A more generalised Bruun rule, involving an equilibrium profile but applied to entire barrier islands, has been proposed (Dean and Maurmeyer, 1983).

A shoreline translation model, incorporating modification of the generalised Bruun rule to accommodate continuity of mass, simulates the migration of the shore as sea level rises (Figure 9.9b and c). It provides a powerful tool to test the sensitivity of shoreline change to variations in other parameters (Cowell and Thom, 1994; Cowell *et al.*, 1995). It is applicable over a longer term than empirically based sediment transport models such as SBEACH (Larson and Kraus, 1989). Morphodynamic responses similar to those that have been interpreted from reconstruction of Holocene barrier sequences have been simulated using the shoreline translation model. For instance, a barrier simulated using a mass continuity procedure can show rollover (Figure 9.9b), whereas in other circumstances it may experience overstepping (Cowell and Thom, 1994). Overstepping appears to have occurred in the case of barriers on the Nova Scotia coast, and these have become more organised through time (Orford *et al.*, 1991; Forbes *et al.*, 1995).

9.4.2 Multidimensional modelling

A major constraint on many morphodynamic models is that they simulate the coast in only two dimensions, whereas it is clear that sediment transport and the mutual co-adjustment of form and process operates in three dimensions (van Rijn, 1993). To monitor and model the coast in three dimensions (3D) remains a challenge that is being addressed as part of several international multidisciplinary initiatives under the title of 'large-scale coastal behaviour'. This represents an expansion from relatively small-scale process studies to increasingly refined mass-balance sediment studies at the time scale of decades and spatial scale of 10 km (Roy *et al.*, 1994; Stive and de Vriend, 1995). This remains small scale in comparison with geomorphological traditions championed by Davis (see Section 1.3.2) and Johnson, but nevertheless has involved significant developments in cross-disciplinary collaboration.

A number of models have been applied to sandy coasts. SBEACH is a cross-shore sediment transport model (Larson and Kraus, 1989). Simulation modelling has been applied to the shoreface using GENESIS (Hanson and Kraus, 1989) and ONELINE (Kamphuis, 2000). GENESIS simulates shoreline planform adjustment using a constant shore profile, solving for conservation of mass by finite difference along a discretised (sectioned) 1D shoreline for a range of individual wave conditions. Attempts have been made to extend a similar approach to three or more dimensions (Davis and Fox, 1972; Yamashita and Tsuchiya, 1992). Shore-profile modelling has been extended to millennial time scales in the shoreface translation model (STM, Figure 9.9b and c) which simulates the transgressive backstepping of barriers on a coast under conditions of rising sea level (Cowell and Thom, 1994; Roy *et al.*, 1994; Cowell *et al.*, 1995). Similar modelling of barrier islands has been undertaken (Forbes and Syvitski, 1994). Recent modelling has also included a simulation of the formation of shore platforms (Trenhaile, 2000, 2001a,b,c).

There are computational constraints on extending models from 1D or 2D to more dimensions. They can be extended to 2.5 dimensions, either in profile (2D-vertical, 2DV) or in planform (2D-horizontal, 2DH). 2DV models concentrate on cross-shore patterns of sediment movement, whereas 2DH models use vertically integrated values to map variations across the nearshore (van Rijn, 1993). These give a quasi 3D capability (Q3D), but to simulate sediment movements genuinely in 3D still requires more computational time than is generally available (Kamphuis, 2000). Initial models (also called sediment transport models) compute sediment-transport rates and bed level changes for one time step based on the topography of the seafloor and hydrodynamic conditions. Dynamic morphological models are iterative models in which the morphological outputs of one time step are input into the next and a sequence is run through time (van Rijn, 1993). Such morphological models offer powerful simulation capabilities where bedload is the principal mode of transport and adjustments to new hydraulic conditions occur rapidly, but they are less suitable for the simulation of suspended sediment transport in which concentration gradients and diffusive processes are significant.

Cell-based simulations are increasingly possible using 2.5D or 3D GIS (Figure 9.10a), combining upscaling from process studies and downscaling from Quaternary evolutionary studies (Cowell and Thom, 1994). For example, passive GIS-based mapping has been extended by computationally intensive dynamic landscape modelling. Coastal wetlands have been modelled, based on a grid of cells each with a characteristic vegetation community that accretes on the basis of internal sedimentary processes and external transfers of water, sediment and nutrients, until habitats switch at threshold values (Sklar *et al.*, 1983). The modelling of wave reshaping of fluvially derived delta deposits was achieved by combining two Fortran programs (WAVE and SEDSIM). The SEDSIM component, based on unidirectional unsteady flow in

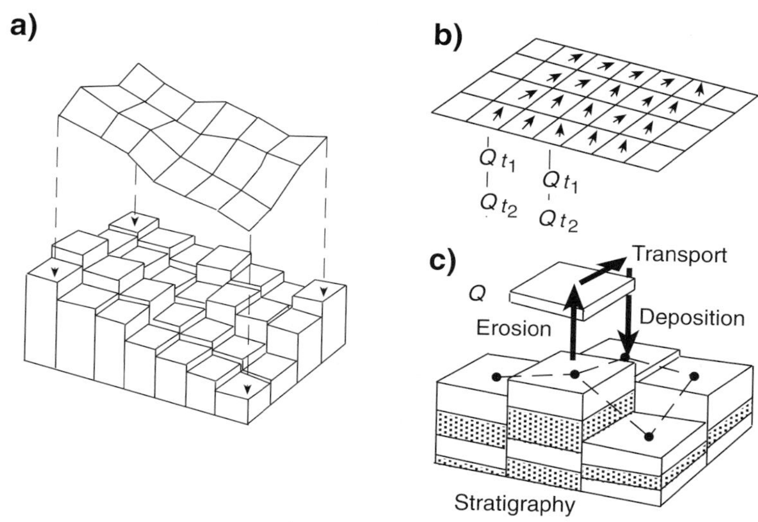

Figure 9.10. Grid cell modelling. (a) Schematic representation of the topography in the coastal zone as a 2-dimensional array of fishnet grid (upper) and 3-dimensional (often termed 2.5D) block diagram (lower). (b) Example of the derivation of subsequent grids, such as vectors for current speed and direction and consequently sediment transport volumes over a series of discrete time steps. (c) Models offer the potential to compute 4-dimensional (or more) arrays, as in this enlargement of several sediment cells, with calculation of erosion, transport, and deposition on a cell-by-cell basis. Where different grain sizes behave differently it may be possible to simulate stratigraphic development through time (based on WAVE/SEDSIM model of Martinez and Harbaugh, 1993).

rivers, has been modified to simulate the coastal and nearshore wave-induced erosion, transport and deposition of sediment by addition of a WAVE suite of procedures (Martinez and Harbaugh, 1989, 1993). It is a hybrid 4D finite-difference/finite-element model with algorithms that redistribute four different grain sizes in response to currents (Figure 9.10). The model simulates incident waves, breaking and surf-zone characteristics (such as radiation stress and longshore currents). It is based on simplified, and empirically tuned, process assumptions but produces a morphological and stratigraphic output at each time step. A series of accounting procedures manage the transportation of sediment, moving it between adjacent cells under continuity constraints. Grid cells are filled and emptied (Figure 9.10b), and may be viewed in terms of accumulation of sediment of different sizes (Figure 9.10c), enabling simulation of depositional strata in balance with topography and incident wave conditions (Martinez and Harbaugh, 1989, 1993). SEDSIM/WAVE has also been used to simulate longshore extension of spits (Raper, 2000). The outputs are not predictions or reconstructions,

Figure 9.11. The mouth of the River Teign at Teignmouth, Devon, southern England. This has been the site of 3D modelling as part of the Coast3D project. The shoals undergo complex cyclical movements in response to coastal processes (photograph South West Water).

but model outcomes under different variables which may be compared with independent outcomes measured in the field. As with the early modelling of spit evolution (King and McCullach, 1971), it demonstrates spit sensitivity to wave direction and nearshore bathymetry (Raper, 2000).

International collaboration in Europe within the Coast3D project has enabled measurement and modelling for several coastal sites in Europe. Modelling of a relatively straight barrier beach and bar coast at Egmond in the Netherlands is regarded as 2.5-dimensional. Sedimentary non-uniformities mean that 2D models are not entirely appropriate, and reasonable consistency has been possible between measurement and modelling, simulating nearshore bar behaviour on this coast at the event scale of individual storms, but rather less accurately simulating response of the subaerial beach (van Rijn, 2001).

A series of studies were undertaken at the mouth of the River Teign in Devon, southern England, as a part of the Coast3D project (Figure

9.11). The overall morphology appears simple and consists of a barrier estuary on which the busy port of Teignmouth is situated. It is similar to the schematic geomorphology shown in Figure 1.1, and used to illustrate several other basic geomorphological concepts throughout this book. The river discharges through an estuary, whose southern shore is controlled by a rocky headland, the Ness. However, the complex interplay of waves, wave-induced currents, littoral drift and strong tidal currents (up to 2 m s^{-1}) through the mouth, together with winter storms, leads to a complex and dynamic 3-dimensional system of shoals at the river mouth. The shoals were shown to undergo cyclic change by studies undertaken by Spratt (1856). The pattern of change of the shoals was suggested to be part of a closed cycle adopting a dynamic equilibrium based on a 10-year period of monitoring (Robinson, 1975). The shoals have to be dredged daily to maintain access to the port. The highly irregular geomorphology at Teignmouth stretches presently available models and will require more complex models to understand fully. Nevertheless, preliminary modelling within the Coast3D project has provided considerable insight into the patterns of behaviour and confirms some of the hypotheses of previous researchers (Soulsby, 2001). However, details of such a system are still too complex to predict (Whitehouse, 2001; Sutherland, 2001).

The value of long records of monitoring is shown where these reveal a pattern. The cut and fill sequence that beach profiling at Moruya reveals would not be apparent to the casual observer but can be seen in 2–3 decades of beach profiles (see Figure 6.18c). Similarly, regular beach profiling identifies a sweep zone within which the beach volume varies, which is a very useful beach management tool, and on the basis of which storm-cut volumes can be defined. Nevertheless, the variability caused by such subtle factors as the long-term ENSO variation will require a period of record considerably longer than the wavelength of the modulation being studied and from a number of sites around the Pacific Basin. Whether or not oscillations like the Pacific Decadal Oscillation influence beach volumes will not be addressed solely from monitoring, but will require multidisciplinary study incorporating paleoenvironmental proxy data as well as modelling.

If the principles on which a system is based are understood, then it may be possible to forecast aspects of its behaviour. Coastal morphodynamic models provide a framework within which the most marked variation and complexity of change at particular sites might be understood, incorporating new elements into the models and refining the domain within which the system operates. Historical and, in some cases, prehistorical records provide further insight into the possibility that the fluctuations are part of a larger oscillatory pattern.

9.5 Summary

Morphodynamic studies of coastal systems offer a framework within which to attempt to link contemporary process studies with longer-term evolutionary trends. Although it is unrealistic to anticipate a high degree of predictive capability, conceptual models narrow the level of uncertainty. The body of knowledge on which this book is based provides the foundation for morphodynamic models of coastal evolution each of which will continue to be subject to revision and refinement. The scientific method should guide selection of the most probable hypotheses, recognising the advantages advised by Douglas Johnston. 'The most for which we may reasonably hope is by correct methods of research to reduce the chances of error to a minimum, and to raise to its maximum the probability of discovering the real causes and relations of things' (Johnson, 1933, pp.492–493). To achieve this he recognised the need to be accurate in observing facts, careful in classifying them, cautious in generalising from them, fertile in inventing hypotheses, ingenious and impartial in testing their validity, skilful in securing their confirmation or revision, and judicial in formulating ultimate interpretations (Johnson, 1933).

Clues to factors relating to cause and effect can be sought in the relative time sequencing of events, as well as in the geographical distribution of landforms in terms of consistency and strength of association of landforms with the processes considered responsible for their formation. Schumm (1991) recommends the following steps: first, assembling all historical information and considering past events (what is it?); second, understanding the processes involved (what controls it?); and third, comparing different settings in space (how general is it?).

There are good reasons for studying past landforms. The Holocene stratigraphy of coastal environments contains a cumulative record of varying process operation including extreme events of infrequent occurrence (Braithwaite *et al.*, 2000), and provides context for observed historical trends and predicted near-future conditions (Blum and Törnqvist, 2000). Historical and paleoecological records provide a library of past experiments (Wasson, 1994), and an accurate perception of environmental history is a prerequisite for valid environmental concepts and future environmental management (Dickinson, 2000).

Coastal morphodynamics are still imperfectly understood, but the concept that there is a time-dependent equilibrium to which coastal landforms adjust has been a useful one in enabling generalisations. Inheritance from preceding conditions and antecedent states, and thresholds and lags have all contributed to the sedimentary or erosional landforms. 'The present provides insights into the past, it influences future geological processes and events, and the past influences both present and future geological processes and events' (Schumm, 1991, p. 27).

The unprecedented technical sophistication of monitoring and modelling capabilities offers opportunities to examine and understand the trajectory of change and to mitigate the most adverse impacts. Multidisciplinary studies, involving multidimensional modelling, enable a greater understanding of the range of variability. The challenge, however, will be to ask the right questions, to formulate new hypotheses and to test them more rigorously. Human society is having an increasing impact on the way in which coasts change, and in order to understand and mitigate that role, it will be important to be able to assess the natural pathway of change on different coasts. There is likely to be a wide range of emotive, social, cultural and political reasons to distort the role that humans have had, or are having, in changing the way the coastal system behaves. These issues are examined in the final chapter.

Chapter 10
Human activities and future coasts

> The real conflict of the beach is not between sea and shore, for theirs is only a lover's quarrel, but between man and nature. On the beach, nature has achieved a dynamic equilibrium that is alien to man and his static sense of equilibrium. (Soucie, 1973, p. 56)

Previous chapters have described coastal landforms and discussed morphodynamic frameworks for interpreting the pattern of adjustments for different coastal types. The range of adjustments is complex and individual coasts change at varying rates and in varying directions. The level of uncertainty about what will happen in the future increases as the time scale increases. Although coastal landforms and the natural processes of erosion and deposition that shape them are the focus of this book, this natural pattern of adjustment is increasingly influenced directly and indirectly by human activities. Many coasts have been substantially modified by local structural and ecological changes brought about intentionally or unintentionally by humans. The impact of human activities can be felt beyond the local scale. Climate change as a result of the enhanced greenhouse effect and the associated threat of accelerated sea-level rise imply human impact on a global scale at an unprecedented rate. These impacts are added to natural pattens of change.

No coast is now likely to be beyond the influence of humans who have become a force 'as powerful as many natural forces of change, stronger than some and sometimes as mindless as any' (Meyer, 1996, p. 2). Future coasts will be increasingly human-dominated (Messerli *et al.*, 2000). How the coast adjusts is not just the subject of scientific enquiry, but is an issue relevant to a large proportion of the world's population. On developed coasts, protection of assets is the major priority and the shoreline can be completely engineered; recreational use of the coast raises a series of additional, and often conflicting, issues. Coastal management is increasingly concerned with sustaining natural and aesthetic coastal resources. It needs to be based on adequate geomorphological understanding of the way coasts operate, integrating studies of process operation with longer-term reconstruction of coastal behaviour, at time scales that are relevant to planning and sustainable coastal resource use.

Earth scientists have generally been reluctant to study human-dominated coasts, having become accustomed to studying change over longer time scales than human generations. They have been dissuaded by the challenges of discriminating cause and effect where human influence has been prominent, or establishing geomorphological principles where earth-moving machinery is the dominant agent of change (Nordstrom, 1994a, 2000). Nevertheless, it is important to understand how coasts react to human-induced changes, and it will become increasingly necessary to incorporate human actions into future morphodynamic models.

This final chapter examines briefly human modifications to the coast. It suggests a framework whereby human activities can be considered either as a perturbation to coastal landforms, as a boundary-condition change, or as an intrinsic component of a coastal system. There is a growing realisation that human activities have had, and continue to have, an impact on the global environment. Natural patterns of climate change are being altered locally and globally by human activity. Sea-level change, which has been shown in previous chapters to be one of the most important boundary conditions to which the coast responds, is examined as an example. It is shown that most coasts will be subject to some change in this and other boundary conditions, whether or not human actions are causing accelerated sea-level rise. Many human societies will need to make adjustments in the face of sea-level rise; many have already adjusted to similar changes, and there are lessons to be learned from study of these examples. The response of the coast to such changes is unpredictable in detail, but greater understanding of how coasts behave can reduce the degree of uncertainty. It is important to recognise that change is ongoing and is to be expected in the future, and to replace the previously widespread view that we live in a stable world where change is neither anticipated nor tolerated.

10.1 Human interaction with the coast

People have made use of coastal resources for a long time. However, there has been recent recognition of the need to balance exploitative use of those resources with a growing philosophical and recreational appreciation of the value of the coast for cultural, aesthetic and educational purposes (Kenchington, 1990). The relationship between human society and the coast has changed. Pre-industrial societies tended to be dominated by nature. The development of coastal cities and more intense use of coastal resources led to the concept of a struggle with nature that dominated much of the 20th century. This was replaced, in the latter part of the 20th century, by themes such as environmental assessment and sustainable development which sustain nature's values (Thom and Harvey, 2000).

Natural coastal ecosystems, such as coral reefs and mangroves, are now highly valued for a series of goods and services that were previously poorly appreciated (Ewel *et al.*, 1998; Moberg and Folke, 1999). For example, intertidal ecosystems protect the coast, attenuating wave energy and ameliorating the impact of storms (Brampton, 1992). Rapid clearing of mangrove for shrimp farm aquaculture has been occurring at an alarming rate in Southeast Asia; the farms rarely appear sustainable for more than a few years and re-establishment of mangroves is difficult (Bird, 1993a; Wolanski *et al.*, 2000). Mining of coral from coral reefs, as has occurred in the Maldives for construction material, leads to rapid depletion of a resource that has taken centuries to build up (Brown and Dunne, 1988). Coral limestone has been taken from around the capital, Malé, for use as infill to extend the heavily over-populated island. On the southern side of the island, a large unsightly breakwater built from concrete tetrapods now provides the protection from large waves formerly given naturally by the reef before it was mined (Figure 10.1).

Figure 10.1. Malé, capital of the Republic of Maldives, is a reef island on the rim of a coral atoll. It has been the site of population concentration and the fringing reef has been mined to extend the land area of the island. Erosion during heavy seas has necessitated construction of a seawall around much of the island and detached breakwaters along the seaward reef flat. This contrasts with natural islands (background) which are protected by reef.

Coastal management now promotes the view that human actions should not destroy non-renewable resources, but should be sustainable. Impacts, such as resource depletion, environmental degradation and pollution, can often be seen to be the result of poor planning rather than being an independent problem (Kay and Alder, 1999). Increasing emphasis has been placed on integrated coastal zone management (ICZM), 'the comprehensive assessment, setting of objectives, planning and management of coastal systems and resources, taking into account traditional, cultural and historical perspectives and conflicting interests and uses' (IPCC, 1994, p. 19). Management decisions often need to be proactive, anticipating rather than responding to an issue (Salomons *et al.*, 1999), and greater emphasis is placed on capacity building, community involvement and the coastal environment that future generations will inherit (Cicin-Sain, 1993; Hildebrand, 1997).

In the face of environmental change, any particular coast can be considered in terms of its vulnerability (the degree to which it is susceptible to, or unable to cope with, changes), its sensitivity (the degree to which it is affected by changes), and its adaptive capacity (the ability of a coast to moderate potential impacts). Geomorphological factors influence each of these, and past landforms play an important role in terms of inheritance. For instance, former mangrove areas, particularly associated with the extensive mid-Holocene mangrove forests, have left a legacy in terms of potential acid sulphate soil development (Lin *et al.*, 1995). Formerly marine organic-rich sediments contain pyrite which oxidises on exposure to release sulphuric acid. These soils need remediation before they can be used for agriculture (Sammut *et al.*, 1996), and can become a problem where land use is inappropriate (see Chapter 7). Deltaic–estuarine plains are often intensively cultivated, particularly for rice, rubber, coffee and other cash crops, and are the location for many of the world's megacities (Nicholls and Leatherman, 1996). Increasingly intensive land use has masked, modified and overwhelmed natural processes in these deltaic–estuarine areas (Viles and Spencer, 1995; French, 1997). For example, in Thailand, drainage canals ('klongs'), groundwater extraction and compaction of sediments beneath Bangkok has led to lowering of the ground surface by 20–60 mm a^{-1} (Nutalaya and Rau, 1981).

10.2 Human activities within the coastal system

It will rarely be possible to entirely disentangle human factors from natural factors. Models of coastal behaviour will increasingly need to integrate theories of social science. Although human endeavour has often been viewed as an end in itself to which the forces of nature must be subservient, there is increasing recognition of the multiple values and uses, and wider acceptance of the need for greater knowledge and better education, concerning coastal systems. In many cases, it may be easier to manage peoples' activities than to manage coasts. Figure 10.2 is a simplified schematic illustration of how human activity can influence a typical coastal system (like that shown in Figure 1.1). Human activity can be viewed as a perturbation, an altered boundary condition, or as an intrinsic adjustment within the coastal system (Phillips, 1991). These are examined in turn in the following sections.

10.2.1 Perturbations

Some human actions resemble a natural perturbation; for example, a sand castle or other sand sculpture built on the beach during low tide will be reworked by the swash, leaving no trace after the tide has covered the beach. These disturbances and subsequent recovery respond like, and generally operate at a similar rate to, a small storm. The beach is in a temporary state of disequilibrium but, as long as the system is non-transient (boundary conditions remain stationary), sand is reworked and redeposited adjusting back towards the pre-disturbance morphology (Figure 10.2a).

Beach nourishment illustrates this on a larger scale. Beach nourishment involves replenishing the volume of sand on an eroding beach. As a method of shore protection it is an alternative to using physical

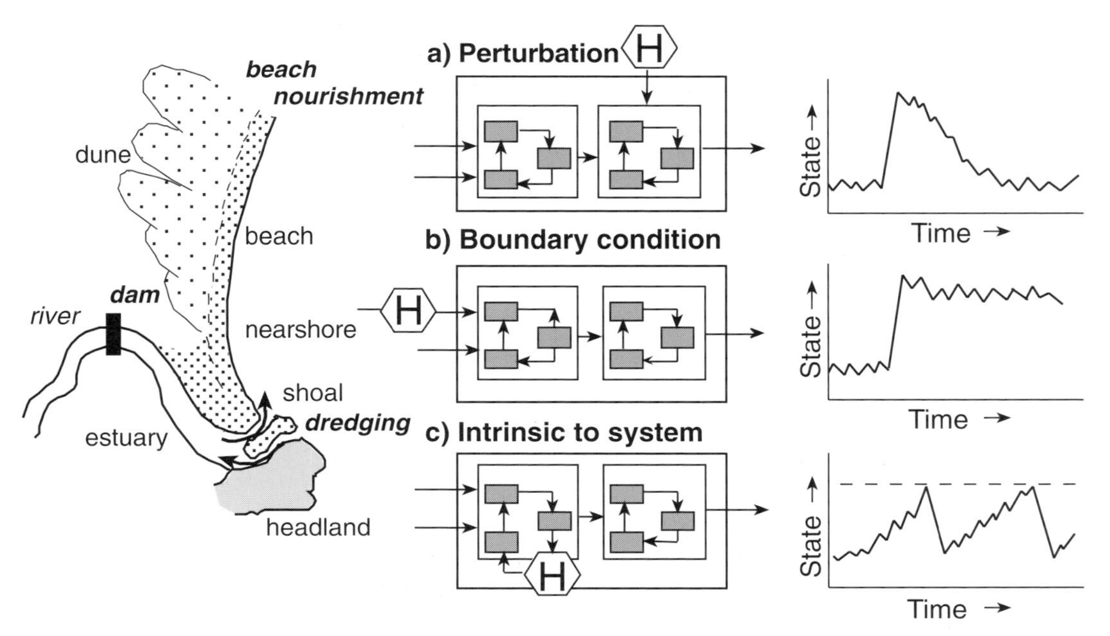

Figure 10.2. Human action (H) and the coastal system: (a) viewed as a perturbation, for instance beach nourishment; (b) altering boundary conditions, for instance damming of a river altering sediment input; and (c) as an intrinsic component of the system, for instance where dredging of a channel shoal is undertaken when navigation is threatened, in response to a threshold water depth. Schematic response of morphology of landforms is shown on the right. Morphology returns to pre-disturbance state in the case of the perturbation; it remains altered where the boundary conditions change, and exhibits repeated negative feedback as a threshold is approached in the case where human action is intrinsic to the system.

structures and is of particular appeal because it looks more natural. Beach nourishment was first undertaken on Coney Island, New York, and more recently on Miami Beach, Florida, which was nourished using 13 000 000 m^3 of sand between 1976 and 1981 (at a cost of $60 000 000). There are now numerous North American and European beaches that have been nourished with sand (Hanson *et al.*, 2002b). The nourishment of beaches means that they are built into a short-term state of disequilibrium which is smoothed out by surf-zone processes and longshore currents in the months after emplacement (Walton, 1994; Larson *et al.*, 1999). The methods of nourishment are largely intuitive; for example, the Dutch approach involves defining the sweep zone, and then overnourishing the beach with an allowance for loss over the lifetime of the project (Work and Rogers, 1998).

A major constraint on beach nourishment is finding a suitable source of sand resembling the natural beach sediment as much as possible; some nourishment schemes involve trucking sand or shingle from the sink at the downdrift end back to the updrift end of a beach (Leafe *et al.*, 1998). Material can be brought from offshore, or from back-barrier or transgressing dune environments, as well as from outside the coastal system, including desert sand and crushed rock (Davidson *et al.*, 1992). There seem to be some advantages to using sediment that is slightly coarser than that on the beach, because this enables a slightly steeper profile that intersects the former beach, rather than requiring that the shoreface also be built up (Dean and Yoo, 1992).

Beaches around the United States that have been nourished rapidly lose sediment and nourishment becomes an ongoing maintenance strategy rather than a solid fix (Bird, 1996). Whatever is, or was, the cause of erosion often remains, particularly if it is a leaky sediment compartment. It is generally perceived that beach nourishment has been unsuccessful if the replenished sand does not stay where it has been put, but it is important to realise that nourishment is a temporary solution, frequently requiring subsequent further replenishment (Pilkey and Wright, 1988).

10.2.2 Boundary conditions

Human impacts can also represent a transient change to boundary conditions. For instance, if a river is dammed, the flow of water and the supply of sediment are altered (Figure 10.2b). Human actions have increased the sediment load of some rivers as a result of clearing of land for agriculture, deforestation and subsequent overgrazing, altering the supply of sediment to the coast (Thomas, 1956). The Huanghe (Yellow) River is experiencing an accelerated rate of delta building, for example, as a result of increased sediment loads resulting from land-use change and forest clearance on the loess plateau (Milliman *et al.*, 1989). Increased sediment supply down rivers in Southeast Asia has resulted from land-use change in catchments (Staub and Esterle, 1993) and locally as a result of open-cast mining (Wright, 1989). Around the Pacific Ocean sediment runoff has had an impact on coral reefs (Cortés *et al.*, 1994; Hopley, 1994), and a major influx of sediment to coastal areas has occurred on Pacific islands (Nunn, 1994). Stratigraphic studies of estuarine systems in northern Northland, New Zealand, for instance, indicate that sedimentation increased by an order of magnitude as a result of Polynesian land clearing and land-use change (Nichol *et al.*, 2000).

Abrupt change has occurred on rivers that have undergone a reduction in discharge and sediment supply as a result of dam construction or other interference. The example of the River Nile (Stanley and Warne, 1993b) is described in Chapter 7, but similar although lesser impacts have occurred on other rivers draining into the Mediterranean Sea (Jimenez and Sanchez-Arcilla, 1997; Pranzini, 2001). Comparable boundary-condition changes occur naturally, but over a longer time scale, as a result of climate change or river diversion. Landforms adjust to this altered forcing, and a sequence of adjustments in processes and forms cascade through the sediment transport system, with erosion of the shoreline at locations where sediment was previously supplied.

Recognition of the interconnectedness of sediment transport pathways in a coastal system provides a sound geomorphological basis for management. Human impact on coastal landforms has involved removal of beaches and dunes where sand and gravel is mined for construction materials. The inappropriateness of extraction directly from such active systems is now widely recognised, and generally illegal, although it still happens (Hesp and Hilton, 1996). Human attempts to halt natural erosion, such as when sources of sediment are sealed through cliff stabilisation programmes, can impede flows and result in erosion elsewhere in the sediment transport system (McKenna *et al.*, 1992; Bray and Hooke, 1997). Sediment removal depletes the littoral sediment budget and disrupts the natural pathways of sediment movement, and too often a static solution is proposed for a dynamic problem. The destruction of the village of Hallsands on the south Devon coast of England, for example, has been shown to have been a consequence of extraction of shingle from Start Bay (Robinson, 1961).

Although traditional use of coastal resources by pre-technical societies involved some measures to sustain their longer-term viability (Johannes, 1978), it is clear that there were major impacts on island and coastal systems. On reefs in Barbados, it appears that disruption of sediment production on one part of the reef can be related to depletion of beach sand elsewhere (Hatcher *et al.*, 1987).

10.2.3 Intrinsic adjustments

Human actions can also be viewed as intrinsic within the system (Figure 10.2c). For instance, dredging is often undertaken where navigation requires a suitable water depth at the mouth of an estuary (for example at Teignmouth, see Figure 9.11). Human response represents a negative feedback similar to the effect that large flood events might have had in flushing an estuary, but activated under a different set of threshold conditions. For instance, when water levels reach a critical elevation in some coastal lagoons in southern New South Wales, they are opened to the sea by breaching the sand barrier, in order to avoid flooding of settlements or other conditions unacceptable to the community. There has been a long awareness that activities such as dredging can have potential impacts on other parts of the system, and social, political and managerial factors serve as additional feedbacks on whether or not such action occurs (Sheail, 2000). Nevertheless, the behaviour of those coastal systems that are influenced or controlled by human action needs to be the subject of more focused research in order to understand better patterns of accretion and erosion on adjacent landforms (Nordstrom, 1987; Cialone and Stauble, 1998).

Many coastal landforms on developed coasts are built or maintained by earth-moving machinery (Nordstrom, 2000). Much of the Dutch coast consists of a constructed dune or dyke built to what is considered the level of the most probable largest storm that is unlikely to be exceeded in a millennium (van der Graaf, 1986; Nordstrom and Arens, 1998). The dunes along the developed coast of New Jersey are rarely more than 1.5 m high; however they are more continuous than they were prior to human construction (Nordstrom, 2000).

Whatever design criteria are used, it is always possible that the next storm exceeds that level. Storm destruction has been experienced along most of the developed coasts of the United States but, surprisingly, the assets are usually rebuilt and on a grander scale than before the disaster (Nordstrom, 2000). Property normally suffers more damage than natural landforms during a storm but the speed of recovery of human societies is potentially quicker than that of natural geomorphological recovery. For example, bulldozers can reshape the beachface and dune in a matter of days following a storm if required, whereas beach recovery normally takes months (see Chapter 6).

Where human actions are intrinsic to a coastal system, they are generally persistent and recurrent and landforms are less free to adjust than they would naturally be. For example, beach and dune can be decoupled through the construction of a physical structure such as a road or boardwalk between them, and interactions such as those implied in Figure 9.2 are no longer feasible. Even where interactions are not completely severed, coastal management initiatives such as dune fences and revegetation can restrict natural processes. There can be recognisable geomorphological patterns to the functioning of artificial coastal landforms. Dune establishment using dune fences, for instance, might involve an initial active phase in which fences trap sand and achieve their purpose, becoming buried during a subsequent phase of vegetated dune development. It might be appropriate to view periodic replenishment of a nourished beach as part of a morphodynamic cycle of the beach. Similarly, solid structures such as a seawall or groyne can also go through phases of equilibration and deterioration over time, eventually needing replacement.

10.3 Tourism and the resort cycle

During the latter part of the 20th century here has been a reassessment of values associated with coastal resource use, now embodied in approaches to coastal management. Although tourism has not been the only factor driving these changes, the coast is now a major tourist destination and tourism has been an important factor. The nature of tourism has changed considerably since the opportunity for travel became more widely available in the 19th century, with resorts developed around railway or steamship termini in Europe but based on road transport in the United States. Air travel has led to increasing internationalisation with major expansions of tourism in areas such as the Mediterranean, Caribbean and southeast Asia.

It has frequently been the case that developers have built hotels in choice spots with little consideration for the geomorphological sensitivity of the site (Wong, 1990; Kamphuis, 2000). In many cases, at European resorts and on selected Australian beaches such as Bondi and Manly beaches, an early development was a promenade, a substantial engineered structure built on the beach or backshore. In North America, it was more common to build a boardwalk above the dunes. Sufficient technology and capital enable development anywhere; for instance, floating hotels have been constructed within coral reef areas, such as the Great Barrier Reef.

In the preceding chapters, models of coastal development have emphasised morphological change of coasts over time. There have been relatively few attempts to view the morphological adjustments that occur as humans make use of the coast. However, one such attempt is shown in Figure 10.3; it integrates an evolutionary model of human use,

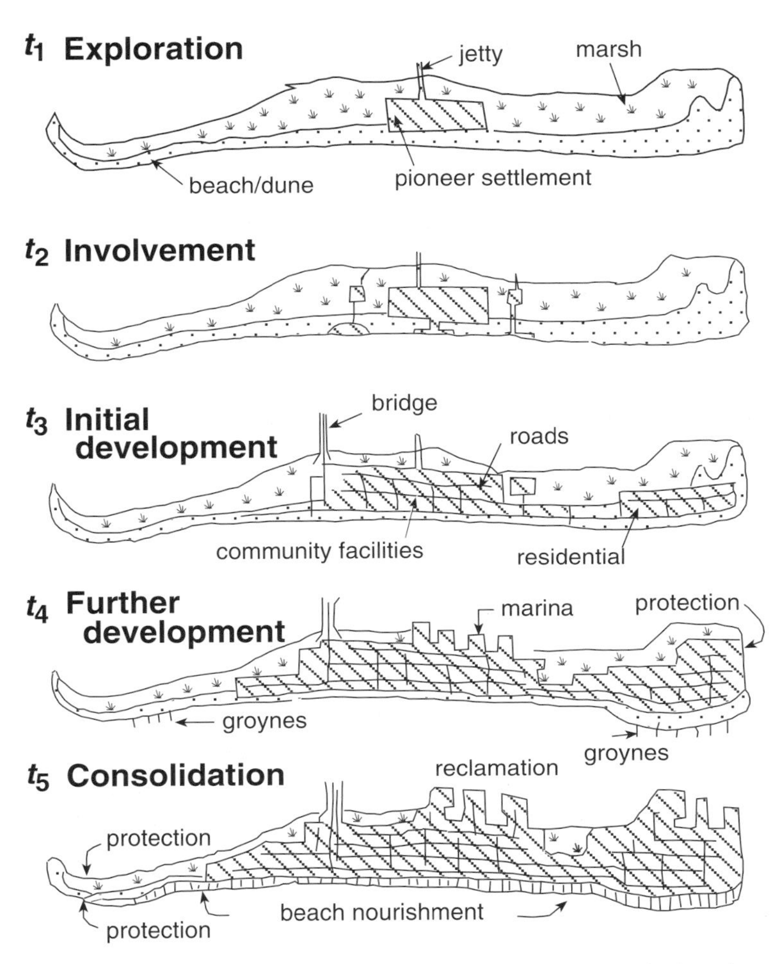

Figure 10.3. Model of barrier island settlement and the resort cycle (based on Meyer-Arendt, 1985 and Nordstrom, 1994a, 2000). The five stages, exploration, involvement, initial development, further development and consolidation are shown to have socioeconomic definition, but to also be expressed in the morphology of the island, with increasing human modification of the shoreline. Grand Isle, a barrier island in southern Louisiana is an example (see Figure 6.25).

with morphological changes of a barrier island. It is based on the concept of the 'resort cycle', also called the tourist area life cycle (Butler, 1980). The resort cycle recognises that a coastal area goes through several stages in relation to its role as a tourist attraction: (i) exploration, during which there are few visitors; (ii) involvement, during which there is some local encouragement of tourism and some tourist response; (iii) development, during which tourism is promoted; (iv) consolidation as carrying capacity is reached; and then (v) stagnation.

These stages have been recognised in the development of several coastal resorts in North America and the West Indies (Meyer-Arendt, 1985; Weaver, 1990; Priestley and Mundet, 1998). Figure 10.3 includes changes in shoreline morphology over time (Nordstrom, 1994a, 2000). Resort expansion leads to increasing pressure to adopt shoreline protection measures, and to extend or 'reclaim' more land.

Tourists and coastal managers now place a higher value on a natural coastline. The growth of ecotourism, and appreciation of the coast for its aesthetic and educational values, means that there is added emphasis on the coast looking natural. Nevertheless, it may be possible to integrate human-induced morphological changes into more traditional morphodynamic models (Nordstrom, 1994b, 2000). Even where there are intensive holiday developments, as along the Gold Coast in Queensland (Figure 10.4), there is an emphasis on a natural looking beach, and beaches are replenished when they undergo excessive erosion (Bird, 1996).

10.4 Global environmental change

There is increased awareness of the nature of global environmental problems, as a result of international research efforts. Assessments by the Intergovernmental Panel on Climate Change (IPCC) and other

Figure 10.4. The Gold Coast of Queensland, showing intensive construction of high-rise holiday accommodation.

international programmes, such as the International Geosphere–Biosphere Program (IGBP), including core projects such as Land–Ocean Interactions in the Coastal Zone (LOICZ), focus research on global issues (Holligan and Reiners, 1993). However, although this trend has encouraged multidisciplinary studies, understanding of the complexity of coastal systems remains at an early stage. The extent to which scientific studies provide background to individual coastal management decisions is highly variable on a case by case basis.

10.4.1 Sea-level rise

The issue of sea-level rise is of particular concern. The rate of sea-level rise, indeed whether there will be accelerated sea-level rise, is still subject to debate and revision (Woodworth, 1990). The behaviour of the coast in response to such changes is even more contentious (Bird, 1993a). The IPCC has suggested that there is likely to be sea-level rise averaging between 0.09 m and 0.88 m by 2100 (Houghton *et al.*, 2001). If sea level is rising as a result of increased greenhouse gases in the atmosphere then that rise will continue at a steady or declining rate for several decades, even if greenhouse-gas emissions can be lowered (Caldeira and Kasting, 1993; Warrick, 1993).

The pattern of sea-level change locally and regionally is still only poorly anticipated, because of many complicating factors (see Chapter 2). The changes to high-tide levels could deviate from that for mean sea level (Woodworth *et al.*, 1991; Dixon and Tawn, 1992). Amongst the other aspects that might change under an enhanced greenhouse effect is the frequency and intensity of storms (Daniels, 1992; Murty, 1993; Evans, 1993), which can also result in higher storm surges (Toppe and Fuhrboter, 1994). Nevertheless, it remains difficult to substantiate that there has been any change in storm frequency (Lighthill *et al.*, 1994; Liu *et al.*, 2001). Whereas short-term records for the eastern United States indicate an increased frequency of storms (Zhang *et al.*, 2000), longer-term records indicate that storms large enough to form gravel ridges on islands on the Great Barrier Reef have had a similar return period for the past five millennia (Hayne and Chappell, 2001; Nott and Hayne, 2001).

For mid ocean in the central Pacific and eastern Indian Oceans there remains a question, based on tide gauges and on coral records, whether the sea is rising (Wyrtki, 1990; Woodroffe and McLean, 1990; Smithers and Woodroffe, 2001). Satellite altimetry suggests that the rate of rise is less than anticipated and is closely related to variations in sea-surface temperature (Cabanes *et al.*, 2001). Individual records can deviate from the regional trend for reasons specific to those stations. For instance,

Hilo on the Big Island of Hawaii records a relative sea-level rise of 3.8 mm a^{-1} as a result of rapid subsidence of this island (see Chapter 2) and anomalous or idiosyncratic records from tide gauges in Japan appear to be associated with seismic events (Baker, 1993).

Future sea-level change along continental margins will also show considerable spatial variability because of tectonic, subsidence, flexural or compactional factors (Gornitz, 1995). No individual coast should be expected to experience the global average sea-level change. Many regional impacts of human-induced climate change relative to natural multi-decadal variability will be undetectable until at least 2050 (Hulme *et al.*, 1999). The natural variability of phenomena such as El Niño will mask global changes and the more subtle but less completely understood variations, such as the Pacific Decadal Oscillation, will add further to the difficulty of discriminating patterns of change (McLean and Tsyban, 2001).

10.4.2 Response of coasts

If sea-level rise does occur as a result of global warming, it will be masked or accentuated by patterns of sea-level change that are already being experienced on many individual coasts. Sea level has changed substantially in the past, and the chapters of this book have emphasised that sea-level changes have been a major boundary condition to which coasts have responded. The postglacial early-Holocene rapid sea-level rise occurred virtually worldwide while humans occupied the coast (see Chapter 2, Figure 2.11) and exceeded in magnitude sea-level rise now attributed to human action (Leinfelder and Seyfried, 1993; Hanebuth *et al.*, 2000). It is also clear that many coastlines in areas remote from former ice sheets experienced a sea level above present during the mid Holocene.

The coastal stratigraphic record provides an archive from which to examine rates and directions of changes in the past and morphodynamic responses and provides partial analogues for present or future conditions. Geomorphological reconstructions extend our understanding of coastal behaviour and assist calibration and evaluation of predictive models. It is necessary to refine some of the stratigraphically based, geochronologically calibrated models of how coasts have responded to past sea-level change to incorporate human influence and to forecast how they might respond to future human-caused steric sea-level changes.

The past is not always the key to the future. For instance, there have been changes in boundary conditions other than sea level, particularly human impacts. However, it is a key to understanding the present which is an outcome of past causes and effects, and this gives greater confidence to the forecasting of future coastal adjustments than would otherwise be possible (Pye and Allen, 2000). It is not a simple matter to extrapolate from site to site; no one type of coast behaves simply in response to sea level, in each case the effect of other boundary conditions, particularly sediment availability and supply, is important (Tooley, 1992). The coast does not remain a passive surface over which the sea rises. Instead it is dynamic, and the natural morphodynamics, described in this book, must be understood to forecast the nature of shoreline response effectively. Simple generalisations such as the Bruun rule (see Figure 9.9) provide only a first-order indication of the sort of response that can be expected, but it is futile to search for universal causes and relationships that will yield accurate estimates of volumes of change or shoreline recession.

Higher sea level will result in many local changes, such as altered wave refraction patterns and energy gradients, and simulations will need to be multidimensional (Pethick, 2001). Forecasting change is likely to be easier on hard rocky coastlines than poorly consolidated soft-rock coastlines, and the dynamic readjustments that will occur on sandy beaches, incorporating responses to frequent high tides and storms, seasonal and longer-term changes, and decadal–century time scales will remain a challenge (Brunsden, 2001). This can be illustrated with reference to coral reefs. Some Holocene reefs have been able to keep pace with sea-level rise of 10 mm a^{-1}. Accelerated sea-level rise could rejuvenate Indo-Pacific reef flats, leading to increased carbonate production across what are often emergent reef flats (Buddemeier and Hopley, 1988). However, it is clear that, although reefs are well adapted to survive natural environmental change, there are other stresses, both natural and human-induced, on modern reefs. Whereas natural stresses are often acute, human-induced stresses are frequently chronic (Brown, 1997a). The synergy between global climate change, as a result of the greenhouse effect, and local anthropogenic stresses severely threatens reefs (Wilkinson, 1996). The issue of more frequent and persistent coral bleaching as a result of an increase in global sea-surface temperatures is of still further concern, and widespread bleaching of reefs in 1998 has caused considerable alarm (Hoegh-Guldberg, 1999; Lough, 2000). Reefs that undergo major mortality or on which coral growth is retarded by bleaching may be less able to adapt to other stresses imposed on them; many have already suffered because of human impacts, especially overfishing, and are far from pristine (Jackson, 1997).

10.4.3 Human adaptation to change on the coast

Whatever the cause of change, humans will need to adapt to it. Four possible strategies have been widely recognised that could be adopted in the face of rising sea level (Titus, 1990a,b): (i) no protection, leading to eventual abandonment; (ii) shoreline protection, whereby seawalls are constructed around the coast; (iii) accommodation, by which adjustments are made to be able to cope with inundation, raising coastal land and buildings vertically; and (iv) managed retreat, which plans for retreat and adopts engineering solutions that recognise natural processes of adjustment. These four patterns of adjustment were initially described in relation to barrier islands, but have subsequently been suggested as options that are available for other shorelines where sea-level rise threatens (Bijlsma *et al.*, 1996). The decision to defend, to nourish, or to retreat is one that is site-specific, depending on the pattern of relative sea-level change, geomorphological setting, sediment availability and erosion, as well as a series of social, economic and political factors (Griggs *et al.*, 1994).

The 'do nothing' option, involving no protection, is a cheap and expedient way to let the coast take care of itself. It involves the abandonment of coastal facilities when they are subject to coastal erosion, and either gradual landward retreat or evacuation and resettlement elsewhere. It is clearly not always feasible; for instance, atoll nations, such as the Maldives (see Figure 10.1), have no hinterland into which to retreat.

Shore protection involving hard engineering structures is already adopted for much of the world's shoreline. Only on those highly developed coasts, such as Long Beach Island, New Jersey, does the land value justify or meet the cost of expensive engineering solutions. It is clear that there will be a range of social, legal, political, as well as economic considerations that will need to be factored into planning in such vulnerable areas (Titus, 1990a; Nordstrom, 1994a).

Shore protection consists of interposing a static structure between the sea and the land to prevent erosion and or flooding, and it has a long history. Modifications were made to natural harbours for navigation and trade; for instance, hard structures were in use 4000 years ago at Mediterranean ports such as Pharos and Tyre. On a geological time scale, the structures do not last; even on the so-called 'engineering' time scale they require frequent maintenance and monitoring. For example, geomorphological processes have overridden former ports such as those around Troy, described in Greek literature, or Ostia near Rome, which have been stranded inland by sediment accumulation (Kraft *et al.*, 1980). Protection of the shore in Italy, England and the Netherlands can be traced back at least to the 6th century. These schemes have often been called reclamation, although few 'reclaimed' areas were previously land (Miossec, 1994). The level of impact has increased substantially as technical ability and methods of construction available to engineers have improved.

a) Vertical wall

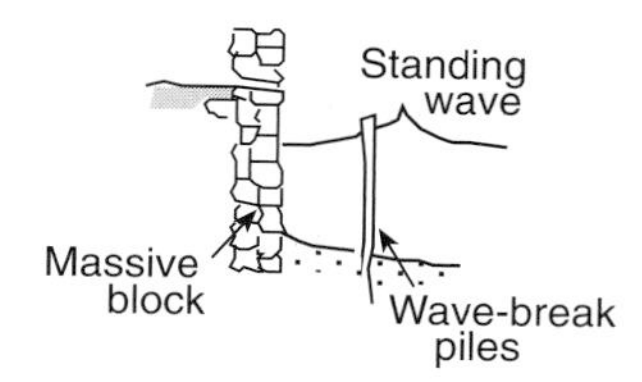

b) Curved concrete wall

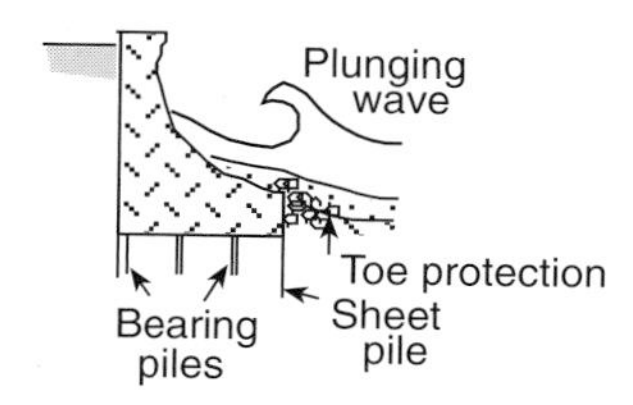

c) Rubble-mound

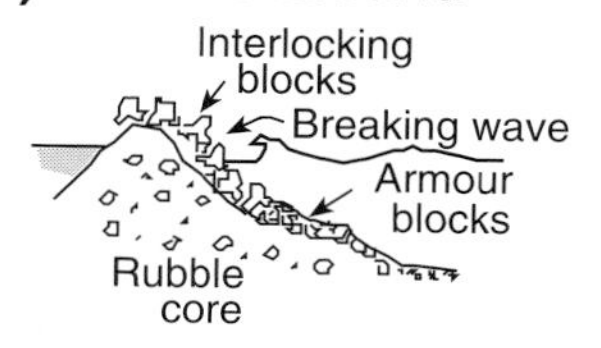

d) Revetment

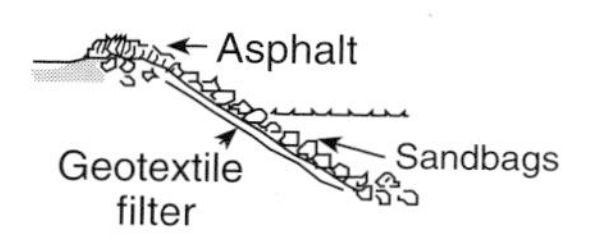

Figure 10.5. Broad types of seawall (based on Carter, 1988). (a) The vertical wall occurs in high-energy setting and reflects wave energy with the potential to set up a standing wave (clapotis); wave-break piles may attenuate some energy if installed. (b) The curved concrete wall causes the plunging wave to break. (c) The rubble-mound is a lower energy option; whereas (d) the revetment would be suitable for the lowest energy setting.

Seawalls

Seawalls became an important protective measure for major coastal cities, where infrastructure must be protected; the seawall at Galveston in Texas was first constructed in the early 20th century; seawall construction in Europe and North America increased considerably after the 1940s (Kraus, 1996). Seawalls are a temporary solution; they are sustainable only as long as funds are available to maintain the protection continually.

A range of seawall types can be envisaged in relation to maximum wave energy (Carter, 1988), their morphology resembling cliff and beach profiles described in previous chapters (Figure 10.5). Vertical seawalls are built in particularly exposed situations or where deep water enables wave energy to reach the shore. These reflect wave energy and under storm conditions standing waves (clapotis) will develop. In some cases piles are placed in front of the wall to lessen wave energy slightly. Curved or stepped seawalls are designed to enable waves to break and to dissipate wave energy. The curve can also prevent the wave overtopping the wall, and provide additional protection for the toe of the wall. A series of mound-type structures are used in lesser energy settings. These comprise rubble mounds and barricades, concrete tetrapods, or 'gabion' mesh bags filled with boulders or cobbles. The least exposed sites involve the lowest-cost technology, bulkheads or revetments of sand bags or

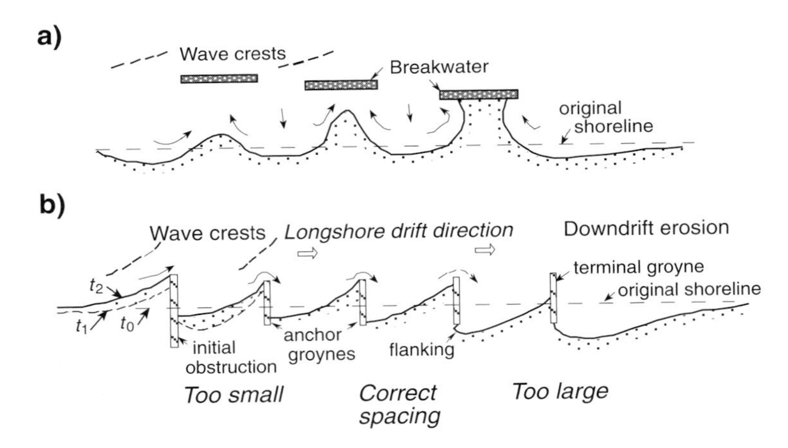

Figure 10.6. Groynes and breakwaters. (a) Breakwaters, note the relationship between detached breakwater distance from shore and the alteration to circulation patterns and development of salients and tombolo (based on Kamphuis, 2000). (b) Spacing of groynes and the gradual chain reaction as groynes fill from updrift to downdrift (based on Inman and Brush, 1973; Carter, 1988).

geotextiles. These serve to armour the shore and impede erosion. Breakwaters perform a similar role to seawalls except that they are primarily designed to reduce wave energy. Detached breakwaters cause refraction (Figure 10.6a) and the beach planform adopts a curved form with salients or tombolos (see Chapter 6) in the lee of breakwaters (McCormick, 1993; King *et al.*, 2000).

Seawalls and breakwaters are generally expensive to construct, and the costs to build protection in the face of sea-level rise would be enormous (Yohe and Schlesinger, 1998). Seawalls are subject to damage, and overtopping by big storms can lead to problems of drainage of water that gets behind them. The wall also serves to encourage erosion of beach deposits from the foot of the wall and can increase longshore sediment transport (Rakha and Kamphuis, 1997; Ruggiero and McDougal, 2001). Some studies have suggested that lowering of the beach can lead to scour beneath or around the wall, often necessitating further construction of a wall in a more seaward position (Carter, 1988). Other studies imply little change in beach morphology as a result of wall construction (Griggs *et al.*, 1994).

Groynes

Where a shoreline is eroding because of longshore transport of sediment, groynes are the most widely adopted solution. These are concrete, metal or wood structures constructed perpendicular or occasionally oblique to the shore. They intercept sediment moved alongshore and the sediment accretes on the updrift side of groynes, while erosion predominates on the downdrift side (Figure 10.6b). Between the groynes the short section of beach can realign itself so that it is more parallel to the prevailing direction of wave approach, which decreases the rate at which sediment can be carried alongshore (Silvester, 1974). Groynes function best in microtidal settings. Length and spacing are critical (see Figure 10.6b); the groynes should not be so short that scouring occurs around the rear (flanking).

A field of groynes, consisting of updrift anchor groynes and a downdrift terminal groyne, slows sediment transport (Carter, 1988). Sand builds up on the updrift side of the first groyne, often with erosion on the downdrift side. Groynes fill up in sequence from updrift to downdrift (see Figure 10.6b). The terminal groyne tends to exacerbate erosion downdrift of it, often requiring still further downdrift groynes (Inman and Brush, 1973). Sediment bypasses groynes when they are all filled (Kamphuis, 2000).

Shore protection

Shore protection is already important where there are extensive low-lying areas that require protection. The city of Venice already experiences extensive flooding; a flood in 1966 inundated St Mark's Square to a depth of 1.5 m. Interference with the mouth of the River Po, diverted during the 16th century, has exacerbated issues of subsidence and compaction in the Venice lagoon (Day *et al.*, 1999). The construction of engineered levées on rivers can have the effect of keeping flow within the banks and depriving the wetlands in interdistributary basins of sediments that would otherwise have built up the plains. This has occurred in the Mississippi River Delta and the city of New Orleans is now largely below sea level and requires constant pumping. The threat of high-tide levels has led to construction of the Thames Barrage at Silvertown, London (Carter, 1988) and similar barrages elsewhere such as on the Nagara River in the Kiso plains of Japan. The Dutch Delta project involving the mouths of the Rhine Scheldt and Maas Rivers is a large example as has been the transformation of the Zuider Zee into the freshwater Ijsselmeer. The Caspian Sea provides an example in which water level rise is faster than present rates of sea-level rise, and provides a further analogy as to how coastal landforms may react (Ignatov *et al.*, 1993; Clauer *et al.*, 2000).

There has been increasing recognition in recent years, called the 'quiet revolution' by Carter (1988), that solid engineering structures in the coastal zone often serve to relocate the problem, and a preference has emerged for soft engineering options, such as beach nourishment, working with nature rather than against it. There are few areas of developed coasts where the shoreline is entirely natural, and many European and North American coasts are now highly artificial after centuries of intervention. Human strategies on the coast have been heavily based on a static engineered response, whereas the coast is in, or strives towards, a dynamic equilibrium. Solid coastal structures are built and persist because they protect expensive infrastructure, but they often relocate the problem downdrift or to another part of the coast. Soft options like beach nourishment, while also being temporary and needing regular replenishment, appear more acceptable, and go some way to restore the natural dynamism of the shoreline. However, in many cases there is a legacy of decisions that were made in the past which have given rise to the present threats to coastal infrastructure and which necessitate immediate shore protection (Doornkamp, 1998). For instance, the seawall and promenade of many coastal cities in Europe represents a highly engineered use of prime seafront space, which might be preferably designated as public open space, parkland and amenities if it were available today. Such open space use might also allow greater flexibility in terms of future land-use change, for instance through managed retreat, in the face of threats of erosion or inundation as a result of sea-level rise. Foredune areas represent a natural reserve which can be called upon in the face of extreme events (see Chapter 6); building on these areas leaves little option but to undertake costly protective measures when extreme events (whether amplified by gradual global change or not) threaten. Managed retreat can comprise 'setbacks', rolling easements and other planning tools including building within a particular design life (Titus, 1998).

The futility of trying to predict future scenarios where there is a large human influence is clearly apparent. Even future climate is to a certain extent a function of what humans choose to make it, with recognition that greenhouse gas emissions influence climate and could be controlled. Future climate and non-climate scenarios are interlinked, and future coasts could to some extent be what planners choose to make them. Modelling options such as managed retreat will require greater incorporation of management goals, values and reflexivity in relation to environmental change (Lorenzoni *et al.*, 2000). There is scope for further 'manufactured landforms' or 'planned adaptation', whereby coastal landforms are built or reshaped. In addition to nourished and replenished beaches, there is increasing need to rehabilitate muddy coasts, with 'managed realignment' (Esselink *et al.*, 1998) and 'foreshore recharge' with fine sediment (Ford *et al.*, 1999). It is likely that under some circumstances entire coral reefs may be built to achieve the adjustments that are sought in the coastal landscape (Pittock, 1999).

Coastal erosion is already widespread, and there are many coasts where exceptional high tides or storm surges result in encroachment on the shore, impinging on human activity (Figure 10.7). If the sea rises, many coasts that are developed with infrastructure along or close to the shoreline will be unable to accommodate erosion, and will experience 'coastal squeeze'. This occurs where the ecological or geomorphological zones that would normally retreat landwards encounter solid structures and are squeezed out. Wetlands, salt marsh and mangroves, and adjacent freshwater wetlands are particularly likely to suffer from this squeeze (Lee, 2001).

10.5 Prospects

The realisation that human society is having impacts on the global environment and the concern about accelerated sea-level changes have served to focus attention on the dynamic changes that occur on the coast. Coastal landforms are dynamic and they are shaped by powerful processes with morphodynamic adjustments in response to extreme events and boundary-condition changes. This book has examined coastal morphology and processes and their interrelationships providing insights into the behaviour of a series of coastal environments. Coastal landforms are complex and unpredictable but there are patterns

Figure 10.7. Recent coastal erosion near Suoi Tien Village, Binh Thuân Province, southern Vietnam (photograph C.V. Murray-Wallace).

to the occurrence of phenomena, events recur, and geomorphology provides a morphodynamic framework for the better management of coastal areas.

Conceptual and simulation models of how the coast evolves provide tools to examine sensitivity to changes in key variables. Coastal scientists must continue to test and refine the premises and practical simplifications on which models are founded, identifying their shortcomings, challenging assumptions, and evaluating outcomes. Coastal resources are important in human use of the coast, but their exploitation needs to be viewed in terms of sustainability and in the context of the other competing uses and values. There are many coasts on which humans exert a direct influence and engineering plays an important role in ensuring that coastal infrastructure remains secure. On other coasts, human influence is indirect, subtle and barely discernible.

Future studies of coastal environments must be multidisciplinary, combining the insights gained from geological and paleoecological reconstructions with outcomes from short-term process studies to increase our understanding of coastal behaviour at the time scales that are important to society. If there is a real desire to move away from a simple hard-engineering approach to the coast, then there is a major challenge for coastal geomorphologists, together with other scientists, to become involved in assessing the probabilities of events occurring and the risk posed by potential hazards. There is a central role for more coastal education with better and wider communication of the results of scientific studies and a broader recognition of the uncertainties involved.

Recognising the role of human factors will require a rigorous understanding of how the natural system adjusts. It will also require awareness and an honest evaluation of a wide range of social, cultural and political issues that obfuscate the role that humans have had, or are having, in changing the way coastal systems operate. More problematic are the many situations where the coastal manager perceives that there has been a change in the coastal system, and endeavours to determine whether the change is part of a natural system progression, whether it is a response to human-induced impact, or whether one has exacerbated the other.

Global environmental change challenges us to look at coasts on time scales that are relevant to society. Major advances in understanding the course of late Quaternary sea-level fluctuations, rigorous analysis of tide-gauge and satellite altimeter records, and more detailed calibration of models, appear to be reducing the magnitude of any future sea-level rise expected as a result of global warming. Coastal geomorphology emphasises the prominent role that past sea-level changes have had in determining the pattern of coastal evolution, and the significance of ongoing relative sea-level changes to the behaviour of the modern coastline. It is clear that there is geographical variability not only in this pattern of relative sea-level adjustment but also in the range of other factors which influence the response of coastal landforms. This variability is still rarely incorporated into globally based predictions. Understanding how coastlines respond to changes in sea level, whether rises or falls and whether natural or human induced, remains a priority for the study of coastal morphodynamics. The challenge for coastal geomorphology is to develop robust morphodynamic models that can be used to assess the impact of changes in boundary conditions, whether associated with sea-level rise, storm incidence, or anthropogenic modification. More may be learned from a theory discarded than by those that appear upheld. It is often only by questioning conventional wisdom that choices can be made between the competing hypotheses. The challenge will be to ask the right questions.

Coastal geomorphology is built on a solid empirical and descriptive foundation. It has been at the forefront of the development of ideas in the earth sciences. Paradigms have shifted and we now have powerful conceptual frameworks within which to understand variability on the coast. The tools available to future coastal scientists promise unprecedented sophistication in terms of research and monitoring at spatial scales and over time frames never before anticipated. It is already possible to generate more detailed morphodynamic models with greater power to simulate the patterns of change that landforms experience. We will be better able to recognise the impact that humans are having, to measure or monitor it and, where appropriate, to take steps to diminish it.

References

Aagaard, T. and Masselink, G., 1999. The surf zone. In: A.D. Short (Editor), *Handbook of Beach and Shoreface Dynamics*. John Wiley & Sons, Chichester, pp. 72–118.

Abbot, R.H., 1997. A first look at Bligh's passage through the Great Barrier Reef. *International Journal of Remote Sensing*, 18: 1203–1206.

Abbs, D.J. and Physick, W.L., 1992. Sea-breeze observations and modelling: a review. *Australian Meteorological Magazine*, 41: 7–19.

Abrahams, A.D. and Oak, H.L., 1975. Shore platform widths between Port Kembla and Durras Lake, New South Wales. *Australian Geographical Studies*, 13: 190–194.

Adam, P., 1981. Australian saltmarshes. *Wetlands*, 1: 8–19.

Adam, P., 1990. *Saltmarsh Ecology*. Cambridge University Press, Cambridge.

Adey, W.H., 1975. The algal ridges and coral reefs of St Croix: their structure and Holocene development. *Atoll Research Bulletin*, 187: 1–67.

Adey, W.H., 1978. Coral reef morphogenesis: a multidimensional model. *Science*, 202: 831–837.

Adey, W.H. and Burke, R.B., 1976. Holocene bioherms (algal ridges and bank-barrier reefs) of the eastern Caribbean. *Geological Society of America Bulletin*, 87: 95–109.

Adey, W.H. and Burke, R.B., 1977. Holocene bioherms of Lesser Antilles: geographic control of development. In: S.H. Frost, M.P. Weiss and J.B. Saunders (Editors), *Reefs and Related Carbonates: Ecology and Sedimentology*. AAPG (American Association of Petroleum Geologists) Studies in Ecology 4, pp. 67–81.

Adjas, A., Masse, J.-P. and Montaggioni, L.F., 1990. Fine-grained carbonates in nearly closed reef environments: Mataiva and Takapoto atolls, Central Pacific Ocean. *Sedimentary Geology*, 67: 115–132.

Agassiz, A., 1899. The islands and coral reefs of Fiji. *Bulletin of the Museum of Comparative Zoology Harvard College*, 33: 1–167.

Aharon, P. and Chappell, J., 1986. Oxygen isotopes, sea level changes, and the temperature history of a coral reef environment in New Guinea over the last 10^5 years. *Palaeogeography, Palaeoclimatology, Palaeoecology*, 56: 337–379.

Ahnert, F., 1960. Estuarine meanders in the Chesapeake Bay area. *Geographical Review*, 50: 390–401.

Airy, G.B., 1845. On tides and waves. *Encyclopedia Metripolitana*, 5: 241–396.

Aitken, M.J., 1990. *Science-based Dating in Archaeology*. Longman, London, 274 pp.

Alexander, C.R., Nittrouer, C.A., Demaster, D.J., Park, Y.-A. and Park, S.-C., 1991. Macrotidal mudflats of the southwestern Korean coast: a model for interpretation of intertidal deposits. *Journal of Sedimentary Petrology*, 61: 805–824.

Alexander, I., Andres, M.S., Braithwaite, C.J.R., Braga, J.C., Cooper, M.J., Davies, P.J., Elderfield, P.J., Gilmour, H., Kay, M.A., Kroon, R.L.F., McKenzie, D., Montaggioni, L.F., Skinner, J.A., Thompson, A., Vasconcelos, R., Webster, C.J. and Wilson, P.A., 2001. New constraints on the origin of the Australian Great Barrier Reef: results from an international project of deep coring. *Geology*, 29: 483–486.

Allen, G.P., Laurier, D. and Thouvenin, J., 1979. Étude sédimentologique du delta de la Mahakam. *Compagnie Français des Pétroles Notes et Mémoires*, 15: 13–156.

Allen, G.P. and Posamentier, H.W., 1993. Sequence stratigraphy and facies model of an incised valley fill: the Gironde estuary, France. *Journal of Sedimentary Petrology*, 63: 378–391.

Allen, J.R.L., 1975. Sediments of the modern Niger delta: a summary and review. In: J.P. Morgan (Editor), *Deltaic Sedimentation: Modern and Ancient*. Society of Economic Palaeontologists and Mineralogists, Tulsa, OK, Special Publication, pp. 138–151.

Allen, J.R.L., 1987a. Late Flandrian shoreline oscillations in the Severn Estuary: the Rumney Formation and its typesite (Cardiff area). *Philosophical Transactions of the Royal Society of London, Series B*, 315: 157–174.

Allen, J.R.L., 1987b. Reworking of muddy intertidal sediments in the Severn Estuary, southwestern U.K. – a preliminary survey. *Sedimentary Geology*, 50: 1–23.

Allen, J.R.L., 1988. Nearshore sediment transport. *Geographical Review*, 78: 148–157.

Allen, J.R.L., 1992. Tidally induced marshes in the Severn Estuary, southwest Britain. In: J.R.L. Allen and K. Pye (Editors), *Saltmarshes: Morphodynamics, Conservation and Engineering Significance*. Cambridge University Press, Cambridge, pp. 123–147.

Allen, J.R.L., 1993. Muddy alluvial coasts of Britain: field criteria for shoreline position and movement in the recent past. *Proceedings of the Geologists Association*, 104: 241–262.

Allen, J.R.L., 1994. Fundamental properties of fluids and their relation to sediment transport processes. In: K. Pye (Editor), *Sediment Transport and Depositional Processes*. Blackwell Science Publications, Oxford, pp. 25–60.

Allen, J.R.L., 1997. Simulation models of salt-marsh morphodynamics: some implications for high-intertidal sediment couplets related to sea-level change. *Sedimentary Geology*, 113: 211–223.

Allen, J.R.L., 2000. Morphodynamics of Holocene salt marshes: a review sketch from the Atlantic and Southern North Sea coasts of Europe. *Quaternary Science Reviews*, 19: 1155–1231.

Allen, J.R.L. and Fulford, M.G., 1996. Late Flandrian coastal change and tidal palaeochannel development at Hills Flats, Severn Estuary (SW Britain). *Journal of the Geological Society of London*, 153: 151–162.

Allen, J.R.L. and Pye, K. (Editors), 1992. *Saltmarshes: Morphodynamics, Conservation and Engineering Significance*. Cambridge University Press, Cambridge, 184 pp.

Alleng, G.P., 1998. Historical development of the Port Royal Mangrove Wetland, Jamaica. *Journal of Coastal Research*, 14: 951–959.

Allison, M.A., 1998. Historical changes in the Ganges–Brahmaputra delta front. *Journal of Coastal Research*, 14: 1269–1275.

Allison, M.A. and Kepple, E.B., 2001. Modern sediment supply to the lower delta plain of the Ganges–Brahmaputra River in Bangladesh. *Geo-Marine Letters*, 21: 66–74.

Allison, M.A., Nittrouer, C.S., Faria, L.E.C., Silveira, O.M. and Mendes, A.C., 1996. Sources and sinks of sediment to the Amazon margin: the Amapa coast. *Geo-Marine Letters*, 16: 36–40.

Allison, M.A., Lee, M.T., Ogston, A.S. and Aller, R.C., 2000. Origin of Amazon mudbanks along the northeastern coast of South America. *Marine Geology*, 163: 241–256.

Allison, R.J., 1989. Rates and mechanisms of change in hard rock coastal cliffs. *Zeitschrift für Geomorphologie, Suppl.-Bd.*, 73: 125–138.

Allison, R.J. and Kimber, O.G., 1998. Modelling failure mechanisms to explain rock slope change along the Island of Purbeck Coast, UK. *Earth Surface Processes and Landforms*, 23: 731–750.

Amos, C.L., 1995. Siliclastic tidal flats. In: G.M.E. Perillo (Editor), *Geomorphology and Sedimentology of Estuaries. Developments in Sedimentology*. Elsevier, Amsterdam, pp. 273–306.

Amos, C.L., Tee, K.T. and Zaitlin, B.A., 1991. The post-glacial evolution of Chignecto Bay, Bay of Fundy, and its modern environment of deposition. *Canadian Society of Petroleum Geologists Memoir*, 16: 59–90.

Andersen, T.J., 2001. Seasonal variation in erodibility of two temperate, microtidal mudflats. *Estuarine, Coastal and Shelf Science*, 53: 1–12.

Anderson, F.E., 1972. Resuspension of estuarine sediments by small amplitude waves. *Journal of Sedimentary Petrology*, 42: 602–607.

Anderson, J.A.R., 1964. The structure and development of the peat swamps of Sarawak and Brunei. *Journal of Tropical Geography*, 18: 7–16.

Anderson, R.C., 1998. Submarine topography of Maldivian atolls suggests a sea level of 130 metres below present at the last glacial maximum. *Coral Reefs*, 17: 339–341.

Anderson, R.S., Densmore, A.L. and Ellis, M.A., 1999. The generation and degradation of marine terraces. *Basin Research*, 11: 7–19.

Andrews, C. and Williams, R.B.G., 2000. Limpet erosion of chalk shore platforms in southeast England. *Earth Surface Processes and Landforms*, 25: 1371–1381.

Andrews, E.C., 1916. Shoreline studies at Botany Bay. *Journal of the Royal Society of New South Wales*, 50: 165–176.

Andrews, J.T., 1970. *A Geomorphological Study of Post-Glacial Uplift with*

Particular Reference to Arctic Canada. Alden and Mowbray, Oxford, 156 pp.

Angulo, R.J. and Lessa, G.C., 1998. The Brazilian sea-level curves: a critical review with emphasis on the curves from the Paranaguá and Cananéia regions. *Marine Geology*, 140: 141–166.

Anthony, E.J., 1989. Chenier plain development in northern Sierra Leone, West Africa. *Marine Geology*, 90: 297–309.

Anthony, E.J., 1991. Beach-ridge plain development: Sherbro Island, Sierra Leone. *Zeitschrift für Geomorphologie, NF Suppl.-Bd.*, 81: 85–98.

Anthony, E.J., 1995. Beach-ridge development and sediment supply: examples from West Africa. *Marine Geology*, 129: 175–186.

Anthony, E.J. and Blivi, A.B., 1999. Morphosedimentary evolution of a delta-sourced, drift-aligned sand barrier-lagoon complex, western Bight of Benin. *Marine Geology*, 158: 161–176.

Arber, E.A.N., 1911. *The Coast Scenery of North Devon*. Kingsmead Reprints, Bath, 261pp.

Arber, M.A., 1949. Cliff profiles of Devon and Cornwall. *Geographical Journal*, 114: 191–197.

Arber, M.A., 1974. The cliffs of north Devon. *Proceedings of the Geologists Association*, 85: 147–157.

Are, F. and Reimnitz, E., 2000. An overview of the Lena River Delta setting: geology, tectonics, geomorphology and hydrology. *Journal of Coastal Research*, 16: 1083–1093.

Arens, S., 1996. Patterns of sand transport on vegetated dunes. *Geomorphology*, 17: 339–350.

Ashley, G.M., 1988. Tidal channel classification for a low mesotidal saltmarsh. *Marine Geology*, 82: 17–32.

Ashton, A., Murray, A.B. and Arnault, O., 2001. Formation of coastline features by large-scale instabilities induced by high-angle waves. *Nature*, 414: 296–300.

Atwater, B.F. and Moore, A.L., 1992. A tsunami about 1000 years ago in Puget Sound, Washington. *Science*, 258: 1614–1617.

Aubrey, D.G., 1979. Seasonal patterns of onshore/offshore sediment movement. *Journal of Geophysical Research*, 85: 3264–3276.

Aubrey, D.G. and Gaines, A.G., 1982. Rapid formation and degradation of barrier spits in areas with low rates of littoral drift. *Marine Geology*, 49: 257–278.

Aucan, J. and Ridd, P.V., 2000. Tidal asymmetry in creeks surrounded by salt-flats and mangroves with small swamp slopes. *Wetlands Ecology and Management*, 8: 223–231.

Augustinus, P.G.E.F., 1989. Cheniers and chenier plains: a general introduction. *Marine Geology*, 90: 219–229.

Augustinus, P.G.E.F., Hazelhoff, L. and Kroon, A., 1989. The chenier coast of Suriname: modern and geological development. *Marine Geology*, 90: 269–281.

Austin, R.M., 1991. Modelling Holocene tides on the NW European continental shelf. *Terra Nova*, 3: 276–288.

Axelsson, V., 1967. The Laitaure delta: a study of deltaic morphology and processes. *Geografiska Annaler*, 49A: 1–127.

Ayukai, T. and Wolanski, E., 1997. Importance of biologically mediated removal of fine sediments from the Fly River plume, Papua New Guinea. *Estuarine, Coastal and Shelf Science*, 44: 629–639.

Azam, M.H. and Mokhtar, N., 2000. An experimental investigation of broken wave over mud bed. *Journal of Coastal Research*, 16: 965–975.

Backshall, D.G., Barnett, J., Davies, P.J., Duncan, D.C., Harvey, N., Hopley, D., Isdale, P.J., Jennings, J.N. and Moss, R., 1979. Drowned dolines – the blue holes of the Pompey Reefs, Great Barrier Reef. *Bureau of Mineral Resources Journal of Australian Geology and Geophysics*, 4: 99–109.

Baeteman, C., Beets, D.J. and Van Strydonck, M., 1999. Tidal crevasse splays as the cause of rapid changes in the rate of aggradation in the Holocene tidal deposits of the Belgian Coastal Plain. *Quaternary International*, 56: 3–13.

Bagnold, R.A., 1940. Beach formation by waves; some model-experiments in a wave tank. *Journal of the Institute of Civil Engineers*, 15: 27–53.

Bagnold, R.A., 1941. *The Physics of Blown Sand and Desert Dunes*. Chapman and Hall, London, 265 pp.

Bagnold, R.A., 1946. Motion of waves in shallow water; interaction between waves and sand bottoms. *Proceedings of the Royal Society of London, Series A*, 187: 1–18.

Bagnold, R.A., 1963. Mechanics of marine sedimentation. In: M.N. Hill (Editor), *The Sea*, Vol. 3. Wiley Interscience, New York, pp. 507–528.

Bagnold, R.A., 1966. An approach to the sediment transport problem from general physics. *United States Geological Survey, Professional Paper*, 422–I.

Bailard, J.A. and Inman, D.L., 1981. An energetics bedload model for a plane sloping beach: local transport. *Journal of Geophysical Research*, 86: 10938–10954.

Baines, G.B.K. and McLean, R.F., 1976a. Resurveys of 1972 hurricane rampart on Funafuti atoll. *Search*, 7: 36–37.

Baines, G.B.K. and McLean, R.F., 1976b. Sequential studies of hurricane bank evolution at Funafuti atoll. *Marine Geology*, 21: M1–M8.

Baines, G.B.K., Beveridge, P.J. and Maragos, J.E., 1974. Storms and island building at Funafuti Atoll, Ellice Islands. *Proceedings of the Second International Coral Reef Symposium*, Vol. 2. Great Barrier-Reef Committee, Brisbane, pp. 485–496.

Bak, P., Tang, C. and Wiesenfeld, K., 1988. Self organized criticality. *Physics Review*, A38: 364–374.

Baker, E.K., Harris, P.T., Short, S.A. and Keene, J.B., 1995. Patterns of sedimentation in the Fly River Delta. In: B.W. Flemming and A. Bartholomä (Editors), *Tidal Signatures in Modern and Ancient Environments*. International Association of Sedimentologists Special Publication, Blackwell Scientific, Oxford, pp. 193–211.

Baker, R.G.V. and Haworth, R.J., 2000. Smooth or oscillating late Holocene sea-level curve? Evidence from palaeo-zoology of fixed biological indicators in east Australia and beyond. *Marine Geology*, 163: 367–386.

Baker, R.G.V., Haworth, R.J. and Flood, P.G., 2001. Warmer or cooler late

Holocene marine palaeoenvironments?: Interpreting southeast Australian and Brazilian sea-level changes using fixed biological indicators and their $\delta^{18}O$ composition. *Palaeogeography, Palaeoclimatology, Palaeoecology*, 168: 249–272.

Baker, T.F., 1993. Absolute sea level measurements, climate change and vertical crustal movements. *Global and Planetary Change*, 8: 149–159.

Baker, V.R., 1977. Stream-channel response to floods with examples from central Texas. *Geological Society of America Bulletin*, 88: 1057–1071.

Baker, V.R., 1994. Geomorphological understanding of floods. *Geomorphology*, 10: 139–156.

Baker, V.R., 1996. Hypotheses and geomorphological reasoning. In: B.L. Rhoads and C.E. Thorn (Editors), *The Scientific Nature of Geomorphology*. John Wiley & Sons, Chichester, pp. 57–85.

Bal, A.A., 1997. Sea caves, relict shore and rock platforms: evidence for the tectonic stability of Banks Peninsula, New Zealand. *New Zealand Journal of Geology and Geophysics*, 40: 299–305.

Balchin, W.G.V., 1941. The raised features of Billfiord and Sassenfiord, West Spitsbergen. *Geographical Journal*, 97: 364–376.

Ball, M.C., 1998. Mangrove species richness in relation to salinity and waterlogging: a case study along the Adelaide River floodplain, northern Australia. *Global Ecology and Biogeography Letters*, 7: 73–82.

Ball, M.M., 1967. Carbonate sand bodies of Florida and the Bahamas. *Journal of Sedimentary Petrology*, 37: 556–591.

Balsillie, J.H., Campbell, K., Coleman, C., Entsminger, L., Glassen, R., Hajishafie, N., Huang, D., Tunsoy, A.F. and Tanner, W.F., 1976. Wave parameter gradients along the wave ray. *Marine Geology*, 22: M17–M21.

Baltzer, F. and Purser, B.H., 1990. Modern alluvial fan and deltaic sedimentation in a foreland tectonic setting: the Lower Mesopotamian Plain and the Arabian Gulf. *Sedimentary Geology*, 67: 175–197.

Bard, E., Hamelin, B. and Fairbanks, R.G., 1990a. U-Th ages obtained by mass spectrometry in corals from Barbados: sea level during the past 130000 years. *Nature*, 346: 456–458.

Bard, E., Hamelin, B., Fairbanks, R.G. and Zindler, A., 1990b. Calibration of the ^{14}C timescale over the past 30000 years using mass spectrometric U-Th ages from Barbados corals. *Nature*, 345: 405–410.

Bard, E., Hamelin, B., Arnold, M., Montaggioni, L., Cabioch, G., Faure, G. and Rougerie, F., 1996. Deglacial sea-level record from Tahiti corals and the timing of global meltwater discharge. *Nature*, 382: 241–244.

Barnes, R.S.K., 1980. *Coastal Lagoons*. Cambridge University Press, Cambridge, 106 pp.

Barnes, R.S.K., 2001. Lagoons. In: J.H. Steele, S.A. Thorpe and K.K. Turekian (Editors), *Encyclopedia of Ocean Sciences*. Academic Press, San Diego, pp. 1427–1438.

Barnett, T.P. and Sutherland, A.J., 1968. A note on an overshoot effect in wind-generated waves. *Journal of Geophysical Research*, 73: 6879–6885.

Barrell, J., 1912. Criteria for the recognition of ancient delta deposits. *Geological Society of America Bulletin*, 23: 377–446.

Barth, M.C. and Titus, J.G., 1984. *Greenhouse Effect and Sea Level Rise: A Challenge for this Generation*. Van Nostrand Reinhold, New York, 325 pp.

Bartholdy, J. and Aagaard, T., 2001. Storm surge effects on a back-barrier tidal flat of the Danish Wadden Sea. *Geo-Marine Letters*, 20: 133–141.

Bartol, I.K., Mann, R. and Luckenbach, M., 1999. Growth and mortality of oysters (*Crassostrea virginica*) on constructed intertidal reefs: effects of tidal height and substrate level. *Journal of Experimental Marine Biology and Ecology*, 237: 157–184.

Barton, M.E. and Coles, B.J., 1984. The characteristics and rates of the various slope degradation processes in the Barton Clay cliffs of Hampshire. *Quarterly Journal of Engineering Geology, London*, 17: 117–136.

Bartrum, J.A., 1916. High water rock platforms: a phase of shoreline erosion. *Transactions of the New Zealand Institute*, 48: 132–134.

Bartrum, J.A., 1926. Abnormal shore platforms. *Journal of Geology*, 34: 793–807.

Bartrum, J.A., 1935. Shore platforms. *Proceedings of the Australian and New Zealand Association for the Advancement of Science*, 22: 135–143.

Bartrum, J.A., 1938. Shore platforms. *Journal of Geomorphology*, 1: 266–278.

Bartrum, J.A. and Turner, F.J., 1928. Pillow lavas, periodotites, and associated rocks from northernmost New Zealand. *Transactions of the New Zealand Institute*, 59: 98–138.

Barua, D.K., 1990. Suspended sediment movement in the estuary of the Ganges–Brahmaputra–Meghna river system. *Marine Geology*, 91: 243–253.

Barua, D.K., 1991. The coastline of Bangladesh: an overview of processes and forms, In: N.C. Kraus, K.J. Gingerida and D.L. Kriebel (Editors), Coastal Sediments '91: American Society of Civil Engineers, New York, pp. 2284–2301.

Barui, N. and Chanda, S., 1992. Late-Quaternary pollen analysis in relation to palaeoecology, biostratigraphy and dating of Calcutta peat. *Proceedings of the Indian National Academy of Science B*, 58: 191–200.

Bascom, W.N., 1951. The relationship between sand size and beach face slope. *Transactions of the American Geophysical Union*, 32: 866–874.

Bascom, W.N., 1964. *Waves and Beaches*. Anchor, New York, 267pp.

Bates, C.C., 1953. Rational theory of delta formation. *Bulletin of the American Association of Petroleum Geologists*, 37: 2119–2162.

Bathurst, R.G.C., 1975. *Carbonate Sediments and their Diagenesis*. Elsevier Scientific Publishing, Amsterdam, 658 pp.

Batiza, R., 2001. Seamounts and off-ridge volcanism. In: J.H. Steele, S.A. Thorpe and K.K. Turekian (Editors), *Encyclopedia of Ocean Sciences*. Academic Press, San Diego, pp. 2696–2708.

Battjes, J.A., 1974. Surf similarity. *Proceedings of the Fourteenth International Conference on Coastal Engineering*, Honolulu, HI, American Society of Civil Engineers, 446–480.

Bauer, B.O., 1991. Aeolian decoupling of beach sediments. *Annals of the Association of American Geographers*, 81: 290–303.

Bauer, B.O. and Greenwood, B., 1988. Surf-zone similarity. *Geographical Review*, 78: 138–147.

Bauer, B.O. and Greenwood, B., 1990. Modification of a linear bar-trough system by a standing edge wave. *Marine Geology*, 92: 177–204.

Bauer, B.O., Sherman, D.J., Nordstron, K.F. and Gares, P.A., 1990. Aeolian transport measurement and prediction across a beach and dune at Catroville, California. In: K.F. Nordstrom, N.P. Psuty and R.W.G. Carter (Editors), *Coastal Dunes: Form and Process*. Wiley, New York, pp. 39–55.

Baumann, R.H., Day, J.W. and Miller, C.A., 1984. Mississippi deltaic wetland survival: sedimentation versus coastal submergence. *Science*, 224: 1093–1095.

Bayliss-Smith, T.P., 1988. The role of hurricanes in the development of reef islands, Ontong Java Atoll, Solomon Islands. *Geographical Journal*, 154: 377–391.

Bayliss-Smith, T.P., Healey, R., Lailey, R., Spencer, T. and Stoddart, D.R., 1979. Tidal flows in salt marsh creeks. *Estuarine, Coastal and Marine Science*, 9: 235–255.

Beck, J.W., Edwards, R.L., Ito, E., Taylor, F.W., Recy, J., Rougerie, F., Joannot, P. and Henin, C., 1992. Sea-surface temperature from coral skeletal strontium/calcium ratios. *Science*, 257: 644–647.

Beeftink, W.G. and Rozema, J., 1988. The nature and functioning of salt marshes. In: W. Salomans, B.L. Bayne, E.K. Duursma and U. Förstner (Editors), *Pollution of the North Sea: An Assessment*. Springer, Berlin, pp. 59–87.

Beets, D.J. and van der Spek, A.J.F., 2000. The Holocene evolution of the barrier and the back-barrier basins of Belgium and the Netherlands as a function of late Weichselian morphology, relative sea-level rise and sediment supply. *Geologie en Mijnbouw*, 79: 3–16.

Belknap, D.F. and Kraft, J.C., 1981. Preservation potential of transgressive coastal lithosomes on the U.S. Atlantic shelf. *Marine Geology*, 42: 429–442.

Bellotti, P., Milli, S., Tortora, P. and Valeri, P., 1995. Physical stratigraphy and sedimentology of the Late Pleistocene–Holocene Tiber Delta depositional sequence. *Sedimentology*, 42: 617–634.

Bellwood, D.R., 1995. Carbonate transport and within-reef patterns of bioerosion and sediment release by parrotfishes (family Scaridae) on the Great Barrier Reef. *Marine Ecology Progress Series*, 117: 127–136.

Belov, A.P., Davies, P. and Williams, A.T., 1999. Mathematical modelling of basal coastal cliff erosion in uniform strata: a theoretical approach. *Journal of Geology*, 107: 99–109.

Belperio, A.P., 1993. Land subsidence and sea level rise in the Port Adelaide estuary: implications for monitoring the greenhouse effect. *Australian Journal of Earth Sciences*, 40: 359–368.

Benavente, J., Gracia, F.J. and Lopez-Aguayo, F., 2000. Empirical model of morphodynamic beachface behaviour for low-energy mesotidal environments. *Marine Geology*, 167: 375–390.

Bennett, R.J. and Chorley, R.J., 1978. *Environmental Systems: Philosophy, Analysis and Control*. Princeton University Press, Princeton, NJ, 624 pp.

Benson, B.E., Grimm, K.A. and Clague, J.J., 1997. Tsunami deposits beneath tidal marshes on northwestern Vancouver Island, British Columbia. *Quaternary Research*, 48: 192–204.

Benum, B.T., Storlazzi, C.D., Seymour, R.J. and Griggs, G.B., 2000. The relationship between incident wave energy and seacliff erosion rates: San Diego County, California. *Journal of Coastal Research*, 16: 1162–1178.

Berendsen, H.J.A. and Stouthamer, E., 2000. Weichselian and Holocene palaeogeography of the Rhine–Meuse delta, The Netherlands. *Palaeogeography, Palaeoclimatology, Palaeoecology*, 161: 311–335.

Berg, N.H., 1983. Field evaluation of some sand transport models. *Earth Surface Processes and Landforms*, 8: 101–114.

Bernard, H.A., Le Blanc, R.J. and Major, C.F., 1962. Recent and Pleistocene geology of southeast Texas. In: E.H. Rainwater and R.P. Zingula (Editors), *Geology of the Gulf Coast and Central Texas: Guidebook of Excursion*. Houston Geological Society, Houston, pp. 175–205.

Beveridge, W.I.B., 1980. *The Seeds of Discovery*. Norton, New York, 130 pp.

Bhattacharya, J.P. and Walker, R.G., 1992. Deltas. In: R.G. Walker and N.P. James (Editors), *Facies Models: Response to Sea Level Change*. Geological Association of Canada, Toronto, pp. 157–177.

Bigarella, J.J., 1965. Sand-ridge structures from Parana coastal plain. *Marine Geology*, 3: 269–278.

Bijlsma, L., Ehler, C.N., Klein, R.J.T., Kulshrestha, S.M., McLean, R.F., Mimura, N., Nicholls, R.J., Nurse, L.A., Pérez Nieto, H., Stakhiv, E.Z., Turner, R.K. and Warrick, R.A., 1996. Coastal zones and small islands. In: R.T. Watson, M.C. Zinyowera and R.H. Moss (Editors), *Climate Change 1995: Impacts, Adaptations and Mitigation*. Cambridge University Press, Cambridge, pp. 289–324.

Bird, E.C.F., 1960. The formation of sand beach-ridges. *Australian Journal of Science*, 22: 349–350.

Bird, E.C.F., 1965. The formation of coastal dunes in the humid tropics: some evidence from North Queensland. *Australian Journal of Science*, 27: 258–259.

Bird, E.C.F., 1967. Coastal lagoons of southeastern Australia. *Australian Geographer*, 8: 365–385.

Bird, E.C.F., 1971. Mangroves as land-builders. *Victorian Naturalist*, 88: 189–197.

Bird, E.C.F., 1973. The evolution of sandy barrier formations on the East Gippsland coast. *Proceedings of the Royal Society of Victoria*, 79: 75–88.

Bird, E.C.F., 1974. Dune stability on Fraser Island. *Queensland Naturalist*, 21: 15–21.

Bird, E.C.F., 1976. *Coasts*, Australian National University Press, Canberra, 282 pp.

Bird, E.C.F., 1985a. *Coastline Changes*. Wiley Interscience, Chichester, 219 pp.

Bird, E.C.F., 1985b. Indonesia. In: E.C.F. Bird and M.L. Schwartz (Editors), *The World's Coastline*. Van Nostrand Reinhold, New York, pp. 879–888.

Bird, E.C.F., 1986. Mangroves and intertidal morphology in Westernport Bay, Victoria, Australia. *Marine Geology*, 69: 251–271.

Bird, E.C.F., 1993a. *Submerging Coasts: The Effects of a Rising Sea Level on Coastal Environments*. John Wiley & Sons, Chichester, 184 pp.

Bird, E.C.F., 1993b. *The Coast of Victoria*. Melbourne University Press, Melbourne, 324 pp.

Bird, E.C.F., 1994. Physical setting and geomorphology of coastal lagoons. In: B.J. Kjerfve (Editor), *Coastal Lagoon Processes*. Elsevier, Amsterdam, pp. 9–39.

Bird, E.C.F. 1996. *Beach Management*. John Wiley & Sons, Chichester, 281 pp.

Bird, E.C.F., 2000. *Coastal Geomorphology: An Introduction*. John Wiley & Sons, Chichester, 322 pp.

Bird, E.C.F. and Dent, O.F., 1966. Shore platforms of the South Coast of New South Wales. *Australian Geographer*, 19: 71–80.

Bird, E.C.F. and Jones, D.J.B., 1988. The origin of foredunes on the coast of Victoria, Australia. *Journal of Coastal Research*, 4: 181–192.

Bird, E.C.F. and Rosengren, N.J., 1987. Coastal cliff management: an example from Black Rock Point, Melbourne, Australia. *Journal of Shoreline Management*, 3: 39–51.

Bird, E.C.F. and Schwartz, M.L., 1985. *The World's Coastline*. Van Nostrand Reinhold, New York, 1071 pp.

Bird, J.B., Richards, A. and Wong, P.P., 1979. Coastal subsystems of western Barbados, West Indies. *Geografiska Annaler*, 61A: 221–236.

Bishop, P. and Cowell, P., 1997. Lithological and drainage network determinants of the character of drowned, embayed coastlines. *Journal of Geology*, 105: 685–699.

Bjerknes, J., 1969. Atmospheric teleconnections from the equatorial Pacific. *Monthly Weather Review*, 97: 163–172.

Black, K.P. and Rosenberg, M.A., 1991. Suspended sediment load at three time scales. In: N.C. Kraus, K.J. Gingerich and D.L. Kriebel (Editors), *Coastal Sediments '91*. American Society of Civil Engineers, New York, pp. 313–327.

Blanchon, P. and Eisenhauer, A., 2001. Multi-stage reef development on Barbados during the last interglaciation. *Quaternary Science Reviews*, 20: 1093–1112.

Blanchon, P. and Jones, B., 1997. Hurricane control on shelf-edge-reef architecture around Grand Cayman. *Sedimentology*, 44: 479–506.

Blanchon, P. and Shaw, J., 1995. Reef drowning during the last deglaciation: Evidence for catastrophic sea-level rise and ice-sheet collapse. *Geology*, 23: 4–8.

Blanchon, P., Jones, B. and Kalbfleisch, W., 1997. Anatomy of a fringing reef around Grand Cayman Island: storm rubble, not coral framework. *Journal of Sedimentary Petrology*, 67: 1–16.

Blasco, F., 1975. *The Mangroves of India*. Institut Français De Pondichery, India, 175 pp.

Blasco, F., Gauquelin, T., Rasolofoharinoro, M., Denis, J., Azipuru, M. and Caldairou, V., 1998. Recent advances in mangrove studies using remote sensing data. *Marine and Freshwater Research*, 49: 287–296.

Bloom, A.L., 1967. Pleistocene shorelines: a new test of isostasy. *Geological Society of America Bulletin*, 78: 1477–1494.

Bloom, A.L., 1970. Paludal stratigraphy of Truk, Ponape, and Kusaie, Eastern Caroline Islands. *Geological Society of America Bulletin*, 81: 1895–1904.

Bloom, A.L., 1977. *Atlas of Sea-Level Curves*. IGCP 61. Cornell University, Ithaca.

Bloom, A.L. and Yonekura, N., 1985. Coastal terraces generated by sea-level change and tectonic uplift. In: M.J. Woldenberg (Editor), *Models in Geomorphology*. Allen & Unwin, Boston, pp. 139–154.

Bloom, A.L., Broecker, W.S., Chappell, J.M.A., Matthews, R.K. and Mesolella, K.J., 1974. Quaternary sea level fluctuations on a tectonic coast, new $^{230}Th/^{234}U$ dates for the Huon Peninsula, New Guinea. *Quaternary Research*, 4: 185–205.

Blum, M.D. and Törnqvist, T.E., 2000. Fluvial responses to climate and sea-level changes: a review. *Sedimentology*, 47 (Suppl. 1): 2–48.

Blum, M.D., Misner, T.J., Collins, E.S., Scott, D.B., Morton, R.A. and Aslan, A., 2001. Middle Holocene sea-level rise and highstand at +2 m, central Texas coast. *Journal of Sedimentary Research*, 71: 581–588.

Blumenstock, D.I., 1961. A report on typhoon effects upon Jaluit Atoll. *Atoll Research Bulletin*, 75: 1–105.

Bondevik, S., Svendsen, J.I. and Mangerud, J., 1998. Distinction between the Storegga tsunami and the Holocene marine transgression in coastal basin deposits of western Norway. *Journal of Quaternary Science*, 13: 529–537.

Boon, J.D. and Byrne, R.J., 1981. On basin hypsometry and the morphodynamic response of coastal inlet systems. *Marine Geology*, 40: 27–48.

Borstad, G., Brown, L., Cross, W., Nallee, M. and Wainwright, P., 1997. Towards a management plan for a tropical reef-lagoon system using airborne multi-spectral imaging and GIS. *Proceedings of the Fourth International Conference on Remote Sensing for Marine and Coastal Environments*: Vol. II, Environmental Research Institute of Michigan, Ann Arbor, pp. 605–610.

Bosence, D.W.J., 1983. The occurrence and ecology of recent rhodiliths: a review. In: T.M. Peryt (Editor), *Coated Grains*. Springer-Verlag, Heidelberg, pp. 225–242.

Boss, S.K. and Neumann, A.C., 1993. Impacts of Hurricane Andrew on carbonate platform environments, northern Great Bahama Bank. *Geology*, 21: 897–900.

Boto, K.G. and Bunt, J.S., 1982. Carbon export from mangroves. In: I.E. Galbally and J.R. Freney (Editors), *The Cycling of Carbon, Nitrogen, Sulfur and Phosphorus in Terrestrial and Aquatic Ecosystems*. Australian Academy of Science, Canberra, pp. 105–110.

Bourman, R.P., 1986. Aeolian sand transport along beaches. *Australian Geographer*, 17: 30–34.

Bourrouilh-Le Jan, F.G. and Talandier, J., 1985. Sédimentation et fracturation de haute énergie en milieu récifal: tsunamis, auragans et cyclones et leurs effets sur la sédimentologie et la géomophologie d'un atoll: motu et hoa, à Rangiroa, Tuamotu, Pacifique SE. *Marine Geology*, 67: 263–333.

Bowen, A.J., 1980. Simple models of nearshore sedimentation: beach profiles and longshore bars. In: S.B. McCann (Editor), *The Coastline of Canada*. Geological Survey of Canada, Ottawa, pp. 1–11.

Bowen, A.J. and Guza, R.T., 1978. Edge waves and surf beat. *Journal of Geophysical Research*, 83: 1913–1920.

Bowen, A.J. and Huntley, D.A., 1984. Waves, longwaves and nearshore morphology. *Marine Geology*, 60: 1–13.

Bowen, A.J. and Inman, D.L., 1966. *Budget of Littoral Sediments in the Vicinity of Point Arguello, California*. US Army Coastal Engineering Research Center, Technical Memorandum, 19.

Bowen, A.J. and Inman, D.L., 1969. Rip currents. 2. Laboratory and field observations. *Journal of Geophysical Research*, 74: 5479–5490.

Bowen, A.J., Inman, D.L. and Simmons, V.P., 1968. Wave 'set-down' and wave set-up. *Journal of Geophysical Research*, 73: 2569–2577.

Bowen, D.Q., 1978. *Quaternary Geology*. Pergamon Press, Oxford, 221 pp.

Bowler, J.M., Hope, G.S., Jennings, J.N., Singh, G. and Walker, D., 1976. Late Quaternary climates of Australia and New Guinea. *Quaternary Research*, 6: 359–394.

Bowman, G.M., 1989. Podzol development in a Holocene chronosequence. I. Moruya Heads, New South Wales. *Australian Journal of Soil Research*, 27: 607–628.

Bowman, H.H.M., 1917. Ecology and physiology of the red mangrove. *Proceedings of the American Philosophical Society*, 56: 589–672.

Boyd, R., Bowen, A.J. and Hall, R.K., 1987. An evolutionary model for transgressive sedimentation on the eastern shore of Nova Scotia. In: D.M. FitzGerald and P.S. Rosen (Editors), *Glaciated Coasts*. Academic Press, London, pp. 87–114.

Boyd, R., Dalrymple, R. and Zaitlin, B.A., 1992. Classification of clastic coastal depositional environments. *Sedimentary Geology*, 80: 139–150.

Bradley, W.C., 1958. Submarine abrasion and wave-cut platforms. *Bulletin of the Geological Society of America*, 69: 967–974.

Braithwaite, C.J.R., Taylor, J.D. and Kennedy, W.J., 1973. The evolution of an atoll: the depositional and erosional history of Aldabra. *Philosophical Transactions of the Royal Society of London, Series B*, 266: 307–340.

Braithwaite, C.J.R., Montaggioni, L.F., Camoin, G.F., Dalmasso, H., Dullo, W.C. and Mangini, A., 2000. Origins and development of Holocene coral reefs: a revisited model based on reef boreholes in the Seychelles, Indian Ocean. *International Journal of Earth Sciences*, 89: 431–445.

Brampton, A.H., 1992. Engineering significance of British saltmarshes. In: J.R.L. Allen and K. Pye (Editors), *Saltmarshes: Morphodynamics, Conservation and Engineering Significance*. Cambridge University Press, Cambridge, pp. 115–122.

Brander, R.W., 1999. Field observations on the morphodynamics of rip currents. *Marine Geology*, 157: 199–218.

Brander, R.W. and Short, A.D., 2000. Morphodynamics of a large-scale rip current system, Muriwai Beach, New Zealand. *Marine Geology*, 165: 27–39.

Bray, M.J., 1997. Episodic shingle supply and the modified development of Chesil Beach, England. *Journal of Coastal Research*, 13: 453–467.

Bray, M.J. and Hooke, J.M., 1997. Prediction of soft-cliff retreat with accelerating sea-level rise. *Journal of Coastal Research*, 13: 453–467.

Bray, M.J., Carter, D.J. and Hooke, J.M., 1995. Littoral cell definition and budgets for central south England. *Journal of Coastal Research*, 11: 381–400.

Bray, T.F. and Carter, C.H., 1992. Physical processes and sedimentary record of a modern, transgressive, lacustrine barrier island. *Marine Geology*, 105: 155–168.

Bremontier, N.T., 1833. Mémoire sur les dunes. *Annales des Ponts et Chausées*, 1: 145–224.

Bretschneider, C.L., 1958. Revisions in wave forecasting: deep and shallow water. *Proceedings Sixth Conference on Coastal Engineering*, American Society of Civil Engineers, pp. 30–67.

Bretz, J.H., 1960. Bermuda: a partially drowned late mature Pleistocene karst. *Geological Society of America Bulletin*, 81: 2523–2524.

Bristow, C.S., Chrostan, P.N. and Bailey, S.D., 2000. The structure and development of foredunes on a locally prograding coast: insights from ground penetrating radar surveys, Norfolk, United Kingdom. *Sedimentology*, 47: 923–944.

Broecker, W.S., 1997. Thermohaline circulation, the Achilles heel of our climate system: will man-made CO_2 upset the current balance? *Science*, 278: 1582–1588.

Broecker, W.S., Thurber, D.L., Goddard, J., Ku, T.-L., Matthews, R.K. and Mesolella, K.J., 1968. Milankovitch hypothesis supported by precise dating of coral reefs and deep-sea sediments. *Science*, 159: 297–300.

Brooke, B., 2001. The distribution of carbonate eolianite. *Earth Science Reviews*, 55: 135–164.

Brooke, B.P., Young, R.W., Bryant, E.A., Murray-Wallace, C.V. and Price, D.M., 1994. A Pleistocene origin for shore platforms along the northern Illawarra coast, New South Wales. *Australian Geographer*, 25: 178–185.

Brown, B.E., 1997a. Adaptations of reef corals to physical environmental stress. *Advances in Marine Biology*, 31: 221–299.

Brown, B.E., 1997b. Coral bleaching: causes and consequences. *Coral Reefs*, 16: S129–S138.

Brown, B.E. and Dunne, R.P., 1988. The environmental impact of coral mining in the Maldives. *Environmental Conservation*, 15: 159–166.

Brown, E.H., 1960. *The Relief and Drainage of Wales*. University of Wales Press, Cardiff, 186 pp.

Brown, S.L., 1998. Sedimentation on a Humber saltmarsh. In: K.S. Black, D.M. Paterson and A. Cramp (Editors), *Sedimentary Processes in the Intertidal Zone*. Geological Society of London, Special Publication, pp. 69–83.

Bruggeman, J.H., van Kessel, A.M., van Rooij, J.M. and Breeman, A.M., 1996. Bioerosion and sediment ingestion by the Caribbean parrotfishes *Scarus vetula* and *Sparisoma viride*: implications of fish size, feeding mode and habitat use. *Marine Ecology Progress Series*, 134: 59–71.

Brunsden, D., 2001. A critical assessment of the sensitivity concept in geomorphology. *Catena*, 42: 99–123.

Brunsden, D. and Jones, D.K.C., 1980. Relative time scales and formative events in coastal landslide systems. *Zeitschrift für Geomorphologie, NF Suppl.-Bd*, 34: 1–19.

Brunsden, D. and Thornes, J.B., 1979. Landscape sensitivity and change. *Transactions of the Institute of British Geographers*, N.S., 4: 463–484.

Bruun, P., 1954. *Coast Erosion and the Development of Beach Profiles*. Beach Erosion Board, US Army Corps of Engineers, Technical Memorandum, 44: 1–79.

Bruun, P., 1962. Sea-level rise as a cause of shore erosion. *American Society of Civil Engineering Proceedings, Journal of Waterways and Harbors Division*, 88: 117–130.

Bruun, P., 1978. *Stability of Coastal Inlets: Theory and Engineering*. Elsevier, Amsterdam, 509 pp.

Bruun, P., 1988. The Bruun rule of erosion by sea-level rise: a discussion on large-scale two- and three-dimensional usages. *Journal of Coastal Research*, 4: 627–648.

Bryant, E.A., 1979. Edge wave and sediment sorting relationships on beach foreshores, Broken Bay. *Search*, 10: 442–443.

Bryant, E.A., 1982. Behaviour of grain size characteristics on reflective and dissipative foreshores, Broken Bay, Australia. *Journal of Sedimentary Petrology*, 52: 431–450.

Bryant, E.A., 1983. Regional sea level, southern oscillation and beach change, New South Wales, Australia. *Nature*, 305: 213–216.

Bryant, E.A., 1988. Storminess and high tide beach change, Stanwell Park, Australia, 1943–1978. *Marine Geology*, 79: 171–187.

Bryant, E.A., 2001. *Tsunami: The Underrated Hazard*. Cambridge University Press, Cambridge, 320 pp.

Bryant, E.A. and Price, D.M., 1997. Late Pleistocene marine chronology of the Gippsland Lakes region, Australia. *Physical Geography*, 18: 318–334.

Bryant, E.A. and Young, R.W., 1996. Bedrock-sculpturing by tsunami, South Coast, New South Wales, Australia. *Journal of Geology*, 104: 565–582.

Bryant, E.A., Young, R.W., Price, D.M. and Short, S.A., 1994. Late Pleistocene dune chronology: near-coastal New South Wales and eastern Australia. *Quaternary Science Reviews*, 13: 209–223.

Bryant, E.A., Young, R.W. and Price, D.W., 1996. Tsunami as a major control on coastal evolution, southeastern Australia. *Journal of Coastal Research*, 12: 831–840.

Bryant, E.A., Young, R.W. and Price, D.M., 1997. Late Pleistocene marine deposition and TL chronology of the New South Wales, Australian coastline. *Zeitschrift für Geomorphologie, NF*, 41: 205–227.

Bryce, S., Larcombe, P. and Ridd, P.V., 1998. The relative importance of landward-directed tidal sediment transport versus freshwater flood events in the Normanby River estuary, Cape York Peninsula, Australia. *Marine Geology*, 149: 55–78.

Buddemeier, R.W. and Hopley, D., 1988. Turn-ons and turn-offs: causes and mechanisms of the initiation and termination of coral reef growth. In: J.H.

Choat *et al.* (Editors), *Proceedings of the Sixth International Coral Reef Symposium*, Townsville, Australia, Vol. 1: 253–261.

Buddemeier, R.W., Smith, S.V. and Kinzie, R.A., 1975. Holocene windward reef-flat history, Enewetak Atoll. *Geological Society of America Bulletin*, 86: 1581–1584.

Büdel, J., 1966. Deltas: a basis of culture and civilization, *Scientific Problems of the Humid Zone Deltas and their Implications: Proceedings of the Dacca Symposium*. UNESCO, Paris, pp. 295–300.

Budetta, P., Galietta, G. and Santo, A., 2000. A methodology for the study of the relation between coastal cliff erosion and the mechanical strength of soils and rock masses. *Engineering Geology*, 56: 243–256.

Bull, W.B., 1985. Correlation of flights of global marine terraces. In: M. Morisawa and J. Hack (Editors), *Tectonic Geomorphology: Proceedings of the Fifteenth Annual Geomorphology Symposium*. George Allen and Unwin, State University of New York at Binghamton, pp. 129–152.

Bunt, J.S., 1996. Mangrove zonation: an explanation of data from seventeen riverine estuaries in tropical Australia. *Annals of Botany*, 78: 333–341.

Bunt, J.S., 1999. Overlap in mangrove species zonal patterns: some methods of analysis. *Mangroves and Salt Marshes*, 3: 155–164.

Bunt, J.S. and Bunt, E.D., 1999. Complexity and variety of zonal pattern in the mangroves of the Hinchinbrook area, northeastern Australia. *Mangroves and Salt Marshes*, 3: 165–176.

Bunt, J.S., Williams, W.T. and Clay, H.J., 1982. River water salinity and the distribution of mangrove species along several rivers in North Queensland. *Australian Journal of Botany*, 30: 401–412.

Butler, R.W., 1980. The concept of a tourist area cycle of evolution: implications for management of resources. *Canadian Geographer*, 24: 5–12.

Cabanes, C., Cazenave, A. and Le Provost, C., 2001. Sea level rise during past 40 years determined from satellite and *in situ* observations. *Science*, 294: 840–842.

Cabioch, G. and Ayliffe, L.K., 2001. Raised coral terraces at Malakula, Vanuatu, southwest Pacific, indicate high sea level during marine isotope stage 3. *Quaternary Research*, 56: 357–365.

Cabioch, G., Camoin, G.F. and Montaggioni, L.F., 1999a. Postglacial growth history of a French Polynesian barrier reef tract, Tahiti, central Pacific. *Sedimentology*, 46: 985–1000.

Cabioch, G., Correge, T., Turpin, L., Castellaro, C. and Recy, J., 1999b. Development patterns of fringing and barrier reefs in New Caledonia (southwest Pacific). *Oceanologica Acta*, 22: 567–578.

Cahoon, D.R. and Lynch, J.C., 1997. Vertical accretion and shallow subsidence in a mangrove forest of southwestern Florida, USA. *Mangroves and Salt Marshes*, 1: 173–186.

Cahoon, D.R., Reed, D.J. and Day, J.W., 1995. Estimating shallow subsidence in microtidal salt marshes of the southeastern United States: Kaye and Barghoorn revisited. *Marine Geology*, 128: 1–9.

Cahoon, D.R., Marin, P.E., Black, B.K. and Lynch, J.C., 2000. A method for measuring vertical accretion, elevation, and compaction of soft, shallow-water sediments. *Journal of Sedimentary Research*, 70: 1250–1253.

Caldeira, K. and Kasting, J.F., 1993. Insensitivity of global warming potentials to carbon dioxide emission scenarios. *Nature*, 366: 251–253.

Calderoni, G., Mazzini, E., Simeoni, U. and Tessari, U., 1999. A new application of system theory to foredune intervention strategies. *Journal of Coastal Research*, 15: 457–470.

Cambers, G., 1976. Temporal scales in coastal erosion systems. *Transactions of the Institute of British Geographers, NS*, 1: 246–256.

Cameron, G.A. and Pritchard, D.W., 1963. Estuaries. In: M.N. Hill (Editor), *The Sea*, Vol. 2. Wiley Interscience, New York, pp. 306–324.

Carannante, G., Esteban, M., Milliman, J.D. and Simone, L., 1988. Carbonate lithofacies as paleolatitude indicators: problems and limitations. *Sedimentary Geology*, 60: 333–346.

Carlton, J.M., 1974. Land-building and stabilization by mangroves. *Environmental Conservation*, 1: 285–294.

Carpenter, K.E., 1997. A critical appraisal of the methodology used in studies of material flux between saltmarshes and coastal waters. In: T.D. Jickells and J. Rae (Editors), *Biogeochemistry of Intertidal Sediments*. Cambridge University Press, Cambridge, pp. 59–83.

Carr, A.P., 1965. Shingle spit and river mouth: short term dynamics. *Transactions of Institute of British Geographers*, 36: 117–129.

Carr, A.P. and Blackley, M.W.L., 1986. The effects and implication of tides and rainfall on the circulation of water within salt marsh sediments. *Limnology and Oceanography*, 31: 266–276.

Carson, R., 1955. *The Edge of the Sea*. Houghton Mifflin, Boston, 238 pp.

Carter, J., 1959. Mangrove succession and coastal change in south-west Malaya. *Transactions of the Institute of British Geographers*, 26: 79–88.

Carter, R.W.G., 1986. The morphodynamics of beach-ridge formation: Magilligan, Northern Ireland. *Marine Geology*, 73: 191–214.

Carter, R.W.G., 1988. *Coastal Environments – An Introduction to the Physical, Ecological and Cultural Systems of Coastlines*. Academic Press, London, 617 pp.

Carter, R.W.G. and Orford, J.D., 1984. Coarse clastic barrier beaches: a discussion of the distinctive dynamic and morphosedimentary characteristics. *Marine Geology*, 60: 377–389.

Carter, R.W.G. and Orford, J.D., 1993. The morphodynamics of coarse clastic beaches and barriers: a short- and long-term perspective. *Journal of Coastal Research*, 15: 158–179.

Carter, R.W.G. and Stone, G.W., 1989. Mechanisms associated with the erosion of sand dune cliffs, Magilligan, Northern Ireland. *Earth Surface Processes and Landforms*, 14: 1–10.

Carter, R.W.G. and Woodroffe, C.D. (Editors), 1994. *Coastal Evolution – Late Quaternary Shoreline Morphodynamics*. Cambridge University Press, Cambridge, 517 pp.

Carter, R.W.G., Johnston, T.W. and Orford, J.D., 1984. Stream outlets through mixed sand and gravel coastal barriers: examples from southeast Ireland. *Zeitschrift für Geomorphologie, NF*, 28: 427–442.

Carter, R.W.G., Johnston, T.W., McKenna, J. and Orford, J.D., 1987. Sea-level, sediment supply and coastal changes: examples from the coast of Ireland. *Progress in Oceanography*, 18: 79–101.

Carter, R.W.G., Forbes, D.L., Jennings, S.C., Orford, J.D., Shaw, J. and Taylor, R.B., 1989. Barrier and lagoon coast evolution under differing relative sea-level regimes: examples from Ireland and Nova Scotia. *Marine Geology*, 88: 221–242.

Carter, R.W.G., Hesp, P.A. and Nordstrom, K.F., 1990a. Erosional landforms in coastal dunes. In: K.F. Nordstrom, N.P. Psuty and R.W.G. Carter (Editors), *Coastal Dunes: Form and Process*. John Wiley & Sons, Chichester, pp. 217–249.

Carter, R.W.G., Jennings, S.C. and Orford, J.D., 1990b. Headland erosion by waves. *Journal of Coastal Research*, 6: 517–529.

Carter, R.W.G., Bauer, B.O., Sherman, D.J., Davidson-Arnott, R.G.D., Gares, P.A., Nordstrom, K.F. and Orford, J.D., 1992. Dune development in the aftermath of stream outlet closure: examples from Ireland and California. In: R.W.G. Carter, T.G.F. Curtis and M.J. Sheehy-Skeffington (Editors), *Coastal Dunes: Geomorphology, Ecology and Management for Conservation*. Balkema, Rotterdam, pp. 57–69.

Cartwright, D.E., 1999. *Tides: A Scientific History*. Cambridge University Press, Cambridge, 292 pp.

Castaing, P. and Guilcher, A., 1995. Geomorphology and sedimentology of rias. In: G.M.E. Perillo (Editor), *Geomorphology and Sedimentology of Estuaries*. Elsevier, Amsterdam, pp. 69–111.

CERC, 1984. *Shore Protection Manual*. Coastal Engineering Research Center, US Corps of Engineers, Vicksburg.

Chagué-Goff, C., Nichol, S.L., Jenkinson, A.V. and Heijnis, H., 2000. Signatures of natural catastrophic events and anthropogenic impact in an estuarine environment, New Zealand. *Marine Geology*, 167: 285–301.

Challinor, J., 1949. A principle in coastal geomorphology. *Geography*, 34: 212–215.

Chamberlin, T.C., 1890. The method of multiple working hypotheses. *Science*, 15: 92–96.

Chandler, J.H. and Brunsden, D., 1995. Steady state behaviour of the Black Ven mudslide: the application of archival analytical photogrammetry to studies of landform change. *Earth Surface Processes and Landforms*, 20: 255–275.

Chapman, D.M., Geary, M., Roy, P.S. and Thom, B.G., 1982. *Coastal Evolution and Coastal Erosion in New South Wales*. Coastal Council of New South Wales, Sydney, 340 pp.

Chapman, V.J., 1944. 1939 Cambridge University Expedition to Jamaica. Part 1. A study of the botanical processes concerned in the development of the Jamaican shoreline. *Journal of the Linnean Society (London), Botany*, 52: 407–447.

Chapman, V.J., 1974. *Salt Marshes and Salt Deserts of the World*. J. Cramer, Lehre, 392 pp.

Chapman, V.J., 1976. *Mangrove Vegetation*. J. Cramer, Germany, 447 pp.

Chapman, V.J. and Ronaldson, J.W., 1958. The mangrove and salt marsh flats of the Auckland Isthmus. *New Zealand Department of Scientific and Industrial Research Bulletin*, 125: 1–79.

Chappell, J., 1974a. Geology of coral terraces, Huon Peninsula, New Guinea: a study of Quaternary tectonic movements and sea-level changes. *Geological Society of America Bulletin*, 85: 553–570.

Chappell, J., 1974b. Late Quaternary glacio- and hydro-isostasy on a layered earth. *Quaternary Research*, 4: 429–440.

Chappell, J., 1980. Coral morphology, diversity and reef growth. *Nature*, 286: 249–252.

Chappell, J., 1983a. Evidence for smoothly falling sea levels relative to north Queensland, Australia, during the past 6000 years. *Nature*, 302: 406–408.

Chappell, J., 1983b. Thresholds and lags in geomorphologic changes. *Australian Geographer*, 15: 357–366.

Chappell, J., 1987. Late Quaternary sea-level changes in the Australian region. In: M.J. Tooley and I. Shennan (Editors), *Sea-Level Changes*. Blackwell, Oxford, pp. 296–331.

Chappell, J., 1993. Contrasting Holocene sedimentary geologies of lower Daly River, northern Australia, and lower Sepik-Ramu, Papua New Guinea. *Sedimentary Geology*, 83: 339–358.

Chappell, J. and Grindrod, J., 1984. Chenier plain formation in Northern Australia. In: B.G. Thom (Editor), *Coastal Geomorphology in Australia*. Academic Press, Sydney, pp. 197–231.

Chappell, J. and Polach, H., 1991. Post glacial sea level rise from a coral record at Huon Peninsula, Papua New Guinea. *Nature*, 349: 147–149.

Chappell, J. and Shackleton, N.J., 1986. Oxygen isotopes and sea level. *Nature*, 324: 137–140.

Chappell, J. and Thom, B.G., 1977. Sea levels and coasts. In: J. Allen, J. Golson and R. Jones (Editors), *Sunda and Sahul: Prehistoric Studies in Southeast Asia, Melanesia and Australia*. Academic Press, London, pp. 275–291.

Chappell, J. and Thom, B.G., 1986. Coastal morphodynamics in north Australia: review and prospect. *Australian Geographical Studies*, 24: 110–127.

Chappell, J. and Veeh, H.H., 1978. Late Quaternary tectonic movements and sea-level changes at Timor and Atauro Island. *Geological Society of America Bulletin*, 89: 356–368.

Chappell, J. and Woodroffe, C.D., 1994. Macrotidal estuaries. In: R.G. Carter and C.D. Woodroffe (Editors), *Coastal Evolution: Late Quaternary Shoreline Morphodynamics*. Cambridge University Press, Cambridge, pp. 187–218.

Chappell, J., Eliot, I., Bradshaw, M.P. and Lonsdale, E., 1979. Experimental control of beach face dynamics by water-table pumping. *Engineering Geology*, 14: 29–41.

Chappell, J., Rhodes, E.G., Thom, B.G. and Wallensky, E., 1982. Hydro-isostasy and the sea-level isobase of 5500 BP in north Queensland, Australia. *Marine Geology*, 49: 81–90.

Chappell, J., Chivas, A., Wallensky, E., Polach, H.A. and Aharon, P., 1983. Holocene palaeo-environment changes, central to north Great Barrier Reef inner zone. *Bureau of Mineral Resources Journal Australian Geology and Geophysics*, 8: 223–235.

Chappell, J., Omura, A., Esat, T., McCulloch, M., Pandolfi, J, Ota, Y. and Pillans, B. 1996. Reconciliation of late Quaternary sea levels derived from coral terraces at Huon Peninsula with deep sea oxygen isotope records. *Earth and Planetary Science Letters*, 141: 227–236.

Chave, K.E., Smith, S.V. and Roy, K.J., 1972. Carbonate production by coral reefs. *Marine Geology*, 12: 123–140.

Chen, J.H., Curran, H.A., White, B. and Wasserburg, G.J., 1991. Precise chronology of the last interglacial period: ^{234}U–^{230}Th data from fossil coral reefs in the Bahamas. *Geological Society of America Bulletin*, 103: 82–97.

Chen, R. and Twilley, R.R., 1998. A gap dynamic model of mangrove forest development along gradients of soil salinity and nutrient resources. *Journal of Ecology*, 86: 37–51.

Cheney, R.E., 2001. Satellite altimetry. In: J.H. Steele, S.A. Thorpe and K.K. Turekian (Editors), *Encyclopedia of Ocean Sciences*. Academic Press, San Diego, pp. 2504–2510.

Cheney, R.E., Douglas, B.C. and Miller, L., 1989. Evaluation of Geosat altimeter data with application to tropical Pacific sea level variability. *Journal of Geophysical Research*, 94: 4737–4747.

Chenhall, B.E., Yassini, I., Depers, A.M., Caitcheon, G., Jones, B.G., Batley, G.E. and Ohmsen, G.S., 1995. Anthropogenic marker evidence for accelerated sedimentation in Lake Illawarra, New South Wales, Australia. *Environmental Geology*, 26: 124–135.

Chepil, W.S., 1945. Dynamics of wind erosion, 3. The transport capacity of the wind. *Soil Science*, 60: 475–480.

Chivas, A., Chappell, J., Polach, H., Pillans, B. and Flood, P., 1986. Radiocarbon evidence for the timing and rate of island development, beach-rock formation and phosphatization at Lady Elliot Island, Queensland, Australia. *Marine Geology*, 69: 273–287.

Choi, D.R. and Ginsburg, R.N., 1982. Siliciclastic foundations of Quaternary reefs in the southernmost Belize Lagoon, British Honduras. *Geological Society of America Bulletin*, 93: 116–126.

Chorley, R.J. and Kennedy, B.A., 1971. *Physical Geography*. Prentice Hall, London, 370 pp.

Chorley, R.J., Schumm, S.C. and Sugden, D.E., 1984. *Geomorphology*. Methuen, London, 605 pp.

Christiansen, T., Wiberg, P.L. and Milligan, T.G., 2000. Flow and sediment transport on a tidal salt marsh surface. *Estuarine, Coastal and Shelf Science*, 50: 315–331.

Christie, M.C., Dyer, K.R. and Turner, P., 1999. Sediment flux and bed level measurements from a macro tidal mudflat. *Estuarine, Coastal and Shelf Science*, 49: 667–688.

Chubb, L.J., 1957. The pattern of some Pacific island chains. *Geological Magazine*, 94: 221–228.

Church, J.A. and Gregory, J.M., 2001. Sea level change. In: J.H. Steele, S.A. Thorpe and K.K. Turekian (Editors), *Encyclopedia of Ocean Sciences*. Academic Press, San Diego, pp. 2599–2604.

Cialone, M.A. and Stauble, D.K., 1998. Historical findings on ebb shoal mining. *Journal of Coastal Research*, 14: 537–563.

Cicin-Sain, B., 1993. Sustainable development and integrated coastal management. *Ocean and Coastal Management*, 21: 11–43.

Cinque, A., De Pippo, T. and Romano, P., 1995. Coastal slope terracing and relative sea-level changes: deductions based on computer simulations. *Earth Surface Processes and Landforms*, 20: 87–103.

Cintron, G., Lugo, A.E., Pool, D.J. and Morris, G., 1978. Mangroves of arid environments in Puerto Rico and adjacent islands. *Biotropica*, 10: 110–121.

Clague, J.J. and Bobrowsky, P.T., 1994a. Evidence for a large earthquake and tsunami 100–400 years ago on western Vancouver Island, British Columbia. *Quaternary Research*, 41: 176–184.

Clague, J.J. and Bobrowsky, P.T., 1994b. Tsunami deposits beneath tidal marshes on Vancouver Island, British Columbia. *Geological Society of America Bulletin*, 106: 1293–1303.

Clague, J.J., Bobrowsky, P.T. and Hamilton, T.S., 1994. A sand sheet deposited by the 1964 Alaska tsunami at Port Alberni, British Columbia. *Estuarine, Coastal and Shelf Science*, 38: 413–421.

Clark, C.D., Ripley, H.T., Green, E.P., Edwards, A.J. and Mumby, P.J., 1997. Mapping and measurement of tropical coastal environments with hyperspectral and high spatial resolution data. *International Journal of Remote Sensing*, 18: 237–242.

Clark, J.A., Farrell, W.E. and Peltier, W.R., 1978. Global change in post glacial sea level: a numerical calculation. *Quaternary Research*, 9: 265–287.

Clark, J.A. and Lingle, C.S., 1979. Predicted relative sea-level changes (18 000 years BP to Present) caused by late-glacial retreat of the Antarctic ice sheet. *Quaternary Research*, 11: 279–298.

Clark, J.S., 1986. Late Holocene vegetation and coastal processes at a Long Island tidal marsh. *Journal of Ecology*, 74: 561–578.

Clark, R.L. and Guppy, J.C., 1988. A transition from mangrove forest to freshwater wetland in the monsoon tropics of Australia. *Journal of Biogeography*, 15: 665–684.

Clarke, D.J. and Eliot, I.G., 1983. Onshore–offshore patterns of sediment exchange in the littoral zone of a sandy beach. *Geological Society of Australia Journal*, 30: 341–351.

Clarke, D.J. and Eliot, I.G., 1988a. Low-frequency changes of sediment volume on the beachface at Warilla Beach, New South Wales, 1975–1985. *Marine Geology*, 79: 189–211.

Clarke, D.J. and Eliot, I.G., 1988b. Low-frequency variation in the seasonal intensity of coastal weather systems and sediment movement on the beachface of a sandy beach. *Marine Geology*, 79: 23–39.

Clarke, L.D. and Hannon, N.J., 1967. The mangrove swamp and salt marsh communities of the Sydney district. I. Vegetation, soils and climate. *Journal of Ecology*, 55: 753–771.

Clarke, P.J. and Kerrigan, R.A., 2000. Do forest gaps influence the population structure and species composition of mangrove stands in northern Australia? *Biotropica*, 32: 642–652.

Clauer, N., Chaudhuri, S., Toulkeridis, T. and Blanc, B., 2000. Fluctuations of Caspian Sea level: beyond climatic variations? *Geology*, 28: 1015–1018.

Clayton, K.M., 1989. Sediment input from the Norfolk cliffs, eastern England – a century of coast protection and its effect. *Journal of Coastal Research*, 5: 433–442.

Clemens, K.E. and Komar, P.D., 1988. Oregon beach-sand compositions produced by the mixing of sediments under a transgressing sea. *Journal of Sedimentary Petrology*, 58: 519–529.

Clements, F.E., 1936. Nature and structure of the climax. *Journal of Ecology*, 24: 252–284.

Clemmensen, L.B., Richardt, N. and Andersen, C., 2001. Holocene sea-level variation and spit development: data from Skagen Odde, Denmark. *The Holocene*, 11: 323–331.

CLIMAP Project Members, 1976. The surface of the ice-age Earth. *Science*, 191: 1131–1137.

Clouard, V. and Bonneville, A., 2001. How many Pacific hotspots are fed by deep-mantle plumes? *Geology*, 29: 695–698.

Coco, G., O'Hare, T.J. and Huntley, D.A., 1999. Beach cusps: a comparison of data and theories for their formation. *Journal of Coastal Research*, 15: 741–749.

Coleman, J.M., 1969. Brahmaputra river; channel processes and sedimentation. *Sedimentary Geology*, 3: 129–239.

Coleman, J.M., 1988. Dynamic changes and processes in the Mississippi River delta. *Geological Society of America Bulletin*, 100: 999–1015.

Coleman, J.M. and Gagliano, S.M., 1964. Cyclic sedimentation in the Mississippi River deltaic plain. *Transactions of the Gulf Coast Association of Geological Societies*, 14: 67–80.

Coleman, J.M., Gagliano, S.M. and Smith, W.G., 1970. *Sedimentation in a Malaysian High Tide Tropical Delta*. Society of Economic Palaeontologists and Mineralogists, Tulsa, OK, Special Publication, 15, pp. 185–197.

Coleman, J.M., Roberts, H.H. and Stone, G.W., 1998. Mississippi River Delta: an overview. *Journal of Coastal Research*, 14: 698–716.

Coleman, J.M. and Wright, L.D., 1978. Sedimentation in an arid macrotidal alluvial river system: Ord River, Western Australia. *Journal of Geology*, 86: 621–642.

Coles, S.M., 1979. Benthic microalgal populations on intertidal sediments and their role as precursors to salt marsh development. In: R.L. Jeffries and A.J. Davy (Editors), *Ecological Processes in Coastal Environments*. Blackwell, pp. 25–41.

Collins, L.B., 1988. Sediments and history of the Rottnest Shelf, southwest Australia: a swell-dominated, non-tropical carbonate margin. *Sedimentary Geology*, 60: 15–49.

Collins, L.B., Zhu, Z.R. and Wyrwoll, K.-H., 1997. Geology of the Houtman

Abrolhos Islands. In: H.L. Vacher and T.M. Quinn (Editors), *Geology and Hydrogeology of Carbonate Islands*. Elsevier, Amsterdam, pp. 811–833.

Collins, M.B., Amos, C.L. and Evans, G., 1981. Observations of some sediment-transport processes over intertidal flats, the Wash, UK. *Special Publications of the International Association of Sedimentologists*, 5: 81–98.

Collinson, J.D. and Thompson, D.B., 1982. *Sedimentary Structures*. George Allen & Unwin, London, 194 pp.

Connell, J.H., 1978. Diversity in tropical rain forests and coral reefs: high diversity of trees and corals is maintained only in a nonequilibrium state. *Science*, 199: 1302–1309.

Cook, P.G., 1986. A review of coastal dunebuilding in eastern Australia. *Australian Geographer*, 17: 133–142.

Cook, P.J. and Mayo, W., 1978. *Sedimentology and Holocene History of a Tropical Estuary (Broad Sound, Queensland)*, Bureau of Mineral Resources Bulletin 170. Australian Government, Canberra, 206 pp.

Cook, P.J. and Polach, H.A., 1973. A chenier sequence at Broad Sound, Queensland, and evidence against a Holocene high sea level. *Marine Geology*, 14: 253–268.

Cook, P.J., Colwell, J.B., Firman, J.B., Lindsay, J.M., Schwebel, D.A. and von de Borch, C.C., 1977. The late Cainozoic sequence of southeast South Australia and Pleistocene sea-level changes. *Bureau of Mineral Resources Journal of Australian Geology and Geophysics*, 2: 81–88.

Cooke, G.A., 1981. Reconstruction of the Holocene coastline of Mesopotamia. *Geoarchaeology*, 2: 15–28.

Cooper, J.A.G., 1993. Sedimentation in a river dominated estuary. *Sedimentology*, 40: 979–1017.

Cooper, J.A.G., 1994. Lagoons and microtidal coasts. In: R.W.G. Carter and C.D. Woodroffe (Editors), *Coastal Evolution: Late Quaternary Shoreline Morphodynamics*. Cambridge University Press, Cambridge, pp. 219–265.

Cooper, J.A.G., 2001. Geomorphological variability among microtidal estuaries from the wave-dominated South African coast. *Geomorphology*, 40: 99–122.

Cooper, J.A.G., McKenna, J. and Orford, J., 1999a. Mesoscale temporal changes to foredunes at Inch Spit, south-west Ireland. *Zeitschrift für Geomorphologie, NF*, 43: 439–461.

Cooper, J.A.G., Wright, I. and Mason, T., 1999b. Geomorphology and sedimentology. In: B. Allanson and D. Baird (Editors), *Estuaries of South Africa*. Cambridge University Press, Cambridge, pp. 5–25.

Cooper, W.S., 1958. *Coastal Sand Dunes of Oregon and Washington*. Geological Society of America Memoir, 72, New York, 169 pp.

Cooper, W.S., 1967. *Coastal Sand Dunes of California*. Geological Society of America Memoir, 104, New York, 131 pp.

Cornaglia, P., 1889. Delle Spiaggie. *Accademia Nazionale dei Lincei, Atti. Cl. Sci. Fis., Mat. e Nat. Mem.*, 5: 284–304.

Cornish, V., 1897. On the formation of sand-dunes. *Geographical Journal*, 9: 278–309.

Cornish, V., 1898. On sea-beaches and sandbanks. *Geographical Journal*, 11: 528–559, 628–647.

Cortés, J., Macintyre, I.G. and Glynn, P.W., 1994. Holocene growth history of an eastern Pacific fringing reef, Punta Islotes, Costa Rica. *Coral Reefs*, 13: 65–73.

Cotton, C.A., 1916. Fault coasts in New Zealand. *Geographical Review*, 1: 20–47.

Cotton, C.A., 1942. Shorelines of transverse deformation. *Journal of Geomorphology*, 5: 45–58.

Cotton, C.A., 1952. The Wellington coast: an essay in coastal classification. *New Zealand Geographer*, 8: 48–62.

Cotton, C.A., 1954. Deductive morphology and genetic classification of coasts. *Science Monthly*, 78: 163–181.

Cotton, C.A., 1956. Rias *sensu stricto* and *sensu lato*. *Geographical Journal*, 122: 360–364.

Cotton, C.A., 1963. Levels of planation of marine benches. *Zeitschrift für Geomorphologie, NF*, 7: 97–111.

Cotton, C.A., 1969a. Marine cliffing according to Darwin's theory. *Transactions of the Royal Society of New Zealand, Geology*, 6: 187–208.

Cotton, C.A., 1969b. The pedestals of oceanic islands. *Geological Society of America Bulletin*, 80: 749–760.

Cotton, C.A., 1974. *Bold Coasts.* A.H. & A.W. Reed, Wellington, 354 pp.

Cowell, P.J. and Thom, B.G., 1994. Morphodynamics of coastal evolution. In: R.W.G. Carter and C.D. Woodroffe (Editors), *Coastal Evolution, Late Quaternary Shoreline Morphodynamics*. Cambridge University Press, Cambridge, pp. 33–86.

Cowell, P.J., Roy, P.S. and Jones, R.A., 1995. Simulation of large-scale coastal change using a morphological behaviour model. *Marine Geology*, 126: 45–61.

Cowell, P.J., Hanslow, D.J. and Meleo, J.F., 1999. The shoreface. In: A.D. Short (Editor), *Handbook of Beach and Shoreface Morphodynamics*. John Wiley & Sons, Chichester, pp. 37–71.

Cox, A. and Hart, B.H., 1986. *Plate Tectonics: How It Works*. Blackwell, Palo Alto, 392 pp.

Craft, C.B., Seneca, E.D. and Broome, S.W., 1993. Vertical accretion in microtidal regularly and irregularly flooded estuarine marshes. *Estuarine, Coastal and Shelf Science*, 37: 371–386.

Craighead, F.C., 1964. Land, mangroves and hurricanes. *Fairchild Tropical Garden Bulletin*, 19: 5–32.

Cramer, W. and Hytteborn, H., 1987. The separation of fluctuation and long-term change in vegetation dynamics of a rising seashore. *Vegetatio*, 69: 157–167.

Crossland, C.J., 1984. Seasonal variations in the rates of calcification and productivity in the coral *Acropora formosa* on a high-latitude reef. *Marine Ecology Progress Series*, 15: 135–140.

Crowley, G.M., 1996. Late Quaternary mangrove distribution in Northern Australia. *Australian Systematic Botany*, 9: 219–225.

Crowley, G.M. and Gagan, M.K., 1995. Holocene evolution of coastal wetlands in wet-tropical northeastern Australia. *The Holocene*, 5: 385–399.

Cundy, A.B., Kortekaas, S., Dewez, T., Stewart, I.S., Collins, P.E.F., Croudace, I.W., Maroukian, H., Papanastassiou, D., Gaki-Papanastassiou, P., Pavlopoulos, K. and Dawson, A., 2000. Coastal wetlands as recorders of earthquake subsidence in the Aegean: a case study of the 1894 Gulf of Atalanti earthquakes, central Greece. *Marine Geology*, 170: 3–26.

Curray, J.R., 1961. Late Quaternary sea level: a discussion. *Geological Society of America Bulletin*, 72: 1707–1712.

Curray, J.R., 1964. Transgressions and regressions. In: R.L. Miller (Editor), *Papers in Marine Geology*. Macmillan, New York, pp. 175–203.

Curray, J.R., Emmel, R.J. and Crampton, P.J.S., 1969. Holocene history of a strand plain, lagoonal coast, Nayarit, Mexico. In: A.A. Castanares and F.B. Phleger (Editors), *Coastal Lagoons – A Symposium*. Universitas Nacional Autonoma de Mexico/UNESCO, Mexico City, pp. 35–43.

Dail, H.J., Merrifield, M.A. and Bevis, M., 2000. Steep beach morphology changes due to energetic wave forcing. *Marine Geology*, 162: 443–458.

Dalongeville, M., 1977. Formes littorales de corrosion dans les roches carbonatées au Liban: étude morphologique. *Mediterranée*, 3: 21–33.

Dalrymple, R.A., 1988. A model for refraction of water waves. *Journal of Waterway, Port, Coastal and Ocean Engineering*, 114: 423–435.

Dalrymple, R.A., Biggs, R.B., Dean, R. and Wang, H., 1986. Bluff recession rates in Chesapeake Bay. *Journal of Waterway, Port, Coastal and Ocean Engineering*, 112: 164–168.

Dalrymple, R.W., Knight, R.J. and Lambiase, J.J., 1978. Bedforms and their hydraulic stability relationships in a tidal environment, Bay of Fundy, Canada. *Nature*, 275: 100–104.

Dalrymple, R.W., Knight, R.J., Zaitlin, B.A. and Middleton, G.V., 1990. Dynamics and facies model of a macrotidal sand-bar complex, Cobequid Bay-Salmon River estuary (Bay of Fundy). *Sedimentology*, 37: 577–612.

Dalrymple, R.W., Zaitlin, B.A. and Boyd, R., 1992. Estuarine facies models: conceptual basis and stratigraphic implications. *Journal of Sedimentary Petrology*, 62: 1130–1146.

Dalrymple, R.W., Boyd, R. and Zaitlin, B.A. (Editors), 1994. *Incised-valley Systems: Origin and Sedimentary Sequences*. SEPM (Society of Economic Palaeontologists and Mineralogists), Tulsa, OK, 391 pp.

Daly, R.A., 1910. Pleistocene glaciation and the coral reef problem. *American Journal of Science*, 30: 297–308.

Daly, R.A., 1915. The glacial-control theory of coral reefs. *Proceedings of the American Academy of Arts and Science*, 51: 155–251.

Daly, R.A., 1925. Pleistocene changes of level. *American Journal of Science*, 10: 281–313.

Daly, R.A., 1934. *The Changing World of the Ice Age*. Yale University Press, New Haven, 271 pp.

Dana, J.D., 1849. Report of the United States Exploring Expedition 1838–1842, *Geology*, Volume 10. C. Sherman, Philadelphia, 759 pp.

Dana, J.D., 1872. *Corals and Coral Islands*. Dodd and Mean, New York. 406 pp.

Daniel, J.R.K., 1989. The chenier plain coastal system of Guyana. *Marine Geology*, 90: 283–287.

Daniels, R.C., 1992. Sea-level rise on the South Carolina coast: two case studies for 2100. *Journal of Coastal Research*, 8: 56–70.

Dankers, N., Binsbergen, M., Zegers, K., Laane, R. and Rutgers van de Loeff, M., 1984. Transportation of water, particulate and dissolved organic and inorganic matter between a salt marsh and the Ems-Dollard Estuary, The Netherlands. *Estuarine, Coastal and Shelf Science*, 19: 143–165.

Darienzo, M.E. and Peterson, C.D., 1990. Episodic tectonic subsidence of Late Holocene salt marshes, northern Oregon central Cascadia margin. *Tectonics*, 9: 1–22.

Darwin, C., 1842. *The Structure and Distribution of Coral Reefs*. Smith Elder and Co., London, 214 pp.

Darwin, C., 1845. *Journal of Researches into the Natural History and Geology of the Countries Visited During the Voyage of H.M.S. Beagle Round the World, Under the Command of Capt. Fitzroy R.N.* John Murray, London, 519 pp.

Darwin, C., 1846. *Geological Observations on Parts of South America*. Smith Elder and Co, London, 279 pp.

Darwin, C., 1851. *Geological Observations on Coral Reefs, Volcanic Islands and on South America Being the Geology of the Voyage of the Beagle, Under the Command of Captain Fitzroy During the Years 1832 to 1836*, 3 vols. Smith Elder and Co, London.

David, T.W.E. and Sweet, G., 1904. The geology of Funafuti. In: The Royal Society (Editor), *The Atoll of Funafuti*. The Royal Society, London, pp. 61–124.

Davidson, A.T., Nicholls, R.J. and Leatherman, S.P., 1992. Beach nourishment as a coastal management tool. *Journal of Coastal Research*, 8: 984–1022.

Davies, J.L., 1957. The importance of cut and fill in the development of sand beach ridges. *Australian Journal of Science*, 20: 105–111.

Davies, J.L., 1958a. Wave refraction and the evolution of shoreline curves. *Geographical Studies*, 5: 1–14.

Davies, J.L., 1958b. Analysis of height variation in sand beach ridges. *Australian Journal of Science*, 21: 51–52.

Davies, J.L., 1974. The coastal sediment compartment. *Australian Geographical Studies*, 12: 139–151.

Davies, J.L., 1980. *Geographical Variation in Coastal Development*. Longman, London, 212 pp.

Davies, K.H., 1983. Amino acid analysis of Pleistocene marine molluscs from the Gower Peninsula. *Nature*, 302: 137–139.

Davies, P. and Williams, A.T., 1991. Sediment supply from solid geology cliffs into the intertidal zone of the Severn Estuary/Inner Bristol Channel, UK. In: M. Elliott and J.-P. Ducrotoy (Editors), *Estuaries and Coasts: Spatial and Temporal Intercomparisons*, ECSA nineteenth Symposium. Olsen & Olsen, Fredensborg, pp. 17–24.

Davies, P.J. and Montaggioni, L.F., 1985. Reef growth and sea-level change: the environmental signature. *Proceedings of the Fifth International Coral Reef*

Congress, Vol. 3 Antenne Museum-Ephe, Moorea, French Polynesia, pp.477–515.

Davies, P.J., Radke, B.M. and Robinson, C.R., 1976. The evolution of One Tree Reef, southern Great Barrier Reef, Queensland. *Bureau of Mineral Resources Journal of Australian Geology and Geophysics*, 1: 231–240.

Davis, C.A., 1910. Salt marsh formation near Boston and its geological significance. *Economic Geology*, 5: 623–639.

Davis, D. and Detro, R., 1980. New Orleans drainage and reclamation – a 200 year problem. *Zeitschrift für Geomorphologie*, 34: 87–96.

Davis, J.H., 1940. The ecology and geologic role of mangroves in Florida. *Papers from Tortugas Laboratory*, 32: 307–412.

Davis, R.A. (Editor), 1994. *Geology of Holocene Barrier Island Systems*. Springer-Verlag, Berlin, 464 pp.

Davis, R.A. and Fox, W.T., 1972. 4-Dimensional model for beach and inner nearshore sedimentation. *Journal of Geology*, 80: 484–493.

Davis, R.A. and Hayes, M.O., 1984. What is a wave-dominated coast? *Marine Geology*, 60: 313–329.

Davis, R.A., Andronaco, M. and Gibeaut, J.C., 1989. Formation and development of a tidal inlet from a washover fan, west-central Florida coast, USA. *Sedimentary Geology*, 65: 87–94.

Davis, W.M., 1896. The outline of Cape Cod. *Proceedings of the American Academy*, 31: 303–332.

Davis, W.M., 1899. The geographical cycle. *Geographical Journal*, 14: 481–504.

Davis, W.M., 1909. *Geographical Essays*. Ginn and Company, Boston, 777 pp.

Davis, W.M., 1928. *The Coral Reef Problem*. Special Publication, 9. American Geographical Society, New York, 596 pp.

Davis, W.M., 1933. Glacial episodes of the Santa Monica Mountains, California. *Geological Society of America Bulletin*, 44: 1041–1133.

Dawson, A.G., 1994. Geomorphological effects of tsunami run-up and backwash. *Geomorphology*, 10: 83–94.

Dawson, A.G., 2000. Tsunami deposits. *Pure and Applied Geophysics*, 157: 875–897.

Dawson, A.G., Matthews, J.A. and Shakesby, R.A., 1987. Rock platform erosion on periglacial shores: a modern analogue for Pleistocene rock platforms in Britain. In: J. Boardman (Editor), *Periglacial Processes and Landforms in Britain and Ireland*. Cambridge University Press, Cambridge, pp. 173–182.

Dawson, A.G., Long, D. and Smith, D.E., 1988. The Storegga Slides: evidence from eastern Scotland for a possible tsunami. *Marine Geology*, 82: 271–276.

Day, J.H., 1980. What is an estuary? *South African Journal of Science*, 76: 198.

Day, J.W., Rybczyk, J., Scarton, F., Rismondo, A., Are, D. and Cecconi, G., 1999. Soil accretionary dynamics, sea-level rise and the survival of wetlands in Venice Lagoon: a field and modelling approach. *Estuarine, Coastal and Shelf Science*, 49: 607–628.

de Beaumont, L.E., 1845. *Leçons de géologie practique. Septième leçon*. Bertrand, Paris, pp. 221–252.

De Jong, J.D., 1977. Dutch tidal flats. *Sedimentary Geology*, 18: 13–23.

de Martonne, E., 1906. La pénéplaine et les côtes Bretonnes. *Annales de Géographie*, 15: 299–328.

de Martonne, E., 1909. *Traité de Géographie Physique.* Armaund Colin, Paris, 920 pp.

de Vriend, H.J., Capobianco, M., Chesher, T., de Swart, H.E., Latteux, B. and Stive, M.J.F., 1993a. Approaches to long-term modelling of coastal morphology: a review. *Coastal Engineering*, 21: 225–269.

de Vriend, H.J., Zyserman, J., Nicholson, J., Roelvink, J.A., Péchon, P. and Southgate, H.N., 1993b. Medium-term 2DH coastal area modelling. *Coastal Engineering*, 21: 193–224.

Dean, R.G., 1973. Heuristic models of sand transport in the surf zone. In: *Engineering Dynamics of the Coastal Zone, Proceedings of the First Australian Conference on Coastal Engineering*. Institution of Engineers, Australia, Sydney, pp. 208–214.

Dean, R.G., 1977. *Equilibrium Beach Profiles: US Atlantic and Gulf Coast Ocean*, University of Delaware, Engineering Report, No 12.

Dean, R.G., 1987. Additional sediment input to the nearshore region. *Shore and Beach*, 55: 76–81.

Dean, R.G., 1990. Beach response to sea level change. In: B. Le Méhauté and D.M. Hanes (Editors), *The Sea*. Wiley, New York, pp. 869–887.

Dean, R.G., 1991. Equilibrium beach profiles: characteristics and applications. *Journal of Coastal Resarch*, 7: 53–84.

Dean, R.G., 1997. Models for barrier island restoration. *Journal of Coastal Research*, 7: 53–84.

Dean, R.G. and Maurmeyer, E.M., 1983. Models for beach profile response. In: P.D. Komar (Editor), *CRC Handbook of Coastal Processes and Erosion*. CRC Press, Boca Raton, pp. 151–165.

Dean, R.G. and Yoo, C., 1992. Beach nourishment performance predictions. *Journal of Waterway, Port, Coastal, and Ocean Engineering*, 118: 567–586.

Dean, R.G., Healy, T.R. and Dommerholt, A.P., 1993. A 'blind-folded' test of equilibrium beach profile concepts with New Zealand data. *Marine Geology*, 109: 253–266.

Decho, A.W., 2000. Microbial biofilms in intertidal systems: an overview. *Continental Shelf Research*, 20: 1257–1273.

DeLaune, R.D., Smith, C.J., Patrick, W.H. and Roberts, H.H., 1987. Rejuvenated marsh and bay-bottom accretion on the rapidly subsiding coastal plain of US Gulf Coast: a second-order effect of the emerging Atchafalaya delta. *Estuarine, Coastal and Shelf Science*, 25: 381–389.

DeLaune, R.D., Patrick, W.H. and Van Breemen, N., 1990. Process governing marsh formation in a rapidly subsiding coastal environment. *Catena*, 17: 277–288.

Delesalle, B., 1985. Mataiva atoll, Tuamotu Archipelago. *Proceedings of the Fifth Coral Reef Congress, Tahiti*, Antenne Museum-Ephe, Moorea, French Polynesia, pp. 269–322.

Denys, L. and Baeteman, C., 1995. Holocene evolution of relative sea level and local mean high water spring tides in Belgium – a first assessment. *Marine Geology*, 124: 1–19.

Devoy, R.J.N., 1979. Holocene sea level changes and vegetational history of the lower Thames Estuary. *Philosophical Transactions of the Royal Society, Series B*, 285: 355–407.

Devoy, R.J.N., 1982. Analysis of the geological evidence for Holocene sea-level movements in southeast England. *Proceedings of the Geological Association*, 93: 65–90.

Dickinson, K.A., Berryhill, H.L. and Holmes, C.W., 1972. Criteria for recognising ancient barrier coastlines. *Society of Economic Palaeontologists and Mineralogists, Special Publication*, 16: 192–214.

Dickinson, W.R., 2000. Changing times: the Holocene legacy. *Environmental History*, 5: 1169–1177.

Dickinson, W.R., 2001. Paleoshoreline record of relative Holocene sea levels on Pacific islands. *Earth Science Reviews*, 55: 191–234.

Dickinson, W.R., Burley, D.V. and Shutler, R., 1999. Holocene paleoshoreline record in Tonga: geomorphic features and archaeological implications. *Journal of Coastal Research*, 15: 682–700.

Dickinson, W.W. and Woolfe, K.J., 1997. An *in situ* transgressive barrier model for the Nelson boulder bank, New Zealand. *Journal of Coastal Research*, 13: 937–952.

Dietz, R.S., 1963. Wave base marine profile of equilibrium, and wave-built terraces, a critical appraisal. *Geological Society of America Bulletin*, 74: 971–990.

Dijkema, K.S., 1987. Geography of salt marshes in Europe. *Zeitschrift für Geomorphologie. NF*, 31: 489–499.

Dijkema, K.S., Bossinade, J.H., Bouwesna, P. and de Glopper, R.J., 1990. Salt marshes in the Netherlands Wadden Sea: rising high-tide levels and accretion enhancement. In: J.J. Benkema (Editor), *Expected Effects of Climatic Change on Marine Coastal Ecosystems*. Kluwer, Dordrecht, pp. 173–188.

Dingler, J.R., Reiss, T.E. and Plant, N.G., 1993. Erosional patterns of the Isles Dernieres, Louisiana, in relation to meteorological influences. *Journal of Coastal Research*, 9: 112–125.

Dionne, J.-C., 1967. Formes de corrosion littorale, côte sud du Saint-Laurent. *Cahiers de Géographie, Québec*, 11: 379–395.

Dionne, J.-C., 1999. Indices de fluctuations mineures du niveau marin relatif à l'Isle-Verte, côte sud de l'estuaire du Saint-Laurent, Québec. *Géographie Physique et Quaternaire*, 53: 277–285.

Dittmann, S., 2000. Zonation of benthic communities in a tropical tidal flat of north-east Australia. *Journal of Sea Research*, 43: 33–51.

Dittmar, T., Lara, R.J. and Kattner, G., 2001. River or mangrove? Tracing major organic matter sources in tropical Brazilian coastal waters. *Marine Chemistry*, 73: 253–271.

Dix, G.R. and Kyser, K., 2000. Punctuated environmental change along the Holocene (<6ka) bank margin of south-west Exuma Sound, Bahamas. *Sedimentology*, 47: 421–434.

Dixon, L.F.J., Barker, R., Bray, M., Farres, P., Hooke, J., Inkpen, R., Merel, A., Payne, D. and Shelford, A., 1998. Analytical photogrammetry for geomorpohlogical research. In: S.N. Lane, K.S. Richards and J.H. Chandler

(Editors), *Landform Monitoring, Modelling and Analysis*. John Wiley & Sons, Chichester, pp. 63–94.

Dixon, M.J. and Tawn, J.A., 1992. Trends in UK extreme sea levels: a spatial approach. *Geophysical Journal International*, 93: 607–616.

Dodge, R.E., Fairbanks, R.G., Benninger, L.K. and Maurrasse, F., 1983. Pleistocene sea levels from raised coral reefs of Haiti. *Science*, 219: 1423–1425.

Dolan, R., 1971. Coastal landforms: crescentic and rhythmic. *Geological Society of America Bulletin*, 82: 177–180.

Dolan, R. and Hayden, B., 1983. Patterns and predictions of shoreline change. In: P.D. Komar (Editor), *CRC Handbook of Coastal Processes and Erosion*. CRC Press, Boca Raton, pp. 123–149.

Dolan, R., Fenster, M.S. and Holme, S.J., 1991. Temporal analysis of shoreline recession and accretion. *Journal of Coastal Research*, 7: 723–744.

Dolan, R., Fenster, M.S. and Holme, S.J., 1992. Spatial analysis of shoreline recession and accretion. *Journal of Coastal Research*, 8: 263–285.

Dominey-Howes, D., Cundy, A. and Croudace, I., 2000. High energy marine flood deposits on Astypalaea Island, Greece: possible evidence for the AD 1956 southern Aegean tsunami. *Marine Geology*, 163: 303–315.

Done, T.J., 1999. Coral community adaptability to environmental change at the scales of regions, reefs and reef zones. *American Zoologist*, 39: 66–79.

Donnelly, J.P., Bryant, S.S., Butler, J., Dowling, J., Fan, L., Hausmann, N., Newby, P., Shuman, B., Stern, J., Westover, K. and Webb, T., 2001a. 700 yr sedimentary record of intense hurricane landfalls in southern New England. *Geological Society of America Bulletin*, 113: 714–727.

Donnelly, J.P., Roll, S., Wengren, M., Butler, J., Lederer, R. and Webb, T., 2001b. Sedimentary evidence of intense hurricane strikes from New Jersey. *Geology*, 29: 615–618.

Donoghue, D.N.M., Thomas, D.C.R. and Zong, Y., 1994. Mapping and monitoring the intertidal zone of the East Coast of England using remote sensing techniques and a coastal monitoring GIS. *Marine Technology Society Journal*, 28: 19–29.

Doornkamp, J.C., 1998. Coastal flooding, global warming and environmental management. *Journal of Environmental Management*, 52: 327–333.

Dott, R.H., 1998. What is unique about geological reasoning? *GSA Today*, 10: 15–18.

Douglas, B.C., 2001. Sea level change in the era of the recording tide gauge. In: B.C. Douglas, M.S. Kearney and S.P. Leatherman (Editors), *Sea Level Rise: History and Consequences*. Academic Press, San Diego, pp. 37–64.

Douglas, B.C., Kearney, M.S. and Leatherman, S.P., 2001. *Sea Level Rise: History and Consequences*. Academic Press, San Diego, 232 pp.

Dredge, L.A. and Nixon, F.M., 1992. Glacial and environmental geology of northeastern Manitoba. *Geological Survey of Canada Memoir*, 432: 1–80.

Dronkers, J., 1986a. Tidal asymmetry and estuarine morphology. *Netherlands Journal of Sea Research*, 20: 117–131.

Dronkers, J., 1986b. Tide-induced residual transport of fine sediment. In: J. van de Kreeke (Editor), *Physics of Shallow Estuaries and Bays. Lecture Notes on Coastal and Estuarine Studies*. Springer-Verlag, Berlin, pp. 228–244.

Duffy, W., Belknap, D.F. and Kelley, J.T., 1989. Morphology and stratigraphy of small barrier-lagoon systems in Maine. *Marine Geology*, 88: 243–262.

Dunbar, G.B., Dickens, G.R. and Carter, R.M., 2000. Sediment flux across the Great Barrier Reef Shelf to the Queensland Trough over the last 300 ky. *Sedimentary Geology*, 133: 49–92.

Dyer, K.R., 1995. Sediment transport processes. In: G.M.E. Perillo (Editor), *Geomorphology and Sedimentology of Estuaries*. Elsevier, New York, pp. 423–449.

Dyer, K.R., 1998. The typology of intertidal mudflats. In: K.S. Black, D.M. Paterson and A. Cramp (Editors), *Sedimentary Processes in the Intertidal Zone*. Geological Society of London, Special Publication, pp. 11–24.

Dyer, K.R., 2001. Estuarine circulation. In: J.H. Steele, S.A. Thorpe and K.K. Turekian (Editors), *Encyclopedia of Ocean Sciences*. Academic Press, San Diego, pp. 846–852.

Dyer, K.R., Christie, M.C., Feates, N., Fennessy, M.J., Pejrup, M. and van der Lee, W., 2000. An investigation into processes influencing the morphodynamics of an intertidal mudflat, the Dollard estuary, the Netherlands: I. Hydrodynamics and suspended sediment. *Estuarine, Coastal and Shelf Science*, 50: 607–625.

Eagleson, P.S. and Dean, R.G., 1961. Wave induced motion of bottom sediment particles. *Transactions of the American Society of Engineers*, 126: 1162–1189.

Earle, M.D., 1975. Extreme wave conditions during Hurricane Camille. *Journal of Geophysical Research*, 80: 377–379.

Easton, W.H. and Olson, E.A., 1976. Radiocarbon profile of Hanauma Reef, Oahu, Hawaii. *Geological Society of America Bulletin*, 87: 711–719.

Ebisemiju, F.S., 1987. An evaluation of factors controlling present rates of shoreline retrogradation in the western Niger delta, Nigeria. *Catena*, 14: 1–12.

Edinger, E.N., Limmon, G.V., Jompa, J., Widjatmoko, W., Heikoop, J.M. and Risk, M.J., 2000. Normal coral growth rates on dying reefs: are coral growth rates good indicators of reef health? *Marine Pollution Bulletin*, 40: 404–425.

Edwards, A.B., 1941. Storm-wave platforms. *Journal of Geomorphology*, 4: 223–236.

Edwards, A.B., 1958. Wave-cut platforms at Yampi Sound, in the Buccaneer Archipelago, W.A. *Journal of the Royal Society of Western Australia*, 41: 17–21.

Edwards, R.L., Chen, J.H., Ku, T.-L. and Wasserburg, G.J., 1987. Precise timing of the Last Interglacial Period from mass spectrometric determination of Thorium-230 in corals. *Science*, 236: 1547–1553.

Egler, F.E., 1952. Southeast saline Everglades vegetation, Florida: and its management. *Vegetatio*, 3: 213–265.

Ehlers, J., Nagorny, K., Schmidt, P., Stieve, B. and Zietlow, K., 1993. Storm surge deposits in North Sea salt marshes dated by ^{134}Cs and ^{137}Cs determination. *Journal of Coastal Research*, 9: 698–701.

Eisenhauer, A., Wasserburg, G.J., Chen, J.H., Bonani, G., Collins, L.B., Zhu, Z.R. and Wyrwoll, K.H., 1993. Holocene sea-level determination relative to the Australian continent: U/Th (TIMS) and ^{14}C (AMS) dating of coral cores from the Abrolhos Islands. *Earth and Planetary Science Letters*, 114: 529–547.

Eiser, W.C. and Kjerfve, B., 1986. Marsh topography and hypsometric characteristics of a South Carolina salt marsh basin. *Estuarine, Coastal and Shelf Science*, 23: 595–605.

Eisma, D., 1998. *Intertidal Deposits: River Mouths, Tidal Flats, and Coastal Lagoons*. CRC Press, Boca Raton, 525 pp.

Eisma, D. and van Bennekom, A.J., 1978. The Zaire River and estuary and the Zaire outflow in the Atlantic Ocean. *Netherlands Journal of Sea Research*, 12: 255–272.

Eisma, D., Augustinus, P.G.E.F. and Alexander, C., 1991. Recent and subrecent changes in the dispersal of Amazon mud. *Netherlands Journal of Sea Research*, 28: 181–192.

Eitner, V., 1996. Geomorphological response of the East Frisian barrier islands to sea-level rise: an investigation of past and future evolution. *Geomorphology*, 15: 57–65.

Ekdale, A.A. and Lewis, D.W., 1993. Sabellariid reefs in Ruby Bay, New Zealand: a modern analogue of *Skolithos* 'Piperock' that is not produced by burrowing activity. *Palaios*, 8: 614–620.

El-Asmar, H.M., 1994. Aeolianite sedimentation along the northwestern coast of Egypt, evidence for Middle to Late Quaternary aridity. *Quaternary Science Reviews*, 13: 699–708.

El-Sabh, M.I., Aung, T.H. and Murty, T.S., 1997. Physical processes in inverse estuarine systems. *Oceanography and Marine Biology, Annual Review*, 35: 1–69.

Eleveld, M.A., Blok, S.T. and Bakx, J.P.G., 2000. Deriving relief of a coastal landscape with aerial video data. *International Journal of Remote Sensing*, 21: 189–195.

Eliot, I., Fuller, M. and Sanderson, P., 1998. Historical development of a foredune plain at Desperate Bay, Western Australia. *Journal of Coastal Research*, 14: 1187–1201.

Ellison, A.M., Mukherjee, B.B. and Karim, A., 2000. Testing patterns of zonation in mangroves: scale dependence and environmental correlates in the Sundarbans of Bangladesh. *Journal of Ecology*, 88: 813–824.

Ellison, J.C. and Stoddart, D.R., 1991. Mangrove ecosystem collapse during predicted sea-level rise: Holocene analogues and implications. *Journal of Coastal Research*, 7: 151–165.

Emery, K.L. and Aubrey, D.G., 1991. *Sea Levels, Land Levels and Tide Gauges*. Springer-Verlag, New York, 237 pp.

Emery, K.O., 1941. Rate of surface retreat of sea cliffs based on dated inscriptions. *Science*, 93: 617–618.

Emery, K.O. and Foster, H.L., 1956. Shoreline nips in tuff at Matsushima, Japan. *American Journal of Science*, 254: 380–385.

Emery, K.O. and Kuhn, G.G., 1980. Erosion of rock shores at La Jolla. *Marine Geology*, 37: 197–208.

Emery, K.O. and Kuhn, G.G., 1982. Sea cliffs, their processes, profiles and classification. *Geological Society of America Bulletin*, 93: 644–654.

Emery, K.O., Tracey, J.I. and Ladd, H.S., 1954. Geology of Bikini and nearby atolls. *United States Geological Survey Professional Paper*, 260–A: 1–265.

Emmel, F.J. and Curray, J.R., 1982. A submerged late Pleistocene delta and other features related to sea level changes in the Malacca Strait. *Marine Geology*, 47: 197–216.

Eronen, M., Glückert, G., Hatakka, L., van de Plassche, O., van der Plicht, J. and Rantala, P., 2001. Rates of Holocene isostatic uplift and relative sea-level lowering of the Baltic in SW Finland based on studies of isolation contacts. *Boreas*, 30: 17–30.

Escoffier, F.F., 1940. The stability of tidal inlets. *Shore and Beach*, 8: 114–115.

Esselink, P., Dijkema, K.S., Reents, S. and Hagerman, G., 1998. Vertical accretion and profile changes in abandoned man-made tidal marshes in the Dollard Estuary, the Netherlands. *Journal of Coastal Research*, 14: 570–582.

Evans, G., 1965. Intertidal flat sediments and their environments of deposition in the Wash. *Geological Society of London Quarterly Journal*, 121: 209–241.

Evans, G., 1975. Intertidal flat deposits of the Wash, western margin of the North Sea. In: R.N. Ginsburg (Editor), *Tidal Deposits*. Springer-Verlag, New York, pp. 13–20.

Evans, J.L., 1993. Sensitivity of tropical cyclone intensity to sea surface temperature. *Journal of Climate*, 6: 1133–1140.

Evans, O.F., 1939. Sorting and transportation of material in the swash and backwash. *Journal of Sedimentary Petrology*, 9: 28–31.

Evans, O.F., 1942. The origin of spits, bars and related structures. *Journal of Geology*, 50: 846–865.

Everitt, J.H., Judd, F.W., Escobar, D.E. and Davis, M.R., 1996. Integration of remote sensing and spatial information technologies for mapping black mangrove on the Texas Gulf coast. *Journal of Coastal Research*, 12: 64–69.

Ewel, K.C., Twilley, R.R. and Ong, J.E., 1998. Different kinds of mangrove forests provide different goods and services. *Global Ecology and Biogeography Letters*, 7: 83–94.

Fagherazzi, S. and Furbish, D.J., 2001. On the shape and widening of salt marsh creeks. *Journal of Geophysical Research*, 106: 991–1003.

Fagherazzi, S., Bortoluzzi, A., Dietrich, W.E., Adami, A., Lanzoni, S., Marani, M. and Rinaldo, A., 1999. Tidal networks, I, automatic network extraction and preliminary scaling features from DTMs. *Water Resources Research*, 35: 3891–3904.

Fairbanks, R.G., 1989. A 17000-year glacio-eustatic sea level record: influence of glacial melting rates on the Younger Dryas event and deep-ocean circulation. *Nature*, 342: 637–642.

Fairbridge, R.W., 1948. Notes on the geomorphology of the Pelsart Group of the Houtman's Abrolhos Islands. *Journal of the Royal Society of Western Australia*, 30: 1–143.

Fairbridge, R.W., 1961. Eustatic changes in sea level. *Physics and Chemistry of the Earth*, 4: 99–185.

Fairbridge, R.W., 1980. The estuary: its definition and geodynamic cycle. In: E. Olausson and I. Cato (Editors), *Chemistry and Biogeochemistry of Estuaries*. Wiley, New York, pp. 1–36.

Fairbridge, R.W. and Johnson, D.L., 1978. Eolianite. In: R.W. Fairbridge and J. Bourgeois (Editors), *The Encyclopedia of Sedimentology*. Dowden Hutchinson and Ross, Stroudsberg, pp. 279–282.

Falkland, A.C. and Woodroffe, C.D., 1997. Geology and hydrogeology of Tarawa and Christmas Island (Kiritimati), Kiribati, Central Pacific. In: H.L. Vacher and T.M. Quinn (Editors), *Geology and Hydrogeology of Carbonate Islands*. Elsevier, Amsterdam, pp. 577–610.

Feng, J. and Zhang, W., 1998. The evolution of the modern Luanhe River delta, north China. *Geomorphology*, 25: 269–278.

Fenneman, N.M., 1902. Development of the profile of equilibrium of the subaqueous shore terrace. *Journal of Geology*, 10: 1–31.

Ferland, M.A., Roy, P.S. and Murray-Wallace, C.V., 1995. Glacial lowstand deposits on the outer continental shelf of southeastern Australia. *Quaternary Research*, 44: 294–299.

Field, M.E. and Duane, D.B., 1976. Post-Pleistocene history of the United States inner continental shelf: significance to origin of barrier islands. *Geological Society of America Bulletin*, 98: 691–702.

Fischer, A.G., 1961. Stratigraphic record of transgressing seas in light of sedimentation on Atlantic coast of New Jersey. *American Association of Petroleum Geologists Bulletin*, 45: 1656–1666.

Fisk, H.N., 1944. *Geological Investigation of the Alluvial Valley of the Lower Mississippi River*. Mississippi River Commission, Vicksburg, MS, 78 pp.

Fisk, H.N., 1959. Padre Island and Laguna Madre mud flats, south coastal Texas. *Coastal Studies Institute, second Geography Conference*, Louisiana State University, Baton Rouge, pp. 103–151.

Fisk, H.N., 1960. Recent Mississippi River sedimentation and peat accumulation. *Comptes Rendus 4th Congress Avanc. Étud. Stratigr. Géol. Carb.*: 187–199.

Fisk, H.N., McFarlan, E., Kolb, C.R. and Wilbert, L.J., 1954. Sedimentary framework of the modern Mississippi Delta. *Journal of Sedimentary Petrology*, 24: 79–99.

Fitt, W.K., Brown, B.E., Warner, M.E. and Dunne, R.P., 2001. Coral bleaching: interpretation of thermal tolerance limits and thermal thresholds in tropical corals. *Coral Reefs*, 20: 51–65.

FitzGerald, D.M. and van Heteren, S., 1999. Classification of paraglacial barrier systems: coastal New England, USA. *Sedimentology*, 46: 1083–1108.

FitzGerald, D.M., Baldwin, C.T., Ibrahim, N.A. and Humphries, S.M., 1992. Sedimentologic and morphologic evolution of a beach-ridge barrier along an indented coast: Buzzards Bay, Massachusetts. In: C.H. Fletcher and J.F. Wehmiller (Editors), *Quaternary Coasts of the United States: Marine and Lacustrine Systems*. SEPM Special Publication, Society of Economic Palaeontologists and Mineralogists, Tulsa, OK, pp. 65–75.

FitzGerald, D.M., Rosen, P.S. and van Heteren, S., 1994. New England barriers. In: R.A. Davis (Editor), *Geology of Holocene Barrier Island Systems*. Springer-Verlag, Berlin, pp. 305–394.

Fjeldskaar, W., 1991. Geoidal-eustatic changes induced by the deglaciation of Fennoscandia. *Quaternary International*, 9: 1–6.

Flather, R.A., 2001. Storm surges. In: J.H. Steele, S.A. Thorpe and K.K. Turekian (Editors), *Encyclopedia of Ocean Sciences*. Academic Press, San Diego, pp. 2882–2892.

Fleming, M., Lin, G. and Sternberg, L.d.S.L., 1990. Influence of mangrove detritus in an estuarine ecosystem. *Bulletin of Marine Science*, 47: 663–669.

Flemming, N.C., 1965. Form and relation to present sea level of Pleistocene marine erosion features. *Journal of Geology*, 73: 799–811.

Fletcher, C.H. and Jones, A.T., 1996. Sea-level highstand recorded in Holocene shoreline deposits on Oahu, Hawaii. *Journal of Sedimentary Research*, 66: 632–641.

Flinn, D., 1997. The role of wave diffraction in the formation of St Ninian's Ayre (Tombolo) in Shetland, Scotland. *Journal of Coastal Research*, 13: 202–208.

Flood, P.G. and Heatwole, H., 1986. Coral Cay instability and species-turnover of plants at Swain Reefs, southern Great Barrier Reef, Australia. *Journal of Coastal Research*, 2: 479–496.

Flood, P.G. and Scoffin, T.P., 1978. Reefal sediments of the northern Great Barrier Reef. *Philosophical Transactions of the Royal Society, London, Series A*, 291: 55–71.

Focke, J.W., 1978a. Limestone cliff morphology and organism distribution on Curaçao (Netherlands Antilles). *Leidse Geologische Mededelingen*, 51: 131–150.

Focke, J.W., 1978b. Limestone cliff morphology on Curaçao (Netherlands Antilles), with special attention to the origin of notches and vermetid/coralline algal surf benches ('cornices', 'trottoirs'). *Zeitschrift für Geomorphologie, NF*, 22: 329–349.

Folk, R.L., 1964. *Petrology of Sedimentary Rocks*. Hemphills, Austin, 182 pp.

Folk, R.L. and Robles, R., 1964. Carbonate sands of Isla Perez, Alacran Reef complex, Yucatan. *Journal of Geology*, 72: 255–292.

Foote, Y.L.M. and Huntley, D.A., 1994. Velocity moments on a macro-tidal intermediate beach. *Coastal Dynamics '94*, American Society of Civil Engineers, New York, pp. 794–808.

Forbes, D.L. and Syvitski, J.P.M., 1994. Paraglacial coasts. In: R.W.G. Carter and C.D. Woodroffe (Editors), *Coastal Evolution: Late Quaternary Shoreline Morphodynamics*. Cambridge University Press, Cambridge, pp. 373–424.

Forbes, D.L. and Taylor, R.B., 1994. Ice in the shore zone and the geomorphology of cold coasts. *Progress in Physical Geography*, 18: 59–89.

Forbes, D.L., Taylor, R.B., Orford, J.D., Carter, R.W.G. and Shaw, J., 1991. Gravel-barrier migration and overstepping. *Marine Geology*, 97: 305–313.

Forbes, D.L., Orford, J.D., Carter, R.W.G., Shaw, J. and Jennings, S.C., 1995. Morphodynamic evolution, self-organisation, and instability of coarse-clastic barriers on paraglacial coasts. *Marine Geology*, 126: 63–85.

Ford, M.A., Cahoon, D.R. and Lynch, J.C., 1999. Restoring marsh elevation in a rapidly subsiding salt marsh by thin-layer deposition of dredged material. *Ecological Engineering*, 12: 189–205.

Fórnos, J.J. and Ahr, W.M., 1997. Temperate carbonates on a modern, low-energy, isolated ramp: the Balearic platform, Spain. *Journal of Sedimentary Research*, 67: 364–373.

Fórnos, J.J., Forteza, V. and Martinez-Taberner, A., 1997. Modern polychaete reefs in Western Mediterranean lagoons: *Ficopotamus enigmaticus* (Fauvel) in the Albufera of Menorca, Balearic Islands. *Palaeogeography, Palaeoclimatology, Palaeoecology*, 128: 175–186.

Fosberg, F.R., 1966. Vegetation as a geological agent in tropical deltas, In: *Scientific Problems of the Humid Tropical Zone Deltas and their Implications: Proceedings of the Dacca Symposium*. UNESCO, Paris, pp. 227–233.

Foster, I.D.L., Dawson, A., Dawson, S., Lees, J.A. and Mansfield, L., 1993. Tsunami sedimentation sequences in the Scilly Isles, southwest England. *Science of Tsunami Hazards*, 11: 35–46.

Fox, W.T., 1985. Modeling coastal environments. In: R.A. Davis (Editor), *Coastal Sedimentary Environments*. Springer-Verlag, New York, pp. 665–705.

Francis, J., 1992. Physical processes in the Rufiji delta and their possible implications on the mangrove ecosystem. *Hydrobiologia*, 247: 173–179.

Frankel, E., 1977. Previous *Acanthaster* aggregations in the Great Barrier Reef. *Proceedings of the Third International Coral Reef Symposium*, Vol. 1. Rosenstiel School of Marine and Atmospheric Sciences, University of Miami, pp. 201–208.

Frazier, D.E., 1967. Recent deltaic deposits of the Mississippi River: their development and chronology. *Transactions of the Gulf Coast Association of Geological Societies*, 17: 287–315.

French, J.R., 1993. Numerical simulation of vertical marsh growth and adjustment to accelerated sea-level rise, North Norfolk, UK. *Earth Surface Processes and Landforms*, 18: 63–81.

French, J.R. and Clifford, N.J., 1992. Characteristics and 'event-structure' of near-bed turbulence in a macrotidal saltmarsh channel. *Estuarine, Coastal and Shelf Science*, 34: 49–69.

French, J.R. and Spencer, T., 1993. Dynamics of sedimentation in a tide-dominated backbarrier salt marsh, Norfolk, UK. *Marine Geology*, 110: 315–331.

French, J.R. and Stoddart, D.R., 1992. Hydrodynamics of salt marsh creek systems: implications for marsh morphological development and material exchange. *Earth Surface Processes and Landforms*, 17: 235–252.

French, J.R., Spencer, T., Murray, A.L. and Arnold, N.S., 1995. Geostatistical analysis of sediment deposition in two small tidal wetlands, Norfolk, UK. *Journal of Coastal Research*, 11: 308–321.

French, P.W., 1997. *Coastal and Estuarine Management*. Routledge, London, 251 pp.

Frey, R.W. and Basan, P.B., 1978. Coastal salt marshes. In: R.A. Davis (Editor), *Coastal Sedimentary Environments*. Springer, New York, pp. 101–169.

Friedrichs, C.T. and Aubrey, D.G., 1994. Tidal propagation in strongly convergent channels. *Journal of Geophysical Research*, 99: 3321–3336.

Frith, C.A., 1983. Lagoonal sedimentation, One Tree Reef, Great Barrier Reef. *Bureau of Mineral Resources Journal of Australian Geology and Geophysics*, 8: 211–221.

Froidefond, J.M., Pujos, M. and Andre, X., 1988. Migration of mud banks and changing coastline in French Guiana. *Marine Geology*, 84: 19–30.

Fujii, S. and Fuji, N., 1967. Postglacial sea level in the Japanese Islands. In: N. Ikebe (Editor), *Sea Level Changes and Crustal Movements of the Pacific During the Pliocene and Post-Pliocene Time. (Journal of Geoscience)* pp. 43–51, Osaka: Osaka City University.

Fujii, S. and Mogi, A., 1970. On coasts and shelves in their mutual relations in Japan during the Quaternary. *Quaternaria*, 12: 155–164.

Funnell, B.M. and Person, I., 1989. Holocene sedimentation on the North Norfolk barrier coast in relation to relative sea-level change. *Journal of Quaternary Science*, 4: 25–36.

Furukawa, K., Wolanski, E. and Mueller, H., 1997. Currents and sediment transport in mangrove forests. *Estuarine, Coastal and Shelf Science*, 44: 301–310.

Gagan, M.K., Ayliffe, L.K., Beck, J.W., Cole, J.E., Druffel, E.R.M., Dunbar, R.B. and Schrag, D.P., 2000. New views of tropical paleoclimates from corals. *Quaternary Science Reviews*, 19: 45–64.

Gagan, M.K., Johnson, D.P. and Crowley, G.M., 1994. Sea level control of stacked late Quaternary coastal sequences, central Great Barrier Reef. *Sedimentology*, 41: 329–351.

Gallagher, B., 1972. Some qualitative aspects of nonlinear wave radiation in a surf zone. *Geophysical Fluid Dynamics*, 3: 347–354.

Gallagher, E., Guza, R.T. and Elgar, S., 1998. Observations of sand bar evolution on a natural beach. *Journal of Geophysical Research*, 103: 3203–3215.

Galloway, W.E., 1975. Process framework for describing the morphologic and stratigraphic evolution of deltaic depositional systems. In: M.L. Broussard (Editor), *Deltas: Models for Exploration*. Houston Geological Society, Houston, pp. 87–98.

Galloway, W.E., 1976. Sediments and stratigraphic framework of the Copper River fan-delta, Alaska. *Journal of Sedimentary Petrology*, 46: 726–737.

Galloway, W.E., 1989. Genetic stratigraphic sequences in basin analysis I. Architecture and genesis of flooding-surface bounded depositional units. *American Association of Petroleum Geologists Bulletin*, 73: 125–142.

Galvin, C.J., 1968. Breaker type classification on three laboratory beaches. *Journal of Geophysical Research*, 73: 3651–3659.

Galvin, C.J., 1972. Wave breaking in shallow water. In: R.E. Meyer (Editor), *Waves on Beaches*. Academic Press, New York, pp. 413–456.

Gao, S. and Collins, M., 1994. Tidal inlet equilibrium, in relation to cross-sectional area and sediment transport patterns. *Estuarine, Coastal and Shelf Science*, 38: 157–172.

Gaposchkin, E.M., 1973. *Smithsonian Standard Earth*. Special Report, 353. Smithsonian Institution Astrophysical Observatory, Cambridge, MA, 388 pp.

Gardiner, J.S., 1931. *Coral Reefs and Atolls*, Macmillan, London, 182 pp.

Gardner, L.R. and Bohn, M., 1980. Geomorphic and hydraulic evolution of tidal creeks on a subsiding beach ridge plain, North Inlet, South Carolina. *Marine Geology*, 34: 791–797.

Gares, P.A., 1992. Topographic changes associated with coastal dune blowouts at Island Beach State Park, New Jersey. *Earth Surface Processes and Landforms*, 17: 589–604.

Gastaldo, R.A. and Huc, A.-Y., 1992. Sediment facies, depositional environments, and distribution of phytoclasts in the recent Mahakam River delta, Kalimantan, Indonesia. *Palaios*, 7: 574–590.

Gehrels, W.R., 1994. Determining relative sea-level change from salt-marsh foraminifera and plant zones on the coast of Maine, USA. *Journal of Coastal Research*, 10: 990–1009.

Gehrels, W.R., 1999. Middle and late Holocene sea-level change in eastern Maine reconstructed from foraminiferal saltmarsh stratigraphy and AMS ^{14}C dates on basal peats. *Quaternary Research*, 52: 350–359.

Gehrels, W.R., Belknap, D.F. and Kelley, J.T., 1996. Integrated high-precision analyses of Holocene relative sea-level changes: lessons from the coast of Maine. *Geological Society of America Bulletin*, 108: 1073–1088.

Geister, J., 1977. The influence of wave exposure on the ecological zonation of Caribbean coral reefs. *Proceedings of the Third International Coral Reef Symposium*, Vol. 1, Rosenstiel School of Marine and Atmospheric Sciences, University of Miami, pp. 23–29.

Geyh, M.A. and Schleicher, H.S., 1990. *Absolute Age Determination*. Springer-Verlag, Berlin, 503 pp.

Gilbert, G.K., 1877. *Report on the Geology of the Henry Mountains*. Government Printing Office, Washington, DC, 160 pp.

Gilbert, G.K., 1885. The topographic features of lake shores. *United States Geological Survey Annual Report*, 5: 75–123.

Gilbert, G.K., 1886. The inculcation of the scientific method by example with an illustration drawn from the Quaternary geology of Utah. *American Journal of Science*, 31: 284–299.

Giles, R.T. and Pilkey, O.H., 1965. Atlantic beach and dune sediments of the southern United States. *Journal of Sedimentary Petrology*, 35: 900–910.

Gill, E.D., 1950. Some unusual shore platforms near Gisborne, North Island, New Zealand. *Transactions and Proceedings of the Royal Society of New Zealand*, 78: 64–68.

Gill, E.D., 1972. The relationship of present shore platforms to past sea levels. *Boreas*, 1: 1–25.

Gill, E.D., 1973. Rate and mode of retrogradation on rocky coasts in Victoria, Australia, and their relationship to sea level changes. *Boreas*, 2: 143–171.

Gill, E.D. and Hopley, D., 1972. Holocene sea levels in eastern Australia – a discussion. *Marine Geology*, 12: 223–233.

Gill, E.D. and Lang, J.G., 1983. Micro-erosion meter measurements of rock wear on the Otway coast of southeast Australia. *Marine Geology*, 52: 141–156.

Ginsburg, R.N., 1975. *Tidal Deposits: A Casebook of Recent Examples and Fossil Counterparts*. Springer-Verlag, New York, 425 pp.

Giosan, L., Bokuniewicz, H., Panin, N. and Postolache, I., 1999. Longshore sediment transport pattern along the Romanian Danube Delta coast. *Journal of Coastal Research*, 15: 859–871.

Gischler, E., Lomando, A.J., Hudson, J.H. and Holmes, C.W., 2000. Last interglacial reef growth beneath Belize barrier and isolated platform reefs. *Geology*, 28: 387–390.

Glynn, P.W., 1984. Widespread coral mortality and the 1982–83 El Niño warming event. *Environmental Conservation*, 11: 133–146.

Glynn, P.W. and Ault, J.S., 2000. A biogeographic analysis and review of the far eastern Pacific coral reef region. *Coral Reefs*, 19: 1–23.

Glynn, P.W., Almodovar, L.R. and Gonzalez, J.C., 1964. Effects of Hurricane Edith on marine life in La Parguera, Puerto Rico. *Caribbean Journal of Science*, 4: 335–345.

Godfrey, P.J., Leatherman, S.P. and Zaremba, R., 1979. A geobotanical approach to classification of barrier beach systems. In: S.P. Leatherman (Editor), *Barrier Islands*. Academic Press, New York, pp. 99–126.

Godwin, H. and Clifford, M.A., 1938. Studies of the postglacial history of British vegetation. I. Origin and stratigraphy of the Fenland deposits near Woodwalton, Hunts. II. Origin and stratigraphy of deposits in southern Fenland. *Philosophical Transactions of the Royal Society of London, Series B*, 229: 323–406.

Godwin, H., Suggate, R.P. and Willis, E.H., 1958. Radiocarbon dating of the eustatic rise in ocean level. *Nature*, 181: 1518–1519.

Goede, A., Harmon, R. and Kiernan, K., 1979. Sea caves of King Island. *Helictite*, 17: 51–64.

Goff, J., Chagué-Goff, C. and Nichol, S., 2001. Palaeotsunami deposits: a New Zealand perspective. *Sedimentary Geology*, 143: 1–6.

Goldsmith, V., 1989. Coastal sand dunes as geomorphic systems. *Proceedings of the Royal Society of Edinburgh, Series B*, 96: 3–15.

González Bonorino, G., Bujalesky, G., Colombo, F. and Ferrero, M., 1999. Holocene coastal paleoenvironments in Atlantic Patagonia, Argentina. *Journal of South American Earth Sciences*, 12: 325–331.

González, M. and Medina, R., 2001. On the application of static equilibrium bay formulations to natural and man-made beaches. *Coastal Engineering*, 43: 209–225.

Good, R.E., Good, N.F. and Frasco, B.R., 1982. A review of primary production and decomposition: dynamics of the belowground marsh component.

In: U.S. Kennedy (Editor), *Estuarine Comparisons*. Academic Press, New York, pp. 139–157.

Goodbred, S.L. and Hine, A.C., 1995. Coastal storm deposition: salt-marsh response to a severe extratropical storm, March 1993, west-central Florida. *Geology*, 23: 679–682.

Goodbred, S.L. and Kuehl, S.A., 2000a. Enormous Ganges–Brahmaputra sediment discharge during strengthened early Holocene monsoon. *Geology*, 28: 1083–1086.

Goodbred, S.L. and Kuehl, S.A., 2000b. The significance of large sediment supply, active tectonism, and eustasy on margin sequence development: Late Quaternary stratigraphy and evolution of the Ganges–Brahmaputra delta. *Sedimentary Geology*, 133: 227–248.

Goodbred, S.L., Wright, E.E. and Hine, A.C., 1998. Sea-level change and storm-surge deposition in a late Holocene Florida salt marsh. *Journal of Sedimentary Research*, 68: 240–252.

Gornitz, V., 1995. Monitoring sea level changes. *Climatic Change*, 31: 515–544.

Gostin, V.A., Belperio, A.P. and Cann, J.H., 1988. The Holocene non-tropical coastal and shelf carbonate province of southern Australia. *Sedimentary Geology*, 60: 51–70.

Gould, H.R. and McFarlan, E., 1959. Geologic history of the chenier plain, southwestern Louisiana. *Transactions of the Gulf Coast Association of Geological Societies*, 9: 261–270.

Gould, S.J., 1989. *Wonderful Life: The Burgess Shale and the Nature of History*. Hutchinson Radius, London, 347 pp.

Gourlay, M.R., 1994. Wave transformation on a coral reef. *Coastal Engineering*, 23: 17–42.

Gourlay, M.R., 1996a. Wave set-up on coral reefs. 1. Set-up and wave-generated flow on an idealised two-dimensional horizontal reef. *Coastal Engineering*, 27: 161–193.

Gourlay, M.R., 1996b. Wave set-up on coral reefs. 2. Set-up on reefs with various profiles. *Coastal Engineering*, 28: 17–55.

Grant, D.R., 1970. Recent coastal submergence of the Maritime Provinces, Canada. *Canadian Journal of Earth Sciences*, 7: 676–698.

Gray, A.J., 1992. Saltmarsh plant ecology: zonation and succession revisited. In: J.R.L. Allen and K. Pye (Editors), *Saltmarshes: Morphodynamics, Conservation and Engineering Significance*. Cambridge University Press, Cambridge, pp. 63–79.

Gray, S.C., Hein, J.R., Hausmann, R. and Radtke, U., 1992. Geochronology and subsurface stratigraphy of Pukapuka and Rakahanga atolls, Cook Islands: Late Quaternary reef growth and sea level history. *Palaeogeography, Palaeoclimatology, Palaeoecology*, 91: 377–394.

Greenstein, B.J., 1989. Mass mortality of the West Indian echinoid *Diadema antillarum* (Echinodermata: Echinoidea): a natural experiment in taphonomy. *Palaios*, 4: 487–492.

Greenwood, B.G. and Davidson-Arnott, R.G.D., 1979. Sedimentation and equilibrium in wave-formed bars: a review and case study. *Canadian Journal of Earth Sciences*, 16: 312–332.

Gregory, J.W., 1913. *The Nature and Origin of Fjords*. John Murray, London, 542 pp.

Grigg, R.W., 1982. Darwin point: a threshold for atoll formation. *Coral Reefs*, 1: 29–34.

Grigg, R.W., 1998. Holocene coral reef accretion in Hawaii: a function of wave exposure and sea level history. *Coral Reefs*, 17: 263–272.

Grigg, R.W. and Epp, D., 1989. Critical depth for the survival of coral islands: effects on Hawaiian Archipelago. *Science*, 243: 638–641.

Griggs, G.B. and Trenhaile, A.S., 1994. Coastal cliffs and platforms. In: R.W.G. Carter and C.D. Woodroffe (Editors), *Coastal Evolution: Late Quaternary Shoreline Morphodynamics*. Cambridge University Press, Cambridge, pp. 425–450.

Griggs, G.R., Tait, J.F. and Corona, W., 1994. The interaction of seawalls and beaches: seven years of monitoring, Monterey Bay, California. *Shore and Beach*, 62: 21–28.

Grindrod, J., 1988. The palynology of Holocene mangrove and saltmarsh sediments, particularly in northern Australia. *Review of Palaeobotany and Palynology*, 55: 229–245.

Grindrod, J. and Rhodes, E.G., 1984. Holocene sea level history of a tropical estuary: Missionary Bay, North Queensland. In: B.G. Thom (Editor), *Coastal Geomorphology in Australia*. Academic Press, Sydney, pp. 151–178.

Grindrod, J., Moss, P. and van der Kaars, S., 1999. Late Quaternary cycles of mangrove development and decline on the north Australian continental shelf. *Journal of Quaternary Science*, 14: 465–470.

Groen, P., 1967. On the residual transport of suspended matter by an alternating tidal current. *Netherlands Journal of Sea Research*, 3: 564–574.

Grossman, E.E. and Fletcher, C.H., 1998. Sea level higher than present 3500 years ago on the northern main Hawaiian Islands. *Geology*, 26: 363–366.

Grossman, E.E., Fletcher, C.H. and Richmond, B.M., 1998. The Holocene sea-level highstand in the equatorial Pacific: analysis of the insular paleosea-level database. *Coral Reefs*, 17: 309–327.

Grottoli, A.G., 2001. Past climate from corals. In: J.H. Steele, S.A. Thorpe and K.K. Turekian (Editors), *Encyclopedia of Ocean Sciences*. Academic Press, San Diego, pp. 2098–2107.

Guilcher, A., 1953. Essai sur la zonation et la distribution des formes littorales de dissolution du calcaire. *Annales de Géographie*, 62: 161–179.

Guilcher, A., 1958. *Coastal and Submarine Morphology*. Methuen, London, 274 pp.

Guilcher, A., 1966. Les grandes falaises et megafalaises des côtes Sud-Ouest et Ouest de l'Irlande. *Annales de Géographie*, 75: 26–38.

Guilcher, A., 1979. Marshes and estuaries in different latitudes. *Interdisciplinary Science Reviews*, 4: 158–168.

Guilcher, A., 1985. Retreating cliffs in the humid tropics: an example from Paraiba, north-eastern Brazil. *Zeitschrift für Geomorphologie*, Suppl.-Bd, 57: 95–103.

Guilcher, A., 1988. *Coral Reef Geomorphology*. John Wiley & Sons, Chichester, 228 pp.

Guilcher, A. and Berthois, L., 1957. Cinq années d'observations sédimentologiques dans quatre esturaires-Temoins de l'ouest de la Bretagne. *Revue de Géomorphologie Dynamique*, 8: 67–86.

Guillou, H., Brousse, R., Gillot, P.Y. and Guille, G., 1993. Geological reconstruction of Fangataufa atoll, South Pacific. *Marine Geology*, 110: 377–391.

Gulliver, F.P., 1896. Cuspate forelands. *Bulletin of the Geological Society of America*, 7: 399–422.

Gulliver, F.P., 1899. Shoreline topography. *Proceedings of the American Academy of Arts and Sciences*, 34: 151–258.

Guo, Q.Z. and Psuty, N.P., 1997. Flood-tide deltaic wetlands – detection of their sequential evolution. *Photogrammetric Engineering and Remote Sensing*, 63: 273–280.

Guppy, H.B., 1888. A recent criticism of the theory of subsidence as affecting coral reefs. *Scottish Geographical Magazine*, 4: 121–137.

Guttenberg, B., 1941. Changes in sea level, postglacial uplift, and mobility of the earth's interior. *Geological Society of America Bulletin*, 52: 721–772.

Guza, R.T. and Inman, D.L., 1975. Edge waves and beach cusps. *Journal of Geophysical Research*, 80: 2997–3012.

Guza, R.T. and Thornton, E.B., 1982. Swash oscillations on a natural beach. *Journal of Geophysical Research*, 87: 483–491.

Hack, J.T. and Goodlett, J.C., 1960. Geomorphology and forest ecology of a mountain region in the central Appalachians. *United States Geological Survey, Professional Paper*, 347.

Hackshaw, C., 1878. On the action of limpets (*Patella*) in sinking pits and abrading the surface of the chalk at Dover. *Journal of the Linnean Society (Zoology)*, 14: 406–411.

Haff, P.K., 1991. Basic physical models in sediment transport, In: N.C. Kraus, K.J. Gingerich and D.L. Kriebel (Editors), *Coastal Sediments '91*. American Society of Civil Engineers, New York, pp. 1–14.

Haff, P.K., 1996. Limitations on predictive modeling in geomorphology. In: B.L. Rhoads and C.E. Thorn (Editors), *The Scientific Nature of Geomorphology*. John Wiley & Sons, Chichester, pp. 337–358.

Hafsten, U., 1983. Shore-level changes in South Norway during the last 13000 years traced by biostratigraphical methods and radiometric dating. *Norsk Geographische Tidsskrift*, 37: 63–79.

Hallam, A. and Wignall, P.B., 1999. Mass extinctions and sea-level changes. *Earth Science Reviews*, 48: 217–250.

Hallemeier, R.J., 1981. A profile zonation for seasonal sand beaches from wave climate. *Coastal Engineering*, 4: 253–277.

Halley, R.B., Shinn, E.A., Hudson, J.H. and Lidz, B., 1977. Recent and relict topography of Boo Bee patch reef, Belize. *Proceedings of Third International Coral Reef Symposium*, Vol. 2. Rosenstiel School of Marine and Atmospheric Sciences, University of Miami, pp. 29–35.

Hallock, P., Hine, A.C., Vargo, G.A., Elrod, J.A. and Jaap, W.C., 1988. Platforms of the Nicaraguan Rise: examples of sensitivity of carbonate sedimentation to excess trophic resources. *Geology*, 16: 1104–1107.

Hanazawa, S., 1940. Micropaleontological studies of drill cores from a deep well in Kita Daito Zima (North Borodino Island). *Jubilee Publication in Commemoration Professor H. Yabe's 60th Birthday*, 2: 755–802.

Hands, E.B., 1983. Erosion of the Great Lakes due to changes in the water level. In: P.D. Komar (Editor), *CRC Handbook of Coastal Processes and Erosion*. CRC Press, Boca Raton, pp. 167–189.

Hanebuth, T., Stattegger, K. and Grootes, P.M., 2000. Rapid flooding of the Sunda Shelf: a late-glacial sea-level record. *Science*, 288: 1033–1035.

Hansom, J.D., 2001. Coastal sensitivity to environmental change: a view from the beach. *Catena*, 42: 291–305.

Hanson, H. and Kraus, N.C., 1989. *GENESIS: Generalized Model for Simulating Shoreline Change*. US Corps of Engineers, Vicksburg, 247 pp.

Hanson, H., Larson, M., Capobianco, M., Dette, H.H., Hamm, L., Lechuga, A., Spanhoff, R., Brampton, A. and Laustrup, C., 2002a. Beach nourishment projects, practices and objectives? A European overview. *Journal of Coastal Research*, in press.

Hanson, H., Larson, M., Steezel, H., Nicholls, R., Capobianco, M., Jimenez, J., Plant, N., Stive, M., Southgate, H. and Aarninkhof, S., 2002b. Modelling of coastal evolution on yearly to decadal time scales. *Journal of Coastal Research*, in press.

Haq, B.U., Hardenbol, J. and Vail, P., 1987. Chronology of fluctuating sea levels since the Triassic. *Science*, 235: 1156–1167.

Hardisty, J., 1990. *Beaches: Form and Process*. Unwin Hyman, London, 324 pp.

Hardisty, J., 1994. Beach and nearshore sediment transport. In: K. Pye (Editor), *Sediment Transport and Depositional Processes*. Blackwell Science Publications, Oxford, pp. 219–255.

Hardisty, J. and Whitehouse, R.J.S., 1988. Evidence for a new sand transport process from experiments on Saharan dunes. *Nature*, 322: 532–534.

Harmelin-Vivien, M.L. and Laboute, P., 1986. Catastrophic impact of hurricanes on atoll outer reef slopes in the Tuamotu (French Polynesia). *Coral Reefs*, 5: 55–62.

Harmsworth, G.C. and Long, S.P., 1986. An assessment of saltmarsh erosion in Essex, England, with reference to the Dengie Peninsula. *Biological Conservation*, 35: 377–387.

Harriott, V.J., Harrison, P.L. and Banks, S.A., 1995. The coral communities of Lord Howe Island. *Marine and Freshwater Research*, 46: 457–465.

Harris, P.T. and Collins, M., 1988. Estimation of annual bedload flux in a macrotidal estuary: Bristol Channel, UK. *Marine Geology*, 83: 237–252.

Harris, P.T. and Collins, M.B., 1991. Sand transport in the Bristol Channel: bedload parting zone or mutually evasive transport pathways? *Marine Geology*, 101: 209–216.

Harrison, W., 1970. Prediction of beach changes. *Progress in Geography*, 2: 208–235.

Haseldonckx, P., 1977. The palynology of a Holocene marginal peat swamp environment in Johore, Malaysia. *Review of Palaeobotany and Palynology*, 24: 227–238.

Haslett, S.K., 2000. *Coastal Systems.* Routledge, New York, 218 pp.

Haslett, S.K. and Curr, R.H.F., 1998. Coastal rock platforms and Quaternary sea-levels in the Baie d'Audierne, Brittany, France. *Zeitschrift für Geomorphologie, NF*, 42: 507–515.

Haslett, S.K., Davies, P., Curr, R.H.F., Davies, C.F.C., Kennington, K., King, C.P. and Margetts, A.J., 1998. Evaluating late-Holocene relative sea-level change in the Somerset Levels, southwest Britain. *The Holocene*, 8: 197–207.

Haslett, S.K., Bryant, E.A. and Curr, R.H.F., 2000. Tracing beach sand provenance and transport using foraminifera: preliminary examples from NW Europe and SE Australia. In: I. Foster (Editor), *Tracers in Geomorphology*. John Wiley & Sons, Chichester, pp. 437–452.

Hasselmann, K., Ross, D.B., Muller, P. and Sell, W., 1976. A parametric wave prediction model. *Journal of Physical Oceanography*, 6: 200–228.

Hatcher, B.G., 1997. Coral reef ecosystems: how much greater is the whole than the sum of the parts? *Coral Reefs*, 16: S77–S91.

Hatcher, B.G., Imberger, J. and Hendry, M.D., 1987. Scaling analysis of coral reef systems: an approach to problems of scale. *Coral Reefs*, 5: 171–181.

Hatton, R.S., DeLaune, R.D. and Patrick, W.H., 1983. Sedimentation, accretion and subsidence in marshes of Barataria Basin, Louisiana. *Limnology and Oceanography*, 28: 494–502.

Hayes, M.O., 1975. Morphology of sand accumulation in estuaries. In: L.E. Cronin (Editor), *Estuarine Research*. Academic Press, New York, pp. 3–22.

Hayes, M.O., 1979. Barrier island morphology as a function of tidal and wave regime. In: S. Leatherman (Editor), *Barrier Islands from the Gulf of St Lawrence to the Gulf of Mexico*. Academic Press, New York, pp. 1–27.

Hayes, M.O., 1994. The Georgia Bight barrier system. In: R.A. Davis (Editor), *Geology of Holocene Barrier Island Systems*. Springer-Verlag, Berlin, pp. 233–304.

Hayne, M. and Chappell, J., 2001. Cyclone frequency during the last 5000 years at Curacao Island, north Queensland, Australia. *Palaeogeography, Palaeoclimatology, Palaeoecology*, 168: 207–219.

Hays, J.D., Imbrie, J. and Shackleton, N.J., 1976. Variations in the earth's orbit: pacemaker of the Ice Ages. *Science*, 194: 1121–1132.

Healey, R.G., Pye, K., Stoddart, D.R. and Bayliss-Smith, T.P., 1981. Velocity variations in salt marsh creeks, Norfolk, England. *Estuarine, Coastal and Shelf Science*, 13: 535–545.

Healy, T.R., 1968. Shore platform morphology on the Whangaparaoa Peninsula, Auckland. *Conference Series, New Zealand Geographical Society*, 5: 163–168.

Hearty, P.J., 1997. Boulder deposits from large waves during the last interglaciation on north Eleuthera Island, Bahamas. *Quaternary Research*, 48: 326–338.

Hearty, P.J., 1998. The geology of Eleuthera Island, Bahamas: a rosetta stone of Quaternary stratigraphy and sea-level history. *Quaternary Science Reviews*, 17: 333–355.

Jones, B. and Hunter, I.G., 1992. Very large boulders on the coast of Grand Cayman: the effects of giant waves on rocky coastlines. *Journal of Coastal Research*, 8: 763–774.

Jones, B. and Hunter, I.G., 1995. Vermetid buildups from Grand Cayman, British West Indies. *Journal of Coastal Research*, 11: 973–983.

Jones, D.G. and Williams, A.T., 1991. Statistical analysis of factors influencing cliff erosion along a section of the west Wales coast, UK. *Earth Surface Processes and Landforms*, 16: 95–111.

Jones, J.R., Cameron, B. and Fisher, J.J., 1993. Analysis of cliff retreat and shoreline erosion: Thompson Island, Massachusetts, USA. *Journal of Coastal Research*, 9: 87–96.

Jones, M.R., 1995. The Torres Reefs, North Queensland, Australia – strong tidal flows a modern control on their growth. *Coral Reefs*, 14: 63–69.

Jongsma, D., 1970. Eustatic sea level changes in the Arafura Sea. *Nature*, 228: 150–151.

Judd, J.W., 1904. General report on the materials sent from Funafuti and the methods of dealing with them. In: *The Atoll of Funafuti*. The Royal Society, London, pp. 167–185.

Jungerius, P.D. and van der Meulen, F., 1989. The development of dune blowouts, as measured with erosion pins and sequential air photos. *Catena*, 16: 369–376.

Jupp, D.L.B., Mayo, K.K., Kuchler, D.A., van Classen, D., Kenchington, R.A. and Guerin, P.R., 1985. Remote sensing for planning and managing the Great Barrier Reef of Australia. *Photogrammetria*, 40: 21–42.

Jurgen, E., 1993. Storm surge deposits in North Sea salt marshes dated by ^{134}Cs and ^{137}Cs determination. *Journal of Coastal Research*, 9: 698–701.

Jutson, J.T., 1939. Shore platforms near Sydney, New South Wales. *Journal of Geomorphology*, 2: 237–250.

Kaczorowski, R.T., 1980. Stratigraphy and coastal processes of the Louisiana chenier plain. In: *Gulf Coast Association of Geological Societies, Field Trip Guide*, 18 October: 1–27.

Kamaluddin, H., 1993. The changing mangrove shorelines in Kuala Kurau, Peninsular Malaysia. *Sedimentary Geology*, 83: 187–197.

Kamphuis, J.W., 1987. Recession rate of glacial till bluffs. *Journal of Waterway, Port, Coastal and Ocean Engineering*, 113: 60–73.

Kamphuis, J.W., 1991. Alongshore sediment transport rate distribution. In: N.C. Kraus, K.J. Ginge and D.L. Kriebel (Editors), *Coastal Sediments '91*, American Society of Civil Engineers, New York, pp. 170–183.

Kamphuis, J.W., 2000. *Introduction to Coastal Engineering and Management*. World Scientific, Singapore, 437 pp.

Kan, H., Hori, N., Kawana, T., Kaigara, T. and Ichikawa, K., 1997. The evolution of a Holocene fringing reef and island: reefal environmental sequence and sea level changes in Tonaki Island, the Central Ryukyus. *Atoll Research Bulletin*, 443: 1–20.

Kana, T.W., Hayter, E.J. and Work, P.A., 1999. Mesoscale sediment transport at southeastern United States tidal inlets: conceptual model applicable to mixed energy settings. *Journal of Coastal Research*, 15: 303–313.

Kastler, J.A. and Wiberg, P.L., 1996. Sedimentation and boundary changes of Virginia salt marshes. *Estuarine, Coastal and Shelf Science*, 42: 683–700.

Kawana, T. and Nakata, T., 1994. Timing of Late Holocene tsunamis originated around the Southern Ryukyu Islands, Japan, deduced from coralline tsunami deposits. *Journal of Geography*, 103: 352–376.

Kay, R. and Alder, J., 1999. *Coastal Planning and Management*. Routledge, London, 375 pp.

Kay, R.C., Eliot, I., Caton, B., Morvell, G. and Waterman, P., 1996. A review of the Intergovernmental Panel on Climate Change's Common Methodology for assessing the vulnerability of coastal areas to sea-level rise. *Coastal Management*, 24: 165–188.

Kayanne, H., Yonekura, N., Ishii, T. and Matsumoto, E., 1988. Geomorphic and geological development of Holocene emerged reefs in Rota and Guam, Mariana Islands. In: N. Yonekura (Editor), *Sea-Level Changes and Tectonics in the Middle Pacific: Report of the HIPAC Project*. Faculty of Science, University of Tokyo, Tokyo, pp. 35–57.

Kaye, C.A., 1959. *Shoreline Features and Quaternary Shoreline Changes, Puerto Rico. United States Geological Survey Professional Paper 317–B*: 1–140.

Kaye, C.A. and Barghoorn, E.S., 1964. Late Quaternary sea-level change and crustal rise at Boston, Massachusetts, with notes on the autocompaction of peat. *Geological Society of America Bulletin*, 75: 63–80.

Ke, X., Evans, G. and Collins, M.B., 1996. Hydrodynamics and sediment dynamics of The Wash embayment, eastern England. *Sedimentology*, 43: 157–174.

Kelletat, D.H., 1995. *Atlas of Coastal Geomorphology and Zonality*. *Journal of Coastal Research*, Special Issue No. 13. Coastal Education & Research Foundation (CERF), Charlottesville, Virginia, 286 pp.

Kelley, J.T., Belknap, D.F., Jacobson, G.L. and Jacobson, H.A., 1988. The morphology and origin of salt marshes along the glaciated coastline of Maine, USA. *Journal of Coastal Research*, 4: 649–665.

Kemp, P.H., 1975. Wave asymmetry in the nearshore zone and beach area. In: J.R. Hails and A.P. Carr (Editors), *Nearshore Sediment Dynamics and Sedimentation*. Wiley Interscience, Chichester, pp. 47–67.

Kench, P.S., 1997. Contemporary sedimentation in the Cocos (Keeling) Islands, Indian Ocean: interpretation using settling velocity analysis. *Sedimentary Geology*, 114: 109–130.

Kench, P.S., 1998. Physical processes in an Indian Ocean atoll. *Coral Reefs*, 17: 155–168.

Kench, P.S., 1999. Geomorphology of Australian estuaries: review and prospect. *Australian Journal of Ecology*, 24: 367–380.

Kench, P.S. and McLean, R.F., 1996. Hydraulic characteristics of bioclastic deposits: new possibilities for environmental interpretation using settling velocity fractions. *Sedimentology*, 43: 561–570.

Kench, P.S. and McLean, R.F., 1997. A comparison of settling and sieve techniques for the analysis of bioclastic sediments. *Sedimentary Geology*, 109: 111–119.

Kenchington, R.A., 1990. *Managing Marine Environments.* Taylor and Francis, New York, 248 pp.

Kenchington, R.A. and Crawford, D., 1993. On the meaning of integration in coastal zone management. *Ocean and Coastal Management*, 21: 109–127.

Kennedy, D.M. and Woodroffe, C.D., 2000. Holocene lagoonal sedimentation at the latitudinal limits of reef growth, Lord Howe Island, Tasman Sea. *Marine Geology*, 169: 287–304.

Kennedy, D.M. and Woodroffe, C.D., 2002. Fringing reef growth and morphology: a review. *Earth Science Reviews*, 57: 257–279.

Kesel, R.H. and Smith, J.S., 1978. Tidal creek and pan formation in intertidal salt marshes, Nigg Bay, Scotland. *Scottish Geographical Magazine*, 94: 159–168.

Kestner, F.J.T., 1970. Cyclic changes in Morecambe Bay. *Geographical Journal*, 136: 85–97.

Kestner, F.J.T., 1975. The loose boundary regime of the Wash. *Geographical Journal*, 141: 388–414.

Keulegan, G.H. and Krumbein, W.C., 1949. Stable configuration of bottom slope in a shallow sea and its bearing on geological processes. *EOS, Transactions of the American Geophysical Union*, 30: 855–861.

Kidson, C., 1963. The growth of sand and shingle spits across estuaries. *Zeitschrift für Geomorphologie, NF*, 7: 1–2.

Kidson, C. and Carr, A.P., 1959. The movement of shingle over the sea bed close inshore. *Geographical Journal*, 125: 380–389.

Kindler, P. and Hearty, P.J., 2000. Elevated marine terraces from Eleuthera (Bahamas) and Bermuda: sedimentological, petrographic and geochronological evidence for important deglaciation events during the middle Pleistocene. *Global and Planetary Change*, 24: 41–58.

King, C.A.M., 1959. *Beaches and Coasts.* First Edition. Edward Arnold, London, 403 pp.

King, C.A.M., 1963. Some problems concerning marine planation and the formation of erosion surfaces. *Transactions of the Institute of British Geographers*, 33: 29–44.

King, C.A.M., 1969. Changes in the spit at Gibraltar Point, Lincolnshire, 1951 to 1969. *East Midlands Geographer*, 5: 19–30.

King, C.A.M., 1972. *Beaches and Coasts.* Second Edition. Arnold, London, 570 pp.

King, C.A.M. and McCullach, M.J., 1971. A simulation model of a complex recurved spit. *Journal of Geology*, 79: 22–36.

King, C.A.M. and Williams, W.W., 1949. The formation and movement of sand bars by wave action. *Journal of Geography*, 113: 70–85.

King, D.M., Cooper, N.J., Morfett, J.C. and Pope, D.J., 2000. Application of offshore breakwaters to the United Kingdom: a case study at Elmer Beach. *Journal of Coastal Research*, 16: 172–187.

King, L.C., 1930. Raised beaches and other features of the south-east coast of the North Island of New Zealand. *Transactions of the New Zealand Institute*, 61: 498–523.

Kirby, J.T. and Dalrymple, R.A., 1986. An approximate model for nonlinear dispersion in monochromatic wave propagation models. *Coastal Engineering*, 9: 545–561.

Kirby, J.T., Dalrymple, R.A. and Liu, P.L., 1981. Modification of edge waves by barred-beach topography. *Coastal Engineering*, 5: 35–49.

Kirby, R., 1992. Effects of sea-level rise on muddy coastal margins. In: D. Prandle (Editor), *Dynamics and Exchanges in Estuaries and the Coastal Zone. Coastal and Estuarine Studies*. American Geophysical Union, Washington, DC, pp. 311–334.

Kirby, R., 2000. Practical implications of tidal flat shape. *Continental Shelf Research*, 20: 1061–1077.

Kirk, R.M., 1977. Rates and forms of erosion in intertidal platforms at Kaikoura Peninsula, South Island, New Zealand. *New Zealand Journal of Geology and Geophysics*, 18: 787–801.

Kirk, R.M., 1980. Mixed sand and gravel beaches: morphology, process and sediments. *Progress in Physical Geography*, 4: 189–210.

Kjerfve, B. and Magill, K.E., 1989. Geographic and hydrodynamic characteristics of shallow coastal lagoons. *Marine Geology*, 88: 187–199.

Klein, G.V., 1985. Intertidal flats and intertidal sand bodies. In: R.A. Davis (Editor), *Coastal Sedimentary Environments*. Springer-Verlag, New York, pp. 187–224.

Kleypas, J.A., 1996. Coral reef development under naturally turbid conditions: fringing reefs near Broad Sound. *Coral Reefs*, 15: 153–167.

Kleypas, J.A. and Hopley, D., 1993. Reef development across a broad continental shelf, southern Great Barrier Reef, Australia. *Proceedings of the Seventh International Coral Reef Symposium, Guam*, Vol. 2. University of Guam Press, Mangilao, 1129–1141.

Kleypas, J.A., McManus, J.W. and Menez, L.A.B., 1999. Environmental limits to coral reef development: where do we draw the line? *American Zoologist*, 39: 146–159.

Knighton, A.D., 1998. *Fluvial Forms and Processes: A New Perspective*. Arnold, London, 383 pp.

Knighton, A.D., Mills, K. and Woodroffe, C.D., 1991. Tidal creek extension and saltwater intrusion in northern Australia. *Geology*, 19: 831–834.

Knowlton, N., 1992. Thresholds and multiple stable states in coral reef community dynamics. *American Zoologist*, 32: 674–682.

Knutson, P.L., 1988. Role of coastal marshes in energy dissipation and shore protection. In: D.D. Hook *et al.* (Editors), *The Ecology and Management of Wetlands*. Croom Helm, London, pp. 161–175.

Kochel, R.C., Kahn, J.H., Dolan, R., Hayden, B.P. and May, P.F., 1985. US mid-Atlantic barrier island geomorphology. *Journal of Coastal Research*, 1: 1–9.

Kolb, C., 1980. Should we permit Mississippi–Atchafalaya diversion? *Transactions of the Gulf Coast Association of Geological Societies*, 30: 145–150.

Kolb, C.R. and van Lopik, J.R., 1966. Depositional environments of the Mississippi River deltaic plain, southeastern Louisiana. In: M.L. Shirley and J.A. Ragsdale (Editors), *Deltas*. Texas Geological Society, Houston, pp. 17–62.

Komar, P.D., 1976. *Beach Processes and Sedimentation*. First Edition. Prentice Hall, New Jersey, 429 pp.

Komar, P.D., 1987. Selective grain entrainment by a current from a bed of mixed sizes: a re-analysis. *Journal of Sedimentary Petrology*, 57: 203–211.

Komar, P.D., 1996. The budget of littoral sediments: concepts and applications. *Shore and Beach*, 64: 18–26.

Komar, P.D., 1998. *Beach Processes and Sedimentation*. Second Edition. Prentice Hall, Upper Saddle River, 544 pp.

Komar, P.D. and Shih, S.-M., 1993. Cliff erosion along the Oregon coast: a tectonic-sea level imprint plus local controls by beach processes. *Journal of Coastal Research*, 9: 747–765.

Komar, P.D. and Wang, C., 1984. Processes of selective grain transport and the formation of placers on beaches. *Journal of Geology*, 92: 629–652.

Kominz, M.A., 2001. Sea level variations over geologic time. In: J.H. Steele, S.A. Thorpe and K.K. Turekian (Editors), *Encyclopedia of Ocean Sciences*. Academic Press, San Diego, pp. 2605–2608.

Konicki, K.M. and Holman, R.A., 2000. The statistics and kinematics of transverse sand bars on an open coast. *Marine Geology*, 169: 69–101.

Kotvojs, F.J. and Cowell, P.J., 1991. Refinement of the Dean profile model for beach design. *Australian Civil Engineering Transactions, Institute of Engineers Australia*, CE33: 9–15.

Kraft, J.C., 1971. Sedimentary facies patterns and geologic history of a Holocene marine transgression. *Geological Society of America Bulletin*, 82: 2131–2158.

Kraft, J.C. and John, C.J., 1979. Lateral and vertical facies relations of transgressive barriers. *The American Association of Petroleum Geologists Bulletin*, 63: 2145–2163.

Kraft, J.C., Kayan, I. and Erol, O., 1980. Geomorphic reconstructions in the environs of ancient Troy. *Science*, 209: 776–782.

Kraus, N.C. (Editor), 1996. *History and Heritage of Coastal Engineering*. American Society of Civil Engineering, New York, 603 pp.

Krogstad, H.E., Wolf, J., Thompson, S.P. and Wyatt, L.R., 1999. Methods of intercomparison of wave measurements. *Coastal Engineering*, 37: 235–257.

Krone, R.B., 1987. A method for simulating historic marsh elevations. *Coastal Sediments '87*, American Society of Civil Engineers, New Orleans, pp. 316–323.

Kroonenberg, S.B., Rusakov, G.V. and Svitoch, A.A., 1997. The wandering of the Volga delta: a response to rapid Caspian sea-level change. *Sedimentary Geology*, 107: 189–209.

Krumbein, W.C., 1944. Shore currents and sand movements on a model beach. *Beach Erosion Board, Technical Memorandum*, 7: 1–44.

Ku, T.-L., Ivanovich, M. and Luo, S., 1990. U-series dating of the last interglacial high sea stands: Barbados revisited. *Quaternary Research*, 33: 129–147.

Kudrass, H.R. and Schlüter, H.U., 1994. Development of cassiterite-bearing

B.C. Patten (Editor), *Systems Analysis and Simulation in Ecology*. Academic Press, New York, pp. 113–145.

Lugo, A.E. and Snedaker, S.C., 1974. The ecology of mangroves. *Annual Review of Ecology and Systematics*, 5: 39–64.

Lundberg, J., Ford, D.C., Schwarcz, H.P., Dickin, A.P. and Li, W.-X., 1990. Dating sea level in caves. *Nature*, 343: 217–218.

L'Yavanc, J. and Bassoullet, P., 1991. Nouvelle approche dans l'étude de la dynamique sédimentaire des estuaires macrotidaux à faible débit fluvial. *Oceanologica Acta, Actes du Colloque International sur l'Environnement des Mers Épicontinentales, Lille, 1990*, 11: 129–136.

Lyell, C., 1832. *Principles of Geology*, 3 vols. Murray, London.

Lynch, J.C., Meriwether, J.R., McKee, B.A., Vera-Herrera, F. and Twilley, R.R., 1989. Recent accretion in mangrove ecosystems based on ^{137}Cs and ^{210}Pb. *Estuaries*, 12: 284–299.

Maa, P.Y. and Mehta, A.J., 1987. Mud erosion by waves: a laboratory study. *Continental Shelf Research*, 7: 1269–1284.

Maa, P.Y., Hsu, T.-W., Tsai, C.H. and Juang, W.J., 2000. Comparison of wave refraction and diffraction models. *Journal of Coastal Research*, 16: 1073–1082.

Macdonald, G.A., Abbott, A.T. and Pererson, F.L., 1970. *Volcanoes in the Sea: The Geology of Hawaii*. University of Hawaii Press, Honolulu, 517 pp.

Macintyre, I.G., 1988. Modern coral reefs of Western Atlantic: new geological perspective. *American Association of Petroleum Geologists Bulletin*, 72: 1360–1369.

Macintyre, I.G. and Glynn, P.W., 1976. Evolution of modern Caribbean fringing reef, Galeta Point, Panama. *American Association of Petroleum Geologists Bulletin*, 60: 1054–1072.

Macintyre, I.G., Littler, M.M. and Littler, D.S., 1995. Holocene history of Tobacco Range, Belize, Central America. *Atoll Research Bulletin*, 430: 1–18.

Macintyre, I.G., Precht, W.F. and Aronson, R.B., 2000. Origin of the Pelican Cays ponds, Belize. *Atoll Research Bulletin*, 466: 1–11.

Mackin, J.H., 1948. Concept of a graded river. *Geological Society of America Bulletin*, 59: 463–511.

Macnae, W., 1966. Mangroves in eastern and southern Australia. *Australian Journal of Botany*, 14: 67–104.

Macnae, W., 1968. A general account of the fauna and flora of mangrove swamps and forests in the Indo-West-Pacific region. *Advances in Marine Biology*, 6: 73–270.

MacNeil, F.S., 1954. The shape of atolls: an inheritance from subaerial forms. *American Journal of Science*, 252: 402–427.

Maiklem, W.R., 1968. Some hydraulic properties of bioclastic carbonate grains. *Sedimentology*, 10: 101–109.

Manabe, S. and Stouffer, R.J., 1994. Multiple-century response of a coupled ocean-atmosphere model to an increase of atmospheric carbon dioxide. *Journal of Climate*, 7: 5–23.

Mann, K.H., 1982. *Ecology of Coastal Waters – A Systems Approach*. Blackwell Scientific Publications, Oxford, 322 pp.

Mantua, N.J., Hare, S.R., Zhang, Y., Wallace, J.M. and Francis, R.C., 1997. A Pacific interdecadal climate oscillation with impacts on salmon production. *Bulletin of the American Meteorological Society*, 78: 1069–1079.
Maragos, J.E., Baines, G.B.K. and Beveridge, P.J., 1973. Tropical cyclone creates a new land formation on Funafuti atoll. *Science*, 181: 1161–1164.
Maritorena, S., 1996. Remote sensing of the water attenuation in coral reefs: a case study in French Polynesia. *International Journal of Remote Sensing*, 17: 155–166.
Marker, M.E., 1976. Aeolianite: Australian and South African deposits compared. *Annals of the South African Museum*, 71: 115–124.
Marshall, J.F. and Davies, P.J., 1978. Skeletal carbonate variation on the continental shelf of eastern Australia. *Bureau of Mineral Resources Journal of Australian Geology and Geophysics*, 3: 85–92.
Marshall, J.F. and Davies, P.J., 1982. Internal structure and Holocene evolution of One Tree Reef, southern Great Barrier Reef. *Coral Reefs*, 1: 21–28.
Marshall, J.F. and Davies, P.J., 1988. *Halimeda* bioherms of the northern Great Barrier Reef. *Coral Reefs*, 6: 139–148.
Marshall, J.F. and Jacobson, G., 1985. Holocene growth of a mid-plate atoll: Tarawa, Kiribati. *Coral Reefs*, 4: 11–17.
Marshall, J.F. and Thom, B.G., 1976. The sea level in the Last Interglacial. *Nature*, 263: 120–121.
Marshall, J.F., Tsuji, Y., Matsuda, H., Davies, P.J., Iryu, Y., Honda, N. and Satoh, Y., 1998. Quaternary and Tertiary subtropical carbonate platform development on the continental margin of southern Queensland, Australia. *Special Publications of the International Association of Sedimentologists*, 25: 163–195.
Martin, L., Flexor, J.-M., Blitzkow, D. and Suguio, K., 1985. Geoid change indications along the Brazilian coast during the last 7000 years. *Proceedings of the Fifth International Coral Reef Congress*, vol. 3: Antenne Museum-Ephe, Moorea, French Polynesia, pp. 85–90.
Martin, R.T., Gadel, F.Y. and Barusseau, J.P., 1981. Holocene evolution of the Canet-St Nazaire lagoon (Golfe du Lion, France) as determined from a study of sediment properties. *Sedimentology*, 28: 823–836.
Martinez, J.O., Gonzalez, J.L., Pilkey, O.H. and Neal, W.J., 1995. Tropical barrier islands of Columbia's Pacific coast. *Journal of Coastal Research*, 11: 432–453.
Martinez, P.A. and Harbaugh, J.W., 1989. Computer simulation of wave and fluvial-dominated nearshore environments. In: V.C. Lakhan and A.S. Trenhaile (Editors), *Applications in Coastal Modeling*. Elsevier, Amsterdam, pp. 297–340.
Martinez, P.A. and Harbaugh, J.W., 1993. *Simulating Nearshore Environments. Computer Methods in the Geosciences*. Pergamon Press, Oxford, 265 pp.
Martinson, D.G., Pisias, N.G., Hays, J.D., Imbrie, J., Moore, T.C. and Shackleton, N.J., 1987. Age dating and the orbital theory of the ice ages: development of a high-resolution 0 to 300000-year chronostratigraphy. *Quaternary Research*, 27: 1–29.
Mason, D.C. and Garg, P.K., 2001. Morphodynamic modelling of intertidal

sediment transport in Morecambe Bay. *Estuarine, Coastal and Shelf Science*, 53: 79–92.

Mason, D.C., Davenport, I.J., Flather, R.A. and Gurney, C., 1998. A digital elevation model of the inter-tidal areas of the Wash, England, produced by the waterline method. *International Journal of Remote Sensing*, 19: 1455–1460.

Mason, D.C., Amin, M., Davenport, I., Flather, R.A., Robinson, G.J. and Smith, J.A., 1999. Measurement of recent intertidal sediment transport in Morecambe Bay using the water-line method. *Estuarine, Coastal and Shelf Science*, 49: 427–456.

Mason, S.J. and Hansom, J.D., 1988. Cliff erosion and its contribution to a sediment budget for part of the Holderness Coast, England. *Shore and Beach*, 56: 30–38.

Massel, S.R. and Done, T.J., 1993. Effects of cyclone waves on massive coral assemblages on the Great Barrier Reef: meteorology, hydrodynamics and demography. *Coral Reefs*, 12: 153–166.

Massel, S.R. and Gourlay, M.R., 2000. On the modelling of wave breaking and set-up on coral reefs. *Coastal Engineering*, 39: 1–27.

Masselink, G., 1998a. Field investigation of wave propagation over a bar and the consequent generation of secondary waves. *Coastal Engineering*, 33: 1–9.

Masselink, G., 1998b. Morphological evolution of beach cusps and associated swash circulation patterns. *Marine Geology*, 146: 93–113.

Masselink, G., 1999. Alongshore variation in beach cusp morphology in a coastal embayment. *Earth Surface Processes and Landforms*, 24: 335–347.

Masselink, G. and Black, K.P., 1995. Magnitude and cross shore distribution of bed return flow measured on natural beaches. *Coastal Engineering*, 25: 165–190.

Masselink, G. and Hegge, B., 1995. Morphodynamics of meso- and macrotidal beaches: examples from central Queensland, Australia. *Marine Geology*, 129: 1–23.

Masselink, G. and Pattiaratchi, C.B., 1998. The effects of sea breeze on beach morphology, surf zone hydrodynamics and sediment resuspension. *Marine Geology*, 146: 115–135.

Masselink, G. and Short, A.D., 1993. The effect of tide range on beach morphodynamics and morphology: a conceptual beach model. *Journal of Coastal Research*, 9: 785–800.

Masselink, G. and Turner, I.L., 1999. The effect of tides on beach morphodynamics. In: A.D. Short (Editor), *Handbook of Beach and Shoreface Morphodynamics*. John Wiley & Sons, Chichester, pp. 204–229.

Masselink, G., Hegge, B.J. and Pattiaratchi, C.B., 1997. Beach cusp morphodynamics. *Earth Surface Processes and Landforms*, 22: 1139–1155.

Masson, D.G., 1996. Catastrophic collapse of the volcanic island of Hierro 15 ka ago and the history of landslides in the Canary Islands. *Geology*, 24: 231–234.

Mastronuzzi, G. and Sansò, P., 2000. Boulders transport by catastrophic waves along the Ionian coast of Apulia (southern Italy). *Marine Geology*, 170: 93–103.

Mathers, S. and Zalasiewicz, J., 1999. Holocene sedimentary architecture of the Red River Delta, Vietnam. *Journal of Coastal Research*, 15: 314–325.

Maul, G.A., Ches, F., Bushnell, M. and Mayer, D.A., 1985. Sea level variation as an indicator of Florida Current volume transport: comparisons with direct measurements. *Science*, 227: 304–307.

Maxwell, W.G.H., 1968. *Atlas of the Great Barrier Reef*. Elsevier, Amsterdam, 258 pp.

May, J.P. and Tanner, W.F., 1973. The littoral power gradient and shoreline changes. In: D.R. Coates (Editor), *Coastal Geomorphology*. Publications in Geomorphology, State University of New York, Binghamton, New York, pp. 43–60.

May, V. and Heeps, C., 1985. The nature and rates of change on chalk coastlines. *Zeitschrift für Geomorphologie, NF*, Suppl.-Bd, 57: 81–94.

McBride, R.A., 1987. Tidal inlet history, morphology and stability, eastern Florida, USA. *Coastal Sediments '87*, American Society of Civil Engineers, New Orleans, pp. 1592–1607.

McBride, R.A., Byrnes, M.R. and Hiland, M.W., 1995. Geomorphic response-type model for barrier coastlines: a regional perspective. *Marine Geology*, 126: 143–159.

McCann, S.B., 1977. Coastal landforms. *Progress in Physical Geography*, 1: 339–344.

McCave, I.N., 1979. Suspended sediment. In: K.R. Dyer (Editor), *Estuarine Hydrography and Sedimentation, A Handbook*. Cambridge University Press, Cambridge, pp. 131–185.

McCormick, M.E., 1993. Equilibrium shoreline response to breakwaters. *Journal of Waterway, Port, Coastal, and Ocean Engineering*, 119: 657–670.

McCowan, J., 1894. On the highest wave of permanent type. *Philosophical Magazine, Series 5*, 38: 351–357.

McDougall, I., 1964. Potassium-argon ages from lavas of the Hawaiian Islands. *Geological Society of America Bulletin*, 75: 107–128.

McDougall, I. and Duncan, R.A., 1980. Linear volcanic chains: recording plate motions? *Tectonophysics*, 63: 275–295.

McDowell, D.M. and O'Connor, B.A., 1977. Hydraulic behavior of estuaries. Wiley, New York, 292 pp.

McGee, W.J., 1890. Encroachment of the sea. *Forum*, 9: 437–449.

McGill, J.T., 1958. Map of coastal landforms of the world. *Geographical Review*, 48: 402–405.

McGreal, W.S., 1979. Marine erosion of glacial sediments from a low-energy cliffline environment near Kilkeel, Northern Ireland. *Marine Geology*, 32: 89–103.

McInnes, K.L. and Hubbert, G.D., 2001. The impact of eastern Australian cut-off lows on coastal sea levels. *Meteorological Applications*, 8: 229–243.

McKee, E.D., 1959. Storm sediments on a Pacific atoll. *Journal of Sedimentary Petrology*, 29: 354–364.

McKee, E.D. and Ward, W.C., 1983. Eolian environment. In: P.A. Scholle, D.G. Bedont and C.H. Moore (Editors), *Carbonate Depositional Environments*.

American Assocation of Petroleum Geologists, Memoir, Tulsa, OK, pp. 132–169.

McKee, K.L. and Faulkner, P.L., 2000. Mangrove peat analysis and reconstruction of vegetation history at the Pelican Cays, Belize. *Atoll Research Bulletin*, 468: 45–58.

McKee, K.L. and Patrick, W.H., 1988. The relationship of smooth cordgrass (*Spartina alterniflora*) to tidal datums: a review. *Estuaries*, 11: 143–151.

McKenna, J., Carter, R.W.G. and Bartlett, D., 1992. Coast erosion in northeast Ireland; Part II. Cliffs and shore platforms. *Irish Geography*, 25: 111–128.

McKenzie, P., 1958. Rip current systems. *Journal of Geology*, 66: 103–113.

McLachlan, A., 1990. The exchange of materials between dune and beach systems. In: K.F. Nordstrom, N.P. Psuty and R.W.G. Carter (Editors), *Coastal Dunes: Form and Process*. John Wiley & Sons, Chichester, pp. 201–215.

McLean, R.F., 1967. Measurement of beach rock erosion by some tropical marine gastropods. *Bulletin of Marine Science*, 17: 551–561.

McLean, R.F., 1978. Recent coastal progradation in New Zealand. In: J.L. Davies and M.A.F. Williams (Editors), *Landform Evolution in Australasia*. Australian National University, Canberra, pp. 168–196.

McLean, R.F., 1984. Coastal landforms: sea-level history and coastal evolution. *Progress in Physical Geography*, 8: 431–440.

McLean, R.F. and Davidson, C.F., 1968. The role of mass movement in shore platform development along the Gisborne coastline, New Zealand. *Earth Science Journal*, 2: 15–25.

McLean, R.F. and Kirk, R.M., 1969. Relationship between grain size, size-sorting, and foreshore slope on mixed sand-shingle beaches. *New Zealand Journal of Geology and Geophysics*, 12: 138–155.

McLean, R.F. and Stoddart, D.R., 1978. Reef island sediments of the northern Great Barrier Reef. *Philosophical Transactions of the Royal Society, London, Series A* 291: 101–117.

McLean, R.F. and Tsyban, A., 2001. Coastal zones and marine ecosystems. In: J.J. McCarthy, O.F. Canziani, N.A. Leary, D.J. Dokken and K.S. White (Editors), *Climate Change 2001: Impacts, Adaptation, and Vulnerability*. Cambridge University Press, Cambridge, pp. 343–379.

McLean, R.F. and Woodroffe, C.D., 1994. Coral atolls. In: R.W.G. Carter and C.D. Woodroffe (Editors), *Coastal Evolution: Late Quaternary Shoreline Morphodynamics*. Cambridge University Press, Cambridge, pp. 267–302.

McManus, J.W., 2001. Coral reefs. In: J.H. Steele, S.A. Thorpe and K.K. Turekian (Editors), *Encyclopedia of Ocean Sciences*. Academic Press, San Diego, pp. 524–534.

McMaster, R., Lachance, T.P. and Ashraf, A., 1970. Continental shelf geomorphic features off Portuguese Guinea, Guinea and Sierra Leone (West Africa). *Marine Geology*, 9: 203–213.

McNinch, J.E. and Luettich, R.A., 2000. Physical processes around a cuspate foreland: implications to the evolution and long-term maintenance of a cape-associated shoal. *Continental Shelf Research*, 20: 2367–2389.

McNutt, M. and Menard, H.W., 1978. Lithospheric flexure and uplifted atolls. *Journal of Geophysical Research*, 83: 1206–1212.

McPherson, J.G., Shanmugam, G. and Moiola, R.J., 1987. Fan-deltas and braid deltas: varieties of coarse-grained deltas. *Geological Society of America Bulletin*, 99: 331–340.

McSaveney, M.J., Goff, J.R., Darby, D.J., Goldsmith, P., Barnett, A., Elliott, S. and Nongkas, M., 2000. The 17 July 1998 tsunami, Papua New Guinea: evidence and initial interpretation. *Marine Geology*, 170: 81–92.

Menard, H.W., 1983. Insular erosion, isostasy and subsidence. *Science*, 220: 913–918.

Menard, H.W., 1986. *Islands*, Scientific American Library. New York, 230 pp.

Mendelssohn, I.A. and McKee, K.L., 1980. Sublethal stresses controlling *Spartina alterniflora* productivity. In: B. Gopal, R.E. Turner, R.G. Wetzel and D.F. Whigham (Editors), *Wetlands: Ecology and Management: Proceedings of the First International Wetlands Conference*. National Institute of Ecology, Jaipur and International Scientific Publications, New Delhi, India, pp. 223–242.

Mesolella, K.J., Matthews, R.K., Broecker, W.S. and Thurber, D.L., 1969. The astronomical theory of climatic change: Barbados data. *Journal of Geology*, 77: 250–274.

Messerli, B., Grosjean, M., Hofer, T., Nunez, L. and Pfister, C., 2000. From nature-dominated to human-dominated environmental changes. *Quaternary Science Reviews*, 19: 459–479.

Meyer, R.E., 1972. *Waves on Beaches and Resulting Sediment Transport*. Academic Press, New York, 462 pp.

Meyer, W.B., 1996. *Human Impact on the Earth*. Cambridge University Press, Cambridge, 253 pp.

Meyer-Arendt, K.J., 1985. The Grand Isle, Louisiana resort cycle. *Annals of Tourism Research*, 12: 449–465.

Miall, A.D., 1996. *The Geology of Stratigraphic Sequences*. Springer-Verlag, Berlin, 433 pp.

Miall, A.D. and Miall, C.E., 2001. Sequence stratigraphy as a scientific enterprise: the evolution and persistence of conflicting paradigms. *Earth Science Reviews*, 54: 321–348.

Michel, D. and Howa, H.L., 1999. Short-term morphodynamic response of a ridge and runnel system on a mesotidal sandy beach. *Journal of Coastal Research*, 15: 428–437.

Mikkelsen, O. and Pejrup, M., 1998. Comparison of flocculated and dispersed suspended sediment in the Dollard estuary. In: K.S. Black, D.M. Paterson and A. Cramp (Editors), *Sedimentary Processes in the Intertidal Zone*. Geological Society of London, Special Publication, pp. 199–209.

Milankovitch, M., 1941. *Kanon der Erdbestrahlung und seine Anwendung auf das Eiszeitenproblem*, 132. Royal Serbian Academy, Special Publication, Belgrade, 633 pp.

Miller, D.L., Smeins, F.E. and Webb, J.W., 1996. Mid-Texas coastal marsh change (1939–1991) as influenced by Lesser Snow Goose herbivory. *Journal of Coastal Research*, 12: 462–476.

Miller, J.L. and Gardner, L.R., 1981. Sheet flow in a salt-marsh basin, North Inlet, South Carolina. *Estuaries*, 4: 234–237.

Miller, R.L. and Zeigler, J.M., 1958. A model relating dynamics and sediment pattern in equilibrium in the region of shoaling waves, breaker zone and foreshore. *Journal of Geology*, 66: 417–441.

Milliman, J.D., 1974. *Marine Carbonates*. Springer-Verlag, Berlin, 375 pp.

Milliman, J.D., 2001. River inputs. In: J.H. Steele, S.A. Thorpe and K.K. Turekian (Editors), *Encyclopedia of Ocean Sciences*. Academic Press, San Diego, pp. 2419–2427.

Milliman, J.D. and Emery, K.O., 1968. Sea levels during the past 35000 years. *Science*, 162: 1121–1123.

Milliman, J.D. and Meade, R.H., 1983. World-wide delivery of river sediment to the oceans. *Journal of Geology*, 91: 1–21.

Milliman, J.D. and Syvitski, J.P.M., 1992. Geomorphic/tectonic control of sediment discharge to the ocean: the importance of small mountainous rivers. *Journal of Geology*, 100: 525–544.

Milliman, J.D., Broadus, J.M. and Gable, F., 1989. Environmental and economic implications of rising sea level and subsiding deltas: the Nile and Bengal examples. *Ambio*, 18: 340–345.

Minoura, K., Imamura, F., Takahashi, T. and Shuto, N., 1997. Sequence of sedimentation process caused by the 1992 Flores tsunami: evidence from Babi Island. *Geology*, 25: 523–526.

Minoura, K., Nakaya, S. and Uchida, M., 1994. Tsunami deposits in a lacustrine sequence of the Sanriku coast, northeast Japan. *Sedimentary Geology*, 89: 25–31.

Miossec, A. (Editor), 1994. Défense des côtes ou protection de l'espace littoral. Quelles perspectives? Université de Nantes Institut de Géographie, 364 pp.

Mitchell, N.C., 1998. Characterising the irregular coastlines of volcanic ocean islands. *Geomorphology*, 23: 1–14.

Mitrovica, J.X. and Davis, J.L., 1995. Present-day post-glacial sea level change far from the Late Pleistocene ice sheets: implications for recent analyses of tide gauge records. *Geophysical Research Letters*, 22: 2529–2532.

Mitrovica, J.X. and Peltier, W.R., 1991. On postglacial geoid subsidence over the equatorial oceans. *Journal of Geophysical Research*, 96: 20053–20071.

Moberg, F. and Folke, C., 1999. Ecological goods and services of coral reef ecosystems. *Ecological Economics*, 29: 215–233.

Moeller, I., Spencer, T. and French, J.R., 1996. Wind wave attenuation over saltmarsh surfaces: preliminary results from Norfolk, England. *Journal of Coastal Research*, 12: 1009–1016.

Mogi, A., Tsuchide, M. and Fukushima, M., 1980. Coastal erosion of the new volcanic island Nisbinoskimo. *Geographical Review of Japan*, 53: 449–462.

Möller, I., Spencer, T., French, J.R., Leggett, D.J. and Dixon, M., 1999. Wave transformation over salt marshes: a field and numerical modelling study from North Norfolk, England. *Estuarine, Coastal and Shelf Science*, 49: 411–426.

Montaggioni, L.F., 1988. Holocene reef growth history in mid-plate high volcanic islands. *Proceedings Sixth International Coral Reef Symposium*, Vol. 3. Sixth Coral Reef Symposium Executive Committee, Townsville, pp. 455–460.

Montaggioni, L.F. and Faure, G., 1997. Response of reef coral communities to sea-level rise: a Holocene model from Mauritius (Western Indian Ocean). *Sedimentology*, 44: 1053–1070.

Montaggioni, L.F. and Pirazzoli, P.A., 1984. The significance of exposed coral conglomerates from French Polynesia (Pacific Ocean) as indications of recent sea-level changes. *Coral Reefs*, 3: 29–42.

Montaggioni, L.F., Cabioch, G., Camoinau, G.F., Bard, E., Ribaud-Laurenti, A., Faure, G., Dejardin, P. and Recy, J., 1997. Continuous record of reef growth over the past 14 k.y. on the mid-Pacific island of Tahiti. *Geology*, 25: 555–558.

Moore, C.H., 1989. *Carbonate Diagenesis and Porosity*. Elsevier, New York, 338 pp.

Moore, D.G., 1954. Origin and development of sea caves. *American Caver National Speleological Society Bulletin*, 16: 71–76.

Moore, G.W. and Moore, J.G., 1988. Large scale bedforms in boulder gravel produced by giant waves in Hawaii. *Geological Society of America Special Publication*, 229: 101–110.

Moore, J.G. and Moore, G.W., 1984. Deposit from a giant wave on the island of Lana'i, Hawaii. *Science*, 226: 1312–1315.

Moore, J.G., Clague, D.A., Holcomb, R.T., Lipman, P.W., Normark, W.R. and Torresan, M.E., 1989. Prodigious submarine landslides on the Hawaiian Ridge. *Journal of Geophysical Research*, 94: 17465–17484.

Moore, J.G., Bryan, W.B. and Ludwig, K.R., 1994a. Chaotic deposition by a giant wave, Molokai, Hawaii. *Geological Society of America Bulletin*, 106: 962–967.

Moore, J.G., Normark, W.R. and Holcomb, R.T., 1994b. Giant Hawaiian underwater landslides. *Science*, 264: 46–47.

Morgan, J.P. and McIntire, W.G., 1959. Quaternary geology of the Bengal Basin, East Pakistan and India. *Geological Society of America Bulletin*, 70: 319–342.

Mörner, N.-A., 1976. Eustasy and geoid changes. *Journal of Geology*, 84: 123–151.

Morton, R.A., 1979. Temporal and spatial variations in shoreline changes and their implications, examples from the Texas Gulf Coast. *Journal of Sedimentary Petrology*, 49: 1101–1112.

Morton, R.A., 1994. Texas barriers. In: R.A. Davis (Editor), *Geology of Holocene Barrier Island Systems*. Springer-Verlag, Berlin, pp. 75–114.

Morton, R.A., Leach, M.P., Paine, J.G. and Cardoza, M.A., 1993. Monitoring beach changes using GPS surveying techniques. *Journal of Coastal Research*, 9: 702–720.

Morton, R.A., Paine, J.G. and Gibeaut, J.C., 1994. Stages and durations of post-storm beach recovery, southeastern Texas coast, USA. *Journal of Coastal Research*, 10: 884–908.

Morton, R.A., Gibeaut, J.C. and Paine, J.G., 1995. Meso-scale transfer of sand during and after storms: implications for prediction of shoreline movement. *Marine Geology*, 126: 161–179.

Morton, R.A., Paine, J.G. and Blum, M.D., 2000a. Responses of stable bay-margin and barrier-island systems to Holocene sea-level highstands, western Gulf of Mexico. *Journal of Sedimentary Research*, 70: 478–490.

Morton, R.A., Ward, G.H. and White, W.A., 2000b. Rates of sediment supply and sea-level rise in a large coastal lagoon. *Marine Geology*, 167: 261–284.

Moslow, T.M. and Heron, S.D., 1979. Quaternary evolution of Core Banks, North Carolina: Cape Lookout to New Drum Inlet. In: S.P. Leatherman (Editor), *Barrier Islands*. Academic Press, New York, pp. 211–246.

Muckersie, C. and Shepherd, M.J., 1995. Dune phases as time-transgressive phenomena, Manawatu, New Zealand. *Quaternary International*, 26: 61–67.

Mudge, B.F., 1858. The salt marsh formations of Lynn. *Essex Institute Proceedings*, 2: 117–119.

Muhs, D.R., Kennedy, G.L. and Rockwell, T.K., 1994. Uranium series ages of marine terrace corals from the Pacific coast of North America and implications for Last-Interglacial sea level history. *Quaternary Research*, 42: 72–87.

Mulrennan, M.E., 1992. Ridge and runnel beach morphodynamics: an example from the central east coast of Ireland. *Journal of Coastal Research*, 8: 906–918.

Mulrennan, M.E. and Woodroffe, C.D., 1998. Holocene development of the lower Mary River plains, Northern Territory, Australia. *The Holocene*, 8: 565–579.

Mumby, P.J., Clark, C.D., Green, E.P. and Edwards, A.J., 1998. Benefits of water column correction and contextual editing for mapping coral reefs. *International Journal of Remote Sensing*, 19: 203–210.

Munk, W.H. and Sargent, M.C., 1948. Adjustment of Bikini Atoll to ocean waves. *Transactions of the American Geophysical Union*, 29: 855–860.

Munk, W.H. and Sargent, M.S., 1954. Adjustment of Bikini Atoll to ocean waves. *United States Geological Survey Professional Paper*, 260–C: 275–280.

Muñóz-Pérez, J.J., Tejedor, L. and Medina, R., 1999. Equilibrium beach profile model for reef-protected beaches. *Journal of Coastal Research*, 15: 950–957.

Murray, A.S. and Roberts, R.G., 1997. Determining the burial time of single grains of quartz using optically stimulated luminescence. *Earth and Planetary Science Letters*, 152: 163–180.

Murray, J., 1889. Structure, origin, and distribution of coral reefs and islands. *Nature*, 39: 424–428.

Murray, J.W. and Hawkins, A.B., 1976. Sediment transport in the Severn Estuary during the past 8000–9000 years. *Journal of the Geological Society of London*, 132: 385–398.

Murray-Wallace, C.V., Belperio, A.P. and Cann, J.H., 1998. Quaternary neotectonism and intra-plate volcanism: the Coorong to Mount Gambier Coastal Plain, southeastern Australia: a review. In: I.S. Stewart and C. Vita-Finzi (Editors), *Coastal Tectonics*. Geological Society, Special Publication, London, pp. 255–267.

Murray-Wallace, C.V., Brooke, B.P., Cann, J.H., Belperio, A.P. and Bourman, R.P., 2001. Whole-rock aminostratigraphy of the Coorong Coastal Plain, South Australia: towards a 1 million year record of sea-level highstands. *Journal of the Geological Society, London*, 158: 111–124.

Murray-Wallace, C.V., Banerjee, D., Bourman, R.P., Olley, J.M. and Brooke, B.P., 2002. Optically stimulated luminescence dating of Holocene relict foredunes, Guichen Bay, South Australia. *Quaternary Science Reviews*, 21: 1077–1086.

Murty, T.S., 1993. Episodic sea level changes in the western Indian Ocean: a review. *Marine Geodesy*, 16: 73–85.

Nairn, R.B. and Southgate, H.N., 1993. Deterministic profile modelling of nearshore processes. Part 2. Sediment transport and beach profile development. *Coastal Engineering*, 19: 57–96.

Nakada, M., 1986. Holocene sea levels in oceanic islands: implications for the rheological structure of the Earth's mantle. *Tectonophysics*, 121: 263–276.

Nanayama, F., Shigeno, K., Satake, K., Shimokawa, K., Koitabashi, S., Miyasaka, S. and Ishii, M., 2000. Sedimentary differences between the 1993 Hokkaido-nansei-oki tsunami and the 1959 Miyakojima typhoon at Taisei, southwestern Hokkaido, northern Japan. *Sedimentary Geology*, 135: 255–264.

Neal, A. and Roberts, C.L., 2000. Applications of ground-penetrating radar (GPR) to sedimentological, geomorphological and geoarchaeological studies in coastal environments. In: K. Pye and J.R.L. Allen (Editors), *Coastal and Estuarine Environments: Sedimentology, Geomorphology and Geoarchaeology*. Geological Society, Special Publication, London, pp. 139–171.

Nemec, W., 1995. The dynamics of deltaic suspension plumes. In: M.N. Oti and G. Postma (Editors), *Geology of Deltas*. A.A. Balkema, Rotterdam, pp. 31–93.

Nemec, W. and Steel, R.J., 1988. *Fan Deltas: Sedimentology and Tectonic Settings*. Blackie and Son Ltd, Glasgow, 444 pp.

Nerem, R.S., 1995. Measuring global mean sea level variations using Topex/Poseidon altimeter data. *Journal of Geophysical Research*, 100: 25135–25151.

Nerem, R.S., 1999. Measuring very low frequency sea level variations using satellite altimeter data. *Global and Planetary Change*, 20: 157–171.

Nerem, R.S. and Mitchum, G.T., 2001. Observation of sea level change from satellite altimetry. In: B.C. Douglas, M.S. Kearney and S.P. Leatherman (Editors), *Sea Level Change: History and Consequences*. Academic Press, San Diego, pp. 121–163.

Neumann, A.C., 1966. Observations on coastal erosion in Bermuda and measurements of the boring rate of the sponge, *Cliona lampa*. *Limnology and Oceanography*, 11: 92–108.

Neumann, A.C., 1972. Quaternary sea level history of Bermuda and the Bahamas, *American Quaternary Association Second National Conference*. AMQUA, University of Miami, Miami, pp. 41–44.

Neumann, A.C. and Hearty, P.J., 1996. Rapid sea-level changes at the close of the Last Interglacial (substage 5e) recorded in Bahamian island geology. *Geology*, 24: 775–778.

Neumann, A.C. and Macintyre, I.G., 1985. Reef response to sea level rise: keep-up, catch-up or give-up. *Proceedings of the Fifth International Coral Reef Congress*, vol. 3: Antenne Museum-Ephe, Moorea, French Polynesia, pp. 105–110.

Newell, N.D., 1961. Recent terraces of tropical limestone shores. *Zeitschrift für Geomorphologie, NF, Suppl-Bd.*, 3: 87–106.

Newell, N.D. and Bloom, A.L., 1970. The reef flat and 'two-meter eustatic terrace' of some Pacific atolls. *Geological Society of America Bulletin*, 81: 1881–1894.

Newell, N.D., Rigby, J.K., Whiteman, A.J. and Bradley, J.S., 1951. Shoal-water geology and environments, eastern Andros Island, Bahamas. *Bulletin of the American Museum of Natural History*, 97: 1–30.

Newman, W.S., Cinquemani, L.J., Pardi, R.R. and Marcus, L.F., 1980. Holocene delevelling of the United States' east coast. In: N.-A. Mörner (Editor), *Earth Rheology, Isostasy and Eustasy*. John Wiley & Sons, Chichester, pp. 449–463.

Nguyen, V.L., Ta, T.K.O. and Tateishi, M., 2000. Late Holocene depositional environments and coastal evolution of the Mekong River Delta, southern Vietnam. *Journal of Asian Earth Sciences*, 18: 427–439.

Nichol, S.L. and Boyd, R., 1993. Morphostratigraphy and facies architecture of sandy barriers along the eastern shore of Nova Scotia. *Marine Geology*, 114: 59–80.

Nichol, S.L., Zaitlin, B.A. and Thom, B.G., 1997. The upper Hawkesbury River, New South Wales, Australia: a Holocene example of an estuarine bayhead delta. *Sedimentology*, 44: 263–286.

Nichol, S.L., Augustinus, P.C., Gregory, M.R., Creese, R. and Horrocks, M., 2000. Geomorphic and sedimentary evidence of human impact on the New Zealand coastal landscape. *Physical Geography*, 21: 109–132.

Nicholls, R.J. and Leatherman, S.P., 1996. Adapting to sea-level rise: relative sea-level trends to 2100 for the United States. *Coastal Management*, 24: 301–324.

Nicholls, R.J., Hoozemans, E. and Marchand, M., 1999. Increasing flood risk and wetland loss due to global sea-level rise: regional and global analyses. *Global Environmental Change*, 9: S69–S88.

Nichols, M.M., 1989. Sediment accumulation rates and relative sea-level rise in lagoons. *Marine Geology*, 88: 201–219.

Nichols, M.M. and Biggs, R.B., 1985. Estuaries. In: R.A.J. Davis (Editor), *Coastal Sedimentary Environments*. Springer-Verlag, New York, pp. 77–186.

Niedoroda, A.W., Swift, D.J.P. and Hopkins, T.S., 1985. The shoreface. In: R.A. Davis (Editor), *Coastal Sedimentary Environments*. Springer-Verlag, New York, pp. 533–624.

Niedoroda, A.W., Reed, C.W., Swift, D.J.P., Arato, H. and Hoyanagi, K., 1995. Modeling shore-normal large-scale coastal evolution. *Marine Geology*, 126: 181–199.

Nielsen, A.F. and Gordon, A.D., 1981. Tidal inlet behavioral analysis. *Proceedings of the Seventeenth International Coastal Engineering Conference*. American Society of Civil Engineers, New York, pp. 2461–2480.

Nielsen, P., 1992. *Coastal Bottom Boundary Layers and Sediment Transport*. World Scientific, Singapore, 324 pp.

Nittrouer, C.A., Kuehl, S.A., DeMaster, D.J. and Kowsmann, R.O., 1986. The deltaic nature of Amazon shelf sedimentation. *Geological Society of America Bulletin*, 97: 444–458.

Nolan, T.J., Kirk, R.M. and Shulmeister, J., 1999. Beach cusp morphology on sand and mixed gravel beaches, South Island, New Zealand. *Marine Geology*, 157: 185–198.

Noller, J.S., Sowers, J.M. and Lettis, W.R. (Editors), 2000. *Quaternary Geochronology: Methods and Application*. American Geophysical Union, Washington, DC, 582 pp.

Nordstrom, K.F., 1987. Predicting shoreline changes at tidal inlets on a developed coast. *Professional Geographer*, 39: 457–465.

Nordstrom, K.F., 1989. Erosion control strategies for bay and estuarine beaches. *Coastal Management*, 17: 25–35.

Nordstrom, K.F., 1994a. Beaches and dunes of human-altered coasts. *Progress in Physical Geography*, 18: 497–516.

Nordstrom, K.F., 1994b. Developed coasts. In: R.W.G. Carter and C.D. Woodroffe (Editors), *Coastal Evolution: Late Quaternary Shoreline Morphodynamics*. Cambridge University Press, Cambridge, pp. 477–509.

Nordstrom, K.F., 2000. *Beaches and Dunes of Developed Coasts*. Cambridge University Press, Cambridge, 338 pp.

Nordstrom, K.F. and Arens, S.M., 1998. The role of human actions in evolution and management of foredunes in the Netherlands and New Jersey, USA. *Journal of Coastal Conservation*, 4: 169–180.

Nordstrom, K.F. and Jackson, N.L., 1992. Two-dimensional change on sandy beaches in meso-tidal estuaries. *Zeitschrift für Geomorphologie*, 36: 465–478.

Nordstrom, K.F. and Roman, C.T. (Editors), 1996. *Estuarine Shores: Evolution, Environments and Human Alterations*. John Wiley & Sons, Chichester, 486 pp.

Norrman, J.O., 1980. Coastal erosion and slope development in Surtsey Island, Iceland. *Zeitschrift für Geomorphologie, NF, Suppl.-Bd*, 34: 20–38.

Nossin, J.J., 1965. Analysis of younger beach ridge deposits in eastern Malaya. *Zeitschrift für Geomorphologie*, 9: 186–208.

Nott, J., 1990. The role of sub-aerial processes in sea cliff retreat: a south east Australian example. *Zeitschrift für Geomorphologie, NF*, 34: 75–85.

Nott, J., 1997. Extremely high-energy wave deposits inside the Great Barrier Reef, Australia: determining the cause – tsunami or tropical cyclone. *Marine Geology*, 141: 193–207.

Nott, J. and Hayne, M., 2001. High frequency of 'super-cyclones' along the Great Barrier Reef over the past 5000 years. *Nature*, 413: 508–512.

Novak, B. and Pedersen, G.K., 2000. Sedimentology, seismic facies and stratigraphy of a Holocene spit-platform complex interpreted from high-resolution shallow seismics, Lysegrund, southern Kattegat, Denmark. *Marine Geology*, 162: 317–335.

Nummedal, D., 1983. Barrier islands. In: P.D. Komar (Editor), *CRC Handbook of Coastal Processes and Erosion*. CRC Press, Boca Raton, pp. 77–121.

Nummedal, D., Oertel, G.F., Hubbard, D.K. and Hine, A.C., 1977. Tidal inlet variability: Cape Hatteras to Cape Canaveral, *Coastal Sediments '77*. American Society of Civil Engineers, Charleston, South Carolina, pp. 543–562.

Nummedal, D., Penland, S., Gerdes, R., Schramm, W., Kahn, J. and Roberts, H., 1980. Geologic response to hurricane impact on low-profile Gulf Coast barriers. *Transactions of the Gulf Coast Association of Geological Societies*, 30: 183–195.

Nunn, P.D., 1990. Coastal processes and landforms of Fiji and their bearing on Holocene sea-level changes in the south and west Pacific. *Journal of Coastal Research*, 6: 279–310.

Nunn, P.D., 1993. Role of *Porolithon* algal-ridge growth in the development of the windward coast of Tongatapu Island, Tonga, south Pacific. *Earth Surface Processes and Landforms*, 18: 427–439.

Nunn, P.D., 1994. *Oceanic Islands*. Blackwell, Oxford, 413 pp.

Nunn, P.D., 1998. *Pacific Island Landscapes: Landscape and Geological Development of Southwest Pacific Islands, Especially Fiji, Samoa and Tonga*. Institute of Pacific Studies, The University of the South Pacific, 318 pp.

Nutalaya, P. and Rau, J.L., 1981. Bangkok: the sinking metropolis. *Episodes*, 4: 3–8.

Oaks, R.Q. and DuBar, J.R. (Editors), 1974. *Post-Miocene Stratigraphy, Central and Southern Atlantic Coastal Plain*. Utah State University Press, Logan, 275 pp.

O'Brien, D.J., Whitehouse, R.J.S. and Cramp, A., 2000. The cyclic development of a macrotidal mudflat on varying timescales. *Continental Shelf Research*, 20: 1593–1619.

O'Brien, M.P., 1931. Estuary tidal prisms related to entrance areas. *Journal of Civil Engineering*, 1: 738–739.

O'Brien, M.P., 1969. Equilibrium flow areas of inlets in sandy coasts. *Journal of Waterways Harbors and Coastal Engineering, American Society of Civil Engineers*, 95: 2261–2280.

Odum, E.W., 1984. Dual-gradient concept of detritus transport and processing in estuaries. *Bulletin of Marine Science*, 35: 510–521.

Oenema, O. and DeLaune, R.D., 1988. Accretion rates in salt marshes in the eastern Scheldt, south-west Netherlands. *Estuarine, Coastal and Shelf Science*, 26: 379–394.

Oertel, G.F., 1979. Barrier island development during the Holocene recession, southeastern United States. In: S.P. Leatherman (Editor), *Barrier Islands*. Academic Press, New York, pp. 273–290.

Oertel, G.F., 2001. Hypsographic, hydro-hypsographic and hydrological analysis of coastal bay environments, Great Machipongo Bay, Virginia. *Journal of Coastal Research*, 17: 775–783.

Oertel, G.F., Wong, G.T.F. and Conway, J.D., 1989. Sediment accumulation at a fringe marsh during transgression, Oyster, Virginia. *Estuaries*, 12: 18–26.

Ollier, C.D., 1984. *Weathering*. Arnold, London, 270 pp.

Ollier, C.D., 1988. *Volcanoes*. Blackwell, Oxford, 228 pp.

Oomkens, E., 1974. Lithofacies relations in the Late Quaternary Niger Delta complex. *Sedimentology*, 21: 195–222.

O'Regan, P.R., 1996. The use of contemporary information technologies for coastal research and management. *Journal of Coastal Research*, 12: 192–204.

Oreskes, N., Shrader-Frechette, K. and Belitz, K., 1994. Verification, validation, and confirmation of numerical models in the earth sciences. *Science*, 263: 641–646.

Orford, J.D. and Carter, R.W.G., 1995. Examination of mesoscale forcing of a swash-aligned gravel barrier from Nova Scotia. *Marine Geology*, 126: 201–211.

Orford, J.D. and Wright, P., 1978. What's in a name? Descriptive or genetic implications of 'ridge and runnel' topography. *Marine Geology*, 28: M1–M8.

Orford, J.D., Carter, R.W.G. and Jennings, S.C., 1991. Coarse clastic barrier environments: evolution and implications for Quaternary sea level interpretations. *Quaternary International*, 9: 87–104.

Orford, J.D., Jennings, S.C. and Forbes, D.L., 2001. Origin, development, reworking and breakdown of gravel-dominated coastal barriers in Atlantic Canada: future scenarios for the British coast. In: J.R. Packham (Editor), *British Shingles*. Otley, Westbury, pp. 23–55.

Orme, A.R., 1962. Abandoned and composite sea cliffs in Britain and Ireland. *Irish Geographer*, 4: 279–291.

Orme, A.R., 1990. The instability of Holocene dunes, the case of the Morro dunes, California. In: K.F. Nordstrom, N.P. Psuty and R.W.G. Carter (Editors), *Coastal Dunes: Form and Process*. John Wiley & Sons, Chichester, pp. 315–338.

Orme, A.R. and Orme, A.J., 1988. Ridge-and-runnel enigma. *Geographical Review*, 78: 169–184.

Orme, G.R., 1973. Aspects of sedimentation in the coral reef environment. In: O.A. Jones and R. Endean (Editors), *The Geology and Biology of Coral Reefs, vol. VI, Geology 2*, pp. 129–182.

Orson, R.A., Panageotou, W. and Leatherman, S.P., 1985. Response of tidal salt marshes of the US Atlantic and Gulf coasts to rising sea levels. *Journal of Coastal Research*, 1: 29–37.

Orson, R.A., Warren, R.S. and Niering, W.A., 1987. Development of a tidal marsh in a New England river valley. *Estuaries*, 10: 20–27.

Orton, G.J. and Reading, H.G., 1993. Variability of deltaic processes in terms of sediment supply, with particular emphasis on grain size. *Sedimentology*, 40: 475–512.

Ostrander, G.K., Armstrong, K.M., Knobbe, E.T., Gerace, D. and Scully, E.P., 2000. Rapid transition in the structure of a coral reef community: the effects of coral bleaching and physical disturbance. *Proceedings of the National Academy of Science*, 97: 5297–5302.

Ota, Y., 1986. Marine terraces as reference surfaces in late Quaternary tectonics studies: examples from the Pacific rim. *Royal Society of New Zealand Bulletin*, 24: 357–375.

Ota, Y. and Kaizuka, S., 1991. Tectonic geomorphology at active plate boundaries: examples from the Pacific rim. *Zeitschrift für Geomorphologie, Suppl.-Bd*, 82: 119–146.

Ota, Y. and Omura, A., 1991. Late Quaternary shorelines in the Japanese islands. *Quaternary Research*, 30: 175–186.

Ota, Y., Berryman, K.R., Hull, A.G., Miyauchi, T. and Iso, N., 1988. Age and height distribution of Holocene transgressive deposits in eastern North Island, New Zealand. *Palaeogeography, Palaeoclimatology, Palaeoecology*, 68: 135–151.

Ota, Y., Hull, A.G., Iso, N., Ikeda, Y., Moriya, I. and Yoshikawa, T., 1992. Holocene marine terraces on the northeast coast of North Island, New Zealand, and their tectonic significance. *New Zealand Journal of Geology and Geophysics*, 35: 273–288.

Otvos, E.G., 1970. Development and migration of barrier islands, northern Gulf of Mexico. *Geological Society of America Bulletin*, 81: 241–246.

Otvos, E.G., 1981. Barrier island formation through nearshore aggradation: stratigraphic and field evidence. *Marine Geology*, 43: 195–243.

Otvos, E.G., 1985. Barrier island genesis – questions of alternatives for the Apalachicola Coast, northeastern Gulf of Mexico. *Journal of Coastal Research*, 1: 267–278.

Otvos, E.G., 1986. Island evolution and 'stepwise retreat': late Holocene transgressive barriers, Mississippi delta coast – limitations of a model. *Marine Geology*, 72: 325–340.

Otvos, E.G., 2000. Beach ridges – definitions and significance. *Geomorphology*, 32: 83–108.

Otvos, E.G., 2001. Assumed Holocene highstands, Gulf of Mexico: basic issues of sedimentary and landform criteria – discussion. *Journal of Sedimentary Research*, 71: 645–647.

Otvos, E.G. and Price, W.A., 1979. Problems of chenier genesis and terminology – an overview. *Marine Geology*, 31: 251–263.

Packham, J.R. and Willis, A.J., 1997. *Ecology of Dunes, Salt Marsh and Shingle*. Chapman and Hall, London, 335 pp.

Paine, J.G., 1993. Subsidence of the Texas coast: inferences from historical and late Pleistocene sea levels. *Tectonophysics*, 222: 445–458.

Pandolfi, J.M., 1996. Limited membership in Pleistocene reef coral assemblages from the Huon Peninsula, Papua New Guinea: constancy during global change. *Paleobiology*, 22: 152–176.

Pandolfi, J.M. and Minchin, P.R., 1995. A comparison of taxonomic composition and diversity between reef coral life and death assemblages in Madang Lagoon, Papua New Guinea. *Palaeogeography, Palaeoclimatology, Palaeoecology*, 119: 321–341.

Parkinson, R.W., 1989. Decelerating Holocene sea-level rise and its influence on southwest Florida coastal evolution: a transgressive/regressive stratigraphy. *Journal of Sedimentary Petrology*, 59: 960–972.

Parsons, B. and Sclater, J.G., 1977. An analysis of the variation of ocean floor bathymetry and heat flow with age. *Journal of Geophysical Research*, 82: 803–827.

Paskoff, R.P., 1978. Sur l'évolution géomorphologique du grand escarpement côtier du désert Chilien. *Géographie Physique et Quaternaire*, 32: 351–360.

Paskoff, R.P., 1991. Likely occurrence of a mega-tsunami in the Middle Pleistocene, near Coquimbo, Chile. *Revista Geologica de Chile*, 18: 87–91.

Patterson, R.T., Hutchinson, I., Guilbault, J.P. and Clague, J.J., 2000. A comparison of the vertical zonation of diatom, foraminifera, and macrophyte assemblages in a coastal marsh: implications for greater paleo-sea level resolution. *Micropaleontology*, 46: 229–244.

Pattiaratchi, C., Hegge, B., Gould, J. and Eliot, I., 1997. Impact of sea-breeze activity on nearshore and foreshore processes in southwestern Australia. *Continental Shelf Research*, 17: 1539–1560.

Paul, A.K., Bandyopadhyay, M.K. and Chowdhury, A., 1987. Morpho-ecological variations in the Sundarban mudflats. *Geographical Review of India*, 49: 1–14.

Pejrup, M., 1988. Suspended sediment transport across a tidal flat. *Marine Geology*, 82: 187–198.

Peltier, W.R., 1988. Global sea level and earth rotation. *Science*, 240: 895–901.

Peltier, W.R., 1999. Global sea level rise and glacial isostatic adjustment. *Global and Planetary Change*, 20: 93–123.

Peltier, W.R., 2001. Global glacial isostatic adjustment and modern instrumental records of relative sea level history. In: B.C. Douglas, M.S. Kearney and S.P. Leatherman (Editors), *Sea Level Rise: History and Consequences*. Academic Press, San Diego, pp. 65–95.

Peltier, W.R. and Tushingham, A.M., 1989. Global sea level rise and the greenhouse effect: might they be connected? *Science*, 244: 806–810.

Penck, A., 1894. *Morphologie der Erdoberfläche Bibliothek der Geographie*, 2 vols. Handbücher Engelhorn, Stuttgart.

Penck, A. and Brückner, E., 1909. *Die Alpen im Eiszeitalter*, 3 vols. G.H. Tauchnitz, Leipzig.

Pendón, J.G., Morales, J.A., Borrego, J., Jimenez, I. and Lopez, M., 1998. Evolution of estuarine facies in a tidal channel environment, SW Spain: evidence for a change from tide- to wave-domination. *Marine Geology*, 147: 43–62.

Penland, S., Boyd, R., Nummedal, D. and Roberts, H., 1981. Deltaic barrier development on the Louisiana coast. *Supplement to Gulf Coast Association of Geological Societies Transactions, Thirty-first Annual Meeting*. Gulf Coast Association of Geological Sciences, Corpus Christi, TX, pp. 471–476.

Penland, S., Boyd, R. and Suter, J.R., 1988. Transgressive depositional systems of the Mississippi Delta plain: a model for barrier shoreline and shelf sand development. *Journal of Sedimentary Petrology*, 58: 932–949.

Penland, S. and Suter, J.R., 1989. The geomorphology of the Mississippi River chenier plain. *Marine Geology*, 90: 231–258.

Penland, S., Suter, J.R. and Boyd, R., 1985. Barrier island arcs along abandoned Mississippi River deltas. *Marine Geology*, 63: 197–233.

Périgaud, C. and Delecluse, P., 1992. Annual sea level variations in the southern tropical Indian Ocean from GEOSAT and shallow-water simulations. *Journal of Geophysical Research*, 97: 20169–20178.

Perillo, G.M.E., 1995. *Geomorphology and Sedimentology of Estuaries*. Developments in Sedimentology, 53. Elsevier, Amsterdam, 471 pp.

Perillo, G.M.E., Ripley, M.D., Piccolo, M.C. and Dyer, K.R., 1996. The formation of tidal creeks in a salt marsh: new evidence from the Loyola Bay salt marsh, Rio Gallegos Estuary, Argentina. *Mangroves and Salt Marshes*, 1: 37–46.

Perrin, C., 1990. Genèse de la morphologie des atolls: le cas de Mururoa (Polynésie Française). *Comptes Rendus, Académie des Sciences. Série II*, 311: 671–678.

Perry, C.T., 2001. Storm-induced coral rubble deposition: Pleistocene records of natural reef disturbance and community response. *Coral Reefs*, 20: 171–183.

Pestrong, R., 1965. The development of drainage patterns on tidal marshes. *Stanford University Publications in Geological Science*, 10: 1–87.

Pethick, J.S., 1980. Salt marsh initiation during the Holocene transgression: the example of the north Norfolk marshes. *Journal of Biogeography*, 7: 1–9.

Pethick, J.S., 1981. Long-term accretion rates on tidal marshes. *Journal of Sedimentary Petrology*, 61: 571–577.

Pethick, J.S., 1984. *An Introduction to Coastal Geomorphology*. Arnold, London, 260 pp.

Pethick, J.S., 1992. Saltmarsh geomorphology. In: J.R.L. Allen and K. Pye (Editors), *Saltmarshes: Morphodynamics, Conservation and Engineering Significance*. Cambridge University Press, Cambridge, pp. 41–62.

Pethick, J.S., 1996. The geomorphology of mudflats. In: K.F. Nordstrom and C.T. Roman (Editors), *Estuarine Shores: Evolution, Environment and Human Alterations*. John Wiley & Sons, Chichester, pp. 185–211.

Pethick, J.S., 2001. Coastal management and sea-level rise. *Catena*, 42: 307–322.

Pethick, J.S., Leggett, D. and Husain, L., 1990. Boundary layers under salt marsh vegetation developed in tidal currents. In: J.B. Thornes (Editor), *Vegetation and Erosion*. John Wiley & Sons, Chichester, pp. 113–124.

Petty, W.H., Delcourt, P.A. and Delcourt, H.R., 1996. Holocene lake-level fluctuations and beach-ridge development along the northern shore of Lake Michigan, USA. *Journal of Paleolimnology*, 15: 147–169.

Philander, S.G., 1990. *El Niño, La Niña, and the Southern Oscillation*. Academic Press, San Diego, 293 pp.

Philander, S.G., 2001. El Niño Southern Oscillation (ENSO) models. In: J.H. Steele, S.A. Thorpe and K.K. Turekian (Editors), *Encyclopedia of Ocean Sciences*. Academic Press, San Diego, pp. 827–832.

Phillips, J.D., 1985. Headland-bay beaches revisited: an example from Sandy Hook, New Jersey. *Marine Geology*, 65: 21–31.

Phillips, J.D., 1991. The human role in earth surface systems: some theoretical considerations. *Geographical Analysis*, 23: 316–331.

Phillips, J.D., 1992. Qualitative chaos in geomorphologic systems, with an example from wetland response to sea-level rise. *Journal of Geology*, 100: 365–374.

Phillips, J.D., 1995. Biogeomorphology and landscape evolution: the problem of scale. *Geomorphology*, 13: 337–347.

Phillips, J.D., 1996. Deterministic complexity, explanation, and predictability in geomorphic systems. In: B.L. Rhoads and C.E. Thorn (Editors), *The Scientific Nature of Geomorphology*. John Wiley & Sons, Chichester, pp. 315–335.

Phillips, J.D., 1999. Event timing and sequence in coastal shoreline erosion: Hurricanes Bertha and Fran and the Neuse Estuary. *Journal of Coastal Research*, 15: 616–623.

Pierce, J.W., 1969. Sediment budget along a barrier island chain. *Sedimentary Geology*, 3: 5–16.

Pilkey, O.H. and Wright, H.L., 1988. Seawalls versus beaches. *Journal of Coastal Research*, 4: 41–64.

Pilkey, O.H., Young, R.S., Riggs, S.R., Smith, A.W.S., Wu, H. and Pilkey, W.D., 1993. The concept of shoreface profile of equilibrium: a critical review. *Journal of Coastal Research*, 9: 255–278.

Pillay, S., Gardner, L.R. and Kjerfve, B., 1992. The effect of cross-sectional velocity and concentration variations on suspended sediment transport rates in tidal creeks. *Estuarine, Coastal and Shelf Science*, 35: 331–345.

Piper, D.J.W. and Panagos, A.G., 1981. Growth patterns of the Acheloos and Evinos deltas, western Greece. *Sedimentary Geology*, 28: 111–132.

Pirazzoli, P.A., 1986a. Secular trends of relative sea level (RSL) changes indicated by tide-gauge records. *Journal of Coastal Research*, Special issue, 1: 1–26.

Pirazzoli, P.A., 1986b. Marine notches. In: O. van de Plassche (Editor), *Sea-Level Research: A Manual for the Collection and Evaluation of Data*. Geobooks, Norwich, pp. 361–400.

Pirazzoli, P.A., 1991. *World Atlas of Holocene Sea-Level Changes*. Elsevier Oceanography Series, 58. Elsevier, Amsterdam, 300 pp.

Pirazzoli, P.A., 1994. Tectonic coasts. In: R.W.G. Carter and C.D. Woodroffe (Editors), *Coastal Evolution: Late Quaternary Shoreline Morphodynamics*. Cambridge University Press, Cambridge, pp. 451–476.

Pirazzoli, P.A., 1996. Sea-level changes: the last 20000 years. John Wiley & Sons, Chichester, 211 pp.

Pirazzoli, P.A. and Montaggioni, L., 1988. The 7000 yr sea-level curve in French Polynesia: geodynamic implications for mid-plate volcanic islands. *Proceedings of the Sixth International Coral Reef Symposium*, Vol. 3 Sixth International Coral Reef Symposium Executive Committee, Townsville, pp. 467–472.

Pirazzoli, P.A., Montaggioni, L.F., Vergnaud-Grazzini, C. and Saliège, J.F., 1987. Late Holocene sea levels and coral reef development in Vahitahi Atoll, eastern Tuamotu Islands, Pacific Ocean. *Marine Geology*, 76: 105–116.

Pirazzoli, P.A., Montaggioni, L.F., Salvat, B. and Faure, G., 1988. Late Holocene sea level indicators from twelve atolls in the central and eastern Tuamotus (Pacific Ocean). *Coral Reefs*, 7: 57–68.

Pirazzoli, P.A., Radtke, U., Hantoro, W.D., Jouannic, C., Hoang, C.T., Causse, C. and Borel Best, M., 1991. Quaternary raised coral-reef terraces on Sumba Island, Indonesia. *Science*, 252: 1834–1836.

Pirazzoli, P.A., Stiros, S.C., Arnold, M., Laborel, J. and Laborel-Deguen, F., 1999. Late Holocene coseismic vertical displacements and tsunami deposits near Kynos, Gulf of Euboea, Central Greece. *Physics and Chemistry of the Earth*, 24: 361–367.

Pitman, J.I., 1985. Thailand. In: E.C.F. Bird and M.L. Schwartz (Editors), *The World's Coastline*. Van Nostrand Reinhold, New York, pp. 771–787.

Pitman, W.C., 1978. Relationship between eustacy and stratigraphic sequences of passive margins. *Geological Society of America Bulletin*, 89: 1369–1403.

Pittock, A.B., 1999. Coral reefs and environmental change: adaptation to what? *American Zoologist*, 39: 110–129.

Pizzuto, J.E. and Schwendt, A.E., 1997. Mathematical modeling of autocompaction of a Holocene transgressive valley-fill deposit, Wolfe Glade, Delaware. *Geology*, 25: 57–60.

Playfair, J., 1802. *Illustrations of the Huttonian Theory of the Earth*. Dover, New York, 528 pp.

Playford, P.E., 1990. Geology of the Shark Bay area. In: P.F. Berry, S.D. Bradshaw and B.R. Wilson (Editors), *Research in Shark Bay: Report to the France-Australe Committee Bicentenary Expedition*, Western Australian Museum, Perth, pp. 13–32.

Playford, P.E., 1997. Geology and hydrogeology of Rottnest Island. In: H.L. Vacher and T.M. Quinn (Editors), *Geology and Hydrogeology of Carbonate Islands*. Elsevier, Amsterdam, pp. 783–810.

Plius, J.L.A., 1992. Relationships between deflation and near surface wind velocity in a coastal dune blowout. *Earth Surface Processes and Landforms*, 17: 663–673.

Popper, K.R., 1968. *The Logic of Scientific Discovery*. Harper and Row, New York, 480 pp.

Posamentier, H.W. and Vail, P.R., 1988. Eustatic controls on clastic deposition II – sequence and systems tracts models. In: C.K. Wilgus *et al.* (Editors), *Sea-level Research: An Integrated Approach*. Society of Economic Palaeontologists and Mineralogists, Tulsa, OK, Special Publication, pp. 109–124.

Postma, G., 1990. An analysis of the variation in delta architecture. *Terra Nova*, 2: 124–130.

Postma, G., 1995. Causes of architectural variation in deltas. In: M.N. Oti and G. Postma (Editors), *Geology of Deltas*. Balkema, Rotterdam, pp. 3–16.

Postma, H., 1954. Hydrography of the Dutch Wadden Sea. *Archives Nederland Zoologie*, 10: 405–511.

Postma, H., 1961. Transport and accumulation of suspended matter in the Dutch Wadden Sea. *Netherlands Journal of Sea Research*, 1: 148–190.

Postma, H., 1967. Sediment transport and sedimentation in the estuarine environment. In: G.H. Lauff (Editor), *Estuaries*. American Association for the Advancement of Science, Washington, pp. 158–179.

Powell, J.W., 1875. *Exploration of the Colorado River of the West and its Tributaries*. Government Printing Office, Washington, DC, 291 pp.

Prager, E.J., 1991. Numerical simulation of circulation in a Caribbean-type backreef lagoon. *Coral Reefs*, 10: 177–182.

Pranzini, E., 2001. Updrift river mouth migration on cuspate deltas: two examples from the coast of Tuscany (Italy). *Geomorphology*, 38: 125–132.

Prêcheur, C., 1960. *Le littoral de la Manche, de Sainte-Adresse à Ault, étude morphologique*, Special issue, Centre National de la Recherche Scientifique et du Conseil Général de la Seine-Maritime, Poitiers, 138 pp.

Price, D.M., Brooke, B.P. and Woodroffe, C.D., 2001. Thermoluminescence dating of aeolianites from Lord Howe Island and south-west Western Australia. *Quaternary Science Reviews*, 20: 841–846.

Price, W.A., 1947. Equilibrium of form and forces in tidal basins of the coast of Texas and Louisiana. *Bulletin of the American Association of Petroleum Geologists*, 31: 1619–1663.

Priestley, G. and Mundet, L., 1998. The post-stagnation phase of the resort cycle. *Annals of Tourism Research*, 25: 85–111.

Prior, D.B. and Bornhold, B.D., 1990. The underwater development of Holocene fan deltas. In: A. Colella and D.B. Prior (Editors), *Coarse-Grained Deltas*. Special Publication of the International Association of Sedimentologists. Blackwell Scientific, Oxford, pp. 75–90.

Prior, D.B. and Renwick, W.H., 1980. Landslide morphology and processes on some coastal slopes in Denmark and France. *Zeitschrift für Geomorphologie, NF, Suppl.-Bd*, 34: 63–86.

Pritchard, D., 1952. Estuarine hydrology. *Advances in Geophysics*, 1: 243–280.

Psuty, N.P., 1965. Beach-ridge development in Tabasco, Mexico. *Annals of the Association of American Geographers*, 55: 112–124.

Psuty, N.P., 1992. Spatial variation in coastal foredune development. In: R.W.G. Carter, T.G.F. Curtis and M.J. Sheehy-Skeffington (Editors), *Coastal Dunes: Geomorphology, Ecology and Management for Conservation*. Balkema, Rotterdam, pp. 3–13.

Psuty, N.P. and Moreira, M.E.S.A., 2000. Holocene sedimentation and sea level rise in the Sado Estuary, Portugal. *Journal of Coastal Research*, 16: 125–138.

Pugh, D.T., 1987. *Tides, Surges and Mean Sea-Level*. John Wiley & Sons, Chichester, 471 pp.

Pugh, D.T., 2001. Tides. In: J.H. Steele, S.A. Thorpe and K.K. Turekian (Editors), *Encyclopedia of Ocean Sciences*. Academic Press, San Diego, pp. 2961–2968.

Purdy, E.G., 1963. Recent calcium carbonate facies of the Great Bahama Bank: 2. Sedimentary facies. *Journal of Geology*, 71: 472–497.

Purdy, E.G., 1974a. Karst-determined facies patterns in British Honduras: Holocene carbonate sedimentation model. *American Association of Petroleum Geologists Bulletin*, 58: 825–855.

Purdy, E.G., 1974b. Reef configurations, cause and effect. In: L.F. Laporte (Editor), *Reefs in Time and Space*. Society of Economic Palaeontologists and Mineralogists, Tulsa, OK, Special Publication, pp. 9–76.

Purdy, E.G. and Winterer, E.L., 2001. Origin of atoll lagoons. *Geological Society of America Bulletin*, 113: 837–854.

Putnam, W.C., 1937. The marine cycle of erosion for a steeply sloping shoreline of emergence. *Journal of Geology*, 45: 844–850.

Pye, K., 1983. Coastal dunes. *Progress in Physical Geography*, 7: 531–557.

Pye, K., 1990. Physical and human influences on coastal dune development between the Ribble and Mersey estuaries, northwest England. In: K.F. Nordstrom, N.P. Psuty and R.W.G. Carter (Editors), *Coastal Dunes: Form and Process*. John Wiley & Sons, Chichester, pp. 339–359.

Pye, K., 1992. Saltmarshes on the barrier coastline of North Norfolk, eastern England. In: J.R.L. Allen and K. Pye (Editors), *Saltmarshes: Morphodynamics, Conservation and Engineering Significance*. Cambridge University Press, Cambridge, pp. 149–178.

Pye, K. and Allen, J.R.L., 2000. Past, present and future interactions, management challenges and research needs in coastal and estuarine environments. In: K. Pye and J.R.L. Allen (Editors), *Coastal and Estuarine Environments: Sedimentology, Geomorphology and Geoarchaeology*. Geological Society, Special Publication, London, pp. 1–4.

Pye, K. and Bowman, G.M., 1984. The Holocene marine transgression as a forcing function in episodic dune activity on the eastern Australian coast. In: B.G. Thom (Editor), *Coastal Geomorphology in Australia*. Academic Press, Sydney, pp. 179–196.

Pye, K. and French, P.W., 1993. *Erosion and Accretion Processes on British Salt Marshes*, 5 vols. Cambridge Environmental Research Consultants, Cambridge.

Pye, K. and Rhodes, E.G., 1985. Holocene development of an episodic transgressive dune barrier, Ramsay Bay, north Queensland, Australia. *Marine Geology*, 64: 189–202.

Pye, K. and Tsoar, H., 1990. *Aeolian Sand and Sand Dunes*. Unwin Hyman, London, 396 pp.

Qinshang, Y., Shiyuan, X. and Xusheng, S., 1989. Holocene cheniers in the Yangtze Delta, China. *Marine Geology*, 90: 337–343.

Quinlan, G., 1985. A numerical model of postglacial relative sea level change near Baffin Island. In: J.T. Andrews (Editor), *Quaternary Environments, Eastern Canadian Arctic, Baffin Island and Western Greenland*. Allen and Unwin, London, pp. 560–584.

Rakha, K.A. and Kamphuis, J.W., 1997. Wave-induced currents in the vicinity of a seawall. *Coastal Engineering*, 30: 23–52.

Rampino, M.R. and Sanders, J.E., 1980. Holocene transgression in south-central Long Island, New York. *Journal of Sedimentary Petrology*, 50: 1063–1080.

Ramsey, E.W. and Laine, S.C., 1997. Comparison of Landsat Thematic Mapper and high resolution photography to identify change in complex coastal wetlands. *Journal of Coastal Research*, 13: 281–292.

Randerson, P.F., 1979. A simulation model of salt-marsh development and plant ecology. In: B. Knights and A.J. Phillips (Editors), *Estuarine and Coastal Land Reclamation and Water Storage*. Saxon House, Farnborough, pp. 48–57.

Ranwell, D.S., 1964. Spartina salt marshes in southern England. II. Rate and seasonal pattern of sediment accretion. *Journal of Ecology*, 52: 79–94.

Raper, J., 2000. *Multidimensional Geographic Information Science*. Taylor & Francis, London, 300 pp.

Raper, J. and Livingstone, D., 1995. Development of a geomorphological spatial model using object-oriented design. *International Journal of Geographical Information Systems*, 9: 359–383.

Raybould, A.F., 1997. The history and ecology of *Spartina anglica* in Poole Harbour. *Proceedings of the Dorset Natural History and Archaeological Society*, 119: 147–158.

Rayleigh, L., 1876. On waves. *Philosophical Magazine, Series 5*, 1: 257–279.

Rayleigh, L., 1877. On progressive waves. *Proceedings of the London Mathematical Society*, 9: 21–26.

Rea, C.C. and Komar, P.D., 1975. Computer simulation models of a hooked beach shoreline configuration. *Journal of Sedimentary Petrology*, 45: 866–872.

Rector, R.L., 1954. Laboratory study of equilibrium profiles of beaches. *Beach Erosion Board Technical Memorandum*, 41: 1–38.

Redfield, A.C., 1967. Postglacial change in sea level in the western North Atlantic Ocean. *Science*, 157: 687–691.

Redfield, A.C., 1972. Development of a New England salt marsh. *Ecological Monographs*, 42: 201–237.

Redfield, A.C. and Rubin, M., 1962. The age of salt marsh peat and its relation to recent changes in sea level at Barnstable, Massachusetts. *Proceedings of the National Academy of Science*, 48: 1728–1735.

Redman, J.B., 1852. On the alluvial formations and the local change of the south coast of England. *Minutes of Proceedings of the Institution of Civil Engineers*, 11: 162–223.

Reed, D.J., 1988. Sediment dynamics and deposition in a retreating coastal salt marsh. *Estuarine, Coastal and Shelf Science*, 26: 67–79.

Reed, D.J., 1989. Patterns of sediment deposition in subsiding coastal salt marshes, Terrebonne Bay, Louisiana: the role of winter storms. *Estuaries*, 12: 222–227.

Reed, D.J., 1990. The impact of sea-level rise on coastal salt marshes. *Progress in Physical Geography*, 14: 465–481.

Reed, D.J., Muir Wood, R. and Best, J., 1988. Earthquakes, rivers and ice: scientific research at Laguna San Rafael, southern Chile. *Geographical Journal*, 154: 392–405.

Reed, D.J., Spencer, T., Murray, A.L., French, J.R. and Leonard, L., 1999. Marsh surface sediment deposition and the role of tidal creeks: implications for created and managed coastal marshes. *Journal of Coastal Conservation*, 5: 81–90.

Reeve, D., Li, B. and Thurston, N., 2001. Eigenfunction analysis of decadal fluctuations in sandbank morphology at Gt Yarmouth. *Journal of Coastal Research*, 17: 371–382.

Reineck, H.E., 1972. Tidal flats. In: J.K. Rigby and W.K. Hamblin (Editors), *Recognition of Ancient Sedimentary Environments*. Society of Economic Palaeontologists and Mineralogists, Tulsa, OK, Special Publication, 16, pp. 146–159.

Ren, M.E. and Shi, Y.L., 1986. Sediment discharge of the Yellow River (China) and its effect on the sedimentation of the Bohai and the Yellow Sea. *Continental Shelf Research*, 6: 785–810.

Reynolds, K.C., 1933. Investigation of wave-action on sea walls by the use of models. *American Geophysical Union Transactions*, 14: 512–516.

Reynolds, O., 1877. On the rate of progression of groups of waves and the rate at which energy is transmitted by waves. *Nature*, 36: 343–344.

Rhoads, B.L. and Thorn, C.E., 1996. Observation in geomorphology. In: B.L. Rhoads and C.E. Thorn (Editors), *The Scientific Nature of Geomorphology*. John Wiley & Sons, Chichester, pp. 21–56.

Rhodes, E.G., 1982. Depositional model for a chenier plain, Gulf of Carpentaria, Australia. *Sedimentology*, 29: 201–221.

Richard, G.A., 1978. Seasonal and environmental variations in sediment accretion in a Long Island salt marsh. *Estuaries*, 1: 29–35.

Richards, F.J., 1934. The salt marshes of the Dovey Estuary. IV. The rates of vertical accretion, horizontal extension and scarp erosion. *Annals of Botany*, 48: 235–259.

Ricketts, P.J., 1992. Current approaches in Geographic Information Systems for coastal management. *Marine Pollution Bulletin*, 25: 82–87.

Riddell, K.J. and Fuller, T.W., 1995. The Spey Bay geomorphological study. *Earth Surface Processes and Landforms*, 20: 671–686.

Riggs, S.R., Cleary, W.J. and Snyder, S.W., 1995. Influence of inherited geologic framework on barrier shoreface morphology and dynamics. *Marine Geology*, 126: 213–234.

Ritchie, W., 1979. Machair development and chronology in the Uists and adjacent islands. *Transactions and Proceedings of the Botanical Society of Edinburgh*, 77B: 107–122.

Ritchie, W., 1992. Scottish landform examples – 4. Coastal parabolic dunes of the Sands of Forvie. *Scottish Geographical Magazine*, 108: 39–44.

Ritchie, W. and Penland, S., 1990. Aeolian sand bodies of the south Louisiana coast. In: K.F. Nordstrom, N.P. Psuty and R.W.G. Carter (Editors), *Coastal Dunes: Form and Process*. John Wiley & Sons, Chichester, pp. 105–127.

Roberts, H.H., 1974. Variability of reefs with regard to changes in wave power around an island. *Proceedings of the Second International Coral Reef Congress*, Vol. 2. Great Barrier Reef Committee, Brisbane, pp. 497–512.

Roberts, H.H., 1983. Shelf margin reef morphology: a clue to major off-shelf sediment transport routes, Grand Cayman Island, West Indies. *Atoll Research Bulletin*, 263: 1–19.

Roberts, H.H., 1997. Dynamic changes of the Holocene Mississippi River delta plain: the delta cycle. *Journal of Coastal Research*, 13: 605–627.

Roberts, H.H., 1998. Delta switching: early reponses to the Atchafalaya River diversion. *Journal of Coastal Research*, 14: 882–899.

Roberts, H.H., Murray, S.P. and Suhayda, J.N., 1975. Physical processes in a fringing reef system. *Journal of Marine Research*, 33: 233–260.

Roberts, H.H., Murray, S.P. and Suhayda, J.N., 1977. Evidence for strong currents and turbulence in a deep coral reef groove. *Limnology and Oceanography*, 22: 152–156.

Roberts, H.H., Rouse, L.J., Walker, N.D. and Hudson, J.H., 1982. Cold water stress in Florida Bay and Northern Bahamas: a product of cold air outbreaks. *Journal of Sedimentary Petrology*, 52: 145–155.

Roberts, H.H., Phipps, C.V. and Effendi, L., 1987. *Halimeda* bioherms of the eastern Java Sea, Indonesia. *Geology*, 15: 371–374.

Roberts, H.H., Wilson, P.A. and Lugo-Fernandez, A., 1992. Biologic and geologic responses to physical processes: examples from modern reef systems of the Caribbean–Atlantic region. *Continental Shelf Research*, 12: 809–834.

Roberts, W., Le Hir, P. and Whitehouse, R.J.S., 2000. Investigation using simple mathematical models of the effect of tidal currents and waves on the profile shape of intertidal mudflats. *Continental Shelf Research*, 20: 1079–1097.

Robinson, A.H.W., 1961. The hydrography of Start Bay and its relationship to beach changes at Hallsands. *Geographical Journal*, 127: 63–77.

Robinson, A.H.W., 1975. Cyclical changes in shoreline development at the entrance to Teignmouth Harbour, Devon, England. In: J. Hails and A. Carr (Editors), *Nearshore Sediment Dynamics and Sedimentation*. John Wiley & Sons, Chichester, pp. 181–200.

Robinson, A.H.W., 1980. Erosion and accretion along part of the Suffolk coast of East Anglia, England. *Marine Geology*, 37: 133–146.

Robinson, D.A. and Jerwood, L.C., 1987. Sub-aerial weathering of chalk shore platforms during harsh winters in southeast England. *Marine Geology*, 77: 1–14.

Robinson, L.A., 1977. Marine erosive processes at the cliff foot. *Marine Geology*, 23: 257–271.

Rodriguez, A.B., Fassell, M.L. and Anderson, J.B., 2001. Variations in shoreface progradation and ravinement along the Texas coast, Gulf of Mexico. *Sedimentology*, 48: 837–853.

Rodriguez, H.N. and Mehta, A.J., 1998. Considerations on wave-induced fluid mud streaming at open coasts. In: K.S. Black, D.M. Paterson and A. Cramp (Editors), *Sedimentary Processes in the Intertidal Zone*. Geological Society of London, Special Publication, pp. 177–186.

Rodriguez, H.N. and Mehta, A.J., 2000. Longshore transport of fine-grained sediment. *Continental Shelf Research*, 20: 1419–1432.

Rodriguez-Navarro, C., Doehne, E. and Sebastian, E., 1999. Origins of honeycomb weathering: the role of salts and wind. *Geological Society of America Bulletin*, 111: 1250–1255.

Roelvink, J.A. and Stive, M.J.F., 1989. Bar-generating cross-shore flow mechanisms on a beach. *Journal of Geophysical Research*, 94: 4785–4800.

Roep, T.B. and van Regteren Altena, J.F., 1988. Paleotidal levels in tidal sediments (3800–3635 BP); compaction, sea level rise and human occupation (3275–2620 BP) at Bovenkarspel, NW Netherlands. In: P.L. de Boer, A. van Gelder and S.D. Nio (Editors), *Tide-Influenced Sedimentary Environments and Facies*. D. Reidel, Dordrecht, pp. 215–231.

Ross, J.C., 1855. Review of the theory of coral formations set forth by Ch. Darwin in his book entitled: Researches in Geology and Natural History. *Natuurkundig Tijdschrift voor Nederlandsch Indië*, 8: 1–43.

Roy, P.S., 1984. New South Wales estuaries: their origin and evolution. In: B.G. Thom (Editor), *Coastal Geomorphology in Australia*. Academic Press, Sydney, pp. 99–121.

Roy, P.S., 1994. Holocene estuary evolution – stratigraphic studies from southeastern Australia. In: R.W. Dalrymple, R. Boyd and B.A. Zaitlin (Editors), *Incised-Valley Systems: Origin and Sedimentary Sequences*. Society of Economic Palaeontologists and Minerologists, Tulsa, OK, Special Publication No. 51. pp. 241–263.

Roy, P.S. and Thom, B.G., 1981. Late Quaternary marine deposition in New South Wales and southern Queensland: an evolutionary model. *Journal of the Geological Society of Australia*, 28: 471–489.

Roy, P.S., Thom, B.G. and Wright, L.D., 1980. Holocene sequences on an embayed high-energy coast: an evolutionary model. *Sedimentary Geology*, 16: 1–9.

Roy, P.S., Cowell, P.J., Ferland, M.A. and Thom, B.G., 1994. Wave-dominated coasts. In: R.W.G. Carter and C.D. Woodroffe (Editors), *Coastal Evolution: Late Quaternary Shoreline Morphodynamics*. Cambridge University Press, Cambridge, pp. 121–186.

Roy, P.S., Williams, R.J., Jones, A.R., Yassini, I., Gibbs, P.J., Coates, B., West, R.J., Scanes, P.R., Hudson, J.P. and Nichol, S., 2001. Structure and function of south-east Australian estuaries. *Estuarine, Coastal and Shelf Science*, 53: 351–384.

Rubin, K.H., Fletcher, C.H. and Sherman, C., 2000. Fossiliferous Lana'i deposits formed by multiple events rather than a single giant tsunami. *Nature*, 408: 675–681.

Rudberg, S., 1967. The cliffed coast of Gotland and the rate of cliff retreat. *Geografiska Annaler*, 49A: 283–298.

Ruessink, B.G. and Terwindt, J.H.J., 2000. The behaviour of nearshore bars on the time scale of years: a conceptual model. *Marine Geology*, 163: 289–302.

Ruessink, B.G., van Enckevort, I.M.J., Kingston, K.S. and Davidson, M.A., 2000. Analysis of observed two- and three-dimensional nearshore bar behaviour. *Marine Geology*, 169: 161–183.

Ruggiero, P. and McDougal, W.G., 2001. An analytical model for the prediction of wave set-up, longshore currents and sediment transport on beaches with seawalls. *Coastal Engineering*, 43: 162–182.

Russell, R.J., 1936. Physiography of lower Mississippi River delta. Reports on the geology of Plaquemine and St Bernard parishes, Louisiana, *Louisiana Department of Conservation, Geological Bulletin*, 8: 3–199.

Russell, R.J., 1940. Quaternary history of Louisiana. *Geological Society of America Bulletin*, 51: 1199–1234.

Russell, R.J., 1942. Geomorphology of the Rhone Delta. *Annals of the Association of American Geographers*, 32: 149–254.

Russell, R.J., 1963. Recent recession of tropical cliffy coasts. *Science*, 139: 9–15.

Russell, R.J., 1967. *River Plains and Sea Coasts*. University of California Press, Berkeley, 173 pp.

Russell, R.J. and Howe, H.V., 1935. Cheniers of Southwestern Louisiana. *Geographical Review*, 25: 449–61.

Russell, R.J. and McIntire, W.G., 1965. Beach cusps. *Geological Society of America Bulletin*, 76: 307–320.

Ruz, M.-H., Héquette, A. and Hill, P.R., 1992. A model of coastal evolution in a transgressed thermokarst topography, Canadian Beaufort Sea. *Marine Geology*, 106: 251–278.

Ruz, M.-H. and Allard, M., 1994. Foredune development along a subarctic emerging coastline, eastern Hudson Bay, Canada. *Marine Geology*, 117: 57–74.

Saad, S., Husain, M.L. and Asano, T., 1999. Sediment accretion of a tropical estuarine mangrove: Kemaman, Terengganu, Malaysia. *Tropics*, 8: 257–266.

Saenger, P. and Siddiqi, N.A., 1993. Land from the sea: the mangrove afforestation program of Bangladesh. *Ocean and Coastal Management*, 20: 23–39.

Saenger, P. and Snedaker, S.C., 1993. Pantropical trends in mangrove above-ground biomass and annual litterfall. *Oecologia*, 96: 293–299.

Sahagian, D.L., Schwartz, F.W. and Jacobs, D.K., 1994. Direct anthropogenic contributions to sea level rise in the twentieth century. *Nature*, 367: 54–57.

Saito, Y., Wei, H., Zhou, Y., Nishimura, A., Sato, Y. and Yokota, S., 2000. Delta progradation and chenier formation in the Huanghe (Yellow River) delta, China. *Journal of Asian Earth Sciences*, 18: 489–497.

Salomons, W., Turner, R.K., de Lacerda, L.D. and Ramachandran, S. (Editors), 1999. *Perspectives on Integrated Coastal Zone Management*. Springer-Verlag, Berlin, 386 pp.

Sammut, J., White, I. and Melville, M.D., 1996. Acidification of an estuarine tributary in eastern Australia due to drainage of acid sulfate soils. *Marine and Freshwater Research*, 47: 669–684.

Samoilov, I.V., 1956. *Die Flussmündungen*. H. Haack, Gotha.

Sanders, N.K., 1968a. The development of Tasmanian shore platforms. PhD Thesis, University of Tasmania.

Sanders, N.K., 1968b. Wave tank experiments on the erosion of rocky coasts. *Papers and Proceedings of the Royal Society of Tasmania*, 102: 11–16.

Sanderson, P.G. and Eliot, I., 1996. Shoreline salients, cuspate forelands and tombolos on the coast of Western Australia. *Journal of Coastal Research*, 12: 761–773.

Sanderson, P.G., Eliot, I., Hegge, B. and Maxwell, S., 2000. Regional variation of coastal morphology in southwestern Australia: a synthesis. *Geomorphology*, 34: 73–88.

Sarre, R., 1989. The morphological significance of vegetation and relief on coastal foredune processes. *Zeitschrift für Geomorphologie, Suppl.-Bd*, 73: 17–31.

Sasaki, T., 1983. Three-dimensional topographic changes on the foreshore zone of sandy beaches. *Science Report, Institute of Geoscience, University of Tsukuba, Section A*, 4: 69–95.

Sauer, J., 1961. *Coastal Plant Geography of Mauritius*. Coastal Studies Series, 5. Louisiana State University Press, Baton Rouge, 153 pp.

Savigear, R.A.G., 1952. Some observations on slope development in South Wales. *Transactions of the Institute of British Geographers*, 18: 31–52.

Savigear, R.A.G., 1962. Some observations on slope development in North Devon and North Cornwall. *Transactions of the Institute of British Geographers*, 31: 23–42.

Saville, T., 1950. Model study of sand transport along an indefinitely long, straight beach. *American Geophysical Union Transactions*, 31: 555–565.

Sayles, R.W., 1931. Bermuda during the ice age. *Proceedings of the American Academy of Arts and Sciences*, 66: 381–467.

Sayre, W.O. and Komar, P.D., 1988. The Jump-Off Joe landslide at Newport, Oregon: history of erosion, development and destruction. *Shore and Beach*, 56: 15–22.

Scheidegger, A.E., 1994. Hazards: singularities in geomorphic systems. *Geomorphology*, 10: 19–25.

Schlager, W., Reijmer, J.J.G. and Droxler, A., 1994. Highstand shedding of carbonate platforms. *Journal of Sedimentary Research*, B64: 270–281.

Schlanger, S.O., 1963. Subsurface geology of Eniwetok Atoll. *United States Geological Survey, Professional Paper*, 260–BB: 991–1066.

Schofield, J.C., 1960. Sea level fluctuations during the last 4000 years as recorded by a chenier plain, Firth of Thames, New Zealand. *New Zealand Journal of Geology and Geophysics*, 3: 467–85.

Schofield, J.C., 1977. Effect of Late Holocene sea-level fall on atoll development. *New Zealand Journal of Geology and Geophysics*, 20: 531–536.

Scholl, D.W., 1964. Recent sedimentary record in mangrove swamps and rise in sea level over the southwestern coast of Florida: Part 1. *Marine Geology*, 1: 344–366.

Scholl, D.W., Craighead, F.C. and Stuiver, M., 1969. Florida submergence curve revised: its relation to coastal sedimentation rates. *Science*, 163: 562–564.

Scholten, M. and Rozema, J., 1990. The competitive ability of *Spartina anglica* on Dutch salt marshes. In: A.J. Gray and P.E.M. Benham (Editors), Spartina anglica *– A Research Review*. Institute of Terrestrial Ecology, Research Publication, HMSO, London, pp. 39–47.

Schoonees, J.S. and Theron, A.K., 1993. Review of the field-data base for longshore transport. *Coastal Engineering*, 19: 1–25.

Schoonees, J.S. and Theron, A.K., 1995. Evaluation of 10 cross-shore sediment transport/morphological models. *Coastal Engineering*, 25: 1–41.

Schou, A., 1945. *Det Marine Forland*, Folia Geografiske Danica, 4.H. Hagerup, Copenhagen, 236 pp.

Schou, A., 1952. Direction determining influence of the wind on shoreline simplification and coastal dunes. *Seventeenth International Geographical Congress*. International Geographical Union, Washington, DC, pp. 370–373.

Schumm, S.A., 1979. Geomorphic thresholds: the concept and its applications. *Transactions of the Institute of British Geographers, NS*, 4: 485–515.

Schumm, S.A., 1991. *To Interpret the Earth: Ten Ways to be Wrong*. Cambridge University Press, Cambridge, 133 pp.

Schumm, S.A., 1993. River response to baselevel change: implications for sequence stratigraphy. *Journal of Geology*, 101: 279–294.

Schumm, S.A. and Lichty, R.W., 1965. Time, space, and causality in geomorphology. *American Journal of Science*, 263: 110–119.

Schwartz, M.L., 1965. Laboratory study of sea-level rise as a cause of shore erosion. *Journal of Geology*, 73: 528–534.

Schwartz, M.L., 1968. The scale of shore erosion. *Journal of Geology*, 76: 508–517.

Schwartz, M.L., 1971. The multiple causality of barrier islands. *Journal of Geology*, 79: 76–92.

Schwimmer, R.A. and Pizzuto, J.E., 2000. A model for the evolution of marsh shorelines. *Journal of Sedimentary Petrology*, 70: 1026–1035.

Scoffin, T.P., 1970. The trapping and binding of subtidal carbonate sediments by marine vegetation in Bimini Lagoon, Bahamas. *Journal of Sedimentary Petrology*, 40: 249–273.

Scoffin, T.P., 1987. *An Introduction to Carbonate Sediments and Rocks*. Chapman & Hall, New York, 274 pp.

Scoffin, T.P., 1993. The geological effects of hurricanes on coral reefs and the interpretation of storm deposits. *Coral Reefs*, 12: 203–221.

Scoffin, T.P. and Le Tissier, M.D.A., 1998. Late Holocene sea level and reef-flat progradation. Phuket, south Thailand. *Coral Reefs*, 17: 273–276.

Scoffin, T.P., Stearn, C.W., Boucher, D., Frydl, P., Hawkins, C.M. and Hunter, I.G., 1980. Calcium carbonate budget of a fringing reef on the west coast of Barbados. Part II. Erosion, sediments and internal structure. *Bulletin of Marine Science*, 32: 457–508.

Scoffin, T.P., Brown, B.E., Dunne, R.P. and Le Tissier, M.D.A., 1997. The controls on growth form of intertidal massive corals, Phuket, south Thailand. *Palaios*, 12: 237–248.

Scott, D.B. and Greenberg, D.A., 1983. Relative sea-level rise and tidal development in the Fundy tidal system. *Canadian Journal of Earth Sciences*, 20: 1554–1564.

Scott, D.B. and Medioli, F.S., 1978. Vertical zonations of marsh foraminifera as accurate indicators of former sea-levels. *Nature*, 272: 528–531.

Scott, D.B. and Medioli, F.S., 1980. Post-glacial emergence curves in the Maritimes determined from marine sediments in raised basins. *Canadian Coastal Conference*, 1980. *Proceedings Associate Committee for Research on*

Shoreline Erosion and Sedimentation, National Research Council, Ottawa, pp. 428–446.

Scott, D.B. and Medioli, F.S., 1986. Foraminifera as sea-level indicators. In: O. van de Plassche (Editor), *Sea-Level Research: A Manual for the Collection and Evaluation of Data*. Geo-Books, Norwich, pp. 435–456.

Scott, G.A.J. and Rotondo, G.M., 1983. A model to explain the differences between Pacific plate island atoll types. *Coral Reefs*, 1: 139–150.

Scruton, P.C., 1960. Delta building and the deltaic sequence. In: F.P. Shepard *et al.* (Editors), *Recent Sediments, Northwest Gulf of Mexico*. American Association of Petroleum Geologists, Tulsa, OK, pp. 82–102.

Selby, M.J., 1985. *Earth's Changing Surface: An Introduction to Geomorphology*. Clarendon Press, Oxford, 607 pp.

Semeniuk, V., 1980a. Quaternary stratigraphy of the tidal flats, King Sound, Western Australia. *Journal of the Royal Society of Western Australia*, 63: 65–78.

Semeniuk, V., 1980b. Mangrove zonation along an eroding coastline in King Sound, northwestern Australia. *Journal of Ecology*, 68: 789–812.

Semeniuk, V., 1981. Long-term erosion of the tidal flats King Sound, northwestern Australia. *Marine Geology*, 43: 21–48.

Semeniuk, V., 1985a. Development of mangrove habitats along ria shorelines in north and northwestern tropical Australia. *Vegetatio*, 60: 3–23.

Semeniuk, V., 1985b. Mangrove environments of Port Darwin, Northern Territory: the physical framework and habitats. *Journal of the Royal Society of Western Australia*, 67: 81–97.

Semeniuk, V., 1986. Terminology for geomorphic units and habitats along the tropical coast of Western Australia. *Journal of Royal Society of Western Australia*, 68: 53–79.

Semeniuk, V., Searle, D.J. and Woods, P.J., 1988. The sedimentology and stratigraphy of a cuspate foreland, southwestern Australia. *Journal of Coastal Research*, 4: 551–564.

Sestini, G., 1992. Implications of climatic change for the Nile Delta. In: L. Jeftic, J.D. Milliman and G. Sestini (Editors), *Climatic Change and the Mediterranean: Environmental and Societal Impacts of Climatic Change and Sea-level Rise in the Mediterranean Region*. Edward Arnold, Sevenoaks, pp. 535–601.

Settlemyre, J.L. and Gardner, L.R., 1977. Suspended sediment flux through a salt marsh drainage basin. *Estuarine, Coastal and Marine Science*, 5: 653–663.

Shackleton, N.J., 1987. Oxygen isotopes, ice volume and sea level. *Quaternary Science Reviews*, 6: 183–190.

Shackleton, N.J., 2000. The 100000-year ice-age cycle identified and found to lag temperature, carbon dioxide, and orbital eccentricity. *Science*, 289: 1897–1902.

Shackleton, N.J. and Opdyke, N.D., 1973. Oxygen isotope and paleomagnetic stratigraphy of equatorial Pacific core V28–238: oxygen isotope temperatures and ice volumes on a 10^5 and 10^6 year scale. *Quaternary Research*, 3: 39–55.

Shaler, N.S., 1896. Beaches and marshes of the Atlantic coast. *National Geographic Monthly*, 1: 157–159.

Shaw, J. and Ceman, J., 1999. Salt-marsh aggradation in response to late-Holocene sea-level rise at Amherst Point, Nova Scotia, Canada. *Geology*, 9: 439–451.

Shaw, J. and Forbes, D.L., 1992. Barriers, barrier platforms and spillover deposits in St. George's Bay, Newfoundland: paraglacial sedimentation on the flanks of a deep coastal basin. *Marine Geology*, 105: 119–140.

Sheail, J., 2000. Dredging the Tyne: an institutional perspective on process management. *Science of the Total Environment*, 251: 139–151.

Shennan, I., 1986a. Flandrian sea-level changes in the Fenland. I: The geographical setting and evidence of relative sea-level changes. *Journal of Quaternary Science*, 1: 119–154.

Shennan, I., 1986b. Flandrian sea-level changes in the Fenland. II: Tendencies of sea-level movement, altitudinal changes, and local and regional factors. *Journal of Quaternary Science*, 1: 155–179.

Shennan, I., 1989. Holocene crustal movements and sea-level changes in Great Britain. *Journal of Quaternary Science*, 4: 77–89.

Shennan, I. and Woodworth, P.L., 1992. A comparison of late Holocene and twentieth-century sea-level trends from the UK and North Sea region. *Geophysical Journal International*, 109: 96–105.

Shennan, I., Innes, J.B., Long, A.J. and Zong, Y., 1994. Late Devensian and Holocene relative sea-level changes at Loch nan Eala, near Arisaig, northwest Scotland. *Journal of Quaternary Science*, 9: 261–283.

Shennan, I., Tooley, M., Green, F., Innes, J., Kennington, K., Lloyd, J. and Rutherford, M., 1999. Sea level, climate change and coastal evolution in Morar, northwest Scotland. *Geologie en Mijnbouw*, 77: 247–262.

Shennan, I., Horton, B., Innes, J., Gehrels, R., Lloyd, J., McArthur, J. and Rutherford, M., 2000a. Late Quaternary sea-level changes, crustal movements and coastal evolution in Northumberland, UK. *Journal of Quaternary Science*, 15: 215–237.

Shennan, I., Lambeck, K., Horton, B., Innes, J., Lloyd, J., McArthur, J., Purcell, T. and Rutherford, M., 2000b. Late Devensian and Holocene records of relative sea-level changes in northwest Scotland and their implications for glacio-hydro-isostatic modelling. *Quaternary Science Reviews*, 19: 1103–1135.

Shennan, I., Lambeck, K., Horton, B., Innes, J., Lloyd, J., McArthur, J. and Rutherford, M., 2000c. Holocene isostasy and relative sea-level changes on the east coast of England. In: I. Shennan and J. Andrews (Editors), *Holocene Land–Ocean Interaction and Environmental Change around the North Sea*. Geological Society, Special Publication, London, pp. 275–298.

Shepard, F.P., 1948. *Submarine Geology*. Harper, New York, 348 pp.

Shepard, F.P., 1963a. *Submarine Geology*. Second edition. Harper & Row, New York, 557 pp.

Shepard, F.P., 1963b. Thirty-five thousand years of sea level. In: T. Clements (Editor), *Essays in Marine Geology in Honor of K.O. Emery*. University of South California Press, Los Angeles, pp. 1–10.

Shepard, F.P., 1976. Coastal classification and changing coastlines. *Geoscience and Man*, 14: 53–64.

Shepard, F.P. and Kuhn, G.G., 1983. History of sea arches and remnant stacks of La Jolla, California. *Marine Geology*, 51: 139–161.

Shepard, F.P. and Lafond, E.C., 1940. Sand movements near the beach in relation to tides and waves. *American Journal of Science*, 238: 272–285.

Shepard, F.P. and Wanless, H.R., 1971. *Our Changing Coastlines*. McGraw-Hill, New York, 579 pp.

Shepard, F.P., MacDonald, G.A. and Cox, D.C., 1950. The tsunami of April 1, 1946. *Bulletin of the Scripps Institution of Oceanography of the University of California*, 5: 391–528.

Shepard, F.P., Curray, J.R., Newman, W.A., Bloom, A.L., Newell, N.D., Tracey, J.I. and Veeh, H.H., 1967. Holocene changes in sea level: evidence in Micronesia. *Science*, 157: 542–544.

Shepherd, M.J., 1987. Sandy beach ridge system profiles as indicators of changing coastal processes. *New Zealand Geographical Society Conference Series*, 14: 106–112.

Shepherd, M.J., 1990. Relict and contemporary foredunes as indicators of coastal processes. In: G. Brierley and J. Chappell (Editors), *Applied Quaternary Studies*, Australian National University, Canberra, pp. 17–24.

Sherman, C.E., Fletcher, C.H. and Rubin, K.H., 1999. Marine and meteoric diagenesis of Pleistocene carbonates from a nearshore submarine terrace, Oahu, Hawaii. *Journal of Sedimentary Research*, 69: 1083–1097.

Sherman, D.J., 1988. Empirical evaluation of longshore-current models. *Geographical Review*, 78: 158–168.

Sherman, D.J., 1995. Problems of scale in the modeling and interpretation of coastal dunes. *Marine Geology*, 124: 339–349.

Sherman, D.J. and Bauer, B.O., 1993. Dynamics of beach–dune systems. *Progress in Physical Geography*, 17: 413–447.

Sherman, D.J. and Hotta, S., 1990. Aeolian sediment transport, theory and measurement. In: K.F. Nordstrom, N.P. Psuty and R.W.G. Carter (Editors), *Coastal Dunes, Form and Process*. John Wiley & Sons, Chichester, pp. 17–37.

Sherman, D.J. and Lyons, W., 1994. Beach-state controls on aeolian sand delivery to coastal dunes. *Physical Geography*, 15: 381–395.

Shi, Z., Hamilton, L.J. and Wolanski, E., 2000. Near-bed currents and suspended sediment transport in saltmarsh canopies. *Journal of Coastal Research*, 16: 909–914.

Shields, A., 1936. *Anwendung der Ähnlichkeits-mechanik und der Turbulenzforschung auf die Geschiebebewlgung*. Preuss Versuchsastalt für Wasserbau und Schiffbau, Heft 26, Berlin.

Shigemura, T., 1980. Tidal prism-throat area relationships of the bays of Japan. *Shore and Beach*, 48: 30–35.

Shinn, E.A., 1963. Spur and groove formation on the Florida reef tract. *Journal of Sedimentary Petrology*, 33: 291–303.

Shinn, E.A., Hudson, J.H., Robbin, D.M. and Lidz, B., 1981. Spurs and grooves revisited: construction versus erosion Looe Key Reef, Florida, *Proceedings of the Fourth International Coral Reef Symposium*, University of the Phillipines, Manila, pp. 475–483.

Shinn, E.A., Hudson, J.H., Halley, R.B., Lidz, B., Robbin, D.M. and Macintyre, I.G., 1982. Geology and sediment accumulation rates at Carrie Bow Cay, Belize. In: K. Rützler and I.G. Macintyre (Editors), *The Atlantic Barrier Reef Ecosystem at Carrie Bow Cay, Belize I: Structure and Communities*. Smithsonian Contributions to Marine Science, Washington, DC, pp. 63–75.

Shinn, E.A., Steinen, R.P., Dill, R.F. and Major, R., 1993. Lime-mud layers in high-energy tidal channels: a record of hurricane deposition. *Geology*, 21: 603–606.

Short, A.D., 1979. Three dimensional beach-stage model. *Journal of Geology*, 87: 553–571.

Short, A.D., 1988. Holocene coastal dune formation in southern Australia. *Sedimentary Geology*, 55: 121–142.

Short, A.D., 1989. Chenier research on the Australian coast. *Marine Geology*, 90: 345–351.

Short, A.D., 1993. *Beaches of the New South Wales Coast*. Australian Beach Safety and Management Program, Sydney, 359 pp.

Short, A.D. (Editor), 1999a. *Handbook of Beach and Shoreface Morphodynamics*. John Wiley & Sons, Chichester, 379 pp.

Short, A.D., 1999b. Wave-dominated beaches. In: A.D. Short (Editor), *Handbook of Beach and Shoreface Morphodynamics*. John Wiley & Sons, Chichester, pp. 173–203.

Short, A.D., 2001. Beaches. In: J.H. Steele, S.A. Thorpe and K.K. Turekian (Editors), *Encylopedia of Ocean Sciences*. Academic Press, San Diego, pp. 245–255.

Short, A.D. and Aagaard, T., 1993. Single and multi-bar beach change models. *Journal of Coastal Research*, Special Issue 15: 141–157.

Short, A.D. and Hesp, P.A., 1982. Wave, beach and dune interactions in south-eastern Australia. *Marine Geology*, 48: 259–284.

Short, A.D. and Masselink, G., 1999. Embayed and structurally controlled beaches. In: A.D. Short (Editor), *Handbook of Beach and Shoreface Morphodynamics*. John Wiley & Sons, Chichester, pp. 230–250.

Short, A.D. and Wright, L.D., 1984. Morphodynamics of high energy beaches: an Australian perspective. In: B.G. Thom (Editor), *Coastal Geomorphology in Australia*. Academic Press, Sydney, pp. 43–68.

Short, A.D., Cowell, P.J., Cadee, M., Hall, W. and van Dijk, B., 1995. Beach rotation and possible relation to the Southern Oscillation. In: T.H. Aung (Editor), *Ocean Atmosphere Pacific Conference*. National Tidal Facility, Adelaide, pp. 329–334.

Shulmeister, J. and Kirk, R.M., 1993. Evolution of a mixed sand and gravel barrier system in north Canterbury, New Zealand, during Holocene sea-level rise and still-stand. *Sedimentary Geology*, 87: 215–235.

Silvester, R., 1960. Stabilization of sedimentary coastlines. *Nature*, 188: 467–469.

Silvester, R., 1974. *Coastal Engineering*. Elsevier, Amsterdam, 338 pp.

Simpson, R. and Riehl, H., 1981. *The Hurricane and its Impact*. Louisiana State University Press, Baton Rouge, 398 pp.

Sissons, J.B., 1974. The Quaternary in Scotland: a review. *Scottish Journal of Geology*, 10: 311–337.

Sissons, J.B., Cullingford, R.A. and Smith, D.E., 1966. Late-glacial and post-glacial shorelines in south-east Scotland. *Transactions of the Institute of British Geographers*, 39: 9–18.

Sklar, F.H., Costanza, R., Day, J.W. and Conner, W.H., 1983. Dynamic simulation of aquatic material flows in an impounded swamp habitat in the Barataria Basin, LA. In: W.K. Lauenroth, G.V. Skogerboe and M. Flug (Editors), *Analysis of Ecological Systems: State-of-the-Art in Ecological Modelling*. Developments in Environmental Modelling 5. Elsevier Science Publishing, New York, pp. 741–749.

Slaymaker, O. and Spencer, T., 1998. *Physical Geography and Global Environmental Change*. Longman, Singapore, 292 pp.

Sloss, L.L., 1991. The tectonic factor in sea level change: a countervailing view. *Journal of Geophysical Research*, 96: 6609–6617.

Smith, A.W.S., 2001. Discussion of: Munoz-Perez, J.J.; Tejedor, L., and Medina, R., 1999. Equilibrium beach profile model for reef protected beaches. *Journal of Coastal Research* 15(4), 950–957. *Journal of Coastal Research*, 17: 241–242.

Smith, D.E., Cullingford, R.A. and Firth, C.R., 2000. Patterns of isostatic land uplift during the Holocene: evidence from mainland Scotland. *The Holocene*, 10: 489–501.

Smith, D.E. and Dawson, A.G. (Editors), 1983. *Shorelines and Isostasy*. Academic Press, London, 387 pp.

Smith, D.E., Cullingford, R.A. and Haggart, B.A., 1985. A major coastal flood during the Holocene in eastern Scotland. *Eiszeitalter und Gegenwart*, 35: 109–118.

Smith, G.M., Spencer, T. and Möller, I., 2000. Visualization of coastal dynamics: Scolt Head island, north Norfolk, England. *Estuarine, Coastal and Shelf Science*, 50: 137–142.

Smith, S.V. and Kinsey, D.W., 1976. Calcium carbonate production, coral reef growth, and sea level change. *Science*, 194: 937–939.

Smithers, S.G. and Woodroffe, C.D., 2000. Microatolls as sea-level indicators on a mid-ocean atoll. *Marine Geology*, 168: 61–78.

Smithers, S.G. and Woodroffe, C.D., 2001. Coral microatolls and 20th century sea level in the eastern Indian Ocean. *Earth and Planetary Science Letters*, 191: 173–184.

Smithers, S.G., Woodroffe, C.D., McLean, R.F. and Wallensky, E., 1993. Lagoonal sedimentation in the Cocos (Keeling) Islands, Indian Ocean, *Proceedings of the Seventh International Coral Reef Symposium*. University of Guam Press, Mangilao, pp. 273–288.

Snedaker, S.C. and Snedaker, J.G. (Editors), 1984. *The Mangrove Ecosystem: Research Methods*. UNESCO, Paris, 251 pp.

So, C.L., 1965. Coastal platforms of the Isle of Thanet, Kent. *Transactions of the Institute of British Geographers*, 37: 147–156.

Solomon, S.M. and Forbes, D.L., 1999. Coastal hazards, and associated management issues on South Pacific islands. *Ocean and Coastal Management*, 42: 523–554.

Somoza, L. and Rey, J., 1991. Holocene fan deltas in a 'ria' morphology: prograding clinoform types and sea-level control. *Cuadernos de Geologia Iberica*, 15: 37–48.

Sonu, C.J. and James, W.R., 1973. A Markov model for beach profile changes. *Journal of Geophysical Research*, 78: 1462–1471.

Sonu, C.J. and van Beek, J.L., 1971. Systematic beach changes on the Outer Banks, North Carolina. *Journal of Geology*, 79: 416–425.

Soucie, G., 1973. Where beaches have been going: into the ocean. *Smithsonian*, 4: 55–61.

Soulsby, R., 2001. *Final Volume of Summary Papers, COAST3D*, Report TR121. HR Wallingford.

Spackman, W., Dolsen, C.P. and Riegel, W., 1966. Phytogenic organic sediments and sedimentary environments in the Everglades–mangrove complex. *Paleontographica Abt. B*, 117: 135–152.

Spalding, M.D., 2001. Mangroves. In: J.H. Steele, S.A. Thorpe and K.K. Turekian (Editors), *Encyclopedia of Ocean Sciences*. Academic Press, San Diego, pp. 1533–1542.

Spenceley, A.P., 1976. Unvegetated saline tidal flats in North Queensland. *Journal of Tropical Geography*, 42: 78–85.

Spenceley, A.P., 1977. The role of pneumatophores in sedimentary processes. *Marine Geology*, 24: M31–M37.

Spenceley, A.P., 1982. Sedimentation patterns in a mangal on Magnetic Island near Townsville, north Queensland, Australia. *Singapore Journal of Tropical Geography*, 3: 100–107.

Spencer, C.D., Plater, A.J. and Long, A.J., 1998. Rapid coastal change during the mid to late Holocene: the record of barrier estuary sedimentation in the Romney Marsh region, southeast England. *The Holocene*, 8: 143–163.

Spencer, T., 1985. Weathering rates on a Caribbean reef limestone: results and implications. *Marine Geology*, 69: 195–201.

Spencer, T., 1988a. Coastal biogeomorphology. In: H.A. Viles (Editor), *Biogeomorphology*. Blackwells, Oxford, pp. 255–318.

Spencer, T., 1988b. Limestone coastal morphology: the biological contribution. *Progress in Physical Geography*, 12: 66–101.

Spencer, T., 1992. Bioerosion and biogeomorphology. In: D.M. John, S.J. Hawkins and J.H. Price (Editors), *Plant–Animal Interactions in the Marine Benthos*. Systematics Association Special Volume. Clarendon Press, Oxford, pp. 493–509.

Spencer, T., Stoddart, D.R. and Woodroffe, C.D., 1987. Island uplift and lithospheric flexure: observations and cautions from the South Pacific. *Zeitschrift für Geomorphologie, NF, Suppl.-Bd.*, 63: 87–102.

Spratt, T., 1856. *An Investigation of Movements of Teignmouth Bar*. John Weale, London.

Sprigg, R.C., 1979. Stranded and submerged sea-beach systems of southeast South Australia and the aeolian desert cycle. *Sedimentary Geology*, 22: 53–96.

Stamp, L.D., 1940. The Irrawaddy River. *Geographical Journal*, 95: 329–356.

Stanley, D.J. and Hait, A.K., 2000. Holocene depositional patterns, neotectonics and Sundarban mangroves in the western Ganges–Brahmaputra Delta. *Journal of Coastal Research*, 16: 26–39.

Stanley, D.J. and Warne, A.G., 1993a. Sea level and initiation of Predynastic culture in the Nile Delta. *Nature*, 363: 435–438.

Stanley, D.J. and Warne, A.G., 1993b. Nile Delta: recent geological evolution and human impact. *Science*, 260: 628–634.

Stanley, D.J. and Warne, A.G., 1994. Worldwide initiation of Holocene marine deltas by deceleration of sea-level rise. *Science*, 265: 228–231.

Stanley, D.J. and Warne, A.G., 1998. Nile Delta in its destructional phase. *Journal of Coastal Research*, 14: 794–825.

Stapor, F.W., 1971. Sediment budgets on a compartmented low-to-moderate energy coast in northwest Florida. *Marine Geology*, 10: M1–M7.

Stapor, F.W., 1975. Holocene beach ridge plain development, northwest Florida. *Zeitschrift für Geomorphologie, NF, Suppl.-Bd.*, 22: 116–144.

Stapor, F.W., Mathews, T.D. and Lindfors-Kearns, F.E., 1991. Barrier-island progradation and Holocene sea-level history in southwest Florida. *Journal of Coastal Research*, 7: 815–838.

Staub, J.R. and Esterle, J.S., 1993. Provenance and sediment dispersal in the Rajang River delta/coastal plain system, Sarawak, East Malaysia. *Sedimentary Geology*, 85: 191–201.

Staub, J.R. and Esterle, J.S., 1994. Peat-accumulating depositional systems of Sarawak, east Malaysia. *Sedimentary Geology*, 89: 91–106.

Stearn, C.W., Scoffin, T.P. and Martindale, W., 1977. Calcium carbonate budget of a fringing reef on the west coast of Barbados. Part I. Zonation and productivity. *Bulletin of Marine Science*, 27: 479–510.

Stearns, H.T., 1935. Shore benches on the island of Oahu, Hawaii. *Geological Society of America Bulletin*, 46: 1467–1482.

Stearns, H.T., 1970. Ages of dunes on Oahu, Hawaii. *Occasional Papers of the Bernice P. Bishop Museum, Honolulu, Hawaii*, 24: 49–72.

Steers, J.A., 1953. *The Sea Coast*. Collins, London, 288 pp.

Steers, J.A. (Editor), 1960. *Scolt Head Island*. W. Heffer & Sons Ltd, Cambridge, 269 pp.

Steers, J.A., 1969. *The Coastline of England and Wales*. Cambridge University Press, Cambridge, 762 pp.

Steers, J.A., 1973. *The Coastline of Scotland*. Cambridge University Press, Cambridge, 335 pp.

Stephenson, W.J., 2000. Shore platforms: a neglected coastal feature? *Progress in Physical Geography*, 24: 311–327.

Stephenson, W.J. and Kirk, R.M., 1996. Measuring erosion rates using the micro-erosion meter: 20 years of data from shore platforms, Kaikoura Peninsula, South Island, New Zealand. *Marine Geology*, 131: 209–218.

Stephenson, W.J. and Kirk, R.M., 1998. Rates and patterns of erosion on intertidal shore platforms, Kaikoura Peninsula, South Island, New Zealand. *Earth Surface Processes and Landforms*, 23: 1071–1085.

Stephenson, W.J. and Kirk, R.M., 2000a. Development of shore platforms on

Kaikoura Peninsula, South Island, New Zealand: I: the role of waves. *Geomorphology*, 32: 21–41.

Stephenson, W.J. and Kirk, R.M., 2000b. Development of shore platforms on Kaikoura Peninsula, South Island, New Zealand: II: the role of subaerial weathering. *Geomorphology*, 32: 43–56.

Stevenson, J.C. and Kearney, M.S., 1996. Shoreline dynamics on the windward and leeward shores of a large temperate estuary. In: K.F. Nordstrom and C.T. Roman (Editors), *Estuarine Shores: Evolution, Environments and Human Alterations*. John Wiley & Sons, Chichester, pp. 233–259.

Stevenson, J.C., Ward, L.G. and Kearney, M.S., 1986. Vertical accretion in marshes with varying rates of sea level rise. In: D. Wolfe (Editor), *Estuarine Variability*. Academic Press, New York, pp. 241–260.

Stirling, C.H., Esat, T.M., Lambeck, K. and McCulloch, M.T., 1998. Timing and duration of the Last Interglacial: evidence for a restricted interval of widespread coral reef growth. *Earth and Planetary Science Letters*, 160: 745–762.

Stirling, C.H., Esat, T.M., McCulloch, M.T. and Lambeck, K., 1995. High-precision U-series dating of corals from Western Australia and implications for the timing and duration of the Last Interglacial. *Earth and Planetary Science Letters*, 135: 115–130.

Stive, M.J.F. and de Vriend, H.J., 1995. Modelling shoreface profile evolution. *Marine Geology*, 126: 235–248.

Stive, M.J.F. and Wind, H.G., 1982. A study of radiation stress and set-up in the nearshore region. *Coastal Engineering*, 6: 1–25.

Stoddart, D.R., 1962. Catastrophic storm effects on the British Honduras reefs and cays. *Nature*, 196: 512–515.

Stoddart, D.R., 1965. The shape of atolls. *Marine Geology*, 3: 369–383.

Stoddart, D.R., 1966. Darwin's impact on geography. *Annals of the American Association of Geographers*, 56: 683–698.

Stoddart, D.R., 1969a. Ecology and morphology of recent coral reefs. *Biological Reviews*, 44: 433–498.

Stoddart, D.R., 1969b. Geomorphology of the Marovo elevated barrier reef, New Georgia. *Philosophical Transactions of the Royal Society of London*, 255: 383–402.

Stoddart, D.R., 1971. Coral reefs and islands and catastrophic storms. In: J.A. Steers (Editor), *Applied Coastal Geomorphology*. Macmillan, London, pp. 155–197.

Stoddart, D.R., 1972. Catastrophic damage to coral reef communities by earthquake. *Nature*, 239: 51–52.

Stoddart, D.R., 1973a. Coral reefs: the last two million years. *Geography*, 58: 313–323.

Stoddart, D.R., 1973b. Coral reefs of the Indian Ocean. In: O.A. Jones and R. Endean (Editors), *Biology and Geology of Coral Reefs vol. I, Geology I*. Academic Press, London, pp. 51–92.

Stoddart, D.R., 1975. Almost-atoll of Aitutaki: geomorphology of reefs and islands. *Atoll Research Bulletin*, 190: 31–57.

Stoddart, D.R., 1980. Mangroves as successional stages, inner reefs of the northern Great Barrier Reef. *Journal of Biogeography*, 7: 269–284.

Stoddart, D.R., 1985. Hurricane effects on coral reefs: conclusions. *Proceedings of the Fifth International Coral Reef Congress*, vol. 3: Antenne Museum-Ephe, Moorea, French Polynesia, pp. 349–350.

Stoddart, D.R. and Steers, J.A., 1977. The nature and origin of coral reef islands. In: O.A. Jones and R. Endean (Editors), *Biology and Geology of Coral Reefs vol. IV, Geology 2*. Academic Press, London, pp. 59–105.

Stoddart, D.R., McLean, R.F. and Hopley, D., 1978. Geomorphology of reef islands, northern Great Barrier Reef. *Philosophical Transactions of the Royal Society London, Series B*, 284: 39–61.

Stoddart, D.R., Reed, D.J. and French, J.R., 1989. Understanding salt-marsh accretion, Scolt Head Island, Norfolk, England. *Estuaries*, 12: 228–236.

Stoddart, D.R., Woodroffe, C.D. and Spencer, T., 1990. Mauke, Mitiaro and Atiu: geomorphology of Makatea islands in the Southern Cooks. *Atoll Research Bulletin*, 341: 1–61.

Stokes, G.G., 1847. On the theory of oscillatory waves. *Transactions of the Cambridge Philosophical Society*, 8: 441–473.

Stone, G.W. and McBride, R.A., 1998. Louisiana barrier islands and their importance in wetland protection: forecasting shoreline change and subsequent response of wave climate. *Journal of Coastal Research*, 14: 900–915.

Stone, J.O., Ballantyne, C.K. and Fifield, L.K., 1998. Exposure dating and validation of periglacial weathering limits, northwest Scotland. *Geology*, 26: 587–590.

Storlazzi, C.D. and Griggs, G.B., 2000. Influence of El Niño–Southern Oscillation (ENSO) events on the evolution of central California's shoreline. *Geological Society of America Bulletin*, 112: 236–249.

Strahler, A.N., 1952a. Dynamic basis of geomorphology. *Bulletin of the Geological Society of America*, 63: 923–938.

Strahler, A.N., 1952b. Hypsometric (area-altitude) analysis of erosional topography. *Bulletin of the Geological Society of America*, 63: 1117–1142.

Strahler, A.N., 1966. Tidal cycle of changes in an equilibrium beach, Sandy Hook, New Jersey. *Journal of Geology*, 74: 247–268.

Streif, H., 1972. The results of stratigraphical and facial investigations in the coastal Holocene of Woltzeten/Ostfriesland, Germany. *Geologiska Föreningens i Stockholm Förhandlingar*, 94: 280–299.

Streif, H., 1989. Barrier islands, tidal flats, and coastal marshes, resulting from a relative rise of sea level in East Frisia on the German North Sea. In: W.J.M. Van der Linden, S.A.P.L. Cloentingh, J.P.K. Kaasschieter and J.A.M. Van de Graaf (Editors), *Proceedings of the Symposium on Coastal Lowlands*. Kluwer, Royal Geological Mining Society, Netherlands, Dordrecht, pp. 213–223.

Strickland, C., 1940. *Deltaic formation with Special Reference to the Hydrographic Processes of the Ganges and the Brahmaputra*. Longmans, Green and Co Ltd, Calcutta, 157 pp.

Stringer, W.J., Groves, J.E. and Olmsted, C., 1988. Landsat determined geographic change. *Photogrammetric Engineering and Remote Sensing*, 54: 347–351.

Stuiver, M., Reimer, P.J., Bard, E., Beck, J.W., Burr, G.S., Hughen, K.A.,

Kromer, B., McCormac, G., van der Plicht, J. and Spurk, M., 1998. Intercal 98 radiocarbon age calibration, 24000–0 cal BP. *Radiocarbon*, 40: 1041–1083.

Stumpf, R.P., 1983. The process of sedimentation on the surface of a salt marsh. *Estuarine, Coastal and Shelf Science*, 17: 495–508.

Suess, E., 1888. *Das Antlitz der Erde*, 3 vols. F. Tempsky, Vienna.

Suk, N.S., Guo, Q. and Psuty, N.P., 1999. Suspended solids flux between salt marsh and adjacent bay: a long-term continuous measurement. *Estuarine, Coastal and Shelf Science*, 49: 61–81.

Sunamura, T., 1983. Processes of sea cliff and platform retreat. In: P.D. Komar (Editor), *CRC Handbook of Coastal Processes and Erosion*. CRC Press, Boca Raton, pp. 233–265.

Sunamura, T., 1984. Quantitative prediction of beach-face slopes. *Geological Society of America Bulletin*, 95: 242–245.

Sunamura, T., 1989. Sandy beach geomorphology elucidated by laboratory modelling. In: V.C. Lakhan and A.S. Trenhaile (Editors), *Applications in Coastal Modeling*. Elsevier, Amsterdam, pp. 159–213.

Sunamura, T., 1991. The elevation of shore platforms: a laboratory approach to the unsolved problem. *Journal of Geology*, 99: 761–766.

Sunamura, T., 1992. *Geomorphology of Rocky Coasts*. John Wiley & Sons, Chichester, 302 pp.

Sunamura, T., 1996. A physical model for the rate of coastal tafoni development. *Journal of Geology*, 104: 741–748.

Sunamura, T. and Horikawa, K., 1977. Sediment budget in Kujykuri coastal area, Japan. *Coastal Sediments '77*, American Society of Civil Engineers, New York, pp. 475–487.

Sunamura, T. and Mizuno, O., 1987. A study on depositional shoreline forms behind an island. *Annual Report, Institute of Geoscience, University of Tsukuba*, 13: 71–73.

Suter, J.R., 1994. Deltaic coasts. In: R.W.G. Carter and C.D. Woodroffe (Editors), *Coastal Evolution: Late Quaternary Shoreline Morphodynamics*. Cambridge University Press, Cambridge, pp. 87–120.

Sutherland, J., 2001. Synthesis of Teignmouth coastal area modelling. In: R. Soulsby (Editor), *Final Volume of Summary Papers, COAST3D*, Report TR121. HR Wallingford, pp. D5.1–D5.5.

Sverdrup, H.U. and Munk, W.H., 1947. *Wind, Sea, and Swell Theory of Relationships Forecasting*. 601, US Department of the Navy Hydrography Office, Washington, DC, 44 pp.

Swain, A., 1989. Beach profile development. In: V.C. Lakhan and A.S. Trenhaile (Editors), *Applications in Coastal Modeling*. Elsevier, Amsterdam, pp. 215–232.

Swan, B., 1979. Sand dunes in the humid tropics: Sri Lanka. *Zeitschrift für Geomorphologie, NF*, 23: 152–171.

Swift, D.J.P., 1970. Quaternary shelves and the return to grade. *Marine Geology*, 8: 5–30.

Swift, D.J.P., 1975. Barrier island genesis: evidence from the central Atlantic Shelf, eastern USA. *Sedimentary Geology*, 14: 1–43.

Swift, D.J.P. and Thorne, J.A., 1991. Sedimentation on continental margins. I. A general model for shelf sedimentation. In: D.J.P. Swift, G.F. Oertel, R.W. Tillman and J.A. Thorne (Editors), *Shelf Sand and Sandstone Bodies: Geometry, Facies and Sequence Stratigraphy*. Special Publication of the International Association of Sedimentologists. Blackwell Scientific Publications, Oxford, pp. 3–31.

Symonds, G., Black, K.P. and Young, I.R., 1995. Wave-driven flow over shallow reefs. *Journal of Geophysical Research*, 100: 2639–2648.

Syvitski, J.P.M. and Shaw, J., 1995. Sedimentology and geomorphology of fjords. In: G.M.E. Perillo (Editor), *Geomorphology and Sedimentology of Estuaries*. Elsevier, Amsterdam, pp. 113–178.

Syvitski, J.P.M., Burrell, D.C. and Skei, J.M., 1987. *Fjords: Processes and Products*. Springer-Verlag, New York, 379 pp.

Szabo, B.J., 1979. Uranium-series age of coral reef growth on Rottnest Island, Western Australia. *Marine Geology*, 29: M11–M15.

Szabo, B.J., Ward, W.C., Weidie, A.E. and Brady, M.J., 1978. Age and magnitude of the late Pleistocene sea-level rise on the eastern Yucatan Peninsula. *Geology*, 6: 713–715.

Szabo, B.J., Tracey, J.I. and Goter, E.R., 1985. Ages of subsurface stratigraphic intervals in the Quaternary of Eniwetak Atoll, Marshall Islands. *Quaternary Research*, 23: 54–61.

Tait, R.J., 1972. Wave set-up on coral reefs. *Journal of Geophysical Research*, 77: 2207–2211.

Talandier, J. and Bourrouilh-Le Jan, F., 1988. High energy sedimentation in French Polynesia: cyclone or tsunami. In: M.I. El-Sabh and T.S. Murty (Editors), *Natural and Man-Made Hazards*. D. Reidel, Dordrecht and Boston, pp. 193–199.

Tanner, W.F., 1958. The equilibrium beach. *American Geophysical Union Transactions*, 39: 889–891.

Tanner, W.F., 1960. Bases for coastal classification. *Southeastern Geology*, 2: 13–22.

Tanner, W.F., 1990. Origin of barrier islands on sandy coasts. *Transactions of the Gulf Coast Association of Geological Societies*, 40: 819–823.

Tanner, W.F., 1993. 8000-year record of sea-level change from grain size parameters: data from beach ridges in Denmark. *The Holocene*, 3: 220–231.

Tanner, W.F., 1995. Origin of beach ridges and swales. *Marine Geology*, 129: 149–161.

Tanner, W.F. and Stapor, F.W., 1971. Tabasco beach-ridge plain: an eroding coast. *Transactions of the Gulf Coast Association of Geological Societies*, 21: 231–232.

Taylor, M. and Stone, G.W., 1996. Beach-ridges: a review. *Journal of Coastal Research*, 12: 612–621.

Teal, J.M., 1962. Energy flow in the salt marsh ecosystem of Georgia. *Ecology*, 43: 614–624.

Teal, J.M., 2001. Salt marshes and mudflats. In: J.H. Steele, S.A. Thorpe and K.K. Turekian (Editors), *Encyclopedia of Ocean Sciences*. Academic Press, San Diego, pp. 2490–2495.

Tejan-Kella, M.S., Chittleborough, D.J., Fitzpatrick, R.W., Thompson, C.H., Prescott, J.R. and Hutton, J.T., 1990. Thermoluminescence dating of coastal sand dunes at Cooloola and North Stradbroke Island, Australia. *Australian Journal of Soil Research*, 28: 465–481.

Terwindt, J.H.J., Augustinus, P.G.E.F., Boersma, J.R. and Hoekstra, P., 1987. Mud discharge, dispersion and deposition in a monsoon-dominated coastal environment. In: N.C. Kraus (Editor), *Coastal Sediments '87: Proceedings of a Specialty Conference on Advances in Understanding of Coastal Sediment Processes*. American Society of Civil Engineers, New Orleans, pp. 1976–1988.

Terzaghi, K., 1962. Stability of steep slopes on hard unweathered rocks. *Geotechnique*, 12: 251–270.

Thieler, E.R. and Danforth, W.W., 1994. Historical shoreline mapping. 2. Application of the digital shoreline mapping and analysis systems (DSMS DSAS) to shoreline change mapping in Puerto Rico. *Journal of Coastal Research*, 10: 600–620.

Thieler, E.R., Pilkey, O.H., Young, R.S., Bush, D.M. and Chai, F., 2000. The use of mathematical models to predict beach behavior for US coastal engineering: a critical review. *Journal of Coastal Research*, 16: 48–70.

Thom, B.G., 1967. Mangrove ecology and deltaic geomorphology: Tabasco, Mexico. *Journal of Ecology*, 55: 301–343.

Thom, B.G., 1973. The dilemma of high interstadial sea levels during the last glaciation. *Progress in Geography*, 5: 167–246.

Thom, B.G., 1978. Coastal sand deposition in southeast Australia during the Holocene. In: J.L. Davies and M.A.F. Williams (Editors), *Landform Evolution in Australasia*. Australian National University Press, Canberra, pp. 197–214.

Thom, B.G., 1982. Mangrove ecology: a geomorphological perspective. In: B.F. Clough (Editor), *Mangrove Ecosystems in Australia, Structure, Function and Management*. Australian National University Press, Canberra, pp. 3–17.

Thom, B.G., 1984a. Sand barriers of eastern Australia: Gippsland – a case study. In: B.G. Thom (Editor), *Coastal Geomorphology in Australia*. Academic Press, Sydney, pp. 233–261.

Thom, B.G., 1984b. Transgressive and regressive stratigraphies of coastal sand barriers in eastern Australia. *Marine Geology*, 56: 137–158.

Thom, B.G. and Chappell, J., 1975. Holocene sea levels relative to Australia. *Search*, 6: 90–93.

Thom, B.G. and Hall, W., 1991. Behaviour of beach profiles during accretion and erosion dominated phases. *Earth Surface Processes and Landforms*, 16: 113–127.

Thom, B.G. and Harvey, N., 2000. Triggers for late twentieth century reform of Australian coastal management. *Australian Geographical Studies*, 38: 275–290.

Thom, B.G. and Roy, P.S., 1985. Relative sea levels and coastal sedimentation in southeast Australia in the Holocene. *Journal of Sedimentary Petrology*, 55: 257–264.

Thom, B.G. and Roy, P.S., 1988. Sea-level rise and climate: lessons from the Holocene. In: G.I. Pearman (Editor), *Greenhouse: Planning for Climatic Change*. CSIRO, Melbourne, pp. 177–188.

Thom, B.G. and Wright, L.D., 1983. Geomorphology of the Purari Delta. In: T. Petr (Editor), *The Purari: Tropical Environment of a High Rainfall River Basin*. Dr. W. Junk, The Hague, pp. 47–65.

Thom, B.G., Hails, J.R., Martin, A.R.H. and Phipps, C.V.G., 1972. Post-glacial sea levels in eastern Australia – a reply. *Marine Geology*, 12: 233–242.

Thom, B.G., Wright, L.D. and Coleman, J.M., 1975. Mangrove ecology and deltaic–estuarine geomorphology, Cambridge Gulf-Ord River, Western Australia. *Journal of Ecology*, 63: 203–222.

Thom, B.G., Orme, G.R. and Polach, H.A., 1978a. Drilling investigations of Bewick and Stapleton Islands. *Philosophical Transactions of the Royal Society, London, Series A*, 291: 37–54.

Thom, B.G., Polach, H.A. and Bowman, G.M., 1978b. *Holocene Age Structure of Coastal Sand Barriers in New South Wales, Australia*. Geography Department, Faculty of Military Studies, University of NSW, Duntroon, Canberra, 86 pp.

Thom, B.G., Bowman, G.M., Gillespie, R., Temple, R. and Barbetti, M., 1981. *Radiocarbon Dating of Holocene Beach-ridge Sequences in Southeast Australia*. Geography Department, Faculty of Military Studies, University of NSW, Duntroon, Canberra, 36 pp.

Thom, B.G., Shepherd, M., Ly, C.K., Roy, P.S., Bowman, G.M. and Hesp, P.A., 1992. *Coastal Geomorphology and Quaternary Geology of the Port Stephens–Myall Lakes Area*. Australian National University Press, Canberra, pp. 1–407.

Thom, B.G., Hesp, P. and Bryant, E., 1994. Last glacial 'coastal' dunes in eastern Australia and implications for landscape stability during the Last Glacial Maximum. *Palaeogeography, Palaeoclimatology, Palaeoecology*, 111: 229–248.

Thom, R.M., 1992. Accretion rates of low intertidal salt marshes in the Pacific northwest. *Wetlands*, 12: 147–156.

Thomas, M.L.H. and Stevens, J., 1991. Communities of constructional lips and cup reef rims in Bermuda. *Coral Reefs*, 9: 225–230.

Thomas, M.S. and Anderson, J.B., 1992. *Eustatic Controls on the Facies Architecture of the Trinity/Sabine Incised Valley System, Texas Continental Shelf*. Society of Economic Palaeontologists and Mineralogists, Tulsa, OK, Special Paper, 52.

Thomas, W.L., 1956. *Man's Role in Changing the Face of the Earth*. University of Chicago Press, Chicago, 1193 pp.

Thompson, C.H., 1981. Podzol chronosequences on coastal dunes of eastern Australia. *Nature*, 291: 59–61.

Thompson, R.W., 1968. Tidal flat sedimentation on the Colorado River delta, northwestern Gulf of California. *Memoir of the Geological Society of America*, 107: 1–133.

Thompson, T.A. and Baedke, S.J., 1995. Beach-ridge development in Lake Michigan: shoreline behavior in response to quasi-periodic lake-level events. *Marine Geology*, 129: 163–174.

Thorne, J.A. and Swift, D.J.P., 1991a. Sedimentation on continental margins, VI: a regime model for depositional sequences, their component system tracts, and bounding surfaces. *Special Publication of the International Association of Sedimentologists*, 14: 189–255.

Thorne, J.A. and Swift, D.J.P., 1991b. Sedimentation on continental margins. II. Application of the regime concept. *Special Publication of the International Association of Sedimentologists*, 14: pp. 33–58.

Thornton, E.B. and Guza, R.T., 1982. Energy saturation and phase speeds measured on natural beaches. *Journal of Geophysical Research*, 87: 9499–9508.

Thornton, E.B. and Guza, R.T., 1983. Transformation of wave height distribution. *Journal of Geophysical Research*, 88: 5925–5938.

Thornton, E.B. and Kim, C.S., 1993. Longshore current and wave height modulation at tidal frequency inside the surf zone. *Journal of Geophysical Research*, 98: 16509–16519.

Thorpe, S.A., 2001. Breaking waves and near-surface turbulence. In: J.H. Steele, S.A. Thorpe and K.K. Turekian (Editors), *Encyclopedia of Ocean Sciences*. Academic Press, San Diego, pp. 349–352.

Thurber, D.L., Broecker, W.S., Blanchard, R.L. and Potratz, H.A., 1965. Uranium-series ages of Pacific atoll coral. *Science*, 149: 55–58.

Titus, J.G. (Editor), 1990a. *Changing Climate and the Coast, Vol 1: Adaptive Responses and their Economic, Environmental, and Institutional Implications*. US Environmental Protection Agency, Miami, 396 pp.

Titus, J.G. (Editor), 1990b. *Changing Climate and the Coast, Vol. II: Western Africa, the Americas, the Mediterranean Basin and the Rest of Europe*. US Environmental Protection Agency, Miami, 508 pp.

Titus, J.G., 1998. Rising seas, coastal erosion, and the takings clause: how to save wetlands and beaches without hurting property owners. *Maryland Law Review*, 57: 1279–1399.

Tjia, H.D., 1996. Sea-level changes in the tectonically stable Malay–Thai peninsula. *Quaternary International*, 31: 95–101.

Todd, T.W., 1968. Dynamic diversion: influence of longshore current-tidal flow interaction on chenier and barrier island plains. *Journal Sedimentary Petrology*, 38: 734–746.

Toldo, E.E., Dillenburg, S.R., Correa, I.C.S. and Almeida, E.S.B., 2000. Holocene sedimentation in Lagoa dos Patos Lagoon, Rio Grande do Sul, Brazil. *Journal of Coastal Research*, 16: 816–822.

Tooley, M., 1992. Sea-level changes and the impacts on coastal lowlands. *Journal of the Society of Fellows*, 7: 36–45.

Toppe, A. and Fuhrboter, A., 1994. Recent anomalies in mean and high tidal water levels at the German North Sea coastline. *Journal of Coastal Research*, 10: 206–209.

Törnqvist, T.E., 1993. Holocene alternation of meandering and anastomosing fluvial systems in the Rhine–Meuse Delta (Central Netherlands) controlled by sea-level rise and subsoil erodibility. *Journal of Sedimentary Petrology*, 63: 683–693.

Törnqvist, T.E., 1994. Middle and late Holocene avulsion history of the River Rhine (Rhine–Meuse delta, Netherlands). *Geology*, 22: 711–714.

Törnqvist, T.E., Kidder, T.R., Autin, W.J., van der Borg, K., de Jong, A.F.M., Klerks, C.J.W., Snijders, E.M.A., Storms, J.E.A., van Dam, R.L. and Wiemann, M.C., 1996. A revised chronology for Mississippi River subdeltas. *Science*, 273: 1693–1696.

Tracey, J.I., 1972. Holocene emergent reefs in the central Pacific. *American Quaternary Association Second Conference*, AMQUA, University of Miami, Miami, pp. 51–52.

Trenberth, K.E., 2001. El Niño Southern Oscillation (ENSO). In: J.H. Steele, S.A. Thorpe and K.K. Turekian (Editors), *Encyclopedia of Ocean Sciences*. Academic Press, San Diego, pp. 815–827.

Trenhaile, A.S., 1974. The geometry of shore platforms in England and Wales. *Transactions of the Institute of British Geographers*, 46: 129–142.

Trenhaile, A.S., 1980. Shore platforms: a neglected coastal feature. *Progress in Physical Geography*, 4: 1–23.

Trenhaile, A.S., 1987. *The Geomorphology of Rock Coasts*. Clarendon Press, Oxford, 384 pp.

Trenhaile, A.S., 1989. Sea level oscillations and the development of rock coasts. In: V.C. Laklan and A.S. Trenhaile (Editors), *Applications in Coastal Modeling*. Elsevier, Amsterdam, pp. 271–295.

Trenhaile, A.S., 1997. *Coastal Dynamics and Landforms*. Clarendon Press, Oxford, 366 pp.

Trenhaile, A.S., 1999. The width of shore platforms in Britain, Canada, and Japan. *Journal of Coastal Research*, 15: 355–364.

Trenhaile, A.S., 2000. Modeling the development of wave-cut shore platforms. *Marine Geology*, 166: 163–178.

Trenhaile, A.S., 2001a. Modeling the effect of late Quaternary interglacial sea levels on wave-cut shore platforms. *Marine Geology*, 172: 205–223.

Trenhaile, A.S., 2001b. Modeling the effect of weathering on the evolution and morphology of shore platforms. *Journal of Coastal Research*, 17: 398–406.

Trenhaile, A.S., 2001c. Modelling the Quaternary evolution of shore platforms and erosional continental shelves. *Earth Surface Processes and Landforms*, 26: 1103–1128.

Trenhaile, A.S., Pepper, D.A., Trenhaile, R.W. and Dalimonte, M., 1998. Stacks and notches at Hopewell Rocks, New Brunswick, Canada. *Earth Surface Processes and Landforms*, 23: 975–988.

Trenhaile, A.S., Alberti, A.P., Cortizas, A.M., Casais, M.C. and Chao, R.B., 1999. Rock coast inheritance: an example from Galicia, northwestern Spain. *Earth Surface Processes and Landforms*, 24: 605–621.

Tricart, J., 1962. Observations de geomorphologie littorale a Mamba Point (Monrovica) Liberia. *Erdkunde*, 16: 49–57.

Tricart, J., 1972. *The Landforms of the Humid Tropics, Forests and Savannas*. Geographies for Advanced Study. Longman, London.

Trichet, J., Repellin, P. and Oustrière, P., 1984. Stratigraphy and subsidence of the Mururoa Atoll (French Polynesia). *Marine Geology*, 56: 241–257.

Trudgill, S.T., 1976. The marine erosion of limestone on Aldabra Atoll, Indian Ocean. *Zeitschrift für Geomorphologie, Suppl.-Bd.*, 26: 164–200.

Trudgill, S.T., 1987. Bioerosion of intertidal limestone, Co Clare, Eire – 3: zonation, process and form. *Marine Geology*, 74: 256–262.

Tsuji, Y., 1993. Tide influenced high energy environments and rhodolith-associated carbonate deposition on the outer shelf and slope off the Miyako Islands, southern Ryukyu Island arc, Japan. *Marine Geology*, 113: 255–271.

Tsujimoto, H., 1987. Dynamic conditions for shore platform initiation. *Science Reports, Institute of Geoscience, University of Tsukuba, Section A*, 8: 45–93.

Tudhope, A.W., 1989. Shallowing-upwards sedimentation in a coral reef lagoon, Great Barrier Reef of Australia. *Journal of Sedimentary Petrology*, 59: 1036–1051.

Tudhope, A.W., Chilcott, C.P., McCulloch, M.T., Cook, E.R., Chappell, J., Ellam, R.M., Lea, D.W., Lough, J.M. and Shimmield, G.B., 2001. Variability in the El Niño–Southern Oscillation through a glacial-interglacial cycle. *Science*, 291: 1511–1517.

Tudhope, A.W. and Risk, M.J., 1985. Rate of dissolution of carbonate sediments by microboring organisms, Davies Reef, Australia. *Journal of Sedimentary Petrology*, 55: 440–447.

Tudhope, A.W. and Scoffin, T.P., 1984. The effects of *Callianassa* bioturbation on the preservation of carbonate grains in Davies Reef Lagoon, Great Barrier Reef, Australia. *Journal of Sedimentary Petrology*, 54: 1091–1096.

Tudhope, A.W. and Scoffin, T.P., 1994. Growth and structure of fringing reefs in a muddy environment: south Thailand. *Journal of Sedimentary Research*, A64: 752–764.

Tushingham, A.M. and Peltier, W.R., 1991. Ice-3G: a new global model of late Pleistocene deglaciation based upon geophysical predictions of post-glacial relative sea level change. *Journal of Geophysical Research*, 96: 4497–4523.

Twidale, C.R. and Campbell, E.M., 1999. Development of a basin, doughnut and font assemblage on a sandstone coast, western Eyre Peninsula, South Australia. *Journal of Coastal Research*, 14: 1385–1394.

Twilley, R.R., 1985. The exchange of organic carbon in basin mangrove forests in a southwest Florida estuary. *Estuarine, Coastal and Shelf Science*, 20: 543–557.

Twilley, R.R., Chen, R.H. and Hargis, T., 1992. Carbon sinks in mangroves and their implications to carbon budget of tropical coastal ecosystems. *Water, Air and Soil Pollution*, 64: 265–288.

Umitsu, M., 1993. Late Quaternary sedimentary environments and landforms in the Ganges Delta. *Sedimentary Geology*, 83: 177–186.

Umitsu, M., 1997. Landforms and floods in the Ganges Delta and coastal lowlands of Bangladesh. *Marine Geodesy*, 20: 77–87.

Umitsu, M., Buman, M., Kawase, K. and Woodroffe, C.D., 2001. Holocene palaeoecology and formation of the Shoalhaven River deltaic–estuarine plains, southeast Australia. *The Holocene*, 11: 407–418.

Uncles, R.J. and Stephens, J.A., 1989. Distribution of suspended sediment at high water in a macrotidal estuary. *Journal of Geophysical Research*, 94: 14395–14405.

Vacher, H.L., Hearty, P.J. and Rowe, M.P., 1995. Stratigraphy of Bermuda: nomenclature, concepts, and status of multiple systems of classification. *Geological Society of America Special Paper*, 300: 271–294.

Vacher, H.L. and Rowe, M.P., 1997. Geology and hydrogeology of Bermuda. In: H.L. Vacher and T.M. Quinn (Editors), *Geology and Hydrogeology of Carbonate Islands*. Elsevier, Amsterdam, pp. 35–90.

Vail, P.R., 1992. The evolution of seismic stratigraphy and the global sea-level curve. *Geological Society of America Memoir*, 180: 83–91.

Vail, P.R. and Hardenbol, J., 1979. Sea-level changes during the Tertiary. *Oceanus*, 22: 71–79.

Vail, P.R., Mitchum, R.M., Todd, R.G., Widmier, J.M., Thompson, S., Songree, J.B., Bubb, J.N. and Hatelid, W.G., 1977. Seismic stratigraphy – applications to hydrocarbon exploration. *American Association of Petroleum Geologists Memoir*, 26: 49–212.

Valentin, H., 1952. *Die Küsten der Erde*, 246. Petermanns Geographische Mitteilungen, Gotha, J. Perthes, 118 pp.

Valentin, H., 1954. Der Landverlust in Holderness, Ostengland, von 1852 bis 1952. *Die Erde*, 6: 296–315.

Vallejo, L.E. and Degroot, R., 1988. Bluff response to wave action. *Engineering Geology*, 26: 1–16.

van Andel, T.H. and Veevers, J.J., 1967. *Morphology and Sediments of the Timor Sea*, Bulletin No. 83. Bureau of Mineral Resources, Geology and Geophysics, Canberra, 172 pp.

van de Plassche, O., 1982. Sea level changes and water-level movements in the Netherlands during the Holocene. *Mededelingen Rijks Geologische Dienst*, 36: 3–93.

van de Plassche, O. (Editor), 1986. *Sea-level Research: A Manual for the Collection and Evaluation of Data*. Geobooks, Norwich, 618 pp.

van de Plassche, O., Van der Borg, K. and De Jong, A.F.M., 1998. Sea level-climate correlation during the past 1400 yr. *Geology*, 26: 319–322.

van der Graaf, J., 1986. Probabilistic design of dunes: an example from the Netherlands. *Coastal Engineering*, 9: 479–500.

van der Molen, J. and de Swart, H.E., 2001. Holocene tidal conditions and tide-induced sand transport in the southern North Sea. *Journal of Geophysical Research*, 106: 9339–9362.

van Dongeren, A.R. and de Vriend, H.J., 1994. A model of morphological behaviour of tidal basins. *Coastal Engineering*, 22: 287–310.

van Eerdt, M.M., 1985. The influence of vegetation on erosion and accretion in salt marshes of the Oosterschelde, The Netherlands. *Vegetatio*, 62: 367–373.

van Heerden, I.L. and Roberts, H.H., 1980. The Atchafalaya Delta: rapid progradation along a traditionally retreating coast (south-central Louisiana). *Zeitschrift für Geomorphologie, Suppl.-Bd.*, 34: 188–201.

van Rijn, L.C., 1993. *Principles of Sediment Transport in Rivers, Estuaries and Coastal Seas*. Aqua Publications, Amsterdam.

van Rijn, L.C., 2001. Comparison of profile models with hydrodynamic and morphodynamic data on time scale of a storm month at Egmond coast, October–November 1998. In: R. Soulsby (Editor), *Final Volume of Summary Papers, COAST3D*, Report TR121. HR Wallingford, pp. B7.1–B7.7.

van Straaten, L.M.J.U., 1954. Radiocarbon datings and changes of sea level at Velzen (Netherlands). *Geologie en Mijnbouw*, 16: 247–253.

van Straaten, L.M.J.U. and Kuenen, P.H., 1957. Accumulation of fine grained sediments in the Dutch Wadden Sea. *Geologie en Mijnbouw*, 19: 329–354.

van Wagoner, J.C., Mitchum, R.M., Campion, K.M. and Rahmanian, V.D., 1990. *Siliclastic Sequence Stratigraphy in Well-logs, Cores, and Outcrops*. American Association of Petroleum Geologists, Methods in Exploration Series, 7, 55 pp.

Van Wellen, E., Chadwick, A.J. and Mason, T., 2000. A review and assessment of longshore sediment transport equations for coarse-grained beaches. *Coastal Engineering*, 40: 243–275.

Vann, J.H., 1959. Landform-vegetation relationships in the Atrato Delta. *Annals of the Association of American Geographers*, 49: 345–360.

Vaughan, T.W., 1909. The geological work of mangroves in southern Florida. *Smithsonian Miscellaneous Collection*, 52: 461–464.

Vaughan, T.W., 1916. Results of investigations of ecology of the Floridian and Bahamian shoal-water corals. *Proceedings of the National Academy of Sciences*, 2: 95–100.

Vaughan, T.W., 1932. Rate of sea cliff recession on the property of the Scripps Institution of Oceanography at La Jolla, California. *Science*, 75: 250.

Vaughn, D.M., Rickman, D. and Huh, O., 1996. Modeling spatial and volumetric changes in the Atchafalaya Delta, Louisiana. *Geocarto International*, 11: 71–80.

Veeh, H.H., 1966. Th^{230}/U^{238} and U^{234}/U^{238} ages of Pleistocene high sea level stand. *Journal of Geophysical Research*, 71: 3379–3386.

Veeh, H.H. and Veevers, J.J., 1970. Sea level at −175 m off the Great Barrier Reef 13600 to 17000 years ago. *Nature*, 226: 536–537.

Vening Meinesz, F.A., Umbgrove, J.H.F. and Kuenen, P.H., 1934. *Gravity Expeditions at Sea, 1923–1932*, Vol. 2. Waltman, Delft, 208 pp.

Vernberg, F.J., 1993. Salt-marsh processes: a review. *Environmental Toxicology and Chemistry*, 12: 2167–2195.

Veron, J.E.N., 1995. *Corals in Space and Time: the Biogeography and Evolution of the Scleractinia*. University of New South Wales Press, Sydney, 331 pp.

Veron, J.E.N. and Minchin, P.R., 1992. Correlations between sea surface temperature, circulation patterns and the distribution of hermatypic corals of Japan. *Continental Shelf Research*, 12: 835–857.

Verstappen, H.T., 1960. On the geomorphology of raised coral reefs and its tectonic significance. *Zeitschrift für Geomorphologie*, 4: 1–28.

Vézina, J.L., Jones, B. and Ford, D., 1999. Sea-level highstands over the last

500000 years: evidence from the Ironshore Formation on Grand Cayman, British West Indies. *Journal of Sedimentary Research*, 69: 317–327.

Viles, H.A., 1988. *Biogeomorphology*. Blackwell, Oxford, 365 pp.

Viles, H.A. and Spencer, T., 1995. *Coastal Problems*. Arnold, London, 350 pp.

Viles, H.A. and Trudgill, S.T., 1984. Long term remeasurements of micro-erosion meter rates, Aldabra Atoll, Indian Ocean. *Earth Surface Processes and Landforms*, 9: 89–94.

Vine, F.J., 2001. Magnetics. In: J.H. Steele, S.A. Thorpe and K.K. Turekian (Editors), *Encyclopedia of Ocean Sciences*. Academic Press, San Diego, pp. 1515–1525.

Visher, G.S., 1969. Grain size distribution and depositional processes. *Journal of Sedimentary Petrology*, 39: 1074–1106.

Vishnu-Mittre and Gupta, H.P., 1972. Pollen analytical study of the Quaternary deposits in the Bengal Basin. *Palaeobotanist*, 18: 297–306.

Volker, A., 1966. Tentative classification and comparison with deltas of other climatic regions, In: *Scientific Problems of the Humid Tropical Zone Deltas and their Implications: Proceedings of the Dacca Symposium*. UNESCO, Paris, pp. 399–408.

von Arx, W.S., 1948. The circulation systems of Bikini and Rongelap Lagoons. *American Geophysical Union Transactions*, 29: 861–870.

von Bertalanffy, L., 1952. *Problems of Life*. Watts and Co, London, 216 pp.

von Bertalanffy, L., 1968. *General Systems Theory: Foundations, Developments, Applications*. Allen Lane Penguin Press, London, 311 pp.

von Richthofen, F., 1886. *Führer für Forschungsreisende*. Jänecke, Hannover, 745 pp.

Vos, P.C. and van Kesteren, W.P., 2000. The long-term evolution of intertidal mudflats in the northern Netherlands during the Holocene: natural and anthropogenic processes. *Continental Shelf Research*, 20: 1687–1710.

Walcott, R.I., 1970. Flexure of the lithosphere at Hawaii. *Tectonophysics*, 9: 435–446.

Walcott, R.I., 1972. Past sea levels, eustasy and deformation of the Earth. *Quaternary Research*, 2: 1–14.

Walker, H.J., 1998. Arctic deltas. *Journal of Coastal Research*, 14: 718–738.

Walker, R.G., 1992. Facies, facies models and modern stratigraphic concepts. In: R.G. Walker and N.P. James (Editors), *Facies Models: Response to Sea Level Change*. Geological Association of Canada, Toronto, pp. 1–14.

Walbran, P.D., 1994. The nature of the pre-Holocene surface, John Brewer Reef, with implications for the interpretation of Holocene reef development. *Marine Geology*, 122: 63–79.

Walton, T.L., 1994. Shoreline solution for tapered beach fill. *Journal of Waterway, Port, Coastal and Ocean Engineering*, 120: 651–655.

Wang, B.-C. and Eisma, D., 1988. Mudflat deposition along the Wenzhou coastal plain in southern Zhejiang, China. In: P.L. de Boer, A. van Gelder and S.D. Nio (Editors), *Tide-influenced Sedimentary Environments and Facies*. D. Reidel, Dordrecht, pp. 265–274.

Wang, F.C., Lu, T. and Sikora, W.B., 1993. Intertidal marsh suspended sediment transport processes, Terrebonne Bay, Louisiana, USA. *Journal of Coastal Research*, 9: 209–220.

Wang, P. and Murray, J.W., 1983. The use of foraminifera as indicators of tidal effects in estuarine deposits. *Marine Geology*, 51: 239–250.

Wang, Y., 1983. The mudflat system of China. *Canadian Journal of Fisheries and Aquatic Science*, 40: 160–171.

Wang, Y. and Ke, X., 1989. Cheniers on the east coastal plain of China. *Marine Geology*, 90: 321–335.

Wang, Z.B., Louters, T. and de Vriend, H.J., 1995. Morphodynamic modelling for a tidal inlet in the Wadden Sea. *Marine Geology*, 126: 289–300.

Wang, Z.-Y. and Liang, Z.-Y., 2000. Dynamic characteristics of the Yellow River mouth. *Earth Surface Processes and Landforms*, 25: 765–782.

Ward, E.M., 1922. *English Coastal Evolution*. Methuen, London, 262 pp.

Ward, J.M., 1967. Studies in ecology on a shell barrier beach. *Vegetatio*, 14: 241–342.

Ward, L.G., 1981. Suspended-material transport in marsh tidal channels, Kiawah Island, South Carolina. *Marine Geology*, 40: 139–154.

Ward, W.T., 1977. Sand movement on Fraser Island: a response to changing climates. *Occasional Papers in Anthropology, Anthropology Museum, University of Queensland*, 8: 113–126.

Warrick, R.A., 1993. Slowing global warming and sea-level rise: the rough road from Rio. *Transactions of the Institute of British Geographers, NS*, 18: 140–148.

Wasson, R.J., 1994. Living with the past: uses of history for understanding landscape change and degradation. *Land Degradation and Rehabilitation*, 5: 79–87.

Wasson, R.J. and Clark, R.L., 1987. *The Quaternary in Australia – Past, Present and Future*. Bureau of Mineral Resources Journal of Australian Geology and Geophysics, Report, 282: 29–34.

Watson, J.G., 1928. Mangrove forests of the Malay Peninsula. *Malayan Forestry Records*, 6: 1–275.

Watts, A.B. and Ribe, N.M., 1984. On geoid heights and flexure of the lithosphere at seamounts. *Journal of Geophysical Research*, 89: 11152–11170.

Watts, A.B. and Zhong, S., 2000. Observations of flexure and the rheology of oceanic lithosphere. *Geophysics Journal International*, 142, 855–875.

Weaver, D.B., 1990. Grand Cayman Island and the resort cycle concept. *Journal of Travel Research*, 29: 9–15.

Weber, J.N. and Woodhead, P.M.J., 1972. Carbonate lagoon and beach sediments of Tarawa atoll, Gilbert Islands. *Atoll Research Bulletin*, 157: 1–28.

Webster, J.M., Davies, P.J. and Konishi, K., 1998. Model of fringing reef development in response to progressive sea level fall over the last 7000 years – (Kikai-jima, Ryukyu Islands, Japan). *Coral Reefs*, 17: 289–308.

Wegener, A., 1915. *Die Entstehung der Kontinente und Ozeane*. Vieweg, Braunschweig, 248 pp.

Wells, J.T., 1983. Dynamics of coastal fluid muds in low-, moderate-, and high-tide-range environments. *Canadian Journal of Fisheries and Aquatic Science*, 40: 130–142.

Wells, J.T., 1995. Tide-dominated estuaries and tidal rivers. In: G.M.E. Perillo (Editor), *Geomorphology and Sedimentology of Estuaries*. Elsevier, Amsterdam, pp. 179–205.

Wells, J.T. and Coleman, J.M., 1981. Physical processes and fine-grained sediment dynamics, coast of Surinam, South America. *Journal of Sedimentary Petrology*, 51: 1053–1068.

Wells, J.T. and Coleman, J.M., 1987. Wetland loss and subdelta life cycle. *Estuarine, Coastal and Shelf Science*, 25: 111–125.

Wentworth, C.K., 1927. Estimates of marine and fluvial erosion in Hawaii. *Journal of Geology*, 25: 117–133.

Wentworth, C.K., 1931. Geology of the Pacific equatorial islands. *B.P. Bishop Museum Occasional Papers*, 9: 1–25.

Wentworth, C.K., 1938. Marine bench-forming processes, water level weathering. *Journal of Geomorphology*, 1: 6–32.

Wentworth, C.K., 1939. Marine bench-forming processes II, solution benching. *Journal of Geomorphology*, 2: 3–25.

Werner, B.T. and Fink, T.M., 1993. Beach cusps as self-organized patterns. *Science*, 260: 968–970.

Wescott, W.A., 1993. Geomorphic thresholds and complex response of fluvial systems – some implications for sequence stratigraphy. *American Association of Petroleum Geologists Bulletin*, 77: 1208–1218.

West, R.C., 1956. Mangrove swamps of the Pacific coast of Columbia. *Annals of the Association of American Geographers*, 46: 98–121.

West, R.C., 1977. Tidal salt-marsh and mangal formations of middle and south America. In: V.J. Chapman (Editor), *Wet Coastal Ecosystems. Ecosystems of the World.* Elsevier, Amsterdam, pp. 193–213.

Westcott, W.A. and Etheridge, F.G., 1980. Fan delta sedimentology and tectonic setting – Yallahs fan delta, southeast Jamaica. *Bulletin of the American Association of Petroleum Geologists*, 64: 374–399.

Wheeler, A.J. and Waller, M.P., 1995. The Holocene lithostratigraphy of fenland, eastern England: a review and suggestions for redefinition. *Geological Magazine*, 132: 223–233.

Whitehouse, R.J.S., 2001. Synthesis of Teignmouth process measurements and interpretation. In: R. Soulsby (Editor), *Final Volume of Summary Papers, COAST3D*, Report TR121. HR Wallingford, pp. C8.1–C8.5.

Whitehouse, R.J.S. and Mitchener, H.J., 1998. Observations of the morphodynamic behaviour of an intertidal mudflat at different timescales. In: K.S. Black, D.M. Paterson and A. Cramp (Editors), *Sedimentary Processes in the Intertidal Zone*. Geological Society of London, Special Publication, pp. 255–271.

Widdows, J., Brinsley, M. and Elliott, M., 1998. Use of *in situ* flume to quantify particle flux (biodeposition rates and sediment erosion) for an intertidal mudflat in relation to changes in current velocity and benthic macrofauna. In: K.S. Black, D.M. Paterson and A. Cramp (Editors), *Sedimentary*

Processes in the Intertidal Zone. Geological Society of London, Special Publication, pp. 85–97.

Wiegel, R.L., 1994. Ocean beach nourishment on the USA Pacific coast. *Shore and Beach*, 62: 11–36.

Wiens, H.J., 1959. Atoll development and morphology. *Annals of the Association of American Geographers*, 49: 31–54.

Wiens, H.J., 1962. Atoll environment and ecology. Yale University Press, New Haven, CT, 532 pp.

Wigley, T.M.L. and Raper, S.C.B., 1992. Implications for climate and sea level of revised IPCC emission scenarios. *Nature*, 357: 293–300.

Wilkinson, C.R., 1996. Global change and coral reefs: impacts on reefs, economies and human cultures. *Global Change Biology*, 2: 547–558.

Williams, A.T., 1973. Problems of beach cusp development. *Journal of Sedimentary Petrology*, 45: 857–866.

Williams, A.T. and Roberts, G.T., 1995. The measurement of pebble impacts and wave action on shore platforms and beaches: the swash force transducer (swashometer). *Marine Geology*, 129: 137–143.

Williams, G.C., 1983. *Hydraulic Characteristics of NSW Estuarine Lake Inlets*, 2 vols. University of New South Wales, Sydney.

Williams, H. and Hutchinson, I., 2000. Stratigraphic and microfossil evidence for late Holocene tsunamis at Swantown Marsh, Whidbey Island, Washington. *Quaternary Research*, 54: 218–227.

Williams, M., Dunkerley, D., De Deckker, P., Kershaw, P. and Chappell, J., 1998. *Quaternary Environments*. Arnold, London, 329 pp.

Williams, W.W., 1956. An east coast survey: some recent changes in the coast of East Anglia. *Geographical Journal*, 122: 317–334.

Williams, W.W., 1960. *Coastal Changes*. Routledge & Kegan Paul, London, 220 pp.

Wilson, J.P. and Burrough, P.A., 1999. Dynamic modelling, geostatistics, and fuzzy classification. *Annals of the Association of American Geographers*, 89: 736–746.

Wilson, J.T., 1963. Pattern of uplifted islands in the main ocean basins. *Science*, 139: 592–594.

Wilson, P.A., Jenkyns, H.C., Elderfield, H. and Larson, R.L., 1998. The paradox of drowned carbonate platforms and the origin of Cretaceous Pacific guyots. *Nature*, 392: 889–894.

Winland, H.D. and Matthews, R.K., 1974. Origin and significance of grapestone, Bahama Islands. *Journal of Sedimentary Petrology*, 44: 921–923.

Winnant, C.D., Inman, D.L. and Nordstrom, C.E., 1975. Description of seasonal beach changes using empirical eigenfunctions. *Journal of Geophysical Research*, 80: 1979–1986.

Witman, J.D., 1992. Physical disturbance and community structure of exposed and protected reefs: a case study from St John, US Virgin Islands. *American Zoologist*, 32: 641–654.

Witter, R.C., Kelsey, H.M. and Hemphill-Haley, E., 2001. Pacific storms, El Niño and tsunamis: competing mechanisms for sand deposition in a coastal marsh, Euchre Creek, Oregon. *Journal of Coastal Research*, 17: 563–583.

Wolanski, E., 1986. An evaporation-driven salinity maximum zone in Australian tropical estuaries. *Estuarine, Coastal and Shelf Science*, 22: 415–424.

Wolanski, E., 1992. Hydrodynamics of mangrove swamps and their coastal waters. *Hydrobiologia*, 247: 141–161.

Wolanski, E., 1994. *Physical Oceanographic Processes of the Great Barrier Reef*. CRC Press, Boca Raton, 194 pp.

Wolanski, E. and Chappell, J., 1996. The response of tropical Australian estuaries to a sea level rise. *Journal of Marine Systems*, 7: 267–279.

Wolanski, E. and Ridd, P., 1986. Tidal mixing and trapping in mangrove swamps. *Estuarine, Coastal and Shelf Science*, 23: 759–771.

Wolanski, E., Jones, M. and Bunt, J.S., 1980. Hydrodynamics of a tidal creek–mangrove swamp system. *Australian Journal of Marine and Freshwater Research*, 31: 431–450.

Wolanski, E., Chappell, J., Ridd, P. and Vertessy, R., 1988. Fluidization of mud in estuaries. *Journal of Geophysical Research*, 93: 2351–2361.

Wolanski, E., Spagnol, S. and Lim, E.B., 1997. The importance of mangrove flocs in sheltering seagrass in turbid coastal waters. *Mangroves and Salt Marshes*, 1: 187–191.

Wolanski, E., Nhan, N.H. and Spagnol, S., 1998. Sediment dynamics during low flow conditions in the Mekong River Estuary, Vietnam. *Journal of Coastal Research*, 14: 472–482.

Wolanski, E., Spagnol, S., Thomas, S., Moore, K., Alongi, D.M., Trott, L. and Davidson, A., 2000. Modelling and visualizing the fate of shrimp pond effluent in a mangrove-fringed tidal creek. *Estuarine, Coastal and Shelf Science*, 50: 85–97.

Wolaver, T.G., Dame, R.F., Spurrier, J.D. and Miller, A.B., 1988. Sediment exchange between a euhaline salt marsh in South Carolina and the adjacent tidal creek. *Journal of Coastal Research*, 4: 17–26.

Wolman, M.G. and Miller, J.P., 1960. Magnitude and frequency of forces in geomorphic processes. *Journal of Geology*, 68: 54–74.

Wong, P.P., 1981. Beach evolution between headland breakwaters. *Shore and Beach*, 49: 3–12.

Wong, P.P., 1990. The geomorphological basis of beach resort sites – some Malaysian examples. *Ocean and Shoreline Management*, 13: 127–147.

Wood, A., 1968. Beach platforms in the chalk of Kent, England. *Zeitschrift für Geomorphologie, NF*, 12: 107–113.

Wood, R., 1999. *Reef Evolution*. Oxford University Press, Oxford, 414 pp.

Wood-Jones, F., 1912. *Coral and Atolls: A History and Description of the Keeling–Cocos Islands, with an Account of their Fauna and Flora, and a Discussion of the Method of Development and Transformation of Coral Structures in General*. Lovell Reeve and Co., London, 392 pp.

Woodroffe, C.D., 1982. Geomorphology and development of mangrove swamps, Grand Cayman Island, West Indies. *Bulletin of Marine Science*, 32: 381–398.

Woodroffe, C.D., 1985. Studies of a mangrove basin, Tuff Crater, New Zealand

III. The flux of organic and inorganic particulate matter. *Estuarine, Coastal and Shelf Science*, 20: 447–462.

Woodroffe, C.D., 1988. Vertical movement of isolated oceanic islands at plate margins: evidence from emergent reefs in Tonga (Pacific Ocean), Cayman Islands (Caribbean Sea) and Christmas Island (Indian Ocean). *Zeitschrift für Geomorphologie, Suppl.-Bd.*, 69: 17–37.

Woodroffe, C.D., 1990. The impact of sea-level rise on mangrove shorelines. *Progress in Physical Geography*, 14: 483–520.

Woodroffe, C.D., 1992. Mangrove sediments and geomorphology. In: D. Alongi and A. Robertson (Editors), *Tropical Mangrove Ecosystems*. Coastal and Estuarine Studies, American Geophysical Union, Washington, DC, pp. 7–41.

Woodroffe, C.D., 1993a. Late Quaternary evolution of coastal and lowland riverine plains of Southeast Asia and northern Australia: an overview. *Sedimentary Geology*, 83: 163–173.

Woodroffe, C.D., 1993b. Morphology and evolution of reef islands in the Maldives, *Proceedings of the Seventh International Coral Reef Symposium*. University of Guam Press, Mangilao, pp. 1217–1226.

Woodroffe, C.D., 1995a. Mangrove vegetation of Tobacco Range and nearby mangroves, central Belize barrier reef. *Atoll Research Bulletin*, 427: 1–35.

Woodroffe, C.D., 1995b. Response of tide-dominated mangrove shorelines in northern Australia to anticipated sea-level rise. *Earth Surface Processes and Landforms*, 20: 65–85.

Woodroffe, C.D., 2000. Deltaic and estuarine environments and their Late Quaternary dynamics on the Sunda and Sahul shelves. *Journal of Asian Earth Sciences*, 18: 393–413.

Woodroffe, C.D., Bardsley, K.N., Ward, P.J. and Hanley, J.R., 1988. Production of mangrove litter in a macrotidal embayment, Darwin Harbour, NT, Australia. *Estuarine, Coastal and Shelf Science*, 26: 581–598.

Woodroffe, C.D., Bryant, E.A., Price, D.M. and Short, S.A., 1992. Quaternary inheritance of coastal landforms, Cobourg Peninsula, Northern Territory. *Australian Geographer*, 23: 101–115.

Woodroffe, C.D. and Chappell, J., 1993. Holocene emergence and evolution of the McArthur River Delta, southwestern Gulf of Carpentaria, Australia. *Sedimentary Geology*, 83: 303–317.

Woodroffe, C.D. and Falkland, A.C., 1997. Geology and hydrogeology of the Cocos (Keeling) Islands, Indian Ocean. In: H.L. Vacher and T.M. Quinn (Editors), *Geology and Hydrogeology of Carbonate Islands*. Elsevier, Amsterdam, pp. 885–908.

Woodroffe, C.D. and Grime, D., 1999. Storm impact and evolution of a mangrove-fringed chenier plain, Shoal Bay, Darwin, Australia. *Marine Geology*, 159: 303–321.

Woodroffe, C.D. and Grindrod, J., 1991. Mangrove biogeography: the role of Quaternary environmental and sea-level fluctuations. *Journal of Biogeography*, 18: 479–492.

Woodroffe, C.D. and McLean, R.F., 1990. Microatolls and recent sea level change on coral atolls. *Nature*, 344: 531–534.

Woodroffe, C.D. and McLean, R.F., 1998. Pleistocene morphology and

Holocene emergence of Christmas (Kiritimati) Island, Pacific Ocean. *Coral Reefs*, 17: 235–248.

Woodroffe, C.D. and Morrison, R.J., 2001. Reef-island accretion and soil development, Makin Island, Kiribati, central Pacific. *Catena*, 44: 245–261.

Woodroffe, C.D., Curtis, R.J. and McLean, R.F., 1983a. Development of a chenier plain, Firth of Thames, New Zealand. *Marine Geology*, 53: 1–22.

Woodroffe, C.D., Stoddart, D.R., Harmon, R.S. and Spencer, T., 1983b. Coastal morphology and late Quaternary history, Cayman Islands, West Indies. *Quaternary Research*, 19: 64–84.

Woodroffe, C.D., Thom, B.G. and Chappell, J., 1985. Development of widespread mangrove swamps in mid-Holocene times in northern Australia. *Nature*, 317: 711–713.

Woodroffe, C.D., Chappell, J., Thom, B.G. and Wallensky, E., 1989. Depositional model of a macrotidal estuary and floodplain, South Alligator River, Northern Australia. *Sedimentology*, 36: 737–756.

Woodroffe, C.D., McLean, R.F., Polach, H. and Wallensky, E., 1990a. Sea level and coral atolls: Late Holocene emergence in the Indian Ocean. *Geology*, 18: 62–66.

Woodroffe, C.D., Stoddart, D.R., Spencer, T., Scoffin, T.P. and Tudhope, A., 1990b. Holocene emergence in the Cook Islands, South Pacific. *Coral Reefs*, 9: 31–39.

Woodroffe, C.D., Short, S., Stoddart, D.R., Spencer, T. and Harmon, R.S., 1991a. Stratigraphy and chronology of Pleistocene reefs in the southern Cook Islands, South Pacific. *Quaternary Research*, 35: 246–263.

Woodroffe, C.D., Veeh, H.H., Falkland, A., McLean, R.F. and Wallensky, E., 1991b. Last interglacial reef and subsidence of the Cocos (Keeling) Islands, Indian Ocean. *Marine Geology*, 96: 137–143.

Woodroffe, C.D., Mulrennan, M.E. and Chappell, J., 1993. Estuarine infill and coastal progradation, southern van Diemen Gulf, northern Australia. *Sedimentary Geology*, 83: 257–275.

Woodroffe, C.D., Murray-Wallace, C.V., Bryant, E.A., Brooke, B., Heijnis, H. and Price, D.M., 1995. Late Quaternary sea-level highstands in the Tasman Sea: evidence from Lord Howe Island. *Marine Geology*, 125: 61–72.

Woodroffe, C.D., McLean, R.F., Smithers, S.G. and Lawson, E., 1999. Atoll reef-island formation and response to sea-level change: West Island, Cocos (Keeling) Islands. *Marine Geology*, 160: 85–104.

Woodroffe, C.D., Kennedy, D.M., Hopley, D., Rasmussen, C.E. and Smithers, S.G., 2000. Holocene reef growth in Torres Strait. *Marine Geology*, 170: 331–346.

Woodwell, G.M., Houghton, R.A., Hall, C.A.S., Whitney, D.E., Moll, R.A. and Juers, D.W., 1979. The flax pond ecosystem study: the annual metabolism and nutrient budgets of a salt marsh. In: R.L. Jeffries and A.J. Davy (Editors), *Ecological Processes in Coastal Environments*. Blackwell, Oxford, pp. 491–511.

Woodworth, P.L., 1990. A search for accelerations in records of European mean sea level. *International Journal of Climatology*, 10: 129–143.

Woodworth, P.L., 1993. A review of recent sea-level research. *Oceanography and Marine Biology Annual Review*, 31: 87–109.

Woodworth, P.L., Shaw, S.M. and Blackman, D.L., 1991. Secular trends in mean tidal range around the British Isles and along the adjacent European coastline. *Geophysical Journal International*, 104: 593–609.

Wooldridge, S.W. and Linton, D.L., 1938. Influence of Pliocene transgression on the geomorphology of south-east England. *Journal of Geomorphology*, 1: 40–54.

Wooldridge, S.W. and Linton, D.L., 1955. *Structure, Surface, and Drainage in South-east England*. Philip, London, 176 pp.

Woolfe, K.J. and Larcombe, P., 1999. Terrigenous sedimentation and coral reef growth: a conceptual framework. *Marine Geology*, 155: 331–345.

Woolfe, K.J., Larcombe, P., Naish, T. and Purdon, R.G., 1998a. Lowstand rivers need not incise the shelf: an example from the Great Barrier Reef, Australia, with implications for sequence stratigraphic models. *Geology*, 26: 75–78.

Woolfe, K.J., Larcombe, P., Orpin, A.R., Purdon, R.G., Michaelsen, P., McIntyre, C.M. and Amjad, N., 1998b. Controls upon inner-shelf sedimentation, Cape York Peninsula, in the region of 12° S. *Australian Journal of Earth Sciences*, 45: 611–621.

Woolnough, S.J., Allen, J.R.L. and Wood, W.L., 1995. An exploratory numerical model of sediment deposition over tidal salt marshes. *Estuarine, Coastal and Shelf Science*, 41: 515–543.

Work, P.A. and Rogers, W.E., 1998. Laboratory study of beach nourishment behaviour. *Journal of Waterway, Port, Coastal, and Ocean Engineering*, 124: 229–237.

Wright, D. and Bartlett, D. (Editors), 1999. *Marine and Coastal Geographical Information Systems*. Taylor and Francis, Philadelphia, 320 pp.

Wright, H.E., 1963. The Late Pleistocene geology of coastal Lebanon. *Quaternaria*, 6: 525–539.

Wright, L.D., 1970. The influence of sediment availability on patterns of beach ridge development in the vicinity of the Shoalhaven River delta, NSW. *Australian Geographer*, 11: 336–348.

Wright, L.D., 1985. River deltas. In: R.A. Davis (Editor), *Coastal Sedimentary Environments*. Springer-Verlag, New York, pp. 1–76.

Wright, L.D., 1989. Dispersal and deposition of river sediments in coastal seas: models from Asia and the tropics. *Netherlands Journal of Sea Research*, 23: 493–500.

Wright, L.D., 1995. *Morphodynamics of Inner Continental Shelves*. CRC Press, Boca Raton, 241 pp.

Wright, L.D. and Coleman, J.M., 1971. The discharge/wave power climate and the morphology of delta coasts. *Association of American Geographers Proceedings*, 3: 186–189.

Wright, L.D. and Coleman, J.M., 1972. River delta morphology: wave climate and the role of the subaqueous profile. *Science*, 176: 282–284.

Wright, L.D. and Coleman, J.M., 1973. Variations in morphology of major river deltas as functions of ocean wave and river discharge regimes. *American Association of Petroleum Geologists Bulletin*, 57: 370–398.

Wright, L.D. and Short, A.D., 1983. Morphodynamics of beaches and surf zones in Australia. In: P.D. Komar (Editor), *CRC Handbook of Coastal Processes and Erosion*. CRC Press, Boca Raton, pp. 35–64.

Wright, L.D. and Short, A.D., 1984. Morphodynamic variability of surf zones and beaches: a synthesis. *Marine Geology*, 56: 93–118.

Wright, L.D. and Thom, B.G., 1977. Coastal depositional landforms: a morphodynamic approach. *Progress in Physical Geography*, 1: 412–459.

Wright, L.D., Coleman, J.M. and Thom, B.G., 1973. Processes of channel development of a high-tide range environment: Cambridge Gulf–Ord River Delta, Western Australia. *Journal of Geology*, 81: 15–41.

Wright, L.D., Coleman, J.M. and Erickson, M.W., 1974. *Analysis of Major River Systems and their Deltas: Morphologic and Process Comparisons*. Technical Report No. 156. Louisiana State University, Louisiana, 114 pp.

Wright, L.D., Chappell, J., Thom, B.G., Bradshaw, M.P. and Cowell, P., 1979. Morphodynamics of reflective and dissipative beach and inshore systems: Southeastern Australia. *Marine Geology*, 32: 105–140.

Wright, L.D., Nielsen, P., Short, A.D. and Green, M.O., 1982. Morphodynamics of a macrotidal beach. *Marine Geology*, 50: 97–128.

Wright, L.D., Yang, Z.-S., Bornhold, B.D., Keller, G.H., Prior, D.B. and Wiseman, W.J., 1986. Hyperpycnal plumes and plume fronts over the Huanghe (Yellow River) delta front. *Geo-Marine Letters*, 9: 97–106.

Wright, L.D., Short, A.D. and Green, M., 1995. Short-term changes in the morphodynamic state of beaches and surf zones: an empirical predictive model. *Marine Geology*, 62: 339–364.

Wright, L.W., 1970. Variation in the level of the cliff/shore platform junction along the south coast of Great Britain. *Marine Geology*, 9: 347–353.

Wright, V.P. and Marriott, S.B., 1993. The sequence stratigraphy of fluvial depositional systems: the role of floodplain sediment storage. *Sedimentary Geology*, 86: 203–210.

Wright, W.B., 1914. *The Quaternary Ice Age*. Macmillan, London, 464 pp.

Wyrtki, K., 1990. Sea level rise: the facts and the future. *Pacific Science*, 44: 1–16.

Xitao, Z., 1989. Cheniers in China: an overview. *Marine Geology*, 90: 311–320.

Yamano, H., Hori, K., Yamaguchi, M., Yamagawa, O. and Ohmura, A., 2001. Highest-latitude coral reef at Iki Island, Japan. *Coral Reefs*, 20: 9–12.

Yamano, H., Kayanne, H., Yonekura, N., Nakamura, H. and Kudo, K., 1998. Water circulation in a fringing reef located in a monsoon area: Kabira Reef, Ishigaki Island, Southwest Japan. *Coral Reefs*, 17: 89–99.

Yamano, H., Miyajima, T. and Koike, I., 2000. Importance of foraminifera for the formation and maintenance of a coral sand cay: Green Island, the Great Barrier Reef, Australia. *Coral Reefs*, 19: 51–58.

Yamashita, T. and Tsuchiya, Y., 1992. Numerical simulation of pocket beach formation. *Proceedings of the Twenty-third Coastal Engineering Conference*, American Society of Civil Engineers: 2556–2566.

Yan, Q., Xu, S. and Shao, X., 1989. Holocene cheniers in the Yangtse delta, China. *Marine Geology*, 90: 311–320.

Yanagi, T., 1999. *Coastal Oceanography*. Kluwer, Dordrecht, 162 pp.

Yang, S.L., 1999. Sedimentation on a growing intertidal island in the Yangtze River mouth. *Estuarine, Coastal and Shelf Science*, 49: 401–410.

Yang, S.L., Ding, P.X. and Chen, S.L., 2001. Changes in progradation rate of the tidal flats at the mouth of the Changjiang (Yangtze) River, China. *Geomorphology*, 38: 167–180.

Yapp, R.H., Johns, D. and Jones, O.T., 1917. The salt marshes of the Dovey Estuary. Part II. The salt marshes. *Journal of Ecology*, 5: 65–103.

Yasso, W.E., 1965. Plan geometry of headland-bay beaches. *Journal of Geology*, 73: 702–714.

Yohe, G.W. and Schlesinger, M.E., 1998. Sea-level change: the expected economic cost of protection or abandonment in the United States. *Climatic Change*, 38: 447–472.

Yokoyama, Y., Lambeck, K., De Deckker, P., Johnston, P. and Fifield, K., 2000. Timing of the Last Glacial Maximum from observed sea-level minima. *Nature*, 406: 713–716.

Yokoyama, Y., De Deckker, P., Lambeck, K., Johnston, P. and Fifield, L.K., 2001. Sea-level at the Last Glacial Maximum: evidence from northwestern Australia to constrain ice volumes for oxygen isotope stage 2. *Palaeogeography, Palaeoclimatology, Palaeoecology*, 165: 281–297.

Yonekura, N. (Editor), 1988. *Sea-Level Changes and Tectonics in the Middle Pacific*. University of Tokyo, Japan, 191 pp.

Yonekura, N. (Editor), 1990. *Sea-Level Changes and Tectonics in the Middle Pacific*. University of Tokyo, Japan, 280 pp.

Yonekura, N., Ishii, T., Saito, Y., Maeda, Y., Matsushima, Y., Matsumoto, E. and Kayanne, H., 1988. Holocene fringing reefs and sea-level change in Mangaia Island, southern Cook Islands. *Palaeogeography, Palaeoclimatology, Palaeoecology*, 68: 177–188.

Yonge, C.M., 1940. The biology of reef-building corals. *Scientific Reports of the Great Barrier Reef Expedition 1928–1929, British Museum (Natural History)*, 1: 353–891.

Yoshikawa, T., 1985. Landform development by tectonics and denudation. In: A. Pitty (Editor), *Themes in Geomorphology*. Croom & Helm, London, pp. 194–210.

Young, I.R., 1989. Wave transformation over coral reefs. *Journal of Geophysical Research*, 94: 9779–9789.

Young, R.W. and Bryant, E.A., 1992. Catastrophic wave erosion on the southeastern coast of Australia: impact of the Lanai tsunamis ca. 105 ka? *Geology*, 20: 199–202.

Young, R.W. and Bryant, E.A., 1993. Coastal rock platforms and ramps of Pleistocene and Tertiary age in southern New South Wales, Australia. *Zeitschrift für Geomorphologie, NF*, 37: 257–272.

Young, R.W. and Bryant, E.A., 1998. Morphology and process on the lateritic coastline near Darwin, northern Australia. *Zeitschrift für Geomorphologie, NF*, 42: 97–107.

Young, R.W., Bryant, E.A., Price, D.M., Wirth, L. and Pease, M., 1993. Theoretical constraints and chronological evidence of Holocene coastal development in central and southern New South Wales, Australia. *Geomorphology*, 7: 317–329.

Young, R.W., Bryant, E.A. and Price, D.M., 1996a. Catastrophic wave (tsunami?) transport of boulders in southern New South Wales, Australia. *Zeitschrift für Geomorphologie, NF*, 40: 191–207.

Young, R.W., White, K.L. and Price, D.M., 1996b. Fluvial deposition on the Shoalhaven Deltaic Plain, Southern New South Wales. *Australian Geographer*, 27: 215–234.

Zarillo, G.A., 1985. Tidal dynamics and substrate response in a salt-marsh estuary. *Marine Geology*, 67: 13–35.

Zeff, M.L., 1988. Sedimentation in a salt marsh-tidal channel system, southern New Jersey. *Marine Geology*, 82: 33–48.

Zeigler, J.M., Tuttle, S.D., Tasha, H.J. and Giese, G.S., 1965. The age and development of the Provincelands hook, outer Cape Cod, Massachusetts. *Limnology and Oceanography*, 10, Supplement: R298–R311.

Zenkovich, V.P., 1946. On the study of shore dynamics. *Akademiya nauk USSR, Institut Okeandogi, Trudy*, 1: 99–112.

Zenkovich, V.P., 1959. On the genesis of cuspate spits along lagoon shores. *Journal of Geology*, 67: 269–277.

Zenkovich, V.P., 1967. *Processes of Coastal Development*. Oliver & Boyd, Edinburgh, 738 pp.

Zeuner, F.E., 1945. *The Pleistocene Period*. Ray Society, London, 322 pp.

Zhang, K., Douglas, B.C. and Leatherman, S.P., 2000. Twentieth-century storm activity along the US east coast. *Journal of Climate*, 13: 1748–1761.

Zingg, A.W., 1952. Wind tunnel studies of the movement of sedimentary material. *Proceedings of the Fifth Hydraulics Conference*. State University of Iowa, Iowa, pp. 111–135.

Zong, Y. and Tooley, M.J., 1999. Evidence of mid-Holocene storm-surge deposits from Morecambe Bay, northwest England: a biostratigraphical approach. *Quaternary International*, 55: 43–50.

Index